Springer Collected Works in Mathematics

For further volumes:
http://www.springer.com/series/11104

Arizona, 1968

Peter D. Lax

Selected Papers I

Editors

Peter Sarnak · Andrew J. Majda

Reprint of the 2005 Edition

Springer

Author
Peter D. Lax
Courant Institute
New York, NY
USA

Editors
Peter Sarnak
Princeton University
Princeton, NJ
USA

Andrew J. Majda
New York University
New York, NY
USA

ISSN 2194-9875
ISBN 978-1-4614-9432-4 (Softcover)
978-0-387-22925-6 (Hardcover)
DOI 10.1007/978-0-387-28148-3
Springer New York Heidelberg Dordrecht London

Library of Congress Control Number: 2012954381

Printed on acid-free paper

Springer is part of Springer Science+Business Media (www.springer.com)

Preface

Peter Lax's research spans many areas of pure and applied mathematics. It ranges from functional analysis, partial differential equations, and numerical methods to conservation laws, integrable systems, and scattering theory. Many of the papers in these volumes have become classics. They are a "must read" for any serious student of these topics, and their impact continues to be felt both explicitly and implicitly in current research. In terms of insight, depth, and breadth, Lax has few equals. The reader of these selecta will quickly appreciate his brilliance as well as his masterful touch. Having this collection of papers in one place allows one to follow the evolution of his ideas and mathematical interests and to appreciate how many of these papers initiated topics that developed a life of their own. It proves that even in today's highly specialized world of mathematics and science, it is still possible to work across disciplines at the very highest level.

The two volumes contain a selection from Lax's published papers. A complete list of these papers to date as well as a list of his books and monographs is provided in the text. The books are an outgrowth of years of research experience and far-reaching insights, an excellent example being his recent text on functional analysis. The selected papers are assembled according to topic and theme. After each paper, or collection of papers, is a commentary placing the paper in context and where relevant, discussing more recent developments. The volumes were edited together with Peter Lax.

We are very grateful to P. Deift, S. Friedland, F. Gesztesy, V. Guillemin, T.Y. Hou, S. Klainerman, H. McKean, B. Parlette and H. Widom for providing their expert commentaries on various papers. Thanks also to J. Heinze, I. Lindemann, and M. Spencer from Springer, who initiated and saw through this "Lax Selecta" project, and to Keisha Grady for her help in preparing the commentaries. The project took longer than was originally anticipated, but we believe that the result has been well worth the wait.

Andy Majda
and
Peter Sarnak
New York
May 2004

Acknowledgments

The papers appearing in these two volumes have all been previously published, and are being reprinted here with the permission of the publishers or editors of the journals and books in which they originally appeared. Below is a list of journals, books, and proceedings, in alphabetical order by publisher, along with the necessary credit lines, where these papers have been previously published. Only papers requiring the publisher's permission have been included, and papers published by Springer-Verlag are not included at all.

Elsevier:

[58] Reprinted from ***Contributions to Nonlinear Functional Analysis***, P. Lax, Shock Waves and Entropy, pp. 603–634, 1971, with permission from Elsevier.

[83] Reprinted from ***Topics in Functional Analysis***, P. Lax and R.S. Phillips, The time delay operator and a related trace formula, pp. 197–215, 1978, with permission from Elsevier.

[84] Reprinted from ***Recent Advances in Numerical Analysis***, P. Lax, Accuracy and resolution in the computation of solutions of linear and nonlinear equations, pp. 107–117, 1978, with permission from Elsevier.

[85] Reprinted from *Linear Algebra and Its Applications*, **20**, P. Lax and A. Lax, On Sums of Squares, 71–75, 1978, with permission from Elsevier.

[98] Reprinted from *Journal of Functional Analysis*, **46**, P. Lax, The asymptotic distribution of lattice points in Euclidean and non-Euclidean spaces, pp. 280–350, 1982, with permission from Elsevier.

American Institute of Physics:

[33] Reprinted from *Journal of Mathematical Physics*, **5**, P. Lax, Development of singularities of solutions of nonlinear hyperbolic partial differential equations, pp. 611–613, 1964, with permission from the American Institute of Physics.

Society for Industrial and Applied Mathematics:

[73] Reprinted from *SIAM Review*, **18**, P. Lax, Almost periodic solutions of the KdV equations, pp. 351–375, 1976, with permission from SIAM.

American Mathematical Society:

[5] Reprinted from *Proceedings of the American Mathematical Society*, **3**, P. Lax, On the existence of Green's functions, pp. 526–531, 1952, with permission from the AMS.

[36] Reprinted from *Proceedings of the American Mathematical Society*, **16**, J.F. Adams, P. Lax, and R.S. Phillips, On matrices whose real linear combinations are nonsingular, pp. 318–322, 1965, with permission from the AMS.

[36]' Reprinted from *Proceedings of the American Mathematical Society*, **17**, P. Lax, Correction to "On matrices whose real linear combinations are nonsingular," pp. 945–947, 1966, with permission from the AMS.

[27] Reprinted from *Bulletin of the American Mathematical Society*, **68**, P. Lax and R.S. Phillips, The wave equation in exterior domans, pp. 47–49, 1962, with permission from the AMS.

[35] Reprinted from *Bulletin of the American Mathematical Society*, **70**, P. Lax and R.S. Phillips, Scattering Theory, pp. 130–142, 1964, with permission from the AMS.

[91] Reprinted from *Bulletin of the American Mathematical Society*, **2**, P. Lax and R.S. Phillips, Scattering theory and automorphic forms, pp. 161–195, 1980, with permission from the AMS.

[99] Reprinted from *Bulletin of the American Mathematical Society*, **4**, P. Lax, The multiplicity of eigenvalues, pp. 213–214, 1982, with permission from the AMS.

[109] Reprinted from *Transactions of the American Mathematical Society*, **289**, P. Lax, Translations Representations for automorphic solutions of the wave equation in non-Euclidean spaces; the case of finite volume, pp. 715–735, 1985, with permission from the AMS.

Princeton University Press:

[9] Reprinted from ***Contributions to the Theory of Partial Differential Equations***, edited by L. Bers, S. Bochner, and F. John, copyright 1954 by Princeton University Press. P. Lax and A. Milgram, Parabolic Equations, pp. 167–190, 1954, reprinted by permission of Princeton University Press.

Contents

PETER D. LAX
List of Publications

1944

[1] Proof of a conjecture of P. Erdös on the derivative of a polynomial, *Bull. Amer. Math. Soc.* **50**, 509–513.

1948

[2] The quotient of exponential polynomials, *Duke Math. J.* **15**, 967–970.

1950

[3] Partial Diffential Equations, Lecture Notes, NYU, IMS (1950–51).

1951

[4] A remark on the method of orthogonal projections, *Comm. Pure Appl. Math.* **4**, 457–464.

1952

[5] On the existence of Green's function, *Proc. Amer. Math. Soc.* **3**, 526–531.

1953

[6] Nonlinear hyperbolic equations, *Comm. Pure Appl. Math.* **6**, 231–258.

1954

[7] Weak solutions of nonlinear equations and their numerical computation, *Comm. Pure Appl. Math.* **7**, 159–194.

[8] Symmetrizable linear transformations, *Comm. Pure Appl. Math.* **7**, 633–648.

[9] (with A. Milgram) Parabolic equations, *Ann. Math. Studies 33* (Princeton) 167–190.

[10] The initial value problem for nonlinear hyperbolic equations, *Ann. Math. Studies 33* (Princeton) 211–229.

1955

[11] Reciprocal extremal problems in function theory, *Comm. Pure Appl. Math.* **8**, 437–454.

[12] On Cauchy's problem for hyperbolic equations and the differentiability of solutions of elliptic equations, *Comm. Pure Appl. Math.* **8**, 615–633.

[13] (with R. Courant) Cauchy's problem for hyperbolic differential equations, *Ann. Mat. Pura Appl.* **40**, 161–166.

1956

[14] A stability theorem for solutions of abstract differential equations, and its application to the study of local behavior of solutions of elliptic equations, *Comm. Pure Appl. Math.* **9**, 747–766.

[15] (with R.D. Richtmyer) Survey of the stability of linear finite difference equations, *Comm. Pure Appl. Math.* **9**, 267-293.

[16] (with R. Courant) The propagation of discontinuities in wave motion, *Proc. Nat. Acad. Sci.* **42**, 872–876.

1957

[17] A Phragmen–Lindelöf theorem in harmonic analysis and its application to some questions in the theory of elliptic equations, *Comm. Pure Appl. Math.* **10**, 361–389.

[18] Hyperbolic systems of conservation laws, II, *Comm. Pure Appl. Math.* **10**, 537–566.

[19] Remarks on the preceding paper, *Comm. Pure Appl. Math.* **10**, 617–622.

[20] Asymptotic solutions of oscillatory initial value problems, *Duke Math. J.* **24**, 627–646.

1958

[21] Differential equations, difference equations and matrix theory, *Comm. Pure Appl. Math.* **11**, 175–194.

1959

[22] Translation invariant spaces, *Acta Math.* **101**, 163-178.

1960

[23] (with B. Wendroff) Systems of conservation laws, *Comm. Pure Appl. Math.* **13**, 217–237.

[24] (with R.S. Phillips) Local boundary conditions for dissipative symmetric linear differential operators, *Comm. Pure Appl. Math.* **13**, 427–455.

1961

[25] Translation invariant spaces, Proc. Int. Symp. on Linear Spaces, Israeli Acad. of Sciences and Humanities, Jerusalem (1960) (Pergamon), 299–307.

[26] On the stability of difference approximations to solutions of hyperbolic equations with variable coefficients, *Comm. Pure Appl. Math.* **14**, 497–520.

1962

[27] (with R.S. Phillips) The wave equation in exterior domains, *Bull. Amer. Math. Soc.* **68**, 47–79.

[28] A procedure for obtaining upper bounds for the eigenvalues of a Hermitian symmetric operator, *Studies in Mathematical Analysis and Related Topics* (Stanford Univ. Press), 199–201.

1963

[29] On the regularity of spectral densities, *Teoriia Veroiatnosteii i ee Prim.* **8**, 337–340.

[30] An inequality for functions of exponential type, *Comm. Pure Appl. Math.* **16**, 241–246.

[31] (With C.S. Morawetz and R. Phillips) Exponential decay of solutions of the wave equation in the exterior of a star-shaped obstacle, *Comm. Pure Appl. Math.* **16**, 477–486.

[32] Survey of stability of difference schemes for solving initial value problems for hyperbolic equations, *Symp. Appl. Math.* **15**, 251–258.

1964

[33] Development of singularities of solutions of nonlinear hyperbolic partial differential equations, *J. Math. Phys.* **5**, 611–613.

[34] (with B. Wendroff) Difference schemes for hyperbolic equations with high order of accuracy, *Comm. Pure Appl. Math.* **17**, 381–398.

[35] (with R.S. Phillips) Scattering theory, *Amer. Math. Soc. Bull.* **70**, 130–142.

1965

[36] (with J.F. Adams and R.S. Phillips) On matrices whose real linear combinations are nonsingular, *Proc. AMS* **16**, 318–322; correction, *ibid.* **17** (1966) 945–947.

[37] Numerical solution of partial differential equations, *Amer. Math. Monthly* **72** II, 74–84.

[38] (with K.O. Friedrichs) Boundary value problems for first order operators, *Comm. Pure Appl. Math.* **18**, 355–388.

1966

[39] (with R.S. Phillips) Analytic properties of the Schrödinger scattering matrix, in *Perturbation Theory and Its Application in Quantum Mechanics*, Proc., Madison, 1965 (John Wiley & Sons), 243–253.

[40] Scattering theory; remarks on the energy theory, and on scattering theory for a geometrical optics model, in *Proc. Conf. on Dispersion Theory*, Cambridge, MA, 1966, M 11, pp. 36–39, 40–42.

[41] (with L. Nirenberg) On stability for difference schemes; a sharp form of Garding's inequality, *Comm. Pure Appl. Math.* **19**, 473–492.

[42] (with J.P. Auffray) Aspects mathématiques de la mécanique de phase, *Acad. Sci. Paris, Compt. Rend.* (B) **263**, 1355–1357.

1967

[43] (with J. Glimm) Decay of solutions of systems of hyperbolic conservation laws, *Bull. AMS* **73**, 105.

[44] Hyperbolic difference equations: a review of the Courant–Friedrichs–Lewy paper in the light of recent developments, *IBM J. Res. Develop.* II(2), 235–238.

[45] (with R.S. Phillips) *Scattering Theory* (Academic Press).

[46] (with R.S. Phillips) The acoustic equation with an indefinite energy form and the Schrödinger equation, *J. Func. Anal.* **1**, 37–83.

[47] (with R.S. Phillips) Scattering theory for transport phenomena, in *Functional Analysis*, Proc. of Conf. at Univ. of Calif., Irvine, 1966, ed. Gelbaum (Thompson Book Co.), 119–130.

[48] (with K.O. Friedrichs) On symmetrizable differential operators, Symp. on Pure Math., Singular Integrals, *AMS* **10**, 128–137.

1968

[49] Integrals of nonlinear equations of evolution and solitary waves, *Comm. Pure Appl. Math.* **21**, 467–490.

1969

[50] Nonlinear partial differential equations and computing, *SIAM Rev.* **11**, 7–19.

[51] (with R.S. Phillips) Decaying modes for the wave equation in the exterior of an obstacle, *Comm. Pure Appl. Math.* **22**, 737–787.

[52] Toeplitz operators, in *Lecture Series in Differential Equations 2*, ed. A.K. Aziz (Van Nostrand), 257–282.

1970

[53] (with J. Glimm) Decay of solutions of systems of nonlinear hyperbolic conservation laws, *Memoirs of the AMS* **101**.

[54] (with R.S. Phillips) The Paley–Wiener theorem for the Radon transform, *Comm. Pure Appl. Math.* **23**, 409–424; Errata *ibid.* **24** (1971) 279.

1971

[55] Nonlinear partial differential equations of evolution, *Proc. Int. Conf. of Mathematicians*, Nice, September 1970 (Gauthier-Villars), 831–840.

[56] (with R.S. Phillips) Scattering theory, *Rocky Mountain J. Math.* **1**, 173–223.

[57] (with H. Brezis, W. Rosenkrantz, and B. Singer) On a degenerate elliptic parabolic equation occurring in the theory of probability, *Comm. Pure Appl. Math.* **24**, 410–415; appendix by P.D. Lax.

[58] Shock waves and entropy, *Contributions to Functional Analysis*, ed. E.H. Zarantonello (Academic Press), 603–634.

[59] Approximation of measure preserving transformations, *Comm. Pure Appl. Math.* **24**, 133–135.

[60] (with K.O. Friedrichs) Systems of conservation equations with a convex extension, *Proc. Nat. Acad. Sci.* **68**, 1686–1688.

[61] (with R.S. Phillips) A logarithmic bound on the location of the poles of the scattering matrix, *Arch. Rational Mech. Anal.* **40**, 268–280.

1972

[62] (with R.S. Phillips) On the scattering frequencies of the Laplace operator for exterior domains, *Comm. Pure Appl. Math.* **25**, 85–101.

[63] The formation and decay of shock waves, *Amer. Math. Monthly* **79**, 227–241.

[64] (with R.S. Phillips) Scattering theory for the acoustic equation in an even number space dimensions, *Ind. U. Math. J.* **22**, 101–134.

[65] Exponential modes of the linear Boltzmann equation, in *The Boltzmann Equation*, ed. F.A. Grunbaum, NYU, CIMS, 111–123.

1973

[66] Hyperbolic systems of conservation laws and the mathematical theory of shock waves, Conf. Board of the Mathematical Sciences, Regional Conf. Series in Appl. Math. (*SIAM*), 11.

[67] The differentiability of Pólya's function, *Adv. Math.* **10**, 456–465.

[68] (with R.S. Phillips) Scattering theory for dissipative hyperbolic systems, *J. Func. Anal.* **14**, 172–235.

1974

[69] Invariant functionals of nonlinear equations of evolution, in *Nonlinear Waves*, eds. S. Leibovich and R. Seebass (Cornell Univ. Press), 291–310.

[70] Applied mathematics and computing, *Symp. on Appl. Math.* (AMS) **20**, 57–66.

[71] Periodic solutions of the KdV equations, in *Nonlinear Wave Motion*, ed. A.C. Newell, Lectures in Appl. Math. **15**, 85–96.

1975

[72] Periodic solutions of the KdV equation, *Comm. Pure Appl. Math.* **28**, 141–188.

[73] Almost periodic behavior of nonlinear waves, *Adv. Math.* **16**, 368–379.

1976

[74] Almost periodic solutions of the KdV equation, *SIAM Rev.* **18**, 351–375.

[75] (with A. Harten, J.M. Hyman, and B. Keyfitz) On finite-difference approximations and entropy conditions for shocks, *Comm. Pure Appl. Math.* **29**, 297–322.

[76] On the factorization of matrix-valued functions, *Comm. Pure Appl. Math.* **29**, 683–688.

[77] (with R.S. Phillips) *Scattering Theory for Automorphic Functions*, Ann. Math. Studies 87 (Princeton Univ. Press and Univ. of Tokyo Press).

[78] (with S. Burstein and A. Lax) *Calculus with Applications and Computing*, Undergrad. Texts in Math. 1 (Springer).

1977

[79] The bomb, Sputnik, computers and European mathematicians, the Bicentenial Tribute to American Mathematics, San Antonio, 1976 (Math. Assoc. of America), 129–135.

[80] (with R.S. Phillips) The scattering of sound waves by an obstacle, *Comm. Pure Appl. Math.* **30**, 195–233.

1978

[81] (with M.S. Mock) The computation of discontinuous solutions of linear hyperbolic equations, *Comm. Pure Appl. Math.* **31**, 423–430.

[82] (with R.S. Phillips) An example of Huygens' principle, *Comm. Pure Appl. Math.* **31**, 415–421.

[83] (with R.S. Phillips) The time delay operator and a related trace formula, *Topics in Functional Analysis*, Adv. Math. Suppl. Studies, 3, eds. I.T. Gohberg and M. Kac (Academic Press), 197–215.

[84] Accuracy and resolution in the computation of solutions of linear and nonlinear equations, *Recent Adv. in Numer. Anal.*, Proc. of Symp., Madison (Academic Press), 107–117.

[85] (with A. Lax) On sums of squares, *Linear Algebra and Its Appl.* **20**, 71–75.

[86] (with R.S. Phillips) Scattering theory for domains with nonsmooth boundaries, *Arch. Rational Mech. Anal.* **68**, 93–98.

[87] Chemical kinetics, in *Lectures on Combustion Theory*, eds. S. Burstein, P.D. Lax, and G.A. Sod, NYU, COO-3077-153, 122–136.

1979

[88] (with R.S. Phillips) Translation representations for the solution of the non-Euclidean wave equation, *Comm. Pure Appl. Math.* **32**, 617–667.

[89] (with C.D. Levermore) The zero dispersion limit for the Korteweg–de Vries equation, *Proc. Natl. Acad. Sci.* **76**, 3602–3606.

[90] Recent methods for computing discontinuous solutions – a review, in *Computing Methods in Applied Sciences and Engineering*, 1977, II; 3rd Int. Symp. IRIA, eds. R. Glowinski and D.L. Lions, Springer Lecture Notes in Physics, **91**, 3–12.

1980

[91] (with R.S. Phillips) Scattering theory for automorphic functions (AMS), *Bull.* **2**, 161–195.

1981

[92] On the notion of hyperbolicity, *Comm. Pure Appl. Math.* **33**, 395–397.

[93] (with A. Harten) A random choice finite difference scheme for hyperbolic conservation laws, *SIAM J. Numer. Anal.* **18**, 289–315.

[94] (with R.S. Phillips) The translation representation theorem, *Integral Equations and Operator Theory* **4**, 416–421.

[95] (with R.S. Phillips) Translation representations for the solution of the non-Euclidean wave equation, II, *Comm. Pure Appl. Math.* **34**, 347–358.

[96] Applied mathematics 1945 to 1975, *Amer. Math. Heritage, Algebra & Appl. Math.*, 95–100.

[97] *Mathematical Analysis and Applications*, Part B, essays dedicated to Laurent Schwartz, ed. L. Nachbin (Academic Press) Advances in Math. Suppl. Studies, 7B, 483–487.

1982

[98] (with R.S. Phillips) The asymptotic distribution of lattice points in Euclidean and non-Euclidean spaces, *J. Func. Anal.* **46**, 280–350.

[99] The multiplicity of eigenvalues, *AMS Bull.*, 213–214.

[100] (with R.S. Phillips) A local Paley–Wiener theorem or the Radon transform of *L*2 functions in a non-Euclidean setting, *Comm. Pure Appl. Math.* **35**, 531–554.

1983

[101] Problems solved and unsolved concerning linear and nonlinear partial differential equations, *Proc. Int. Cong. Mathematicians*, 1 (North-Holland), 119–137.

[102] (with A. Harten and B. van Leer) On upstream differencing and Godunovo-type schemes for hyperbolic conservation laws, *SIAM Rev.* **25**, 35–61.

[103] (with C.D. Levermore) The small dispersion limit of the Korteweg–de Vries equation, *Comm. Pure Appl. Math.* **36**, I 253–290, II 571–593, III 809–930.

1984

[104] On a class of high resolution total-variation-stable finite difference schemes, Ami Harten with Appendix by Peter D. Lax, *SIAM* **21**, 1–23.

[105] (with R.S. Phillips) Translation representations for automorphic solutions of the wave equation in non-Euclidean spaces, I, II, III. *Comm. Pure Appl. Math.* **37**, I 303–328, II 779–813, III **38** (1985), 179–207.

[106] *Shock Waves, Increase of Entropy and Loss of Information, Seminar on Non-linear Partial Differential Equations*, ed. S.S. Chern (Math. Sci. Res. Inst. Publ.), 129–171.

1985

[107] *Large Scale Computing in Science, Engineering and Mathematics*, Rome.

[108] (with R.J. Leveque and C.S. Peskin) Solution of a two-dimensional cochlea model using transform techniques, *SIAM J. Appl. Math.* **45**, 450–464.

[109] (with R. Phillips) Translation representations for automorphic solutions of the wave equation in non-Euclidean spaces; the case of finite volume, *Trans. AMS* **289**, 715–735.

[110] (with P. Constantin and A. Majda) A simple one-dimensional model for the three-dimensional vorticity equation, *Comm. Pure Appl. Math.* **38**, 715–724.

1986

[111] On dispersive difference schemes, 1985, Kruskal Symposium, *Physica* **18D**, 250–255.

[112] Mathematics and computing, *J. Stat. Phys.* **43**, 749–756.

[113] (with A. Jameson) Conditions for the construction of multipoint total variation diminishing difference schemes, *Appl. Numer. Math.* **2**, 335–345.

[114] Mathematics and its applications, *Math. Intelligencer* **8**, 14–17.

[115] Hyperbolic systems of conservation laws in several space variables, *Current Topics in Partial Differential Equations*, papers dedicated to Segeru Mizohata (Tokyo Press), 327–341.

1987

[116] *The Soul of Mathematics*, Studies in Mathematics and Its Applications 16, Patterns and Waves (North-Holland).

1988

[117] Oscillatory solutions of partial differential and difference equations, Mathematics Applied to Science (Academic Press), 155–170.

[118] (with R.J. Leveque and C. Peskin) Solution of a two-dimensional cochlea model with fluid viscosity, *SIAM J. Appl. Math.* **48**, 191–213.

[119] The flowering of applied mathematics in America, AMS Centennial Celebration Proc., 455-466; *SIAM Rev.* **31**, 65–75.

[120] (with J. Goodman) On dispersive difference schemes I, *Comm. Pure Appl. Math.* **41**, 591–613.

1989

[121] Science and computing, *Proc. IEEE* **77**.

[122] Writing mathematics well, Leonard Gillman review, *Amer. Math. Monthly* **96**, 380–381.

[123] From cardinals to chaos: Reflections on the life and legacy of Stanislaw Ulam, reviewed in *Phys. Today* **42**, 69–72; *Bull. AMS* **22** (1990) 304–310, *St. Petersburg Math. J.* **4** (1993) 629–632.

[124] Deterministic turbulence, *Symmetry in Nature*, Volume in Honor of Luigi A. Radicati di Brozolo II (Scuola Normale Superiore), 485–490.

1990

[125] Remembering John von Neumann, *Proc. Symp. Pure Math.* 50.

[126] The ergodic character of sequences of pedal triangles, *Amer. Math. Monthly* **97**, 377–381.

1991

[127] Deterministic analogues of turbulence, *Comm. Pure Appl. Math.* **44**, 1047–1055.

[128] (with T. Hou) Dispersive approximations in fluid dynamics, *Comm. Pure Appl. Math.* **44**.

1992

[129] (with R.S. Phillips) Translation representation for automorphic solutions of the wave equation in non-Euclidean spaces. IV. *Comm. Pure Appl. Math.* **45** (1992), no. 2, 179–201.

1993

[130] (with C.D. Levermore and S. Venakides) The generation and propagation of oscillations in dispersive IVP's and their limiting behavior, in *Important Developments in Soliton Theory 1980–1990*, eds. T. Fokas and V.E. Zakharov (Springer-Verlag).

[131] The existence of eigenvalues of integral operators, in Honor of C. Foias, ed. R. Temam, *Ind. Univ. Math. J.* **42**, 889–991.

1994

[132] Trace formulas for the Schrödinger operator, *Comm. Pure Appl. Math.* **47**, 503–512.

[133] Cornelius Lanczos and the Hungarian phenomenon in science and mathematics, *Proc. Lanczos Centennary Conf.* (N.C. State University Press).

1995

[134] Computational fluid dynamics at the Courant Institute 1-5, *Computational Fluid Dynamics Review*, eds. M. Hafez and K. Oshima (John Wiley & Sons).

[135] A short path to the shortest path, *Amer. Math. Monthly* **102**, 158–159.

1996

[136] Outline of a theory of the KdV equation, Lecture Notes in Mathematics, *Recent Mathematical Methods in Nonlinear Wave Propagation* (Springer) **1640**, 70–102.

[137] (with Xu-Dong Liu) Positive schemes for solving multidimensional hyperbolic conservation laws, *Comp. Fluid Dynamics J.* **5**, 133–156.

[138] *The Old Days: A Century of Mathematical Meetings* (AMS), 281–283.

[139] (with A. Harten, C.D. Levermore, and W.J. Morokoff) Convex entropies and hyperbolicity for general Euler equations, *SIAM J. Numer. Anal.*

1997

[140] *Linear Algebra*, Pure and Applied Math. Series (Wiley-Interscience).

1998

[141] (with Xu-Dong Liu) Solution of the two-dimensional Riemann problem of gas dynamics by positive schemes, *SIAM J. Sci. Comput.* **19**, 319–340.

[142] Jean Leray and Partial Differential Equations, Introduction to Volume II, *Selected Papers of Jean Leray* (Springer-Verlag).

[143] On the discriminant of real symmetric matrices, *Comm. Pure Appl. Math.* **51**, 1387–1396.

[144] The beginning of applied mathematics after the Second World War, *Quart. Appl. Math.* **56**, 607–615.

1999

[145] A mathematician who lived for mathematics, Book Review, *Review, Phys. Today*, 69–70.

[146] The mathematical heritage of Otto Toeplitz, in *Otto Toeplitz*, Bonner Mathematische Schriften, 319, 85–100.

[147] Mathematics and computing, in *Useful Knowledge*, ed. A.G. Bearn (Amer. Philos. Soc.), 23–44.

[148] Change of variables in multiple integrals, *Amer. Math. Monthly* **105**, 497–501.

2000

[149] Mathematics and computing, in IMU, Mathematics: Frontiers and Perspectives (AMS), 417–432.

2001

[150] Change of variables in multiple integrals II, *Amer. Math. Monthly* **108**, 115–119.

[151] On the accuracy of Glimm's scheme, *Math. Appl. Anal.*, **7**, 473–478.

[152] The Radon transform and translation representation, *J. Evol. Equ.* **1**, 311–323.

2002

[153] *Functional Analysis*, Pure and Applied Mathematics Series (Wiley-Interscience).

[154] Jaques-Louis Lions, International Scientist (SMAI Journal MATAPLI), to appear.

[155] Richard Courant, *National Academy of Sciences*, Biographical Memoirs, **82**.

[156] Jürgen Moser, *Ergod. Th. & Dynam. Sys.* **22**, 1337–1342.

2003

[157] *John von Neumann: The Early Years, the Years at Los Alamos and the Road to Computing*, to appear.

[158] (with G. Francsics) A fundamental domain for the Picard modular group in C^2, *ESI* **1273**, 1–18.

PART I

Partial Differential Equations

ON THE EXISTENCE OF GREEN'S FUNCTION

PETER D. LAX

In this note we shall present a very short proof of the existence of Green's function for Laplace's equation for any domain with sufficiently smooth boundary in any number of independent variables. The proof is based on the continuous dependence of solutions of Laplace's equation on their boundary values. It is a modification of a proof given by Paul Garabedian, see [1]; the difference between the two approaches is that whereas Garabedian operates with a representation of harmonic functions in terms of their boundary data which he obtains by a variational argument, in our argument only the linear and bounded dependence of the solution on the boundary values figures.

1. In this section we shall treat the somewhat simpler two-dimensional case.

We consider a bounded domain D whose boundary C consists of a finite number of smooth curves (i.e., curves with continuous tangents).

B is the Banach space of all continuous functions defined on C, normed by the maximum norm.

B' is the submanifold of those elements of B for which the boundary value problem can be solved.[1]

Received by the editors December 13, 1951.

[1] It is easy to show that B' is closed, but this is not necessary for the argument.

Let P be any arbitrary point of D. Associated with P is a linear functional L_P defined over B':

$$L_P[\phi'] = h(P)$$

where h is the harmonic function whose value on C is ϕ'. L_P is clearly a linear functional, and by the maximum principle it is bounded, its bound being exactly one.

By the Hahn-Banach extension theorem, see [2], L_P can be extended as a bounded linear functional to the whole space B; and we imagine it so extended.[2]

Consider then the following two-parameter family of elements in B:

$$\psi(\xi, \eta) = \psi(Q) = \log |z - Q|, \qquad z \in C;$$

the point $Q = (\xi, \eta)$ may lie anywhere off the boundary C.

Clearly $\psi(\xi, \eta)$ has derivatives of second (in fact of all) order with respect to ξ and η, and is harmonic with respect to Q:

$$\psi_{\xi\xi} + \psi_{\eta\eta} = 0.$$

We construct the function

$$k_P(Q) = L_P[\psi(Q)].$$

The operations of differentiating with respect to ξ, η and applying the bounded linear functional L_P commute. Therefore $k_P(Q)$ is a harmonic function, i.e., harmonic in each of the components of the complement of C.

Let Q lie outside of D; the element $\psi(Q)$ lies in B', being the boundary value of the regular harmonic function $\log |z-Q|$. So

$$k_P(Q) = L_P[\psi(Q)] = \log | P - Q | \qquad \text{for } Q \text{ not in } D.$$

This explicit formula shows that $k_P(Q)$ is continuous up to the boundary in the exterior components. Next we shall show that the same is true as Q approaches the boundary from the interior, and in fact we shall show that $k_P(Q)$ is continuous across C. This would mean that $k_P(Q)$ in D is a harmonic function continuous up to the boundary and is equal to $\log |P-Q|$ there, i.e., $k_P(Q)$ is the regular part of Green's function.

As to the continuity of $k_P(Q)$ across C: Let Q be any point in D, Q' its mirror image with respect to the tangent at the nearest boundary point. If Q is near enough to the boundary, Q' will lie outside of D; furthermore on account of the smoothness of C the quotient

[2] It should be pointed out that since B is a separable Banach space, the extension can be accomplished without recourse to transfinite induction.

$$\frac{|z-Q|}{|z-Q'|}$$

tends to one uniformly for all z on C as the distance of Q from the boundary shrinks to zero. Hence

$$\max_{z \text{ on } C} \left| \log \frac{|z-Q|}{|z-Q'|} \right| = \|\psi(Q) - \psi(Q')\|$$

tends to zero as Q approaches C. Since L_P is a *bounded* linear functional,

$$k_P(Q) - k_P(Q') = L_P[\psi(Q) - \psi(Q')]$$

also tends to zero as Q approaches C.

This completes the proof for the plane; now we pass on to higher dimensions. The scheme of things remains the same; there is, however, some trouble in showing that $k_P(Q)$ is continuous across C for it is no longer true that

$$\psi(Q) - \psi(Q') = \frac{1}{|z-Q|^{n-2}} - \frac{1}{|z-Q'|^{n-2}}$$

tends to zero uniformly for all z on C as Q approaches the boundary. This difficulty will be overcome by using a more refined property of L_P than mere boundedness, namely, its monotonicity. According to the maximum theorem $L_P[\phi']$ is monotone over B', i.e., $L_P[\phi'] \geqq L_P[\phi'']$ if $\phi' \geqq \phi''$.[3] What we shall show is that L_P can be extended to B without losing this property of monotonicity. The proof of this relies on the full statement of the Hahn-Banach theorem:

Let $L[\phi]$ be a bounded linear functional defined on the submanifold B', $N(\phi)$ a subadditive, positive homogeneous functional defined over the whole space B which exceeds L over B':

$$L[\phi] \leqq N(\phi) \qquad \text{over } B';$$

Then there exists an extension of L to the full space which stays below N:

(1) $$L[\phi] \leqq N[\phi] \qquad \text{over } B.$$

Our choice of $N[\phi]$ is as follows:

(2) $$N[\phi] = \text{lub}\, L_P[\phi''], \qquad \phi'' \in B', \qquad \phi'' \leqq \phi.$$

---

[3] The symbol $\phi' \geqq \phi''$ is defined to mean that $\phi' - \phi''$ is a non-negative function on C.

$N[\phi]$, ~~being the lub of linear functionals~~, is subadditive. It is clearly positive-homogeneous.

Since L_P is monotone over B', $L_P = N$ over B'. So, according to the Hahn-Banach theorem, $L_P[\phi]$ can be extended to the full space, linearly, so that it never exceeds $N[\phi]$.

To verify that L_P thus extended is monotone it is sufficient to show that $L_P[\phi]$ is nonpositive if ϕ is nonpositive on C. Now $N[\phi]$ itself is monotone, as is easily seen from its definition (2); for nonpositive ϕ, the handy sequence of inequalities

$$L_P[\phi] \leqq N[\phi] \leqq N[0] = 0$$

holds and constitutes a proof of the result desired.

We define as before

$$\psi(Q) = \frac{1}{|z - Q|^{n-2}}$$

and construct $k_P(Q) = L_P[\psi(Q)]$. $k_P(Q)$ is harmonic in D; what remains to be shown is that it is equal to $|Q-P|^{-n+2}$ on the boundary of D.

We have to impose a further restriction on the boundary: that through each boundary point we can pass two spheres of radius d, one lying entirely within D, the other outside of D, with the same constant d for all boundary points.[4]

Let Q be some point of D, R the nearest boundary point, and K' and K'' the interior and exterior tangent spheres through R. Let Q' and Q'' denote the inverted image of Q with respect to K' and K''. If Q is close enough to the boundary, Q' and Q'' are in the exterior.

Inverse points have the well known property that the quotient $|z-Q|/|z-Q'|$ has some constant value s' for all points z on the circumference of the circle K'. The value of s' approaches unity as Q tends to the boundary since the radius of K' does not change. Furthermore, if z is any point outside of K', $|z-Q|/|z-Q'|$ is greater than s'. This is in particular the case for z lying on C:

$$s' \leqq \frac{|z - Q|}{|z - Q'|} \qquad \text{for } z \text{ on } C$$

which implies that

$$\frac{s'^{n-2}}{|z - Q|^{n-2}} \leqq \frac{1}{|z - Q'|^{n-2}} \qquad \text{for } z \text{ on } C.$$

---

[4] This kind of restriction on domains is familiar in potential theory.

We rewrite the above as

$$\psi(Q) \leqq \frac{1}{s'^{n-2}}\psi(Q'). \tag{3}$$

The monotone character of L_P permits us to apply L_P to both sides of (3) and retain the inequality:

$$L_P[\psi(Q)] \leqq \frac{1}{s'^{n-2}} L_P[\psi(Q')].$$

By definition of the function k_P this can be written as

$$k_P(Q) \leqq \frac{1}{s'^{n-2}} k_P(Q').$$

As in the two-dimensional case $k_P(Q')$ can be evaluated explicitly since Q' is not in D:

$$k_P(Q') = \frac{1}{|P - Q'|^{n-2}}\cdot$$

So our inequality becomes

$$k_P(Q) \leqq \frac{1}{s'^{n-2}} \frac{1}{|P - Q'|^{n-2}}.$$

The reverse inequality holds for Q'':

$$\frac{1}{s''^{n-2}} \frac{1}{|P - Q''|^{n-2}} \leqq k_P(Q).$$

As Q tends to some boundary point R, so does Q' and Q'', and s' as well as s'' tend to one. Thus the upper and lower bounds for $k_P(Q)$ both tend to $1/|P-R|^{n-2}$, and $k_P(Q)$ itself cannot do otherwise.

I should like to mention that C. Miranda [3] has made use of the extension theorem for this problem in a different manner; his reasoning is slightly more delicate but then it gives more: the solvability of the boundary value problem for all continuous boundary data, not just the existence of Green's function. A similar reasoning plays a role in his treatment [4] of the first boundary value problem for the biharmonic equation; in this case of course the bounded dependence of the solution on the boundary data lies much deeper.

G. Fichera has made use of an equivalent of the Hahn-Banach theorem in his investigations on boundary value problems, see [5] and [6], and so has J. Deny in [7]. M. Brelot, in [8], utilized the possibility of monotonic extension of monotonic linear transformations to prove that the generalized solution, in Wiener's sense, of the first boundary value problem for Laplace's equation can be characterized as the least (resp. greatest) harmonic majorant (resp. minorant) of harmonic functions which are continuous up to the boundary and whose boundary values are less (resp. greater) than the prescribed data.

Bibliography

1. P. R. Garabedian, *The classes L_p and conformal mapping*, Trans. Amer. Math. Soc. vol. 69 (1950) pp. 392–415.

2. S. Banach, *Théorie des opérations linéaires*, Warsaw, 1932.

3. C. Miranda, *Sul principio di Dirichlet per le funzioni armoniche*, Atti della Accademia Nazionale dei Lincei. Rendiconti. Classe di Scienze Fisiche, Matematiche e Naturali (8) vol. 3 (1947) pp. 55–59.

4. ———, *Formule di maggiorazione e teorema di esistenza per le funzioni biharmoniche di due variabli*, Giornale di Matematiche di Battaglini (4) vol. 2 (1948) pp. 97-118.

5. G. Fichera, *Teoremi di completezza sulla frontiera di una dominio per taluni sistemi di funzioni*, Atti della Accademia Nazionale dei Lincei. Rendiconti. Classe di Scienze Fisiche, Matematiche e Naturali (8) vol. 3 (1947) pp. 502–507.

6. ———, *Teoremi di competezza sulla frontiera di un dominio per taluni sistemi di funzioni*, Annali di Matematica Pura ed Applicata (4) vol. 27, pp. 1–28.

7. J. Deny, *Sur l'approximation des fonctions harmoniques*, Bull. Soc. Math. France vol. 73 (1945) pp. 71–73.

8. M. Brelot, *Remarque sur le prolongement fonctionnel linéaire et le problème de Dirichlet.*

New York University

IX. PARABOLIC EQUATIONS

P. D. Lax and A. N. Milgram

§1. INTRODUCTION

This paper is about the initial-boundary value problem for the parabolic equation.

$$u_t = - Lu \tag{1.1}$$

where L is a $2p^{th}$ order elliptic differential operator

$$Lu = (-1)^p \sum_{\nu=0}^{2p} \sum_{i_1,\ldots,i_\nu=1}^{n} a_{i_1 \ldots i_\nu} \frac{\partial^\nu}{\partial x_{i_1} \ldots \partial x_{i_\nu}} u \tag{1.2}$$

The coefficients a may depend on the space variables but not on time; they are supposed to have continuous derivatives of at least p^{th} order

The value of u at $t = 0$ is prescribed on a bounded domain G of the Euclidean space E^n; on the boundary of G the function u and all its derivatives up to order $p - 1$ are prescribed to be zero.

We shall prove in this note that this problem has a unique solution. Our proof is an application of a theorem of Hille (see [11] and [13] and Yosida (see [23]) on unbounded operators which are infinitesimal generators of semigroups (i.e., for which the exponential function e^{At} can be defined). This theorem has been applied by Hille and Yosida to the case where L is a second order elliptic operator or a system of such (see [12], [24], [25]). The application presented here for the higher order case is made possible by the recently developed theory of higher order elliptic operators (see Gårding [10] and Browder [1] and [2]) in particular by Gårding's lemma, which asserts that such operators are bounded from below for functions satisfying the first boundary condition.

The operator (1.2), as written there, applies only to $2p$ times differentiable functions (which are required to satisfy the boundary condition); it has to be extended before the Hille-Yosida theorem can be applied to it. This is accomplished here by generalizing the Friedrichs extension of symmetric half-bounded operators to the non-symmetric case. The Friedrichs extension (see [4], [5] and [19]), we recall, assigns a unique self-adjoint extension to every halfbounded symmetric operator by means of the quadratic form induced by the operator; it has been used by Friedrichs to discuss formally self-adjoint second order elliptic operators. Both Browder and Gårding make use of it in treating the first boundary value problem for formally self-adjoint higher order elliptic operators.

It is easy to verify that the extended operator $\tilde{L}$ satisfies the hypotheses of the Hille-Yosida theorem, but all we can conclude is that the generalized equation $u_t = -\tilde{L}u$ has a unique solution with prescribed initial value. In section 5 we show however that these generalized solutions are genuine ones, at least if the coefficients of L are sufficiently differentiable. In the case of constant coefficients we show this by use of the fundamental solution of the parabolic equation (1.1), whose properties were investigated by Ladyzhenskaya (see [17]) and P. C. Rosenbloom (see [20]). For variable coefficients the differentiability properties can be deduced from the differentiability properties of solutions of elliptic equations; such theorems were obtained recently by Gårding [10], who refers to L. Schwartz, also by Browder, who uses the fundamental singularity (see [1]), by F. John who uses the method of spherical means (see [14], [15] and [16]), and by K. O. Friedrichs who employs estimates for the L_2 norms of higher derivatives and the mollifier (see [8]). In this paper we employ Friedrichs' version of these differentiability theorems.

The extension of unsymmetric operators is presented in section 2; some further properties of this extension (which are not needed in the rest of this paper) are discussed in section 4. The Hille-Yosida theorem is described in section 3, and differentiability properties of the solution in section 5.

The initial-boundary value problem for the parabolic equation (1.1) has also been treated by F. Browder by means of eigenfunction expansions (see [3]).

Our thanks are due to the Office of Naval Research and the Office of Ordnance Research for their support.

§2. POSITIVE BILINEAR FUNCTIONALS IN HILBERT SPACE AND ASSOCIATED OPERATORS

In this section we describe the Friedrichs extension of a nonsymmetric positive definite operator. The extension is based on the theory of linear transformations induced by bilinear forms.

The following theorem is a mild generalization of the Fréchet-Riesz Theorem on the representation of bounded linear functionals in Hilbert space.

THEOREM 2.1. Let H be a real Hilbert space, and $B(x,y)$ a (not necessarily symmetric) bilinear functional which is

a) bounded, i.e.,

$$|B(x,y)| \leqq C' \|x\| \cdot \|y\|$$

and

b) positive definite in the sense that there exists a positive constant C such that

$$C \|x\|^2 \leqq B(x,x)$$

for all x in H.

If a fixed element is substituted for either argument of $B(x,y)$, B becomes a bounded linear functional of the other. We claim that all linear functionals can be obtained in this way, i.e., to each bounded linear functional l defined over H there corresponds two unique elements x_l, x_l^* in H such that

$$l(x) \equiv B(x_l,x) \equiv B(x,x_l^*)$$

for all x in H.

PROOF. Let V be the subset of H consisting of those elements y to which there corresponds an element z in H for which

$$B(z,x) \equiv (y,x)$$

for all x in H.

We note that z is unique. For suppose that z and $\overline{z}$ both satisfy this condition; then $B(z,x) = B(\overline{z},x)$ for all x in H, that is $B(z-\overline{z},x) = 0$ for all x, in particular for $x = z-\overline{z}$. Because of the positiveness of B we can conclude that $\|z-\overline{z}\| = 0$, i.e., $z = \overline{z}$.

This reasoning shows us at the same time that the dependence of z on y is <u>bounded</u> with bound $1/C$. For

$$C\|z\|^2 \leqq B(z, z) = (y, z) \leqq \|y\|\|z\|$$

Our aim is to show that all elements of H belong to V. Clearly, V is a linear subspace; furthermore from the bounded dependence of z on

y and the continuity of the bilinear functional B it follows that the linear subspace V is <u>closed</u>. If V were not equal to all of H there would exist an element $z^o \neq 0$ orthogonal to all of V. Consider the linear functional $B(z^o,x)$. Since this is a bounded linear functional, by the Fréchet-Riesz Theorem there exists y such that $B(z^o,x) \equiv (y,x)$. Hence y lies in V. But taking x to be z^o in this relation yields $B(z^o,z^o) = (y,z^o)$, and this is zero, because of the orthogonality of z_o to V; it follows that V is all of H.

The above reasoning shows that every ordinary scalar product can be represented as a scalar product with respect to B; since according to the Riesz-Fréchet Theorem all bounded linear functionals can be so represented this proves Theorem 2.1.

COROLLARY 2.1. If U is a proper closed linear submanifold in H, there exists x_U and x_U^* in H such that

$$B(x_U,x) \equiv B(x,x_U^*) = 0$$

for all x in U.

PROOF. Let $\overline{x}_U$ be orthogonal to U. By Theorem 2.1, there exist x_U, x_U^* such that

$$B(x_U,x) \equiv B(x,x_U^*) \equiv (\overline{x}_U,x)$$

In the applications to follow, this representation Theorem 2.1 will be applied in the case where H is a subset of another Hilbert space H_o.

We denote by (x,y), $(x,y)_o$ and $\|x\|$, $\|x\|_o$ the respective inner products and norms in H and H_o. In addition we suppose that

(2.1) $\quad H$ is a dense subset of H_o

(2.2) $\quad$ there exists a constant k such that

$$\|x\| \geq k \|x\|_o$$

for all x in H.

We shall suppose as before that we are given a bilinear functional $B(x,y)$ defined on H, which, in terms of the metric in H, is bounded and positive. We shall show that this bilinear form induces two closed transformations, which are adjoints of each other:

THEOREM 2.2. There exist two linear transformations E and E^* with domains D and D^* dense in H which satisfy the relations

$$B(x,y) = (Ex,y)_o \tag{2.3}$$

for all x in D and y in H and

$$B(x,y) = (x,E^*y)_o \tag{2.4}$$

for all x in D^* and y in H.

The range of E and E^* is H_0 and E and E^* have single valued inverses S and S^* which are bounded as mappings of H_0 into H:

$$\|Sz\| \leqq k_1 \|z\|_o \tag{2.5}$$

they therefore are, a fortiore, bounded as mappings of H_0 into H_0:

$$\|Sz\|_0 \leqq k_2 \|z\|_0$$

PROOF. Each element z in H_0 induces a linear functional $l(x) = (z,x)_0$ in H. By assumption (2.2) about the relation of the norms in H and H_0, it follows that $(z,x)_0$ is a bounded linear functional in H. By Theorem 2.1 there exists a unique element y in H such that $B(y,x) = (z,x)_0$ for all x in H. Denote the dependence of y on z by $y = Sz$. Clearly Sz is linear and sends all of H_0 into a subset D of H. Suppose D were not dense in H. Then $\overline{D}$ is a closed linear manifold in H, and by Corollary 2.1 there exists an element x_D^* such that $B(x,x_D^*) = 0$ for all x in $\overline{D}$. But Sx_D^* lies in D, and so

$$0 = B(Sx_D^*,x_D^*) = (x_D^*,x_D^*)_0 = \|x_D^*\|_0^2$$

Hence $x_D^* = 0$ and $\overline{D} = H$.

Next we show that S has a single valued inverse. For let z and v be two elements of H_0 for which $Sz = Sv$; it follows by the definition of S that $(z,x)_0 = (v,x)_0$ for all x in H. Thus $z - v$ is orthogonal in the sense of scalar product in H_0 to a dense subset, from which we conclude $z = v$. This proves that S has a single valued inverse; call this inverse E. Clearly it satisfies (2.3).

The boundedness of S follows from the positive definite character of B, the relation of the norms in H and H_0 and the Schwarz inequality:

$$C \|Sz\|^2 \leq B(Sz,Sz) = (z,Sz)_o \leq \|z\|_o \|E^{-1}z\|_o$$
$$\leq \|z\|_o 1/k \|Sz\|$$

This proves the boundedness of S as mapping of H_o into H with bound $(kC)^{-1}$.

Our next theorem is concerned with possible modifications of B which do not disturb its boundedness, positiveness, or even the domain of its associated operator.

THEOREM 2.3. Let N be a linear operator with domain H and range a subset of H_o. Suppose also that

a) $|(Nx,y)_o| < M \|x\| \|y\|$ for x,y in H, and that

b) the bilinear functional $B_N(x,y) = B(x,y) + (Nx,y)_o$ is positive definite over H.

Then the operators E and E_N associated with the bilinear forms B and B_N have the same domain, and $E_N = E + N$.

PROOF. $B_N(x,y)$ is clearly bounded. Hence, B_N determines an operator E_N and domain D_N in H whose inverse S_N satisfies $B_N(S_N z,x) \equiv (z,x)_o$ for all x in H and z in H_o. Therefore, for all x in H,

$$B(S_N z,x) + (NS_N z,x)_o \equiv (z,x)_o$$

or

$$B(S_N z,x) = (z-NS_N z,x)_o$$

By definition of S as inverse of E, this asserts that

$$S_N z = S(z-NS_N z) \tag{2.7}$$

This shows that $S_N z$ lies in D, i.e., D_N is contained in D. Denote $S_N z$ in (2.7) by w; apply the operator E to both sides:

$$Ew = E_N w - Nw \tag{2.8}$$

(2.7) holds for all z in H_o, and thus (2.8) holds for all w in D_N.

Reversing the role of B and B_N we conclude that $D \subset D_N$; consequently $D = D_N$.

Let L be an unbounded operator in H_o, with domain D_o dense in H_o. We shall call L positive definite if it satisfies these two conditions:

$$(2.9) \qquad (Lx,x)_o \geqq k(x,x)_o$$

$$(2.10) \qquad (Lx,y)_o \leqq C'(Lx,x)_o^{1/2} (Ly,y)_o^{1/2}$$

The term "positive definite" is slightly misleading since it refers only to condition (2.9); "positive definite and not too unsymmetric" would be more appropriate but it is too long a phrase. We shall give a procedure which associates to each positive definite operator L an extension called its Friedrichs extension and denoted by $\tilde{L}$, and which has a bounded inverse $\tilde{L}^{-1}$ whose norm is at most k^{-1}.

Denote $(Lx,x)_o^{1/2}$ by $\|x\|$; this is a Hilbert norm over D_o (the scalar product (x,y) is $1/2(Lx,y)_o + 1/2(x,Ly)_o$). D_o can be completed[1] within H_o to a Hilbert space H under the norm $\|x\|$. $(Lx,y)_o$ is, by assumption (2.10), a bounded bilinear functional over D_o and so can be extended by closure to a bilinear functional $B(x,y)$ over H; $B(x,y)$ satisfies the hypotheses of Theorem (2.2) (with the constant C equal to one) and so there exists a linear transformation E with domain D uniquely characterized by the requirement that $B(x,y) = (Ex,y)_o$ for all y in H and all x in D. For x in D_o, $Ex = Lx$ clearly satisfies this requirement; so E is an extension of L, and we call it[2] $\tilde{L}$. The aforementioned properties of L follow from Theorem 2.2.

---

[1] Denote by $\overline{D}_o$ the closure of D_o under the new norm. Since every Cauchy sequence in the new norm is, by assumption 2.9, a Cauchy sequence in the old norm, there is a natural homomorphism of $\overline{D}_o$ into H_o; Friedrichs has shown, see [5], that the kernel of this homomorphism consists of the zero element, so that $\overline{D}_o$ can be regarded as a subspace H of H_o. If one operates merely with a positive definite bilinear form $B(x,y)$ over D_o, not necessarily one induced by an operator L, the kernel of the homomorphism may contain nonzero elements. Nevertheless this does not interfere with the construction of the induced operator $\tilde{L}$ (and $\tilde{L}^*$).

[2] If we operate with a bilinear form $B(x,y)$ over D_o not induced by an operator and if the mapping T of $\overline{D}_o$ into H_o is not one-to-one, we proceed as follows to construct $\tilde{L}$: $(z,Ty)_o$ is, for any fixed z in H_o, a bounded linear functional of y over $\overline{D}_o$. So by Theorem 2.1 it can be represented as $B(x,y)$ over $\overline{D}_o$; $x = Sz$ is a bounded transformation of H_o into $\overline{D}_o$, so TS is a bounded mapping of H_o into H_o. It is easy to show that this mapping has a single valued inverse, which we call $\tilde{L}$.

REMARK. $(\tilde{L}x,x)$ is equal to $\|x\|^2$.

THEOREM 2.4. Suppose that L_0 is positive definite over D_0, and that N is an operator defined over D_0. Put $L = L_0 + N$ and assume that these conditions are satisfied:

a) $(Lx,x)_0 \geqq c(L_o x,x)_o$ in D_0

b) $\|Nx\|_0 \leqq M \|x\| = M(x,L_0 x)_0$

Then $\tilde{L}$ has the same domain as $\tilde{L}_0$, and is equal to $\tilde{L}_0 + \bar{N}$, where $\bar{N}$ is the extension of N to D by closure.

PROOF. Conditions a) and b) imply that L is positive definite and that the norm induced by it is equivalent to the norm induced by L_0. So the spaces H associated with L and L_0 are the same, and L has a Friedrichs extension $\tilde{L}$. That $\tilde{L}$ is equal to $\tilde{L}_0 + \bar{N}$ follows immediately from Theorem 2.3 after we observe that condition b) enables us to extend N to H by closure.

We have now all the results that are needed for the application of the Hille-Yosida theorem. Theorem 2.4 is needed only in the special case when N is bounded by the H_0 norm, in fact when N is just a constant multiple of the identity.

§3. THE HILLE-YOSIDA THEOREM

Let L be a positive definite operator with lower bound k; if λ is a positive number, $L + \lambda I$ is positive definite with lower bound $k + \lambda$. By the fundamental property of the Friedrichs extension, $\|\widetilde{(L+\lambda I)}^{-1}\|$ is less than $(k+\lambda)^{-1}$; by Theorem 2.4, taking N as λI, $\widetilde{(L+\lambda I)}$ is equal to $\tilde{L} + \lambda I$. These facts can be stated in this form: The transformation $A = -\tilde{L}$ satisfies the conditions

a) The positive real axis belongs to the resolvent set of A.

b) $\|(A-\lambda I)^{-1}\|_0 \leqq (k+\lambda)^{-1}$ for positive λ.

According to the theorem of Hille-Yosida, these are just the conditions for A to be the infinitesimal generator of a strongly continuous, one-parametric semigroup of transformations which reduce to the identity of $t = 0$. More precisely:

If A satisfies conditions a) and b), there exists a one-parameter family of operators $E(t)$,

$0 \leqq t < \infty$ with these properties:

(i) It forms a semigroup, $E(t_1+t_2) = E(t_1) \cdot E(t_2)$.

(ii) It is strongly continuous for $0 \leqq t < \infty$

(iii) $E(0) = I$, $\|E(t)\| \leqq e^{-kt}$

(iv) $\lim_{h \to o} \frac{E(h)-I}{h} x = Ax$ for every x in the domain of A

(v) $E(t)$ commutes with $(\mu - A)^{-1}$ for all positive μ.

Note that (v) implies that $E(t)$ maps the domain of A into itself.

All this can be summarized by saying that $x(t) = E(t)x_o$ is the solution of the equation

$$\frac{dx(t)}{dt} = Ax(t)$$

with initial value x_o (at least if x_o lies in the domain of A). To make this paper complete we give here Yosida's elegant proof:[3]

PROOF. Define the operator A_λ as $\lambda A(\lambda - A)^{-1}$. The decomposition $A_\lambda = -\lambda + \lambda^2(\lambda - A)^{-1}$ shows that A_λ is bounded. We claim that if x belongs to the domain of A, $A_\lambda x$ tends to Ax as $\lambda \longrightarrow \infty$. To see this, write $Ax = y$; $A_\lambda x$ is equal to $\lambda(\lambda - A)^{-1}y$. By condition b) we see that $\|A_\lambda x\|_o \leqq \|y\|_o$. If y belongs to the domain of A, we can write $A_\lambda x = \lambda(\lambda - A)^{-1}y = y + (\lambda - A)^{-1}Ay$; for fixed y, the second term tends to zero by condition b) and so $A_\lambda x \longrightarrow y$ for y in the domain of A and hence, because of the bounded dependence of $A_\lambda x$ on y, for all y.

So far only a fraction of the information in b) was used, namely that $\|(A - \lambda I)^{-1}\|$ is $O(\lambda^{-1})$. For the next step, however, the full strength of b) is needed. We construct $E_\lambda(t) = \exp. tA_\lambda$ and we need to know that the norm of $E_\lambda(t)$ is uniformly bounded.

---

[3] See [23]. If L is symmetric, the one-parameter family of operators $E(t)$ can be constructed with the aid of the functional calculus for the bounded symmetric transformation $\tilde{L}^{-1}$; $E(t)$ is namely $\exp(-t/\tilde{L}^{-1})$. Altogether the problem of solving the initial value problem for the equation $U_t = AU$ can be regarded as part of the problem of developing a functional calculus for the operator A.

$$E_\lambda(t) = \exp.\ tA_\lambda = \exp.\ t\lambda A(\lambda - A)^{-1} =$$

$$= \exp.\ t\left\{-\lambda + \lambda^2(\lambda - A)^{-1}\right\} = \exp.(-t\lambda)\ \exp.\ t\lambda^2(\lambda - A)^{-1}$$

So $\| E_\lambda(t) \|_0 \leqq \exp.(-t\lambda)\ \exp.\ t\lambda^2\ \| (A-\lambda)^{-1} \| \leqq \exp.(-tk\lambda(k+\lambda)^{-1})$. Next we show that $E_\lambda(t)$ converges to some operator $E(t)$ in the strong sense as λ tends to ∞. For

$$(3.1)\quad \begin{aligned} E_\lambda(t) - E_\mu(t) &= \exp.\ tA_\lambda - \exp.\ tA_\mu = \exp.[stA_\lambda + (1-s)tA_\mu]\Big|_{s=0}^{s=1} = \\ &= \int_0^1 \exp.[stA_\lambda + (1-s)tA_\mu] \cdot t[A_\lambda - A_\mu]ds \end{aligned}$$

The first factor under the integral on the right is less than one in norm (this can be shown by the same analysis that led to the inequality $\| E_\lambda \| \leqq 1$). The second factor, $[A_\lambda - A_\mu]$ applied to an element x in the domain of A tends to zero as λ and μ tend to ∞, and so $E_\lambda(t)x - E_\mu(t)x$ tends to zero for such x. Since the norm of $E_\lambda(t) - E_\mu(t)$ is uniformly bounded (at most 2), this proves the convergence of $E_\lambda(t)x$ to some limit for every x. Call this limit $E(t)x$; we claim that it has properties (i) - (v). Properties (i), (iii) follow immediately from the analogous properties of E_λ. Property (iv) follows from the integral equation

$$E(t)x = x + \int_0^t E(t)Ax\ dt$$

valid for all x in the domain of A. This integral relation is an immediate consequence of its analogue

$$E_\lambda(t)x = x + \int_0^t E_\lambda(t)A_\lambda x\ dt$$

Differentiability over the domain of A implies of course continuity over the domain of A and since $E(t)$ is uniformly bounded, this proves the strong continuity of $E(t)$.

It is not hard to show that if A is any operator satisfying the hypotheses a) and b) then there exists only <u>one</u> one-parameter family $E(t)$ satisfying conditions (i)-(v). The uniqueness proof is particularly simple in our case where $-A$ is positive definite, for this implies that for any solution $x(t)$ of

$$\frac{dx(t)}{dt} = Ax(t)$$

the norm of $x(t)$ is a decreasing function of t, and therefore two solutions of this equation with the same initial values are identical. This shows that $E(t)x$ is uniquely determined for x in the domain of A, and thus, $E(t)$ being bounded, for all x.

§4. EFFECTIVE CALCULATION OF THE FRIEDRICHS EXTENSION

In this section we shall give an effective method for computing the Friedrichs extension of a positive definite operator L which has an adjoint over its domain of definition D (and the operators occurring in most applications do have adjoints). The method relies on an extension of Theorem 2.4 to the case where, instead of requiring N to satisfy condition b), we merely require

c) $$|(Nx,y)_0| \leq M \|x\| \cdot \|y\|$$

The positive definiteness of L is still assured and the norm induced by L is equivalent to the norm induced by L_0. So the closure of D under the norm induced either by L or L_0 is the same space H; but the domains of $\tilde{L}$ and $\tilde{L}_0$ need not be the same. Nevertheless it is possible even under this weaker hypothesis to relate $\tilde{L}$ to $\tilde{L}_0$. Purely formally, we can write

$$L = L_0 + N = L_0(I+L_0^{-1}N) = L_0(I+T) \tag{4.1}$$

We shall prove that (4.1) is literally true provided that L and L_0 are replaced by their Friedrichs extensions $\tilde{L}$ and $\tilde{L}_o$:

$$\tilde{L} = \tilde{L}_o(I+T) \tag{4.2}$$

and if T is defined as the completion of the transformation $\tilde{L}_0^{-1}N$ over H. We start first by proving that $\tilde{L}_0^{-1}N$, defined over D_0, is bounded in the norm of H. To see this put $\tilde{L}_0^{-1}Nx = y$, which can be rewritten as $\tilde{L}_0 y = Nx$ and take the scalar product of both sides with y:

$$(y,\tilde{L}_0 y)_0 = (y,Nx)_0$$

The left side is, by definition, equal to $\|y\|^2$; the right side is, by assumption c), less than $M\|x\| \ \|y\|$, and this proves the boundedness of $\tilde{L}_0^{-1}N$, with bound M.

Next we show that (4.2) holds for all elements x in the domain of $\tilde{L}$. Let x be such an element, with $\tilde{L}x = y$, i.e., $B(x,z) = (y,z)_0$

for all z in H. $B(x,z)$ is defined by a limiting process: $B(x,z) = \lim(Lx_i,z)_0$, where x_i is a sequence of elements in D_0 tending toward x in the sense of the norm of H. The following sequence of equations uses the fact that $\tilde{L}_0$ is an extension of L_0 and has an inverse defined over H_0:

$$(Lx_i,z)_0 = (L_0x_i+Nx_i,z)_0 = (L_0x_i,z)_0 + (Nx_i,z) =$$

$$= (\tilde{L}_0x_i,z)_0 + (\tilde{L}_0\tilde{L}_0^{-1}Nx_i,z)_0 = (\tilde{L}_0(I+T)x_i,z)_0 = B_0((I+T)x_i,z)$$

As i tends to infinity, the left term tends to $B(x,z)$ which is equal to $(y,z)_0$; the term on the right tends to $B_0((I+T)x,z)$ because of the boundedness of T and B_0. So $B_0((I+T)x,z) = (y,z)_0$ which, by definition of $\tilde{L}_0$, means that $(I+T)x$ belongs to the domain of $\tilde{L}_0$, and $\tilde{L}_0(I+T)x$ is y; this is precisely (4.2). The same argument can be used to show that every element in the domain of $\tilde{L}_0(I+T)$ belongs to the domain of $\tilde{L}$; this completes the proof of the validity of (4.2).

In a practical application it is the inverse of $\tilde{L}$ rather than $\tilde{L}$ itself that we are interested in computing, since this is the operator which relates the solution of the problem to the given data. Therein lies the usefulness of (4.2): for, if we choose L_0 so that both $\tilde{L}_0^{-1}$ and $(I+T)^{-1}$ can be computed, $\tilde{L}^{-1}$ (being equal to $(I+T)^{-1}\tilde{L}_0^{-1}$) can also be computed. Such a choice of L_0 is possible if L has an adjoint over D_0 (i.e., an operator L^* such that $(Lx,y)_0 = (x,L^*y)_0$ for x,y in D_0), namely the <u>symmetric part</u> of L:

$$L_0 = \frac{L + L^*}{2}$$

L_0 is positive definite since $(x,L_0x)_0 = \frac{1}{2}(x,Lx + L^*x)_0 = \frac{1}{2}(x,L^*x)_0 + \frac{1}{2}(Lx,x)_0 = (Lx,x)_0$ and so if L satisfies (2.9), so does L_0. (2.10) is satisfied by L_0 with constant C equal to one (Schwarz inequality). Denote by N the unsymmetric part of L: $N = 1/2(L-L^*)$. Since the quadratic forms induced by L and L_0 are bounded, the quadratic form induced by N is also bounded:

$$(4.3) \qquad |(Nx,y)_0| \leqq M\|x\| \cdot \|y\|$$

The classical <u>variational</u> principle enables one to compute $\tilde{L}_0^{-1}$ effectively; the same is true of T, since T is the product of $\tilde{L}_0^{-1}$ and N. We shall prove now that $(I+T)^{-1}$ exists[4] and can be approximated

[4] In [1], Browder proves that if L is an elliptic operator, $I+T$ has an inverse, by showing that a) T is completely continuous (this follows from Rellich's lemma) b) $I+T$ annihilates no element of H. By the Riesz theory, a) and b) imply that $I+T$ has an inverse. Our proof does not depend on the complete continuity of T.

uniformly to any degree of accuracy by <u>polynomials</u> in T, and so $(I+T)^{-1}$ too can be computed effectively.

T is an <u>antisymmetric</u> transformation over H; this follows from the definition of scalar product in H, the symmetry of L_o and the antisymmetry of N with respect to the scalar product of H_o. Therefore $S = iT$ is a symmetric bounded transformation over H, (or rather the extension of H obtained by adjoining complex scalars). This shows right away that $I + T = I - iS$ has an inverse; this inverse can be computed with the aid of the operational calculus for bounded symmetric operators. According to this operational calculus if $f(\lambda)$ is any given continuous function over the spectrum of S, and $f_n(\lambda)$ is a sequence of polynomials tending uniformly to $f(\lambda)$ over the spectrum of S, then $f_n(S)$ tends to $f(S)$; in fact the norm of the difference of $f(S)-f_n(S)$ is not greater than the absolute value of the deviation of $f_n(\lambda)$ from $f(\lambda)$ over the spectrum of S.

As we have already shown, the norm of our operator $S = iT$ is at most M, where M is the bound of the bilinear form $(Nx,y)_o$ (in fact, it is easy to show that the norm of S -- and thus its spectral radius -- is equal to M). So to compute $(I+iS)^{-1}$ we have to approximate $f(\lambda) = (1+i\lambda)^{-1}$ uniformly by polynomials $f_n(\lambda)$ over the interval $-M < \lambda < M$. If M is less than one, this can be done by the power series expansion of $(1+i\lambda)^{-1}$ around $\lambda = 0$; for M greater than one some other method must be used.

Since the real part of $(1+i\lambda)$ is an even function of λ and its imaginary part odd, the coefficients of the even powers in $f_n(\lambda)$ will be real, those of the odd powers of λ pure imaginary; so when $f_n(S)$ is written as a polynomial in T, $f_n(S) = f_n(+iT)$, all coefficients will be real.

What we have proved so far is that $\tilde{L}^{-1} = (I+T)^{-1}\tilde{L}_o^{-1}$ can be approximated by an expression of the form $p(T)\tilde{L}_o^{-1}$, where $p(T)$ is a polynomial in T which can be picked so that $\|(I+T)^{-1}-p(T)\|$ is arbitrarily small say $< \epsilon$. (Note that since $\|(I+T)^{-1}\| = 1$, $\|p(T)\| \leqq 1 + \epsilon$.) According to Theorem 2.2 and inequality (2.5), $\tilde{L}_o^{-1}$ as transformation of H_o into H is bounded, its bound being equal to k^{-1}, the reciprocal of the lower bound (2.9) of the operator L_o. So if x is any element of H_o, $p(T)\tilde{L}_o^{-1}x$ differs from $\tilde{L}^{-1}x$ in the sense of the norm of H by at most $\epsilon k^{-1}\|x\|_o$.

For sake of completeness we state the variational principle for positive definite symmetric operators:

Let x be any element of H_o, and let y vary in the domain of $\tilde{L}_o$. Consider the function $V(y) = (\tilde{L}_o y,y)_o - 2(x,y)_o$; it reaches its minimum for $y = \tilde{L}_o^{-1}x$. Denote the value of this minimum by d. If y' is any other element in the domain of $\tilde{L}_o$, the deviation of $V(y')$ from d measures the distance of y' from y in the sense of the norm of H:

$$\| y'-y \|^2 = V(y') - d$$

This shows that a sequence of elements y_1 approaches y if and only if $V(y_1)$ approaches d. This is the basis of the Rayleigh-Ritz procedure which does effectively produce a sequence of elements of D_o tending to y in the sense of the norm of H (note that in order to estimate the distance of y_1 from y, we need a lower bound for d).

Suppose x is any given element of H_o and we wish to evaluate $\tilde{L}^{-1}x$ to some desired degree of accuracy. According to what was said before $\tilde{L}^{-1}x$ differs from $p(T)\tilde{L}_o^{-1}x$ by less than $\epsilon k^{-1} \| x \|_o$. To evaluate $p(T)\tilde{L}_o^{-1}x$, we find by the Rayleigh-Ritz method an approximation to $\tilde{L}_o^{-1}x$, i.e., an element y' of D_o whose distance from $\tilde{L}_o^{-1}x$ in the sense of the H norm is less than η. $p(T)\tilde{L}_o^{-1}x$ differs from $p(T)y'$ by less than $\|p(T)\| \cdot \eta \leq (1+\epsilon)\eta$ To evaluate $p(T)y'$ to any desired degree of accuracy, we have to be able to evaluate terms of the form $T^n y'$ to any desired degree of accuracy; we shall show how to evaluate Ty': Since y' lies in D_o, the original domain of definition of $T = \tilde{L}_o^{-1}N$, we simply form $Ny' = z$ and approximate $\tilde{L}_o^{-1}z$ by the Rayleigh-Ritz method to an arbitrary degree of accuracy by an element of D_o.

We close this section by stating a result on positive definite operators with an adjoint; this result will be used in Section 5:

> If L is a positive definite operator defined over D_o with an adjoint L^* over D_o, then $\tilde{L}$ and $\tilde{L}^*$ are the adjoints of each other.

This result is a corollary of Theorem 2.2. Actually, the only part of it needed in the next section is that $\tilde{L}$ and $\tilde{L}^*$ are adjoint to each other, i.e., $(\tilde{L}x,y) = (x,\tilde{L}^*y)$ for all x in D and all y in D_o. This property of $\tilde{L}$ is sometimes expressed by saying that $\tilde{L}$ is a <u>weak</u> extension of L.

§5. DIFFERENTIABILITY PROPERTIES OF SOLUTIONS

The $2p^{th}$ order operator L of (1.2) is called <u>elliptic</u> if the $2p^{th}$ order form

$$\sum_{i_1,\ldots,i_{2p}=1}^{n} a_{i_1\ldots i_{2p}} \xi_{i_1}\ldots\xi_{i_{2p}} \tag{5.1}$$

does not vanish (say it is positive) for any choice of the variables ξ except $\xi_1 = \ldots \xi_n = 0$. For the domain D_0 of this operator we take all 2p times differentiable functions which vanish outside of a given bounded domain G. Denote by $(u,v)_0$ and $\| u \|_0$ the L_2 scalar product and norm of the function u over G and by $\| u \|_k$ the sum of the L_2 norms of u and its first k derivatives. We define H_0 as the closure of the space of continuous square integrable functions u over G under the norm $\| u \|_0$; D_0 is a dense subspace of H_0. For elliptic operators whose highest order coefficients are constant, L. Gårding has proved by the use of the Fourier transform (see [9]) that if L is augmented by a suitably large multiple of the identity, the resulting operator $L + t_0 I$ is positive definite; specifically that these inequalities hold:

$$(5.2)\qquad \begin{aligned} (Lu+t_0 u,u)_0 &\geqq h \; \| u \|_p^2 \\ (Lu+t_0 u,v)_0 &\leqq C' \; \| u \|_p \; \| v \|_p \end{aligned}$$

In [10], Gårding has announced the same result for elliptic operators with variable coefficients; it can be deduced from the result for the constant coefficient case with the aid of a well-known estimate for $\| u \|_k$, $k < p$, in terms of $\| u \|_0$ and $\| u \|_p$, and a sufficiently fine partition of unity. (See for instance F. Browder, [2] or J. Leray, [18]; Leray treats the more general case of strongly elliptic systems.)

Since, for the purpose of solving the parabolic equation

$$\frac{\partial u}{\partial t} = - Lu$$

the addition of a constant multiple of the identity to L is irrelevant, we may as well assume that already L itself satisfies (5.2):

$$(5.2')\qquad \begin{aligned} (Lu,u)_0 &\geqq h \; \| u \|_p^2 \\ (Lu,v)_0 &\leqq C' \; \| u \|_p \; \| v \|_p \end{aligned}$$

which implies that L satisfies (2.9) and (2.10), i.e., L is a positive definite operator.

According to sections 2 and 3, our elliptic operator L has a Friedrichs extension $\tilde{L}$, and $-\tilde{L} = A$ is the infinitesimal generator of a semigroup $E(t)$, i.e., $E(t)u_0 = u(t)$ satisfies $du(t)/dt = -\tilde{L}u(t)$, $u(0) = u_0$ for every u_0 in the domain of A.

$u(t)$ is a generalized solution of our parabolic equation with initial value u_0. We shall prove now that $u(t)$ is a genuine solution; i.e., that $u(t)$ has 2p continuous derivatives with respect to the space variables and continuous derivatives of all orders with respect to t at every point of the domain D and all time $t > 0$, and that it satisfies the partial differential equation $u_t = - Lu$ there, provided that the coefficients of the operator L are sufficiently differentiable. Further differentiability of the coefficients implies further differentiability of $u(t)$ with respect to the space variables. We shall prove these statements for any initial function in H_0 and not just for elements of the domain of A; furthermore we shall show that if the initial function u_0 is sufficiently differentiable, $u(t)$ approaches u_0 as t tends to zero not only in the L_2 sense but pointwise.

It should be pointed out that from the relation $u_t = - \tilde{L}u$ and differentiability properties of $u(t)$ we can deduce that the partial differential equation $u_t = - Lu$ is satisfied, on the basis of this simple result:

> If u is a function with continuous derivatives up to order $2p$ which belongs to the domain of $\tilde{L}$, then $\tilde{L}u$ is indeed equal to Lu, L being interpreted as the originally given differential operator.

This is not an immediate consequence of the definition of $\tilde{L}$, since $\tilde{L}$ is the Friedrichs extension of the operators L as defined over D_0, D_0 being the set of $2p$ times differentiable functions <u>which satisfy the boundary conditions</u>. Nevertheless the result is true as stated; its proof is based on the fact, noted at the end of section 4, that $\tilde{L}$ is a weak extension of L. Denote namely $\tilde{L}u$ by v, Lu by v', and take w as some function with $2p$ continuous derivatives in D which vanishes outside some closed subset of D. w belongs to the domain of L^* (in order that L^* exist the coefficients of the terms of order ν, $\nu = 0,1,\ldots,2p$ have to be ν-p times differentiable) and since $\tilde{L}$ is a weak extension, $(v,w) = (\tilde{L}u,w) = (u,L^*w)$. On the other hand, by Green's formula, $(v',w) = (Lu,w)$ is also equal to (u,L^*w); this shows that v' is square integrable over D and equal to v.

Our proof of the differentiability properties of $u(t)$ is based on a differentiability theorem of K. O. Friedrichs (see [8]) for solution of elliptic equations. The reduction is via this lemma:

> LEMMA 5.1. For t positive, $E(t)$ has derivatives of arbitrary order with respect to t, and $d^mE(t)/dt^m$ is equal to $A^mE(t)$, $m = 1,2,\ldots,$. The differentiation with respect to t is meant in the uniform topology.

A consequence of Lemma 5.1 is that $E(t)$ transforms every element of H_o into an element to which the operator A can be applied an arbitrary number of times; since it follows from the above mentioned theorem of Friedrichs that such elements are as differentiable as the coefficients of L permit them to be, this proves the differentiability of $E(t)u_o$ with respect to the space variables. To be precise:

FRIEDRICHS THEOREM. Let u be an element in the domain of A and denote Au by v. If v has square integrable derivatives up to order S in every closed subset of G, then u has square integrable derivatives up to order $2p + S$ in every closed subset of G, and so -- by Sobolev's lemma (see [22])-- continuous derivatives of order $2p + S - [n/2] - 1$ in G. The coefficients of L have to have continuous derivatives up to order $p + S$.

COROLLARY. If u lies in the domain of A^m, u has continuous derivatives up to the order $2mp - [n/2] - 1$ in G, provided that the coefficients of L have continuous derivatives up to the order $(2m-1)p$.

Combining this corollary with Lemma 5.1 we conclude that if the coefficients of L have $p + [n/2] + 1$ continuous derivatives, all elements of the form $E(t)u$ have continuous derivatives up to order $2p$ with respect to the space derivatives, and if the coefficients have derivatives of arbitrary order, so do all elements of the form $E(t)u$.

To prove the pointwise differentiability of $E(t)u$ with respect to t, we need the full statement of Friedrichs theorem; we shall state it as a corollary to the corollary:

COROLLARY². A^{-m} not only transforms every element u of H_o into a function with $2mp - [n/2] - 1$ continuous derivatives but the magnitude of these derivatives is bounded by the L_2 norm of u, uniformly in every closed subset of G.

Choose m_o so large that $2m_op - [n/2] - 1$ is non-negative; by the second corollary A^{-m_o} maps every sequence of elements of H_o convergent in the norm of H_o into a sequence of functions uniformly convergent in every closed subset of G.

Let u_0 be any element of H_0, t_1 any positive value: Choose t_2 as some smaller positive number; using the semigroup property of $E(t)$, the fact that $E(t_2)u_0$ lies in the domain of A^{m_0} (Lemma 5.1) and that A^{-m_0} and $E(t)$ commute (property (v)) we can write:

$$E(t_1)u_0 = E(t_1-t_2)\,E(t_2)u_0 = E(t_1-t_2)A^{-m_0}A^{m_0}\,E(t_2)u_0 =$$

$$= A^{-m_0}\,E(t_1-t_2)v_0$$

where v_0 is an abbreviation for $A^{m_0}E(t_2)u_0$. The same process applied to $t_1 + \delta$ in place of t_1 yields a similar identity; subtracting the two and dividing the difference by δ we obtain

$$(5.3)\qquad \frac{E(t_1+\delta)u_0 - E(t_1)u_0}{\delta} = A^{-m_0}\,\frac{E(t_1+\delta-t_2)v_0 - E(t_1-t_2)v_0}{\delta}$$

$E(\tau)v_0$ is differentiable at $\tau = t_1 - t_2$, and so the function to the right of A^{-m_0} on the right side of (5.3) tends to a limit in the sense of H_0, as δ tends to zero. According to what was said before, the images of these functions under the transformation A^{-m_0} converge uniformly in every closed subset of G; this proves the pointwise differentiability of $E(t)u_0$ at $t = t_1$.

By an entirely analogous reasoning one can show that all higher t derivatives of $E(t)u_0$ exist pointwise; no further differentiability conditions need be imposed on the coefficients of L. This type of argument also shows that if u_0 is sufficiently smooth, i.e., belongs to the domain of A^m, $E(t)u_0$ tends to u_0 as t tends to zero not only in the sense of the norm of H_0 but pointwise as well.

Now we turn to the proof of Lemma 5.1; it rests on a contour integral representation for $E(t)$. This powerful method for studying semigroups has been developed by E. Hille (see [11]).

We start with the remark that if A is an operator that satisfies the hypotheses of the Hille-Yosida theorem, then not only the positive real axis but the whole right half plane $\operatorname{Re}\{\mu\} > 0$ belongs to the resolvent set of A and the norm of the resolvent, $\|(A-\mu I)^{-1}\|_0$, is bounded by $[\operatorname{Re}\{\mu\}]^{-1}$ there.[5] On the strength of this an integral representation can be derived for $E(t)$ (or at least for $A^{-1}E(t)$) but not one from which the differentiability of $E(t)$ with respect to t could be deduced. To deduce that we need to know more about the resolvent

[5] This can be deduced from the resolvent equation $A - \mu I = (A-\lambda I)\left\{I + (\lambda+\mu)(A-\lambda I)^{-1}\right\}$, λ large positive.

set[6] of the operator A:

> LEMMA 5.2. If L is a $2p^{th}$ order elliptic operator, $\tilde{L}$ its Friedrichs extension, and $A = -\tilde{L}$, then all complex numbers $\mu = \alpha + i\beta$ in the left half plane for which
>
> $$|\beta| \geqq \text{const} |\alpha|^{\frac{2p-1}{2p}}$$
>
> holds belongs to the resolvent set, and the norm of the resolvent at such a point is bounded by
>
> $$\|(A-\mu I)^{-1}\|_0 \leqq \text{const} |\beta|^{-1}$$

A proof of this lemma will be given in a forthcoming paper by J. Berkowitz and P. D. Lax; it is based on the Gårding inequality, a well-known inequality[7] for $\|u\|_k$, $k < p$, in terms of $\|u\|_p$ and $\|u\|_0$ and the Friedrichs extension (in a slightly more general setting than was needed in section 2).

The contour integral representation for $E(t)$ is

$$E(t) = (2\pi i)^{-1} \int (\mu - A)^{-1} e^{\mu t} d\mu \tag{5.4}$$

the contour being the curve

$$|\beta| = \text{const} |\alpha|^{\frac{2p-1}{p}} \qquad -\infty < \alpha \leqq 0$$

(or any other contours in the left half plane). For positive t the factor $e^{\mu t}$ tends to zero exponentially as $\mu \longrightarrow \infty$, while the first factor $(\mu - A)^{-1}$ is

$$O\left(|\mu|^{-\frac{2p-1}{p}}\right)$$

---

[6] Of course if L is self-adjoint all complex numbers with nonzero imaginary parts belong to the resolvent set, and $\|(A-\mu I)^{-1}\| \leq |\beta|^{-1}$. But in this case no integral representation is needed to deduce the differentiability of $E(t)$ with respect to t; this follows from the functional calculus for the bounded symmetric transformation $\tilde{L}^{-1}$.

[7] $\|u\|_k^2 \leqq C\, a^{p/(p-k)} \|u\|_0^2 + \frac{1}{a} \|u\|_p^2$, where C is a constant depending on the indices p, k and the domain; and a is any positive number.

so the integral on the right of (5.4) converges in the uniform topology. If we differentiate the right side of (5.4) with respect to t under the integral sign any number of times, the resulting integral still converges in the uniform topology; this shows that the right side of (5.4) has derivatives of arbitrary order with respect to t.

Denote the object on the right side of (5.4) by $F(t)$ temporarily. We shall prove that $F(t) \equiv E(t)$ by showing that

a) $$\frac{dF(t)}{dt} = AF(t)$$

b) $F(t)u_o \longrightarrow u_o$ as $t \longrightarrow 0$ for all u_o in the domain of A.

According to the remarks at the end of section 3, a) and b) imply that $F(t)u_o$ is equal to $E(t)u_o$ for all t and all u_o in the domain of A and so, by continuity, for all u_o in H_o.

To prove a), we note that the operator A can be applied to the integrand in (5.4) and that the resulting integral converges absolutely. Since A is a closed operator this implies that

$$AF(t) = (2\pi i)^{-1} \int A(\mu - A)^{-1} e^{\mu t} \, d\mu$$

On the other hand, dF/dt is given by

$$\frac{d\,F(t)}{dt} = (2\pi i)^{-1} \int u(\mu - A)^{-1} e^{\mu t} \, d\mu$$

The difference of these two is $(2\pi i)^{-1} \int e^{\mu t} \, d\mu$, and the value of this integral is zero, as may be seen by shifting the contour of integration to the left. This shows that dF/dt is equal to AF; the same type of reasoning shows that d^mF/dt^m is equal to A^mF.

We turn now to the proof of b). Let u_o be any element in the domain of A; choose for λ_o any complex number to the <u>right</u> of the contour of integration in (5.4), say $\lambda_o = 1$; denote $(A - \lambda_o)u_o$ by v_o.

$$F(t)u_o = F(t)(A - \lambda_o)^{-1} v_o =$$

$$(2\pi i)^{-1} \int (A - \lambda_o)^{-1} A - (\mu - A)^{-1} v_o e^{\mu t} \, d\mu$$

We claim that the right side is equal to

$$(2\pi i)^{-1} \int (\mu - \lambda_o)^{-1} (A - \mu)^{-1} v_o \, e^{\mu t} \, d\mu$$

This is most easily verified by showing that the difference between the two expressions which we claim to be equal,

$$(2\pi i)^{-1} \int (A-\lambda_o)^{-1} (\mu-\lambda_o)^{-1} e^{\mu t} d\mu$$

is zero. Since the only singularity of this integral is at $\mu = \lambda_o$, we may shift the contour to the left without altering the value of the integral but we see that, on the new contour, the integral is arbitrarily small.

The representation

$$(2\pi i)^{-1} \int (\mu-\lambda_o)^{-1} (A-\mu)^{-1} v_o e^{\mu t} d\mu = F(t)u_o$$

was derived in order to determine the limit of $F(t)u_o$ as t tends to zero. For that purpose note that the integral converges absolutely and uniformly for all t, including $t = 0$, since the first factor is $O(|\mu|^{-1})$ near infinity, the second factor - according to Lemma 5.2 -

$$O\left(|\mu|^{-\frac{2p-1}{2p}}\right)$$

and the third, $e^{\mu t}$, never exceeds one in absolute value. This justifies passage to the limit under the integral sign, and so

$$\lim_{t \to o} F(t)u_o = (2\pi i)^{-1} \int (\mu-\lambda_o)^{-1}(A-\mu)^{-1} v_o \, d\mu$$

To evaluate this remaining integral we make a closed path out of the contour by adjoining the arc of the circle $|\mu| = R$ which is to the right of the contour, and throwing away that portion of the old contour which lies outside of this circle. The value of the integral along the new arc and the discarded arc is small if R is large; therefore this change makes no difference in the value of the integral, which can now be put equal to the sum of the residues inside. The only singularity inside is at $\mu = \lambda_o$, and the residue there is $(A-\lambda)^{-1}v_o = u_o$. Q.E.D.

We should like to mention that it is possible to prove differentiability theorems for generalized solutions of a parabolic equation directly, without the aid of differentiability theorems for solutions of elliptic equations. We have carried out such a direct verification in case the coefficients of L are constant, using the fundamental solution $K(x,t)$ for our parabolic equation.[8] The fundamental solution in this case can be written down explicitly; denoting by $L^*(\xi)$ the form

[8] Yosida, in [26], uses such a method to prove the differentiability of generalized solutions of the diffusion equation, even in the case of variable coefficients.

$$L^*(\xi) = \sum_{\nu=0}^{2p} \sum_{i_1,\ldots,i_\nu=1}^{n} (-1)^{p-\mu} a_{i_1,\ldots,i_\nu} \xi_{i_1},\ldots,\xi_{i_\nu}$$

we have

$$K(x,t) = \int_{E_n} \exp\left\{2\pi i\xi\cdot x - tL^*(2\pi i\xi)\right\} d\xi$$

The integral defining $K(x,t)$ converges absolutely and uniformly for $T \geqq t_0 > 0$. Various properties of $K(x,t)$ were investigated by Ladyzhenskaya, [17] and P. C. Rosenbloom, [20]. The ones we need to know are these:

a) $K(x,t)$ satisfies $\frac{\partial K}{\partial t} = -L^*K$ for $t > 0$.

b) $K(x,t)$ has derivatives of arbitrary order with respect to x and t for $t > 0$; as t tends to zero, K and its derivatives remain bounded (and in fact tend to zero) uniformly outside of any sphere $\| x \| = r$.

c) $u(y,t) = (K(x-y,t),u)_0$ tends to $u(y)$ in the sense of the norm of H_0 as $t \longrightarrow 0$.

Let x_0 be any point of G, $\psi(x)$ some function identically equal to one in some neighborhood N of x_0, zero outside of some closed subset of G and $2p$ times differentiable. Let u_0 be some element in the domain of $\tilde{L}$, so that $E(\tau)u_0 = u(\tau)$ satisfies $u_\tau = -\tilde{L}u$. Take the scalar product of both sides with $\psi(x)K(x-y,t-\tau)$:

$$(u_\tau, \psi K) = (-\tilde{L}u, \psi K)$$

Since $\tilde{L}$ is a weak extension, and ψK lies in the domain of L^*, the right side can be written as $(u,-L^*\psi K)$; the left side can be written as $(u,\psi K)_\tau - (u,\psi K_\tau)$ which is equal to $(u,\psi K)_\tau + (u,\psi K_t)$, since K is a function of $t-\tau$. Integrating both sides with respect to τ from zero to $t-\delta$, we get:

$$\left(u(t-\delta), \psi(x)K(x-y,\delta)\right)_o = \tag{5.5}$$

$$\left(u, \psi(x)K(x-y,t)\right)_0 - \int_0^{t-\delta} (u, \psi K_t + L^*\psi K)_0 \, d\tau$$

The factor ψ is $\equiv 1$ in N; so by property a) of K, $\psi K_t + L^*\psi K = K_t + L^*K$ is zero in N. This shows that the space integration in the second term of (5.5) is extended only over $G - N$; so, by property b) of K, the function on the right side of (5.5) is an infinitely differentiable function of y and t and its derivatives remain uniformly bounded as δ tends to zero (i.e., uniformly with respect to δ, not the order of the derivative) if y is restricted to some closed subset of N. So the limit of the right side as $\delta \longrightarrow 0$ as function of y, y in N, is an infinitely differentiable function. The limit of the left side, regarded as element of H_o, is by property c) of K equal to $\Psi u(t)$. This shows that $u(t)$ is an infinitely differentiable function in N, i.e., (N being arbitrary) in all of G.

The methods of section 5 can be used to derive analogous results in case u is a vector of functions and L is a strongly elliptic matrix differential operator.

BIBLIOGRAPHY

[1] BROWDER, F. E., "The Dirichlet problem for linear elliptic equations of arbitrary even order with variable coefficients," Proceedings of the National Academy of Sciences, 38 (1952), 230-235.

[2] BROWDER, F. E., "The Dirichlet and vibration problems for linear elliptic differential equations of arbitrary order," Proceedings of the National Academy of Sciences, 38 (1952), 741-747.

[3] BROWDER, F. E., "Linear parabolic equations of arbitrary order," abstract 605, Bulletin of the American Mathematical Society, 58 (1952), 632.

[4] FREUDENTHAL, H., "Ueber die Friedrichssche Fortsetzung halbbeschränkter Hermitescher Operatoren," Proceedings of the Koninklijke Nederlandsche Akademie van Wetenschappen, 39 (1936), 832-833.

[5] FRIEDRICHS, K. O., "Spectraltheorie halbbeschränkter Operatoren und Anwendung auf die Spektralzerlegung von Differzialoperatoren I, II," Mathematische Annalen, 109 (1934), 465-486, 685-713.

[6] FRIEDRICHS, K. O., "On differential operators in Hilbert space," American Journal of Mathematics 61 (1939), 523-544.

[7] FRIEDRICHS, K. O., "The identity of weak and strong extensions of differential operators," Transactions of the American Mathematical Society 55 (1944), 132-151.

[8] FRIEDRICHS, K. O., "Differentiability of solutions of elliptic operators," Communications on Pure and Applied Mathematics, 6, 3.

[9] GARDING, L., "Dirichlet's problem and the vibration problem for linear elliptic partial differential equations with constant coefficients," Proceedings of the symposium on spectral theory and differential problems, Stillwater, Oklahoma, (1951), 291-301.

[10] GARDING, L., "Le problème de Dirichlet pour les équations aux dérivées partielles elliptiques linéaires dans des domaines bornes," Comptes rendus des séances de l'Académie des Sciences, 233 (1951), 1554-1556.

[11] HILLE, E., Functional Analysis and Semi-Groups, American Mathematical Society Colloquium Publication, New York, (1948).

[12] HILLE, E., "On the integration problem for Fokker-Planck's equation," 10th Congress of Scandinavian Mathematicians, Trondheim (1950).

[13] HILLE, E., "On the differentiability of semi-group operators," Acta Scientiarum Mathematicarum, 12, part B (1950), 19-26.

[14] JOHN, F., "Derivatives of continuous weak solutions of linear elliptic equations," Communication on Pure and Applied Mathematics, 6, 3.

[15] JOHN, F., "General properties of solutions of linear elliptic partial differential equations," Proceedings of the symposium on spectral theory and differential problems, Stillwater, Oklahoma (1951).

[16] JOHN, F., "Derivatives of solutions of linear partial differential equations," this Study.

[17] LADYZENSKAYA, O. A., "On the uniqueness of the solution of the Cauchy problem for a linear parabolic equation," Matematiceskii Sbornik, 27, 69 (1950), 175-184.

[18] LERAY, J., "Hyperbolic equations with variable coefficients," lectures delivered at the Institute for Advanced Study, Princeton, Fall, (1952).

[19] NAGY, B. v. Sz., Spectraldarstellung linearer Transformationen des Hilbertschen Raumes, Springer, Berlin (1942).

[20] ROSENBLOOM, P. C., "Linear equations of parabolic type with constant coefficients," this Study.

[21] ROTHE, E., "Ueber die Wärmeleitungsgleichung mit nichtkonstanten Koeffizienten im räumlichen Falle I, II," Mathematische Annalen 104 (1931), 340-354, 354-362.

[22] SOBOLEV, S., "Sur quelques évaluations concernant les familles des fonctions ayant des dérivées à carré intégrables, Akademiya Nauk USSR, Doklady, N.S. (Comptes Rendus) (1938), 279-282.

[23] YOSIDA, K., "On the differentiability and the representation of one-parameter semigroups of linear operators," Journal of the Mathematical Society of Japan, 1 (1948), 15-21.

[24] YOSIDA, K., "Integration of Fokker-Planck's equation in a compact Riemann space," Arkiv for Matematik, 1 (1949), 71-75.

[25] YOSIDA, K., "Integration of the Fokker-Planck equation with boundary conditions," Journal of the Mathematical Society of Japan, 3 (1951).

[26] YOSIDA, K., "On the integration of diffusion equations in Riemannian spaces," Proceedings of the American Mathematical Society, 3 (1952), 864-874.

[27] VISCHIK, M. I., "The method of orthogonal and direct decomposition in the theory of elliptic differential equations," (Russian), Matemakiceskii Sbornik, 25 (1949), 189-234.

[28] WEYL, H., "The method of orthogonal projection in potential theory," Duke Mathematical Journal, 7 (1940), 414-444.

COMMUNICATIONS ON PURE AND APPLIED MATHEMATICS, VOL. VIII, 615–633 (1955)

On Cauchy's Problem for Hyperbolic Equations and the Differentiability of Solutions of Elliptic Equations*

By PETER D. LAX

This paper is about solutions of *linear partial differential equations with variable coefficients.* Two different problems will be treated, with the same method. The first problem is the Cauchy problem for *symmetric hyperbolic systems of first order*, a class of differential equations introduced and studied by Friedrichs, see [43]. This class, which will be discussed at the beginning of Section 1, includes all equations of physics describing reversible time-dependent phenomena, such as Maxwell's equations (and therefore also second order hyperbolic equations), the equations of compressible flow, of the motion of elastic bodies, of magneto-hydrodynamics and others. The Cauchy problem will be solved with the aid of *energy integral* inequalities, developed for this purpose by Friedrichs. Such energy inequalities have been employed previously by H. Weber [22], Hadamard [23], Zaremba [24], Rubinowicz [25], Hans Lewy to derive various uniqueness theorems, and by Courant-Friedrichs-Lewy [1], Friedrichs [43], Schauder [27], Frankl [29], Petrovskii [30] and Leray [32] to derive existence theorems. Leray in particular was able to extend the energy method in a remarkably elegant way to the most general kind of hyperbolic equation (see also Gårding, [40].)

In all these treatments the energy inequality is used to show that the solution at some later time depends *boundedly* on the initial values in an appropriate norm; the appropriate norm in the present case is just the L_2 norm and—more generally—the s-fold Dirichlet integral. To derive an existence theorem however one needs, in addition to the a priori energy estimates, auxiliary constructions. Schauder, in [27], has observed that the Cauchy-Kowalewski procedure can be used as such a constructive step; his method was taken over by both Petrovskii and Leray. In his work the constructive steps are *projections*. In Friedrichs' approach the solution is also constructed by projections; but to show that the object constructed is smooth if the data determining it are smooth, he needs an approximation by solutions of difference equations. In my method the projection is performed in an appropriate space, so that the solution arises fully outfitted with all its existing derivatives. The appropriate space is the space of functions with negative norms, described in

*The work for this paper was supported by the Office of Naval Research, United States Navy, under contract N6ori-201 T. O. No. 1.

Section 2 of this paper. To be able to operate in this space, one has to show that the solution depends boundedly on the initial data in the sense of these negative norms. This result too turns out to be a consequence of the energy inequality, if one employs the inverse of the s-fold iterated Laplace operator. The other type of construction used in this paper is the inversion of the s-fold Laplacean; it is a particularly simple construction, since it has to be carried out in the *periodic case* only.

In Section 1, I show that the Cauchy problem is entirely equivalent to the problem of finding periodic solutions; this problem is then solved in Section 3. No doubt, the Cauchy problem can be treated directly with the methods of this paper, but going through the periodic problem allows for especially smooth sailing.

The combined result of Sections 1 and 3 is the *Existence Theorem*: If for a linear symmetric hyperbolic system, whose coefficients are sufficiently smooth functions of the independent variables, the prescribed initial functions have square integrable partial derivatives of order s, then there exists a unique solution with the same number of square integrable partial derivatives.

The interest in such an existence theorem is intensified by Sobolev's lemma, which asserts that a function with s square integrable derivatives has a (lesser) number of *continuous* derivatives. The existence theorem is, of course, not new, being the same as Friedrichs'; only its proof is new, in the aspects given.

The second part of this paper, Section 4, contains a theorem on the *differentiability of weak solutions* of an elliptic equation $Lu = v$ of order $2m$ asserting that if v has square integrable derivatives up to a certain order, say j, in some domain, then u has square integrable derivatives of order $j + 2m$. Such differentiability theorems are needed in the modern version of the direct methods of the Calculus of Variations applied to boundary value problems, where the solution is constructed by orthogonal projection or an equivalent of it. The solutions constructed in this way have to satisfy the differential equation in a weak sense only. There are several distinct ways of proving the differentiability of weak solutions. One method employs the fundamental singularity of the equation under discussion. The prototype of such a proof is given in Weyl's famous paper on orthogonal projection, [37]; it has been extended to the most general elliptic equations with variable coefficients by Browder [17], Gårding [12], L. Schwartz [11], Vishik [10], making use of Fritz John's construction of the fundamental solution. Another method employed by Fritz John is the method of spherical means, [18]. A third one invented by Friedrichs [42], see also [19], employs mollifiers and a priori estimates of higher derivatives; Louis Nirenberg shows in his paper (in this issue) that mollifiers can be replaced by difference quotients.

The treatment employed in this paper uses Friedrichs' approach but dispenses with mollifiers or difference quotients. The basic idea, expressed somewhat differently in Section 4, is: if u is a weak solution of $Lu = v$, then it is also a weak solution of the higher order equation $\Delta^s Lu = w$, where w stands for $\Delta^s v$

By orthogonal projection, or the like, one can construct a solution $\tilde{u}$ of $\Delta^s Lu = w$ which has square integrable derivatives of order $s + m$. If one could identify $\tilde{u}$ with the originally given u, the objective would be attained. This much of the idea is related to one proposed by Courant in [39]. The difficulty with this approach is that $\tilde{u}$ will be equal to u only if it is made to satisfy appropriate boundary conditions, and this isn't a simple matter. This point doesn't come up, however, if the underlying domain has no boundary; therefore we operate in the simplest of such domains, a period parallelogram. Then the only tools needed to carry out such a program are the a priori inequality of Gårding, and the inversion of the iterated Laplacean, a particularly simple task in the periodic case. To pass from the periodic to the nonperiodic case, we simply introduce $u' = \phi u$, where ϕ is equal to one in some neighborhood of a preassigned point, and zero outside of some larger neighborhood. The function u' satisfies an appropriate differential equation, and can be regarded as periodic with respect to some parallelogram which contains the support of ϕ. Repeated application of the differentiability theorem for periodic solutions and of the above construction gives then the differentiability theorem for nonperiodic solutions.

In the differentiability theorem derived in this paper the functions u and v may be, just as in the work of Laurent Schwartz, merely *generalized functions*, and the degree of differentiability of u and v may be *negative* (this is explained more precisely in the text). The result was expressed in this form not for the sheer gain of generality but because it is of distinct advantage in the reduction of the nonperiodic to the periodic case.

During the writing of this paper I had many conversations with L. Nirenberg and K. O. Friedrichs on various aspects of the differentiability of solutions. These have been a great help to me, and I wish to thank them.

1. *The Cauchy Problem for Symmetric Hyperbolic Systems and its Relation to the Periodic Problem*

Let a denote a vector of m functions over the Euclidean space, A_i and B m-by-m matrices which are assumed to be *smooth* functions. The first order operator

$$Lu = A_i u_{x^i} + Bu$$

is called *symmetric* if the matrices A_i are symmetric matrices.[1] It is called *hyperbolic* if there exists a *space-like hyperplane* at each point, i.e. one whose normal $(\xi_1, \cdots, \xi_n)$ has the property that the matrix

$$\xi_1 A_1 + \cdots + \xi_n A_n$$

is *positive definite.* In what follows we shall assume that the hyperplanes $x^1 =$ const. are always space-like, i.e. that the matrix A_1 is positive definite. We

[1]Or simultaneously symmetrizable by the same similarity transformation.

may, without loss of generality, take A_i to be the *unit matrix*; for, A_i can be written as $T'T$, and the change of variables $Tu = w$ and multiplication of the operator by $(T')^{-1}$ leads to a symmetric operator where the coefficient of the x^1 derivative is the identity. Since in many physical applications the coordinate x^1 is time, we shall hereafter denote it by t and refer to the others as *space variables*:

$$Lu = u_t + A_i u_{x^i} + Bu.$$

We shall assume that the coefficients A_i, B, (and the functions $u(x, t)$ to which L shall be applied) are *periodic* with respect to all the space coordinates.[2] We can always achieve this state by altering the coefficients and initial data outside a sufficiently large compact set. Since signals propagate at a finite speed, this has no influence on the local Cauchy problem.

More precisely, we rely on the following known

UNIQUENESS THEOREM. *The value of the solution u of $Lu = v$ at a given point depends on the value of the coefficients of L, v and initial values only in a finite portion of space.*

The precise delineation of the domain of dependence is in terms of characteristic surfaces. The theorem is due to Holmgren, and holds for any equation for which the Cauchy problem can be solved for a dense set of data, such as equations with analytic coefficients, regardless of type, and (as remarked by Petrovskii) hyperbolic equations.

Similarly, we alter the coefficients if necessary at some sufficiently far-away time so as to make them periodic with respect to time as well. We denote the time period by T.

Let $\{u\}$ denote the totality of smooth solutions of the homogeneous equation

$$Lu = 0$$

and denote by ϕ and ψ the initial and final values of u respectively: $u(0) = \phi$, $u(T) = \psi$. The relation of ϕ to ψ is linear, and we denote it by S (the solution operator):

$$\psi = S\phi.$$

Friedrichs has shown that the operator S is *bounded* in the square integral norm. We reproduce his reasoning:

Denote by $N(t)$ the square integral of a solution u at time t over a period-parallelogram: $N(t) = (u(t), u(t))$. Next, we compute $\dot{N}(t) = dN/dt$, making use of the differential equation $Lu = 0$,

$$\dot{N} = 2(u, u_t) = -2(u, A_i u_{x^i} + Bu). \tag{1.1}$$

Shift over the matrix A_i and integrate by parts:

$$\dot{N} = 2(A_i u_{x^i}, u) + 2([A_{i,i} - B]u, u). \tag{1.2}$$

[2]This useful device is due to Petrovskii, [30].

Adding these two expressions for $\dot{N}$ we get a third one which involves no derivatives of u, and can therefore be estimated by the square integral of u:

$$\dot{N} \leqq KN,$$

which, as is well known, implies

$$N(t) \leqq N(0)e^{Kt}.$$

This expresses the fact that the solution operator S is bounded with respect to the L_2 norm, in fact $\| S \| \leqq e^{KT/2}$.

We define the norm $\| \phi \|_s$ as the square root of the integral of the sum of the squares of ϕ and all derivatives of ϕ up to order s; the underlying domain, for functions of the space variables, is the period parallelogram, for functions of space and time, the cylinder $0 \leq t \leq T$ erected over the period parallelogram. The completion of the space of smooth functions under the s-norm is denoted[3] by H_s ; H_s is a Hilbert space. We shall sometimes refer to its elements as functions with s strong derivatives.

The partial derivatives of a solution u of $Lu = 0$ can be thought of as satisfying a symmetric hyperbolic system of differential equations, namely of the system obtained from $Lu = 0$ by differentiating it. Applying the a priori estimate just derived to this enlarged system we conclude:

The solution operator S is bounded with respect to the s-norm.

The transformation $u' = e^{-\lambda t}u$ maps solutions of $Lu = 0$ into solutions of $L'u' = 0$, where the operator L' differs from L merely by the addition of λI to B, the coefficient of u. S', the solution operator associated with L' is just $e^{-\lambda t}S$. Therefore choosing λ large enough we can accomplish that it has norm less than, say, $\frac{1}{2}$ with respect to any specified norm $\| \phi \|_s$. From now on we assume that such a transformation had already been performed.

Our aim is to prove the existence theorem formulated in the introduction:

EXISTENCE THEOREM. *Let L be a symmetric hyperbolic operator with sufficiently differentiable coefficients, ϕ an initial function over the period parallelogram with s strong derivatives, and v a function over the cylinder $0 \leq t \leq T$ with s strong derivatives. The initial value problem*

$$Lu = v, \qquad u(0) = \phi$$

has a unique solution u in the cylinder with s strong derivatives over every period parallelogram $t = const.$

In this section we shall reduce this existence theorem to another one about periodic solutions:

EXISTENCE THEOREM ABOUT PERIODIC SOLUTIONS. *Let L be a symmetric operator with sufficiently differentiable periodic coefficients and assume that the*

[3] A more detailed discussion of the space H_s is given in Section 2.

coefficient B of u is sufficiently positive. Let v be a function over the period parallelogram which has s periodic strong derivatives. Then the equation

$$Lu = v$$

has a unique periodic solution u with s periodic strong derivatives.

COROLLARY TO BOTH EXISTENCE THEOREMS. *It follows by Sobelov's theorem that if the data ϕ and v have a sufficient number of strong derivatives, the corresponding solution will have a sufficient (lesser) number of continuous derivatives.*

We show first (although this is irrelevant in the logical structure of this paper) that the existence of periodic solutions follows from the existence of the solution to the Cauchy problem. Denote by $\bar{S}$ the closure of the solution operator S with respect to the s-norm; assuming that the Cauchy problem can be solved, it follows that the domain of $\bar{S}$ is all of H_s. To find the periodic solution u of $Lu = v$, write u as $u_1 + u_2$, where u_1 is the solution of $Lu_1 = v$ with initial value zero, and u_2 is a solution of the homogeneous equation with initial value ϕ yet to be determined so that $u_1 + u_2$ is periodic:

$$u_1(0) + u_2(0) = u_1(T) + u_2(T).$$

Denoting $u_1(T)$ by ψ this can be written as

$$\phi - \bar{S}\phi = \psi.$$

Since $\bar{S}$ has norm less than one, this equation clearly has a solution, given by the Neumann series.

To pass from the periodic problem to the Cauchy problem is just as simple. It is sufficient to consider the homogeneous problem, since we have at least one solution of the inhomogeneous equation at our disposal, namely the periodic one.

We shall show first:

LEMMA. *If ψ is any smooth function, then there exists a ϕ in the domain of S such that*

$$\phi - S\phi = \psi.$$

To show this, we shall construct an auxiliary function u_1 with these properties:

i) $u_1(0) = 0, u_1(T) = \psi$,

ii) $Lu_1 = v$ is a smooth periodic function.

For small positive t, we define $u_1(t)$ as identically zero; that makes $v = Lu_1$ identically zero for small t. So, if v, and a certain number, say k, of its derivatives are to be periodic with period T, we must have v and its first k derivatives with respect to t zero at T. We therefore define $u_1(t)$ in the range $T - \epsilon < t \leqq T$ by the formula

$$u_1(t) = \psi + \psi_1(t - T) + \psi_2 \frac{(t - T)^2}{2} + \cdots + \psi_{k+1} \frac{(t - T)^{k+1}}{(k + 1)!},$$

where ψ is the given smooth function, and $\psi_1, \psi_2, \cdots, \psi_{k+1}$ are determined by the requirement that v and its first kt derivatives be zero at $t = T$. The successive determination of the $\psi_1, \psi_2, \cdots, \psi_{k+1}$ is done as in the Cauchy-Kowalewski calculation. Finally we connect the two ends of u_1 in any way, e.g. by defining $u_1(t)$ as

$$h(t)\left\{\psi + \psi_1(t - T) + \cdots + \psi_{k+1} \frac{(t - T)^{k+1}}{(k + 1)!}\right\},$$

where $h(t)$ is a function equal to one in the range $T - \epsilon \leqq t \leqq T$, zero in the range $0 \leqq t \leqq \epsilon$ and infinitely differentiable in between.

Having constructed u_1, we denote u_2 as the solution of the periodic problem

$$Lu_2 = v.$$

According to the corollary to the periodic existence theorem, which we assumed to hold, u_2 is a smooth function. The function $u = u_2 - u_1$ is then a smooth solution of the homogeneous equation; the difference between its value at $t = 0$ and $t = T$ is ψ. So, denoting $u(0)$ by ϕ, we have

$$\phi - S\phi = \psi. \tag{1.3}$$

This completes the proof of the lemma. Since the norm of S is less than one half, we have

$$\tfrac{1}{2} \| \phi \|_s \leqq \| \phi - S\phi \|_s = \| \psi \|_s ,$$

so that the dependence of ϕ on ψ is continuous.

Denoting by $\overline{S}$ the closure of S, we can conclude from this

LEMMA′. *If ψ is any element of H_s, there exists an element ϕ in the domain of $\overline{S}$ such that*

$$\phi - \overline{S}\phi = \psi. \tag{1.3 ′}$$

We claim that this implies that the domain of $\overline{S}$ is all of H_s. For, suppose it were not so; then, by the projection theorem, there would be an element ϕ_0 orthogonal (in the sense of the scalar product of H_s) to the domain of $\overline{S}$. Take ψ to be ψ_0 in the relation (1.3′) with some $\phi = \phi_0$ and take the scalar product of both sides with ϕ_0; that of the right side is zero, and that of the left not less than $\frac{1}{2} \| \phi_0 \|_s$, since $\overline{S}$ has norm less than $\frac{1}{2}$. Thus $\| \phi_0 \|_s$ is zero, and so, consequently, is $\| \psi_0 \|_s$. This completes the proof.

We have shown that the domain of $\overline{S}$ is H_s, and thus the Cauchy problem can be solved for all ϕ in H_s.

2. *Positive and Negative Norms*

Throughout this paper we shall use this language and notation: a function, or vector of functions, over a domain of Euclidean space is called *smooth* if it has a sufficient number of derivatives for the purpose at hand. For most part in this paper the underlying domain will be a period-parallelogram.

The L_2 scalar product of two smooth functions u, w shall be denoted, as usual, by (u, w). For any non-negative integer s we define the *s-norm* of a smooth function u, in the customary way, as the square root of the s-fold Dirichlet integral

$$\| u \|_s^2 = \sum_{|p| \leq s} (D^p u, D^p u). \tag{2.1}$$

The *completion* of the smooth functions under the s-norm will be denoted by H_s. Since the s-norm is not less than the zero-norm, every Cauchy sequence in the s-norm is a fortiori a Cauchy sequence in the zero-norm. Thus there is a mapping of H_s into H_0, and furthermore to each function u in H_s we can assign functions in H_0 as its partial derivatives up to order s so that the norm relation (2.1), and the formula for integration by parts for smooth, compactly carried functions holds. Functions in H_s shall be referred to as "having finite s-norm" or "having s strong derivatives".

Next we show, following Friedrichs, that the mapping of H_s into H_0 is one-to-one. In other words, if not all of the partial derivatives of u of order up to s are zero, then the square integral of u cannot be zero. For, let u be a function with finite s-norm whose square integral is zero; we aim to show that then $\| u \|_s$ is also zero. Let Du be any of the first partial derivatives of u; Du is square integrable, and the formula for integration by parts holds:

$$(Du, a) = -(u, Da)$$

for every smooth test function a with compact support. Here, and hereafter, the parentheses denote the ordinary L_2 scalar product over the underlying domain. If the square integral of u is zero, the right side is zero; thus the left side must be zero too:

$$(Du, a) = 0.$$

But since the set of smooth functions a with compact support is dense in the space of square integrable functions, it follows that the square integral of Du is zero. Continuing in this fashion, we can show that all partial derivatives of u up to order s have zero square integrals, and therefore the s-norm of u is zero.

For any element b of H_0, the expression $(w, b) = l(w)$ is a bounded linear functional of w with respect to the zero norm, and therefore a fortiori with respect to the s-norm. Denote its *norm* by the symbol[4] $\| b \|_{-s}$:

[4]Leray, in [32], also introduces negative norms, in a somewhat different fashion but amounting to the same thing. The applications in the present paper are different from his.

$$\| b \|_{-s} = \sup_w \frac{(w, b)}{\| w \|_s}. \tag{2.2}$$

These linear functionals form a complete set; i.e., their simultaneous vanishing for a fixed element u of H_s takes place only for the zero element.

The space H_s is a Hilbert space, therefore *reflexive*; according to a classical theorem, a *complete set* of linear functionals over such a space is a *dense set* of linear functionals. So the linear functionals $(w, b) = l(w)$ over H_s are dense.

Now according to a classical theorem of normed linear spaces (the projection theorem for a Hilbert space, the Hahn-Banach theorem for a Banach space), the *norm* of an element w is equal to $\sup_l | l(w) | / \| l \|$, the supremum being taken over any *dense* set of bounded linear functionals. In our case, we have just shown that the linear functionals (w, b), b in H_0, constitute a dense set of linear functionals, thus

$$\| w \|_s = \sup_b \frac{(w, b)}{\| b \|_{-s}}; \tag{2.3}$$

i.e., the relation between positive and negative norm is reciprocal.

Denote the *closure* of the set of square integrable functions under the minus s-norm by H_{-s}. H_{-s} is isometrically isomorphic to the space of bounded linear functionals over H_s, therefore H_{-s} is a Hilbert space. To every element b of H_{-s} there is associated a linear functional $l(w)$ over H_s; we shall denote this linear functional by (w, b). Its norm is $\| b \|_{-s}$:

$$(w, b) \leqq \| b \|_{-s} \| w \|_s. \tag{2.4}$$

We shall refer to (2.4) as the generalized Schwarz inequality. Please observe that the elements b of H_{-s} are *generalized functions* (distributions) in the sense of Laurent Schwartz, since their scalar product with functions w possessing s strong derivatives is defined.

For fixed w in H_s, (w, b) is a bounded linear functional of b in H_{-s}. We assert now that, conversely, these are *all* the bounded linear functionals over H_{-s}.

REPRESENTATION THEOREM. *Every bounded linear functional $m(b)$ over H_{-s} can be represented as*

$$m(b) = (w, b)$$

where w belongs to H_s.

Proof: The norm of the linear functional $m(b) = (w, b)$ is, according to (2.3), $\| w \|_s$; therefore, since the space H_s is *complete*, the space of linear functionals of the above form is *closed*. Next we show that the space of linear functionals of the above form constitutes a *complete* set, i.e., that $(w, b) = 0$ for all w in H_s implies that b is zero. Since H_{-s} is reflexive (in fact a Hilbert space), it would follow that these constitute *all* linear functionals.

There remains to be shown that $(b, w) = 0$ for all w in H_s implies that b is zero. But, by definition, the *norm* of (w, b) *as a linear functional of* w is $\| b \|_{-s}$; thus $(b, w) \equiv 0$ implies $\| b \|_{-s} = 0$.

As we remarked at the beginning, the main result of this section is the representation theorem. We shall use it to construct functions w with s strong derivatives, as *representers* of linear functionals bounded in the minus s norm.

3. *The Periodic Problem for Symmetric Systems*

Let Lu denote a *symmetric* first order operator:

$$Lu = A_i u_{x^i} + Bu; \tag{3.1}$$

i.e., u is a vector of m unknown functions, A_i and B m-by-m matrices which are smooth periodic functions of the independent variables, the A_i symmetric. In addition we assume that B is *sufficiently positive*, i.e. that it is the sum of a matrix B_0 and some sufficiently large multiple of the identity.

It is no longer necessary to take L hyperbolic (although this was necessary to make L sufficiently positive by an exponential transformation). Throughout this section, the underlying domain is the period-parallelogram.

The content of this section is this

EXISTENCE THEOREM. *Let v be a function with s strong derivatives, $s \geqq 1$; then the equation $Lu = v$ has a solution u with s strong derivatives.*

The proof of this theorem will be based on an a priori estimate for the adjoint operator L^*:

A PRIORI INEQUALITY. *Let a be a smooth vector, and define b as L^*a. Then*

$$\| a \|_{-s} \leqq \| b \|_{-s} \tag{3.2}$$

for $s \geqq 0$.

Remark: This implies that the correspondence between a and b is one-to-one, i.e. that the operator L^* has a *unique* inverse.

Proof: We shall first handle the case $s = 0$. The inequality (3.2) is a fairly exact analogue of the inequality described in the first section, expressing the boundedness of the solution operator. The proof is based on the *boundedness* of the operator, L^* *from below*, i.e. the so called *energy inequality*:

$$(a, a) \leqq (a, L^*a).$$

The a priori estimate (3.2) for $s = 0$ follows by taking the scalar product of both sides of $L^*a = b$ with a, applying the energy inequality to the left, the Schwarz inequality to the right.

If L is symmetric with B sufficiently positive, so is L^*; we shall therefore employ the same notation for L^* as for L (namely (3.1)). Take now

$$(a, L^*a) = (a, A_i a_{x^i} + Ba). \tag{3.3}$$

Since the A_i are symmetric, this can be written as

$$(a, L^*a) = (A_i a, a_{x_i} + Ba).$$

Integrating by parts we transform this into[5]

$$(3.4) \quad \begin{aligned} (a, L^*a) &= \left(-\frac{\partial}{\partial x_i} A_i a, a\right) + (a, Ba) \\ &= -(A_i a_{x_i}, a) + \left(a, \left[B - \frac{\partial}{\partial x_i} A_i\right] a\right). \end{aligned}$$

Adding the two expressions (3.3) and (3.4) for (a, L^*a) we get

$$(3.5) \qquad (a, L^*a) = \left(a, \left[B - \frac{1}{2}\frac{\partial}{\partial x_i} A_i\right] a\right).$$

Since B was assumed sufficiently positive, the right side of (3.5) is indeed greater than (a, a). This proves the energy inequality.

To prove the a priori inequality for positive s, we construct a smooth function c related to a by

$$(3.6) \qquad \Delta^s c = a.$$

Here Δc denotes c minus the Laplacean of c, so that Δ^s is a *positive definite* elliptic operator:

$$\| c \|_s^2 \leqq (c, \Delta^s c).$$

We can construct c and prove its smoothness either by using Fourier series, or by orthogonal projection and mollifiers à la Friedrichs. The latter method for proving differentiability is particularly simple in the periodic case and for an operator with constant coefficients.

The operator $L^*\Delta^s$ is bounded from below:

$$(3.7) \qquad \| c \|_s^2 \leqq (c, L^*\Delta^s c);$$

this can be deduced, because of the positive definiteness of Δ^s, as in the case $s = 0$. Now take the scalar product of both sides of $L^*\Delta^s c = b$ with c, apply the energy inequality (3.7) on the left and the generalized Schwarz inequality (2.4) on the right:

$$\| c \|_s^2 \leqq (c, L^*\Delta^s c) = (c, b) \leqq \| c \|_s \, \| b \|_{-s},$$

thus

$$\| c \|_s \leqq \| b \|_{-s}.$$

To complete the derivation of the a priori estimate, we have to show that $\| a \|_{-s} \leqq \| c \|_s$. We do this as follows: By definition,

$$\| a \|_{-s} = \sup \frac{(a, w)}{\| w \|_s};$$

[5]There are no boundary terms because of periodicity.

from (3.6) we have

$$(3.8) \qquad \frac{(a, w)}{\| w \|_s} = \frac{(\Delta^s c, w)}{\| w \|_s};$$

shifting[6] s differentiations onto w and applying the Schwarz inequality we know that the expression on the right in (3.8) does not exceed

$$\frac{\| c \|_s \, \| w \|_s}{\| w \|_s} = \| c \|_s \,.$$

Thus

$$\| a \|_{-s} \leqq \| c \|_s \,.$$

Remark: Of course, an a priori estimate (3.2) holds for positive s also as consequence of the energy inequality for $\Delta^s L^*$.

Proof of the existence theorem: Let a be any smooth function, and b related to it by $L^*a = b$. The a priori estimate (3.2) for $s = 0$ shows that each b comes from at most one function a. Consider the linear functional $l(a)$ defined as (v, a). If v has s strong derivatives, $l(a)$ is, according to the generalized Schwarz inequality (2.4), a bounded linear functional of a in the minus s-norm. In view of the one-to-one correspondence between a and b, $l(a)$ can be regarded as a linear functional $l'(b)$ of b as well, and in view of the a priori estimate, it will be bounded by $\| b \|_{-s}$. The domain of definition of $l'(b)$ is the linear space of functions b of the form L^*a. Imagine $l'(b)$ extended (boundedly) to all of H_{-s}. According to the *representation theorem* of Section 2, $l'(b)$ can be represented as (u, b), where u is a function with s strong derivatives. We claim that this u is a solution of our differential equation

$$Lu = v.$$

Certainly, u is a weak solution:

$$l'(b) = (v, a) = (u, b)$$

for all smooth functions a related to b by $b = L^*a$. Since u has s strong derivatives and L is a first order operator, we can, for $s \geqq 1$, integrate (u, b) by parts, turning it into (Lu, a). Thus $(v, a) = (Lu, a)$ for all smooth a, and since the smooth a form a dense subset among the square integrable functions, we must have $Lu = v$.

Remark: It is easy to show, although this is not needed in the foregoing, that $l'(b)$ is *densely defined*, i.e. that the set of elements b of the form L^*a is dense in the minus s-norm, s *positive*. We prove this by showing that all bounded linear functionals $l(b)$ over H_{-s} which vanish for all b of the form L^*a vanish identically. According to the representation theorem, $l(b)$ can be represented as (u, b), where u has s strong derivatives. So we have

$$l(b) = (u, b) = (u, L^*a) = 0$$

[6]Again, there are no boundary terms on account of periodicity.

for all smooth a. Again, integration by parts gives us $Lu = 0$; since the operator L has a *unique* inverse u is zero, i.e. $l(b) \equiv 0$.

In his treatment of symmetric hyperbolic systems, Friedrichs gives a direct proof (with the aid of mollifiers) that $b = L^*a$ is dense among all *square integrable* functions ($s = 0$). This of course can also be deduced a posteriori from an adequate existence theory.

4. *Differentiability of Weak Solutions of Elliptic Equations*

Let L be an elliptic operator of order $2m$ with sufficiently differentiable coefficients, and let u be a weak solution of $Lu = v$ in some domain D; i.e.

$$(L^*a, u) = (a, v) \tag{4.1}$$

for all smooth functions a which are zero in a boundary strip. Here u and v may be generalized functions in the sense of Laurent Schwartz, i.e. they are only assumed to have a finite k-norm for some value of k, possibly very large negative. The differentiability theorem we wish to prove states:

DIFFERENTIABILITY THEOREM. *Let L be an elliptic operator of order $2m$ with sufficiently smooth coefficients, u a weak solution of $Lu = v$ in some domain D over which v belongs to H_{t-m}. Then ϕu belongs to H_{t+m} for every smooth factor ϕ which vanishes outside of a compact subset of D.*

Remark: If $t + m$ is nonnegative, this implies that u belongs to H_{t+m} with respect to any compact subset of D.

We shall prove the differentiability theorem first for the case where D is a period parallelogram. The result then is the following

DIFFERENTIABILITY THEOREM. *If $\| v \|_{t-m}$ is finite, then $\| u \|_{t+m}$ is also finite.*

Another way of stating this result is: The operator L maps H_{t-m} onto H_{t+m}.

We shall show first how to use the theorem in the periodic case to handle non-periodic solutions.

Let ϕ be a smooth function which is zero in a boundary strip of D but identically one in some domain D'. Denote the function ϕu by u'; u' is a weak solution of

$$Lu' = \phi v + Nu,$$

where N is a differential operator of *degree at most* $2m - 1$, whose coefficients are zero in the boundary strip. Denote $\phi v + Nu$ by v'.

Draw a parallelogram in Euclidean space containing the domain D, and imagine the coefficients of L so altered inside this parallelogram but outside the domain D that they become periodic but lose none of their ample differentiability properties. Denote the altered L by L'; since u' and v' vanish outside the domain D, they may both be imagined to be periodic functions satisfying

$$L'u' = v'.$$

Applying the differentiability theorem for the periodic case we can conclude that u' has $2m$ more strong derivatives than v'; although v' depends on derivatives of u, it depends only on derivatives of order *less* than $2m$, so there is a net gain of at least one derivative.

Next, let ϕ_2 be a smooth function which is zero outside of D, and identically one in some subdomain D'' of D'. Denote $\phi_2 u$ by u''; u'' is a weak solution of $Lu'' = \phi_2 v + N'u$, where N' is a differential operator of degree at most $2m - 1$ and has coefficients which are zero outside of D'. Since in D' the functions u and u' are identical, we can replace $N'u$ by $N'u'$, an expression with one higher degree of differentiability. This provides a "bootstrap" with which u can pull itself up to the required degree of differentiability.

The proof of the differentiability theorem in the periodic case is based on a *calculus criterion* for a function u to have finite s-norm. The criterion itself is a corollary of the representation theorem of Section 2 for bounded linear functionals in H_{-s}.

CRITERION OF DIFFERENTIABILITY. *Let u be a function in H_k, k any positive or negative integer, and let s denote an integer greater than k. Let $\{b\}$ be a linear space of elements in H_{-k} which are dense there. Assume now that*

$$(b, u) \leqq \text{const.}\ \| b \|_{-s} \tag{4.2}$$

for all elements b[7] *in this linear subspace. Conclusion: u belongs to H_s.*

Proof: The condition is clearly necessary. To show its sufficiency, consider the linear functional $l(b) = (b, u)$ as defined on the given linear space. The domain of $l(b)$, being dense in H_{-k}, is a fortiori dense in H_{-s}; according to (4.2) l is bounded in H_{-s}, so by closure it can be extended to the whole space. According to the representation theorem, the closure of $l(b)$ can be represented as (b, w), where w belongs to H_s. We claim that w is equal to u. For the linear functional $l'(b) = (b, u - w)$ is zero for a set of b dense in H_{-k}, and thus for all b in H_{-k}. But then the norm of $l'(b)$, which is $\| u - w \|_k$ must be zero; i.e., $u = w$.

This criterion is tailor-made for showing the differentiability of weak solutions of differential equations. Weak solutions are characterized namely by the relation (4.1)

$$(L^*a, u) = (a, v),$$

valid for all smooth functions a, L^* being the adjoint of L. To apply the criterion we need the following theorems:

A PRIORI ESTIMATE: *Let a be a smooth function, and define b as L^*a. Then*

[7]Since k is less than s, $\| b \|_{-s}$ is finite for every b in H_{-k}.

(4.3) $$\| a \|_{m-t} \leqq \| b \|_{-(m+t)}$$

for any integer t, positive, negative or zero.

DENSITY THEOREM. *The range of L^*a for smooth a is dense in H_k for any integer k, positive, negative or zero.*

From these two results and our criterion for differentiability we can deduce our differentiability theorem for weak solutions of $Lu = v$. For, assume that $\| v \|_{t-m}$ is finite for some integer t. Let a denote any smooth test function; define b as $b = L^*a$ and denote (b, u) by $l(b)$. By definition of weak solutions, $l(b) = (b, u) = (a, v)$. Since $\| v \|_{t-m}$ is finite, $(a, v) \leqq \| a \|_{m-t} \| v \|_{t-m}$. By the a priori estimate (4.3), $\| a \|_{m-t} \leqq \| b \|_{-(m+t)}$ and thus

(4.4) $$l(b) \leqq \text{const.} \| b \|_{-(m+t)} .$$

The weak solution u is assumed to have finite k norm for some (possibly very large negative) value of k. According to our density theorem, the set of elements b of form L^*a, for which therefore inequality (4.4) holds, is dense in the minus k norm. Hence according to our criterion of differentiability (with $s = m + t$), u has $m + t$ strong derivatives.

There remain the a priori estimates and the density theorem to be proved. Both are consequences of Gårding's inequality, plus a simple construction. Gårding's inequality states namely that scalar elliptic operators L of order $2m$ are *bounded from* below; i.e.,

(4.5) $$\| a \|_n \leqq (a, La) + T(a, a)$$

for all[8] smooth functions a, with some sufficiently large constant T. We may assume without loss of generality that L itself is already bounded, i.e. that (4.5) holds with $T = 0$:

(4.5)′ $$\| a \|_m^2 \leqq (a, La).$$

(Otherwise one may replace the operator Lu by $L'u = Lu + Tu$; the differentiability theorem for weak solutions of L follows from that for L' via a bootstrap argument.)

The expression (a, La') is a bounded bilinear functional of a and a' in the m-norm, and shall be abbreviated as $Q(a, a')$. Gårding's inequality can then be stated for all functions a with finite m-norm:

(4.5)″ $$\| a \|_m^2 \leqq Q(a, a).$$

Our a priori inequality (4.3) for $t = 0$ is an immediate consequence of Gårding's inequality (4.5) applied to the operator L^*. For, take the scalar product of both sides of $L^*a = b$ with a, apply Gårding's inequality on the left and the generalized Schwarz inequality (2.4) on the right:

$$\| a \|_m^2 \leqq (a, L^*a) = (a, b) \leqq \| a \|_m \| b \|_{-m} .$$

[8]This is for a compact domain, such as a period parallelogram. In the noncompact case Gårding's inequality holds in general only for functions which vanish in a boundary strip.

Dividing both sides by $\| a \|_m$ yields (4.3) with t equal to zero. For t negative, $-t = s$, take the scalar product of both sides of $L^*a = b$ with $\Delta^s a$, where Δ denotes the same operator as in Section 3. We get

$$(\Delta^s a, L^*a) = (a, \Delta^s L^*a) = (\Delta^s a, b). \tag{4.6}$$

Since $\Delta^s L^*$ is an elliptic operator of order $2m + 2s$, Gårding's inequality applies (we again assume $\Delta^s L^*$ positive definite):

$$\| a \|_{m+s}^2 \leqq (a, \Delta^s L^* a).$$

On the other hand, after shifting s differentiations onto b in $(\Delta^s a, b)$ we obtain the estimate:

$$(\Delta^s a, b) \leqq \| D^s a \|_m \, \| D^s b \|_{-m} \leqq \| a \|_{m+s} \, \| b \|_{s-m} .$$

Substituting these last two inequalities into the identity (4.6), we get the a priori estimate (4.3).

For t positive, we construct (see Section 3) a smooth function c which is related to a by

$$\Delta^t c = a.$$

Substitute this into the relation between a and b; we obtain

$$L^* \Delta^t c = b.$$

Take the scalar product of both sides with c. Since $L^*\Delta^t$ is an elliptic operator of order $2m + 2t$, Gårding's inequality applies:

$$\| c \|_{t+m}^2 \leqq (c, L^*\Delta^t c) = (c, b) \leqq \| c \|_{t+m} \, \| b \|_{-(t+m)} .$$

Since $\| a \|_{m-t} \leqq \| c \|_{m+t}$ (see Section 3), we have the desired inequality.

We turn now to the density theorem. We shall prove first that the range of L^*a for smooth a is dense in the minus m-norm (and therefore a fortiori is dense in any k-norm for k less than $-m$). For, assume to the contrary that $L^*a = b$ is not dense in H_{-m}. Then there is a bounded linear functional $l(b)$ over H_{-m}, not identically zero, which vanishes for all b of the above form. According to the *representation theorem* of Section 2, all linear functionals over H_{-m} can be represented as (b, w), with w having m strong derivatives, and the norm of this linear functional is $\| w \|_m$. So presumably we have

$$(L^*a, w) = 0$$

for all smooth a. Rewrite this as

$$Q(a, w) = 0.$$

Since Q is continuous in $\| a \|_m$, $Q(a, w)$ vanishes not only for all smooth a but for all a in H_m, and in particular for $a = w$:

$$Q(w, w) = 0.$$

But according to Gårding's inequality this implies that $\| w \|_m$ is zero, i.e. that the linear functional $l(b)$ has zero norm and so is identically zero.

Next we prove the density of the range of L^*a in any k-norm, $k > -m$. Set $k = -m + s$ and write

$$\Delta^s L^* a = \Delta^s b = c.$$

Since $\Delta^s L^*$ is an elliptic operator of order $2m + 2s$, its range is, by the foregoing, dense in H_{-m-s}. We claim that from this it follows that the set of $b = L^*a$ is dense in H_{-m+s}. We shall show namely that any *smooth* b_0 in H_{-m+s} can be approximated in the $(-m + s)$-norm by elements b of the form L^*a. Since the smooth elements b_0 of H_{-m+s} are dense, this will do.

Let b_0 be a smooth function and denote $\Delta^s b_0$ by c_0. Since the elements c of the form $\Delta^s L^* a = \Delta^s b$ are dense in H_{-m-s}, there will be an element c of this form such that $\| c - c_0 \|_{-(m+s)}$ is arbitrarily small. But according to our a priori inequality applied to the operator Δ^s, the quantity

$$\| b - b_0 \|_{-m+s} \leqq \| c - c_0 \|_{-m-s}$$

and hence is also arbitrarily small.

Bibliography

[1] Courant, R., Friedrichs, K., and Lewy, H., *Über die partiellen Differenzengleichungen der mathematischen Physik*, Math. Ann., Vol. 100, No. 1/2, 1928, pp. 32–74.

[2] Sobolev, S., *Sur quelques évaluations concernant les familles des fonctions ayant des dérivées à carré intégrable*, C. R. Acad. Sci. S.S.S.R., N. S. L. 1936, pp. 279–282.

[3] Courant, R. and Hilbert, D., *Methoden der mathematischen Physik*, Vol. II, Chapt. VII, Springer, Berlin, 1937.

[4] Sobolev, S., *On a theorem of functional analysis*, Mat. Sbornik, N.S. 4, 1938, pp. 471–497.

[5] Friedrichs, K. O., *On differential operators in Hilbert spaces*, Amer. J. Math., Vol. LXI, No. 2, 1939, pp. 523–544.

[6] Petrowsky, I. G., *Sur l'analyticité des solutions des systèmes d'équations différentielles*, Recueil mathématique, N.S. (Mat. Sbornik), Vol. 5, 1939, pp. 3–70.

[7] Friedrichs, K. O., *The identity of weak and strong extensions of differential operators*, Trans. Amer. Math. Soc., Vol. 55, No. 1, 1944, pp. 132–151.

[8] Friedrichs, K. O., *A theorem of Lichtenstein*, Duke Math. J., Vol. 14, No. 1, 1947, pp. 67–82.

[9] Van Hove, L., (A) *Sur l'extension de la condition de Legendre du calcul des variations aux intégrales multiples à plusieurs fonctions inconnues*, Indagationes Math., Vol. 7, No. 1, 1947, pp. 3–8.
(B) *Sur le signe de la variation seconde des intégrales multiples à plusieurs fonctions inconnues*, Acad. Roy. Belgique, Cl. Sci., Mém., Vol. 24, No. 5, 1949.

[10] Vishik, M. I., *The method of orthogonal and direct decomposition in the theory of elliptic differential equations*, Mat. Sbornik, N.S. 25, 1949, pp. 189–234.

[11] Schwartz, L., *Théorie des distributions*, Paris, Hermann, 1950–1951.

[12] Gårding, L., *Le problème de Dirichlet pour les équations aux dérivées partielles élliptiques homogènes à coéfficients constants*, C. R. Acad. Sci. Paris, Vol. 230, 1950, pp. 1030–1032.

[13] Vishik, M. I., *On strongly elliptic systems of differential equations*, (A) Akad. Nauk, S.S.S.R., Doklady, Vol. 74, 1950, pp. 881–884.
(B) Mat. Sbornik, N.S. 29, 1951, pp. 615–676.

[14] John, F., *General properties of solutions of linear elliptic partial differential equations*, Proc. of the Symposium on Spectral Theory and Differential Problems, Stillwater, Oklahoma, 1951, Chap. III.

[15] Gårding, L., *Le problème de Dirichlet pour les équations aux dérivées partielles élliptiques linéaires dans des domaines bornés*, C. R. Acad. Sci. Paris, Vol. 233, 1951, pp. 1554–1556.

[16] Milgram, A. and Rosenbloom, P. C., *Harmonic forms and heat conduction: Part I, Closed Riemannian manifolds; Part II, Heat distribution on complexes and approximation theory*, Proc. Nat. Acad. Sci., Vol. 37, 1951, pp. 180–184, pp. 435–438.

[17] Browder, F. E., (A) *The Dirichlet problem for linear elliptic equations of arbitrary even order with variable coefficients*, Proc. Nat. Acad. Sci., Vol. 38, No. 3, 1952, pp. 230–235.
(B) *The Dirichlet and vibration problems for linear elliptic differential equations of arbitrary order*, Vol. 38, No. 8, 1952, pp. 741–747.
(C) *Assumption of boundary values and the Green's function in the Dirichlet problem for the general linear elliptic equation*, Vol. 39, No. 3, 1953, pp. 179–184.
(D) *Linear parabolic differential equations of arbitrary order; general boundary-value problems for elliptic equations*, Vol. 39, No. 3, 1953, pp. 185–190.

[18] John, F., *Derivatives of continuous weak solutions of linear elliptic equations*, Comm. Pure Appl. Math., Vol. VI, No. 3, 1953, pp. 327–335.

[19] Browder, F. E., *Strongly elliptic systems of differential equations*, Contributions to the theory of partial differential equations, Annals of Mathematics Studies, No. 33, Princeton Univ. Press, 1954, pp. 15–51.

[20] Milgram, A. and Lax, P. D., *Parabolic equations*, Contributions to the theory of partial differential equations, Annals of Mathematics Studies, No. 33, Princeton Univ. Press, 1954, pp. 167–190.

[21] Morrey, C. B., *Second order elliptic systems of differential equations*, Contributions to the theory of partial differential equations, Annals of Mathematics Studies, No. 33, Princeton Univ. Press, 1954, pp. 101–159.

[22] Weber, H., *Die partiellen Differentialgleichungen der mathematischen Physik nach Riemann's Vorlesungen in 4-ter Auflgae neu bearbeitet*, Braunschweig, Friedrich Vieweg, 1900, Vol. 1, p. 390.

[23] Hadamard, J., *Sur l'intégrale résiduelle*, Bull. Soc. Math. France, Vol. 28, 1900, pp. 69–90.

[24] Zaremba, S., *Sopra un teorema d'unicità relativo alla equazione delle onde sferiche*, Rend. Accad. Naz. Lincei, Ser. 5, Vol. 24, 1915, pp. 904–908.

[25] Rubinowicz, A., (A) *Hertstellung von Lösungen gemischter Randwertprobleme bei hyperbolischen Differentialgleichungen zweiter Ordnung durch Zusammenstückelung aus Lösungen einfacherer gemischter Randwertaufgaben*, Monatsh. Math. Phys., Vol. 30, 1920, p. 65.
(B) *Über die Eindeutigkeit der Lösung der Maxwellschen Gleichungen*, Phys. Zeit., Vol. 27, 1926, p. 707.
(C) *Zur Integration der Wellengleichung auf Riemannschen Flächen*, Math. Ann., Vol. 96, 1927, pp. 648–687.

[26] Friedrichs, K., and Lewy, H., (A) *Über die Eindeutigkeit und das Abhängigkeitsgebiet der Lösungen beim Anfangswertproblem linearer hyperbolischer Differentialgleichungen*, Math. Ann., Vol. 98, 1927, pp. 192–204.
(B) *Über fortsetzbare Anfangsbedingungen bei hyperbolischen Differentialgleichungen in drei Veränderlichen*, Nachr. Ges. Wiss. Göttingen, No. 26, 1932, pp. 135–143.

[27] Schauder, J., *Das Anfangswertproblem einer quasilinearen hyperbolischen Differentialgleichung zweiter Ordnung in beliebiger Anzahl von unabhangigen Veränderlichen*, Fund. Math., Vol. 24, 1935, pp. 213–246.

[28] Sobolev, E. L., (A) *Diffractsia vol i na rimanovykh poverkhnostakh*, Akad. Nauk, SSSR, Trudy Mat. Inst. V. A. Steklova, Vol. 9, 1935, pp. 39–105.
(B) *Méthode nouvelle à résoudre le problème de Cauchy pour les équations linéaires hyperboliques normales*, Rec. Math. (Mat. Sbornik), N. S. 1(43), 1936, pp. 39–72.

[29] Frankl, F., *Über das Anfangswertproblem für lineare und nichtlineare hyperbolische partielle Differentialgleichungen zweiter Ordnung*, Rec. Math. (Mat. Sbornik), N. S. 2(44), 1937, pp. 793–814.

[30] Petrovskii, I., *Über das Cauchysche Problem für Systeme von partiellen Differentialgleichungen*, Rec. Math. (Mat. Sbornik), N. S. 2(44), 1937, pp. 814–868.

[31] Sobolev, S. L., *Sur la théorie des équations hyperboliques aux dérivées partielles*, Rec. Math. (Mat. Sbornik), N. S. 5(47), 1939, pp. 71–99.

[32] Leray, J., (A) *Lectures on hyperbolic equations with variable coefficients*, Princeton, Inst. for Adv. Stud., Fall, 1952.
(B) *On linear hyperbolic differential equations with variable coefficients on vector spaces*, Annals of Mathematics Studies, No. 33, Princeton Univ. Press, 1954, pp. 201–210.

[33] Holmgren, E., *Über Systeme von linearen partiellen Differentialgleichungen*, Öfvers. Kongl. Vetens.-Akad. Förh., Vol. 58, 1901, pp. 91–103.

[34] Hadamard, J., *Lectures on Cauchy's problem in linear partial differential equations*, Yale Univ. Press, 1923, or *Le Problème de Cauchy et les équations aux dérivées partielles linéaires hyperboliques*, Hermann, Paris, 1932.

[35] Riesz, M., *L'intégrale de Riemann-Liouville et le problème de Cauchy*, Acta Math., Vol. 81, 1948, pp. 1–222.

[36] Bureau, F., *Quelques questions de Géométrie suggérées par la théorie des équations aux dérivées partielles totalement hyperboliques*, Colloque de Géométrie Algébrique, Liege, 1949.

[37] Weyl, H., *The method of orthogonal projection in potential theory*, Duke Math. J., Vol. 7, 1940, pp. 411–444.

[38] Gårding, L., *On a lemma by H. Weyl*, Kungl. Fys. Sällskapets i Lund Förh., Vol. 20, 1950, pp. 1–4.

[39] Courant, R., *Variational methods for the solution of problems of equilibrium and vibrations*, Bull. Amer. Math. Soc., Vol. 49, 1943, pp. 1–23.

[40] Gårding, L., *L'inégalité de Friedrichs et Lewy pour les équations hyperboliques linéaires d'ordre supérieur*, C. R. Acad. Sci. Paris, Vol. 239, 1954, pp. 849–850.

[41] Nirenberg, L., *Remarks on strongly elliptic partial differential equations*, appearing in this issue.

[42] Friedrichs, K. O., *Differentiability of solutions of elliptic differential equations*, Comm. Pure and Appl. Math., Vol. 6, 1953, pp. 299–326.

[43] Friedrichs, K. O., *Symmetric hyperbolic linear differential equations*, Comm. Pure and Appl. Math., Vol. 7, 1954, pp. 345–392.

Received May 6, 1955.

Reprinted from the Proceedings of the NATIONAL ACADEMY OF SCIENCES,
Vol. 42, No. 11, pp. 872–876. November, 1956.

THE PROPAGATION OF DISCONTINUITIES IN WAVE MOTION*,†

BY R. COURANT AND P. D. LAX

NEW YORK UNIVERSITY

Communicated September 4, 1956

Discontinuities of "waves," i.e., of solutions of hyperbolic systems of partial differential equations, can occur only along characteristic surfaces; moreover, on these surfaces the discontinuities spread along bicharacteristics ("rays") on which they satisfy ordinary differential equations.

These facts are well known for the classical equations of wave propagation. In this note they will be presented for general hyperbolic systems of k linear equations of first order for a vector function $u(x^0, x^1, \ldots, x^n)$ with k components; the first variable x^0 may be identified with time t. By analyzing the possible discontinuities we shall show that discontinuities in the initial data on a spacelike manifold are propagated along characteristics.

THEOREM. *Assume that*

$$Mu = \sum_{\nu=0}^{n} A^\nu u_{x^\nu} + Bu = 0$$

is a hyperbolic system of k linear equations with k unknowns, and that the hyperplanes $x^0 =$ *const. are spacelike, i.e., the matrix* A^0 *is the identity, and that all linear com-*

binations $\sum \xi_\nu A^\nu$ *of the matrices* A^ν *with real coefficients* ξ_ν *have real and distinct eigenvalues. In addition, assume that* A^ν *and* B *are sufficiently differentiable functions of the independent variables. On the spacelike manifold* $x^0 = 0$ *initial data are assigned:*

$$u(0, x^1, \ldots, x^n) = f(x^1, \ldots, x^n),$$

which have on either side of a sufficiently smooth $(n - 1)$*-dimensional manifold* Γ *continuous derivatives of sufficiently high order that suffer jump discontinuities across* Γ*. Then the solution* u *of* $Mu = 0$ *with these initial values has continuous partial derivatives of sufficiently high order everywhere except on the k characteristic surfaces issuing from* Γ*; across these, the partial derivatives of the solution have jump discontinuities.*

This discontinuous solution is understood as a "weak solution" u, i.e., as one which satisfies the relation

$$\int uM^*w\, dx = 0 \tag{1}$$

for all smooth test functions w with compact support; M^* is the adjoint of M.

1. *Analysis of Discontinuities.*—Let u be a piecewise smooth weak solution with discontinuity along a smooth surface C. Integration by parts of equation (1) leads to the condition

$$\sum \phi_\nu A^\nu(u) = 0 \tag{2}$$

for the jump (u) across C, where ϕ_ν are the direction cosines of the normal to the surface of discontinuity. If (u) is not zero, the matrix $A = \sum \phi_\nu A^\nu$ must be singular, i.e., the carrier of the discontinuity must be a characteristic surface and the jump (u) must be annihilated by A. Since we have assumed that A has the rank $k - 1$, we have

$$(u) = \sigma R, \tag{3}$$

where R is the normalized right null vector of A, σ a scalar as yet unrestricted.

Denoting the jump in the partial derivatives of u by (u_{x^ν}), we conclude, since the differential equation $Mu = 0$ is satisfied on both sides, that

$$\sum A^\nu(u_{x^\nu}) + B(u) = 0.$$

We multiply this relation by the left null vector L of A and get

$$\sum LA^\nu(u_{x^\nu}) + LB(u) = 0. \tag{4}$$

Since $\sum \phi_\nu LA^\nu = \sum L\phi_\nu A^\nu = LA = 0$, equation (4) is independent of the jump in the normal derivative. On the other hand, the jump in the tangential derivative is equal to the tangential derivative in the jump in u. So equation (4) can be rewritten as

$$\sum LA^\nu(u)_{x^\nu} + LB(u) = 0. \tag{4'}$$

Substitution of expression (3) for (u) into equation (4′) yields

$$\sum LA^\nu R\cdot\sigma_{x^\nu} + (LA^\nu R_{x^\nu} + LB)\sigma = 0, \tag{5}$$

a single first-order linear differential equation for σ. If the surface of discontinuity is regarded as the level surface of a function ϕ, the characteristic condition $\Delta =$

det $\sum \phi_{x^\nu} A^\nu = 0$ may be regarded as a nonlinear partial differential equation for ϕ; it is easy to show that the direction of differentiation in equation (5) is the *bicharacteristic* direction relative to this equation.

There are k distinct characteristic hypersurfaces through a given $(n - 1)$-dimensional manifold Γ in the initial hyperplane; across each, the jump in (u) must be parallel to the corresponding null vector R. On Γ these null vectors are the eigenvectors of $\sum_{\nu=1}^{n} \gamma_\nu A^\nu$, where γ_ν, $\nu = 1, \ldots, n$, are the direction numbers of the normal to Γ in the hyperplane $x_0 = 0$. They form a complete set, and therefore the jump in the initial values along Γ can be uniquely decomposed as

$$(f) = \sum \sigma_0^j R_j.$$

The quantities σ_0^j are the values on Γ of the scalars σ in equation (3); these initial values and the differential equation (5) together determine the value of σ uniquely on each characteristic surface.

The magnitude of the jump of all higher derivatives of u can be determined recursively, by the condition of continuity for the higher derivatives of Mu across a discontinuity surface. The jumps of the higher derivatives satisfy inhomogeneous ordinary differential equations along the bicharacteristics.

The magnitude of the discontinuities of a solution along the characteristic surfaces issuing from the intersection of an initial and boundary surface can be calculated in a similar fashion in terms of the prescribed data.

2. *Construction of Solution of a Discontinuous Initial-Value Problem.*—Let $v(x^0, x^1, \ldots, x^n)$ be an auxiliary function which has continuous partial derivatives of a sufficiently high order l in each of the $k + 1$ wedges cut out of the half-space $x^0 > 0$ by the k characteristic surfaces through Γ, so chosen that the jumps of v and of its partial derivatives across the characteristic surfaces are equal to the jumps of the unknown solution u of the discontinuous initial-value problem calculated a priori by the method of section 1. The function $\psi = Mv$ has, then, partial derivatives of order $l - 1$; the method for calculating the jumps guarantees that ψ and its partial derivatives are continuous across the characteristic surfaces.

Introduce $u - v = w$ as a new unknown; it has to satisfy the inhomogeneous equation

$$Mw = -\psi$$

with a sufficiently differentiable right-hand side. The initial values of w,

$$w(0, x^1, \ldots, x^n) = f(x^1, \ldots, x^n) - v(0, x^1, \ldots, x^n),$$

are, by construction of v, sufficiently differentiable. Therefore, by known theorems,[1] such a solution w exists and is sufficiently differentiable. This completes the proof of the theorem.

It is easy to verify that if the initial vector f has piecewise square-integrable derivatives of order m, the solution has piecewise square-integrable derivatives of order $m - 2$. Likewise, if the coefficients are analytic, f piecewise analytic, and Γ an analytic curve, the solution u is piecewise analytic.

The above method for constructing the solution shows clearly that if the coefficients A^ν, B depend on a number of parameters in a sufficiently differentiable manner, so will the solution of a fixed discontinuous initial-value problem.

3. *Generalized Huygens Principle.*—So far we have shown that jump discontinuities in the initial values cause jump discontinuities in the solution along the characteristics issuing from the position of the discontinuity but do not affect the differentiability of the solution elsewhere. It turns out that the effects of other kinds of singularities in the data are likewise propagated only along the bicharacteristics. This property of wave propagation may be called the

GENERALIZED HUYGENS PRINCIPLE. *The differentiability properties of a solution u at a point P, time T, depend only on the differentiability properties of the initial data $u(0, x) = f(x)$ in the neighborhood of those points x which lie on the bicharacteristics issuing from (T, P).*

This result could be derived by approximating singular data by a sequence of data with a finite number of jumps. But the result can be read off from a representation formula for the solution derived with the aid of a Riemann function and Radon's formula for representing functions in terms of their plane averages.[2]

Assume that the number of space variables is 3. Apply Green's formula to a pair of solutions u and v of $Mu = 0$ and $M^*v = 0$ in the slab $0 \leq t \leq T$. Assuming that one of them, say u, vanishes for $|x|$ large, we get

$$\iiint u(T, x)\, v(T, x)\, dx = \iiint u(0, x)\, v(0, x)\, dx. \tag{6}$$

Denote the initial value of u by f; choose the value of v on $t = T$ to be as follows:

$$v(T, x) = \begin{cases} 1 & \text{for } (x - P)\cdot\omega > 0, \\ 0 & \text{for } (x - P)\cdot\omega < 0, \end{cases}$$

P some space point, ω an arbitrary unit vector in space. This function v is the solution of a discontinuous backward initial-value problem for M^*; according to our theory, the solution is piecewise smooth and depends smoothly on the parameters P and ω. So we have

$$\iiint_{(x - P)\cdot\omega \geq 0} u(T, x)\, dx = \iiint f(x)\, v(x, P, \omega)\, dx, \tag{6'}$$

where $v(x, P, \omega)$ denotes the value of v on $t = 0$. Differentiate equation (6′) with respect to that component of P which is perpendicular to ω. We get

$$\iint_{(x - p)\cdot\omega = 0} u(T, x)\, dx = \iiint f(x)\, v_p\, dx + \sum \iint_{C_j} f(x)\, (v)_j\, dS. \tag{7}$$

Here $C_j = C_j(P, \omega)$ is the intersection of the j-th characteristic surface through $(x - P)\cdot\omega = 0$, $t = T$, with the initial hyperplane $t = 0$; v_p denotes the derivative of v with respect to $P\cdot\omega$, and $(v)_j$ is the jump of v across C_j. Integrate equation (7) with respect to ω over the unit sphere. We get

$$\iiint u(T, x)\, \frac{dx}{r} = \iiint f(x)\, k(x, P)\, dx, \tag{8}$$

where r denotes distance from the point P. It is not very difficult to show that the kernel k is a smooth function of both variables except at points where x lies on the envelope of the surfaces $C_j(P, \omega)$. The set of these points x is clearly the set lying on the bicharacteristics through P. Applying the Laplacian with respect to P to equation (8), we get on the left $u(P)$, on the right an expression which is a smooth function of P, provided that f vanishes in a neighborhood of the crucial point set. This completes the proof of the generalized Huygens principle. If u satisfies an inhomogeneous relation $Mu = g$, the above reasoning shows that the differentiability of $u(P)$ is governed by the differentiability of g only in the neighborhood of the bicharacteristics issuing from P.

* Dedicated to Carl Ludwig Siegel in Göttingen on his sixtieth birthday.

† A detailed paper will appear in *Comm. Pure and Appl. Math.* The laws of propagation of discontinuities along bicharacteristics, in particular for differential equations of second order, have often been discussed in the literature. (See, e.g., R. Courant and D. Hilbert, *Methods of Mathematical Physics*, Vol. II, chaps. V, VI, and J. B. Keller, "Geometrical Acoustics I. The Theory of Weak Shock Waves," *J. Appl. Phys.*, **25**, 938–947, 1954.)

[1] See K. O. Friedrichs, *Comm. Pure and Appl. Math.*, **7**, 345–393, 1954, for symmetric hyperbolic systems, and J. Leray, *Lectures on Hyperbolic Equations with Variable Coefficients* (Princeton, N.J.: Institute for Advanced Study, 1952), for general hyperbolic systems.

[2] Radon's formula was used in the solution of Cauchy's problem for equations with constant coefficients by R. Courant and A. Lax, *Comm. Pure and Appl. Math.*, Vol. **8**, 1955. For diverse application see Fritz John, *Plane Waves and Spherical Means* (New York: Interscience Publishers, Inc., 1956).

ASYMPTOTIC SOLUTIONS OF OSCILLATORY INITIAL VALUE PROBLEMS

By Peter D. Lax

Introduction. In this paper we investigate the asymptotic behavior of solutions of linear partial differential equations with highly oscillatory initial values. We consider an operator M (a first order matrix operator throughout this paper), and a solution $u(x^1, \cdots, x^m, t) = u(x, t)$ of the equation

$$Mu = 0$$

with initial values

$$u(x, 0) = \phi(x)$$

where $\phi(x)$ depends on a parameter ξ in an oscillatory manner

$$\phi(x) = e^{i\xi l(x)}\psi(x),$$

with $l(x)$ some given real-valued function.

In §1, we shall construct a formal asymptotic series for u:

$$u \sim e^{i\xi l(x,t)}\left\{v_0 + \frac{1}{\xi}v_1 + \cdots\right\}$$

and show that, if the hyperplane $t = 0$ is space-like for the operator M, our expansion has an asymptotic validity. This is potentially useful in calculating asymptotic diffraction patterns (see [10], [11], and [14] for a detailed discussion); namely, the problem of diffraction can be formulated as finding the steady state of a solution with oscillatory initial values.

In §2 we use the formal asymptotic series to show that, if the hyperplane $t = 0$ is *not* space-like for an operator M with analytic coefficients, then the initial value problem is incorrectly posed. The method used in this paper is a generalization of the method which Hadamard used to obtain this result for operators with constant coefficients.

In §3, we study the manner in which solutions of hyperbolic equations depend on their initial values. This is done by using the asymptotic expansion to construct, in §4, by Fourier synthesis, an approximate influence function; i.e. we find—modulo a smooth function—the kernel G of the integral operator expressing the value of a solution of a hyperbolic equation at any point P in terms of its values at points Q of the initial manifold. A different construction of the influence function, based on an expression of the δ-function in terms of

Received May 13, 1957. Presented in part to the American Mathematical Society, Cambridge meeting, October, 1956. This work was supported by the Office of Naval Research, United States Navy, under Contract N6ori-201, T.O. 1. Reproduction in whole or in part permitted for any purpose of the United States Government.

plane waves, has been given by R. Courant and the author in [3]. Both constructions show that the influence function $G(P, Q)$ is a smooth function except when P and Q lie on the same bicharacteristic. This is equivalent to the generalized Huygens' principle: the differentiability properties of solutions at a point P depend on the differentiability properties of the initial data only at those points of the initial manifold which lie on a bicharacteristic issuing from P.

A more detailed exposition of both methods will be given in a forthcoming joint paper in the Communications for Pure and Applied Mathematics.

I wish to thank Avron Douglis for a number of helpful suggestions concerning the presentation of this paper.

1. Formal construction of asymptotic series. Let M be a linear first order matrix operator

$$Mu = u_t + \sum A_j u_{x^j} + Bu \tag{1.1}$$

acting on vector-valued functions u with n components; here A_j and B are n by n matrices which are functions of $x = (x^1, x^2, \cdots, x^m)$ and t.

We shall abbreviate the array A_j by A; likewise, we shall denote the gradient $\{u_{x^j}\}$ by u_x. With this notation we can write M as follows:

$$Mu = u_t + A \cdot u_x + Bu. \tag{1.1'}$$

We wish to construct formal asymptotic solutions

$$\tilde{u} \sim e^{i\xi l(x,t)}\left\{v_0 + \frac{v_1}{\xi} + \cdots\right\} \tag{1.2}$$

of $Mu = 0$. Here l is a scalar, and the v_n are vector-valued functions.

If we substitute (1.2) into (1.1), we get

$$M\tilde{u} = i\xi e^{i\xi l}[l_t + A \cdot l_x] \sum_{n=0}^{\infty} \frac{v_n}{\xi^n} + e^{i\xi l} \sum_{n=0}^{\infty} \frac{Mv_n}{\xi^n};$$

the coefficient of ξ is

$$ie^{i\xi l}[l_t + A \cdot l_x]v_0 \tag{1.3}$$

and that of ξ^{-n}, $n \geq 0$, is

$$ie^{i\xi l}[l_t + A \cdot l_x]v_{n+1} + e^{i\xi l}Mv_n . \tag{1.4}$$

We require that these quantities vanish. For the expression (1.3), this means that the matrix $l_t + A \cdot l_x$ annihilates the vector v_0. For $v_0 \neq 0$, this can be the case if and only if the matrix $l_t + A \cdot l_x$ is singular.

We assume from now on that at each point x, t, the matrix $p \cdot A$ has n *distinct* eigenvalues for all real values of the parameters p_j under discussion. Then in any simply connected domain of x, t, p-space, we can define n functions $\lambda_\nu(x, t, p_1, p_2, \cdots, p_m)$ $(\nu = 1, 2, \cdots, n)$ as the eigenvalues of the matrix $p \cdot A$. (If the eigenvalues are distinct and real, then n functions λ_ν can be defined as single-valued functions even in a multiply connected domain.) These

λ_ν depend analytically on the elements of $p \cdot A$; therefore, they are analytic functions of the p's and possess the same differentiability properties as functions of x and t as the coefficients of A_i.

The condition that the matrix $l_t + A \cdot l_x$ be singular means that $-l_t$ is an eigenvalue of $A \cdot l_x$ and is expressed by the following differential equation for the "phase function" l:

$$l_t + \lambda(x, t, l_{x^1}, l_{x^2}, \cdots, l_{x^m}) = 0, \tag{1.5}$$

where λ is any one of the n functions λ_ν defined above. Equation (1.5) is the analogue of the classical eikonal equation.

DEFINITION 1.1. *The hyperplane $t = 0$ is called space-like for an operator M if the eigenvalues λ_ν of the matrix $p \cdot A$ are real for all real p_i.*

If λ is real-valued, the classical method of characteristics in the theory of first order equations (see e.g. Courant-Hilbert, Vol. II, Chapter II) enables us to construct solutions of (1.5) with prescribed values at $t = 0$, provided that λ is a twice differentiable function of all its arguments. We have seen that λ is analytic in its last m arguments and if we assume from now on that, whenever the λ are real (hyperbolic case), A_i is twice differentiable in x, t, then λ is twice differentiable.

Since the eikonal equation is nonlinear, the solution l will not, in general, exist for all t. This places a limitation on this method.

If λ is a complex-valued function, then the only general existence theorem for a solution l of (1.5) is Cauchy-Kowalewski's theorem. In that case, we assume that the A_i (and thereby the functions λ) are analytic functions of x and t, and we obtain a complex-valued solution l.

From the definition of the function λ, it follows that λ is homogeneous of degree one in the variables p, so that Euler's identity

$$\lambda = p \cdot \lambda_p$$

holds. We substitute it into (1.5):

$$l_t + l_x \cdot \lambda_p = 0. \tag{1.6}$$

Recalling that the direction of the characteristic curves of (1.5) is

$$\dot{t} = 1, \qquad \dot{x}^i = \lambda_{p_i}, \tag{1.7}$$

we conclude from (1.6) that the phase function l is constant along the characteristics of (1.5).

Equation (1.5) is the characteristic equation for the operator M; the characteristics of (1.5) are the bicharacteristics of M. Thus we have proved

LEMMA 1.1. *The phase function l is constant along the bicharacteristics of M.*

Denote by

$$R = R_\nu(x, t, p_1, \cdots, p_m) \qquad (\nu = 1, 2, \cdots, n)$$

the right eigenvector of the matrix $p \cdot A$ corresponding to the eigenvalue λ_ν, normalized in some convenient fashion. Now we choose a particular solution l of (1.5), put it into (1.3) and (1.4) and note that in order that (1.3) vanish, v_0 must be a scalar multiple of R:

$$v_0 = \sigma R, \tag{1.8}$$

where σ is a function of x and t. In order to make (1.4) vanish for $n = 0$, v_1 and v_0 must satisfy

$$[l_t + l_x + Al_x]v_1 - iMv_0 = 0. \tag{1.9}$$

This is a linear equation for v_1 with a singular matrix and hence has a solution if and only if the inhomogeneous term iMv_0 is orthogonal to the left nullvector L of the matrix $(l_t + l_x + Al_x)$:

$$LMv_0 = 0$$

or, by (1.8),

$$\begin{aligned} LM[\sigma R] &= L\{\sigma_t R + A \cdot \sigma_x R + B\sigma R + \sigma R_t + \sigma A \cdot R_x\} \\ &= L\{\sigma_t + A \cdot \sigma_x R + \sigma M[R]\} = 0. \end{aligned}$$

Assume that L and R were so normalized that $LR = 1$. Set

$$LA_j R = a_j , \qquad LM[R] = b;$$

then the above relation can be written as

$$\sigma_t + a \cdot \sigma_x + b\sigma = 0, \tag{1.10}$$

i.e. as a first order differential equation for σ in which, as we shall see below, the direction of differentiation is that of the bicharacteristic of M: Differentiate the identity

$$p \cdot AR = \lambda R$$

with respect to p_μ, obtaining

$$A_\mu R + p \cdot AR_{p_\mu} = \lambda_{p_\mu} R + \lambda R_{p_\mu} .$$

Multiply this relation by the left eigenvector L of the matrix $p \cdot A$. We get

$$LA_\mu R = \lambda_{p_\mu}$$

which, according to (1.7), is indeed the bicharacteristic direction.

If the hyperplane $t = 0$ is space-like, then the quantities λ, l, R and L are real-valued, and so are therefore the coefficients a_j and b of equation (1.10). In this case, we can construct a solution σ of (1.10) with prescribed initial values by the method of characteristics.

If the hyperplane $t = 0$ is not space-like, i.e. if λ is complex-valued then the coefficients a_j and b will be complex-valued, and we rely on the Cauchy-Kowalewski theorem to construct a solution σ of (1.10) with prescribed analytic initial values.

Having thus determined a solution σ of (1.10) we determine the vector v_1 modulo an arbitrary scalar multiple of R. When we try to determine v_2 so that (1.4) vanishes for $n = 1$, we shall again find that this scalar multiple of R must satisfy a differential equation similar to (1.10) and so on. This is the method of determining successively the terms v_i in our asymptotic series (1.2). Since each step involves differentiation of the previous terms, we have to require, in order to carry out N steps, that the coefficients A_i of M have a sufficient number of continuous derivatives with respect to x and t.

The functions l and σ were determined as solutions of certain differential equations; their initial values can be assigned arbitrarily. We wish to take advantage of this fact in order to construct asymptotic solutions

$$\tilde{u} = e^{i\xi l(x,t)}\left\{v_0 + \frac{1}{\xi}v_1 + \cdots\right\}$$

of $Mu = 0$ with prescribed initial values

$$u(x, 0) = \phi(x)$$

of the form

$$\phi(x) = e^{i\xi l(x)}\psi(x). \tag{1.11}$$

We could equally well consider more general initial functions

$$\tilde{\varphi}(x) = e^{i\xi l}\left\{\psi_0 + \frac{\psi_1}{\xi} + \cdots\right\}.$$

When we determine the phase function $l(x, t)$ from the differential equation (1.5), we prescribe its initial value $l(x, 0)$ so that it coincides with $l(x)$ in equation (1.11).

We represent the vector $\psi(x)$ as a linear combination

$$\psi(x) = \sum_{j=1}^{n} \sigma_j(x)R_j(x) \tag{1.12}$$

of all the right eigenvectors R_j of the matrix

$$l_x \cdot A(x, 0).$$

This is possible since the eigenvalues of the matrix $p \cdot A$ were assumed to be simple, and therefore, the set of eigenvectors is complete. The functions $\sigma_j(x)$ determined by (1.12) furnish the initial values $\sigma(x, 0)$ for the solutions $\sigma(x, t)$ of equation (1.10). These $\sigma(x, t)$ are the coefficients which appear in (1.8). The full asymptotic expansion itself is the sum of n terms, each of the form (1.2).

2. Cauchy's problem for a non-space-like hyperplane. Let M be a first order operator of the form (1.1). Let D be a domain on the hyperplane $t = 0$ and G a domain of x, t-space containing D. Let $\phi(x)$ be a function defined in D, and $f(x, t)$ one defined on G; Cauchy's problem is to find a function $u(x, t)$ in G with continuous first derivatives which is equal to ϕ on D and which satisfies the equation $Mu = f$. If a unique solution exists for any ϕ, f belonging to a pair of linear spaces $\{\phi\}$ and $\{f\}$, Cauchy's problem for the operator M is said to be *properly posed* in G with respect to these spaces.

DEFINITION. A domain G in x, t-space containing a domain D on the hyperplane $t = 0$ is called a *domain of determinacy* of D if the only solution in G of the equation $Mu = 0$ with $\phi(x) \equiv 0$ in D is $u \equiv 0$.

We shall assume throughout this section that the coefficients of M are analytic functions of x and t. For such equations Fritz John's extension of Holmgren's uniqueness theorem, see [9], holds. According to this theorem, *$u(x, t)$ is uniquely determined by ϕ and f provided that G is a domain which can be obtained from D by a smooth deformation which is never characteristic anywhere.* From here on, G denotes such a domain of determinacy.

The linear spaces $\{\phi\}$ and $\{f\}$ shall be taken as the space C^∞ of all infinitely differentiable functions on D and G respectively, with the following customary topology, see Schwartz, [17]: The sequence ϕ_k tends to zero if all sequences of derivatives $D^l\phi_k$ tend to zero in D, uniformly for fixed l. The corresponding topology is taken in the space $\{f\}$.

If Cauchy's problem is properly posed, then for every pair ϕ, f, there is defined a solution

$$u = S(\phi, f) \tag{2.1}$$

where S is a mapping of $C^\infty(D) \times C^\infty(G)$ into the space $C^1(G)$ of once differentiable functions u over G. S is linear and has, as Banach has observed (see [1; 44–47]), a *closed graph*; i.e., if ϕ_k tends to ϕ, f_k to f and $u_k = S(\phi_k, f_k)$ to u in the respective topologies, then u is equal to $S(\phi, f)$. This is easy to show: $u = \lim u_k$ in the C^1 topology and $f^k \to f$ together imply that u is a solution of $Mu = f$, while $\phi_k = u_k(x, 0) \to \phi$ implies that $u(x, 0) = \phi$. But since G is a domain of determinacy of D, we can apply Holmgren's uniqueness theorem and conclude that there is just one solution, namely $S(\phi, f)$.

Moreover, Banach observed that C^∞ is a metric space and therefore the operator S, being an everywhere defined linear operator with a closed graph mapping a metric space into a metric space, is necessarily continuous (closed graph theorem). So we have

LEMMA 2.1. *In a properly posed Cauchy problem, the solution depends continuously on the data ϕ and f in the $C^\infty \times C^\infty$ topology.*

From the definition of convergence in the C^∞ topology it follows that there exists an integer k such that u depends boundedly on the $C^k \times C^k$ topology. So, we have

LEMMA 2.1′. *In a properly posed Cauchy problem, the solution is a continuous function of the data in the* $C^k \times C^k$ *topology for some* k.

The significance of this property in cases where the data ϕ and f are obtained by physical measurement has been emphasized by Hadamard.

THEOREM 2.1 *Let* M *denote a linear, first order operator*

$$Mu = u_t + A \cdot u_x + Bu,$$

whose coefficients A_i *and* B *are analytic functions of* x *and* t. *Assume that the hyperplane* $t = 0$ *is not space-like for* M *at* $x = x_0$, i.e. *that some linear combination* $p \cdot A$ *with real coefficients* p_i *has a non-real simple eigenvalue at* $x = x_0$, $t = 0$. *Then Cauchy's problem for* M *is incorrectly posed over any domain* D *on the hyperplane* $t = 0$ *containing the point* x_0.

According to Lemma 2.1′, to show that the initial value problem for a given operator M is *incorrectly* posed on a domain D, it suffices to display, for any value of k and any domain G in x, t-space, a one-parameter family of data ϕ_ξ and f_ξ which remain bounded in the C^k topology over D and G respectively, but such that as ξ tends to infinity, the corresponding solutions u_ξ of $Mu_\xi = f_\xi$, $u_\xi(x, 0) = \phi_\xi$ tend to infinity in the C^1 topology over the domain G. We shall use the finite section of our formal asymptotic series for this purpose.

We consider the case that the hyperplane $t = 0$ is *not* space-like on D for the operator M, i.e. that there exists a set of real numbers $p_1, p_2, \cdots, p_k$ such that the matrix $p \cdot A$ has a *non-real* eigenvalue at some point $(x_0, 0)$ of D. Denote this eigenvalue by $\lambda(x, t, p_1, p_2, \cdots, p_k)$.

Choose the initial phase linear in x:

$$l(x, 0) = p \cdot x \tag{2.2}$$

and let $l(x, t)$ be the solution of the eikonal equation (1.5). Since $\operatorname{Im} \{\lambda(x_0, 0, p_1, \cdots, p_k)\} \neq 0$, the imaginary part of l does not vanish for $x = x_0$, t sufficiently small but not zero.

Let u_N be the N-th section of the formal series (1.2):

$$u_N = e^{i\xi l}\left\{v_0 + \frac{v_1}{\xi} + \cdots + \frac{v_N}{\xi^N}\right\}.$$

The asymptotic series (1.2) was so constructed that

$$Mu_N = e^{i\xi l}\,\frac{Mv_N}{\xi^N}.$$

Let G be any domain in the x, t-space containing D and only points sufficiently close to D. Denote by m the maximum of the imaginary part of l in $\bar{G}$:

$$m = \max_{(x,t)\in\bar{G}} \operatorname{Im} l(x, t)$$

Since the imaginary part of l changes sign at $t = 0$, m is positive.

Abbreviate Mu_N by f_ξ and $u_N(x, 0)$ by ϕ_ξ. The C^k norm of quantities over G and D respectively behave, as ξ tends to infinity, as follows:

$$| f_\xi |_k = O(e^{\xi m}\xi^{k-N})$$

$$| \phi_\xi |_k = O(\xi^k);$$

however, the asymptotic behavior of the function $u_\xi = u_N$ itself, at the point P^* where the maximum of Im $l(x, t)$ is reached, is as follows:

$$u_\xi(P^*) = e^{\xi m}v_0(P^*)[1 + O(\xi^{-1})]. \tag{2.3}$$

For t small enough the function v_0 does not vanish anywhere (see §1). Taking $N > k$, we see from (2.3) that $u_\xi(P^*)$ tends to infinity faster than the C^k norm of f and ϕ. This completes the proof of Theorem 2.1.

One can speak of solving Cauchy's problem in a one-sided neighborhood of D. This too is incorrectly posed for a hyperplane which is not space-like. The same technique for constructing counter examples will also work here if care is taken to choose the proper sign for ξ.

Recalling the definition of a properly posed problem, Theorem 2.1 can be stated as

THEOREM 2.1′. *Let M be an operator as in Theorem 2.1, G any domain in x, t-space containing the point $(x_0, 0)$. Let $C^\infty(G)$ denote the space of infinitely differentiable functions in G* (i.e. *the space of admissible f for the equation $Mu = f$*) *and let $C_0^1(G)$ denote the subspace of those once-differentiable functions* (i.e. *of* C^1) *which vanish on $t = 0$. Then the image of C_0^1 under the differential operator M does not contain $C^\infty(G)$*; i.e., *there exists at least one function f in C^∞ for which the equation $Mu = f$ has no solution.*

So far, we have called Cauchy's problem correctly posed if a unique solution exists in a common domain. Now, we shall show that, even if we permitted the domain of existence to vary with the data, this dependence would turn out to be illusory.

THEOREM 2.2. *If the initial value problem*

$$Mu = 0, \qquad u(x, 0) = \phi(x), \qquad x \text{ in } D \tag{2.1′}$$

has, for every infinitely differentiable ϕ, a solution u in some domain G_ϕ of x, t-space, then there is a common domain of existence for all solutions.

Proof. First we show that there is a common domain of existence containing any subdomain D' whose closure is contained in the domain D of the hyperplane $t = 0$. Let G_n be a sequence of domains of determinacy of D' such that the distance of any point of G_n to D' is less than, say, $1/n$. Clearly, any domain of x, t-space which contains D contains all G_n for n sufficiently large.

Denote by A_n the set of all those data ϕ for which the corresponding solution u of (2.1′) exists in G_n, $n = 1, 2, \cdots$. The A_n are linear subspaces of C^∞ and their union is, by assumption, the whole space C^∞. By the category principle, A_n is of second category for $n = N$ large enough.

Denote by S_N the operator which assigns to the data ϕ in A_N the corresponding solution u in G_N (see (2.1)). As before, this operator has a closed graph; in other words, if a sequence of data ϕ_k in A_N tends to a limit ϕ in C^∞ and the corresponding solutions u_k tend to a limit u in the respective topologies, then the limit ϕ belongs to A_N and $u = S_N(\phi)$ is the corresponding solution. Since A_N is a subspace of second category, it follows from an extension of the closed graph theorem (Theorem 2.13.7 of Hille, [8A; 28]), that the operator S_N defined over A_N is *bounded*, and thus can be extended to the whole space by closure

From the geometry of domains of determinacy, it follows that the intersection of G_K and G_H is again a domain of determinacy. Therefore the solutions u_K and u_H defined in G_K and G_H agree in the intersection of their domains of definition.

Let $\{D_K'\}$ be a collection of open sets whose union is D and $\{G_k\}$ the corresponding domains of existence. We denote the union of the $\{G_k\}$ by G. Given any function ϕ in $C^\infty(D)$, we can define a solution u of (2.4) by putting $u = u_k$ in G_K. This solution is unique in G since G, being the union of domains of determinacy of D, is a domain of determinacy. This completes the proof of Theorem 2.2.

The situation for equations with nonanalytic coefficients is as follows: There are examples of equations which have nonzero solutions with zero Cauchy data. If we exclude these, it is easy to extend Theorem 2.1 to equations whose coefficients are merely infinitely differentiable, not necessarily analytic. As before, we have to construct for any value of k and any domain G of x, t-space containing a given domain D of the initial plane a one-parameter family of functions u_ξ such that as ξ tends to infinity, the ratio of the maximum of $| u_\xi |$ over G to the C^k norm of $u_\xi(z, 0)$ over D and of Mu_ξ over G tends to infinity.

We consider again the eikonal equation (1.5); if the coefficients of the original equation are not analytic functions of x and t, neither is λ, and therefore equation (1.5) may no longer possess a solution with initial values (2.2). But at any rate we can construct, for any given integer r, a function $l = l_r$ such that $l_r(x, 0)$ is equal to the prescribed initial value (2.2) and which satisfies the eikonal equation modulo a quantity which is $O(t^r)$. Such a "near solution" is e.g. furnished by the r-th section of the formal Taylor series in t computed from (1.5) and (2.2).

Having determined $l = l_r$ in this fashion, determine $v_0, v_1, \cdots, v_N$ as before. Denote by u_N the N-th section of the asymptotic series just constructed:

$$u_N = e^{i\xi l}\left\{v_0 + \frac{v_1}{\xi} + \cdots + \frac{v_N}{\xi}\right\}.$$

We observe that in Mu_N the coefficient of all powers of ξ greater than $-N$ are $O(t^r)$. So we have

$$| Mu_N | = O[(\xi t^r + 1/\xi^N) \mid e^{i\xi l} \mid]$$

and, likewise,

$$| Mu_N |_k = O[(\xi^k t^{r-k} + \xi^{k-N}) \, | e^{i\xi l} |].$$

The Taylor expansion for l gives

$$l = ct + O(t^2), \quad c = c(x).$$

From the differential equation (1.5) we have $c = -\lambda(x, 0, p)$. Since we have assumed that the hyperplane $t = 0$ is not space-like at (x_0, p), it follows that $c(x_0)$ has a nonzero, say positive, imaginary part γ. Denote by $P = P_\xi$ the point $(x_0, K \log \xi/\xi)$; the modulus of $e^{i\xi l}$—and therefore that of u_N—is of the order $\xi^{K\gamma}$ at this point:

$$(2.5) \qquad | u_N(P) | = \xi^{K\gamma} \, | v_0 | \, [1 + O(\xi^{-1})].$$

Denote by $G' = G'_\xi$ a subdomain of G which contains the point P_ξ on its boundary and is contained in the slab $0 \leq t \leq K \log \xi/\xi$. In G we have

$$| e^{i\xi l} | = O(\xi^{K\gamma}).$$

From this and (2.4) we conclude that in G'

$$(2.4') \qquad | Mu_N |_k = O(\xi^{K\gamma+k+1-r} + \xi^{K\gamma+k-N}).$$

For the initial values we have as before

$$(2.6) \qquad | \phi |_k = O(\xi^k).$$

Define f_ξ in G as follows: $f_\xi = Mu_N$ in G', and is so continued outside that its C^k norm is increased at most by a constant factor. Such a smooth continuation of f_ξ from G' to G is possible. So as a consequence of (2.4′) we have in G:

$$(2.7) \qquad | f_\xi |_k = O(\xi^{K\gamma+k+1-r} + \xi^{K\gamma+k-N}).$$

Denote by u_ξ the solution in G of the initial value problem

$$u_\xi(x, 0) = \phi_\xi(x) \quad \text{on } D$$

$$Mu_\xi = f_\xi \qquad \text{in } G.$$

Assuming that the value of u on D and of Mu_ξ in G' uniquely determines u_ξ in G', we conclude that u_ξ equals u_N in G'; in particular we have from (2.5) that

$$(2.5') \qquad u_\xi(P_\xi) = u_N(P_\xi) = \xi^{K\gamma} \, | v_0 | \, [1 + O(\xi^{-1})].$$

We choose now K, r and N so that K_γ, $r - 1$ and N are greater than k. Comparing (2.5′) with (2.6) and (2.7) we see that $u_\xi(P_\xi)$ tends to infinity faster than the C^k norms of ϕ_ξ and f_ξ.

3. Generalized Huygens' principle. Throughout this section, M shall denote a first order operator of the form (1.1) with the following properties:

a) The coefficients of M are sufficiently smooth functions of x and t.

b) M is hyperbolic and the hyperplanes t = constant are space-like (see Definition 1.1).

We introduce the following familiar norms for smooth functions ϕ of x with compact support

$$|| \phi ||_l^2 = \int \sum_{s \leq l} | D^s \phi |^2 \, dx,$$

the summation being over all partial derivatives $D^s\phi$ of ϕ of order $s \leq l$.

Following Laurent Schwartz, we introduce the dual of these norms with respect to the L_2 scalar product:

$$|| \phi ||_{-l} = \sup_{\psi} \frac{| (\phi, \psi) |}{|| \psi ||_l},$$

where (ϕ, ψ) abbreviates the L_2 scalar product in x-space

$$(\phi, \psi) = \int \phi \psi \, dx.$$

The closure of the set of smooth functions with compact support with respect to these norms is called the space of functions with plus l-norm and minus l-norm respectively, and denoted by H_l and H_{-l}.

For smooth ϕ and ψ with compact support we have by definition the *Schwartz inequality*:

$$| (\phi, \psi) | \leq || \phi ||_{-l} \, || \psi ||_l .$$

This shows that the scalar product (ϕ, ψ) is a continuous bilinear functional; thus it can be extended by continuity to any ϕ in H_{-l}, ψ in H_l.

The quantity $|| \psi ||_{-l}$ is by definition the norm of the linear functional

$$l(\phi) = (\phi, \psi)$$

with respect to the l-norm. Since these linear functionals are complete we conclude:

FIRST REPRESENTATION THEOREM. *Every bounded linear functional $l(\phi)$ over H_l can be expressed in the form*

$$l(\phi) = (\phi, \psi), \qquad \psi \in H_{-l} .$$

It is not hard to show (see e.g. [17] or [12] for a proof) that this relation between H_l and H_{-l} is dual:

SECOND REPRESENTATION THEOREM. *Every bounded linear functional $l(\phi)$ over H_{-l} can be represented in the form*

$$l(\phi) = (\phi, \psi), \qquad \psi \in H_l .$$

In this section we shall study the manner of dependence of solutions of $Mu = 0$

on their values on the initial hyperplane $t = 0$. At any given point P, the value of a solution u is a linear functional of the initial values $u(x, 0) = \phi$:

$$u(P) = L(\phi).$$

The fundamental theorem in the theory of hyperbolic equations is the so-called *energy inequality*, which asserts that the solution at a later time t depends boundedly on its initial value in the l-norm. For second order hyperbolic equations this was derived by Friedrichs and Lewy, [5], for symmetric hyperbolic systems by Friedrichs, [4] and Lax [12] and for general hyperbolic equations by Petrowsky, [15], Leray [13] and Gårding [6]. We state now

LEMMA 3.1. *On any hyperplane* $t =$ constant, *the l-norm of a solution is bounded by a constant multiple of its l-norm on the initial hyperplane; l is any positive or negative integer.*

According to Sobolev's lemma, the value of a function at any point in x-space is, for l large enough positive, bounded by a constant times its l-norm; how large l must be taken depends on the number of space dimensions.

From these two facts, we can conclude that for l large enough positive, $u(P)$ *depends boundedly on the initial data* ϕ *in the l-norm*:

$$| L(\phi) | = | u(P) | \leq \text{constant } \| u(T) \|_l \leq \text{constant } \| \phi \|_l ;$$

here and hereafter, T denotes the t-coordinate of the point P. According to the first representation theorem, such a bounded linear functional L can be represented as an integral

$$u(P) = \int G(x, P)\phi(x)\, dx = (G, \phi). \tag{3.1}$$

The kernel G is called the *influence function* and has finite minus l-norm.

As is well known, the value of the solution at P depends on a portion only of the initial data, namely on those contained in the so-called domain of dependence of the point P. This is precisely the support of the influence function G.

The classical uniqueness theorem (see [2], Vol. II, Chapter VI, and the literature quoted there) in the theory of hyperbolic equations gives the following description of the domain of dependence in case P is close enough to the initial hyperplane:

LEMMA 3.2. *The domain of dependence of P is contained in the outermost ray-conoid issuing from P*, i.e. *the conoid formed by the outermost bicharacteristic rays issuing from P.*

For a very special class of equations *Huygens' Principle* is valid, i.e. the domain of dependence of P consists entirely of those points x^* of the initial hyperplane which lie on a bicharacteristic issuing from P. For such equations the influence function $G(x, P)$ is zero at all other points. In a previous publication, [3], R. Courant and I have determined approximately (i.e. modulo a smooth function) the influence function of an arbitrary hyperbolic equation. It turns out that G, although in general not zero at all points other than x^*, is nevertheless a smooth function there:

THEOREM 3.1. *The influence function $G(x, P)$ is a smooth function of x and P whenever x does not lie on a bicharacteristic issuing from P.*

This property of the influence function is equivalent to the following intrinsic property of solutions:

THEOREM 3.1′. *Let u be a smooth solution of $Mu = 0$ whose initial value $\phi(x) = u(x, 0)$ is zero in an open set 0 containing all points x^* which lie on a bicharacteristic issuing from P. Then for all positive integers k and p the following inequality holds:*

$$(3.1') \qquad | D^k u(P) | \leq \text{constant} \, \| \phi \|_{-p}.$$

Here $D^k u$ stands for any partial derivative of order k of u. The constant depends solely on k, p, the open set 0 and the operator M.

Theorem 3.1′ remains valid for *limits* of smooth solutions in the sense of any minus l-norm. These can be characterized as weak solutions of $Mu = 0$ belonging to H_{-l}, i.e. elements of H_{-l} which satisfy $[u, M^*w] = 0$ for all smooth w with compact support. Here the square brackets denote the L_2 scalar product over x, t-space.

COROLLARY. *Theorem 3.1′ is valid for weak solutions.*

Inequality (3.1′) of Theorem 3.1′ can be derived from Theorem 3.1 by differentiating u, as given by (3.1), k times and estimating the resulting expression for $D^k u(P)$ by the Schwartz inequality. Conversely, the smoothness of $G(x, P)$ follows from inequality (3.1′) with the aid of the second representation theorem stated at the beginning of this section for linear functions bounded in the negative norm.

Let ϕ be an initial function which is smooth in an open set 0 of the kind described in Theorem 3.1′. This ϕ can be written as the sum of two functions ϕ_1 and ϕ_2, one smooth everywhere, the other vanishing in 0. According to the principle of superposition, the solution u with initial value ϕ is the sum of the solutions u_1 and u_2 with initial values ϕ_1 and ϕ_2. According to the standard differentiability theorems, u_1 is a smooth function everywhere, and according to the corollary of Theorem 3.1′, u_2 is a smooth function in a neighborhood of P. Thus we have the

Generalized Huygens' Principle. The differentiability properties of a solution in the neighborhood of a point P depend on the differentiability properties of its initial values $\phi(x)$ only in a neighborhood 0 of those points x^* that can be connected to P by a bicharacteristic.

4. Fourier synthesis. In this section we shall present a new method for determining approximately the influence function G. This gives another proof of Theorems 3.1, 3.1′, and the generalized Huygens' principle.

Like the previous one in [3], the present method has this defect: it is applicable only if the point P does not lie too far from the initial hyperplane i.e. before any caustics develop. Therefore both proofs of the generalized Huygens' principle are restricted to this case.

Green's formula, applied in the slab $0 \leq t \leq T$ to a pair of functions u, v one of which vanishes for $| x |$ large, reads as follows:

$$\int u(x, t)v(x, t)\, dx \Big|_0^T = \iint \{uM^*v - vMu\}\, dx\, dt,$$

where M^* is the operator adjoint to M. Suppose that u is a solution of $Mu = 0$ and that v is a matrix solution of $M^*v = 0$ (i.e. each column of v is annihilated by M^*), and that the value of v at $t = T$ is

$$v(x, T) = I\,\delta(x),$$

where I is the unit matrix and δ the Dirac δ-function. Denoting as before the initial value of u by ϕ, Green's formula reads

$$u(P) = \int \phi(x)v(x, 0)\, dx, \tag{4.1}$$

where P denotes the point $(0, T)$. Thus the influence function appears as the value at $t = 0$ of a solution of a backward initial value problem for the adjoint equation; the prescribed initial value on the hyperplane $t = T$ is a δ-function.

In Hadamard's classical method of finding fundamental solutions for second order equations, [8], this auxiliary function v is constructed directly. In Marcel Riesz' method, [16], a one-parameter family of auxiliary functions is constructed and the representation formula is derived by analytic continuation. In [3], this backward initial value problem for M^* was solved by representing the δ function as a superposition of one-dimensional δ-functions. Here we shall synthesize the δ-function from its Fourier components.

Let $v = v(x, t; \xi, w)$ be the matrix solution of $M^*v = 0$ whose value on $t = T$ is

$$v(x, T; \xi, \omega) = Ie^{i\xi x\cdot\omega}, \tag{4.2}$$

where ξ is any positive number and ω a real vector in the dual of x-space. For this v Green's formula yields an expression

$$\int u(x, T)e^{i\xi x\cdot\omega}\, dx = \int u\phi(x)v(x, 0)\, dx = U(\xi, \omega) \tag{4.3}$$

for the Fourier transform $U(\xi, \omega)$ (with respect to the space variables) of the function $u(x, T)$ in terms of the initial data for u.

The value of the function u and the values of all its spatial partial derivatives at any point of the hyperplane $t = T$ can be expressed in terms of $U(\xi, \omega)$ by inverting the Fourier transform. Employing e.g. Abel summation we get the formulae

$$u(0, T) = \lim_{\alpha \to 0} \int_0^\infty \int_{|\omega|=1} U(\xi, \omega) e^{-\alpha\xi} \xi^{m-1} \, d\omega \, d\xi \tag{4.4}$$

and

$$D^k u(0, T) = \lim_{\alpha \to 0} \int_0^\infty \int_{|\omega|=1} U(\xi, \omega) p(\omega) e^{-\alpha\xi} \xi^{k+m-1} \, d\omega \, d\xi; \tag{4.5}$$

here D^k stands for any k-th order partial derivative with respect to the space variables, and p is defined by

$$D^k e^{i\xi x \cdot \omega} = p(\omega) \xi^k e^{i\xi x \cdot \omega}.$$

Substituting the expression (4.3) for U into formulae (4.4) and (4.5), we get an expression for $u(0, T)$ and $D^k u(0, T)$ in terms of the initial data:

$$u(0, T) = \int \phi(x) G(x) \, dx \tag{4.6}$$

and

$$D^k u(0, T) = \int \phi(x) G^k(x) \, dx, \tag{4.7}$$

where the kernels G and G^k are given by

$$G(x) = \lim_{\alpha \to 0} \int_0^\infty \int_{|\omega|=1} v(x, 0; \xi, \omega) e^{-\alpha\xi} \xi^{m-1} \, d\omega \, d\xi \tag{4.8}$$

and

$$G^k(x) = \lim_{\alpha \to 0} \int_0^\infty \int_{|\omega|=1} v(x, 0; \xi, \omega) p(\omega) e^{-\alpha\xi} \xi^{m-1} \, d\omega \, d\xi, \tag{4.9}$$

v denoting the solution of $M^* v = 0$ with initial value (4.2).

If M (and M^*) is an operator with constant coefficients, this auxiliary function v can be written down explicitly as an exponential function. For an operator with variable coefficients we have an asymptotic expansion for v in the reciprocal of the frequency ξ, and this leads to an approximate determination of the influence function.

We separate the frequency range in the integrals (4.4) and (4.5) into a low region, $0 \leq \xi \leq 1$, and a high region, $1 \leq \xi$. We get

$$u(0, T) = \lim_{\alpha \to 0} \int_1^\infty \int_{|\omega|=1} U(\xi, \omega) e^{-\alpha\xi} \xi^{m-1} \, d\omega \, d\xi + E, \tag{4.4'}$$

where E denotes the contribution of the low frequencies; it can be written in the form

$$E = \int E(x) u(x, T) \, dx, \tag{4.10}$$

where the kernel $E(x)$,

$$E(x) = I \int_0^1 \int_{|\omega|=1} e^{i\xi x\cdot\omega} \xi^{m-1}\, d\omega\, d\xi,$$

is an entire analytic function of x. An analogous formula (4.5′) can be derived for $D^k u(0, T)$.

Let v_N denote the N-th partial sum of the asymptotic series (see §1) for the solution of $M^*v = 0$ with prescribed values (4.2) at $t = T$. This function v_N is a finite sum of the form

$$v_N = e^{i\xi l} \sum_{n=0}^{N} \frac{w_n}{\xi^n} \tag{4.11}$$

which takes on the prescribed value (4.2) at $t = T$, and

$$M^*v_N = \frac{e^{i\xi l} M^* w_N}{\xi^N}; \tag{4.12}$$

l and w_N are infinitely differentiable with respect to ω; as functions of x and t, their degree of smoothness depends on the degree of smoothness of the coefficients of M.

Green's formula (4.3) gives

$$U(\xi, \omega) = \int u(x, T) e^{i\xi x\cdot\omega}\, dx = \int \phi(x) v_N(x, 0)\, dx + \int_0^T \int u(x, t) M^* v_N\, dx\, dt. \tag{4.3$_N$}$$

Substitute the expression (4.3$_N$) for U into formulae (4.4′) and (4.5′); we get the following representation formula:

$$u(0, T) = \int \phi(x) G_N(x) + E + K_N \tag{4.6$_N$}$$

where the kernel G_N is equal to

$$G_N = \lim_{\alpha\to 0} \int_1^\infty \int_{|\omega|=1} v_N(x, 0) e^{-\alpha\xi} \xi^{m-1}\, d\omega\, d\xi. \tag{4.8$_N$}$$

The term K_N comes from the last term in (4.3$_N$) and is equal to

$$K_N = \iint_0^T u(x, t) K_N(x, t)\, dx\, dt \tag{4.13}$$

where the kernel $K_N(x, t)$ is given by

$$K_N(x, t) = \int_1^\infty \int_{|\omega|=1} M^* v_N \xi^{m-1}\, d\omega\, d\xi.$$

Since M^*v_N is equal to (4.12), $K_N(x, t)$ has $N - m - 1$ continuous derivatives with respect to x and t, i.e. is a very smooth function if N is large.

A similar formula (4.7_N) can be derived for $D^k u(0, T)$.

LEMMA 4.1. *The difference $G - G_N$ of the influence function G and the approximate one G_N can be made arbitrarily smooth for all x provided N is taken sufficiently large. Similarly, G^k and G_N^k (given by (4.9) and (4.9_N)) differ by a smooth function of x.*

Proof. The difference of the two functionals

$$\int \phi(x)G(x)\,dx$$

and

$$\int \phi(x)G_N(x)\,dx$$

is, according to (4.6_N), equal to $E + K_N$. E and K_N are integral expressions (4.10) and (4.13) with smooth kernels for N large enough. So we have from the Schwartz inequality for arbitrary l:

$$| E | \leq \| u(T) \|_{-l} \, \| E \|_l$$

and

$$| K_N | \leq \int_0^T \| u(t) \|_{-l} \, \| K_N(t) \|_l \, dt.$$

The l-norm of K_N is finite for N large enough. According to the energy inequality stated at the beginning of this section, $\| u(t) \|_{-l}$ is bounded by constant $\| u(0) \|_{-l}$. So these inequalities can be rewritten as follows:

$$| E | \leq \text{constant} \, \| \phi \|_{-l} \, ,$$

$$| K_N | \leq \text{constant} \, \| \phi \|_{-l} \, ,$$

i.e. E and K_N are *bounded linear functionals* of ϕ in the minus l-norm. According to the second representation theorem of §3, such a function is a scalar product of ϕ with a kernel which has a finite plus l-norm. Thus we have shown that $G - G_N$, has finite l-norm and therefore, by Sobolev's lemma, it has $l - m/2$ continuous derivatives, where l can be made large if N is large.

The smoothness of $G^k - G_N^k$ can be proved by a similar argument.

LEMMA 4.2. *The approximate influence functions G_N and G_N^k are smooth functions of x if x does not lie on a bicharacteristic issuing from P.*

Clearly, our main Theorem 3.1 follows from Lemmas 4.1 and 4.2. We note that the equivalent Theorem 3.1′ can be derived directly from Lemma 4.2 on the basis of formulae (4.6_N) and (4.7_N).

We turn now to the proof of Lemma 4.2. Substituting expression (4.11) for

v into formulae (4.8_N) and (4.9_N) for G_N and G_N^k gives a representation of these kernels as sums of terms of the form

$$\lim_{\alpha\to 0}\int_1^\infty\int_{|\omega|=1}\frac{e^{i\xi l-\alpha\xi}}{\xi^\mu}\,w\,d\omega\,d\xi, \tag{4.15}$$

l and w smooth functions of x and ω.

The derivative with respect to x of an expression of the form (4.15) is again a sum of expressions of the same form. Therefore to show that (4.15) is a smooth function of x at all points not on a bicharacteristic issuing from $(0, T)$, it is sufficient to show that the limit (4.15) exists for each value of μ.

For $\mu \geq 2$ we can pass to the limit $\alpha = 0$ under the integral sign; the limit is an integrable function and therefore bounded. We turn now to the case $\mu \leq 2$.

By a partition of unity on the unit sphere $|\omega| = 1$, the integral (4.15) can be written as the sum of expressions of the form

$$\lim_{\alpha\to 0}\int_1^\infty\int_{|\omega|=1}\frac{e^{i\xi l-\alpha\xi}}{\xi^\mu}\,q\,d\omega\,d\xi, \tag{4.15'}$$

where q is a smooth function which vanishes outside of a small neighborhood on the sphere $|\omega| = 1$. We shall specify below how small the support of q is to be taken.

We pick a positive number ϵ, also to be specified below, and for each fixed x divide the expressions (4.15′) into two classes:

The first class contains those expressions where $|l(x, \omega)| > \epsilon$ *for all* ω *in the support of* q.

The second class contains those expressions where $|l(x, \omega)| \leq \epsilon$ *at least at one point on the support of* q.

The limit

$$\lim_{\alpha\to 0}\int_1^\infty\frac{e^{i\xi l-\alpha\xi}}{\xi^\mu}\,d\xi$$

is a polynomial in $1/l$ for $\mu \leq 0$; for $\mu = 1$ it is a function of l which is bounded (even analytic) except at $l = 0$. This shows that expressions in the first class are bounded by a quantity that depends only on ϵ, μ and the maximum absolute value of q.

We turn now to estimating the expressions in the second class. We shall integrate by parts with respect to ω; for this we need the following lemma:

LEMMA 4.3. *Let* x_0 *be a point on the initial hyperplane which cannot be connected to* $(0, T)$ *by a bicharacteristic, and let* ω_0 *be a point on the unit sphere such that* $l(x_0, 0, \omega_0)$ *is zero. Then at such a point the surface gradient of* l *on* $|\omega| = 1$ *is not zero.*

In a neighborhood of such a point ω_0, the function l can be introduced as one of the coordinates on the surface of the sphere $|\omega| = 1$. Since the second derivatives of l are bounded, the size of this neighborhood depends only on the lower bound for the gradient of l. Clearly, it follows from Lemma 4.3 that on any closed set of pairs (x_0, ω_0) of the above kind, the surface gradient of l has a *uniform* lower bound.

Given any positive δ, we can choose ϵ so small that whenever $|l(x, 0, \omega)| < \epsilon$, there is a point ω_0 within a δ-neighborhood of ω where $l(x, 0, \omega_0)$ vanishes. Combining these facts we obtain the following

COROLLARY TO LEMMA 4.3. *Let S be a closed set of points x_0 of the above kind. It is possible to partition the sphere $|\omega| = 1$ into a finite number of neighborhoods and choose a positive number ϵ such that whenever for some x_0 in S the value of $|l(x_0, 0, \omega)|$ is less than ϵ at some point of one of these neighborhoods, $l(x_0, 0, \omega)$ can be introduced by a smooth change of variables as one of the coordinates in this neighborhood.*

Proof of Lemma 4.3. The function $l(x, 0, \omega)$ is defined as the value at $t = 0$ of the solution of the nonlinear equation

$$l_t + \lambda^*(x, t, l_{x^i}) = 0 \tag{1.5*}$$

whose value at $t = T$ is prescribed to be

$$l(x, T) = x \cdot \omega. \tag{4.16}$$

Denote the gradient of l with respect to ω by $\dot{l}$. Differentiate (1.5*) with respect to ω; we get

$$\dot{l}_t + \lambda^*_i \dot{l}_{x^i} = 0, \tag{4.17}$$

where $\lambda^*_{x^i}$ abbreviates $\partial\lambda^*/\partial x^i$. The left side of (4.17) is the derivative of $\dot{l}$ in a bicharacteristic direction. Therefore (4.17) asserts that $\dot{l}$ is constant along bicharacteristics. Since according to (4.16) $\dot{l}$ is equal to x on the hyperplane $t = T$, we conclude that $\dot{l}$ is different from zero except along bicharacteristics issuing from the point $(0, T)$.

The function l is homogeneous of first order in ω. Therefore if $l(x_0, 0, \omega)$ is zero at $\omega = \omega_0$, so is the derivative of l with respect to ω in the *radial* direction. Since $\dot{l}$ is not zero, neither is the *surface gradient* of l along the sphere $|\omega| = 1$. This completes the proof of Lemma 4.3.

Take now the integral in (4.15′) in the second case, i.e. $|l| \leq \epsilon$ at some point of the support of q. Imagine ϵ chosen so small and the partition of unity so fine that we can appeal to the corollary of Lemma 4.3, i e. imagine l as one of the coordinates on the sphere $|\omega| = 1$ in a neighborhood containing the support of q. Integration by parts n times with respect to l gives

$$\int \frac{e^{i\xi l - \alpha\xi}}{\xi^\mu} q(\omega)\, d\omega = (i)^{-n} \int \frac{e^{i\xi l - \alpha\xi}}{\xi^{\mu+n}} \frac{\partial^n q}{\partial l^n}\, d\omega. \tag{4.18}$$

Take n so large that $n + \mu$ is equal to 2; substitute (4.18) into (4.15′) and let

α tend to zero under the integral sign. The resulting integral is clearly bounded, uniformly for all x in question. This completes the proof of Lemma 4.3.

REFERENCES

1. S. BANACH, *Théorie des Opérations Linéaires*, Warsaw, 1932.
2. R. COURANT AND D. HILBERT, *Methoden der mathematischen Physik*, I-II, Springer, Berlin, 1937.
3. R. COURANT AND P. D. LAX, *The propagation of discontinuities in wave motion*, Proceedings of the National Academy of Sciences, vol. 42(1956), pp. 872–876.
4. K. O. FRIEDRICHS, *Symmetric hyperbolic linear differential equations*, Communications on Pure and Applied Mathematics, vol. 7(1954), pp. 345–392.
5. K. FRIEDRICHS AND H. LEWY, *Über fortsetzbare Anfangsbedingungen bei hyperbolischen Differentialgleichungen in drei Veranderlichen*, Nachrichten von der Gesellschaft der Wissenschaften zu Göttingen, 1932, pp. 135–143.
6. L. GÅRDING, *L'inégalité de Friedrichs et Lewy pour les équations hyperboliques linéaires d'ordre supérieur*, Comptes Rendus de l'Academie des Sciences, Paris, vol. 239(1954), pp. 849–850.
7. L. GÅRDING, *Solution directe du Problème de Cauchy pour les équations hyperboliques*, Collogues Internationaux LXXI sur la Théorie des Équations aux Dérivées Partielles, Nancy, 1956, pp. 71–90.
8. J. HADAMARD, *Le Problème de Cauchy*, Hermann, Paris, 1932.

8A. E. HILLE, *Functional Analysis and Semi-Groups*, American Mathematical Society Colloquium Publications, vol. 31, 1948.

9. FRITZ JOHN, *On linear partial differential equations with analytic coefficients*, Communications on Pure and Applied Mathematics, vol. 2(1949), pp. 209–253.
10. J. B. KELLER, R. M. LEWIS AND B. D. SECKLER, *Asymptotic solution of some diffraction problems*, Communications on Pure and Applied Mathematics, vol. 9(1956), pp. 207–265.
11. M. KLINE, *Asymptotic solution of linear hyperbolic partial differential equations*, Journal of Rational Mechanics and Analysis, vol. 3(1954), pp. 315–342.
12. P. D. LAX, *On Cauchy's problem for hyperbolic equations and the differentiability of solutions of elliptic equations*, Communications on Pure and Applied Mathematics, vol. 8(1955), pp. 615–633.
13. J. LERAY, *Lecture Notes on Hyperbolic Equations*, Princeton, 1952.
14. R. K. LÜNEBURG, *Propagation of Electromagnetic Waves*, Lecture Notes, New York University, 1948.
15. I. G. PETROWSKY, *Sur l'analyticité des solutions des systèmes d'équations différentielles*, Recueil Mathématique, N. S. (Matematiceskii Sbornik), vol. 47(1939), pp. 3–70.
16. M. RIESZ, *L'intégrale de Riemann-Liouville et le problème de Cauchy*, Acta Mathematica, vol. 81(1949), pp. 1-223.
17. L. SCHWARTZ, *Théorie des Distributions*, I-II, Paris, 1950–1951.

NEW YORK UNIVERSITY

Reprinted from
Printed in U.S.A.
JOURNAL OF MATHEMATICAL PHYSICS VOLUME 5, NUMBER 5 MAY 1964

Development of Singularities of Solutions of Nonlinear Hyperbolic Partial Differential Equations*

PETER D. LAX
Courant Institute of Mathematical Sciences, New York University, New York, New York
(Received 10 January 1964)

In a recent paper Zabusky has given an accurate estimate of the time interval in which solutions of the nonlinear string equation $y_{tt} = c^2(1 + \epsilon y_x)y_{xx}$ exist. A previous numerical study of solutions of this equation disclosed an anomaly in the partition of energy among the various modes; Zabusky's estimate shows that at the time when the anomaly was observed the solution does not exist. The proof of Zabusky uses the hodograph method; in this note we give a much simpler derivation of the same result based on an estimate given some years ago by the author.

1. PRELIMINARY LEMMAS ABOUT ORDINARY DIFFERENTIAL EQUATIONS

OUR estimates are based on two simple and well-known results concerning solutions of quadratic ordinary differential equations:

Theorem 1. Let $z(t)$ be the solution of the initial-value problem

$$dz/dt = a(t)z^2, \qquad z(0) = m \tag{1.1}$$

in the interval $(0, T)$. *Suppose that the function $a(t)$ satisfies the inequality*

$$0 < A < a(t), \qquad 0 \leq t \leq T,$$

and suppose that m is positive; then

$$T < (mA)^{-1}. \tag{1.2}$$

Theorem 2. Suppose that $a(t)$ satisfies the inequality

$$|a(t)| < B;$$

* The work presented in this paper is supported by the U. S. Atomic Energy Commission Computing and Applied Mathematics Center, Courant Institute of Mathematical Sciences, New York University, under Contract AT(30-1)-1480.

then the initial value problem (1) *has a solution for* $|t| < |mB|^{-1}$.

Proof: Let $z_0(t)$ be the solution of the comparison equation

$$dz_0/dt = Az_0^2, \qquad z_0(0) = m.$$

Since A is a lower bound for $a(t)$, it follows easily that $z_0(t)$ is a lower bound for $z(t)$ for all positive t. Since $z_0 = m/(1 - mAt) \to \infty$ at $t = (mA)^{-1}$, it follows that $z(t)$ cannot exist beyond this time.

The proof of Theorem 2 is similar: we note that the solution z_1 of

$$dz_1/dt = Bz_1^2, \qquad z_1(0) = |m|$$

is an upper bound for $|z(t)|$ for all positive t.

2. QUASILINEAR SYSTEMS FOR TWO UNKNOWNS

The following estimates were derived in Ref. 1.

Consider a system of two first-order partial differential equations:

$$\begin{aligned} u_t + au_x + bv_x &= 0, \\ v_t + cu_x + dv_x &= 0, \end{aligned} \tag{2.1}$$

a, b, c, d being functions of u and v. Suppose that this system is hyperbolic, i.e., that the matrix

$$\begin{bmatrix} a & b \\ c & d \end{bmatrix}$$

has real and distinct eigenvalues λ and μ for all relevant values of u and v.

Let (l_1, l_2) be the left eigenvector of the above matrix corresponding to the eigenvalue λ. Multiply the first equation in (2.1) by l_1, the second by l_2 and add; we obtain the characteristic equation

$$l_1u' + l_2v' = 0, \tag{2.2}$$

where

$$' = \partial/\partial t + \lambda(\partial/\partial x).$$

Let ϕ be an integrating factor for (2.2), i.e., a function ϕ of u and v such that ϕl_1 and ϕl_2 become the u and v derivatives, respectively, of some function $r(u, v)$. Multiply (2.2) by ϕ; we get

$$r' = r_t + \lambda r_x = 0. \tag{2.3}$$

For the other eigenvalue we get a similar equation

$$\grave{s} = 0, \tag{2.4}$$

where

$$\grave{} = \partial/\partial t + \mu(\partial/\partial x). \tag{2.5}$$

The functions r and s are called Riemann invariants; Eqs. (2.3) and (2.4) express the fact that they remain constant along their respective characteristics.

Differentiate (2.3) with respect to x:

$$r_{tx} + \lambda r_{xx} + \lambda_r r_x^2 + \lambda_s s_x r_x = 0. \tag{2.6}$$

From (2.4) and (2.5) we have

$$0 = \grave{s} = s' - (\lambda - \mu)s_x,$$

so

$$s_x = s'/(\lambda - \mu). \tag{2.7}$$

Substitute (2.7) into (2.6) and abbreviate r_x by w; we get

$$w' + \lambda_r w^2 + [\lambda_s/(\lambda - \mu)]s'w = 0. \tag{2.8}$$

Denote by h a function of r and s which satisfies

$$h_s = \lambda_s/(\lambda - \mu).$$

Using (2.3) we have

$$h' = h_r r' + h_s s' = [\lambda_s/(\lambda - \mu)]s'.$$

Substituting this into (2.8) gives

$$w' + \lambda_r w^2 + h'w = 0.$$

Multiplying by e^h and abbreviating $e^h w$ by z gives

$$z' + e^{-h}\lambda_r z^2 = 0, \tag{2.9}$$

an equation of the form (1.1), with

$$a = -e^{-h}\lambda_r.$$

We make now the additional assumption that λ_r *is nonzero* for the relevant values of r and s; this amounts to requiring that the system (2.1) is genuinely nonlinear.

Consider bounded initial values for r and s; since r and s are constant along characteristics, it follows that r and s stay between the same bounds for all time. The quantity $|\lambda_r e^{-h}|$ has then a lower bound A for the relevant values of r and s. Suppose that λ_r is negative, and denote by m the maximum of the initial value of z. Then, according to Theorem 1, solutions with such initial values cannot exist beyond $t = (Am)^{-1}$.

Denote by B the supremum of $|\lambda_r e^{-h}|$, and by m the maximum of $|z|$. According to the proof of Theorem 2 we can place an *a priori* limitation on $|z|$ valid for all values of t less than $(Bm)^{-1}$; this gives an *a priori* estimate for $|r_x|$. We can get a similar estimate for $|s_x|$ in a similar time interval. According to the theory of first-order quasilinear hyperbolic equations[2], solutions to initial-value

[1] Unpublished note.

[2] R. Courant and D. Hilbert, *Methods of Mathematical Physics* (Interscience Publishers, Inc., New York, 1962), Vol. II, Chap. V.

problems exist as long as one can place an *a priori* limitation on the magnitude of their first derivatives.

To summarize: Using theorems 1 and 2 we can place upper and lower bounds on the time interval in which the solution of a given initial-value problem for (2.1) exists.

For arbitrary initial values, these bounds are far from sharp. There is, however, one case in which these bounds are asymptotically correct; when the initial values differ little from a constant r_0, s_0. It follows then that r and s are nearly constant for all time, and so we have the following bounds:

$$e^{-h(0)}\lambda_r(0) - \epsilon < e^{-h}\lambda_r < e^{-h(0)}\lambda_r(0) + \epsilon,$$

$$(e^{h(0)} - \epsilon)\max r_x(0) < \max z(0) < [e^{h(0)} + \epsilon]\max r_x(0),$$

where $a(0)$ denotes the value of $a(r, s)$ at r_0, s_0, and $r_x(0)$ the initial value of r_x. So according to Theorems 1 and 2, the time T_{crit} beyond which a solution cannot be continued is given asymptotically by the smaller of the two numbers

$$[-\lambda_r(0)\max r_x(0)]^{-1}, \quad [-\mu_s(0)\max s_x(0)]^{-1}. \tag{2.10}$$

(Here the sign of r and s is so chosen that λ_r and μ_s are both negative.

It would be interesting and useful to derive such estimates for solutions of systems of equations for more than two variables.

3. NONLINEAR SECOND-ORDER HYPERBOLIC EQUATIONS

In this section we shall rederive the result of Zabusky[3] about the equation studied by Fermi, Pasta and Ulam.[4]

We shall apply now the foregoing theory to the second-order equation

$$y_{tt} = K^2(y_x)y_{xx}. \tag{3.1}$$

We prescribe initial values in the interval $(0, L)$:

$$y(x, 0) = y_0(x), \qquad y_t(x, 0) = 0, \qquad 0 \le x \le L.$$

At the end points we require y to be fixed:

$$y(0, t) = y(L, t) = 0.$$

This mixed initial-boundary-value problem can be converted into a pure initial-value problem by extending y_0 to be an odd function in $(-L, L)$ and further extending it periodically with period $2L$.

We make a first-order system out of (3.1) by introducing

$$y_x = u, \qquad y_t = v$$

as unknowns.,

$$u_t = v_x,$$
$$v_t = K^2u_x.$$

Multiply the first equation by K and add to (subtract from) the second. We get

$$v' + Ku' = 0 \quad \text{and} \quad v^{\backprime} - Ku^{\backprime} = 0, \tag{3.2}$$

where $\lambda = K$, $\mu = -K$. Since K is a function of u alone, Eqs. (3.2) are exact; the Riemann invariants are

$$r = v + L(u), \qquad s = v - L(u),$$

where

$$dL/du = K.$$

To compute λ_r we write

$$r - s = 2L(u)$$

and differentiate it with respect to r:

$$1 = 2L_u u_r = 2Ku_r.$$

Combining this with

$$\lambda_r = K_r = K_u u_r$$

gives

$$\lambda_r = K_u/2K. \tag{3.3}$$

Differentiate r with respect to x:

$$r_x = v_x + L_u u_x = v_x + Ku_x.$$

Since initially v is zero and $u = y_x$,

$$\max r_x(0) \simeq K(0)\max y(0)_{xx}. \tag{3.4}$$

So, by (2.10), the time beyond which the solution cannot be continued is

$$T_{\text{crit}} \simeq 2[K_u(0)\max y_{xx}(0)]^{-1}. \tag{3.5}$$

In Ref. 3 K is taken to be $c(1 + \epsilon u)^{\frac{1}{2}}$ and $y_0(x) = a\sin(\pi x/L)$. Then $K_u(0) = \frac{1}{2}\epsilon c$; so by (3.5),

$$T_{\text{crit}} \simeq 4L^2/ca\epsilon\pi^2.$$

The period of vibration of the linearized system is $2L/c$, so the critical number of vibrations is

$$2L/a\epsilon\pi^2. \tag{3.6}$$

This agrees with Eq. (5.21) of Ref. 3.

The above formula is asymptotically valid for large values (3.6). Taking $\epsilon = 1$, this means that the maximum displacement a should be small compared to the length L of the string.

In the calculation in Ref. 4, the values chosen correspond to: $L = 1$, $a = 1$, $c = 1$, $\epsilon = \frac{1}{64}$. Formula (3.6) indicates that breakdown occurs after ~ 13.0 cycles.

[3] N. J. Zabusky, J. Math. Phys. 3, 1028 (1962).
[4] E. Fermi, J. Pasta, and S. Ulam, "Studies of Nonlinear Problems I, Los Alamos Sci. Lab. Rept. LA 1940 (1955).

COMMUNICATIONS ON PURE AND APPLIED MATHEMATICS, VOL. XIX, NO. 4, 473–492 (1966)

On Stability for Difference Schemes; a Sharp Form of Gårding's Inequality*

P. D. LAX AND L. NIRENBERG

To MARCEL RIESZ, *on his 80-th birthday.*

Introduction

In this paper inequalities are derived for difference and pseudo-differential operators; we show that if the symbol of such an operator satisfies a certain inequality, then the operator itself satisfies an analogous inequality, modulo a lower order error term. This pattern of inference is typical of many results concerning these operators.

In Section 1 we show how the inequality for difference operators can be employed to derive two new stability criteria for difference schemes used in solving numerically the initial value problem for hyperbolic equations. Unlike most earlier criteria our conditions do not require the difference scheme to be *dissipative*. That is not to say that dissipation has no significance for stability; it certainly is important for the stability of nonlinear schemes, and also plays a subtle role in certain theorems of Kreiss on the stability of linear schemes, see [7] and [8].

The main inequality for difference operators—Theorem 1.1, proved in Section 2—is an analogue for difference schemes of a sharp form of Gårding's inequality for differential operators; a proof of the sharp form of the inequality for pseudo-differential operators acting on scalar valued functions is contained in the recent important paper [3] by Hörmander. (The usual form of the inequality is given in Kohn, Nirenberg [5], and in Hörmander [2]; these papers contain an exposition of the theory of such operators, and other references may be found there.) In Section 3 we prove the sharp inequality for vector functions. Our proof makes use of an ingenious partition of unity introduced by Hörmander in [3], but is otherwise different from Hörmander's. An analogous partition of unity is used in our study of difference operators.

We believe that the sharp form of Gårding's inequality of Section 3 will find application.

---

* This paper represents results obtained under Grant AF-AFOSR-684-64 with the Air Force Office of Scientific Research, and the AEC Computing and Applied Mathematics Center, under Contract AT(30-1)-1480 with the U.S. Atomic Energy Commission.

1. Difference Operator

In this section we consider difference operators P_δ depending on a parameter $\delta > 0$ in the following fashion:

$$P_\delta = \sum_\alpha p_\alpha(x) T^\alpha, \tag{1.1}$$

where α is a multi-index $\alpha = (\alpha_1, \cdots, \alpha_n)$, the α_j being integers, and T^α the shift operator

$$(T^\alpha u)(x) = u(x + \alpha\delta).$$

The coefficients $p_\alpha(x)$ are $m \times m$ matrix functions of x. We denote the scalar product and norm in $\mathbb{C}^m$ by $u \cdot v = \Sigma u_j \bar{v}_j$ and $|u|$, respectively. The norm of an $m \times m$ matrix p is defined as its operator norm in $\mathbb{C}^m$ and is denoted by $|p|$; the conjugate transpose of a matrix is denoted by p^*.

The operator P_δ acts on functions $u(x)$ which are defined for all x in E^n and whose values lie in $\mathbb{C}^m$. These functions form a Hilbert space under the norm

$$\|u\|^2 = \int |u(x)|^2 \, dx.$$

The corresponding scalar product is denoted by (u, v). We denote by $\|P_\delta\|$ the operator norm,

$$\|P_\delta\| = \sup_u \|P_\delta u\| / \|u\|,$$

by P_δ^* the Hilbert space adjoint of P_δ, and by $\mathcal{R}e\, P_\delta$ its hermitian part,

$$\mathcal{R}e\, P_\delta = \tfrac{1}{2}(P_\delta^* + P_\delta).$$

The natural domain for difference operators is the space of functions defined on a lattice. It is easy to show that boundedness and positivity of difference operators over lattice functions of class l_2 are equivalent to these properties of the operators over $L_2(E^n)$.

The *symbol*[1] p of the one parameter family P_δ given by (1.1) is defined as the following function of x and ξ, periodic in $\xi = (\xi_1, \cdots, \xi_n)$ of period 2π in each ξ_j:

$$p(x, \xi) = \sum p_\alpha(x) e^{i\alpha\xi}. \tag{1.2}$$

In what follows we have to impose some differentiability conditions on the coefficients $p_\alpha(x)$ and also require that they tend to zero at some definite rate. These conditions can be conveniently expressed in terms of certain norms for the symbol p which we now define.

For any scalar, vector or matrix valued function $f(x)$ we denote by $|f|_k$ the maximum of the norm of f and the norms of its first k partial derivatives. With

[1] p is also called amplification matrix.

$|\alpha|^2 = \Sigma\alpha_j^2$ we define the k,l-norm of p as

$$|p|_{k,l} = \sum |p_\alpha|_k(1 + |\alpha|)^l .$$

We denote by $C_{k,l}$ the class of p with finite k,l-norm.

Our main result is the following inequality:

THEOREM 1.1. *Let P_δ be a one-parameter family of difference operators of the form* (1.1) *whose symbol $p(x, \xi)$ belongs to the classes $C_{0,2}$ and $C_{2,0}$. Suppose further that the symbol p is a hermitian and non-negative matrix for every x and ξ:*

(1.3) $$p(x, \xi) \geqq 0 ;$$

then the operator P_δ satisfies the inequality

(1.4) $$\mathcal{R}e\, P_\delta \geqq -K\delta$$

for all δ and some constant K.

The proof will be given in Section 2; in this section we deduce some corollaries which are then utilized to establish stability criteria for difference schemes. We start by stating some simple and well known rules about functions of difference operators and their symbols.

LEMMA 1.1. *Let A_δ, B_δ be one-parameter families of difference operators with symbols $a(x, \xi)$, $b(x, \xi)$, respectively. Denote the product ab by c, and the one-parameter family of operators with symbol c by C_δ. Then*

$$\|A_\delta\| \leqq |a|_{0,0} ,$$

$$\|A_\delta B_\delta - C_\delta\| \leqq \delta\, |a|_{0,1}\, |b|_{1,0} ,$$

$$\|A_\delta^* - A_\delta'\| \leqq \delta\, |a|_{1,1} .$$

Here A_δ^* is the L_2 adjoint of A_δ, and A_δ' is the difference operator with $a^*(x, \xi)$ as symbol. Furthermore,

$$|c|_{0,2} \leqq |a|_{0,2}\, |b|_{0,2} ,$$

$$|c|_{2,0} \leqq 4|a|_{2,0}\, |b|_{2,0} .$$

For completeness a proof of the lemma is sketched at the end of this section.

We state now

COROLLARY 1.1. *Let S_δ be a one-parameter family of difference operators whose symbol $s(x, \xi)$ is of class $C_{2,0}$ and of class $C_{0,2}$. Suppose that the norm of s does not exceed* 1:

(1.5) $$|s(x, \xi)| \leqq 1 \qquad \textit{for all} \qquad x, \xi .$$

Then the operator S_δ is bounded by

(1.6) $$\|S_\delta\| \leqq 1 + K\delta$$

for all δ and some constant K.

Proof: We have to show that $\|S_\delta u\|^2 \leqq (1 + K\delta)\, \|u\|^2$ for all functions u,

$$\|S_\delta u\|^2 = (S_\delta u, S_\delta u) = (u, S_\delta^* S_\delta u)\ .$$

Since $C_{0,2} \cap C_{2,0} \subset C_{1,1}$, we easily verify with the aid of Lemma 1.1, that $S_\delta^* S_\delta$ differs by $O(\delta)$ from the operator whose symbol is s^*s. Defining the symbol p as $p = 1 - s^*s$, and denoting by P_δ the corresponding operator, we have

$$\|u\|^2 - \|S_\delta u\|^2 = (u, [1 - S_\delta^* S_\delta]u) = (u, P_\delta u) + O(\delta)\, \|u\|^2\ . \tag{1.7}$$

It follows from assumption (1.5) that the symbol p is non-negative and from the last part of Lemma 1.1 that it is in $C_{0,2} \cap C_{2,0}$; therefore by the main theorem, $(u, \mathcal{R}e\, P_\delta u) = \mathcal{R}e\, (u, P_\delta u)$ is $\geqq O(\delta)\, \|u\|^2$. Since the left side of (1.7) is real, this implies the desired inequality.

Next we deduce from Corollary 1.1 the following stability criterion.

THEOREM 1.2. *Consider the difference scheme*

$$v(t + \delta) = S_\delta v(t)\ , \tag{1.8}$$

where S_δ is a difference operator of the form (1.1) *which may depend on t as well. Suppose that the norm of the symbol $s = s(t, x, \xi)$ is $\leqq 1$ for all t, x, ξ and that, for all t, $|s|_{0,2} + |s|_{2,0}$ is bounded by a fixed constant. Then the difference scheme is stable in the sense that*

$$\|v(T)\| \leqq M(T)\, \|v(0)\| \tag{1.9}$$

for all solutions of the difference scheme (1.8), *where M is a function of T but is independent of δ.*

Proof: It follows from Corollary 1.1 that $\|S_\delta\| \leqq 1 + K\delta$, K independent of T. Therefore, $\|v(t + \delta)\| \leqq (1 + K\delta)\, \|v(t)\|$, and so inductively

$$\|v(n\delta)\| \leqq (1 + K\delta)^n\, \|v(0)\|\ .$$

Setting $n\delta = T$ this is the desired inequality (1.9), with $M = (1 + K\delta)^n \leqq e^{K\delta n} = e^{KT}$ indeed independent of δ.

To state the next corollary and the stability theorem which follows from it we need the concept of *numerical range.*

DEFINITION. *The numerical range of an operator A in Hilbert space, denoted by $w(A)$, is the set of all complex numbers of the form*

$$(Ae, e)\ ,$$

e any complex vector of unit length. The numerical range of a matrix is defined analogously.

COROLLARY 1.2. *Suppose that the numerical range of the symbol s is contained in the unit disk for all x and ξ:*

$$|w(s)| \leqq 1\ . \tag{1.10}$$

Then the numerical range of the operator S_δ *is contained in a disk of radius* $1 + O(\delta)$:

$$|w(S_\delta)| \leqq 1 + O(\delta) . \tag{1.11}$$

Proof: (1.11) means that, for all vector functions u,

$$|(S_\delta u, u)| \leqq [1 + O(\delta)] \, \|u\|^2 .$$

This can also be expressed by saying that for all complex numbers z, $|z| = 1$, and all u,

$$\mathcal{R}e \, z(S_\delta u, u) \leqq (u, u) + O(\delta) \, \|u\|^2$$

which is the same as

$$\mathcal{R}e \, (1 - zS_\delta u, u) \geqq O(\delta) \, \|u\|^2 . \tag{1.12}$$

The symbol of $1 - zS_\delta$ is $1 - zs$; it follows from the hypothesis (1.10) about the numerical range of s that $1 - zs$ has non-negative real part. Therefore, according to the main theorem, the real part of the corresponding operator $1 - zS_\delta$ is greater than $O(\delta)$; this proves inequality (1.12).

Our second stability criterion is

THEOREM 1.3. *Consider difference schemes*

$$v(t + \delta) = S_\delta v(t) ,$$

where the difference operator S_δ *is independent of* t. *Suppose that the symbol* s *belongs to* $C_{0,2} \cap C_{2,0}$ *and satisfies*

$$|w(s)| \leqq 1 \tag{1.13}$$

for all x *and* ξ. *Then the scheme is stable.*

Proof: According to Corollary 1.2 it follows from (1.13) that

$$|w(S_\delta)| \leqq 1 + K\delta . \tag{1.14}$$

Next we use the following theorem[2] of C. A. Berger [1]:

If an operator A *satisfies* $|w(A)| \leqq r$, *then* $|w(A^n)| \leqq r^n$.

Applying this to the operator $A = S_\delta$ and to $r = 1 + K\delta$, we conclude from (1.14) that

$$|w(S_\delta^n)| \leqq (1 + K\delta)^n \leqq e^{K\delta n} .$$

We also use the following elementary observation:

If an operator B *satisfies* $|w(B)| \leqq R$, *then* $\|B\| \leqq 2R$.

(To see this note that $|w(B)| \leqq R$ implies that the norm of both the real and imaginary part of B is $\leqq R$.) Applying this result to $B = S_\delta^n$ and $R = e^{K\delta n}$ we

[2] It was conjectured by Halmos; another proof has been given by C. Pearcy [10], and a more general result recently by T. Kato [4].

conclude that

$$\|S_\delta^n\| \leqq 2e^{K\delta n}. \tag{1.15}$$

Since S_δ is independent of t, the operator relating $v(n\delta)$ to $v(0)$ is S_δ^n. Denoting $n\delta$ by T we find therefore from (1.15) that

$$\|v(T)\| \leqq 2e^{KT} \|v(0)\|,$$

which is stability as defined in (1.9).

The significance of the stability criterion, Theorem 1.2, is that there is a class of difference schemes, investigated in [9], which satisfies the criterion (1.13) concerning the numerical range of the symbol.

It would be interesting to investigate the stability of time dependent difference schemes satisfying condition (1.13); this would involve estimating the numerical range of a product

$$\prod_1^n S_\delta(\nu\delta)$$

under the assumption that each $s(t)$ satisfies (1.13) and that $s(t)$ varies Lipschitz continuously with t.

Remark. If the inequality (1.3) imposed on the symbol is satisfied in the *strict* sense for all x and ξ, then inequality (1.4) for P_δ follows rather easily. For, denote by r the positive square root of p; according to the functional calculus for matrices, r can be expressed by a contour integral,

$$r = \sqrt{p} = \frac{1}{2\pi i}\int (\lambda - p)^{-1}\lambda^{1/2}\, d\lambda,$$

along any contour surrounding the spectrum of p and excluding the origin. If p is positive definite for all x and ξ, then there is such a contour; this shows that r has the same differentiability properties as a function of x and ξ as p. From the relations

$$r^* = r, \qquad r^2 = p,$$

it follows by virtue of Lemma 1.1 that $R_\delta^* R_\delta$ differs from P_δ by $O(\delta)$, so that

$$\begin{aligned}(P_\delta u, u) &= (R_\delta^* R_\delta u, u) + O(\delta)\,\|u\|^2 \\ &= \|R_\delta u\|^2 + O(\delta)\,\|u\|^2 \geqq O(\delta)\,\|u\|^2,\end{aligned}$$

as asserted.

In the application of the main theorem to the first stability criterion we had to deal with an operator P_δ with symbol p of the form $1 - s^*s$ where s was the symbol of the one-step operator. Now one-step operations preserve constants, from which it follows that their symbol s equals 1 for $\xi = 0$; therefore, $p = 1 - s^*s$ is zero for $\xi = 0$, i.e., in this case p is *not* positive definite for all x and ξ. Thus the above remark is inapplicable to stability theory. The following result however, proved in [8], is relevant for stability theory:

If $p(x, \xi)$ is positive definite for all x and all $\xi \neq 0 \pmod{2\pi}$, and if near $\xi = 0$ the symbol has a zero of precisely order $2r$ in the sense that the inequality

$$c_1 |\xi|^{2r} \leqq p(x, \xi) \leqq c_2 |\xi|^{2r} \tag{1.16}$$

holds with some positive constants c_1 and c_2, then $\mathscr{R}e\, P_\delta \geqq O(\delta)$.

The novelty of our main theorem is to remove entirely the condition (1.16), at the expense of requiring p to be twice, not once, differentiable as a function of x.

Recently R. Phillips and L. Sarason have shown that one may define the square root of a positive semidefinite matrix, depending smoothly on parameters, so that the square root is Lipschitz continuous with respect to these parameters. However, this is not sufficient smoothness to enable us to prove Theorem 1.1.

We conclude this section with a brief description of the proof of Lemma 1.1. As in the theory of pseudo-differential operators, [5], one can develop the calculus of the algebra of difference operators.

Proof of Lemma 1.1: We first observe that the operator $A_\delta = \sum_\alpha a_\alpha T^\alpha$ is equivalent to an integral operator with kernel

$$k(x, y) = \sum_\alpha a_\alpha(x) \bar{\delta}(y - \delta\alpha - x)\,,$$

where $\bar{\delta}$ is the delta function. Since

$$\int |k(x, y)|\, dy \leqq |a|_{0,0}\,, \qquad \int |k(x, y)|\, dx \leqq |a|_{0,0}\,,$$

the first inequality of the lemma follows.

For the operator $C_\delta = \Sigma\, c_\alpha T^\alpha$, we have

$$c_\alpha = \sum_\beta a_\beta(x) b_{\alpha-\beta}(x)$$

and the last two inequalities of the lemma follow with the aid of the simple inequality $1 + |\alpha| \leqq (1 + |\beta|)(1 + |\alpha - \beta|)$. In addition, if

$$A_\delta B_\delta = \sum t_\alpha(x) T^\alpha\,,$$

then

$$t_\alpha(x) = \sum_\beta a_\beta(x) b_{\alpha-\beta}(x + \delta\beta)\,.$$

Hence

$$\begin{aligned} \sum_\alpha \sup_x |t_\alpha(x) - c_\alpha(x)| &\leqq \sum_\alpha \sum_\beta \sup_x |a_\beta(x)|\, |b_{\alpha-\beta}(x + \delta\beta) - b_{\alpha-\beta}(x)| \\ &\leqq \delta \sum_\alpha \sum_\beta |\beta|\, |a_\beta|_0\, |b_{\alpha-\beta}|_1 \leqq \delta\, |a|_{0,1}\, |b|_{1,0}\,, \end{aligned}$$

and the second inequality of Lemma 1.1 results—with the aid of the first.

Finally, to verify the third inequality, we have

$$A_\delta^* = \sum (a_{-\alpha}^*(x + \delta\alpha) - a_{-\alpha}^*(x))\, T^\alpha ,$$

and the result follows—again with the aid of the first inequality.

2. Proof of the Main Theorem

The main theorem asserts that if the symbol P is non-negative, then for all vector functions u

$$\mathscr{R}e\,(P_\delta u, u) \geqq -K\delta\, \|u\|^2 . \tag{2.1}$$

First we localize the problem, i.e., show that it is sufficient to prove it for functions u whose support is of size $O(\sqrt{\delta})$.

Let ϕ_j be a Gårding type partition of unity:

$$\sum \phi_j^2(x) \equiv 1 \tag{2.2}$$

so formed that the support of each ϕ_k has diameter $\leqq c\sqrt{\delta}$, c some constant to be fixed later, and such that each point is contained in the supports of at most a fixed number (independent of δ) of supports of the ϕ_k. The ϕ_k can be chosen so that $|\phi_k|_1 \leqq C/\sqrt{\delta}$, i.e.,

$$|\phi_k(x) - \phi_k(y)| \leqq \frac{C}{\sqrt{\delta}}\, |x - y| . \tag{2.3}$$

We show next, using a device of Hörmander, that $(P_\delta u, u)$ and $\sum_k (P_\delta \phi_k u, \phi_k u)$ differ by $O(\delta)\, \|u\|^2$. For, by definition,

$$(P_\delta u, u) = \int \sum_\alpha p_\alpha(x) u(x + \alpha\delta) \cdot u(x)\, dx ,$$

and hence

$$\begin{aligned} &(P_\delta u, u) - \sum (P_\delta \phi_k u, \phi_k u) \\ &\quad = \int \sum_\alpha p_\alpha(x) u(x + \alpha\delta) \cdot u(x) (1 - \sum_k \phi_k(x + \alpha\delta)\phi_k(x))\, dx . \end{aligned} \tag{2.4}$$

Summing with respect to k first and using the identity (2.2), we see that

$$1 - \sum_k \phi_k(x + \alpha\delta)\phi_k(x) \equiv \tfrac{1}{2} \sum_k (\phi_k(x + \alpha\delta) - \phi_k(x))^2 .$$

Using inequality (2.3) and the fact that for each point y there are only a *finite* number of k such that $\phi_k(y) \neq 0$, we see that

$$\sum_k (\phi_k(x + \alpha\delta) - \phi_k(x))^2 \leqq C\, |\alpha|^2\, \delta .$$

Substituting this into (2.4), and using the first inequality of Lemma 1.1, we see that

$$|(P_\delta u, u) - \sum (P_\delta \phi_k u, \phi_k u)| \leqq C^2 \, |p|_{0,2} \, \delta \, \|u\|^2 , \tag{2.5}$$

as asserted.

Denote $\phi_k u$ by u_k. Suppose that we have proved the main inequality (2.1) for the function u_k:

$$(P_\delta u_k , u_k) \geqq -K\delta \, \|u_k\|^2 ; \tag{2.6}$$

summing with respect to k and using the inequality (2.5) as well as the identity

$$\sum \|u_k\|^2 = \|u\|^2 ,$$

which is a consequence of (2.2), we obtain

$$\mathcal{R}e \, (P_\delta u, u) \geqq -K_1\delta \, \|u\|^2 ,$$

i.e., the asserted inequality (2.1) for u.

For the proof of the main theorem for functions with small support we need the following two lemmas:

LEMMA 2.1. *Let P_δ be a one-parameter family of difference operators of the form* (1.1) *with constant coefficients whose symbol $p(\xi)$ is hermitian. Let ϕ be a real scalar function, with Lipschitz constant K. Then for all functions u*

$$|\mathcal{R}e \, (\phi P_\delta u, \phi u) - \mathcal{R}e \, (P_\delta \phi u, \phi u)| \leqq \tfrac{1}{2} \, |p|_{0,2} \, K^2\delta^2 \, \|u\|^2 . \tag{2.7}$$

Proof: Since $p(\xi)$ is hermitian, $p_{-\alpha} = p_\alpha^*$. Now

$$I = \mathcal{R}e \, [(\phi P_\delta u, \phi u) - (P_\delta \phi u, \phi u)]$$

$$= \mathcal{R}e \int \sum [p_\alpha u(x + \delta\alpha) \cdot u(x)](\phi(x) - \phi(x + \delta\alpha))\phi(x) \, dx .$$

Replacing x by $x - \delta\alpha$ as variable of integration and α by $-\alpha$, we find, since $p_{-\alpha} = p_\alpha^*$, that

$$I = \mathcal{R}e \int \sum [p_\alpha^* u(x) \cdot u(x + \delta\alpha)](\phi(x + \delta\alpha) - \phi(x))\phi(x + \delta\alpha) \, dx .$$

The terms in the square brackets of the last two expressions have equal real parts; hence adding the expressions we find that

$$I = \tfrac{1}{2} \, \mathcal{R}e \int \sum [p_\alpha u(x + \delta\alpha) \cdot u(x)](\phi(x) - \phi(x + \delta\alpha))^2 \, dx .$$

Thus

$$|I| \leqq \tfrac{1}{2}\delta^2 K^2 \int \sum |\alpha|^2 \, |p_\alpha| \, |u(x + \delta\alpha)| \, |u(x)| \, dx$$

$$\leqq \tfrac{1}{2} \, |p|_{0,2} \, \delta^2 K^2 \, \|u\|^2 ,$$

as asserted.

Remark. A similar inequality holds for difference operators with *variable* coefficients, with the term $\frac{1}{2}|p|_{1,2}\,\delta^2 K \max|\phi| \cdot \|u\|^2$ added to the upper bound. This result is essentially a special case of Lemma 3.1 of [6].

LEMMA 2.2. *Let A and B be two hermitian $m \times m$ matrices such that, for all vectors $v \in \mathbb{C}^m$,*

$$|Bv \cdot v| \leqq Av \cdot v\,.$$

Then for any vectors v, w the following inequality holds:

$$|Bv \cdot w| \leqq (Av \cdot v)^{1/2}(Aw \cdot w)^{1/2}\,. \tag{2.8}$$

Proof: By assumption,

$$(A + B)(v + w) \cdot (v + w) \geqq 0\,,$$
$$(A - B)(v - w) \cdot (v - w) \geqq 0\,.$$

Add and divide by 2:

$$Av \cdot v + 2\,\mathcal{R}e\,Bv \cdot w + Aw \cdot w \geqq 0\,. \tag{2.9}$$

Multiplying v and w by arbitrary real numbers s, t, turns (2.9) into a non-negative quadratic form in s and t, and its discriminant must therefore be non-negative. This fact, after multiplying one of the vectors by a convenient complex scalar of unit modulus, yields (2.8).

We turn to the proof of the main inequality (2.1) for the functions u_k. We shall denote the function u_k by v; by construction, the support of v has diameter $c\sqrt{\delta}$. To simplify writing we take the support of v to be centered around the origin $x = 0$. So

$$(P_\delta v, v) = \int_{|x| \leqq c\sqrt{\delta}} \sum p_\alpha(x) v(x + \alpha\delta) \cdot v(x)\, dx\,.$$

The coefficients p_α were assumed to be twice differentiable; therefore by Taylor's theorem (here $p_\alpha^l = \partial p_\alpha / \partial x_l$),

$$p_\alpha(x) = p_\alpha(0) + \sum x_l p_\alpha^l(0) + O(|p_\alpha|_2\,|x|^2)\,.$$

In particular, for $|x| \leqq c\sqrt{\delta}$,

$$|p_\alpha(x) - p_\alpha(0) - \sum x_l p^l(0)| \leqq O(\delta)\,|p_\alpha|_2\,.$$

Denoting by P_δ^0 and P_δ^l the operators

$$P_\delta^0 = \sum p_\alpha(0)\,T^\alpha\,, \qquad P_\delta^l = \sum p_\alpha^l(0)\,T^\alpha\,,$$

we see with the aid of Lemma 1.1 that, for all v whose support lies in $|x| \leqq c\sqrt{\delta}$,

$$(P_\delta v, v) = ((P_\delta^0 + \sum x_l P_\delta^l)v, v) + \delta\, O(|p|_{2,0})\,\|v\|^2\,.$$

This shows that in order to prove (2.5) it suffices to show that the quadratic form Q defined by

$$(P_\delta^0 v, v) + \sum \mathcal{R}e\,(x_l P_\delta^l v, v) = Q \tag{2.10}$$

is bounded from below by $O(\delta)\,\|v\|^2$. To estimate Q we make use of the fact that P_δ^0 and P_δ^l are operators with constant coefficients: by Parseval's theorem,

$$(P_\delta^0 v, v) = \int p^0(\xi)\hat{v}(\xi)\cdot\hat{v}(\xi)\,d\xi\,, \tag{2.11}$$

where $\hat{v}(\xi)$ is the Fourier transform

$$\hat{v}(\xi) = (2\pi)^{-n/2}\int e^{-ix\cdot\xi}v(x)\,dx\,.$$

Similarly, denoting $x_l v$ by v^l we get

$$(x_l P_\delta^l v, v) = (P_\delta^l v, v^l) = \int p^l(\xi)\hat{v}(\xi)\cdot\hat{v}^l(\xi)\,d\xi\,. \tag{2.12}$$

At this point we remark that if we were dealing with scalar valued functions, we could complete the proof quite easily with the aid of this simple inequality for non-negative functions $p(x,\xi)$ whose second derivatives with respect to x is bounded by K:

$$|p^l(0,\xi)|^2 \leqq 2Kp(0,\xi)\,.$$

Combining the above inequality with a standard one, we get

$$|p^l\hat{v}\cdot\hat{v}^l| \leqq \frac{1}{4Kn}\,|p^l\hat{v}|^2 + nK\,|\hat{v}^l|^2 \leqq \frac{1}{2n}\,p^0\,|\hat{v}|^2 + nK\,|\hat{v}^l|^2$$

so that

$$Q \geqq \frac{1}{2}\int p^0(\xi)\hat{v}(\xi)\cdot\hat{v}(\xi)\,d\xi - nK\sum\,(v^l, v^l)\,.$$

The first term on the right is non-negative while the second is $O(\delta)\,\|v\|^2$, since $v^l = x_l v$ and $|x_l| \leqq c\sqrt{\delta}$ on the support of v.

The general case is however more difficult. In Taylor's formula

$$p(x,\xi) = p^0(\xi) + \sum x_l p^l(\xi) + O(|x|^2)$$

set $x_1 = x_2 = \cdots x_{l-1} = x_{l+1} = \cdots = x_n = 0$, $x_l = \pm\sqrt{\delta}$; using the fact that the matrix $p(x,\xi)$ is assumed non-negative, we get

$$p^0(\xi) \pm \sqrt{\delta}\,p^l(\xi) \geqq O(\delta)\,.$$

Adding a term of order δ to $p^0(\xi)$ alters Q only by an amount $O(\delta)\,\|v\|^2$; by such means we can eliminate the term $O(\delta)$ from the above inequality and have instead

$$p^0(\xi) \pm \sqrt{\delta}\,p^l(\xi) \geqq 0\,. \tag{2.13}$$

Inequality (2.13) shows that the hypothesis of Lemma 2.2 with $A = p^0$, $B = \sqrt{\delta}\,p^l$ is satisfied. Applying the lemma with the vectors $\hat{v}(\xi)$, $\hat{v}^l(\xi)$, we find from (2.8) that

$$\sqrt{\delta}\,|p^l\hat{v}\cdot\hat{v}^l| \leqq (p^0\hat{v}\cdot\hat{v})^{1/2}(p^0\hat{v}^l, \hat{v}^l)^{1/2}. \tag{2.14}$$

Using the inequality

$$\sqrt{\alpha\beta} \leqq \frac{k}{2}\alpha + \frac{1}{2k}\beta$$

with $\alpha = p^0\hat{v}\cdot\hat{v}$, $\beta = p^0\hat{v}^l\cdot\hat{v}^l$ and $k = \sqrt{\delta}/n$, we obtain from (2.14) that, for any ξ,

$$|p^l\hat{v}\cdot\hat{v}^l| \leqq \frac{1}{2n}p^0\hat{v}\cdot\hat{v} + \frac{n}{2\delta}p^0\hat{v}^l\cdot\hat{v}^l. \tag{2.15}$$

Integrating this relation with respect to ξ and using once more the Parseval relation to express the right side in terms of v and v^l, we get with the help of the identity (2.12) the inequality

$$|(P^l_\delta v, v^l)| \leqq \frac{1}{2n}(P^0_\delta v, v) + \frac{n}{2\delta}(P^0_\delta v^l, v^l).$$

Using this estimate for the second term in (2.10), we get the following lower bound for Q:

$$\begin{aligned} Q &\geqq (P^0_\delta v, v) - \sum_l \left\{\frac{1}{2n}(P^0_\delta v, v) + \frac{n}{2\delta}(P^0_\delta v^l, v^l)\right\} \\ &= \tfrac{1}{2}(P^0_\delta v, v) - \frac{n}{2\delta}\sum (P^0_\delta v^l, v^l). \end{aligned} \tag{2.16}$$

By definition, $v^l = x_l v$; according to Lemma 2.1, with $\phi = x_l$,

$$(P^0_\delta v^l, v^l) = (P^0_\delta x_l v, x_l v) = (P^0_\delta v, x_l^2 v) + O(\delta^2)\,\|v\|^2.$$

Substituting this into (2.16) we find, setting $\sum x_l^2 = r^2$,

$$\begin{aligned} Q &\geqq \tfrac{1}{2}(P^0_\delta v, v) - \tfrac{1}{2}\mathscr{Re}\left(P^0_\delta, \frac{nr^2}{\delta}v\right) + O(\delta)\,\|v\|^2 \\ &= \tfrac{1}{2}\mathscr{Re}\left(P^0_\delta v, \left(1 - \frac{nr^2}{\delta}\right)v\right) + O(\delta)\,\|v\|^2. \end{aligned} \tag{2.17}$$

We recall that the function v is zero for $r \geqq c\sqrt{\delta}$; we choose now the constant c to be $1/\sqrt{2n}$ and introduce the auxiliary function ϕ defined as

$$\phi(r) = \begin{cases} \sqrt{1 - \dfrac{nr^2}{\delta}} & \text{for} \quad r < c\sqrt{\delta}, \\ \sqrt{1/2} & \text{for} \quad r \geqq c\sqrt{\delta}. \end{cases}$$

We rewrite the right side of (2.17) in terms of ϕ, obtaining

$$Q \geqq \tfrac{1}{2}\mathscr{Re}\,(P^0_\delta v, \phi^2 v) + O(\delta)\,\|v\|^2. \tag{2.18}$$

As defined above, ϕ satisfies a Lipschitz condition with constant $K = O(1/\sqrt{\delta})$; hence, by Lemma 2.1 again,

$$\mathcal{R}e\,(P_\delta^0 v, \phi^2 v) = \mathcal{R}e\,(P_\delta^0 \phi v, \phi v) + O(K^2\delta^2)\,\|v\|^2 .$$

The first term on the right when expressed in the Fourier representation is visibly non-negative; the second term is of order δ. Substituting into (2.17), we obtain therefore the desired estimate

$$Q \geqq O(\delta)\,\|v\|^2 .$$

This completes the proof of the main theorem.

3. Sharp Gårding Inequality for Pseudo-Differential Operators

Gårding's inequality refers to differential operators of the form $A = a(x, D)$ acting on vectors $u(x)$; here $a(x, \xi)$ is an $n \times n$ hermitian matrix whose elements are polynomials in ξ. If $a(x, \xi)$ is, say, homogeneous in ξ of degree r, reasonably smooth, and is positive semi-definite, then Gårding's inequality may be expressed in the form: for every $\varepsilon > 0$, there is a constant $K(\varepsilon)$ such that

$$\mathcal{R}e\,(Au, u) \geqq -\varepsilon\,\|u\|_{r/2}^2 - K(\varepsilon)\,\|u\|_{(r-1)/2}^2 .$$

Here we are using the familiar H_s norms, s real,

$$\|u\|_s^2 = \int (1 + |\xi|^2)^s\,|\hat{u}(\xi)|^2\,d\xi .$$

The sharp Gårding inequality states that one may take $\varepsilon = 0$ in Gårding's inequality for some constant K:

$$\mathcal{R}e\,(Au, u) \geqq -K\,\|u\|_{(r-1)/2}^2 . \tag{3.1}$$

In this section we shall prove (3.1) for pseudo-differential operators acting on vector valued functions $u(x)$ in R^n. As mentioned earlier, Hörmander [3] proved the result in the scalar case. Our proof for that case is simpler but it also makes use of a particular partition of unity given in [3].

As in [5] and [3], we consider, for functions $a(x, \xi)$ satisfying some degree of smoothness for $\xi \neq 0$ and behaving as though they were homogeneous in ξ of degree r for $|\xi| > 1$, the pseudo-differential operator given by

$$Au(x) = (2\pi)^{-n/2} \int_{R^n} e^{ix\cdot\xi} a(x, \xi)\hat{u}(\xi)\,d\xi . \tag{3.2}$$

We prove that if the matrix $a(x, \xi)$ is hermitian positive semi-definite, then (3.1) holds.

For convenience we shall consider only the case $r = 0$. The function $a(x, \xi)$ is assumed to have a limit $a(\infty, \xi)$ as $x \to \infty$ (see however Theorem 3.2), and is supposed to belong to certain classes $\mathscr{C}_{k,l}$ analogous to $C_{k,l}$ of Section 1. These are defined in terms of the Fourier transform $\hat{a}'(\eta, \xi)$ of $a'(x, \xi) = a(x, \xi) - a(\infty, \xi)$ with respect to x:

$$\hat{a}'(\eta, \xi) = (2\pi)^{-n/2} \int e^{-ix\cdot\eta} a'(x, \xi)\, dx .$$

For a function of ξ set

$$|f|_k = \sum_{0 \leqq |\beta| \leqq k} \sup\, (1 + |\xi|)^{\beta}\, |D_\xi^\beta f| .$$

Define the k, l-norm of $a(x, \xi)$ as

$$|a|_{k,l} = |a(\infty, \xi)|_k + (2\pi)^{-n/2} \int |\hat{a}'(\eta, \cdot)|_k\, (1 + |\eta|)^l\, d\eta \tag{3.3}$$

and denote by $\mathscr{C}_{k,l}$ the class of a with finite k, l-norm.

THEOREM 3.1. *Let $a(x, \xi)$ be hermitian positive semi-definite belonging to $\mathscr{C}_{0,2} \cap \mathscr{C}_{2,0}$. Then the operator A given by* (3.2) *satisfies*

$$\mathscr{R}e\,(Au, u) \geqq -K\, \|u\|_{-1/2}^2 \tag{3.4}$$

for some constant Kf.

We shall make use of the partition of unity in ξ-space described on pages 141–142 of [3]:

$$\sum_0^\infty \psi_k^2(\xi) = 1 ,$$

where each ξ is contained in at most a fixed (independent of ξ) number of supports of the ψ_k, and $\xi = 0$ is contained only in supp ψ_0.

The $\psi_k \in C_0^\infty(R^n)$ satisfy for all α

$$(1 + |\xi|)^{|\alpha|} \sum |D^\alpha \psi_k(\xi)|^2 \leqq C_\alpha \tag{3.5}$$

and, for some constant C,

$$|\xi - \eta| \leqq C(1 + |\xi|)^{1/2} \quad \text{if} \quad \xi, \eta \in \operatorname{supp} \psi_k ,\ \ k = 0, 1, \cdots , \tag{3.6}$$

$$\sum |\psi_k(\xi) - \psi_k(\eta)|^2 \leqq C \frac{|\xi - \eta|^2}{(1 + |\xi|)^{1/2}(1 + |\eta|)^{1/2}} . \tag{3.7}$$

In fact, if $|\xi - \eta| \leqq \frac{1}{2}(1 + |\xi|)$, (3.7) follows from (3.5) for $\alpha = 1$, and otherwise (3.7) holds trivially.

Proof of Theorem 3.1: Since it is very similar to the proof of Theorem 1.1 we shall not carry out all the computations. For convenience we shall suppose that $a(\infty, \xi) = 0$, the extra terms due to $a(\infty, \xi)$ being handled in a similar way.

We first iocalize in ξ-space by setting $u_k = \psi_k(D)u$, i.e.,

$$\hat{u}_k(\xi) = \psi_k(\xi)\hat{u}(\xi) ,$$

and assert, as in Section 1.3 of [3], that (Au, u) differs from $\Sigma\,(Au_k, u_k)$ by $O(\|u\|^2_{-1/2})$. For by definition,

$$A\hat{u}(\xi) = (2\pi)^{-n/2} \int \hat{a}'(\xi - \eta, \eta)\hat{u}(\eta)\, d\eta\,.$$

Hence

$$\begin{aligned} I &= (Au, u) - \sum (Au_k, u_k) \\ &= (2\pi)^{-n/2} \iint \hat{a}'(\xi - \eta, \eta)\hat{u}(\eta) \cdot \hat{u}(\xi)(1 - \sum \psi_k(\eta)\psi_k(\xi))\, d\xi\, d\eta \\ &= \tfrac{1}{2}(2\pi)^{-n/2} \iint \hat{a}'(\xi - \eta, \eta)\hat{u}(\eta) \cdot \hat{u}(\xi) \sum |\psi_k(\eta) - \psi_k(\xi)|^2\, d\xi\, d\eta\,, \end{aligned}$$

as in the derivation of (2.5). Applying (3.7) we see that

$$\begin{aligned} |I| &\leqq \tfrac{1}{2}(2\pi)^{-n/2} \iint |\hat{a}'(\xi - \eta, \eta)\hat{u}(\eta) \cdot \hat{u}(\xi)|\, C\, |\xi - \eta|^2 (1 + |\xi|)^{-1/2} \\ &\qquad \times (1 + |\eta|)^{-1/2}\, d\xi\, d\eta \\ &\leqq \tfrac{1}{2}\, |a|_{0,2}\, \|u\|^2_{-1/2} \end{aligned}$$

—the desired inequality.

Since, now, $\Sigma\, \|u_k\|^2_{-1/2} = \|u\|^2_{-1/2}$, we have only to prove inequality (3.4) for the individual functions u_k. We shall use an analogue of Lemma 2.1.

LEMMA 3.1. *Let the symbol $p(x)$ be a hermitian matrix independent of ξ and belonging to $\mathscr{C}_{0,2}$. If $\phi(\xi)$ is a real scalar function with Lipschitz constant K on the support of u, then*

$$(3.8)\qquad |\mathscr{R}e\, (\phi(D)pu, \phi(D)u) - \mathscr{R}e\, (p\phi(D)u, \phi(D)u)| \leqq \tfrac{1}{2}\, |p|_{0,2}\, K^2\, \|u\|_0^2\,.$$

Here we are using the general notation $\phi(D)$ to represent the operator given by

$$\widehat{\phi(D)u}(\xi) = \phi(\xi)\hat{u}(\xi)\,.$$

To prove (3.8) we start with the identity (we may assume $p(\infty) = 0$)

$$\begin{aligned} &\mathscr{R}e\, (\phi(D)pu, \phi(D)u) - \mathscr{R}e\, (p\phi(D)u, \phi(D)u) \\ &\qquad = (2\pi)^{-n/2}\, \mathscr{R}e \iint \hat{p}(\xi - \eta)\hat{u}(\eta) \cdot \hat{u}(\xi)(\phi(\xi) - \phi(\eta))\phi(\xi)\, d\xi\, d\eta\,, \end{aligned}$$

and using the hermitian property $\hat{p}(\xi) = \hat{p}^*(-\xi)$ we interchange the variables of integration and find, as in the proof of Lemma 2.1, that this equals

$$\tfrac{1}{2}(2\pi)^{-n/2}\, \mathscr{R}e \iint \hat{p}(\xi - \eta)\hat{u}(\eta) \cdot \hat{u}(\xi)(\phi(\xi) - \phi(\eta))^2\, d\xi\, d\eta\,,$$

from which (3.8) follows.

We now indicate the proof of the main inequality (3.4) for any function u_k which we denote by v; (3.4) holds trivially for u_0 so we may suppose $k > 0$, i.e.,

$0 \notin \operatorname{supp} \hat{v}$. We shall use C to denote various constants (independent of k). We have

$$(3.9) \qquad (Av, v) = (2\pi)^{-n/2} \iint \hat{a}(\xi - \eta, \eta)\hat{v}(\eta) \cdot \hat{v}(\xi)\, d\xi\, d\eta\,.$$

Let τ be a point in the support of $\hat{v}(\eta)$. According to (3.6) for $\mu, \nu \in \operatorname{supp} \hat{v}$,

$$1 + |\mu| \leqq 1 + |\nu| + |\nu - \mu| \leqq 1 + |\nu| + \tfrac{1}{2}(1 + |\mu|) + C\,;$$

hence

$$(3.10) \qquad 1 + |\mu| \leqq C(1 + |\nu|) \qquad \text{for} \qquad \mu, \nu \in \operatorname{supp} \hat{v}\,.$$

It follows that

$$(3.10)' \qquad (1 + |\tau|)^{-1}\, \|v\|_0^2 \leqq C\, \|v\|_{-1/2}^2\,.$$

Now by Taylor's theorem, setting

$$\hat{a}^l(\xi - \eta, \tau) = \frac{\partial}{\partial \tau_l}\, \hat{a}(\xi - \eta, \tau)\,,$$

one finds with the aid of the theorem of the mean and (3.10) that, for $\eta \in \operatorname{supp} \psi_k$,

$$|\hat{a}(\xi - \eta, \eta) - \hat{a}(\xi - \eta, \tau) - \sum_{l=1}^{n} (\eta_l - \tau_l)\hat{a}^l(\xi - \eta, \tau)|$$

$$\leqq C\, |\eta - \tau|^2\, |\hat{a}(\xi - \eta, \cdot)|_2\, (1 + |\eta|)^{-2}$$
$$\leqq C\, |a(\xi - \eta, \cdot)|_2\, (1 + |\eta|)^{-1}\,,$$

in virtue of (3.6); from (3.10) it follows that this is

$$\leqq C\, |\hat{a}(\xi - \eta, \cdot)|_2\, (1 + |\eta|)^{-1/2}(1 + |\xi|)^{-1/2}\,.$$

Inserting this into (3.9) we find, on setting $a(x, \tau) = p(x)$, $a^l(x, \tau) = p^l(x)$ and $(D_l - \tau_l)v = v^l$, so that $\hat{v}^l = (\xi_l - \tau_l)\hat{v}$,

$$|(Av, v) - (p(x)v, v) - \sum (p^l(x)v^l, v)| \leqq C\, |a|_{2,0}\, \|v\|_{-1/2}^2\,.$$

Thus to prove (3.4) for v it suffices to show that the quadratic form

$$Q = (pv, v) + \mathcal{R}e \sum (p^l v^l, v)$$

satisfies

$$(3.11) \qquad Q \geqq -C\, \|v\|_{-1/2}^2\,.$$

(We remark that, as on page 483, if v were scalar valued the proof could easily be completed now with the aid of the inequality $p^l(x)^2 \leqq$ constant $p(x)$.)

Now, for any vector ξ the matrix $a(x, \tau + \xi)$ is hermitian positive semi-definite. Hence for c a fixed constant to be chosen later (independent of k) and any $l = 1, \cdots, n$, we have, using the theorem of the mean, and (3.10): for ξ a vector in the direction of $\pm\xi_l$ of magnitude $c(1+|\tau|)^{1/2}$,

$$a(x,\tau) \pm c(1+|\tau|)^{1/2}a^l(x,\tau) + C(1+|\tau|)\frac{|a(x,\cdot)|_2}{(1+|\tau|)^2} \geqq 0\,.$$

We claim that adding a term [constant $\cdot\,(1+|\tau|)^{-1}$ times the identity matrix] to $p(x)$ does not change anything in a substantial way; for by (3.10)′, the additional contribution to (pv, v) is of the order $C\,\|v\|^2_{-1/2}$. Thus we may suppose that such a term has been added so that we have for each l

$$p(x) \pm c(1+|\tau|)^{1/2}p^l(x) \geqq 0\,.$$

We are therefore in a position to apply Lemma 2.2 and conclude that

$$c(1+|\tau|)^{1/2}\,|(p^l v^l, v)| \leqq (pv, v)^{1/2}(pv^l, v^l)\,,$$

or

$$|(p^l v^l, v)| \leqq \frac{1}{2n}(pv, v) + \frac{n}{2c^2}\frac{1}{1+|\tau|}(pv^l, v^l)\,.$$

Hence

$$Q \geqq \tfrac{1}{2}(pv, v) - \frac{n}{2c^2}\frac{1}{1+|\tau|}\sum(pv^l, v^l)\,. \tag{3.12}$$

Following our argument in Section 2 we apply Lemma 3.1 and infer that

$$\sum(pv^l, v^l) = \mathscr{Re}\sum(pv, (D_l-\tau)^2 v) + O(\|v\|_0^2)\,.$$

Inserting this in (3.12) and using (3.10)′ we find

$$Q \geqq \tfrac{1}{2}(pv, v) - \frac{n}{2c^2}\frac{1}{1+|\tau|}\,\mathscr{Re}\,(pv, |D-\tau|^2 v) - \frac{nC}{2c^2}\|v\|^2_{-1/2}\,,$$

or

$$2Q \geqq \mathscr{Re}\left(pv, \left(1 - \frac{n}{2c^2}\frac{|D-\tau|^2}{1+|\tau|}\right)v\right) + O(\|v\|^2_{-1/2})\,. \tag{3.13}$$

Now on the support of $\hat{v}(\xi)$, according to (3.6),

$$\frac{|\xi-\tau|^2}{1+|\tau|} \leqq C^2\,;$$

consequently we choose the constant $c = C\sqrt{n}$ so that

$$1 - \frac{n}{2c^2}\frac{|\xi-\tau|^2}{1+|\tau|} \geqq \frac{1}{2} \quad \text{on} \quad \operatorname{supp}\hat{v}\,.$$

Set

$$\phi(\xi) = \begin{cases} \left(1 - \dfrac{n}{2c^2}\dfrac{|\xi-\tau|^2}{1+|\tau|}\right)^{1/2} & \text{for} \quad |\xi-\tau| \leqq C(1+|\tau|)^{1/2}\,, \\ \sqrt{1/2} & \text{otherwise}\,. \end{cases}$$

Then (3.13) reads

(3.14) $$2Q \geqq \mathscr{Re}\,(pv, \phi^2(D)v) + O(\|v\|^2_{-1/2})\,.$$

The function $\phi(\xi)$ satisfies a Lipschitz condition with constant $K = O((1 + |\tau|)^{-1/2})$ and therefore, applying Lemma 3.1 once more, we find

$$\mathscr{Re}\,(pv, \phi^2(D)v) = \mathscr{Re}\,(p\phi(D)v, \phi(D)v) + \frac{1}{1 + |\tau|}\,O(\|v\|^2_0)\,.$$

The first term on the right is clearly non-negative, while according to (3.10)′ the second is $O(\|v\|^2_{-1/2})$. Insertion of this into (3.14) yields the desired inequality (3.4).

The proof of Theorem 3.1 is complete.

As a consequence of Theorem 3.1 we have the analogue of Corollary 1.1:

COROLLARY 3.1. *Let $a(x, \xi)$ be homogeneous in ξ of degree zero and belong to $\mathscr{C}_{0,2} \cap \mathscr{C}_{2,0}$. Then the operator A given by* (3.2) *satisfies*

(3.15) $$\|Au\|^2_0 \leqq \max_{\substack{|\xi|=1 \\ x}} |a(x, \xi)|^2\, \|u\|^2_0 + K\, \|u\|^2_{-1/2}\,.$$

Hörmander has pointed out to us that inequality (3.4) may be extended also to operators whose symbols $a(x, \xi)$ do not have a limit as $\xi \to 0$ so long as $a(x, \xi)$ has sufficiently many bounded derivatives. We shall present such a result here, using a slight modification of the argument in Theorem 3.5 of his paper [11]; this paper is concerned with non-homogeneous symbols.

In order to keep the argument brief it is convenient to consider a symbol $p(x, \xi)$ which is a hermitian positive semi-definite matrix, and behaves in some sense as though it were homogeneous of order 1 in ξ for $|\xi| \geqq 1$. To be precise, we shall suppose that for $x \in R^n$, $\xi \in R^n$,

(3.16) $$|D^\alpha_x D^\beta_\xi p(x, \xi)| \leqq K_0(1 + |\xi|)^{1-|\beta|} \quad \text{for} \quad |\alpha + \beta| \leqq n + 3\,,$$

and we consider the operator defined by

(3.17) $$(Pu)(x) = (2\pi)^{-n/2} \int e^{ix\cdot\xi} p(x, \xi)\hat{u}(\xi)\, d\xi\,.$$

THEOREM 3.2. *If p satisfies* (3.16), *then*

(3.18) $$\mathscr{Re}\,(Pu, u) \geqq -K\,\|u\|^2_0$$

for some constant K.

Proof: (i) Consider first the case that $p(x, \xi)$ has compact support in x, say that p vanishes for $|x - y| > 10$ with some $y \in R^n$. Let A denote the operator given by (3.2) with $a(x, \xi) = p(x, \xi)(1 + |\xi|^2)^{-1/2}$. It is easily verified (see e.g. Section 3 of [5]) that $P - A(1 + |D|)$ is a continuous map of L_2 into itself with norm bounded by a constant times K_0. Since also $[A, (1 + |D|)^{1/2}](1 + |D|)^{1/2}$, where [,] denotes the commutator, is a bounded map of L_2 into itself (see Section 5 of [5]) we have

$$\mathcal{R}e\,(Pu, u) = \mathcal{R}e\,(A(1 + |D|)^{1/2}u, (1 + |D|)^{1/2}u) + O(\|u\|_0^2),$$

and inequality (3.18) follows from Theorem 3.1—with a constant K which is equal to a constant times K_0; for, it is easily seen that $p(x, \xi/|\xi|)$ satisfies the conditions of Theorem 3.1.

(ii) Turning to the general case, we introduce a locally finite partition of unity:

$$1 \equiv \sum_0^\infty \psi_j(x),$$

where the ψ_j are non-negative C^∞ functions, each of which is a translate of ψ_0, and such that ψ_j has support in a unit ball B_j with center x_j. Let $\phi_j \in C_0^\infty$ satisfy $0 \leqq \phi_j \leqq 1$, $\phi_j \equiv 1$, on a ball of radius 2 concentric with B_j and have support in a concentric ball of radius 3; furthermore, assume that, for all x,

$$\sum_j \phi_j^2(x) \leqq C, \tag{3.19}$$

for some fixed constant C.

Then since $\phi_j\psi_j = \psi_j$, we have

$$\begin{aligned}(Pu, u) &= \sum (\psi_j Pu, \phi_j u)\\ &= \sum (\psi_j P\phi_j u, \phi_j u) + \sum (\psi_j P(1 - \phi_j)u, \phi_j u).\end{aligned}$$

According to part (i) above we have, for K a constant independent of j,

$$\begin{aligned}\mathcal{R}e \sum (\psi_j P\phi_j u, \phi_j u) &\geqq -K \sum \|\phi_j u\|_0^2 = -K \int \sum \phi_j^2 \,|u|^2\, dx\\ &\geqq -CK\,\|u\|_0^2,\end{aligned}$$

by (3.19). Hence to complete the proof of (3.18) it suffices to show that

$$|\sum (\psi_j P(1 - \phi_j)u, \phi_j u)| \leqq \text{constant} \cdot \|u\|_0^2. \tag{3.20}$$

Since $\Sigma\, \|\phi_j u\|_0^2 \leqq C\, \|u\|_0^2$, we need only show that

$$\sum \|\psi_j P(1 - \phi_j)u\|_0^2 \leqq \text{constant} \cdot \|u\|_0^2. \tag{3.21}$$

If $K(x, z)$ is the inverse Fourier transform of $p(x, \xi)$ with respect to ξ, then

$$P(1 - \phi_j)u = \int K(x, x - z)(1 - \phi_j)(z)u(z)\, dz \qquad \text{for} \qquad |x - x_j| < 1.$$

Now $z^\alpha K(x, z)$ is the inverse Fourier transform of $(-D_\xi)^\alpha p(x, \xi)$ which is an integrable function of ξ for $|\alpha| > n + 1$, and it follows that the distribution $K(x, z)$ is a continuous function when $z \neq 0$; furthermore, for some constant c, and $k(z) = (1 + |z|)^{-n-1} \in L_1$,

$$|K(x, z)| \leqq ck(z) \qquad \text{for} \qquad |z| > 1.$$

Hence

$$|P(1 - \phi_j)u| \leqq c \int |k(x - z)|\, |u(z)|\, dz \qquad \text{for} \qquad |x - x_j| < 1,$$

and consequently if $k * u = \int k(x - z)u(z)\, dz$,

$$\sum \|\psi_j P(1 - \phi_j)u\|_0^2 \leqq \sum c^2 \int \psi_j^2 \, |k * u|^2 \, dx$$
$$\leqq c^2 C^2 \, \|k * u\|_0^2 \leqq \text{constant} \cdot \|u\|_0^2$$

—the proof is complete.

Corollary 3.1 may also be extended to symbols $a(x, \xi)$ which need not have limits as $x \to \infty$. We conclude with an illustration.

THEOREM 3.2′. *Let $a(x, \xi)$ be defined for $x \in R^n$, $\xi \in R^n$, be homogeneous in ξ of degree $\frac{1}{2}$ for $|\xi| > 1$, and satisfy*

$$|D_x^\alpha D_\xi^\beta a(x, \xi)| \leqq K_0(1 + |\xi|)^{1/2 - |\beta|} \qquad \textit{for} \qquad |\alpha + \beta| \leqq n + 3 \, .$$

Then the corresponding operator A defined by (3.2) *satisfies*

$$\|Au\|_0^2 \leqq \sup_{\substack{|\xi|=1 \\ x}} |a(x, \xi)|^2 \, \|u\|_{1/2}^2 + K \, \|u\|_0^2 \, .$$

The theorem is proved as above by decomposing $\|Au\|_0^2$ as

$$(Au, Au) = \sum (\psi_j A(1 - \phi_j)u, A(1 - \phi_j)u) + 2 \sum \mathcal{Re} \, (\psi_j A(1 - \phi_j)u, \phi_j A \phi_j u)$$
$$+ \sum (\psi_j A \phi_j u, A \phi_j u) \, .$$

The estimates for the first two sums are obtained similarly to the estimate (3.20), while the last sum is estimated with the aid of Theorem 3.2.

Bibliography

[1] Berger, C. A., *A strange dilation theorem*, Amer. Math. Soc. Notices, Abstract No. 625-152, Vol. 12, 1965, p. 590.

[2] Hörmander, L., *Pseudo-differential operators*, Comm. Pure Appl. Math., Vol. 18, 1965, pp. 501–517.

[3] Hörmander, L., *Pseudo-differential operators and non-elliptic boundary problems*, Annals Math., Vol. 83, 1966, pp. 129–209.

[4] Kato, T., *Some mapping theorems for the numerical range*, Proc. Japan Acad., Vol. 41, 1965, pp. 652–655.

[5] Kohn, J. J., and Nirenberg, L., *An algebra of pseudo-differential operators*, Comm. Pure Appl. Math., Vol. 18, 1965, pp. 269–305.

[6] Kohn, J. J., and Nirenberg, L., *Non-coercive boundary value problems*, Comm. Pure Appl. Math., Vol. 18, 1965, pp. 443–492.

[7] Kreiss, H. O., *On difference approximations of the descriptive type for hyperbolic differential equations*, Comm. Pure Appl. Math., Vol. 17, 1964, pp. 335–353.

[8] Lax, P. D., and Wendroff, B., *On the stability of difference schemes with variable coefficients*, Comm. Pure Appl. Math., Vol. 15, 1962, pp. 363–371.

[9] Lax, P. D., and Wendroff, B., *Difference schemes for hyperbolic equations with high order of accuracy*, Comm. Pure Appl. Math., Vol. 17, 1964, pp. 381–398.

[10] Pearcy, C., *A short proof of the Halmos inequality*, University of Michigan, to appear.

[11] Hörmander, L., *Pseudo-differential operators and hypoelliptic equations*, Proc. Symposium in Singular Integral Operators, Amer. Math. Soc., to appear.

Received May, 1966.

An Example of Huygens' Principle*

PETER D. LAX AND RALPH S. PHILLIPS
Courant Institute *Stanford University*

In this note we discuss Huygens' principle for the spherical wave equation:

(1) $$u_{tt} = \Delta_S u - \left(\frac{n-1}{2}\right)^2 u = L_n u,$$

where Δ_S is the Laplace–Beltrami operator on the n-dimensional unit sphere, n an odd integer.

We solve (1) by expansion in eigenfunctions of L_n:

(2) $$L_n h_j = \lambda_j h_j\ ;$$

then

(3) $$u(\omega, t) = \sum (a_j \exp\{\sqrt{\lambda_j}t\} + b_j \exp\{-\sqrt{\lambda_j}t\})h_j\,.$$

The coefficients a_j and b_j are determined by the Cauchy data:

(4) $$u(\omega, 0) = \sum (a_j + b_j)h_j\,,$$

$(4)_t$ $$u_t(\omega, 0) = \sum \sqrt{\lambda_j}\,(a_j - b_j)h_j\,.$$

The eigenfunctions h_j of Δ_S are the spherical harmonics. To determine λ_j we use the polar coordinate representation of the Euclidian Laplace operator in $n+1$ variables:

$$\Delta_{n+1} = \partial_r^2 + \frac{n}{r}\,\partial_r + \frac{1}{r^2}\,\Delta_S\,.$$

Let h_j be a spherical harmonic of order j; then $H_j = r^j h_j(\omega)$ is a harmonic function of $n+1$ variables, so that

$$0 = \Delta_{n+1} H_j = r^{j-2}(j(j-1) + nj + \Delta_S)h_j\,.$$

* This work was supported by the National Science Foundation partially under Grant No. MCS76-07039 at the Courant Institute and partially under MCS77-04908 at Stanford University. Reproduction in whole or in part is permitted for any purpose of the United States Government.

Communications on Pure and Applied Mathematics, Vol. XXXI, 415–421 (1978)
 0010–3640/78/0031–0415$01.00

Thus

$$(5)\quad L_n h_j = \Delta_S h_j - \left(\frac{n-1}{2}\right)^2 h_j = -\left[j^2 + (n-1)j + \left(\frac{n-1}{2}\right)^2\right] h_j = -\left(j + \frac{n-1}{2}\right)^2 h_j .$$

Substituting this into (3) we get

$$u(\omega, t) = \sum \left(a_j \exp\left\{i\left(j + \frac{n-1}{2}\right)t\right\}\right) + b_j \exp\left\{-i\left(j + \frac{n-1}{2}\right)t\right\} h_j .$$

Now set $t = \pi$; since n is odd,

$$(6)\quad u(\omega, \pi) = (-1)^{(n-1)/2} \sum (-1)^j (a_j + b_j) h_j .$$

Similarly,

$$(6)_t\quad u_t(\omega, \pi) = (-1)^{(n-1)/2} \sum (-1)^j \sqrt{\lambda_j}\,(a_j - b_j) h_j .$$

Since h_j is a homogeneous polynomial of degree j,

$$h_j(-\omega) = (-1)^j h_j(\omega) .$$

Using this in (6), $(6)_t$ and comparing the result with (4), $(4)_t$, we deduce that

$$(7)\quad \begin{aligned} u(\omega, \pi) &= (-1)^{(n-1)/2} u(-\omega, 0) , \\ u_t(\omega, \pi) &= (-1)^{(n-1)/2} u_t(-\omega, 0) . \end{aligned}$$

If the initial data are supported in a ball of radius ε about a point ω_0 of S_n, then by (7) the data at time π will be supported in a ball of radius ε around the antipodal point $-\omega_0$. Let t be some value between 0 and π. Then, since signals propagate at most with speed 1, it follows that the support of $u(\omega, t)$ and $u_t(\omega, t)$ is contained in the ball of radius $t + \varepsilon$ around ω_0. Since signals also propagate backwards with speed at most 1, it follows that $u(\omega, t)$, $u_t(\omega, t)$ are supported in the ball of radius $(1-t) + \varepsilon$ around $-\omega_0$. Thus the data at time t are supported in the spherical strip consisting of points whose distance from ω_0 is greater than $t - \varepsilon$ but less than $t + \varepsilon$. Since ε is arbitrary, we have proved Huygens' principle:

If the support of the initial data of u contains no point whose distance from ω_0 is t, then

$$u(\omega_0, t) = 0 .$$

There are as many ways of proving this result as there are methods for solving a second-order hyperbolic equation. We show now that locally[1] the spherical wave equation can be transformed by a change of variables into the Euclidean wave equation, for which Huygens' principle is known to hold. Since completing this research we learned of the as yet unpublished work of Bent Ørsted [2] in which this same transformation is obtained in the course of an investigation of conformally invariant differential equations.

We introduce stereographic coordinates on the sphere:

$$(8) \qquad x^2+z^2=1\,, \qquad x=(x_1,\cdots,x_n)\,,$$

by

$$(9) \qquad x=\frac{2\xi}{1+\rho^2}\,, \qquad z=\frac{1-\rho^2}{1+\rho^2}\,,$$

where $\rho^2=|\xi|^2=\xi_1^2+\cdots+\xi_n^2$. It follows from (9) that

$$(10) \qquad ds^2=dx^2+dz^2=l^2\,d\xi^2\,,$$

where

$$(11) \qquad l=\frac{2}{1+\rho^2}\,.$$

The mapping of the sphere onto ξ-space is conformal; the Laplace–Beltrami operator in the ξ-space can be expressed in terms of the factor l as

$$\Delta_S=\frac{1}{l^n}\sum \partial_{\xi_j} l^{n-2}\,\partial_{\xi_j}=\frac{1}{l^2}\Delta_\xi+\frac{n-2}{l^3}\sum l_{\xi_j}\,\partial_{\xi_j}\,.$$

Using the definition (11) of l we get

$$(12) \quad L_n=\Delta_S-\left(\frac{n-1}{2}\right)^2=\left(\frac{1+\rho^2}{2}\right)^2\Delta_\xi-(n-2)\frac{(1+\rho^2)}{2}\sum \xi_j\,\partial_{\xi_j}-\left(\frac{n-1}{2}\right)^2.$$

Next we define a mapping to ξ, t-space from X, T-space, $X=(X_1,\cdots,X_n)$,

[1] The spherical wave equation can not be *globally* transformed into the Euclidean wave equation. This follows from the fact that such a transformation would diffeomorphically map the characteristic cones of the Euclidean wave equation onto the "periodic" characteristic cones of the spherical wave equation and this is obviously impossible.

as follows: Let

$$R=|X|=\left(\sum X_j^2\right)^{1/2},$$

(13) $$t=\arctan c, \qquad c=\frac{T^2-R^2-1}{2T},$$

(14) $$\rho=(b^2+1)^{1/2}-b, \qquad b=\frac{T^2-R^2+1}{2R},$$

(15) $$\xi=\rho\frac{X}{R}.$$

Define the function $f=f(X,T)$ by

(16) $$f=R^{(1-n)/2}\left(\frac{1+\rho^2}{\rho}\right)^{(1-n)/2}.$$

Suppose $u(\xi,t)$ is a solution of (1) in stereographic coordinates. Define $v(X,T)$ by

(17) $$v(X,T)=f(X,T)u(\xi,t),$$

ξ, t functions of X, T, via (13)–(15). Then $v(X,T)$ *satisfies the Euclidean wave equation*

(18) $$v_{TT}-\Delta_X v=0,$$

The verification consists of a calculation based on the chain rule, according to which if v is related to u by (17), then

(19) $$v_{TT}-\Delta_X v=fMu+Au_t+\sum B_j u_{\xi_j}+Cv,$$

where

(20) $$M=(t_T^2-t_X\cdot t_X)\,\partial_t^2+2\sum(t_T\xi_{jT}-t_X\cdot\xi_{jX})\,\partial_t\,\partial_{\xi_j}+\sum(\xi_{iT}\xi_{jT}-\xi_{iX}\cdot\xi_{jX})\,\partial_{\xi_i}\,\partial_{\xi_j},$$

(21) $$A=2t_Tf_T-2t_X\cdot f_X+(t_{TT}-\Delta_X t)f,$$

(22) $$B_j=2\xi_{jT}f_T-2\xi_{jX}\cdot f_X+(\xi_{jTT}-\Delta_X\xi_j)f,$$

(23) $$C=f_{TT}-\Delta_X f.$$

Comparing the right-hand side of (12) with that of (19) we see that (18) is implied by (1) if and only if the following relations hold:

$$(24)\qquad M=(t_T^2-t_X\cdot t_X)\left(\partial_t^2-\left(\frac{1+\rho^2}{2}\right)^2\Delta_\xi\right),$$

$$(25)\qquad A=0\,,$$

$$(26)\qquad B_j=-f(t_T^2-t_X\cdot t_X)(n-2)\frac{1+\rho^2}{2}\xi_j\,,$$

$$(27)\qquad C=-f(t_T^2-t_X\cdot t_X)\left(\frac{n-1}{2}\right)^2.$$

The verification of these relations is straightforward; we give now an indication how we arrived at the change of variables (13)–(17).

Denote by H the image in X, T-space of the points $(\omega, 0)$, ω in some neighborhood of S_n. Consider all $(n-1)$-dimensional spheres S_{n-1} in this neighborhood of S_n. Denote the center of such an S_{n-1} by p and its radius by r; $(S_{n-1}, 0)$ can be characterized as the intersection of the characteristic cones issuing from the points (p, r) and $(p, -r)$. Since characteristic cones appear as the singular support of solutions, it follows that such a change of variables must carry the characteristic cones of the spherical wave equation into the characteristic cones of the Euclidean wave equation. Consequently the intersection of two characteristic cones is carried into the intersection of two characteristic cones. Thus $(S_{n-1}, 0)$ is carried into the intersection of two right circular cones, which is an $(n-1)$-dimensional ellipsoid, lying in an n-dimensional hyperplane. This shows that the surface H in X, T-space has the property that its intersections with hyperplanes are $(n-1)$-dimensional ellipsoids. It is easy to show that the only such surfaces are ellipsoids and hyperboloids. We choose H to be the hyperboloid of two sheets:

$$(28)\qquad T^2-X^2=1\,.$$

To determine the mapping of this hyperboloid into S_n we make use of relation (24), which implies that this mapping is conformal with respect to the Lorentz metrics dX^2-dT^2 and $((1+\rho^2)/2)^2\,d\xi^2-dt^2$. Restricting the mapping to a neighborhood of $\xi=0$ at $t=0$ results in a conformal mapping of this neighborhood with the metric $d\xi^2$ into the hyperboloid (28) with the Lorentz metric. Such mappings exist; for instance,

$$(29)\qquad X=\frac{2\xi}{1-\rho^2},\qquad T=\frac{1+\rho^2}{1-\rho^2},\qquad \rho=|\xi|\,,$$

is of this type, since clearly (28) is satisfied and

$$dX^2 - dT^2 = \frac{4}{(1-\rho^2)^2}\, d\xi^2\,,$$

which proves conformality.

Any other conformal map of a neighborhood of $\xi = 0$ into the hyperboloid (28) can be obtained by composing the mapping (29) with a conformal map of ξ-space onto itself, keeping the origin fixed. We make the "Ansatz" that the mapping in question is given by (29) itself. Note that when (29) is composed with (9) it maps the upper hemisphere of (8) onto the upper sheet of (28).

To determine the mapping of X, T-space into S_n, t, take any point X, T inside the cone $T^2 - X^2 > 0$. The characteristic cone through X, T consists of points of the form

$$(30) \qquad X + sE\,, \qquad T + s\,,$$

E an arbitrary unit vector, s any number. The intersection of the cone (30) with the hyperboloid (28) consists of those points (30) for which

$$(31) \qquad s = \frac{1 + R^2 - T^2}{2(T - X \cdot E)}\,.$$

This set is an $(n-1)$-dimensional ellipsoid. The mapping (29) takes this ellipsoid into a set in ξ-space which is the image under the mapping (9) of an $(n-1)$ sphere S_{n-1} in S_n. A straightforward calculation shows that the center of the sphere S_{n-1} corresponds under (9) to the point ξ defined by (15) and (14); the radius of this sphere is equal to t as defined by (13).

Since singularities propagate along characteristics, it follows that there exists a solution $u(\xi, t)$ of (1) whose singularity lies along the cone with vertex at (ξ, t) as defined above. This cone intersects $t = 0$ in the image of S_{n-1} under (9). The singularity of $v(X, T)$ defined by (17) lies along the cone which contains the image of $(S_{n-1}, 0)$ under (29); that cone is given by (30). In particular, the vertices of the two cones, (ξ, t) and (X, T), correspond to each other. This leads to the transformation (13)–(15).

A straightforward calculation shows that with ξ, t defined by (13)–(15), relations (20), (24) hold. Relation (21), (25) is a first-order linear partial differential equation for f which is satisfied by all f of the form

$$f(X, T) = R^{(1-n)/2}\, g(\rho)\,,$$

g an arbitrary function. For f of this form equations (22), (26) are satisfied when

$$g(\rho)=\left(\frac{1+\rho^2}{\rho}\right)^{(1-n)/2}.$$

This completes the determination of f; a calculation shows that relation (23), (27) is also satisfied. This completes the derivation of the change of variables.

We remark that the transformation (13)–(15) carries the separated solutions

$$\exp\{\pm i(j+1)t\}h_j(\omega)$$

of the 3-dimensional spherical wave equation into special solutions $v_j^{\pm}(X, T)$ of the Euclidean wave equations. These special solutions may be useful in some propagation problems; at worst they make a suitable problem for an old-fashioned Tripos Examination.

We remark that the non-Euclidean wave equation

$$u_{tt}=\Delta_{\mathrm{NE}}u+\left(\frac{n-1}{2}\right)^2 u, \tag{32}$$

where Δ_{NE} is the Laplace–Beltrami operator under the metric

$$\left(\frac{2}{1-\rho^2}\right)^2 d\xi^2,$$

also can be transformed to the Euclidean wave equation by a change of variables. In this case the change of variables is far simpler than in the spherical case and works in the large; for $n=2$ it is given in our monograph [1], page 11. On page 237 we observe that for $n=3$ Huygens' principle holds.

We conclude by calling attention to a Huygens' principle discovered by M. Semenov–Tian-Shansky for the hyperbolic system of equations that he associates with the invariant operators of a symmetric space, see [3].

Bibliography

[1] Lax, P. D., and Phillips, R. S., *Scattering Theory for Automorphic Functions*, Annals of Mathematics Studies, No. 87, Princeton University Press, 1976.

[2] Ørsted, B., *Conformally invariant differential equations and projective geometry*, to appear in the Journal of Functional Analysis.

[3] Semenov–Tian-Shansky, M., *Harmonic analysis on Riemannian symmetric spaces of negative curvature and scattering theory*, Izvestiya Akademii Nauk S.S.S.R., Math. Series, Vol. 40, 1976, pp. 562–592.

Received December, 1977.

A Simple One-dimensional Model for the Three-dimensional Vorticity Equation

P. CONSTANTIN
University of Chicago

P. D. LAX
Courant Institute

AND

A. MAJDA
Princeton University

Abstract

A simple qualitative one-dimensional model for the 3-D vorticity equation of incompressible fluid flow is developed. This simple model is solved exactly; despite its simplicity, this equation retains several of the most important structural features in the vorticity equations and its solutions exhibit some of the phenomena observed in numerical computations for breakdown for the 3-D Euler equations.

1. Introduction

In regions far away from boundaries, the physical mechanism of vortex stretching is an important factor responsible for the complexity of incompressible fluid flow. In two space dimensions, where vortex stretching does not occur, the conservation of vorticity leads to the global existence of smooth solutions for the incompressible Euler equations. In three space dimensions, where vortex stretching is a prominent effect, it is an outstanding unsolved problem of mathematical fluid dynamics to determine whether solutions of the Euler equations develop singularities in finite time. This problem is important from the physical point of view because the existence of such singularities signifies the onset of turbulence in high Reynolds number flows and the structure of this conceivable singularity has direct bearing on the inertial cascade in such turbulent flows (see [7]). The possible breakdown of solutions and the structure of singularitites has been studied recently through a wide range of ingenious numerical methods by many authors (see [2], [3], [4], [8], [10]). Several recent theorems support the link between vortex stretching, breakdown for the Euler equations, and the onset of

Communications on Pure and Applied Mathematics, Vol. XXXVIII 715–724 (1985)
 CCC 0010-3640/85/060715–10$04.00

turbulence. In [1], the authors proved that the only way in which smooth solutions of the Euler equations can become singular is that the vorticity become infinite in finite time in a precise fashion; i.e., vortex stretching is the controlling mechanism for breakdown. Another recent theorem (see [5]) establishes that on any closed interval of time, where solutions of the Euler equations remain smooth, the Navier-Stokes equations have a unique smooth regular solution for sufficiently high Reynolds numbers.

In this paper, a simple qualitative one-dimensional mathematical model for the 3-D vorticity equation is developed. Despite its simplicity, this equation retains several of the most important structural features in the vorticity equation, and its solutions exhibit some of the phenomena observed in numerical computations for breakdown of the 3-D Euler equations; a detailed discussion is given at the end of this paper. One great advantage of the simple model which we present is that it can be integrated explicitly. Since the numerical computation of solutions forming singularities involves many subtle issues, this simplified model provides a class of elementary unambiguous test problems for the numerical methods used in studying the breakdown for the 3-D Euler equations. This work is in progress and will be described elsewhere.

2. Heuristic Derivation of the Model Vorticity Equation

The Euler equations for the velocity $v = {}^t(v_1, v_2, v_3)$ and scalar pressure p are given by

$$\begin{aligned} &\frac{Dv}{Dt} = -\nabla p, \qquad x \in \mathbb{R}^3,\ t > 0, \\ (1) \qquad &\operatorname{div} v = 0, \\ &v(x,0) = v_0(x), \end{aligned}$$

where D/Dt is the convective derivative, $D/Dt = \partial/\partial t + \Sigma_{i=1}^3 v_j \partial/\partial x_j$. With $\omega = \nabla \times v$, the vorticity, the Euler equations can be written in the equivalent form

$$\begin{aligned} &\frac{D\omega}{Dt} = \omega \cdot \nabla v, \\ (2.A) \qquad & \\ &\omega(x,0) = \omega_0(x) = \nabla \times v_0, \end{aligned}$$

where the velocity v is determined by the vorticity ω from the equations

$$\operatorname{div} v = 0, \qquad \operatorname{curl} v = \omega,$$

resulting in the familiar Biot-Savart formula,

$$(2.B) \qquad v(x,t) = -\frac{1}{4\pi} \int \frac{(x-y)}{|x-y|^3} \times \omega(y,t)\, dy.$$

The vector $\omega = \operatorname{curl} v$ belongs to the nullspace of the antisymmetric part of the matrix ∇v. Therefore, the term $\omega \cdot \nabla v$ can be replaced by $D\omega$, where the deformation matrix D is the symmetric part of ∇v:

$$D \doteq \tfrac{1}{2}(\nabla v + {}^T\nabla v) = (D_{ij}).$$

Formula (2.B) can be differentiated to express D as a strongly singular integral operator acting on ω:

$$\text{(2.C)} \qquad D(\omega) = (D_{ij}(\omega)) = \sum_{l=1}^{3} \text{P.V.} \int \mathscr{D}_{ij}^{l}(x-y)\,\omega_l(y)\,dy;$$

here $\mathscr{D}_{ij}^{l}(\lambda\vec{x}) = \lambda^{-3}\mathscr{D}_{ij}^{l}(\vec{x})$ for $\vec{x} \neq 0$ and the mean of $\mathscr{D}_{ij}^{l}$ over the unit sphere vanishes. The explicit formulae for the kernels $\mathscr{D}_{ij}^{l}$ are not needed in the developments below. Through the formulae in (2.B) and (2.C) we obtain an integro-differential equation for the vorticity above which is equivalent to the Euler equations in (1),

$$\text{(3)} \qquad \frac{D\omega}{Dt} = D(\omega)\omega, \qquad x \in \mathbb{R}^3,\ t > 0,$$
$$\omega(x,0) = \omega_0(x).$$

For completeness, we remark here that with v defined by (2.B) every smooth solution of (2.A) automatically satisfies $(D/Dt)\operatorname{div}\omega = 0$. Since $\operatorname{div}\omega_0 = 0$, the fact that $\operatorname{div}\omega = 0$ automatically guarantees that the velocity v from (2.B) satisfies both $\operatorname{div} v = 0$ and $\nabla \times v = \omega$.

In two space dimensions, $D(\omega)\omega \equiv 0$, and vorticity is conserved, i.e., $D\omega/Dt = 0$. In three dimensions, the matrix $D(\omega)$ is a symmetric matrix with $\operatorname{tr} D = 0$ and vortex-stretching occurs when ω roughly aligns with an eigenvector of $D(\omega)$ corresponding to a positive eigenvalue. Thus, essential differences in fluid behavior in two and three space dimensions are manifested through the appearance of the term $D(\omega)\omega$ on the right-hand side of (3).

The reformulation of the Euler equations in (3) and the above comments motivate the qualitative model which we present next. The matrix-valued function D depends linearly on the function ω; the operator relating ω to $D\omega$ is a linear singular integral operator that commutes with translation, i.e., it is given by the convolution of ω with a kernel homogeneous of degree -3 and with mean value on the unit sphere equal to zero. In the one space dimension, there is only one such operator, the Hilbert transform,

$$\text{(4)} \qquad H(\omega) = \frac{1}{\pi}\text{P.V.}\int \frac{\omega(y)}{(x-y)}\,dy.$$

The quadratic term $H(\omega)\omega$ is a scalar one-dimensional analogue of the vortex stretching term $D(\omega)\omega$. We replace the convective derivative D/Dt by $\partial/\partial t$ in order to have a one-dimensional incompressible flow and arrive at the *model vorticity equation*,

$$\frac{\partial \omega}{\partial t} = H(\omega)\omega, \qquad \omega(x,0) = \omega_0(x). \tag{5}$$

For the Euler equations, the velocity is determined from the vorticity by convolution with a mildly singular kernel, homogeneous of degree $1 - N$, and the analogue of the velocity for the model is defined within a constant by such a convolution, i.e.,

$$v = \int_{-\infty}^{x} \omega(y,t)\,dy. \tag{6}$$

Since the Hilbert transform is a skew-symmetric operator,

$$\int_{-\infty}^{\infty} H(\omega)\omega\,dy = (H\omega, \omega) = 0.$$

Integrating (5) with respect to y on $\mathbb{R}$ shows then that all smooth solutions of (5) that decay sufficiently rapidly as $|y| \to \infty$ satisfy for all t

$$\frac{d}{dt}\int_{-\infty}^{\infty} \omega(y,t)\,dy = 0. \tag{7}$$

Thus, if $\omega_0(x)$ is the derivative of a function vanishing for $|x| \to \infty$, smooth solutions of (5) also retain this property for $t > 0$. Many studies for the Euler equations concentrate on periodic fluid flow; there is an obvious analogue of the Hilbert transform in (4) on the circle and the periodic model vorticity equation can be defined as in (5). In this case, the mean of ω per period is conserved and v is defined unambiguously by

$$v = \int_{x_0}^{x} \omega(y,t)\,dy$$

provided that the initial data $\omega_0(x)$ satisfies

$$\int_{x_0}^{x_0+p} \omega_0(y)\,dy = 0$$

with p the period.

3. Integration of the Model Vorticity Equation and Explicit Breakdown of Solutions

The nonlinear equation in (5) is well posed in many standard function spaces, for example, $H^1(\mathbb{R})$, the Sobolev space of functions which are square integrable with square integrable first derivative. The local existence and uniqueness follows from the fact that $H^1(\mathbb{R})$ is a Banach algebra of continuous functions and the Hilbert transform maps $H^1(\mathbb{R})$ continuously into itself so that standard existence and uniqueness results for Lipschitz nonlinear ordinary differential equations in Banach space apply. We have the following explicit solution formula for the model equation in (5):

THEOREM. *Suppose $\omega_0(x)$ is a smooth function decaying sufficiently rapidly as $|x| \to \infty$ ($\omega_0 \in H^1(\mathbb{R})$ suffices). Then the solution to the model vorticity equation in (5) is given explicitly by*

$$\omega(x,t) = \frac{4\omega_0(x)}{(2 - tH\omega_0(x))^2 + t^2\omega_0^2(x)}. \tag{8}$$

Remark 1. The equation in (5) can also be defined on any smooth closed curve Γ, using the Hilbert transform H for that curve Γ. The explicit formulae and method of proof given below are exactly the same with this modification—in particular, when Γ is a circle. However, the constraint $\int_\Gamma \omega\, dz = 0$ is no longer automatically preserved by solutions of (5) except when Γ is the circle or the real line.

Remark 2. The proof also provides an explicit expression for $H(\omega)$,

$$(H\omega)(x,t) = \frac{2H\omega_0(x)(2 - tH\omega_0(x)) - 2t\omega_0^2(x)}{(2 - tH\omega_0(x))^2 + t^2\omega_0^2(x)}. \tag{9}$$

The formula in (8) immediately yields the following:

COROLLARY 1. (Breakdown of smooth solutions for the model vorticity equation). *The smooth solution to the differential equation in* (5) *blows up in finite time if and only if the set Z defined by*

$$Z = \{x | \omega_0(x) = 0 \quad and \quad H\omega_0(x) > 0\} \tag{10}$$

is not empty. In this case, $\omega(x,t)$ becomes infinite as $t \uparrow T$, where the blow-up time is given explicitly by $T = 2/M$ with $M = \sup\{H\omega_0)_+(x) | \omega_0(x) = 0\}$.

Proof of Theorem 1: To display the generality of the proof, we shall use the following identities for the Hilbert transform on the line which are also valid for any closed curve Γ (see [9]):

$$H(Hf) = +f, \tag{11.A}$$

$$H(fg) = fHg + gHf + H(Hf \cdot Hg). \tag{11.B}$$

From these identities it follows that

$$H(fHf) = \tfrac{1}{2}\left((Hf)^2 - f^2\right). \tag{12}$$

By applying H to the model vorticity equation and using (12), we obtain an equation satisfied by $(H\omega)(x,t)$:

$$\frac{\partial}{\partial t} H\omega = \tfrac{1}{2}\left((H\omega)^2 - \omega^2\right). \tag{13}$$

We introduce the quantity

$$z(x,t) = H\omega(x,t) + i\omega(x,t) \tag{14}$$

and by combining (5) and (13), we see that $z(x,t)$ satisfies the local equation

$$\frac{\partial z}{\partial t}(x,t) = \tfrac{1}{2} z^2(x,t) \tag{15}$$

with the explicit solution

$$z(x,t) = \frac{z_0(x)}{1 - \frac{1}{2} t z_0(x)}. \tag{16}$$

Formulas (8) and (9) are the real and imaginary parts of (16).

Formula (16) defines $z(x,t)$ as an analytic function in $\mathcal{I}m\, x < 0$. It is well known that the Hilbert transform on the line can be interpreted in terms of complex-valued functions z on the real axis which are boundary values of analytic functions in the lower half-plane that tend to zero sufficiently fast at infinity. The Hilbert transform relates the imaginary part of such a function to its real part on the real axis. Thus a function of the form

$$z = H\omega + i\omega$$

is always the boundary value of a function analytic in the lower half-plane. The identity in (11.A) is then merely the observation that if z is analytic in the lower half-plane, so is iz. The identity (11.B) is the observation that if z and w are analytic in the lower half-plane, so is their product $z \cdot w$. Equation (5) is then the imaginary part of (15); by analyticity (15) holds.

Next, we give an instructive explicit example in the 2π-periodic case.

EXAMPLE. We choose $\omega_0(x) = \cos(x)$ so that $H(\omega_0(x)) = \sin(x)$ and compute that

$$\omega(x, 2t) = \frac{\cos(x)}{1 + t^2 - 2t\sin(x)} \tag{17}$$

and

$$v(x, 2t) = \int_0^x \omega(x', 2t)\,dx' = (2t)^{-1}\log(1 + t^2 - 2t\sin(x)).$$

In this specific example, the breakdown time is $T = 2$ and, as $t \nearrow T$, $\omega(x,t)$ develops a non-integrable local singularity like $1/x$ near $x = 0$. There are two interesting facets to this breakdown process. First,

$$\int_{-\pi}^{\pi} |\omega(x,t)|^p\,dx \nearrow \infty \quad \text{as} \quad t \nearrow T \tag{18.A}$$

for any fixed p with $1 \leqq p < +\infty$. Also, there are finite constants M_p such that

$$\int_{-\pi}^{\pi} |v(x,t)|^p\,dx \leqq M_p \quad \text{as} \quad t \nearrow T \tag{18.B}$$

for any p with $1 \leqq p < \infty$. In particular, the kinetic energy of v remains uniformly bounded as t approaches the breakdown time T. The behavior in the above example is typical for solutions of the model vorticity equation as the following corollary of the theorem indicates:

COROLLARY 2. *Given the initial data $\omega_0(x)$ for the model vorticity equation, assume that the points x_0 with $\omega_0(x_0) = 0$ and defining the breakdown time T are simple zeroes of $\omega_0(x)$. Then as $t \nearrow T$ with T the breakdown time, $\omega(x,t)$ and $v(x,t)$ have the same properties as given in* (18.A) *and* (18.B).

We omit the proof since it is a straightforward but tedious calculation using the explicit solution formulae. It is easy to show that if the initial data ω_0 satisfies the assumption of Corollary 2, then, for $T < t < T + \tau$ with T the breakdown time and τ small enough, the analytic function

$$1 - \tfrac{1}{2}tz_0(x)$$

has a zero in the lower half-plane $\mathcal{I}m\, x < 0$. Thus for such t the function $z(x,t)$ defined by (16) has a pole in the lower half-plane, and therefore its imaginary part given by the formula (8) is *not* related to its real part by the Hilbert transform. In particular, for such t, formula (9) does *not* hold and therefore $\omega(x,t)$ as given in (8) does *not* continue the solution of (5) for $T < t < T + \tau$.

4. Qualitative Comparison of Solutions for the Model Vorticity Equation and the 3-D Euler Equations

First, we recall that $H(\omega)$ in the model vorticity equation has the analogous role as the deformation matrix in the 3-D Euler equations. With this identification, the qualitative fact that blow-up for solutions of the model vorticity equation occurs only at points where $H\omega$ has positive sign is reminiscent of the fact that vorticity for solutions of the 3-D Euler equations increases when it roughly aligns with eigenvectors of the deformation matrix with positive eigenvalues. Below, in discussing properties regarding the conjectured breakdown of solutions for the 3-D Euler equations, we refer to information which can be extracted from various numerical experiments (see [2], [3], [4]).

Smooth solutions of the 3-D Euler equations have the following well-known elementary properties:

SCALE INVARIANCE: If $v(x,t)$ satisfies (1), then for constants λ, α,

$$v_{\lambda,\alpha}(x,t) = \lambda v(\lambda^{\alpha}x, \lambda^{1+\alpha}t) \quad \text{also satisfies} \quad (1).$$

CONSERVATION OF ENERGY: For solutions of (1),

$$\int |v(x,t)|^2\,dx = \int |v_0(x)|^2\,dx \quad \text{for all} \quad t.$$

The reader can easily verify that the function $v(x,t)$ defined in (6) through solutions of the model vorticity equation has the same scale invariant properties for $x \in \mathbb{R}^1$ as for solutions of the 3-D Euler equations with $x \in \mathbb{R}^3$. We mention this explicitly here because some of the numerical methods for studying the blow-up of solutions for the 3-D Euler equations exploit this scale invariance in the numerical algorithm (see [3]). The functions $v(x,t)$ associated with solutions of the model vorticity equation do not satisfy conservation of L^2 norm. However, the computations reported in [3], [4] suggest that for the 3-D Euler equations, as t approaches the conjectured breakdown time T, the vorticity satisfies

$$\int |\omega|^p(x,t)\,dx \nearrow \infty \quad \text{as} \quad t \nearrow T \quad \text{for any} \quad 1 \leqq p < +\infty,$$

while the kinetic energy of $v(x,t)$ remains constant as $t \nearrow T$. Corollary 2 establishes that typical solutions of the model vorticity equation behave in an analogous fashion as the breakdown time is approached. Chorin also reports in

[3], [4], the results of two different numerical procedures which indicate that

(19) the set of breakdown points is a set of Lebesgue measure zero in R^3 with Hausdorff dimension $\sim$ 2.5.

Such a relation was first suggested by Benoit Mandelbrot. For typical solutions of the model vorticity equation, according to Corollary 1, the set of conceivable breakdown points is also a set of Lebesgue measure zero contained in the zero set of $\omega_0(x)$ and typically consists of a finite number of points. In the inviscid calculations for the 3-D Euler equations in [2], it is reported that the deformation matrix $D(\omega)$ becomes large on open sets where ω vanishes. In the model vorticity equation, it is easy to construct explicit examples in which $H(\omega)$ becomes arbitrarily large on an open set where ω vanishes—in fact, an earlier non-constructive breakdown argument for special initial data by Constantin [6] directly exploits this property.

We end this section by remarking that the properties of solutions of the model vorticity equation described in Corollary 2 and also referred to below (16) are *never* satisfied for solutions which blow up for the local scalar quadratic equation,

$$\frac{\partial\omega}{\partial t} = \omega^2,$$

$$\omega(x,0) = \omega_0(x),$$

as the reader can easily verify. Of course, this quadratic equation arises from the characteristic form of the equation for $\omega = u_{\tilde{x}}$, where u satisfies the local quadratic equation, $u_t - uu_{\tilde{x}} = 0$.

Acknowledgment. The authors thank Sergiu Klainerman for his perceptive comments and active interest during the course of this work.

The work of the first author was performed while he was a visiting member at the Courant Institute. The work of the second author was partially supported by Dept. of Energy under contract DE-AC02-76ER03077, that of the third was partially supported by National Science Foundation Grant DMS84-0223.

Bibliography

[1] Beale J. T., Kato, T., and Majda, A., *Remarks on the breakdown of smooth solutions for the* 3-D *Euler equations*, Comm. Math. Phys., 94, 1984, pp. 61–66.

[2] Brachet, M. E., et al., *Small-scale structure of the Taylor–Green vortex*, J. Fluid Mech., 130, 1983, pp. 411–452.

[3] Chorin, A., *Estimates of intermittency, spectra, and blow-up in developed turbulence*, Comm. Pure Appl. Math., 34, 1981, pp. 853–866.

[4] Chorin, A., *The evolution of a turbulent vortex*, Comm. Math. Phys., 83, 1982, pp. 517–535.

[5] Constantin, P., *Note on loss of regularity for solutions of the* 3-*D incompressible Euler and related equations*, (in preparation).

[6] Constantin, P., *Blow-up for a non-local evolution equation*, M.S.R.I. 038-84-6, Berkeley, California, July 1984.
[7] Frisch, U., *Fully developed turbulence and singularities* in *Proc. Les Hauches Summer School* 1981, North Holland, Amsterdam, 1984.
[8] Morf, R., Orszag, S., and Frisch, U., *Spontaneous singularity in three-dimensional incompressible flow*, Phys. Rev. Lett. 44, 1980, pp. 572–575.
[9] Mushkelishvili, N. T., *Singular Integral Equations*, P. Noordhoff, Groningen, 1953.
[10] Siggia, E., *Collapse and amplification of a vortex filament*, preprint, May 1984.

Received May, 1985.

Commentary on Part I

5

This paper presents a very short proof, based on the Hahn–Banach theorem, of the existence of Green's function of the Laplace operator for any domain in space whose boundary is twice differentiable.

In the plane the method works for once-differentiable boundaries. If the boundary is twice differentiable, then a refinement of the argument presented in the paper shows that at any point near the boundary, Green's function is $O(d)$, where d is distance to the boundary. It follows from Poisson's formula for harmonic functions that near the boundary, the first derivatives of Green's function G are bounded. By the Cauchy–Riemann equations the same is true of the harmonic conjugate H of G. It follows that H is Lipschitz continuous up to the boundary.

For a simply connected domain, $\exp(G + iH)$ is a single-valued analytic function. It maps the boundary into the unit circle, and the singular point q onto the origin; no other point is mapped into the origin. It follows that the degree of this mapping is 1 inside the unit circle, and 0 outside. Therefore, it is the mapping of the domain onto the unit disk.

Garabedian and Shiffman have extended this methodology to more general equations; see [1].

References

[1] Garabedian, P.; Shiffman, M. On solution of partial differential equations by the Hahn–Banach theorem, *Trans. AMS* **76** (1954) 288–299.

P.D. Lax

9

The most interesting part of this paper is Theorem 2.1, called usually the Lax–Milgram lemma. The story of this result is as follows: Arthur Milgram was an excellent topologist at the University of Minnesota. In the early nineteen fifties he visited the Courant Institute for a year to learn some analysis. We became friends, and he asked me for a problem to work on. I explained to him how variational arguments can be used to extend self-adjoint operators that are bounded from below, but that there is no known method for dealing with operators that are not symmetric. After some thought he came up with this lemma.

In this paper we show how to use the lemma to generalize the Friedrichs extension to nonsymmetric operators. We apply this to scalar elliptic operators of arbitrary order, and then use the Hille–Yosida theorem to solve the associated parabolic equation.

1

In my book on functional analysis I observe that the second hypothesis of the L-M lemma, boundedness from below, can be weakened to requiring merely $c\|x\| \leq |B(x,x)|$ for some positive constant c. This makes it possible to use the lemma in complex Hilbert space.

P.D. Lax

12

In this paper the negative Sobolev norm $\|u\|_{-s}$ is defined by duality, and a priori estimates in the negative norm are derived for solutions of symmetric hyperbolic equations. The initial value problem is shown to be equivalent to a periodic problem in both space and time for an inhomogeneous equation, which is easily solved using the a priori inequalities.

The interior differentiability of solutions of elliptic equations is first reduced to the periodic case, and then derived recursively using the a priori inequalities.

P.D. Lax

16

The main result of this paper is a proof of the generalized Huygens principle for solutions of linear hyperbolic equations with smooth coefficients in n space variables.

The differentiability properties of a solution u at a point P, T in space-time depend only on the differentiability properties of the initial data $f(x) = u(x,0)$ at those points x that lie on the bicharacteristics issuing from P, T.

The proof is based on first showing that if the initial values of a solution of such an equation are smooth except on a smooth surface of codimension one, across which they and their partial derivatives have jump discontinuities, then the corresponding solution, in the sense of distributions, is smooth, except along the characteristic surfaces issuing from the initial surface of discontinuity; along these the solution and its partial derivatives have jump discontinuities.

To obtain the general result from this special case, Green's formula for a pair of solutions of the equation and its adjoint is applied to u and a solution v of the adjoint equation whose values at time T are zero at all points x where $(x - P)\omega < 0$, and one at all points x where $(x - P) > 0$. Integration with respect to ω gives a weighted average of $u(x, T)$ around the point P, from which the value of u at P can be determined by differentiation, that is, by inversion of the Radon transform.

The proof is carried out for first-order systems and three space dimensions. Another proof of this result is given in the next paper.

The most important extension of the generalized Huygens principle is by Richard Melrose and Michael Taylor to solutions of mixed initial–boundary value problems.

This subject, the propagation of singularities, becomes more precise by formulating it in the microlocal setting, and by using the concept of the wave front set introduced by Hörmander.

References

[1] Melrose, R. Microlocal parametrices for diffractive boundary value problems. *Duke Math. J.* **42** (1975), 605–635.

[2] Taylor, M. Grazing rays and reflection of singularities of solutions to wave equations. *CPAM* **29** (1976), 1–38.

P.D. Lax

20, 16

In the paper, "Asymptotic solutions of oscillatory initial value problems," Lax constructs asymptotic solutions for first-order hyperbolic systems on $\mathbb{R}^m \times \mathbb{R}$ of the form

$$Mu = \frac{\partial u}{\partial t} + \sum_{i=1}^{m} A_i \frac{\partial u}{\partial x_i} + Bu\,, \tag{$*$}$$

the A_i's and B being $n \times n$ matrices whose entries are C^∞ functions of x and t, and $u(x,0)$ being a function of the form, $\phi(x) = (\phi_1(x), \dots, \phi_n(x))$, where

$$\phi_j(x) = e^{i\ell(x)/\hbar}\psi_j(x)\,. \tag{$**$}$$

(I've inserted the $\hbar$ because I will say a few words below about the similarities between what Lax is doing here and what is now a days known as semiclassical methods in the theory of partial differential equations.) Lax posits a solution of ($*$) of the form

$$u(x,t) \sim e^{i\ell(x,t)/\hbar} \sum_{i=0}^{\infty} v_i(x,t)\hbar^i \tag{$***$}$$

and shows that $u(x,t)$ satisfies ($*$) modulo an error term of order $O(\hbar^\infty)$ if and only if $\ell(x,t)$ satisfies the eikonal equation

$$\frac{\partial \ell}{\partial t} = \lambda\left(x, t, \frac{\partial \ell}{\partial x}\right),$$

where $\lambda = \lambda(x,t,\xi)$ is an eigenvalue of the matrix

$$A(t,x,\xi) = \sum \xi_i A_i(t,\chi)\,,$$

3

and shows that if ℓ satisfies this equation, the v_i's can be determined by solving inductively a sequence of transport equations along the bicharacteristics of $\lambda(x,t,\xi)$. He then writes the delta function $\delta(x)$ as a superposition of functions of the form (**) and uses the results above to construct a fundamental solution of (*) modulo smoothing operators. Finally, he makes use of properties of this fundamental solution to deduce propagation of singularities results for (*). Fifteen years after this was written, Hörmander and Duistermaat microlocalized the approach developed here and were able to get around some of the problems it poses, for instance, that of solving the eikonal equation and integrating bicharacteristics flow "downstairs" on x-space rather than "upstairs" on (x,ξ)-space. Moreover, the notion of "wave-front set," i.e., of singularities of distributions living "upstairs" on (x,ξ)-space, simplified considerably the propagation of singularities results that I alluded to above. Nevertheless, it is fascinating to see many of the features of the Fourier integral operator calculus occurring here for the first time (in particular, the explicit formulas in §4 for the fundamental solution of (*)!) It is also (to jump decades ahead) fascinating to see the emphasis placed here on solutions of PDEs of the form (***), i.e., on what we would now call *semiclassical solutions.* (This paper is the second of two papers in solutions of the system (*). An earlier paper of Lax and Courant, "The propagation of discontinuities in wave motion," gives an alternative construction of a fundamental solution modulo smoothing for (*) that involved decomposing the delta function into plane waves rather than onto a superposition of functions of the form (**).)

V. Guillemin

33

It is, of course, well known that smooth classical solutions of one-dimensional quasilinear hyperbolic systems of the form

$$\partial_t u + A(u)u_x = 0,$$

with $u = (u^1, u^2, \ldots, u^N)$ and $A(u)$ an $N \times N$ matrix with real and distinct eigenvalues (i.e., the condition of strict hyperbolicity), are supposed to develop singularities in finite time. This is very easy to see for scalar equations, such as Burger, but it is not so easy for systems. The first rigorous mathematical results to demonstrate the breakdown of general solutions to a 2×2 system are due to O.A. Oleinik [6] and P. Lax [3].

Their results rely on the fact that 2×2 strictly hyperbolic systems, with real eigenvalues $\lambda_1 < \lambda_2$, possess Riemann invariants l, r that effectively decouple the original system for $u = (u_1, u_2)$ into

$$\partial_t l + \lambda_1(u)\partial_x l = 0, \qquad \partial_t r + \lambda_2(u)\partial_x r = 0.$$

Observe, however, that the system remains weakly coupled through $\lambda_1(u), \lambda_2(u)$. In [3], Lax shows, in a characteristically elegant way, that if the so-called *Lax genuine nonlinearity condition* is satisfied, the coupling through u can be shown to be unimportant; i.e., solutions form singularities in finite time in the same fashion as scalar equations do. Note that the

4

arguments of [3] apply not only to compactly supported data but also to Dirichlet, Neumann, and other boundary-type conditions.

The main open questions in the wake of [3] were these:

(1) What is the mechanism of formation of singularities for general $N \times N$ systems of conservation laws that satisfy Lax's genuine nonlinearity condition when Riemann invariants may not exist?

(2) Is the condition of genuine nonlinearity indispensable. Important examples, such as the equation of a nonlinear vibrating string, do not satisfy the genuine nonlinearity condition. Can singularities form in that case, and if so, what is the mechanism?

(3) What happens in higher dimensions?

The third question is by far the most important, but it is beyond the scope of this presentation. I will restrict the discussion below to one-dimensional hyperbolic systems.

The first question was largely settled in a very important paper by F. John; see [1]. The main new idea of the paper was to decouple not the original system, which is impossible in general due to the lack of N linearly independent Riemann invariants, but rather the system obtained by differentiating $u = u^1, u^2, \ldots, u^N$. Using an appropriate *normal* form for the differentiated system and Lax's genuine nonlinearity condition, F. John was able to prove that all compactly supported initial data form singularities in finite time. John's result was extended somewhat by T.P. Liu [5], which allows systems that have both genuine nonlinear and linearly degenerate waves.

The second open question was addressed in a paper by Klainerman and Majda, [2] concerning the equation of a vibrating string. By introducing Riemann invariants the authors are led to study the same type of l, r system as that studied by Lax in [3]. Because of lack of genuine nonlinearity, the argument in [3] breaks down. By a careful study of interactions between the equations satisfied by l, r the authors are able nevertheless to conclude that all solutions of the Dirichlet and Neumann mixed boundary conditions break down in finite time. The case in which the initial data has compact support is much easier, and the argument in that case can be extended to general classes of strictly hyperbolic equations in the spirit of [1]; see [4] for a discussion of this and other relevant results.

References

[1] John, F. Formation of singularities in one dimensional nonlinear wave propagation, *Comm. Pure Appl. Math.* **27**, 1974, pp. 377–405.

[2] Klainerman, S.; Majda, A. Formation of singularities for wave equations including the nonlinear vibrating string, *Comm. Pure Appl. Math.* **33** (1980), pp. 241–263.

[3] Lax, P.D. Development of singularities of solutions of nonlinear hyperbolic partial differential equations, *J. Math. Physics* **5**, 1964, pp. 611–613.

[4] Oleink, O. A. Discontinuous solutions of nonlinear differential equations, *Uspekhi Mat. Nauk*, **12** (1957), no. 3 (75), pp. 3–73; English translation in *Amer. Math. Soc. Transl.* (2), **26** (1963), pp. 95–172.

[5] Ta-tsien, L.; Zhou Y.; De-Xing, K. Weak linear degeneracy and global classical solutions for general quasilinear hyperbolic systems, *Comm. P.D.E.*, **1994**, pp. 1263–1317.

[6] Tai-ping, L. Development of singularities in the nonlinear waves for quasilinear hyperbolic partial differential equations, *J. Diff. Eqs.* **33** (1979), pp. 92–111.

S. Klainerman

41

Garding's inequality for differential (and pseudodifferential) operators A of order $2r$ says that if the symbol of A is reasonably smooth and positive, then the Hermitian part of A is bounded from below as follows:

$$Re(Au, u) > -k\|u\|^2_{(r-1)/2}, \qquad k \text{ some positive number.}$$

Hörmander has shown for scalar-valued operators that this inequality holds even if the symbol of A is merely nonnegative. In this paper we extend this result to matrix-valued operators, and also to difference operators of the form

$$P_\delta = \sum p_\alpha(x) T^\alpha,$$

where α is a multi-index, T is the shift operator

$$(Tu)(x) = u(x+\delta),$$

and δ is a positive parameter that tends to zero. We show that if the matrix symbol

$$p(x,\xi) = \sum p_\alpha(x) e^{i\alpha\xi}$$

is Hermitian and nonnegative, then P_δ satisfies the inequality

$$Re(P_\delta u, u) > -k\delta\|u\|^2, \qquad k \text{ some positive number.}$$

A useful corollary is that if the numerical range of the symbol p is less than or equal to 1, then the numerical range of the operator P_δ is less than $1 + O(\delta)$. According to the Halmos–Berger–Pearcy theorem it follows that the numerical range of P_δ^n is bounded by $1 + O(\delta)^n = \exp kt$. It follows that the norm of P_δ^n is less than $2(\exp kt)$. Since Wendroff and Lax have shown that the numerical range of the symbol of the Lax–Wendroff scheme is less than or equal to 1, provided that a Courant–Friedrichs–Lewy-type condition is satisfied, the stability of the L-W scheme follows. A further extension of Hörmander's theorem has been given by Fefferman and Phong.

6

References

[1] Fefferman, C.; Phong, D. The uncertainty principle and sharp Garding inequalities, *CPAM* **34** (1981), 285–331.

[2] Halmos, P.R. *A Hilbert space problem book*, Springer, NY 1982.

[3] Lax, P.D.; Wendroff, B. Difference schemes for hyperbolic equations with high order of accuracy, *CPAM* **17** (1964), 381–398.

P.D. Lax

82

In this paper we show by a simple argument that Huygens' principle holds for the renormalized spherical wave equation on k-dimensional unit spheres, k odd:

$$u - Lu = 0,\ L = \Delta_s - (k-1)^2/4,$$

where Δ_s is the Laplace–Beltrami operator on the unit sphere. Then we show, by a somewhat lengthy calculation, that there is a transformation that maps any spherical cap onto a ball in $\mathbb{R}^k$, and a multiplier m such that mu satisfies the Euclidean wave equation.

P.D. Lax

110

This paper is part of a research program designed to give insight into one of the outstanding unsolved problems of classical mathematical physics: Do smooth solutions with finite energy for the incompressible three-dimensional Euler equations develop singularities in finite time? A current perspective as well as many other important research contributions to this problem are discussed in Chapter 5 of the recent book *Vorticity and Incompressible Flow*, by A. Majda and A. Bertozzi, Cambridge University Press, 2002.

The story of the genesis of this research paper gives interesting insight into Peter Lax's research talents and scientific friendships. The one-dimensional model was invented by Andy Majda in a Berkeley, California, cafe while he was thinking about simpler models with features of vortex stretching in the three-dimensional Euler equations. Peter Constantin arrived in Berkeley a few days later with a glowing recommendation from Ciprian Foias based on Constantin's student days in Romania and his subsequent visit with Ciprian at Indiana University. Majda showed Constantin the model and gave him the problem of finite-time singularity formation in the model, which they solved rapidly through a sequence of nonlinear estimates. A few weeks later, Majda was traveling to Paris and spent a few days on the way "chez Lax" on Central Park West. He told Peter Lax about the one-dimensional model; Peter's reply was something like this: "How amusing! When I think of the Hilbert transform, I think of analytic continuation and complex analysis." Majda returned from Europe a few weeks later, and the detailed exact solution procedure of the paper using complex analysis was supplied in a shorthand letter from Peter Lax from his Loon Lake summerhouse.

A. Majda

8

PART II

Difference Approximations to PDE

COMMUNICATIONS ON PURE AND APPLIED MATHEMATICS, VOL. IX, 267–293 (1956)

Survey of the Stability of Linear Finite Difference Equations*

P. D. LAX and R. D. RICHTMYER

PART I

AN EQUIVALENCE THEOREM

1. Introduction

Beginning with the discovery by Courant, Friedrichs and Lewy [1] of the conditional stability of certain finite difference approximations to partial differential equations, the subject of stability has been variously discussed in the literature (see bibliography at end). The present paper is concerned with the numerical solution of initial value problems by finite difference methods, generally for a finite time interval, by a sequence of calculations with increasingly finer mesh. Thus if t is the time variable and Δt its increment, we are concerned with limits as $\Delta t \to 0$ for fixed t, not with limits as $t \to \infty$ for fixed Δt (although often the stability considerations are similar). The basic question is whether the solution converges to the true solution of the initial value problem as the mesh is refined. The term *stability*, as usually understood, refers to a property of the finite difference equations, or rather of the above mentioned sequence of finite difference equations with increasingly finer mesh. We shall give a definition of stability in terms of the uniform boundedness of a certain set of operators and then show that under suitable circumstances, for linear initial value problems, stability is necessary and sufficient for convergence in a certain uniform sense for arbitrary initial data. The circumstances are first that a certain consistency condition must be satisfied which essentially insures that the difference equations approximate the differential equations under study, rather than for example some other differential equations, and secondly that the initial value problem be properly posed, in a sense to be defined later.

We shall not be concerned with rounding errors, and in fact assume that all arithmetic steps are carried out with infinite precision. But it will

* The work for this paper was done under Contract AT-(30–1)–1480 of the Atomic Energy Commission.

be evident to the reader that there is an intimate connection between stability and practicality of the equations from the point of view of the growth and amplification of rounding errors. Indeed, O'Brien, Hyman and Kaplan [8] defined stability in terms of the growth of rounding errors. However, we have a slight preference for the definition given below, because it emphasizes that stability still has to be considered, even if rounding errors are negligible, unless, of course, the initial data are chosen with diabolical care so as to be exactly free of those components that would be unduly amplified if they were present.

The basic notions will be spelled out in considerable detail below in an attempt to motivate the definitions given and to justify the approach via the theory of linear operators in Banach space. We shall then give the usual definition of a properly posed initial value problem, define the consistency of a finite difference approximation, define the stability of a sequence of finite difference equations, and prove the equivalence theorem.

2. The Function Space of an Initial-Value Problem

In the solution of an initial-value problem the time variable t plays a special role. An instantaneous state of the physical system is described by one or more functions of certain other variables which we shall call space variables. At any stage of a machine- or hand-calculation one has at hand a numerical representation (e.g. in tabular form) of these functions, that is, of the state of the system at some time t. As time goes on, the state of the system changes according to certain differential or integro-differential equations. It is convenient to think of these functions, for a fixed t, as an element or point in a function space $\mathscr{B}$ and to denote them by a single symbol u.

The initial-value problems under consideration are linear and we suppose $\mathscr{B}$ to be linear also. This may force us to accept as elements of $\mathscr{B}$ some functions not having direct significance as states of a physical system, e.g., functions having negative values for inherently positive quantities like temperature and particle density. But it is convenient to admit such functions as representing *generalized* states of the system, and also to admit complex valued functions. If sums and differences of elements of $\mathscr{B}$ are defined in the obvious manner by sums and differences of the corresponding functions, and if multiplication of an element of $\mathscr{B}$ by a number is defined in the equally obvious manner as multiplication of the corresponding functions by that number, it is clear that $\mathscr{B}$ is a linear vector space.

For a discussion of approximation and errors, one needs a measure of the difference of two states u and v, and it is clear that this measure should have the properties of a norm of the element $w = u - v$; we therefore denote this quantity by $\| w \|$ and suppose that $\mathscr{B}$ is a Banach space. The specific choice of norm may vary from one application to another; in many cases it can be identified with energy. Our assumption that $\mathscr{B}$ is *complete* with respect to the norm plays an important role in the equivalence theorem of Section 8.

3. The Initial Value Problem

Let A denote a linear operator that transforms the element u into the element Au by spatial differentiations, matrix-vector multiplications and the like. The initial value problem is to find a one-parameter set of elements $u(t)$ such that

$$\text{(1)} \qquad \frac{d}{dt} u(t) = Au(t), \qquad 0 \leqq t \leqq T,$$

$$\text{(2)} \qquad u(0) = u_0$$

where u_0 represents a preassigned initial state of the system.

Systems involving higher order derivatives with respect to t can be put into the above form in the usual way by introducing the lower order derivatives as further unknown functions.

All the general considerations in the present discussion apply as well when the operator A depends explicitly on t, and in fact were originally presented in that generality[1], but in the interest of simplicity of the formulas we discuss here only the case of an operator A not depending on the parameter t.

If there are boundary conditions in the problem, it is assumed that they are linear homogeneous and are taken care of by restricting the domain of A to functions satisfying the conditions.

By a *genuine solution* of (1) we mean a one-parameter set $u(t)$ such that first, $u(t)$ is in the domain of A for $0 \leqq t \leqq T$ and secondly

$$\text{(3)} \quad \text{as } \tau \to 0, \quad \left\| \frac{u(t+\tau)-u(t)}{\tau} - Au(t) \right\| \to 0 \text{ uniformly in } t, \quad 0 \leqq t \leqq T.$$

If we pick an element u_0 not in the domain of A (e.g., if A is a differential operator and the functions represented by u_0 are nondifferentiable at one or more points), we obviously cannot find a genuine solution satisfying (2), but we assume that u_0 can always be approximated, as closely as one desires, by an element u_0' for which a unique genuine solution exists. That

[1] P. D. Lax, Seminar, New York University, January 1954.

is, if we define an operator $E_0(t)$ — really a one-parameter family of operators — so that

$$u(t) = E_0(t)u(0), \qquad 0 \leqq t \leqq T,$$

for any genuine solution of (1) depending uniquely on $u(0)$, we assume that *the domain of* $E_0(t)$ *is dense in* $\mathscr{B}$.

It is also desirable that the solution depend continuously on the initial data. If we alter the initial date u_0 by addition of v_0, we want to guarantee that the alteration of the solution is small if v_0 is small, i.e., that there should be a constant K such that

$$\| E_0(t)v_0 \| \leqq K \| v_0 \|, \qquad 0 \leqq t \leqq T.$$

We therefore assume that *the operators* $E_0(t)$ *are uniformly bounded, for* $0 \leqq t \leqq T$.

The foregoing assumptions characterize a *properly posed* problem. For such a problem, $E_0(t)$ has a bounded linear extension $E(t)$ whose domain is the entire space $\mathscr{B}$ and whose bound is the same as that of $E_0(t)$, because a bounded linear operator with a dense domain can always be so extended. Then, for arbitrary u_0, the one-parameter set of elements of $\mathscr{B}$, $u(t)$, given by

$$u(t) = E(t)u_0$$

is interpreted as a generalized solution of the initial value problem (1), (2).

4. Finite Difference Approximations

When an approximate solution is obtained by finite difference methods, the time variable t, in the first place, assumes discrete values $t = t^0, t^1, \cdots, t^n, \cdots$, where $t^n = n\Delta t$, and correspondingly, one deals with a discrete sequence $u^0, u^1, \cdots, u^n, \cdots$, of states of the physical system.

In the second place, the space variables are also discrete so that the functions describing a state of the system are specified only at the points of a lattice or net of values of the space variables. However, we may still regard such a specification (although imperfect) as represented by a point in the same function space $\mathscr{B}$, by adopting some rule for specifying function values between the points of the space lattice, for example linear interpolation. Such a rule, if chosen with reasonable care, will not interfere with the linearity or boundedness of the operators dealt with. (Some authors, such as L. V. Kantorovitch [5] prefer to represent the sates u^n in a different Banach space $\mathscr{B}'$, and to establish suitable homomorphisms between $\mathscr{B}$ and $\mathscr{B}'$.)

The finite difference equations are:

$$u^{n+1} = B(\Delta t, \Delta x, \Delta y, \cdots)u^n, \tag{4}$$

where u^n is (it is hoped) an approximation to $u(t^n)$, and B denotes a linear finite difference operator which depends, as indicated, on the size of the time increment Δt and on the sizes of the space increments $\Delta x, \Delta y, \cdots$.

Contrary to possible appearance, this formulation is not restricted to explicit difference systems. If the system is implicit, the operator B will contain the inverse of a (possibly infinite) matrix, but for present purposes it is not necessary to suppose that B can be easily written in explicit form. Whatever the calculation procedure may be which leads to u^{n+1} when u^n is known, it results in a transformation in $\mathscr{B}$ and this transformation is denoted by B.

We do assume, however, that the calculation procedure is a definite one which can be applied to any function u^n and that the result u^{n+1} depends linearly and continuously on u^n, as is clearly the case for any reasonable scheme. In other words, for any fixed Δt, Δx and Δy, B is a bounded linear transformation whose domain is the whole Banach space.

The concepts of stability and convergence with which we deal here suppose an infinite sequence of calculations with increasingly finer mesh. We assume relations

$$\Delta x = g_1(\Delta t),$$
$$\Delta y = g_2(\Delta t),$$
$$\cdots\cdots\cdots$$

which tell how the space increments approach zero as the time increment goes to zero along the sequence, and we set

$$B(\Delta t,\ g_1(\Delta t),\ g_2(\Delta t), \cdots) = C(\Delta t),$$

so that

$$u^{n+1} = C(\Delta t)u^n. \tag{5}$$

5. The Consistency Condition

Since

$$\frac{u^{n+1} - u^n}{\Delta t}$$

is to be an approximation to the time derivative,

$$\frac{C(\Delta t)u - u}{\Delta t}$$

must be an approximation, in some sense, to Au. We cannot expect this to be true for all u in $\mathscr{B}$, because in general Au is not even defined for all u in $\mathscr{B}$. But we want it to be true for nearly all u that can appear in a genuine solution of the initial value problem; and for any particular genuine solution we want the approximation to be uniformly good for all t in $0 \leqq t \leqq T$. Specifically, we shall call the family of operators $C(\Delta t)$ a *consistent approximation* for the initial value problem, if for some class U of genuine solutions it is true that, for any $u(t)$ in this class,

$$\lim_{\Delta t \to 0} \left\| \left\{ \frac{C(\Delta t) - I}{\Delta t} - A \right\} u(t) \right\| = 0 \text{ uniformly in } t, \quad 0 \leqq t \leqq T, \tag{6}$$

provided that the class U is sufficiently wide and that its initial elements $u(0)$ are dense in $\mathscr{B}$. (6) is called the *consistency condition.*

In applications, A is usually a differential and C a difference operator in the space variables. To verify the consistency condition (6), Au has to be compared to $(C(\Delta t)-I)u/\Delta t$; to carry out this comparison expand each term in $C(\Delta t)u$ into a finite Taylor series (take two or three terms, depending on the order of the differential operator A), obtaining a differential operator. The error in replacing Cu by such a differential expression can be estimated, by Taylor's theorem, for sufficiently smooth functions. Therefore the comparison can be carried out for all sufficiently smooth solutions, and it is well known that the smooth solutions are dense among all solutions.

6. Convergence

Operating n times on u_0 with $C(\Delta t)$ gives $u^n = C(\Delta t)^n u_0$ which, it is hoped, approximates $u(n\Delta t)$. Since $u(t) = E(t)u_0$, we therefore make the following definition: the family of operators $C(\Delta t)$ provides a *convergent approximation* for the initial value problem if for any u_0 in $\mathscr{B}$ and for any sequences $\Delta_j t$, n_j such that $\Delta_j t$ tends to zero and $n_j \Delta_j t \to t$ where $0 \leqq t \leqq T$ then

$$\left\| \left\{ C(\Delta_j t) \right\}^{n_j} u_0 - E(t)u_0 \right\| \to 0, \qquad 0 \leqq t \leqq T. \tag{7}$$

Note that we require (7) to hold for every u_0 in $\mathscr{B}$ if $C(\Delta t)$ is to be called a convergent approximation.

7. Stability

In a sequence of calculations with $\Delta_j t \to 0$, if each calculation is carried from $t = 0$ to $t \approx T$, the operators which are used are those belonging to the set

$$\{C(\Delta_j t)\}^n, \qquad \begin{array}{l} j = 1, 2, 3, \cdots \\ 0 \leqq n\Delta_j t < T \text{ for each } j, \end{array} \tag{8}$$

all applied to u_0. The idea of stability is that there should be a limit to the extent to which any component of an initial function can be amplified in the numerical procedure. Therefore the approximation $C(\Delta, t)$ is said to be *stable* if the operators of the above set are uniformly bounded. Note that we make no reference here to the differential equation whose solution is desired so that stability, as defined, is a property solely of a sequence of difference equation systems.

In practice the bound of $\{C(\Delta t)\}^n$ is generally a continuous function of Δt in some interval, $0 < \Delta t \leqq \tau$, so that we may equivalently define the approximation $C(\Delta t)$ to be stable if for some $\tau > 0$, the set of operators

$$(9) \qquad \{C(\Delta t)\}^n, \qquad \begin{array}{c} 0 < \Delta t \leqq \tau \\ 0 \leqq n\Delta t \leqq T \end{array}$$

is uniformly bounded.

8. The Equivalence Theorem

Given a properly posed initial value problem (1), (2) *and a finite difference approximation* $C(\Delta t)$ *to it that satisfies the consistency condition, stability is a necessary and sufficient condition that* $C(\Delta t)$ *be a convergent approximation.*

According to the definition of Section 6, this involves convergence for an arbitrary initial element u_0. In principle, an unstable scheme can sometimes give convergence for special initial elements. (Such schemes are not generally very useful in practise, because the initial data seldom have the required properties, and even if they do, round-off errors are likely to perturb the calculation enough to throw it into a neighboring divergent situation.)

We now prove the first part of the theorem: a convergent scheme is necessarily stable.

We start by showing that for a convergent scheme, the set of elements

$$(10) \qquad C^n(\Delta t)u_0, \qquad n\Delta t \leqq T$$

are bounded for each fixed u_0 in $\mathscr{B}$. For, assume to the contrary that for a sequence n_j, $\Delta_j t$, $n_j\Delta_j t \leqq T$, the norms of the elements $C^{n_j}(\Delta_j t)u_0$ tend to infinity. Select a subsequence such that $n_j\Delta_j t$ tends to some limit t; since the scheme was assumed convergent, $C^{n_j}(\Delta_j t)u_0$ would have to tend to $E(t)u_0$, which it couldn't if it were unbounded.

We now appeal to the principle of uniform boundedness, which says that if each operator L of a set is bounded and if there exists a function $K(u)$ such that $\| Lu \| \leqq K(u)$ for all L in the set and all u in $\mathscr{B}$, then the set is uniformly bounded. Applying this to the present case, we see that the set (8) is uniformly bounded, and the approximation is stable.

To prove that, conversely, stability implies convergence, let $u(t)=E(t)u_0$ be a genuine solution belonging to the set U referred to in the definition of consistency. Then, for any positive ε,

$$\left\| \left\{ \frac{C(\Delta t)-I}{\Delta t} - A \right\} u(t) \right\| < \frac{\varepsilon}{2}, \qquad 0 \leqq t \leqq T,$$

for sufficiently small Δt. Also, from the definition of a genuine solution,

$$\left\| \left\{ \frac{E(\Delta t)-I}{\Delta t} - A \right\} u(t) \right\| < \frac{\varepsilon}{2}, \qquad 0 \leqq t \leqq T,$$

for sufficiently small Δt, so that by the triangle inequality,

$$\| \{C(\Delta t) - E(\Delta t)\} u(t) \| < \varepsilon \Delta t, \qquad 0 \leqq t \leqq T, \tag{11}$$

for sufficiently small Δt. This last inequality might have been taken as the basis of the definition of consistency, but the definition given in Section 5 is preferred for practical applications because it involves the operator A rather than the generally unknown solution operator $E(t)$. Set

$$\begin{aligned} \psi_j &= [\{C(\Delta_j t)\}^{n_j} - E(n_j \Delta_j t)] u_0 \\ &= \sum_{0}^{n_j-1}{}_{(k)} \{C(\Delta_j t)\}^k [C(\Delta_j t) - E(\Delta_j t)] E((n_j - 1 - k)\Delta_j t) u_0 . \end{aligned}$$

The equality of the second and third members of this equation results from cancellation of all except the first and last terms of the third member, when written out in full. The norm of ψ_j can be estimated by use of inequality (11) with the help of the triangle inequality:

$$\| \psi_j \| < K \sum_{0}^{n_j-1}{}_{(k)} \varepsilon \Delta_j t = K \varepsilon n_j \Delta_j t < K \varepsilon T,$$

for sufficiently small $\Delta_j t$, where K denotes the uniform bound of the set (8). Therefore, since ε was arbitrary,

$$\| \psi_j \| \to 0 \text{ as } \Delta_j t \to 0. \tag{12}$$

Now suppose that $n_j \Delta_j t \to t$ as $j \to \infty$, where t is a number in the interval $(0, T)$. The difference $\{E(n_j \Delta_j t) - E(t)\} u_0$ may be written in either of two ways, depending on which of the two arguments $n_j \Delta_j t$ and t is the larger, that is, as

$$(E(s) - I) E(t') u_0 \text{ if } s = n_j \Delta_j t - t \geqq 0, \qquad t' = t,$$

or as

$$-(E(s) - I) E(t') u_0 \text{ if } s = t - n_j \Delta_j t > 0, \qquad t' = n_j \Delta_j t.$$

(The reason for making the distinction is that the solution operator $E(t)$

is generally defined only for non-negative arguments.) In either case,

$$\| \{E(n_j \Delta_j t) - E(t)\} u_0 \| < K_E \| (E(s) - I) u_0 \|$$

which goes to zero as $s \to 0$ and therefore as $j \to \infty$. Thus, combining this result with (12),

$$\| [\{C(\Delta_j t)\}^{n_j} - E(t)] u_0 \| \to 0 \text{ as } j \to \infty \tag{13}$$

for any u_0 which can be the initial element of a genuine solution of the class U. But these initial elements are dense in $\mathscr{B}$, so that if u is any element of $\mathscr{B}$ there is a sequence $u_1, u_2, \cdots$ converging to u, each u_i the initial element of a genuine solution for which (13) holds. Then

$$[\{C(\Delta_j t)\}^{n_j} - E(t)] u = [\{C(\Delta_j t)\}^{n_j} - E(t)] u_m + \{C(\Delta_j t)\}^{n_j} (u - u_m) + E(t)(u - u_m).$$

The last two terms on the right of this equation can be made as small as one pleases by choosing m sufficiently large, on account of uniform boundedness of the operators C^n and of $E(t)$. Then the first term on the right can be made as small as one pleases by choosing $\Delta_j t$ sufficiently small. Therefore the left member of the above equation goes to zero as $j \to \infty$. Since u was arbitrary, it is now established that $C(\Delta t)$ is a convergent approximation as defined in Section 6, and the equivalence theorem is established.

The above sufficiency proof is an operator-theoretic analogue of Fritz John's result relating the uniform boundedness of the values of the approximate solution to convergence in the maximum norm.

PART II

PARTIAL DIFFERENTIAL EQUATIONS WITH CONSTANT COEFFICIENTS

9. Introduction

Here the stability requirement as defined in Part I, and whose significance is indicated by the equivalence theorem, is applied to a special class of linear initial value problems — those of partial differential equations with constant coefficients and with auxiliary conditions permitting the use of Fourier series or integrals. If the space variables are restricted to a finite domain and the boundary conditions are of such a nature that they can be represented as a periodicity condition, Fourier series are used. If the domain is infinite, but the functions are quadratically integrable, Fourier integrals are used, via Plancherel's theorem. Combinations are also possible, in which some of the space variables have finite domain and others are unlimited. All these cases lead to exactly the same results, and our discussion will be based on Fourier series.

We shall be dealing with the following Banach space $\mathscr{B}$: if p is the number of functions used to describe a state of the physical system, and d is the number of space variables, a point in $\mathscr{B}$ represents a p-vector function defined in a d-dimensional unit cube (or rectangular parallelopiped). We suppose that these functions are in L^2 over this cube and that the square of the Banach norm is given by equation (14).

The advantage of this norm over, for example, the maximum norm is that the Parseval equation then shows that the Fourier transform establishes a norm-preserving isomorphism between $\mathscr{B}$ and the space $\mathscr{B}'$ of the Fourier coefficients. The stability requirement takes on a particularly simple form in $\mathscr{B}'$ leading immediately to the Von Neumann condition as a necessary condition for stability.

Of course, the choice of a norm is restricted by the nature of the problem; i.e., the solution operators $E(t)$ have to be bounded with respect to the norm. In most problems of mathematical physics, the L^2 norm can be used.

We give several sufficient conditions for stability; these are mostly of the nature of an auxiliary condition under which the Von Neumann condition is also sufficient for stability.

One may perhaps surmise that in all practical cases (including problems with variable coefficients,[2] and even nonlinear problems) the Von Neumann condition is both necessary and sufficient for stability. Such a surmise has often been made (so far apparently without misfortune) by people who have to make actual calculations, and one can construct a good bit of heuristic evidence for it. But the purpose of the present discussion is to discuss only certain cases that can be treated rigorously. A few simple applications will be given.

10. Notation; Fourier Series

Let $\mathbf{x} = (x_1, x_2, \cdots, x_d)$ be a vector (vectors will be denoted by bold face type) whose components $x_1, x_2, \cdots, x_d$ are the space variables of the problem. Suppose that the functions with which we deal are periodic with periods $L_1, L_2, \cdots L_d$ in the space variables. Consider a series

$$\sum_{(\mathbf{k})} c(\mathbf{k}) e^{i\mathbf{k}\cdot\mathbf{x}}$$

[2]Fritz John has succeeded in proving in his important paper [13] that for parabolic equations a mildly strengthened form of Von Neumann's condition is sufficient for stability even for operators with variable coefficients. A similar result for a certain class of hyperbolic equations with variable coefficients has been obtained by Peter Lax [14].

where $\mathbf{k}$ is a d-component vector whose components are $2\pi l_1/L_1, \cdots, 2\pi l_d/L_d$ and where the summation is understood to be over all such vectors obtained by letting $l_1, l_2, \cdots, l_d$ run independently over all positive and negative integers and where $c(\mathbf{k})$ is a complex-valued function defined on the lattice of these vectors. This is a general trigonometrical series with the periodicity described above. In our applications there are p functions of the x's; we treat them as the components of a p-component vector $\mathbf{f}(\mathbf{x})$. For any vector $\mathbf{y}$ we denote by $|\mathbf{y}|$ the square root of the sum of the squares of the absolute values of the components. Therefore, if $\mathbf{f}(\mathbf{x})$ can be expanded as

$$\mathbf{f}(\mathbf{x}) = \sum_{(k)} \mathbf{c}(\mathbf{k}) e^{i\mathbf{k}\cdot\mathbf{x}},$$

the Parseval equation is

$$\frac{1}{V}\int_0^{L_1} dx_1 \cdots \int_0^{L_d} dx_d \, |\mathbf{f}(\mathbf{x})|^2 = \sum_{(\mathbf{k})} |\mathbf{c}(\mathbf{k})|^2 \tag{14}$$

where $V = L_1 L_2 \cdots L_d$.

Any periodic $\mathbf{f}(\mathbf{x})$ for which the left member of (14) exists will be called an element of $\mathscr{B}$ and the square root of that member will be called its norm. Similarly, any set of coefficients for which the right member of (14) exists will be called an element of $\mathscr{B}'$ and the square root of that member will be called its norm. Then the Fischer-Riesz theorem says that $\mathscr{B}$ is a complete space and the Riesz-Fischer theorem says that there is a one-to-one correspondence between elements of $\mathscr{B}$ and of $\mathscr{B}'$, if we adopt the usual agreement that functions $\mathbf{f}(\mathbf{x})$ which differ only on a set of measure zero are regarded as identical — this agreement is reasonable, because the corresponding states of the physical system would be physically indistinguishable. The Parseval equation (14) shows that the correspondence between $\mathscr{B}$ and $\mathscr{B}'$ is norm-preserving. Statements of convergence, boundedness and the like can be taken over directly from $\mathscr{B}$ to $\mathscr{B}'$ or from $\mathscr{B}'$ to $\mathscr{B}$.

11. Properly Posed Problems

The general linear differential operator with constant coefficients can be otained formally by taking a function $D(\mathbf{k})$ or $D(k_1, k_2, \cdots, k_d)$ which is a $p \times p$ matrix whose elements are polynomials in $k_1, k_2, \cdots, k_d$, and substituting $\partial/\partial x_1$ for k_1, $\partial/\partial x_2$ for k_2, etc. If A is such an operator and we apply it to the element $\mathbf{v}e^{i\mathbf{k}\cdot\mathbf{x}}$ where $\mathbf{v}$ is a constant vector, the result is simply the product of this element and $D(i\mathbf{k})$. Therefore the solution of the initial value problem

$$\frac{\partial}{\partial t}\mathbf{u}(\mathbf{x}, t) = A\mathbf{u}(\mathbf{x}, t), \tag{15}$$

(16) $$\mathbf{u}(\mathbf{x}, 0) = \mathbf{u}_0(\mathbf{x})$$

is

(17) $$\mathbf{u}(\mathbf{x}, t) = \sum_{(\mathbf{k})} e^{i\mathbf{k}\cdot\mathbf{x}} e^{tD(i\mathbf{k})} \mathbf{v}_0(\mathbf{k})$$

where

(18) $$\mathbf{v}_0(\mathbf{k}) = \frac{1}{V} \int_0^{L_1} dx_1 \cdots \int_0^{L_d} dx_d \, \mathbf{u}_0(\mathbf{x}) e^{-i\mathbf{k}\cdot\mathbf{x}}.$$

The first requirement for a properly posed problem (see Section 3), namely that the domain of the solution operator be dense in $\mathscr{B}$, is automatically satisfied for the problems considered here, because the above solution (equations (17) and (18)) is certainly valid whenever the initial element $\mathbf{u}_0(x)$ is a trigonometric polynomial and the trigonometric polynomials are dense in $\mathscr{B}$.

The second requirement for a properly posed problem takes the form that $\| e^{tD(i\mathbf{k})} \|$ should be a bounded function of $\mathbf{k}$ and that the bound should be uniform in t. (It should be obvious to the reader that if M is a $p \times p$ matrix and we write $\| M \|$ we mean the bound of the transformation corresponding to M in a p-dimensional vector space with complex Euclidean norm.) Whether this condition is satisfied must usually be investigated separately in each case.

12. Finite Difference Equations

Just as the differential operator A is represented, in the space $\mathscr{B}'$, by the matrix $D(i\mathbf{k})$, the finite-difference operator $B(\Delta t, \Delta\mathbf{x})$ will be represented in $\mathscr{B}'$ by a matrix $G(\Delta t, \Delta\mathbf{x}, \mathbf{k})$ whose elements are functions of the components of $\mathbf{k}$ as well as of the parameters Δt, $\Delta\mathbf{x}$. One reason for making the Fourier transformation is that the elements of $G(\Delta t, \Delta\mathbf{x}, \mathbf{k})$ can generally be found easily, even though $B(\Delta t, \Delta\mathbf{x})$ represents an implicit system of difference equations.

Each difference equation equates to zero a certain linear combination of the components of $\mathbf{u}^n$ and of $\mathbf{u}^{n+1}$ at a group of neighboring points of the net used for the numerical work. Specifially, let this group of points be referred to a particular point of the group with coordinates $x_1, x_2, \cdots x_d$ so that a typical neighbor of this point in the group has coordinates $x_1 + \beta_1 \Delta x_1, \cdots, x_d + \beta_d \Delta x_d$ where $\beta_1, \cdots, \beta_d$ are integers. The difference equations can then be written in the form

(19) $$\sum_{(\beta_1, \cdots, \beta_d)} [A(\beta_1, \cdots, \beta_d)\mathbf{u}^{n+1}(x_1 + \beta_1 \Delta x_1, \cdots, x_d + \beta_d \Delta x_d) + B(\beta_1, \cdots, \beta_d)\mathbf{u}^n(x_1 + \beta_1 \Delta x_1, \cdots, x_d + \beta_d \Delta x_d)] = 0,$$

where A and B are $p \times p$ matrices whose elements depend on the β_i and on Δt and the Δx_i but not on t (i.e. n) or the x_i themselves. The summation is over a finite number of neighbors — that is over a finite number of sets of values of $\beta_1, \cdots, \beta_d$.

This system is in general implicit, because in the numerical work the unknowns are the values of the components of $\mathbf{u}^{n+1}$ at the various net points, and each equation contains generally several of the unknowns. We assume, however, that the system is such that if $\mathbf{u}^n(\mathbf{x})$ is given as any element of $\mathscr{B}$, then $\mathbf{u}^{n+1}(\mathbf{x})$ is uniquely determined by the difference equations (19) and the periodicity requirement.

If the Fourier series

$$\mathbf{u}^n(\mathbf{x}) = \sum_{(\mathbf{k})} \mathbf{v}^n(\mathbf{k}) e^{i\mathbf{k}\cdot\mathbf{x}} \tag{20}$$

and a similar one for $\mathbf{u}^{n+1}(\mathbf{x})$ are substituted into (19), the typical term contains a factor

$$\exp \{i[k_1(x_1 + \beta_1 \Delta x_1) + \cdots + k_d(x_d + \beta_d \Delta x_d)]\}$$

from which we cancel out the common part $e^{i\mathbf{k}\cdot\mathbf{x}}$ from all the terms of the equation. What is left can be written as

$$H_1 \mathbf{v}^{n+1}(\mathbf{k}) + H_2 \mathbf{v}^n(\mathbf{k}) = 0$$

where H_1 is an abbreviation for the matrix

$$\sum_{(\beta_1, \cdots, \beta_d)} A(\beta_1, \cdots, \beta_d) \exp \{i[k_1 \beta_1 \Delta x_1 + \cdots + k_d \beta_d \Delta x_d]\}$$

and H_2 is similar. The solvability assumption made in the preceding paragraph is tantamount to the assumption that H_1 has an inverse. Therefore we can write

$$\mathbf{v}^{n+1}(\mathbf{k}) = G\mathbf{v}^n(\mathbf{k}) \tag{21}$$

where the matrix G is given by

$$G = G(\Delta t, \Delta \mathbf{x}, \mathbf{k}) = -H_1^{-1} H_2. \tag{22}$$

G will be called the *amplification matrix*: it is the representation in $\mathscr{B}'$ of the operator $B(\Delta t, \Delta \mathbf{x})$. Therefore the stability requirement is that if the manner of refinement of the mesh is given by $\Delta \mathbf{x} = \mathbf{g}(\Delta t)$, the set of matrices

$$\{G(\Delta_j t, \mathbf{g}(\Delta_j t), \mathbf{k})\}^n, \qquad j = 1, 2, \cdots, \quad 0 \leq n\Delta_j t \leq T \tag{23}$$

should be uniformly bounded for all $\mathbf{k}$ with real components, the bound being uniform in $\mathbf{k}$. As in Part I of this paper $\Delta_j t$ is a sequence tending to zero as $j \to \infty$ and corresponding to each j there is a net or grid of space points such that each $g_1(\Delta_j t) = \Delta x_1$, $g_2(\Delta_j t) = \Delta x_2$, etc. also tends to zero.

The problem of stability is thus reduced to that of finding estimates for the bounds of powers of the amplification matrix G.

13. The Von Neumann (Necessary) Condition for Stability

A lower limit for the bounds of the powers of G is easily given. Let the eigenvalues of G be $\lambda^{(1)}, \lambda^{(2)}, \cdots, \lambda^{(p)}$ (not assumed real, not assumed distinct). The spectral radius of G is

$$r_1 = r_1(\Delta t, \Delta \mathbf{x}, \mathbf{k}) = \mathrm{Max}_{(i)} \, |\lambda^{(i)}|.$$

Suppose the λ's so ordered that $|\lambda^{(1)}| = r_1$ and let $v^{(1)}$ be an eigenvector corresponding to eigenvalue $\lambda^{(1)}$. Then

$$||G|| = \mathrm{Max}_{(\mathbf{v})} \frac{||G\mathbf{v}||}{||\mathbf{v}||} \geqq \frac{||G\mathbf{v}^{(1)}||}{||\mathbf{v}^{(1)}||} = r_1,$$

or generally the spectral radius is a lower bound for the bound of a matrix. If G is raised to any power, each of its eigenvalues gets raised to the same power, and therefore the spectral radius G^n is r_1^n. Therefore

$$||G^n|| \geqq r_1^n.$$

We call

$$R_1 = R_1(\Delta t) = \mathrm{Max}_{(\mathbf{k})} \, r_1(\Delta t, \mathbf{g}(\Delta t), \mathbf{k}),$$

where the maximum is with respect to all $\mathbf{k}$ with real components. The stability requirement of uniform boundedness of the set (23) implies that for some K,

$$\{R_1(\Delta t)\}^n \leqq K, \qquad \begin{array}{l} \Delta t > 0, \\ 0 \leqq n\Delta t \leqq T, \end{array}$$

but this is equivalent to the condition

$$R_1(\Delta t) \leqq 1 + O(\Delta t) \tag{24}$$

where $O(\Delta t)$ denotes a quantity bounded by a constant times Δt. This is the Von Neumann necessary condition for stability.

14. A Sufficient Condition for Stability

Let the eigenvalues of G^*G be denoted by $\mu^{(1)}, \cdots, \mu^{(p)}$. The bound of G is

$$r_2 = r_2(\Delta t, \Delta \mathbf{x}, \mathbf{k}) = ||G|| = \mathrm{Max}_{(i)} \, |\mu^{(i)}|^{1/2}.$$

Since $||G^n|| \leqq ||G||^n$, if we call

$$R_2 = R_2(\Delta t) = \mathrm{Max}_{(\mathbf{k})} \, r_2(\Delta t, \mathbf{g}(\Delta t), \mathbf{k}),$$

the stability requirement is satisfied provided there is a K such that

$$\{R_2(\Delta t)\}^n \leqq K, \qquad \begin{array}{l} \Delta t > 0, \\ 0 \leqq n\Delta t \leqq T. \end{array}$$

but this is equivalent to the condition

$$R_2(\Delta t) \leqq 1 + O(\Delta t). \tag{25}$$

Consequently, we have

THEOREM 1. *Condition* (25) *is sufficient for stability.*

If G is a normal matrix (i.e., one that commutes with its Hermitian conjugate), the eigenvalues of G^*G are just the squares of the absolute values of the eigenvalues of G (because G^* and G can be reduced to diagonal form by the same unitary transformation), so that $R_1(\Delta t) = R_2(\Delta t)$. Therefore, we can state the

COROLLARY. *If G is a normal matrix, the Von Neumann condition* (24) *is sufficient as well as necessary for stability.*

15. A Second Sufficient Condition for Stability

As noted in the corollary, the case in which G is a normal matrix is an important special case, and in that case there is a complete orthogonal set of eigenvectors of G. Even if G is not normal, there *may* be a complete set of linearly independent eigenvectors (not generally orthogonal). A stability condition will now be given for such cases.

Let $\phi^{(1)}, \cdots, \phi^{(p)}$ denote a set of normalized, linearly independent eigenvectors of G. Let T be the matrix having these eigenvectors as columns, so that $T_{ij} = \phi_j^{(i)}$; and let Δ denote the determinant of T. T provides a similarity transformation (not in general unitary) that diagonalizes G. That is,

$$G = T^{-1} \begin{pmatrix} \lambda^{(1)} & \cdot & 0 \\ & \cdot & \\ 0 & & \lambda^{(p)} \end{pmatrix} T,$$

and therefore

$$G^n = T^{-1} \begin{pmatrix} \lambda^{(1)} & \cdot & 0 \\ & \cdot & \\ 0 & & \lambda^{(p)} \end{pmatrix}^n T. \tag{26}$$

The inverse of T has elements given by

$$(T^{-1})_{ij} = \frac{\text{algebraic cofactor of } T_{ji}}{\Delta}.$$

If the columns (or rows) of any determinant are regarded as a set of vectors, the absolute value of the determinant does not exceed the product of the lengths of the vectors (corresponding to the interpretation of the determinant as the volume of a multidimensional parallelopiped of which the vectors form a set of coterminous edges). Each column of the cofactor mentioned above consists of $p-1$ of the components of a normalized eigenvector of G and hence has length less than or equal to 1. Consequently,

$$|(T^{-1})_{ij}| \leqq \frac{1}{|\Delta|}.$$

Clearly, the absolute value of an element of T cannot exceed 1, so from (26),

$$|(G^n)_{ij}| \leqq \frac{p^2}{|\Delta|} r_1^n,$$

where the factor p^2 comes from the fact that there are p^2 terms in the expansion of the matrix product (26). Since the bound of a $p \times p$ matrix does not exceed p times its absolutely largest element,

$$\| G^n \| \leqq \frac{p^3}{|\Delta|} r_1^n.$$

The determinant Δ of course is a function of Δt and $\mathbf{k}$, but if it is bounded away from zero, we can replace $|\Delta|$ by its greatest lower bound in the above inequality and use the same reasoning that led to (25) in Section 14, to prove

THEOREM 2. *If there is a constant a such that $|\Delta| > a > 0$ for all real $\mathbf{k}$ and all sufficiently small Δt, where Δ is the determinant of the normalized eigenvectors of the amplification matrix $G(\Delta t, \mathbf{g}(\Delta t), \mathbf{k})$, the Von Neumann condition* (24) *is sufficient as well as necessary for stability.*

16. A Third Sufficient Condition for Stability

In some cases of practical importance the determinant Δ vanishes for certain values of $\mathbf{k}$ so that a different criterion must be found. To find one, we start from Schur's theorem that any square matrix A can be reduced to triangular form by a unitary transformation

$$B = U^*AU$$

where B is the triangular matrix:

$$B = \begin{bmatrix} \lambda^{(1)} & B_{12} & B_{13} \cdots & B_{1p} \\ 0 & \lambda^{(2)} & B_{23} \cdots & B_{2p} \\ 0 & 0 & \lambda^{(3)} \cdots & B_{3p} \\ \cdot & \cdot & \cdot & \cdot \\ \cdot & \cdot & \cdot & \cdot \\ \cdot & \cdot & \cdot & \cdot \\ 0 & 0 & 0 \cdots & \lambda^{(p)} \end{bmatrix}$$

whose diagonal elements are the eigenvalues of A and such that $B_{ij} = 0$

for $i > j$. Since no element of U can exceed 1 in absolute value

$$\text{Max} \mid B_{ij} \mid \leqq p^2 \text{Max} \mid A_{ij} \mid . \tag{27}$$

The general element of the n-th power of B has the form

$$(B^n)_{ij} = \sum B_{ik_1} B_{k_1 k_2} \cdots B_{k_{n-1} j} \tag{28}$$

summed over the indices $k_1, k_2, \ldots, k_{n-1}$, arranged in every possible way, provided

$$i \leqq k_1 \leqq k_2 \leqq \cdots \leqq k_{n-1} \leqq j.$$

No matter how large n is, at most $j - i$ of the factors in the above product can be off-diagonal elements of B. This will enable us to obtain a satisfactory bound for B^n by imposing restrictions on the diagonal elements only. We assume that

$$\underset{i>1}{\text{Max}} \mid \lambda^{(i)} \mid = \gamma < 1, \tag{29}$$

and call

$$\text{Max} (\mid \lambda^{(1)} \mid, 1) = \lambda^* .$$

We focus our attention temporarily on those factors of the typical product in (28) which are off-diagonal elements of B, disregarding the diagonal elements occurring in the product. The number of such factors can be r where $0 \leqq r \leqq j - i$; let N_r^{j-i} be the number of distinct ways of choosing these r factors from among the off-diagonal elements of B, taking into account the chain rule for subscripts in matrix multiplication. (Except for the trivial cases in which $j - i = 0$ or $r = 0$, N_r^{j-i} is just the binomial coefficient $\binom{j-i-1}{r-1}$).

Having chosen the r off-diagonal elements, we consider the various ways in which they can be combined with diagonal elements to make the general typical product in (28) with the factors in the order shown there. One arrangement is with the off-diagonal elements crowded together at the right side of the product and preceded by a suitable power of $\lambda^{(i)}$ on their left. Other arrangements can be obtained from this one by decreasing the power of $\lambda^{(i)}$ and inserting suitable diagonal elements in positions to the right of the leftmost off-diagonal element. The first such factor can be inserted in any one of r positions, the next in any one of $r + 1$ positions, the next in any one of $r + 2$ positions, and so forth. But the order of insertion is irrelevant, so the number of distinct ways in which q such factors can be inserted is

$$\frac{r(r+1)(r+2)\cdots(r+q-1)}{q!}.$$

The inserted factors are all eigenvalues with index greater than i (because of their positions in the product) hence with index greater than 1. Therefore, by (29), the inserted factors, when multiplied together, are bounded by γ^q. Now the series

$$\sum_{0}^{\infty}{}_{(q)} \frac{r(r+1)\cdots(r+q-1)}{q!}\gamma^q$$

being a hypergeometric series, is convergent to a finite limit because $\gamma < 1$. Let that limit be denoted by $F_r(\gamma)$. The power of $\lambda^{(i)}$ occurring in the product is in any case bounded by $(\lambda^*)^n$, so, finally

$$|(B^n)_{ij}| \leqq (\lambda^*)^n \sum_{1}^{j-1}{}_{(r)} N_r^{j-1} (\operatorname*{Max}_{s,t} |B_{st}|)^r F_r(\gamma).$$

This expansion is clearly maximized by taking $j-1 = p-1$. Then, since the bound of a matrix does not exceed p times the absolute value of its largest element, and using (27) and the fact that the bound is invariant under a unitary transformation, we find

$$||A^n|| \leqq (\lambda^*)^n p \sum_{1}^{p-1}{}_{(r)} N_r^{p-1} (p^2 \operatorname*{Max}_{s,t} |A_{st}|)^r F_r(\gamma).$$

To apply this result to the stability problem, we interpret A as the amplification matrix $G(\Delta t, \Delta \mathbf{x}, \mathbf{k})$. The factor $(\lambda^*)^n$ is then bounded for the set (23) if the Von Neumann condition is satisfied, and the other factors in the above expression are bounded if $\gamma < 1$ and the A_{st} are bounded. The following theorem results:

THEOREM 3. *If the elements of the amplification matrix $G(\Delta t, \mathbf{g}(\Delta t), \mathbf{k})$ are bounded functions of $\mathbf{k}$ and Δt for all real $\mathbf{k}$ and all sufficiently small positive Δt, and if there is a constant γ such that*

$$|\gamma^{(i)}(\Delta t, \mathbf{g}(\Delta t), \mathbf{k})| \leqq \gamma < 1, \qquad i = 2, 3, \cdots, p,$$

then the Von Neumann condition (24) is sufficient as well as necessary for stability.

Roughly speaking, one eigenvalue is permitted to get up to 1, or even $1 + \theta(\Delta t)$ provided the bound of the others is less than 1.

17. The Wave Equation

As a first example to illustrate the foregoing ideas we consider the wave equation

$$\frac{\partial^2 \psi}{\partial t^2} - c^2 \frac{\partial^2 \psi}{\partial x^2} = 0.$$

A satisfactory formulation is obtained by making the further definition

$w = c\,\partial\psi/\partial x$, whereupon the equations become

$$\begin{aligned} \frac{\partial\phi}{\partial t} &= c\frac{\partial w}{\partial x}, \\ \frac{\partial w}{\partial t} &= c\frac{\partial\phi}{\partial x}, \end{aligned} \tag{31}$$

and we now have a properly posed problem. In this case the square of the norm is the energy of the wave motion and by conservation of energy the solution operator is bounded with bound unity.

The first choice of the finite difference equations that we wish to consider is

$$\begin{aligned} \phi^{n+1}(x) &= \phi^n(x) + \frac{c\Delta t}{\Delta x}\left[w^n\left(x+\frac{\Delta x}{2}\right) - w^n\left(x-\frac{\Delta x}{2}\right)\right], \\ w^{n+1}(x) &= w^n(x) + \frac{c\Delta t}{\Delta x}\left[\phi^n\left(x+\frac{\Delta x}{2}\right) - \phi^n\left(x-\frac{\Delta x}{2}\right)\right]. \end{aligned} \tag{32}$$

The amplification matrix is

$$G(\Delta t,\ \Delta x,\ k) = \begin{bmatrix} 1 & 2i\,\dfrac{c\Delta t}{\Delta x}\sin\dfrac{k\Delta x}{2} \\ 2i\,\dfrac{c\Delta t}{\Delta x}\sin\dfrac{k\Delta x}{2} & 1 \end{bmatrix}$$

as found by substituting a Fourier term

$$\begin{pmatrix} a^n \\ b^n \end{pmatrix} e^{ikx}$$

for $\begin{pmatrix} \phi^n \\ w^n \end{pmatrix}$ into (32) and solving for $\begin{pmatrix} a^{n+1} \\ b^{n+1} \end{pmatrix}$ in terms of $\begin{pmatrix} a^n \\ b^n \end{pmatrix}$ as

$$\begin{pmatrix} a^{n+1} \\ b^{n+1} \end{pmatrix} = G\begin{pmatrix} a^n \\ b^n \end{pmatrix}.$$

The quantity $R_1 = R_1(\Delta t)$ defined in Section 13 and appearing in the Von Neumann condition, namely the maximum with respect to k of the spectral radius of G, is

$$R_1 = \sqrt{1 + 4\left(\frac{c\Delta t}{\Delta x}\right)^2}$$

and we reach the well known conclusion that the difference equations (32) are unstable, at least if $c\Delta t/\Delta x$ is kept at any constant value as Δt and $\Delta x \to 0$.

In this example the Von Neumann condition could be satisfied by making Δt and $\Delta x \to 0$ in such a way that $\Delta t/(\Delta x)^2$ is constant, but there is not much point in pursuing this lead because we all know perfectly well that there is a much better difference scheme than (32).

According to the scheme usual in fluid dynamical calculations, the differential equations (21) are approximated by

$$\tag{33} \begin{aligned} \phi^{n+1}(x) &= \phi^n(x) + \frac{c\Delta t}{\Delta x}\left[w^n\left(x+\frac{\Delta x}{2}\right) - w^n\left(x-\frac{\Delta x}{2}\right)\right], \\ w^{n+1}(x) &= w^n(x) + \frac{c\Delta t}{\Delta x}\left[\phi^{n+1}\left(x+\frac{\Delta x}{2}\right) - \phi^{n+1}\left(x-\frac{\Delta x}{2}\right)\right]. \end{aligned}$$

This scheme differs from (32) only in the superscripts on ϕ in the second equation. (The equations would have a more centered look if we had used the notation $\phi^{n+1/2}$ and $\phi^{n-1/2}$ in place of ϕ^{n+1} and ϕ^n). The amplification matrix is

$$\tag{34} G = \begin{pmatrix} 1 & ia \\ ia & 1-a^2 \end{pmatrix}$$

where a is an abbreviation for $\dfrac{2c\Delta t}{\Delta x}\sin\dfrac{k\Delta x}{2}$,

and
$$G^*G = \begin{pmatrix} 1+a^2 & ia^3 \\ -ia^3 & 1-a^2+a^4 \end{pmatrix}.$$

The characteristic equations of these matrices are

$$\tag{35} \lambda^2 - (2-a^2)\lambda + 1 = 0$$

and

$$\tag{36} \mu^2 - (2+a^4)\mu + 1 = 0.$$

For each of these characteristic equations the product of the roots is 1. In (35) the sum of the roots is $2 - a^2$; consequently the roots lie on the unit circle if $a^2 \leqq 4$. In (36) the roots are real. We find, for the quantities $R_1(\Delta t)$ and $R_2(\Delta t)$ introduced in Sections 13 and 14

$$R_1(\Delta t) \begin{cases} = 1 \text{ if } \dfrac{c\Delta t}{\Delta x} \leqq 1, \\ > 1 \text{ if } \dfrac{c\Delta t}{\Delta x} > 1, \end{cases}$$

$$R_2(\Delta t) = \sqrt{1 + 8\left(\frac{c\Delta t}{\Delta x}\right)^4 + 4\left(\frac{c\Delta t}{\Delta x}\right)^2\sqrt{1+4\left(\frac{c\Delta t}{\Delta x}\right)^4}}.$$

The Von Neumann condition is satisfied for $c\Delta t/\Delta x \leqq 1$ but not for any other fixed value of $c\Delta t/\Delta x$.

The sufficient condition for stability given in Section 14, namely

$R_2(\Delta t) \leq 1 + O(\Delta t)$, requires $\Delta t = O((\Delta x)^2)$ as Δt, $\Delta x \to 0$, which is much more stringent than the Von Neumann condition. But the sufficient condition given in Section 15 gives what we want. The normalized eigenvectors of the matrix (34) are easily found, and their determinant has the absolute value

$$|\Delta| = \sqrt{1 - \left(\frac{c\Delta t}{\Delta x} \sin \frac{k\Delta x}{2}\right)^2}.$$

This is bounded away from zero if $c\Delta t < \Delta x$. We arrive thus at the conclusion, first stated in the Courant-Friedrichs-Lewy paper, that equations (33) are stable if $c\Delta t/\Delta x =$ constant < 1 but not if $c\Delta t/\Delta x =$ constant > 1. The case $c\Delta t/\Delta x = 1$ (stable according to Courant, Friedrichs and Lewy) is not handled by our method.

Lastly, we consider the implicit system

$$\begin{aligned}
\phi^{n+1}(x) &= \phi^n(x) + \frac{c\Delta t}{2\Delta x}\left[w^n\left(x + \frac{\Delta x}{2}\right)\right. \\
&\qquad \left. + w^{n+1}\left(x + \frac{\Delta x}{2}\right) - w^n\left(x - \frac{\Delta x}{2}\right) - w^{n+1}\left(x - \frac{\Delta x}{2}\right)\right], \\
w^{n+1}(x) &= w^n(x) + \frac{c\Delta t}{2\Delta x}\left[\phi^n\left(x + \frac{\Delta x}{2}\right)\right. \\
&\qquad \left. + \phi^{n+1}\left(x + \frac{\Delta x}{2}\right) - \phi^n\left(x - \frac{\Delta x}{2}\right) - \phi^{n+1}\left(x - \frac{\Delta x}{2}\right)\right]
\end{aligned} \tag{37}$$

as approximation to the differential equations (31). (Equations of this type have been used, for example, by Arthur Carson of Los Alamos in studies of the dynamics of stellar interiors.) The amplification matrix is

$$G = \begin{bmatrix} \dfrac{1 - a^2/4}{1 + a^2/4} & \dfrac{ia}{1 + a^2/4} \\[2ex] \dfrac{ia}{1 + a^2/4} & \dfrac{1 - a^2/4}{1 + a^2/4} \end{bmatrix}$$

and G^*G is the unit matrix. The criterion of Section 14 is always satisfied and the equations (37) are stable as $\Delta t \to 0$, $\Delta x \to 0$, no matter what are the relative rates at which Δt and Δx approach zero.

18. Diffusion Equation; Two Level Formulas

Consider the equation

$$\frac{\partial u}{\partial t} = A\frac{\partial^2 u}{\partial x^2} + 2B\frac{\partial^2 u}{\partial x \partial y} + C\frac{\partial^2 u}{\partial y^2} \tag{38}$$

where the quadratic form (A, B, C) is required to be positive definite; this makes the differential equation parabolic. In consequence of this requirement, the differential equation provides a properly posed initial value problem; for example in the space $\mathscr{B}$ of functions $u(x, y)$ in L^2 over a rectangle in the x, y-plane.

We consider in this section a certain class of difference equations in which $u^n(x, y)$ and $u^{n+1}(x, y)$ are connected directly. (In the two following sections we will consider some schemes in which u^n, u^{n+1} and u^{n-1} all appear in the same equation; these will be referred to as three-level formulas.)

Introduce the following abbreviations:

$$\begin{aligned} &u^n_{ij} && \text{for } u^n(x, y),\\ &u^n_{i+1\,j} && \text{for } u^n(x+\Delta x, y),\\ &u^n_{i\,j+1} && \text{for } u^n(x, y+\Delta y), \text{ etc.,}\end{aligned}$$

and

$$\begin{aligned}\Phi^n_{ij} \text{ for } \Bigg[A\, & \frac{u^n_{i+1\,j} - 2u^n_{ij} + u^n_{i-1\,j}}{(\Delta x)^2} \\ & + 2B\frac{u^n_{i+1\,j+1} - u^n_{i-1\,j+1} - u^n_{i+1\,j-1} + u^n_{i-1\,j-1}}{4\Delta x \Delta y} \\ & + C\frac{u^n_{i\,j+1} - 2u^n_{ij} + u^n_{i\,j-1}}{(\Delta y)^2}\Bigg].\end{aligned}$$

The class of finite difference equations we wish to consider is

$$u^{n+1}_{ij} - u^n_{ij} = \Delta t[\theta\Phi^{n+1}_{ij} + (1-\theta)\Phi^n_{ij}], \tag{39}$$

where θ is a non-negative constant. The choice $\theta = 0$ gives the usual explicit system and the choices $\theta = \frac{1}{2}$, $\theta = 1$ give the two favorite implicit systems.

(As is well known, if $B = C = 0$, so that the problem reduces to that of one space variable, the implicit equations can be readily solved by a simple algorithm. For two or more space variables it is wise to solve the implicit equations approximately by a relaxation technique; this is of course much more labor than required, per cycle, by the explicit equations, but it is nevertheless worthwhile, in some cases, to use the implicit equations provided the relaxation is done by some method like the extrapolated Liebmann method.)

Since there is only one dependent variable, the amplification matrix has just one element:

$$G(\Delta t, \Delta x, \Delta y, k_x, k_y) = \frac{1 + (1-\theta)W}{1 - \theta W},$$

where

$$V = \Delta t \left[\frac{2A}{(\Delta x)^2} (\cos k_x \Delta x - 1) - \frac{2B}{\Delta x \Delta y} \sin k_x \Delta x \sin k_y \Delta y + \frac{2C}{(\Delta y)^2} (\cos k_y \Delta y - 1) \right]$$

From the positive definite character of the quadratic form (A, B, C), it follows, after a little calculation, that

$$-4\,\Delta t \left[\frac{A}{(\Delta x)^2} + \frac{C}{(\Delta y)^2} \right] \leqq W \leqq 0.$$

The expression $\dfrac{1 + (1-\theta)W}{1 - \theta W}$ is an increasing function of W in the above interval and has the value 1 at $W = 0$. Therefore we will have $|G| \leqq 1$ if this expression is $\geqq -1$ when W has its most negative value. From this the Von Neumann condition is found to be:

1) if $\frac{1}{2} \leqq \theta$, no restriction on the way Δt, Δx, Δy go to zero,

2) if $0 \leqq \theta \leqq \frac{1}{2}$ and if we suppose $\Delta t/(\Delta x)^2$ and $\Delta t/(\Delta y)^2$ kept constant as $\Delta t, \Delta x, \Delta y \to 0$, then

$$2\Delta t \left[\frac{A}{(\Delta x)^2} + \frac{C}{(\Delta y)^2} \right] \leqq \frac{1}{1 - 2\theta}. \tag{40}$$

Since the matrix G has only one element, G commutes with G^* (in this particular example $G = G^*$), so that the Von Neumann condition is sufficient as well as necessary for stability.

This example can be generalized in various ways. For example one may include lower order terms, $D\dfrac{\partial u}{\partial x} + E\dfrac{\partial u}{\partial y} + Fu$, where D, E, F are constants, in the differential equation (38), and investigate their influence on stability. This is easily done because G is still a one-element matrix, but we omit details. It is found that for any reasonable manner of treating these terms in the finite difference equation, the stability condition is the same as before, except that sometimes the sign $\leqq$ in (40) has to be replaced by $<$. We may note, however, that if $F > 0$ it is important to have the Von Neumann condition in the form $R_1(\Delta t) \leqq 1 + O(\Delta t)$ rather than merely $R_1(\Delta t) \leqq 1$, because there are then generally true solutions of the differential equation which increase exponentially as t increases, and clearly we cannot expect (nor do we wish) to exclude such solutions from the numerical work.

19. The Du Fort-Frankel Equations

Du Fort and Frankel [12] have approximated the diffusion equation

$$\frac{\partial u}{\partial t} = \sigma \frac{\partial^2 u}{\partial x^2} \qquad (\sigma = \text{constant} > 0) \tag{41}$$

by the difference equation

$$(42)\quad u^{n+1}(x)-u^{n-1}(x)=\frac{2\sigma\Delta t}{(\Delta x)^2}[u^n(x+\Delta x)-u^{n+1}(x)-u^{n-1}(x)+u^n(x-\Delta x)].$$

This system is of interest for two reasons, the first having to do with consistency and the second with stability. It is readily verified that the consistency condition of Section 5 is satisfied if and only if $\Delta t/\Delta x \to 0$ as Δt, $\Delta x \to 0$. In fact, if $\Delta t/\Delta x \to \mu$ where μ is a constant, it is clear that the difference equation (42) approximates the differential equation

$$\frac{\partial u}{\partial t}=\sigma\frac{\partial^2 u}{\partial x^2}-\mu^2\frac{\partial^2 u}{\partial t^2}$$

rather than (41). Just how large values of $\Delta t/\Delta x$ can be tolerated in practice is of course not settled by our argument.

To write (42) in the form required by our general theory, we must introduce another dependent variable, say $\phi(x)$, as follows:

$$u^{n+1}(x)=\phi^n(x)+\frac{2\sigma\Delta t}{(\Delta x)^2}[u^n(x+\Delta x)-u^{n+1}(x)-\phi^n(x)+u^n(x-\Delta x)],$$

$$\phi^{n+1}(x)=u^n(x).$$

The amplification matrix is

$$(43)\qquad G(\Delta t,\Delta x,k)=\begin{pmatrix}\frac{2\gamma}{1+\gamma}\cos k\Delta x & \frac{1-\gamma}{1+\gamma}\\ 1 & 0\end{pmatrix}$$

where $\gamma=\frac{2\sigma\Delta t}{(\Delta x)^2}$. The characteristic values of G are

$$(44)\qquad \lambda=\frac{\gamma\cos k\Delta x\pm\sqrt{1-\gamma^2\sin^2 k\Delta x}}{1+\gamma},$$

and it is readily found that a) the Von Neumann condition is always satisfied, and in fact $R_1(\Delta t)=1$, b) for any fixed value of γ, $R_2(\Delta t)=$ constant >1 so that condition (25) of Section 14 is of no use, c) the determinant Δ of the normalized eigenvectors of G vanishes when $\gamma\sin k\,\Delta x=1$, so that the condition (Theorem 2) of Section 15 is of no use, d) the criterion of Section 16 is satisfied because the lesser of the roots (44) is always bounded, in absolute value, by $\left|\frac{\gamma-1}{\gamma+1}\right|^{1/2}$, so that the Von Neumann condition is again sufficient as well as necessary for stability.

Therefore, the Du Fort-Frankel equations are always stable, but the time increment must be limited on account of the consistency condition.

20. Positive Operators

In this section we present a sufficient condition, due to Friedrichs [4], for the stability of certain difference schemes. The schemes considered are explicit two level schemes for vector-valued unknowns; i.e., the value of the approximate solution at time $t + h$ and position x is expressed as a linear combination of its computed values at the time t:

$$\mathbf{u}(t + \Delta t) = \sum_r c_r \mathbf{u}(\mathbf{x} + \Delta t \mathbf{d}_r) = C(\Delta t)\mathbf{u}. \tag{45}$$

In Friedrichs' theory the displacement vectors $\mathbf{d}_r$ (finite in number) need not lie on a lattice. The coefficient matrices c_r are functions of $\mathbf{x}$, and are required to satisfy the following conditions:

i) $\sum_r c_r(\mathbf{x}) = I$ (the identity matrix),
ii) $c_r(\mathbf{x})$ is symmetric and positive definite,
iii) $c_r(\mathbf{x})$ is a Lipschitz continuous function of the vector variable $\mathbf{x}$.

Conclusion: The difference scheme (45) is stable.

We reproduce Friedrichs' proof, and show that the norm of $C(\Delta t)$ with respect to the L^2 norm over the whole space, is bounded by

$$|C(\Delta t)| \leqq 1 + A\Delta t, \tag{46}$$

where the constant A depends on the Lipschitz constant of the c_r, and on the number of coefficients. Since stability means the uniform boundedness of $|C^n(\Delta t)|$, $n\Delta t \leqq T$, and $|C^n(\Delta t)| \leqq |C(\Delta t)|^n$, the estimate (46) implies stability.

To estimate $\|C\|$, Friedrichs uses this characterization of the norm of an operator:

$$\|C\| = \text{Sup}\,(\mathbf{u}, C\mathbf{v}),$$
$$\|\mathbf{u}\| = \|\mathbf{v}\| = 1,$$

where the bracket denotes the L^2 scalar product over the period parallelogram:

$$(\mathbf{u}, C\mathbf{v}) = \int \sum \mathbf{u}'(\mathbf{x})c_r(\mathbf{x})\mathbf{v}(\mathbf{x} + \Delta t \mathbf{d}_r)d\mathbf{x}. \tag{47}$$

It follows from this characterization of the norm of C that any upper bound for $(\mathbf{u}, C\mathbf{v})$ valid for all vectors $\mathbf{u}$ and $\mathbf{v}$ of unit length is an upper bound for $\|C\|$. We shall find an upper bound for $(\mathbf{u}, C\mathbf{v})$ from (47). Since the matrices c were assumed to be positive, we can apply the Schwarz inequality to the terms $\mathbf{u}'c\mathbf{v}$ in the integrand. We get, after throwing in the inequality about the arithmetic and the geometric mean,

$$\mathbf{u}'c\mathbf{v} \leqq \tfrac{1}{2}\mathbf{u}'c\mathbf{u} + \tfrac{1}{2}\mathbf{v}'c\mathbf{v}.$$

Substituting this into the integrand in (47) we have the following inequality:

$$(48)\quad (\mathbf{u}, C\mathbf{v}) \leqq \tfrac{1}{2}\sum \int \mathbf{u}'(\mathbf{x})c_r(\mathbf{x})\mathbf{u}(\mathbf{x}) + \tfrac{1}{2}\sum \int \mathbf{v}'(\mathbf{x}+\Delta t\mathbf{d}_r)c_r(\mathbf{x})\mathbf{v}'(\mathbf{x}+\Delta t\mathbf{d}_r).$$

The first term on the right in (48) is, on account of the requirement that $\Sigma\, c_r(\mathbf{x})$ is the identity, just $\frac{1}{2}(\mathbf{u}, \mathbf{u})$ which is $\frac{1}{2}$, since $\mathbf{u}$ has unit norm. In the second group of terms introduce $\mathbf{x}' = \mathbf{x} + \Delta t\mathbf{d}_r$ as new independent variable; we obtain

$$\tfrac{1}{2}\sum \int \mathbf{v}'(\mathbf{x}')c_r(\mathbf{x}' - \Delta t\mathbf{d}_r)\mathbf{v}(\mathbf{x}').$$

If in the above expression we replace $c_r(\mathbf{x}' - \Delta t\mathbf{d}_r)$ by $c_r(\mathbf{x}')$, the error committed is at most a constant times Δt, on account of the assumed Lipschitz continuity of the coefficients c. Imagine such a substitution performed, and treat the resulting group of terms the same way as the first group of terms. This way we find that the value of the second group of terms is at most $1/2 +$ const. Δt, and have the desired $1 +$ const. Δt estimate for the whole expression (48).

Such symmetric positive difference operators come up in difference approximations to solutions of symmetric hyperbolic systems, i.e., equations of the form

$$(49)\qquad \mathbf{u}_t + a_k\mathbf{u}_{x^k} + b\mathbf{u} = 0,$$

where the coefficients a_k are symmetric matrices. A majority of the equations of mathematical physics which describe reversible phenomena are of this form; the general theory of such equations has been developed by Friedrichs (loc. cit), where he gives a recipe for associating a positive symmetric operator to each symmetric hyperbolic operator. We give here such a recipe:

Take for the displacement vectors $\mathbf{d}_r$ the $2d$ vectors

$$\mathbf{d}_r = \pm\,(0,\ 0, \cdots \lambda_r,\ 0, \cdots, 0), \qquad r = 1, \cdots, d.$$

Here the λ_r are arbitrary constants, the side lengths of a rectangular lattice in $\mathbf{x}$-space. Replace the $\mathbf{x}$-space derivative $\mathbf{u}_{x^k}$ in the differential equation (49) by centered difference quotients between $\mathbf{x} + \Delta t\mathbf{d}_k$ and $\mathbf{x} - \Delta t\mathbf{d}_k$, and the time derivative by the forward difference quotient $\mathbf{u}(\mathbf{x}, t + \Delta t) - \bar{\mathbf{u}}(\mathbf{x}, t)$ where $\bar{\mathbf{u}}$ is the weighted average $\frac{1}{2}\alpha_k\mathbf{u}(\mathbf{x} \pm h\mathbf{d}_k, t)$, the sum of the weights α being one. The resulting difference equation can be solved for $\mathbf{u}(\mathbf{x}, t + \Delta t)$:

$$u(x, t + \Delta t) = Cu(x, t) = \sum c_{\pm r}u(x \pm \lambda_r\Delta t),$$

where[3]

$$c_{\pm r} = \tfrac{1}{2}\,\{\alpha_r I \mp \lambda_r^{-1} a_r\}.$$

[3]For the sake of simplicity we have taken b, the coefficient of $\mathbf{u}$ in (49), to be zero.

Clearly, if the α_r are fixed positive constants, the coefficients $c_{\pm r}$ can be made positive definite by taking λ_r large enough. Of course in practice it is the space mesh that stays constant and Δt is made small enough.

Bibliography

[1] Courant, R., Friedrichs, K. O., and Lewy, H., *Über die partiellen Differenzengleichungen der mathematischen Physik*, Math. Ann., Vol. 100, 1928, pp. 32–74.

[2] Courant, R., Isaacson, E., and Rees, M., *On the solution of nonlinear hyperbolic differential equations by finite differences*, Comm. Pure Appl. Math., Vol. 5, 1952, pp. 243–255.

[3] Eddy, R. P., *Stability in the numerical solution of initial value problems in partial differential equations*, Naval Ordnance Laboratory Memorandum 10232.

[4] Friedrichs, K. O., *Symmetric hyperbolic linear differential equations*, Comm. Pure Appl. Math., Vol. 7, 1954, pp. 345–392.

[5] Kantorovitch, L. V., *Functional analysis and applied mathematics*, Uspehi Matem. Nauk, Vol. 3, 1948, p. 89; Bureau of Standards Report 1509, 1952.

[6] Keller, J. B., and Lax, P. D., *The initial and mixed initial and boundary value problems for hyperbolic systems*, Los Alamos Report No. 1210, 1951.

[7] Laasonen, *Über eine Methode zur Lösung der Wärmeleitungsgleichung*, Acta Math., Vol. 81, 1949, pp. 309–317.

[8] O'Brien, G. G., Hyman, M. A., and Kaplan, S., *A study of the numerical solution of partial differential equations*, J. Math. Physics, Vol. 29, 1951, pp. 223–251.

[9] Lewy, H., *On the convergence of solutions of difference equations*, in *Studies and Essays*, Courant Anniversary Volume, Interscience Publishers, New York, 1948, pp.211–214.

[10] Von Neumann, J., and Richtmyer, R. D., *A method for the numerical calculation of hydrodynamic shocks*, J. Appl. Physics, Vol. 21, 1950, pp. 232–237.

[11] Thomas, L. H., *Stability of partial differential equations*, Symposium on Theoretical Compressible Flow, U. S. Naval Ordnance Laboratory, 1949, NOLR 1132, 1950.

[12] Du Fort, E. C., and Frankel, S. P., *Stability conditions in the numerical treatment of parabolic differential equations*, Math. Tables and Other Aids to Computation, Vol. 7, 1953, pp. 135–152.

[13] John, F., *On integration of parabolic equations by difference methods*, Comm. Pure Appl. Math., Vol. 5, 1952, pp. 155–211.

[14] Lax, P. D., *Difference approximation to solutions of linear differential equations – an operator theoretical approach*, Symposium on Partial Differential Equations, Berkeley, Summer 1955; Report of the University of Kansas Mathematics Department (to appear).

Received October, 1955.

COMMUNICATIONS ON PURE AND APPLIED MATHEMATICS, VOL. XIV, 497–520 (1961)

K. O. Friedrichs anniversary issue

On the Stability of Difference Approximations to Solutions of Hyperbolic Equations With Variable Coefficients*

PETER D. LAX

In solving time dependent problems for hyperbolic differential equations with difference methods, approximate solutions at time t are computed as $S_\Delta^n\phi$, $t = n\Delta$, ϕ the prescribed initial data, S_Δ a difference operator on a lattice of mesh size proportional to Δ.

Difference operators with variable coefficients are special but typical examples of convolution operators with variable coefficients, i.e. operators of the form

$$(S_\Delta u)(x) = \int c_\Delta(x; y)u(x-y)dy;$$

c_Δ depends on the parameter $\Delta = \Delta t$ in the following familiar fashion:

$$c_\Delta = \frac{1}{\Delta^k}\, c(x; \frac{y}{\Delta})$$

cdy is a signed measure[1], with total measure one, depending on x as a parameter. It is well known and easy to show, see [13], that the approximate solutions $S_\Delta^n\phi$ will converge to the exact solution at time $t = n\Delta$ for all square integrable initial functions ϕ if and only if the difference equation is stable in the following sense: the L_2 norms of the operators S_Δ^n are uniformly bounded in the range $n\Delta \leq 1$. In this paper we investigate conditions for stability.

If c is independent of x, then S_Δ is a bona fide convolution operator and therefore its properties can be studied by Fourier transformation:

$$\widehat{(S_\Delta u)}(\xi) = C\left(\frac{\xi}{\Delta}\right) \hat{u}(\xi),$$

*The work presented in this paper was supported under Contract AT(30-1)-1480 with the U. S. Atomic Energy Commission. Reproduction in whole or in part is permitted for any purpose of the United States Government.

[1]In the body of the paper we shall be dealing with the case of matrix valued measures, and vector valued functions; but for sake of simplicity in this introduction, we shall speak only of the scalar case.

where $C(\xi)$ is the Fourier-Stieltjes transform of c. Clearly there is stability if and only if

$$|C(\xi)| \leqq 1$$

for all real ξ. Note that for $\xi = 0$ the sign of equality holds, since $C(0) = \int c dy = 1$.

In the case of variable coefficients we form $C(x; \xi)$, the Fourier-Stieltjes transform of c whith respect to y. It is easy to show that, if the condition

$$(1) \qquad |C(x; \xi)| \leqq 1$$

is violated for any value of x and ξ, then there is instability. Von Neumann has posed the converse problem: if (1) holds for all x and ξ, is the difference operator stable? In this paper we prove that, under some additional conditions which in the case of one space variable are only mildly restrictive, the norm of the operator S_Δ is

$$||S_\Delta|| = 1+O(\Delta);$$

clearly this implies stability. One of the additional conditions is that $C(x; \xi)$ should depend Lipschitz continuously on x:

$$|C(x)-C(x')| \leqq \text{const. } |x-x'|,$$

where the norm on the left is the maximum in the whole ξ-space of C and its derivatives with respect to ξ up to some order.

The proof uses so-called local energy inequalities; this method has been used by Friedrichs to show the stability of difference operators with non-negative measure c, see [3] or [8]. Subsequent applications of the method have been given by Lees in [10, 11]. Here we show that the energy method works if and only if $1-|C|^2$ can be decomposed as a linear combination of squares with non-negative coefficients:

$$P \equiv 1-|C|^2 = \sum w_j(x)|D_j(x; \xi)|^2,$$

where both w_j and D_j depend Lipschitz continuously on x. Since $P = 1-|C|^2$ is assumed to be non-positive, we are led to the problem: what non-negative functions can be so decomposed? The difficulty in obtaining such a decomposition is caused by the vanishing of P at $\xi = 0$. In the case of one space variable it can be easily shown that if the order of zero of P at $\xi = 0$ changes at most by 2 locally, then P can be written in the above form. For several variables, and if P has a zero of high order, such a decomposition is not always possible; but in such cases it is sometimes possible to prove stability with respect to some other suitably chosen norm.

The theory is carried out for difference operators, i.e. when the support of C lies on a lattice. Then C is a periodic function of ξ; the methods of this paper apply equally well to the case where C becomes sufficiently small for ξ large.

In Section 1 energy inequalities for solutions of symmetric hyperbolic equations are derived. In Section 2 the basic notions of difference approximations are given. Section 3 contains a derivation of the main inequality by the energy method for one space variable; the case of several variables is treated in Section 4.

1. Symmetric Hyperbolic Differential Equations

Consider any linear first order symmetric hyperbolic system of partial differential equations, i.e. an equation of the form

$$u_t = Gu, \tag{1.1}$$

where

$$G = \sum A_j \frac{\partial}{\partial x^j} + B; \tag{1.2}$$

A_j are real symmetric matrices, u a vector variable. We assume that A_j depends on x and t and has bounded first order partial derivatives.

Denote by G^* the adjoint of the operator G with respect to the L_2 scalar product; since A_j was assumed symmetric, we have

$$G^* = -\sum \frac{\partial}{\partial x^j} A_j + B^* = -\sum A_j \frac{\partial}{\partial x^j} - \sum A_{j,j} + B^*,$$

where $A_{j,j}$ abbreviates $\partial A_j / \partial x^j$. Hence

$$G + G^* = B + B^* - \sum A_{j,j}. \tag{1.3}$$

Since we have assumed that A_j has bounded first derivatives, $G^* + G$ *is a bounded operator*; such an operator G may be called "almost antisymmetric".

In what follows we shall denote the L_2 scalar product with respect to the space variables of vector valued functions by parenthesis (,). Let u be a solution of (1.1) and denote by $E(t)$ the quantity

$$E(t) = (u, u).$$

Using (1.1) and the boundedness of $G+G^*$ we get

$$\begin{aligned} \frac{dE}{dt} &= (u_t, u) + (u, u_t) = (Gu, u) + (u, Gu) \\ &= (u, G+G^* u) \leqq M(u, u) = ME(t); \end{aligned}$$

here M is an upper bound for the norm of $G+G^*$. From the differential inequality

$$\frac{dE}{dt} \leqq ME(t)$$

one can easily deduce that

$$E(t) \leqq e^{Mt} E(0). \tag{1.4}$$

The quantity $E(t)$ is called the energy at time t; for the equations of mathematical physics, such as the Maxwell equations E is the total energy contained in the system at time t. Inequality (1.4) is called an *energy inequality*; such inequalities play a central role in the theory of hyperbolic equations.

If the operator G is strictly antisymmetric, i.e. if

$$B+B^* - \sum A_{j,j} = 0,$$

then $E(t)$ is a constant, i.e. energy is conserved. In the special case of constant coefficients this can also be demonstrated by the use of the Fourier transformation. Denote by U the Fourier transform of u, by ξ the dual variable. Then (1.1), (1.2) become

$$U_t = i(\sum A_j \xi_j) U.$$

Solve this ordinary differential equation:

$$U = \exp\{it(\sum A_j \xi_j)\} U(0);$$

since the exponential of an antisymmetric matrix is unitary, we have that

$$|U(\xi, t)|^2$$

is time independent. Integrating with respect to ξ and using Parseval's relation we deduce anew the law of conservation of energy.

2. Difference Equations and their Degree of Accuracy

We turn now to the study of difference approximations to differential equations of the form (1.1). We shall investigate so-called two-level equations, i.e. equations which relate the values of functions at two different values of time, t and $t+\Delta$. We solve for the value of u at time $t+\Delta$ in terms of its value at t:

$$u(t+\Delta) = S\, u(t), \tag{2.1}$$

where S is an operator which depends on Δ. For purposes of numerical analysis S is taken to be an operator whose value depends on the values of u

on a lattice of points. To make equation (2.1) homogeneous we take the size of the lattice to be proportional to Δ. For sake of simplicity of notation we shall take this lattice to be Δ times the lattice of integers and write

$$(2.2)\qquad (Su)(x) = \sum_j C_j(x, \Delta)u(x-\Delta j),$$

where j is a multiindex and the coefficients C_j are matrices.

We shall assume—and this condition is actually satisfied for all operators used in practice—that the coefficients $C_j(x, \Delta)$ are smooth functions of Δ:

$$(2.3)\qquad C_j(x, \Delta) = C_j(x)+\Delta C_j^{(1)}(x)+ \cdots +O(\Delta^q).$$

$C_j(x) = C_j(x, 0)$ is called the principal part of $C_j(x, \Delta)$; the operator S_0,

$$(2.2'')\qquad (S_0u)(x) = \sum C_j(x)u(x-\Delta j),$$

is called the principal part of S.

The Fourier transform of $\{C_j(x)\}$ with respect to j is called the *amplification factor* of the operator S_0 and is denoted by $C(\xi, x)$:

$$(2.4)\qquad C(x; \xi) = \sum C_j(x)e^{ij\xi}.$$

If S is the sum of a finite number of terms, i.e. if all but a finite number of the coefficients C_j on the right in (2.2) are zero, equation (2.1) is called *explicit*. In this case the amplification factor is a trigonometric polynomial. We shall consider here both explicit and implicit equations; in the latter case we shall assume that the coefficients C_j tend to zero exponentially for $|j|$ large. This condition is satisfied by all difference equations of practical importance; it assures that the amplification factor is analytic for ξ real.

DEFINITION. *The degree of accuracy of the difference equation* (2.1) *is the largest number p such that all solutions of the differential equation* (1.1) *which have continuous derivatives up to order $p+1$ satisfy the difference equation* (2.1) *with an error of magnitude*

$$O(\Delta^{p+1}).$$

In making this definition we assume tacitly that it is not vacuous, i.e. that the coefficients of (1.1) are so smooth that (1.1) possesses $(p+1)$-times differentiable solutions.

With the aid of Taylor's formula we can easily derive an algebraic expression for the degree of accuracy of any difference equation. We have

$$u(t+\Delta) = \sum_{n\leq p} \frac{\Delta^n}{n!}\, u_{t^n}+O(\Delta^{p+1}).$$

If u satisfies the differential equation $u_t = Gu$ we have

$$u_{t^n} = G^n u + \text{lower order terms};$$

so

$$u(t+\Delta) = \sum_{n \leq p} \frac{\Delta^n}{n!} G^n u + \cdots + O(\Delta^{p+1}). \tag{2.5}$$

On the other hand, expanding by Taylor's formula the terms on the right in (2.2) we get

$$(Su)(x) = \sum_{j,\,|q| \leq p} \Delta^{|q|} \frac{(-j)^q C_j}{q!} D^q u + O(\Delta^{p+1}), \tag{2.6}$$

q a multiindex. According to the definition of degree of accuracy, (2.5) and (2.6) differ by $O(\Delta^{p+1})$; hence the coefficients of Δ^n on the right in (2.5) and (2.6) must be the same for $n = 0, 1, \cdots, p$:

$$G^n = \sum_{j,\,|q|=n} (-j)^q C_j(x) D^q + \text{lower order terms}, \tag{2.7}$$

where $C_j(x)$ is the principal part of the coefficient $C_j(x, \Delta)$. The principal part of the right side of (2.7) can be expressed conveniently in terms of the amplification factor C defined in (2.4); the coefficient of D^q, $|q| = n$, is

$$i^q \frac{\partial^q}{\partial \xi^q} C \quad \text{at} \quad \xi = 0.$$

Similarly, the coefficient of D^q, $|q| = n$, on the left in (2.7) is

$$(-i)^q \frac{\partial^q}{\partial \xi^q} e^{i\xi A} \quad \text{at} \quad \xi = 0.$$

This proves

THEOREM 2.1. *If the difference equation (2.1), (2.2) approximates the differential equation (1.1) with p degrees of accuracy, then for ξ small*

$$C(x; \xi) = e^{-i\xi A(x)} + O(|\xi|^{p+1}) \tag{2.8}$$

for all x.

Similar algebraic conditions can be given for the higher order terms of the operator S.

THEOREM 2.2. *If the coefficients of S_0 are symmetric non-negative matrices, then the order of accuracy of (2.1) is equal to one, except possilby in the very special case when the differential equation (1.1) can be split up into single equations with integer coefficients, coupled only through terms of zero order.*

Proof: The order of accuracy has to be at least one; applying theorem (2.1) with $p = 1$ we get

$$\sum C_j = I, \quad \sum j_k C_j = -A_k, \tag{2.9}$$

Denote by B_k the matrix

$$B_k = \sum j_k^2 C_j.$$

Since C_j is non-negative, we have for arbitrary vectors u and v the Schwarz inequality

$$(C_j u, v)^2 \leqq (C_j u, u)(C_j v, v).$$

Thus we have (omitting the subscript k)

$$(Au, v)^2 = \{\sum j(C_j u, v)\}^2 \leqq \{\sum j(C_j u, u)^{1/2}(C_j v, v)^{1/2}\}^2.$$

By the Schwarz inequality the last expression on the right is not greater than

$$\sum j^2(C_j u, u) \sum (C_j v, v) = (Bu, u)(v, v);$$

here we have used the definition of B and the first identity in (2.9). Putting $v = Au$ we get

$$(A_k u, A_k u) = (A_k^2 u, u) \leqq (B_k u, u).$$

We claim that, except in the special situation described in Theorem (2.1), strict inequality holds for some u. This will show that $A_k^2 \neq B_k$. Since

$$\frac{\partial^2}{\partial \xi_k^2} C(\xi) = -B_k \quad \text{and} \quad \frac{\partial^2}{\partial \xi_k^2} e^{-i\xi A} = -A_k^2,$$

we can conclude from Theorem (2.1) that the accuracy of the approximation is not of second degree.

In order that the sign of equality hold in the first application of the Schwarz inequality, u and v must, for every j, be proportional, modulo the nullspace of C_j:

$$v_k = A_k u = a_{k,j} u + n_j, \tag{2.10}$$

where $C_j n_j = 0$. Furthermore the factor of proportionality must be of the same sign for all j.

In order that the sign of equality hold in the second application of the Schwarz inequality, the quantities $j_k^2(C_j u, u)$ and $(C_j v_k, v_k)$ must be proportional. Substituting the above expression for v_k and using the fact that $a_{k,j}$ has the same sign for all j we conclude that $a_{k,j} = \text{const.}\, j_k$, where the constant may depend on k. Applying C_j to (2.10) gives

$$C_j A_k u = \text{const.}\, j_k C_j u.$$

Summing with respect to j and using the identities (2.9), we deduce that the value of the constant is -1; so

$$C_j A_k = -j_k C_j.$$

Repeated application of the above identity gives

$$C_j A_k A_h = j_k j_h C_j,$$

and more generally

$$C_j P(A) = P(-j) C_j$$

for any polynomial P in several variables. Summing with respect to j gives

$$P(A) = \sum P(-j) C_j .$$

From this formula we deduce:

i) the matrices A_k commute,

ii) C_j are mutually orthogonal projections,

iii) the eigenvalues of A_k are integers.

From i) it follows that the matrices A_j can be diagonalized simultaneously; this transformation then splits the system (1.1) into single equations whose coefficients, according to iii), are integers. This completes the proof of Theorem 2.1.

The main problem concerning difference approximations to differential equations is the convergence of solutions of the former to solutions of the latter. The question of convergence turns out to hinge on the validity of an analogue of the energy inequality, called in this context stability. More precisely:

DEFINITION. *The difference equation* (2.1) *is stable if the L_2 norms of the powers of S are bounded, i.e. if there exists a constant M such that*

$$||S^n(\Delta)|| \leqq M \tag{2.11}$$

for all n and Δ in the range $n\Delta \leqq 1$.

The following result is well known and easy to derive, see e.g. [13]:

THEOREM. *Let u be any sufficiently differentiable solution of the differential equation* (1.1), *u_Δ the solution of the difference equation* (2.1) *such that $u_\Delta = u$ at $t = 0$. Assume that the difference equation* (2.1) *is stable. Then for any value of t within a given range, say $t \leqq 1$,*

$$||u(t) - u_\Delta(t)|| \leqq \text{const. } \Delta^p,$$

where p is the degree of accuracy of the approximation and where the value of the constant depends on the magnitude of the derivatives of u but not on Δ.

This theorem shows that stability implies convergence and that the overall error is of the same order in Δ as the accumulated local truncation error. Conversely, stability comes near to being a necessary condition for convergence. Accordingly we shall discuss conditions for a difference equation to be stable. We consider first the case where the coefficients C_j are independent of x and Δ. Then the operator S is a convolution operator and can be studied by Fourier transformation. More precisely, if U is the Fourier transform of u, then the Fourier transform of Su is $C(\Delta\xi)U(\xi)$, where C is the amplification factor introduced earlier. The Fourier transform of $S^n u$ is $C^n U$; since the Fourier transformation preserves the L_2 norm, we see that the stability of the difference equation depends on the uniform boundedness of the powers of the matrix $C(\xi)$. An obvious necessary condition is embodied in the criterion due to von Neumann:

Necessary for the stability of the difference equation (2.1) *is that for any real* ξ *the eigenvalues of the amplification factor* $C(\xi)$ *not exceed* 1 *in absolute value.*

A slightly strengthened form of this criterion is sufficient for stability:

THEOREM 2.3. *Suppose that the coefficients of the operator* S *are independent of* x *and* Δ, *and suppose that the amplification factor* C *of* S *satisfies the following condition:*

For each real ξ *there exists a matrix* $N(\xi)$ *such that, for all real* ξ, $C_N = N^{-1}CN$ *has norm not greater than one, i.e. satisfies*

$$C_N^* C_N \leqq I, \tag{2.12}$$

and that the norms of both N *and* N^{-1} *are bounded by some constant* m *uniformly for all* ξ. *Then the difference equation is stable.*

Proof: The Fourier transform of $S^n u$ is

$$C^n U = N C_N^n N^{-1} U.$$

Therefore, by our assumption about the norm of C_N, N and N^{-1}, we get

$$||S^n u|| = ||C^n U|| \leqq m^2 ||U|| = m^2 ||u||;$$

this is stability.

Necessary and sufficient conditions for stability have been given by Kreiss [6].

Condition (2.12) implies of course that the eigenvalues $\lambda(\xi)$ or $C(\xi)$ satisfy the condition

$$|\lambda(\xi)| \leqq 1$$

for all real ξ. Since $C(\xi)$ is an analytic function of ξ, so is $\lambda(\xi)$, with possible branchpoints. Hence, either $|\lambda(\xi)| = 1$ for all real ξ, or $|\lambda(\xi)| < 1$ at all but a denumerable set of real ξ. The first circumstance takes place only in a few cases of interest. Turning to the second case, given any positive ε we decompose $U(\xi)$ as follows:

$$U = U_1 + U_2,$$

where $|\lambda(\xi)| \leqq \theta < 1$ on the support of U_1 and $||U_2|| < \varepsilon$. By a standard theorem on matrices, if the spectral radius of $C(\xi)$ is less than θ, then for any positive δ, $||C^n(\xi)|| \leqq \text{const. } |\theta+\delta|^n$ for n large enough. This shows that $||S^n u_1||$ tends to zero as n tends to infinity; by Theorem 2.3,

$$||S^n u_2|| \leqq m^2 ||u_2|| < m^2 \varepsilon.$$

Therefore we conclude that

$$||S^n u|| \to 0$$

for every u. This, as pointed out by Friedrichs, is in sharp contrast to differential equations with constant coefficients; there the L_2 norm of solutions is constant for all time.

3. Stability of Difference Equations With Variable Coefficients in One Space Variable

THEOREM 3.1. *The difference equation* (2.1) *is stable if the amplification factor of the operator S in* (2.2) *satisfies the following conditions:*

i) *C has bounded difference quotients with respect to x, i.e.*

$$||C(x)-C(x')||_A \leqq \text{const. } |x-x'|, \tag{3.1}$$

where the norm $||C||_A$ is defined as the maximum of the norm of $C(\xi)$ with respect to ξ in some strip around the real ξ-axis.

ii) *All eigenvalues $\lambda = \lambda(x;\xi)$ of $C(x;\xi)$ satisfy*

$$|\lambda| < 1$$

for all x and all real $\xi \neq 0 \bmod 2\pi$.

iii) *For ξ near* 0,

$$C^*C = I - Q(x)\xi^{2q} + O(\xi^{2q+1}), \tag{3.2}$$

where q is an integer independent of x, and Q is positive definite for all values of x.

Conditions ii) and iii) are mildly strengthened forms of the von Neumann condition. To gain insight into the meaning of the constancy of q appearing in (3.2), we note that if $p = p(x)$ is the *local* degree of accuracy, i.e. the largest integer such that for given x

$$C(x;\xi) = e^{-iA(x)\xi} + O(\xi^{p+1}),$$

then we have

$$C^*C = I + O(\xi^{p+1}).$$

Thus if p is odd, $2q \geqq p+1$, if p is even, $2q \geqq p+2$, and in general the sign of equality would be expected to hold. In that case the constancy of q expresses the fact that, as x varies, the local degree of accuracy changes at most by one.

In the proof of Theorem **3.1** we need the following lemma, whose proof is postponed till the end of the section:

LEMMA **3.1**. *Under assumptions* ii) *and* iii) *there exists a matrix valued function* $M = M(x;\xi)$, *infinitely differentiable with respect to* x *and* ξ, *such that*

$$\text{(3.3)}\qquad \begin{aligned} &\text{a) } M \text{ is postive definite,}\\ &\text{b) } C^*MC \leqq M,\\ &\text{c) } M - C^*MC = Q'(x)\xi^{2q} + O(\xi^{2q+1}) \end{aligned}$$

for ξ *small,* Q' *positive definite.*

For the sake of simplicity we take first the case when Lemma **3.1** is true with $M \equiv I$. At the end of the section we shall indicate the modifications necessary to treat the case when M is $\neq I$.

By assumptions (**3.2**) and (**3.3**), with $M \equiv I$, we can write

$$\text{(3.4)}\qquad I - C^*C = |e^{i\xi}-1|^{2q}P(x,\xi),$$

where P is positive definite for all x and ξ. Let E be a matrix whose absolute value square is P:

$$\text{(3.5)}\qquad E^*E = P.$$

We may e.g. choose E as the positive square root of P, although we shall indicate later why other choices may be preferable. Define D as

$$D = (e^{i\xi}-1)^q E;$$

then (**3.4**) can be rewritten as

$$\text{(3.6)}\qquad C^*C + D^*D = I.$$

We define now the matrix valued function $K = K(x;\xi,\eta)$ as

$$\text{(3.7)}\qquad K = C^*(x;\xi)C(x;\eta) + D^*(x;\xi)D(x;\eta).$$

D depends analytically on C and C^* and therefore is analytic in ξ and Lipschitz continuous in x. So is therefore K; in particular K satisfies a Lipschitz condition with respect to the following norm:

i)

$$\text{(3.8)}\qquad \|K(x) - K(x')\|_5 \leqq \text{const. } |x - x'|,$$

where $\|K(x)\|_5$ denotes the sum of the L_1 norms of the partial derivatives of K up to the second order in ξ, third order in η.

(**3.7**) gives K explicitly as the sum of two degenerate positive kernels. Therefore we deduce the following inequality:

ii) For every x and every square integrable periodic function $U(\xi)$

$$|\int C(\xi)U(\xi)d\xi|^2 \leqq \iint U^*(\xi)K(\xi,\eta)U(\eta)d\eta. \tag{3.9}$$

iii) Setting $\xi=\eta$ in (3.7) we obtain by (3.6) the identity

$$K(x;\xi,\xi)\equiv I. \tag{3.10}$$

Expand K into a Fourier series with respect to ξ and η:

$$K(x;\xi,\eta)=\sum K_{l,m}(x)e^{i(m\eta-l\xi)}. \tag{3.11}$$

Properties i), ii), iii) of K can be expressed in terms of the Fourier coefficients of K as follows:

i)

$$|K_{l,m}(x)-K_{l,m}(x')| \leqq \frac{\text{const. } |x-x'|}{|l|^2|m|^3}; \tag{3.8'}$$

this inequality is obtained by performing, in the integral expression for the Fourier coefficients of $K(x)-K(x')$, two integrations by parts with respect to ξ and three with respect to η, and using (3.8) to estimate the resulting integral.

We turn now to (3.9); we express both sides in terms of the Fourier coefficients of U, C and K. Putting

$$U(\xi)=\sum w_j e^{-ij\xi}$$

we get:

ii) For every x and every square summable sequence of vectors $\{w_j\}$,

$$|\sum C_j(x)w_j|^2 \leqq \sum w_l^* K_{l,m}(x)w_m. \tag{3.9'}$$

We turn now to (3.10); we expand both sides in Fourier series and equate coefficients; we obtain the identities

$$\sum K_{k-r,k-s}(x) = \delta_{r,s}I, \tag{3.10'}$$

where $\delta_{r,s}$ is the Kronecker delta.

Let u be a vector-valued function of x; denote its values at the lattice points $x=k\Delta$ by u_k, those of S_0u by v_k:

$$u_k=u(k\Delta), \qquad v_k=(S_0u)(k\Delta).$$

Substitute these abbreviations into formula (2.2″); we have

$$v_k=\sum C_j(\Delta k)u_{k-j}.$$

Applying (3.9′) to $w_j=u_{k-j}$ we get the "local energy inequality"

$$|v_k|^2 \leqq \sum u_{k-l}^* K_{l,m}(k\Delta)u_{k-m}. \tag{3.12}$$

Sum with respect to k and in the triple sum on the right introduce $k-l=r$ and $k-m=s$ as new indices of summation:

$$\sum|v_k|^2 \leqq \sum u_r^* K_{k-r,k-s}(k\Delta)u_s. \tag{3.13}$$

Rewrite the right side of (3.13) as

$$\sum u_r^* K_{k-r,k-s}(r\Delta)u_s + \sum u_r^* \{K(k\Delta)-K(r\Delta)\}_{k-r,k-s} u_s. \tag{3.14}$$

In the first term, sum with respect to k first; using identity (3.10′) we find that the first sum is equal to $\sum|u_r|^2$; so we get from (3.13)

$$\sum |v_k|^2 \leqq \sum |u_r|^2 + \text{leftover term}, \tag{3.15}$$

where the leftover term denotes the second term in (3.14). Using inequality (3.8′) we get the following upper bound for the leftover term:

$$O(\Delta)\sum \frac{1}{(k-r)^2(k-s)^2}|u_r||u_s|.$$

By the Schwarz inequality,

$$2|u_r||u_s| \leqq |u_r|^2+|u_s|^2;$$

so we have as upper bound for the leftover term

$$O(\Delta)\sum \frac{1}{(k-r)^2(k-s)^2}(|u_r|^2+|u_s|^2).$$

In the term containing $|u_r|$ we carry out the summation first with respect to k and s, in the term containing $|u_s|$ first with respect to k and r. Using $k-r=l$ and $k-s=m$ as new indices of summation we obtain

$$O(\Delta)\sum |u_r|^2$$

as an upper bound for the leftover term. Hence (3.15) can be written as

$$\sum |v_k|^2 \leqq (1+O(\Delta))\sum |u_r|^2$$

Let y denote any number between 0 and Δ. We define

$$u_k(y) = u(k\Delta+y), \qquad v_k(y) = (S_0 u)(k\Delta+y).$$

As before we can derive the inequality

$$\sum|v_k(y)|^2 \leqq (1+O(\Delta))\sum |u_k(y)|^2.$$

Integrate this inequality with respect to y from 0 to Δ; we obtain

$$||S_0 u||^2 \leqq (1+O(\Delta))||u||^2.$$

This inequality asserts that the norm of the operator S_0 is bounded by

$1+O(\Delta)$. Since S differs from its principal part by terms of order $O(\Delta)$, we have by the triangle inequality

$$\|S\| \leqq \|S_0\| + O(\Delta) = 1 + O(\Delta). \tag{3.16}$$

If u is any smooth function independent of Δ, then

$$Su = u + O(\Delta);$$

this shows that in (3.16) the sign of equality holds:

$$\|S\| = 1 + O(\Delta).$$

From (3.16) we conclude further that

$$\|S^n\| \leqq \|S\|^n \leqq \{1+O(\Delta)\}^n = e^{nO(\Delta)},$$

showing that for $n\Delta \leqq 1$ the norms of S^n remain uniformly bounded. This is the stability asserted in Theorem 3.1.

We illustrate the theory on the example of the scalar differential equation

$$u_t = au_x + bu,$$

a, b functions of x and t. We shall set up an explicit three-point difference approximation which is accurate to second order. A three-point explicit scheme is one in which the coefficients C_j are zero except for $j = 0, \pm 1$; so the amplification factor $C(\xi)$ of such an equation is a trigonometric polynomial of first degree. According to Theorem 2.1, for second order accuracy we must have

$$C(\xi) = e^{-ia\xi} + O(\xi^3).$$

This determines uniquely the coefficients of C; a brief calculation gives

$$C(\xi) = 1 - ia \sin \xi - a^2(1 - \cos \xi).$$

Another brief calculation gives

$$1 - |C(\xi)|^2 = (a^2 - a^4)(1 - \cos \xi)^2;$$

this is non-negative if and only if $|a| \leqq 1$. In this case we put

$$K(\xi, \eta) = \bar{C}(\xi)C(\eta) + (a^2 - a^4)(1 - \cos \xi)(1 - \cos \eta).$$

A straightforward calculation gives

$$\{K_{lm}\} = \begin{pmatrix} \frac{a^2(1-a)}{2}, & \frac{a(a^2-1)}{2}, & 0 \\ \frac{a(a^2-1)}{2}, & 1-a^2, & \frac{a(1-a^2)}{2} \\ 0, & \frac{a(1-a^2)}{2}, & \frac{a^2(1+a)}{2} \end{pmatrix}.$$

Hence the local energy inequality is

$$|v_k|^2 \leqq \frac{a^2(1-a)}{2} u_{k-1}^2 + (1-a^2)u_k^2 + \frac{a^2(1+a)}{2} u_{k+1}^2$$
$$+a(a^2-1)[u_{k-1}u_k - u_k u_{k+1}].$$

Before proceeding further, some comments are due:

1) Consider the case of explicit difference equations, i.e. where the operator S involves the values of u at x and at N lattice points to the right and to the left of x. The amplification factor is then a trigonometric polynomial of degree N; thus $I-C^*C$ is a non-negative trigonometric polynomial of degree $2N$, and P one of degree $2N-2q$. Hence by the Fejér-Riesz theorem, and its extension to the matrix case by Rosenblatt [4], E can be chosen as a trigonometric polynomial of degree $N-q$. Thus D, and thereby K, is a trigonometric polynomial of degree N in ξ and η separately. This means that in the local energy inequality (3.12) *only those values of u appear on the right which also appear on the left.* This feature might be useful in extending the method to mixed initial and boundary value problems.

2) The role of assumption iii) is to assure that D, the square root of $I-C^*C$, depends smoothly on ξ and Lipschitz continuously on x. That some such condition is necessary is shown by the example of the scalar function

$$p(x;\xi) = x^2+\xi^2;$$

its square root is not a sufficiently smooth function of ξ.

3) The function p given above is not itself a square of a smooth function but it is a sum of squares. Such a representation for $I-C^*C$ is of course quite sufficient for our purposes and can be obtained under conditions which are milder than iii). In the scalar case we have the following result:

THEOREM 3.2. *Let $p(x;\xi)$ be a smooth function of ξ, Lipschitz continuous in x which is positive for $\xi \neq 0$. Let $2q(x)$ denote the order of zero of p at $\xi = 0$. Suppose that each point x has a neighborhood in which $q(x)$ changes at most by 1. Then p can be written as a linear combination of squares with positive weights:*

$$p = \sum m_k d_k^2,$$

where d_k are smooth functions of ξ, m_k are independent of ξ and both d and m depend Lipschitz continuously on x.

Sketch of proof: We shall construct such a representation locally; then a representation in the large can be obtained by employing a partition of unity. Let x_0 be a given point; if $q(x_0)$ is zero, then in a small enough neighborhood of x_0, p is positive and can be represented as the square of its square root. Take now the case $q(x_0) = 1$; for x near enough to x_0 we define $\xi(x)$ to be that value of ξ for which

$$p(x; \xi)$$

reaches its minimum. Since the first derivative vanishes at the minimum point, we have

$$p_\xi(x; \xi) = 0 \quad \text{for} \quad \xi = \xi(x).$$

Since we have assumed that $q(x_0)$ equals one, p has a zero of precisely the second order at x_0; so $p_{\xi\xi}(x_0; 0) \neq 0$. By the implicit function theorem, $\xi(x)$ depends in the same manner on x as p_ξ, i.e. Lipschitz continuously. We define now

$$m(x) = p(x; \xi(x)).$$

By construction, $p(x; \xi) - m(x)$ has a zero at $\xi = \xi(x)$. Since $p - m$ is non-negative, this zero is at least double; since $p_{\xi\xi}(x_0; 0) \neq 0$, for x near enough to x_0, this zero has multiplicity not greater than 2. Hence

$$p(x; \xi) - m(x) = (\xi - \xi(x))^2 p_1(x, \xi),$$

where p_1 is a positive function. Setting

$$d = (\xi - \xi(x)) p_1^{1/2},$$

we obtain the desired representation

$$p = m + d^2.$$

The case $q(x_0) > 1$ can be reduced to the case $q(x_0) = 1$ as follows: By assumption $q(x) \geq q(x_0) - 1$ for x near x_0; this shows that p is divisible by $\xi^{2q(x_0)-2}$ for x near x_0. We can apply now the previous argument to $p/\xi^{2q(x_0)-2}$ in place of p.

It would be worth while to determine the class of non-negative functions which can be written as the linear combination of squares with positive weights. The condition given in Theorem 3.2 is sufficient but not necessary.

4) We shall show now that, conversely, *if a local energy inequality (3.9′) holds, then $I - C^*C$ can be represented as a linear combination of squares with positive weights.* For: inequality (3.9′) asserts that

$$\sum w_l^* K_{lm} w_m - |\sum C_j w_j|^2$$

is a non-negative hermitean form. Then, according to the theory of hermitean forms [2], it can be written as a linear combination of the squares of absolute values of linear forms $L_k(w)$ with positive weights m_k:

$$\sum m_k |L_k(w)|^2.$$

Set now [3] $w_j = e^{ij\xi} a$, a a fixed vector; we get

$$\sum (a, K_{lm} a) e^{i(m-l)\xi} - |C(\xi) a|^2 = \sum |D_k(\xi) a|^2,$$

where $D_k(\xi)$ denotes $L_k(e^{ij\xi})$. By identity (3.10′) the first term on the left equals $|a|^2$; so, since a is an arbitrary vector, we conclude that

$$I - C^* C = \sum m_k D_k^* D_k$$

as asserted.

We turn now to the proof of Lemma 3.1. According to a well-known theorem, every matrix can be transformed by a similarity transformation into upper triangular form; furthermore, the elements off the main diagonal can be made arbitrarily small. Since the elements on the main diagonal are the eigenvalues, this result can be expressed as follows:

LEMMA 3.2. *Given any matrix C and any positive number ε, there exists a matrix N such that*

$$|NCN^{-1}| < r + \varepsilon,$$

where $r = r(C)$ denotes the absolute value of the largest eigenvalue of C.

The norm of the adjoint is the same as the norm of the matrix itself, and the norm of a product is not greater than the product of the norms. So we get

$$N^{*-1} C^* N^* N C N^{-1} \leqq (r+\varepsilon)^2 I.$$

Multiply by N^* on the left, N on the right; this preserves the inequality. Using the abbreviation

$$N^* N = M$$

(note that M is positive definite) we can write the resulting inequality in the form

$$C^* M C \leqq (r+\varepsilon)^2 M.$$

We now take C to be $C(x; t; \xi)$. According to assumption ii) of Theorem 3.1,

[2] Since both $K_{l,m}$ and C_j tend to zero fast enough, this hermitean form is completely continuous.

[3] Since $\sum |w_j|^2$ diverges, we have to perform a passage to the limit; this however is easy since $K_{l,m}$ and C_j tend to zero sufficiently rapidly.

for $\xi \neq 0, r(C)$ is less than one; so for each x and $\xi \neq 0$ there exists a positive definite matrix M such that

$$C^*MC < M.$$

By continuity this inequality holds in some neighborhood of the point in question. According to assumption iii) such an inequality (with the $\leqq$ sign) holds in the neighborhood of $\xi = 0$ with $M = I$. We apply now a device due to Friedrichs[4]. By the Heine-Borel theorem, a finite number of these neighborhoods cover any compact portion of the Cartesian product of ξ, x, t-space. Construct an infinitely differentiable partition of unity: $\sum p_j \equiv 1$ such that each p_j is non-negative and so that the support of each p_j is contained in some neighborhood of the above covering. Then for each j there is an inequality

$$C^*M_jC \leqq M_j$$

which holds on the support of p_j. Multiply this inequality by p_j and sum with respect to j; we get

$$C^*MC \leqq M,$$

where M denotes $\sum p_jM_j$. This inequality holds everywhere and M is positive definite everywhere. This completes the proof of Lemma 3.1.

This proof, and thus Lemma 3.1, is valid in any number of dimensions.

We shall indicate now the changes necessary in the proof of Theorem 3.1 when $I-C^*C$ is no longer non-negative but $M-C^*MC$ is, for some appropriate M. We shall again prove that the norm of the operator S_0 is bounded by $1+O(\Delta)$, no longer with respect to the L_2 norm but with respect to an M norm, denoted by $||u||_M$ and defined as follows:

$$||u||_M^2 = \int_0^\Delta \sum u^*(y+h\Delta)M_{h-k}(k\Delta)u(y+k\Delta)dy;$$

here $M_p(x)$ is the p-th Fourier coefficient of $M(x;\ \xi)$. As the first step we show

LEMMA 3.3. *The M norm and the I norm are equivalent.*

Proof: We consider first the case when M is independent of x. Then the M norm can be written as

$$\iint u^*(x)m(x-y)u(y)\,dx\,dy,$$

where

$$m(z) = \sum_p M_p\delta((z-p)\Delta).$$

[4] Personal communication.

is the Fourier inverse of $M(\xi)$. The integral with respect to y is a convolution; so we can write the double integral as

$$\int u^*(m * u)dx.$$

Let $U(\xi)$ denote the Fourier transform of $u(x)$; the Fourier transform of m is $M(\Delta\xi)$. Thus by Parseval's identity we can write the integral above as

$$\int U^*(\xi)M(\Delta\xi)U(\xi)d\xi.$$

Since M is positive and continuous, it is bounded from above and below by positive constants. This shows that the M norm of u is equivalent to $||U||$ which, by Parseval's identity, is equal to $||u||$.

To extend this result to the case of variable coefficients we write u as $\sum u_j$, where the support of u_j are small disjoint intervals: this breakup of u is chosen once and for all, independently of Δ. Then

$$||u||_M^2 = \sum (u_i, u_j)_M.$$

For $i \neq j$ the scalar product (u_i, u_j) tends to zero as Δ tends to zero since the Fourier coefficients of M tend to zero fast. Next we set

$$(u_j, u_j)_M = (u_j, u_j)_{M_j} + \text{leftover term},$$

where $M_j(\xi) = M(x_j, \xi)$, x_j a point in the support of u_j. Since M is a continuous function of x and since its Fourier coefficients tend to zero sufficiently rapidly, the leftover term is less than $\varepsilon||u||^2$, ε small, if the support of u_j is small enough and Δ is small enough. We have already proved Lemma 3.3 for M_j; summing we obtain it for M.

From the explicit form for the M norm, when M is independent of x, we can deduce

COROLLARY I. *As Δ tends to zero, $||u||_M$ tends to $||u||$.*

COROLLARY II.

$$||u||_M - ||u||_{M'} \leq d(M, M')||u||,$$

where

$$d(M, M') = \max_{\xi} |M(\xi) - M'(\xi)|.$$

By the technique described above, one can extend these results to M which depends on x.

We proceed now in the proof of Theorem **3.1** as before; we find functions D_k and weights m_k such that

$$C^*MC + \sum m_k D_k^* D_k = M.$$

We define the kernel K as

$$K(x;\xi,\eta) = C^*(x;\xi)M(x;\xi)C(x;\eta) + \sum m_k(x)D_k^*(x;\xi)D(x;\eta).$$

As before, we obtain an identity and an inequality for K which can be expressed in terms of the Fourier coefficients of K and C. This can be used to obtain a quadratic inequality between $\{u_k\}$ and $\{v_k\}$; this time there will be leftover terms on both sides but, as before, these will be $O(\Delta)||u||^2$ and $O(\Delta)||v||^2$, respectively. So we deduce that

$$||S||_M \leqq 1+O(\Delta).$$

In the time independent case we proceed as before. If S depends on t, the amplification factor C and M also depend on t; but since the time dependence of M is sufficiently smooth, we have by Corollary II

$$||u||_{M(t\pm\Delta)} \leqq ||u||_{M(t)} + O(\Delta)||u||.$$

Using this inequality and the inequality

$$||S(t)||_{M(t)} \leqq 1+O(\Delta)$$

derived before we can show recursively that the norm of the operator product

$$S(1)S(1-\Delta)\cdots S(\Delta)S(0)$$

is uniformly bounded for all Δ. This completes the proof of Theorem 3.1.

It should be pointed out that in order to prove the stability of difference approximations to parabolic equations estimates of the following kind are needed:

$$||S_\Delta|| \leqq 1+O(\Delta^2),$$

where S is an operator as before; its characteristic form satisfies the additional condition

$$C_x(x;0) = 0.$$

To derive such estimates by the techniques of this section the difference $K(k\Delta)-K(r\Delta)$ in (3.14) has to be approximated by two terms of the Taylor series, and the relation

$$K_x(x;0,\eta) = 0$$

must be used.

4. Stability of Difference Equations With Variable Coefficients in Several Space Variables

The method of the previous section yields

THEOREM 4.1. *A difference equation is stable if in addition to conditions* i) *and* ii) *of Theorem* 3.1 *the following condition is satisfied:*

iv) *For every x and for $|\xi|$ small enough, $I-C^*C$ can be written as a linear combination of squares with positive weights, i.e. in the form*

$$I-C^*C = \sum_k D_k^* W_k D_k, \tag{4.1}$$

where D_k are smooth matrix valued functions of ξ, W_k are non-negative matrices independent of ξ, and both D and W depend Lipschitz continuously on x.

Clearly, if condition iv) is fulfilled, then $I-C^*C$ must be non-negative. On the other hand, if the coefficients C_j of S are non-negative, then we do have a representation of the form (4.1) with

$$D_k = C(\xi)-e^{ik\xi}, \quad W_k = C_k,$$

as may be easily verified using the definition of C,

$$C(\xi) = \sum C_k e^{ik\xi},$$

and the identity

$$C(0) = \sum C_k = I.$$

In this case $K \equiv C(\eta-\xi)$. Of course this application of Theorem 4.1 is merely a paraphrase of Friedrichs theorem on the stability of difference equations with non-negative coefficients.

If $C^*C \equiv I$, then Theorem 4.1 is trivially applicable. In this case the method of proof yields more than an inequality; we obtain—moduli leftover terms—an energy identity. Such an energy identity has been used by Courant, Friedrichs and Levy in their pioneering study [2] of the symmetric difference equation for the wave equation. For another application see Wendroff [15].

To determine the scope of Theorem 4.1 we have to investigate the class of non-negative functions P which can be represented as linear combinations of squares with non-negative coefficients. Let $Q(\xi)$ be the leading term of P, of order, say, $2q$. Then, if a representation of the form (4.1) holds, the terms of order $2q$ must be the same on both sides of (4.1):

$$Q(\xi) = \sum R_k^*(\xi) W_k R_k(\xi), \tag{4.2}$$

where R_k is the leading term of D_k, of order q.

What non-negative forms can be represented as sums of squares? We shall discuss first the *scalar* case; there it is known that

i) quadratic forms in all variables,
ii) quartic forms in three variables,
iii) forms of any order in two variables,

can be so represented, but that in any other combination of degree and number of variables there exist positive definite forms which cannot be represented as a sum of squares.

Take now one of the three cases above; we shall show that, if Q is positive definite, then not only Q but P itself can be written as a sum of squares. We shall carry out the construction in case iii) [5]:

$$P = Q + \sum_{n=0}^{2q+1} \xi^{2q-n+1}\eta^n R_n(\xi, \eta).$$

We rewrite this as

$$Q + \sum_{m=0}^{q} (\xi^{q-m}\eta^m)^2(\xi R'_m + \eta R''_m);$$

this is equal to

$$Q - \sum \varepsilon(\xi^{q-m}\eta^m)^2 + \sum(\xi^{q-m}\eta^m)^2(\varepsilon + \xi R'_m + \eta R''_m). \tag{4.3}$$

Since Q is assumed to be positive definite, ε can be chosen small enough and positive so that the sum of the first two terms in (4.3) is positive definite; then according to iii) it can be written as a sum of squares. The third term in (4.3) is a sum of products in which the first factor is already a square and the second factor is positive at $\xi = \eta = 0$. Therefore the square root of the second factor is real and smooth for ξ, η small, and depends Lipschitz continuously on x. This completes the proof.

It should be pointed out that if P is a trigonometric polynomial which can be represented as a sum of squares, P need not be representable as a sum of squares of trigonometric polynomials. Thus it may happen that for some difference equations there exists local energy inequalities of the form (3.12), but there exists no energy inequality in which only those terms appear on the right which also appear on the left.

To deal with situations which do not come under one of the above cases we have to introduce a more general norm:

$$||u||_M^2 = \int_{|y| \leq \Delta} u(y+\Delta k) M_{k-h} u(y+\Delta h) dy \tag{4.4}$$

and try to show that the M norm of S is bounded by $1+O(\Delta)$. According to the result described at the end of Section 3, S is so bounded if we can represent

$$P_M = M - C^* MC \tag{4.5}$$

as a sum of squares, where M is the function whose Fourier coefficients are the quantities M_p appearing in formula (4.4). In the scalar case, (4.5) gives

[5] In case i) a theorem of Morse asserts that P is the sum of n squares, n equal to the number of variables.

$$P_M = MP.$$

This shows that if the weight function M is positive at $\xi = 0$, then the leading term of P_M is merely a constant multiple of the leading term of P and so nothing has been gained. Therefore M must be chosen so that $M(0) = 0$; within this class the choice $M = P$ obviously makes P_M a square. The remaining question is: what kind of a norm is (4.4) if M vanishes at $\xi = 0$? In the cases of constant coefficients this can be read off the formula derived for $\|u\|_M$ in the proof of Lemma 3.3:

$$\|u\|_M^2 = \int U^*(\xi) M(\Delta\xi) U(\xi).$$

Clearly

$$\lim_{\Delta \to 0} \Delta^{-2q} \|u\|_M^2 = \int U^* Q(\xi) U(\xi), \tag{4.6}$$

where Q, the leading term of M, is homogeneous of order $2q$. If Q is positive definite, the quantity on the right of (4.6) is equivalent to the q norm of u, i.e. to the quantity

$$\|u\|_q^2 = \sum_{|l|=q} \int |D^l u|^2.$$

Thus the energy inequality which we can derive for these difference equations is the extension of the energy inequality with respect to the q-norm for solutions of the differential equation. In carrying out the details of the estimate we have to assume that the partial derivatives of the coefficients C_j up to order q are bounded.

We turn now to the matrix case; here already the crudest case, i.e. when $I - C^*C$ is quadratic, presents a difficulty. For, it is not known how to write a matrix quadratic form as a sum of squares of linear matrix forms. The writer has verified that for 2×2 matrices and for two variables this can always be done but he (and all his friends whom he has asked) doubt that this is possible in general.

We point out that according to Rosenblatt's theorem all non-negative matrix forms in two variables can be written as a sum of squares.

Bibliography

[1] Artin, E., *Über die Zerlegung definiter Funktionen in Quadrate*, Abh. Math. Sem. Univ. Hamburg, Vol. 5, 1926, pp. 100–115.

[2] Courant, R., Friedrichs, K. O., and Lewy, H., *Über die partiellen Differenzialgleichungen der mathematischen Physik*, Math. Ann., Vol. 100, 1928, pp. 32–74.

[3] Friedrichs, K. O., *Symmetric hyperbolic partial differential equations*, Comm. Pure Appl. Math., Vol. 7, 1954, pp. 345–392.

[4] Grenander, U., and Szego, G., *Toeplitz forms and their applications*, Berkeley, 1958.
[5] John, F., *On the integration of parabolic equations by difference methods*, Comm. Pure Appl. Math., Vol. 5, 1952, pp. 155–211.
[6] Kreiss, H. O., *Über die approximative Lösung von linearen partiellen Differenzialgleichungen mit Hilfe von Differenzengleichungen*, Trans. Roy. Inst. Tech., Stockholm, No. 128, 1958.
[7] Kreiss, H. O., *Über die Lösung des Cauchyproblems*, Acta Math. Vol. 101, 1959, pp. 180–199.
[8] Lax, P. D., and Richtmyer, R. D., *Survey of stability of linear finite difference equations*, Comm. Pure Appl. Math., Vol. 9, 1956, pp. 267–293.
[9] Lax, P. D., *The scope of the energy method*, Bull. Amer. Math. Soc., Vol. 66, 1960, pp. 32–35.
[10] Lees, M., *Energy inequalities for the solution of differential equations*, Trans. Amer. Math. Soc., Vol. 94, 1960, pp. 58–73.
[11] Lees, M., *Von Neumann difference approximation to hyperbolic equations*, Pacific J. Math., Vol. 10, 1960, pp. 213–222.
[12] Lees, M., *A priori estimates for the solutions of difference approximations to parabolic partial differential equations*, Duke J. Math., Vol. 27, 1960, pp. 297–312.
[13] Richtmyer, R. D., *Difference Methods for Initial Value Problems*, Interscience Tracts in Pure and Appl. Math., Vol. 4, Interscience Publishers, New York, 1957.
[14] Rosenblatt, M., *A multidimensional prediction problem*, Ark. Mat., Vol. 3, 1958, pp. 407–424.
[15] Wendroff, B., *On central difference equations for hyperbolic systems*, J. Soc. Indust. Appl. Math., Vol. 8, 1960, pp. 549–555.

Received February, 1961.

The Computation of Discontinuous Solutions of Linear Hyperbolic Equations*

MICHAEL S. MOCK
Rutgers University
AND
PETER D. LAX
Courant Institute

Let L be a linear hyperbolic operator in any number of variables with C^∞ coefficients. As is well known, a solution u of $Lu = 0$ which has C^∞ initial data is C^∞ for all time. Let L_h be a difference approximation to L that is stable, and accurate of order ν. Denote by U the solution of $L_h U = 0$ whose initial values agree with those of u on the lattice points, h denoting the mesh width of the lattice. According to the basic theory of difference approximations, at all times t at which U is available, and in any fixed range $0 \leqq t \leqq T$,

$$|u(t) - U(t)| = O(h^\nu). \tag{1}$$

Consider piecewise C^∞ initial data whose discontinuities occur across C^∞ surfaces. It is known that solutions u of $Lu = 0$ with such initial data are themselves piecewise C^∞, and their discontinuities occur across characteristic surfaces issuing from the discontinuity surface of the initial data. What happens when such a solution is approximated by a solution of $L_h U = 0$? Does U differ from u by $O(h^\nu)$ in those regions where u is C^∞, or have the discontinuities hopelessly polluted the approximate solutions even at smooth regions between discontinuities? In a recent paper [2], Majda and Osher have shown that, for second order accurate schemes applied to hyperbolic equations in one space variables and with constant coefficients, U differs from u in smooth regions by $O(h^2)$, provided that at lattice points of discontinuity the initial value of U is taken as the average of the values of u on the two sides of the discontinuity. In this note we show that, for a scheme of any order ν, the *moments* of U approximate those of u with accuracy $O(h^\nu)$, provided that the initial data of U are *pre-processed* appropriately near the discontinuities; this is true for equations in any number of variables, and with variable coefficients. We show further that we can, by *post-processing* the approximate solution, recover the exact solution, as well as its derivatives,

* This work was supported in part by the U.S. Department of Energy Contract EY-76-C-02-3077*000 at the Courant Mathematics and Computing Laboratory, New York University.

Communications on Pure and Applied Mathematics, Vol. XXXI, 423–430 (1978)

0010-3640/78/0031-0423$01.00

with an error $O(h^{\nu-\delta})$, δ as small as we wish, at all points, no matter how close to the discontinuity. The idea of comparing the moments of a discontinuous solution with the moments of its approximation comes from [3]; there it is used also in studying discontinuous solutions of nonlinear conservation laws.

We need the following quadrature result, which goes back to the 18-th century, see [1]:

LEMMA 1. *Let f be any C^∞ function on R_+, with bounded support. Given any positive integer ν, there exists a quadrature formula accurate of order ν of the form*

$$\int_0^\infty f(x)\,dx = h\sum_0^\infty w_j f(jh) + O(h^\nu)\,, \tag{2}$$

where the weights w_j depend on ν, but

$$w_j = 1 \quad \text{for} \quad j \geqq \nu\,. \tag{3}$$

EXAMPLE. For $\nu = 4$ we have

$$w_0 = \tfrac{3}{8}\,,\quad w_1 = \tfrac{7}{6}\,,\quad w_2 = \tfrac{23}{24}\,,\quad w_j = 1 \quad \text{for} \quad j > 2\,.$$

Using (2) twice we get this

COROLLARY. *Let f be a piecewise C^∞ function, the discontinuity occurring at $x = 0$. Let ν be any integer; then*

$$\int_{-\infty}^\infty f(x)\,dx = h\sum w'_j f(jh) + O(h^\nu)\,, \tag{2$'$}$$

where

$$w'_0 = 2w_0\,,\quad w'_j = w_{|j|} \quad \text{for} \quad j \neq 0\,, \tag{4}$$

and

$$f(0) = \frac{f(0-)+f(0+)}{2}\,.$$

Similar formulas can be obtained in case the discontinuity of f occurs between the mesh points, say at $x = h\theta$, $|\theta| < 1$, and for the integrals of discontinuous functions of several variables whose discontinuities occur along smooth surfaces.

Let L be a first-order hyperbolic matrix operator

(5) $$L=\partial_t+\sum A_j\,\partial_j+B\,, \qquad \partial_j=\frac{\partial}{\partial x_j};$$

the coefficients A_j and B are C^∞ functions of x and t; for simplicity we take them, and all solutions, to be real.

Denote the L_2 scalar product with respect to the x variables by

(6) $$(u,v)=\int u(x)\cdot v(x)\,dx\,.$$

Denote by L^* the adjoint of L; suppose u and v satisfy

(7) $$Lu=0\,, \qquad L^*v=0\,,$$

and one of them vanishes for $|x|$ large. Then, by Green's formula in the slab $0\leqq t\leqq T$,

(8) $$(u(T),v(T))=(u(0),v(0))\,.$$

Let L_h be a two-level, forward difference approximation:

(9) $$L_h=D_t^+ +\sum S_jT^j\,, \qquad T^j \text{ translation by } jh\,,$$

Let U be a lattice function that satisfies $L_hU=0$, i.e.,

(10) $$U_k^{n+1}=\sum C_jU_{k-j}^n\,, \qquad C_j=C_j^n(k)\,.$$

Let V be another lattice function; multiplying (10) by V_k^{n+1} and summing gives

(11) $$\begin{aligned}\sum_k U_k^{n+1}\cdot V_k^{n+1}&=\sum_{k,j}C_jU_{k-j}^n\cdot V_k^{n+1}\\ &=\sum_k U_k^n\cdot\sum_j C_j^*(k+j)V_{k+j}^{n+1}\,.\end{aligned}$$

Suppose V satisfies the two-level backward adjoint equation

$$L_h^*V=0$$

defined to be

(10)* $$V_k^n=\sum C_j^*(k+j)V_{k+j}^{n+1}\,.$$

Then (11) can be rewritten as

$$(U^{n+1}, V^{n+1})_h = (U^n, V^n)_h, \tag{12}$$

where $(\ ,\ )_h$ denotes the lattice scalar product

$$(U, V)_h = h \sum U_k \cdot V_k. \tag{13}$$

Note that, for u, v in C_0^∞,

$$(u, v)_h = (u, v) + O(h^\nu) \tag{14}$$

for any ν.

We draw two conclusions from (12):

(a) If L_h approximates L to order ν, then L_h^* approximates L^* also to order ν.

(b) For all N,

$$(U^N, V^N) = (U^0, V^0). \tag{15}$$

Proof: (a) L_h approximates L to order ν if

$$U^1 = u(h) + O(h^{\nu+1}), \tag{16}$$

where $u(t)$ is any C^∞ solution of $Lu = 0$ and U the solution $L_h U = 0$ with the same initial data as u:

$$U^0 = u(0). \tag{17}$$

Let $v(t)$ be any C^∞ solution of $L^* v = 0$, V a solution of $L_h^* V = 0$ which equals v at $t = h$:

$$V^1 = v(h). \tag{18}$$

Then by (10)* we can determine V^0:

$$V^0 = v(0) + h^{p+1} E + O(h^{p+2}), \tag{19}$$

where p is the order to which L_h^* approximates L^*, and E is the leading truncation error. It follows from (16) and (18) using (14), that

$$(U^1, V^1)_h = (u(h), v(h)) + O(h^{\nu+1}); \tag{20}$$

on the other hand, by (17) and (19),

$$(U^0, V^0)_h = (u(0), v(0)) + h^{p+1}(u(0), E(0)) + O(h^{p+2}). \tag{21}$$

By (12) the left sides of (20) and (21) are equal, and by (8) the first terms on the right are equal; from this we conclude that $p = \nu$.

(b) Relation (15) follows by summing (12) from 0 to $N-1$. It follows that if L_h is stable, so is L_h^*.

We turn now to discontinuous solutions; for the sake of simplicity we take the number of space dimensions to be 1. Suppose u is a piecewise C^∞ solution of $Lu = 0$ whose initial data contain a single discontinuity at, say, $x = 0$. Let U be a solution of $L_h U = 0$ whose initial data are related to those of u as follows:

$$\begin{aligned} U^0(jh) &= w_j' u(jh, 0), \qquad j \neq 0, \\ U^0(0) &= w_0'(u(0-, 0) + u(0+, 0))/2. \end{aligned} \tag{22}$$

where w_j' are the weights (4) entering formula (2)′. Let $\phi(x)$ be an arbitrary C_0^∞ function; denote by v the solution of

$$L^* v = 0, \qquad v(x, T) = \phi(x). \tag{23}$$

Let V be the solution of

$$L_h^* V = 0, \qquad V^N(jh) = \phi(jh), \tag{23$_h$}$$

where N is a time step corresponding to $t = T$. Since v is C^∞ and since L_h^* is accurate of order ν, it follows that if L_h^* is stable, then

$$V^0(jh) = v(jh, 0) + O(h^\nu). \tag{24}$$

We take, in the quadrature formula (2)′, $f = u(x, 0)\, v(x, 0)$; this is a piecewise C^∞ function, with a discontinuity at $x = 0$, therefore

$$(u(0), v(0)) = \int f(x)\, dx = h \sum w_j' u(jh, 0)\, v(jh, 0) + O(h^\nu). \tag{25}$$

Using the definition (22) of $U_h^0(jh)$, the error estimate (24) and the definition (13) of scalar products for lattice functions, we see from (25) that

$$(u(0), v(0)) = (U^0, V^0)_h + O(h^\nu). \tag{26}$$

Now using (8) and (15) we deduce from (26) that

$$(u(T), v(T)) = (U^N, V^N)_h + O(h^\nu).$$

According to (23) and $(23)_h$, both $v(T)$ and V^N were chosen to be ϕ; therefore the relation above can be rewritten as

$$(u(T), \phi) = (U^N, \phi)_h + O(h^\nu). \tag{27}$$

We summarize:

THEOREM 1. *Let L_h be a stable two-level difference operator that approximates L to order ν. Let u be a solution of $Lu=0$ whose initial data are piecewise C^∞ with a discontinuity at $x=0$. Let U be the solution of $L_h U=0$ whose initial data are related to those of U by formula (22). Then at any later time T the moments of u and U with any C_0^∞ function ϕ differ by $O(h^\nu)$.*

We show now how to use the weak error estimate (27) to deduce pointwise estimates. To this end we need to know the dependence of the error term in (27) on ϕ; since the error term comes from (24), it is of order of the size of the $(\nu+1)$-st derivative of v. Thus

$$(u(T), \phi) = (U^N, \phi)_h + ch^\nu, \tag{27'}$$

where

$$c = O(|\phi|_{\nu+1}), \tag{28}$$

the bars denoting the maximum norm of $\partial_x^{\nu+1}\phi$; for more space variables more derivatives of ϕ are needed in (28). Let $s(x)$ be an auxiliary function whose support is contained in the interval $(-1, 1)$, satisfying

$$\int s(x)\,dx = 1, \qquad \int x^l s(x)\,dx = 0, \qquad l = 1, \cdots, p-1, \tag{29}$$

p an arbitrary integer. We set

$$\phi(x) = \frac{1}{\varepsilon} s\left(\frac{x-y}{\varepsilon}\right). \tag{30}$$

For any function g that is C^∞ in $(y-\varepsilon, y+\varepsilon)$

$$g(y) = \int g(x)\phi(x)\,dx + O(\varepsilon^p).$$

So if the interval $(y-\varepsilon, y+\varepsilon)$ is free of discontinuities of u at time T,

$$\int u(x, T)\phi(x)\, dx = u(y, T) + O(\varepsilon^p).$$

Comparing this with (27), gives

$$u(y, T) = (U^N, \phi)_h + O(|\phi|_{\nu+1})h^\nu + O(\varepsilon^p). \tag{31}$$

From the definition of ϕ in (30) we see that

$$|\phi|_{\nu+1} = O\left(\frac{1}{\varepsilon^{\nu+2}}\right);$$

thus

$$u(y, T) = (U^N, \phi)_h + O\left(\frac{h^\nu}{\varepsilon^{\nu+2}}\right) + O(\varepsilon^p). \tag{31$'$}$$

We choose ε so that the two error terms are of the same order:

$$\frac{h^\nu}{\varepsilon^{\nu+2}} = \varepsilon^p, \qquad \varepsilon = h^{\nu/(\nu+2+p)}. \tag{32}$$

With this choice of ε,

$$u(y, T) = (U^N, \phi)_h + O(h^{\nu p/(\nu+p+2)}). \tag{33}$$

Taking p large enough we get

THEOREM 2. *Choose ϕ of the form* (30), (29); *formula* (33) *recovers $u(y, T)$ with accuracy as close to order ν as we wish if p is taken large enough.*

The same technique yields this

COROLLARY. *Any derivative of u can be recovered with an accuracy as close to order ν as desired.*

The points (y, T) at which u and its derivatives can be recovered by formula (33) and its analogue are subject to the restriction that the interval $(y-\varepsilon, y+\varepsilon)$ be free of discontinuities of u at time T. If, however, we choose the support of s to lie in $(0, 1)$ or $(-1, 0)$, this can be replaced by the one-sided restriction that u be free of discontinuities at time T in either $(y-\varepsilon, y)$ or $(y, y+\varepsilon)$. This allows one to get accurate pointwise estimates right up to the discontinuity.

Bibliography

[1] Davis, P., and Rabinowitz, P., *Methods of Numerical Integration*, Academic Press, New York, 1974.

[2] Majda, A., and Osher, S., *Propagation of error into regions of smoothness for accurate difference approximations to hyperbolic equations*, Comm. Pure Appl. Math., Vol. 30, 1977, pp. 671–706.

[3] Mock, M. S., *A difference scheme employing fourth order viscosity to enforce an entropy inequality*, Proc. Bat-Sheva Seminar on Finite Elements for Nonlinear Problems, Tel-Aviv University, 1977, to appear.

Received December, 1977.

Accuracy and Resolution in the Computation of Solutions of Linear and Nonlinear Equations

Peter D. Lax

Let L be a linear, hyperbolic partial differential operator, with C^∞ coefficients, say a first order system:

$$L = \partial_t + \sum A_j \partial_j + B \ , \quad \partial_j = \frac{\partial}{\partial x_j} \ , \tag{1}$$

A_j and B being C^∞ matrix valued functions of x and t of order $m \times m$. We shall not define hyperbolicity but merely use the well-known result that for solutions of $Lu = 0$ the value of $u(t)$ depends boundedly on $u(0)$ in the L_2 norm, and consequently in all the Sobolev norms. In particular, if $u(x,0)$ is C^∞, so is $u(x,t)$.

We shall consider approximations to solutions u of $Lu = 0$ by solutions U of difference equations $L_h U = 0$. We take U to be defined at discrete times $0, h, 2h, \ldots$ on a rectangular spatial lattice; the time increment h is taken to be proportional to the meshsize of the lattice; for sake of simplicity we take the meshsize to be equal to h in all directions. Since L is of first order in t, it is natural to take L_h to be two-level, i.e. of the form

$$L_h = D_t^+ + \frac{1}{h} \sum S_j \, T^j \ , \tag{2}$$

D_t^+ being forward divided difference in t of steplength h, and T^j is translation in x by jh; the matrix coefficients S_j are C^∞ functions of x,t and h. As defined by (2), L_h is explicit in the positive t direction, i.e. given U^0 solutions U of $L_h U = 0$ are uniquely determined at $t = h$, and consequently at all positive integer

ISBN 0-12-208360-1

multiples of h, L_h is called stable forward, if in any given range of $t > 0$, i.e. for all $0 < nh \leq T$, U^n depends boundedly on U^0, uniformly for all h.

We say that L_h approximates L of order ν if for every C^∞ solution u of $Lu = 0$,

$$L_h(u) = O(h^\nu) \tag{3}$$

It is a basic result of the theory of difference approximations that if L_h is stable and approximates L of order ν, then for all C^∞ solutions u and for all h

$$|u(t) - U^N| = O(h^\nu) , \quad t = Nh , \tag{4}$$

where $U^0 = u(0)$, $L_h U = 0$. This shows that the higher the order of approximation, the smaller the error, for h small enough.

In many interesting problems we are presented with piecewise C^∞ initial data whose discontinuities occur along C^∞ surfaces. According to the theory of hyperbolic equations, solutions of $Lu = 0$ with such initial data are themselves piecewise C^∞, their discontinuities occurring across characteristic surfaces issuing from the discontinuity surfaces of the initial data. How good are difference approximations to discontinuous solutions? Consider the model equation

$$u_t + a\, u_x = 0 , \tag{5}$$

and the difference equation obtained by replacing ∂_t by forward, ∂_x by backward difference quotients. The resulting equation can be written as

$$U_k^{n+1} = a\, U_{k-1}^n + (1-a)\, U_k^n . \tag{6}$$

It is easy to show that (6) approximates (5) of first order, and that (6) is stable if and only if $0 \leq a \leq 1$. Suppose the initial data are piecewise constants, say

$$u(x,0) = \begin{cases} 0 & \text{for } x < 0 \\ 1 & \text{for } x > 0 \end{cases} , \quad U_k^0 = \begin{cases} 0 & \text{for } k < 0 \\ 1 & \text{for } k \geq 0 \end{cases} . \tag{7}$$

The solution of (5) with initial values (7) is

$$u(x,t) = \begin{cases} 0 & \text{for } x < at \\ 1 & \text{for } x > at . \end{cases} \tag{8}$$

The solution of (6) has a more complicated structure; we indicate it schematically so:

$$U_k^n \simeq \begin{cases} 0 & \text{for } k < an - W(n) \\ 1 & \text{for } k > an + W(n) ; \end{cases} \tag{9}$$

U_k^n changes gradually from near 0 to near 1 as k goes from $an-W(n)$ to $an+W(n)$. The width $2W(n)$ of this transition region is $O(\sqrt{n})$.

What happens when the first order scheme (6) is replaced by one of higher order? Since the accuracy of higher order schemes is due to small truncation error, and since the truncation error is of form $h^\nu E(u)$, where E is a differential operator of order $\nu+1$, it follows that the truncation error will be large around the discontinuity. Here we are in for a surprise; if U is the solution of $L_h U = 0$ with initial data (7), and L_h approximates L of order ν, then schematically U can be described by (9), the width $2W(n)$ of the transition region is however $O\left(n^{\frac{1}{\nu+1}}\right)$; i.e. the higher the order of accuracy, the narrower the transition region. The same, of course, is true for solutions with arbitrary discontinuous initial data.

At points away from the discontinuities the solution is C^∞, so there the truncation error is small; in these regions it is reasonable to use difference approximations of high order accuracy, except for the danger that the large truncation error at the discontinuities propagates into the smooth region. Majda and Osher, [8], have shown that indeed even in smooth regions $|u - U|$ is $O(h)$ in general; they have further shown that this discrepancy can be reduced to $O(h^2)$ by the simple expedient of taking the initial data of U not as in (9), but by taking for the value of U^0 at the point $k = 0$ of discontinuity the arithmetic mean:

$$U_0^0 = \frac{1}{2}\,(u(-0,0) + u(+0,0)) . \tag{10}$$

Mock and Lax have shown in [8] that accuracy of order ν is regained if one defines the initial values of U as

$$U_k^0 = w_k u(kh,0) , \tag{11}$$

where the weights w_k are $=1$ for $|k| \geq \nu$, and are derived from the Gregory-Newton quadrature formula for $|k| < \nu$. It is indicated in [9] what is the appropriate analogue of (11) for discontinuous initial value problems for functions of several space variables. We shall not repeat the derivation in [9]; the basic idea is to look at moments of the solution; i.e. weighted integrals of the form

$$M(t) = \int u(x,t)\, m(x)\, dx \,. \tag{12}$$

It is easy to show that even for discontinuous solutions u, the moment $M(t)$ is a C^∞ function of t, provided that the weight $m(x)$ is C^∞. For if we write L in the form $L = \partial_t - G$, where G is a linear differential operator in the x variables then $L\,u = u_t - Gu = 0$ implies that

$$\partial_t^n u = G^n u \,,$$

so that

$$\partial_t^n M = \int (\partial_t^n u)\, m dx = \int (G^n u) m \, dx = \int u\, G^{*n} m \, dx$$

where G^* is the adjoint of G. This shows that M is C^n for any n. The analysis in [9] shows that if the initial values of U are chosen according to the recipe (11), then the approximate moment

$$M_h^n = h \sum U_k^n m_k \tag{13}$$

differs from the exact moment $M(t)$, $t = nh$, by $O(h^\nu)$.

We turn now to nonlinear hyperbolic equations in conservation form, i.e. systems of equations

$$u_t + \operatorname{div} f(u) = 0 \,, \tag{14}$$

u a vector valued function of x,t, and f a vector valued function of u. For one space variable (14) reads

$$u_t + f(u)_x = 0 \,. \tag{15}$$

According to the theory of hyperbolic conservation laws, see e.g., [6], solutions of systems of the form (14), (15) are in general discontinuous. The discontinuities, called shocks, need not be present in the initial values but arise spontaneously; their speed of propagations is governed by the Rankine-Hugoniot jump relation

$$s = \frac{f^j(u_-) - f^j(u_+)}{u_-^j - u_+^j} \tag{16}$$

where u^j and f^j stand for any one of the components of the vectors u and f, and $u_\pm$, $f_\pm$ denote the values of u and f on either side of the discontinuity.

We show now that, in contrast to the linear case, the moments of discontinuous solutions of (15) are not C^∞, in fact not even C^2. This can be seen by looking at solutions of single conservation laws that at $t < 0$ contain two shocks travelling with speeds s_1 and s_2, which at time $t = 0$ collide at $x = 0$ and coalesce thereafter into a single shock propagating with speed s_3; between the shocks the solution is constant:

For $t < 0$

$$u(x,t) = \begin{cases} a & \text{for} \quad x < s_1 t \\ b & \text{for} \quad s_1 t < x < s_2 t \\ c & \text{for} \quad s_2 t < x \; . \end{cases} \tag{17$_-$}$$

For $t > 0$

$$u(x,t) = \begin{cases} a & \text{for} \quad x < s_3 t \\ \\ c & \text{for} \quad s_3 t < x \end{cases} \tag{17$_+$}$$

Clearly $u(x,t)$ defined by $(17)_\pm$ is continuous, in t, at $t = 0$. In order to satisfy (15), the jump relation (16) must be satisfied at all discontinuities:

$$s_1 = \frac{f(a) - f(b)}{a-b}, \qquad s_2 = \frac{f(b) - f(c)}{b-c}$$

$$s_3 = \frac{f(a) - f(c)}{a-c} \quad . \tag{18}$$

Let m be any C^∞ function, and define M by (12); then, using (15) and integrating by parts

$$M_t = \int u_t \, m \, dx = -\int f_x \, m \, dx = \int f m_x \, dx \; .$$

Using the definition $(17)_{\pm}$, for $t < 0$

$$M_t = \int_{-\infty}^{s_1 t} f(a) m_x dx + \int_{s_1 t}^{s_2 t} f(b) m_x dx + \int_{s_2 t}^{\infty} f(c) m_x dx = \tag{19$_-$}$$

$$= f(a)m(s_1 t) + f(b)[m(s_2 t) - m(s_1 t)] - f(c)m(s_2 t).$$

Similarly, for $t > 0$

$$M_t = f(a)m(s_3 t) - f(c)m(s_3 t) . \tag{19$_+$}$$

We can verify that M_t is continuous at $t = 0$. Differentiating $(19)_{\pm}$ and setting $t = 0$ we get

$$M_{tt}(-0) = s_1 f(a) + (s_2 - s_1) f(b) - s_2 f(c)$$

$$M_{tt}(+0) = s_3 f(a) - s_3 f(c) ;$$

where we assumed that $m_x(0) = 1$. It is easy to verify that, in general, $M_{tt}(-0) \neq M_{tt}(+0)$, even under the restriction (18). This shows that there are intrinsic difficulties in constructing difference schemes that would even yield moments of discontinuous solutions of nonlinear conservation laws with order of accuracy higher than first. An analysis of other difficulties, and a possible partial cure, are contained in a forthcoming article of Michael Mock.

That it is more difficult to construct accurate approximations of discontinuous solutions of nonlinear equations than of linear equations is hardly surprising. We show now that in some respects it is easier to construct them. Replace in equation (15) u_t by forward and f_x by backward difference quotients; we get

$$U_k^{n+1} = U_k^n + f(U_{k-1}^n) - f(U_k^n) . \tag{20}$$

It is easy to show that this difference scheme is stable if $0 \leq \frac{df}{du} \leq 1$. It is strongly indicated by numerical experiments, and has been shown by Jennings rigorously in [5], that if f is concave then the solution of (20) with initial values (7) is for h small described by

$$U_k^n \simeq w((k-sn)h), \tag{21}$$

where $w(x)$ is a function that tends, exponentially, to 0 as $x \to -\infty$, to 1 as $x \to +\infty$,

$$s = f(0) - f(1)$$

is the shock speed of the exact solution of (15) with initial values (7):

$$u(x,t) = \begin{cases} 0 & \text{for} \quad x < st \\ 1 & \text{for} \quad x > st \end{cases} \tag{22}$$

The result of Jennings is in fact considerably more general than this. The important fact is that (21) is a far better approximation to (22) than (9) is to (8); the width of the transition region in (21) from near 0 to near 1 is O(1), whereas in (9) that width is $O(n^{\frac{1}{\nu+1}})$! We give now a possible theoretical explanation why, as evidenced above, it might be easier to compute solutions of nonlinear equations than of linear ones. Consider a set D of initial data of interest; typically such a set might be a unit ball with respect to some Sobolev norm; this norm defines a distance in D. We denote by S the set of solutions at time t of $Lu = 0$ whose initial values belong to D. As remarked at the beginning, the mapping from D to S is bounded with respect to the norm and, since t is reversible, so is its inverse; in many important cases D and S are isometric. Denote now by D_h the projection of data in D onto data defined on a discrete lattice of mesh width h, e.g. by the formula

$$U_j = \frac{1}{h} \int_{(j-1/2)h}^{(j+1/2)h} u(x)\,dx \tag{23}$$

is one space dimension. (23) carries a ball in Sobolev space onto a ball in discrete Sobolev space defined for functions on the mesh. Define S_h to be the set filled out by U_h^n, $nh = t$, where $L_h U = 0$ and U^0 belongs to D_h. To study how good an approximation solutions of $L_h U = 0$ are to solutions of $Lu = 0$ with initial data in D, we have to obtain a uniform estimate of the distance of corresponding points of S and S_h. Before comparing them we apply the mapping (23) to S, obtaining the set S^h of functions on the mesh. The <u>approximation error</u> is defined by

$$\max_{D_h} |s^h - s_h| = \delta \tag{24}$$

where s^h is the projection via (23) of the exact solution corresponding to the initial data u_0 in D, and s_h is the approximate solution with initial data obtained from u_0 by the projection (23).

We show now how to obtain lower bounds for the approximation error δ by using some notions from information theory. We recall the definition of <u>ε-capacity</u> $C(M,\varepsilon)$ of a set M in a metric space:

$C(M,\varepsilon)$ = largest number of points in M whose distance from each other is $\geq \varepsilon$.

A related notion is the <u>ε-entropy</u> $E(M,\varepsilon)$ of M:

$E(M,\varepsilon)$ = smallest number of ε-balls that cover M; the centers of the balls need not belong to M but may lie in a metric extension of M.

Both ε-capacity and ε-entropy measure the amount of information contained in M.

<u>Theorem:</u> Denote by δ the approximation error defined by (24). Then

$$C(S^h, 3\delta) \leq C(S_h,\delta), \tag{25}$$

and

$$E(S^h,2\delta) \leq E(S_h,\delta). \tag{26}$$

<u>Proof:</u> By definition of C there exist $C = C(S^h,3\delta)$ points $s^1,\dots,s^C$ in S^h such that

$$|s^i - s^j| \geq 3\delta, \quad i \neq j, \tag{27}$$

By (24) the corresponding elements $s_1,\dots,s_C$ of S_h satisfy

$$|s^j - s_j| \leq \delta. \tag{28}$$

By the triangle inequality and (27), (28),

$$\begin{aligned} |s_i - s_j| &\geq |s^i - s^j| - |s^i - s_i| - |s^j - s_j|, \\ &\geq \delta. \end{aligned}$$

Thus S_h contains $C(S^h,3\delta)$ points whose distance from each other exceeds δ; this proves (25).

Similarly, by definition of E there exist $E = E(S_h,\delta)$ points $u_1,\dots,u_E$ such that every point s_h of S_h lies within δ of one of the u_j:

$$|s_h - u_j| \leq \delta.$$

By (29), $|s^h - s_h| \leq \delta$, so by the triangle inequality

$$|s^h - u_j| \leq |s^h - s_h| + |s_h - u_j| \leq 2\delta.$$

Thus S^h can be covered by $E(S_h,\delta)$ balls of radius 2δ; this implies (26).

It was remarked at the beginning that if L is a linear hyperbolic operator, the mapping linking $u(0)$ to $u(t)$ is bounded. Since time is reversible, the inverse mapping is likewise bounded. This shows that $C(D,\varepsilon)$ and $C(S,\varepsilon)$ are comparable quantities, as are $E(D,\varepsilon)$ and $E(S,\varepsilon)$. It follows that $C(D_h,\varepsilon)$ and $C(S^h,\varepsilon)$ are likewise comparable. The forward stability of L_h means that the mapping linking D_h to S_h is bounded. But since L_h is generally unstable backwards, it is plausible to deduce (and true generally) that $C(S_h,\varepsilon)$ is very much smaller than $C(D_h,\varepsilon) \simeq C(S^h,\varepsilon)$. It can be shown that, in general, the higher the order of accuracy of L_h , the larger the ε-capacity (and ε-entropy) of S_h . If we have lower bounds for $C(S^h,\varepsilon)$ or $E(S^h,\varepsilon)$ and upper bounds for $C(S_h,\varepsilon)$ or $E(S_h,\varepsilon)$, then using (25) or (26) we can get a lower bound on the approximation error δ .

Roughly speaking we shall say that an approximation method has high resolution if $C(S_h,\varepsilon)$ and $C(S^h,\varepsilon)$ are comparable, and low resolution if $C(S_h,\varepsilon)$ is very much smaller than $C(S^h,\varepsilon)$. We have shown above that a method with low resolution cannot be very accurate; the converse does not follow, i.e. a method with high resolution need not be highly accurate. But at least it furnishes approximations that contain enough information from which a better approximation may be extracted by a post-processing, hopefully at not too high an expense. Even if that isn't so, a method with high resolution is more likely to preserve qualitative features of solutions, such as number of maxima and minima, which in some cases is all we want to know.

We turn now to nonlinear conservation laws; here time is decidedly not reversible; on the contrary, here the mapping relating initial values to values at time $t > 0$ is a compact mapping. For single conservation laws, with f concave or convex, this follows from the explicit formula for solutions given in [5]; for 2×2 systems this follows from the estimates given in [3]; for general systems this compactness remains an intriguing conjecture. It follows from this conjecture that in the nonlinear case $C(S^h, \varepsilon)$ is much smaller than $C(D_h, \varepsilon)$, and therefore the construction of high resolution methods is easier than in the linear case. It is in this sense that approximating solutions of nonlinear initial value problems is easier than approximating solutions of linear ones.

We conclude by observing that Glimm's method, [2], which recently has been explored by Chorin as a practical one, see [1], is one of very high resolution, since it neither creates nor destroys waves. In [4] Glimm and Marchesin developed an accurate version of Glimm's method.

REFERENCES

1. Chorin, A., Random Choice Solution of Hyperbolic Systems, J. Comp. Physics, Vol. 22, 1976, 517-533.
2. Glimm, J., Solutions in the large for nonlinear hyperbolic systems of equations, Comm. Pure Appl. Math. 18 , 1965 , 697-715.
3. Glimm, J., and Lax, P. D., Decay of solutions of nonlinear hyperbolic conservation laws, Mem. Amer. Math. Soc., 101, 1970.
4. Glimm, J., and Marchesin, D., A random numerical scheme for one dimensional fluid flow with high order of accuracy, preprint to appear.
5. Jennings, G., Discrete travelling waves, Comm. Pure Appl. Math., 26, 1973, 25-37.
6. Lax, P. D., Weak solutions of nonlinear hyperbolic equations and their numerical computation, Comm. Pure Appl. Math., 7, 1954, 159-193.
7. Lax, P. D., Hyperbolic systems of conservation laws and the mathematical theory of shock waves, Vol. 11, Regional Conference Series in Appl. Math, 1973, SIAM, Philadelphia.
8. Majda, A. and Osher, S., Propagation of Error into Regions of Smoothness for Accurate Difference Approximations to Hyperbolic Equations, Comm. Pure Appl. Math., Vol. XXX, 1977, 671-706.

9. Mock, M. and Lax, P. D., The Computation of Discontinuous Solutions of Linear Hyperbolic Equations, to appear.

This work was supported in part by the U. S. Department of Energy Contract EY-76-C-02-307700 at the Courant Mathematics and Computing Laboratory, New York University.

Courant Institute of
Mathematical Sciences
New York University
New York, NY 10012

Commentary on Part II

15

The main result of this paper is the equivalence theorem: a difference scheme converges to a solution of a differential equation for all initial data with finite norm if and only if it is stable with respect to that norm. The necessity was stated because in those early days some people thought, mistakenly, that it is the accumulation and amplification of rounding errors that makes an unstable scheme diverge.

The second part of the paper applies this result to a number of cases in which stability is easily established.

P.D. Lax

26

Proving the stability of a difference scheme is the discrete equivalent of deriving an a priori estimate for solutions of a differential equation; the proof in the discrete case is usually more delicate than its continuum counterpart. In both cases, equations with constant coefficients can be handled by Fourier analysis; the harder part is to prove stability in case the scheme is stable if the coefficients are frozen. This is accomplished here under mild additional conditions. The technique used is to express nonnegative matrix-valued functions as sums of squares; hence an abiding interest in sums of squares.

References

[1] Kreiss, H.O. Über die Lösung des Cauchy Problems für linear partielle Differentialgleichungen mit Hilfe von Differenzengleichungen, *Acta Math.* **101** (1959), 179–199.

[2] Lax, P.D.; Wendroff, B. Difference schemes for hyperbolic equations with high order of accuracy, *CPAM* **17** (1964), 381–398.

P.D. Lax

[81]

The problem tackled here is how to use a high-order-accurate difference equation to compute to the same order of accuracy, solutions whose initial data are only piecewise smooth. The answer is that the initial data have to be preprocessed near the discontinuities by using the the Newton–Coates quadrature formulas. The approximate solution has to be postprocessed to attain the desired accuracy pointwise.

Unfortunately, no numerical examples are given.

P.D. Lax

Here too the focus is the accurate computation of solutions of hyperbolic equations that are piecewise smooth. Unfortunately, the device used by Lax–Mock fails in the nonlinear case. This is not surprising; the solution of nonlinear equations ought to be more difficult than that of linear ones. On the other hand, it has been observed that solutions of nonlinear conservation laws that contain shocks are easier to compute than solutions of similar linear initial value problems, for there is less to compute. In this paper an attempt is made to quantify this statement.

Consider a set D of initial data of interest, for instance the unit ball in the L^1 norm. Denote by S the set of solutions at time t of $Lu = 0$ whose initial values belong to D. Project the initial data in D onto piecewise constant functions on a lattice of mesh width h by averaging over intervals of length h; denote this set of piecewise constant functions by D^h. Denote by S^h the solution at time $t = nh$ of the difference equation $L^h U = 0$ whose initial values lie in D^h. The approximation error is the norm of the difference between corresponding points of S and S^h. The maximum of this norm is a measure of the accuracy of the difference equation in question.

We define the resolving power of the difference equation in terms of the entropy and capacity, in the sense of information theory, of the set S^h, and derive comparisons between accuracy and resolving power. The results presented are speculative; recently, Golse and DeLellis succeeded in estimating the entropy of the set of solutions of a single convex conservation law.

References

[1] DeLellis, C.; Golse, F. A quantitative compactness estimate for scalar conservation laws, to appear in CPAM, 2005.

P.D. Lax

PART III

HYPERBOLIC SYSTEMS OF CONSERVATION LAWS

COMMUNICATIONS ON PURE AND APPLIED MATHEMATICS, VOL. VII, 159–193 (1954)

Weak Solutions of Nonlinear Hyperbolic Equations and Their Numerical Computation

By PETER D. LAX
New York University

Introduction

This paper describes a finite difference scheme for the calculation of *time dependent* one-dimensional compressible fluid flows containing *strong shocks.* This method is closely related to one proposed by J. von Neumann (see [12]) and modified more recently by him and R. D. Richtmyer (see [13]), inasmuch as the path of the shock is *not* regarded as an interior boundary.* The novel feature of the method described here is the use of the *conservation form* of the hydrodynamic equations and, to a lesser extent, the particular way of differencing the equations.

Although the method was designed to deal with hydrodynamic problems, it can be used to construct solutions of discontinuous initial value problems for any hyperbolic system of first order nonlinear conservation laws (to be defined below) in any number of space variables. The evidence for the convergence of the method is a number of calculations carried out on high speed computing machines that show every sign of convergence. Although the flows calculated so far all belong to a somewhat special class, I fully believe that the method will reproduce the most general type of flow. The question of *accuracy* of the approximate solution with a given mesh-size, specifically the detrimental effect of *contact discontinuities* on accuracy, is discussed at the end of Section 1.

In addition to the numerical evidence, I succeeded in proving the convergence of the scheme for arbitrary bounded measurable initial data, for the single conservation law

$$u_t - [\log (a + be^{-u})]_x = 0,$$

a and b being arbitrary positive constants (or even functions depending on x and t). The proof, modeled after a procedure of E. Hopf, see [8], will be published in a separate note.

For the discussion of the difference scheme presented here I found it useful to develop the theory of weak solutions of nonlinear conservation laws a little more systematically than customary. The theory and the numerical evidence supporting it is presented in Section 1. Section 2 contains some remarks, plus

**Added in proof*: See also Ludford, Polachek and Seeger, [18], who employ a linear viscosity term but one which is artificially large. The difference scheme employed in [18] is implicit and is solved by iteration. The difference equations are centered in time.

one numerical example, about mixed initial and boundary value problems for conservation laws. Section 3 discusses the manner in which the approximate solutions computed by our difference scheme approach steady state solutions. Section 4 discusses irreversibility and Section 5 the manner of dependence of weak solutions on the initial data. Section 6 describes how the finite difference scheme would be applied to problems with more space variables and in particular to the equations of compressible flow in Eulerian variables.

1. *Theory of Weak Solutions*

We start with the definition of weak solutions applicable to a certain class of nonlinear hyperbolic systems. We take the number of space variables to be one and consider hyperbolic systems of first order equations, i.e. systems of the form

$$U_t + AU_x + B = 0. \tag{1}$$

U here is a column vector of unknown functions, A and B are matrix and vector coefficients depending on x, t and U. We shall consider systems in which the x-derivatives are perfect derivatives, i.e. AU_x is equal to F_x (plus possibly a vector function of x, t, U), F being a (in general nonlinear) vector function of x, t, U. Such equations,

$$U_t + F_x + B = 0, \tag{2}$$

are called *conservation laws.*

U is called a *weak solution* of equation (2) with initial value Φ if the integral relation

$$\iint \{W_t U + W_x F - WB\}\, dx\, dt + \int W(x, 0)\Phi(x)\, dx = 0, \tag{3}$$

obtained by multiplying (1) by W on the left, integrating the resulting equation and integrating by parts, holds for every *test vector* W which has continuous first derivatives and which vanishes outside of some bounded set.

Clearly, a genuine solution is a weak solution, and conversely: a weak solution with continuous first derivatives is a genuine solution. Weak solutions need not be differentiable; if U_1 and U_2 are two genuine solutions of (2) whose domains of definition in the x,t-plane are separated by a smooth curve, the two taken together will constitute a weak solution if and only if the slope τ of the separating curve and the value of U_1 and U_2 along the curve satisfy the condition

$$\frac{1}{\tau}(U_1 - U_2) = F(U_1) - F(U_2).$$

For the conservation laws of mass, momentum and energy, this relation embodies precisely the Rankine-Hugoniot shock conditions.

These facts about weak solutions are basic:

(a) *The class of weak solutions associated with a given system of equations depends on the form in which the equations are written.*

By this we mean: Suppose that new unknowns V can be introduced as certain nonlinear functions of the old:

$$V = H(U), \tag{4}$$

so that when the differential equation (2) is rewritten in terms of V and solved for V_t, each of the resulting equations is again a conservation law

$$V_t + G_x + C = 0. \tag{2'}$$

Weak solutions of (2′) can be defined the same way as before. But if U_0 is a weak solution of (2), $V_0 = H(U_0)$ is in general *not* a weak solution of (2′); in fact, it can be shown (at least in some special cases) that $V_0 = H(U_0)$ is a weak solution of (2′) if and only if U_0 is a genuine solution of (2) ("genuine" here means Lipschitz continuous); this result holds of course only if the transformation (4) is genuinely nonlinear.

This noninvariance is not paradoxical. Whenever the concept of solution is generalized one has to sacrifice some properties of the original notion for the sake of saving others.

This dependence on the form is especially relevant for the equations of unsteady compressible flow which contain *four* conservation laws: conservation of mass, momentum, energy and entropy. Any one of these equations can be deduced from the other three, that is, a genuine solution of any three is necessarily a genuine solution of the fourth. But a weak solution of any three is in general, not a weak solution of the fourth. In physical problems one is looking for weak solutions of the first three equations.

(b) *Weak solutions cannot be obtained as limits of genuine solutions.*

(c) *The initial values do not in general determine a unique weak solution.*

There are many known examples of different weak solutions (even infinitely many) of the same equation with the same initial data.[1] This shows that the initial value problem for weak solutions is not a meaningful one—unless some additional principle is given which selects a unique weak solution for each initial value problem. However, if we believe that our mathematical model does describe an aspect of the physical world, then there is indeed assigned to each initial function a unique weak solution, namely the one that occurs in nature. The problem is to characterize mathematically this physically relevant solution.

---

[1]For example, the functions

$$u(x, t) = \begin{cases} 0 \text{ for } x < 0 \\ x/t \text{ for } 0 < x < t \\ 1 \text{ for } t < x \end{cases} \quad \text{and} \quad u(x, t) = \begin{cases} 0 \text{ for } 2x < t \\ 1 \text{ for } 2x > t \end{cases}$$

are both weak solution of the equation $u_t + (\frac{1}{2}u^2)_x = 0$ with initial value $\varphi(x) = 0$ for $x < 0$, $\varphi(x) = 1$ for $x > 0$.

First of all, we exclude all solutions where entropy of a particle has been decreased. It is not clear, however, whether this insures the uniqueness of the solution of the initial value problem especially if there are several space variables but even in the case of one space variable.** Some additional principle is needed to pick out a unique solution, such as:

(a) The weak solutions occurring in nature are limits of viscous flows.

(b) The weak solutions occurring in nature must be stable.

It is commonly believed that (a) does characterize uniquely the solutions occurring in nature. But whether the same is true of postulate (b) is seriously doubted by some.

We shall describe now in some detail the viscosity method, the notion of stability and the relation of the two.

Enlarge equation (2) by the additional term λU_{xx} on the right, obtaining a nonlinear parabolic system

$$U_t + F_x + B = \lambda U_{xx} . \tag{5}$$

The initial value problem $U(x, 0) = \Phi(x)$ can be solved for a fairly wide class of initial vectors Φ; it is commonly believed that the solution exists for a range of t which is independent of λ, and that if the initial vector is kept fixed and λ taken smaller and smaller, the corresponding solutions $U_\lambda(x, t)$ converge boundedly, almost everywhere in the strip $0 \leq t \leq T$ to a limit $U(x, t)$. This has been demonstrated so far only for a single equation,[2] $u_t + uu_x = 0$, by E. Hopf*** (see [8]), admitting all *bounded measurable* functions as initial values. For the hydrodynamic case only the convergence of *steady state* solutions[3] of the viscous flow equations to steady state weak solutions of the equations of ideal flow have been investigated.[4]

Granting the validity of the conjectures about the parabolic equation it is an easy matter to show, just as Hopf has shown for the equation considered by him, that the limit $U(x, t)$ of $U_\lambda(x, t)$ is a weak solution of equation (2). This follows from the fact that U, being a genuine solution of (5) is a weak solution of (5) as well; i.e. if we multiply (5) by any twice differentiable test vector W and integrate by parts we obtain the integral relation

---

***Added in proof*: Recently, Germain and Bader, see [16], have found an analogue of the entropy condition for the equation $u_t + (u^2/2)_x = 0$ and have succeeded in showing that this and the jump condition characterizes a unique solution of any initial value problem.

[2]The nonlinear parabolic equation $u_t + uu_x = \lambda u_{xx}$ was first considered by Burgers, see [8].

***See also J. H. Cole, [17].

[3]A steady state solution is one that depends only on a particular linear combination of x and t.

[4]See Becker, [1], Thomas [14], Gilbarg [6], Grad [7], and Courant and Friedrichs [2], pp. 134-138.

$$(6) \qquad -\iint \{W_t U_\lambda + W_x F(U_\lambda) - WB\}\, dx\, dt - \int W(x, 0)\Phi(x)\, dx = \lambda \iint W_{xx} U_\lambda\, dx\, dt.$$

Keep Φ and W fixed and let λ tend to zero. The left side of (6) approaches the left side of (3) and the right side of (6) tends to zero. This proves that U satisfies (3) for all twice differentiable test vectors W (and therefore à fortiore for all once differentiable ones too).

Observe that it was crucial for this argument that U be a *strong* limit of the sequence U_λ, i.e. that $\iint | U_\lambda - U |$ over any bounded set of the x,t-plane tend to zero. For if U_λ converges to U in the *weak* sense only, the sequence $F(U_\lambda)$ will converge in the weak sense but *not* to $F(U)$. This phenomenon can be expressed concisely:

A weak limit of weak solutions is not a weak solution unless it is also a strong limit.

A precise statement and proof of this, in the case of a single conservation law, is given in Section 5.

There are a number of different ways of introducing a viscosity term; the way equation (5) does it is perhaps the simplest, although when applied to the equations of compressible flow it does not exactly correspond to the action of viscosity or heat conduction. All these methods are expected to produce the same weak solution in the limit. The aim of this paper is to describe a different type of limiting process, a finite difference scheme and to show—by mathematical reasoning, plausibility arguments and numerical evidence—that this scheme furnishes the experimentally observed flows.

The difference scheme is as follows:

Replace the space derivatives F_x by the *symmetric* difference quotients $[F(U(x + \Delta x, t)) - F(U(x - \Delta x, t))]/2\Delta x$, the time derivative U_t by the *forward* difference quotient $[U(x, t + \Delta t) - \overline{U}(x, t)]/\Delta t$, where $\overline{U}(x, t)$ is an abbreviation for the average[5] of U at $(x + \Delta x, t)$ and $(x - \Delta x, t)$. If $U(x,0)$ is known, we can determine $U(x, t)$ for all values of t which are integer multiples of Δt; in particular the value of $U(x, 0)$ at the lattice point $x = m\Delta x$, $m = 0, \pm 2, \cdots$ determines $U(x, t)$ at all points of the staggered lattice $t = n\Delta t$, $n = 0, 1, 2 \cdots$; $x = m\Delta x$, $m + (-1)^n = 0, \pm 2, \pm 4, \cdots$.

Denote by Δ a particular choice of mesh width, $\Delta = (\Delta t, \Delta x)$ and by U_Δ the corresponding solution of the finite difference scheme with initial value Φ. $U_\Delta(x, t)$ is defined for all values of t which are integer multiples of Δt, and for sake of convenience we might as well put $U_\Delta(x, t)$ equal to $U_\Delta(x, \nu\Delta t)$ for $\nu\Delta t \le t < (\nu + 1)\Delta t$.

I conjecture that if Δt and Δx tend to zero so that the classical Courant-Friedrichs-Lewy stability criterion is observed; i.e., the domain of dependence of

[5]If we replace $\overline{U}(x, t)$ by $U(x, t)$, the resulting finite difference scheme is unconditionally unstable.

a point with respect to the differential equation always stays within the domain of dependence of this point with respect to the difference equation, then the functions $U_\Delta(x, t)$ remain uniformly bounded and converge almost everywhere to a limit $U(x, t)$ for a wide class of initial vectors. Should this conjecture be true, it is an easy matter to show that the limit $U(x, t)$ is a weak solution of the original system of equations. The proof goes the same way as in the viscosity method, summation by parts replacing integration by parts. And, just as in that case, it is crucial for the argument that U be a *strong* limit of U_Δ ; if U were a weak limit only, it would in general not be a weak solution.

Very likely this conjecture holds only for systems of conservation laws which, in addition to being hyperbolic, satisfy some additional condition, possibly in the large, see e.g. Weyl's paper [15].

I succeeded in proving this conjecture for the single equation

$$u_t - [\log (a + be^{-u})]_x = 0.$$

Details of the proof will be published in a separate note; I would like to mention however that in the course of proving the convergence, I obtained a fairly explicit formula relating $u(x, t)$ to its initial values $\varphi(x)$. This formula is strikingly similar to the formula Hopf obtained for the solution of the equation $u_t + (u^2/2)_x = 0$ by the linear viscosity method and suggests this

CONJECTURE: Both the linear viscosity method and the finite difference scheme described above, when applied to any single homogeneous first order conservation law

$$u_t + [f(u)]_x = 0, \qquad f'' < 0,$$

and arbitrary bounded measurable initial data $u(x, 0) = \varphi(x)$, converge to the same limit $u(x, t)$ given by the explicit formula

$$u(x, t) = g\left(\frac{x - y_0}{t}\right)$$

where $y_0 = y_0(x, t)$ is that value of y which maximizes

$$\int_0^y \varphi(s)\, ds + tG\left(\frac{x - y}{t}\right).$$

The function $g(s)$ is defined as the inverse of $f'(u) = df(u)/du$, i.e. $f'(g(s)) = s$; $G(s)$ is defined as the integral of $g(s)$: $dG/ds = g$. The maximum problem defining y_0 has a unique solution for almost all x and t so that $u(x, t)$ is well defined for almost all x and t. A similar result holds if "f" is positive.

Next we turn to the principle of stability formulated as a

CONJECTURE: *Among all functions $U = S(\Phi)$, which assign to each vector Φ a weak solution U with initial value Φ there exists one which is continuous in some reasonable topology and it is the only one which is continuous in any reasonable topology.*[6]

[6]It is understood that whenever an initial value problem has a genuine solution, $S(\Phi)$ has to coincide with it and that the semigroup property is satisfied by the solutions furnished by S.

This conjecture is a bit vague since it does not define reasonable topology, nor does it specify the domain of initial vectors Φ. Maybe further investigations will indicate what the proper choice for these undefined objects is.

The first part of the conjecture, asserting the existence of a continuous assignment, is the classical principle that in a physical problem the solution must depend continuously on the data.

That there is only *one* way of assigning $U = S(\Phi)$ continuously was proposed merely as a possible explanation of why various types of apparently different limiting procedures, such as the various viscosity methods and finite difference schemes, pick out the same weak solution. Because of the systematic nature of these procedures one would expect that the solutions picked out by each of them do depend continuously on the initial data. (This can be proved rigorously in simple cases; see Section 5.) If, therefore, there is only one way of assigning a weak solution continuously to each initial vector, it would follow that these various limiting procedures do lead to the same weak solution. On the other hand, the reason why these various limiting procedures lead to the same result could easily be some type of dissipative mechanism common to all of them. In this connection it should be pointed out that if the parabolic equation (5) is approximated by the standard finite difference scheme (centered space derivatives, forward time derivative) and if Δx, Δt and λ are let to approach zero *simultaneously*, keeping $\Delta t/\Delta x$ constant and λ equal to $(\Delta x)^2/2\Delta t$, we arrive at the finite difference scheme proposed in this paper. The von Neumann-Richtmyer method is explicitly based on such a simultaneous carrying out of two limiting procedures.

We present now a partial list of experimental calculations carried out for the equation $u_t + (u^2/2)_x = 0$, $u_t + (u^3/3)_x = 0$, and the hydrodynamic equations both in Eulerian and Lagrangean coordinates.

In the first group of calculations the initial data were picked to have a constant value Φ_i for x negative and another constant value Φ_f for x positive. This choice of the initial data leads to homogeneous problems; i.e., the solution depends only on x/t. This has the advantage that advancing in time in the finite difference scheme has the same effect as refining the mesh. The computational plan was to keep on grinding out time cycles until it becomes evident whether the method converges, diverges to infinity or oscillates. All calculations performed so far converged and quite rapidly at that. Unless stated otherwise, the difference scheme used was the one described before.

(1) *Equation*: $u_t + (u^2/2)_x = 0.$

Initial function: $\varphi_i = 1, \quad \varphi_f = 0.$

$$\Delta t/\Delta x = 1;$$

$$\textit{Exact solution: } u(x, t) = \begin{cases} 1 \text{ for } x/t < 1/2 \\ 0 \text{ for } x/t > 1/2. \end{cases}$$

TABLE I

$n = 44$		$n = 48$	
k	u	k	u
17	1.00000	19	1.00000
19	.99548	21	.99548
21	.76818	23	.76817
23	.21061	25	.21061
25	.02343	27	.02344
27	.00210	29	.00210
29	.00018	31	.00018

TABLE II

Rarefraction wave, $n = 48$, $\Delta t/\Delta x = 1$

k	u
47	.92695
45	.88187
43	.83994
41	.79948
39	.7599
37	.7209
35	.6825
33	.6444
31	.6066
29	.5692
27	.5321
25	.4954
23	.4590
21	.4229
19	.3873
17	.3523
15	.3177
13	.2839
11	.2509
9	.2189
7	.1881
5	.1587
3	.1310
1	.1055
−1	.0823
−3	.0619
−5	.0447
−7	.0306
−9	.0198
−11	.0120

Values of the solution after 44 and 48 time cycles are listed in Table I. The values of u for k less than 17 are equal to one within five figures, those for k greater than 31 are zero within five figures. Notice that there is a very rapid transition from $u = 1$ to $u = 0$ around $k = 22$ for $n = 44$, and $k = 24$ for $n = 48$; this corresponds closely to the exact solution which has a sharp discontinuity along the line $x = 2t$.

(2) *Equation*: same.

Initial function: $\varphi_l = 0, \quad \varphi_r = 1.$

$$\Delta t/\Delta x = 1.$$

Exact solution: $$u(x, t) = \begin{cases} 0 & \text{for } x < 0, \\ x/t & \text{for } 0 < x < t, \\ 1 & \text{for } t < x. \end{cases}$$

Values of the calculated solution after 48 time cycles are listed in Table II, and plotted in Figure 1. The dashed line in the graph is the exact solution.

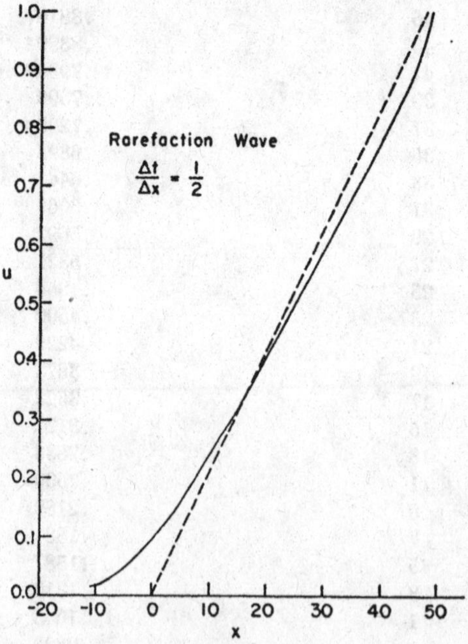

FIGURE 1

(3) *Equation*: same.

Initial function: same as in (2).

$$\Delta t/\Delta x = 1/2.$$

TABLE III

Rarefaction wave, $n = 63$, $\Delta t/\Delta x = 1/2$

k	u
64	.8553
60	.8170
56	.7758
52	.7322
48	.6869
44	.6405
40	.5933
36	.5457
32	.4980
28	.4506
24	.4039
20	.3580
16	.3134
12	.2704
8	.2295
4	.1911
0	.1555
−4	.1234
−8	.0949
−12	.0706
−16	.0505
−20	.0345
−24	.0225
−28	.0139

Values of calculated solution after 63 time cycle are listed in Table III and plotted in Figure 2. The dashed line in the graph is the exact solution.

(4) *Equation*: $u_t + (u^3/3)_x = 0$.

Initial function: $\varphi_l = 1, \quad \varphi_r = 0.$

$$\Delta t/\Delta x = 1.$$

Difference scheme: $u_k^{n+1} = u_k^n - 1/3[(u_k^n)^3 - (u_{k-1}^n)^3](\Delta t/\Delta x)$.

Exact solution: $u(x, t) = \begin{cases} 1 & \text{for } x/t < 1/3, \\ 0 & \text{for } x/t > 1/3. \end{cases}$

Values of the calculated solution after 25, 26 and 27 time cycles are listed in Table IV. The values of u, for x to the left of the range listed, are equal to one within five figures, to the right of the range listed, equal to zero within five

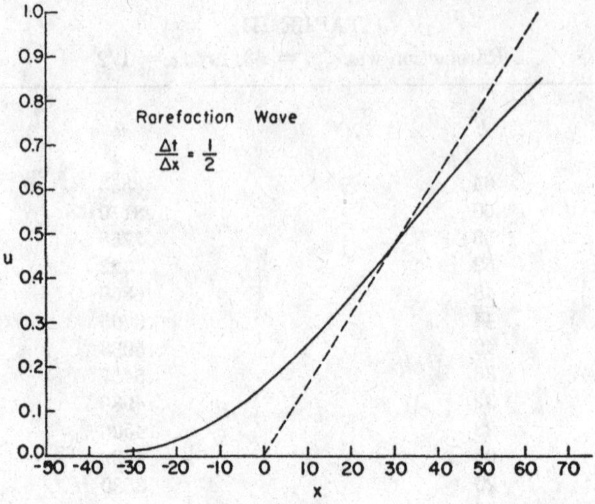

FIGURE 2

figures. The transition from $u = 1$ to $u = 0$ takes place very closely to where theory predicts it at $k = n/3$.

(5) *Equation*: Hydrodynamic equations in Eulerian form

$$\rho_t + (u\rho)_x = 0,$$

$$(u\rho)_t + (u^2\rho + p)_x = 0,$$

$$\left(\rho e + \frac{u^2\rho}{2}\right)_t + \left(\rho e u + \frac{u^3\rho}{2} + up\right)_x = 0.$$

Here ρ, u, p and e denote density, velocity, pressure and internal energy per unit mass. The equation of state expresses e as function of p and ρ; e.g. for an ideal gas e is $p/\rho(\gamma - 1)$.

In the computations the quantities ρ, $u\rho = m$ and $\rho(e + u^2/2) = E$, mass,

TABLE IV

$n = 25$		$n = 26$		$n = 27$	
k	u	k	u	k	u
6	1.00000				
7	.89330	7	.99989	7	1.00000
8	.10670	8	.65283	8	.98902
9	.00000	9	.01396	9	.34391
10		10	.00000	10	.00041

momentum and total energy per unit volume were used as dependent variables. In terms of these the equations are:

$$\rho_t + m_x = 0,$$

$$m_t + \left[(\gamma - 1)E + \frac{3-\gamma}{2}\frac{m^2}{\rho}\right]_x = 0,$$

$$E_t + \left[\gamma \frac{m}{\rho} E - \frac{\gamma - 1}{2}\frac{m^3}{\rho^2}\right]_x = 0.$$

Initial Vector: $u_i = 2, \quad p_i = 50, \quad \rho_i = 50,$

$u_f = 0, \quad p_f = 0, \quad \rho_f = 10,$

$\gamma = 1.5,$

$\dfrac{\Delta t}{\Delta x} = 0.25.$

TABLE V

$n = 49$

ρ	$u/8$	p	k
4998	2500	4997	−6
4996	2501	4994	−4
4993	2501	4990	−2
4988	2503	4983	0
4981	2505	4972	2
4971	2508	4958	4
4959	2512	4941	6
4946	2515	4922	8
4931	2519	4904	10
4915	2522	4890	12
4898	2524	4882	14
4878	2524	4880	16
4852	2523	4887	18
4818	2520	4898	20
4777	2517	4913	22
4734	2513	4926	24
4695	2508	4928	26
4648	2492	4880	28
4497	2431	4620	30
3919	2196	3672	32
2622	1524	1753	34
1424	0490	0289	36
1047	0042	0012	38
1002	0001	0000	40
1000	0000	0000	42

$$\textit{Exact solution: } U = \begin{cases} U_i & \text{for } x/t > 2.5, \\ U_f & \text{for } x/t > 2.5. \end{cases}$$

Values of the calculated solution after 49 time cycles are given in Table V. There is a rapid transition from one state to the other around $k = 31$; this gives

TABLE VI

$n = 99$

ρ	$u/8$	p	k
4996	2501	4995	0
4994	2501	4992	2
4992	2502	4988	4
4988	2503	4983	6
4984	2504	4976	8
4979	2506	4969	10
4973	2508	4959	12
4966	2510	4949	14
4959	2512	4939	16
4952	2514	4929	18
4946	2516	4921	20
4942	2517	4914	22
4938	2518	4910	24
4937	2518	4909	26
4936	2518	4911	28
4937	2517	4915	30
4937	2515	4922	32
4935	2514	4931	34
4931	2512	4940	36
4923	2510	4949	38
4910	2508	4957	40
4893	2507	4965	42
4872	2505	4972	44
4850	2504	4977	46
4829	2503	4981	48
4814	2502	4984	50
4806	2502	4987	52
4809	2501	4987	54
4819	2499	4983	56
4824	2493	4956	58
4777	2469	4840	60
4508	2369	4388	62
3646	2027	3094	64
2208	1207	1150	66
1250	0278	0131	68
1024	0019	0005	70
1001	0000	0000	72
1000	0000	0000	74

124/49 = 2.48 for the speed of propagation of discontinuity, in pretty good agreement with the theoretically calculated value of 2.5.

Values of the calculated solution after 99 time cycles are given in Table VI. There is a rapid transition from one state to another around $k = 62$, giving as speed of propagation 248/99 = 2.50.

(6) *Equation*: Hydrodynamic equations in Eulerian form.

Initial vector: $u_i = 1, \quad p_i = 50, \quad \rho_i = 50,$

$$u_f = 0, \quad p_i = 0, \quad \rho_f = 10,$$

$$\gamma = 1.5,$$

$$\frac{\Delta t}{\Delta x} = 0.25.$$

TABLE VII

$n = 49$

ρ	$u/8$	p	k
4995	1251	4992	−20
4989	1253	4983	−18
4977	1256	4966	−16
4957	1263	4936	−14
4924	1273	4887	−12
4874	1288	4814	−10
4804	1309	4713	−8
4712	1338	4581	−6
4598	1373	4421	−4
4464	1416	4237	−2
4314	1464	4034	0
4154	1517	3822	2
3990	1572	3609	4
3831	1628	3405	6
3685	1682	3218	8
3561	1731	3055	10
3469	1772	2921	12
3415	1804	2816	14
3405	1824	2733	16
3431	1828	2655	18
3461	1806	2539	20
3412	1731	2302	22
3138	1554	1835	24
2541	1212	1127	26
1795	0704	0444	28
1270	0244	0095	30
1062	0047	0012	32
1010	0006	0001	34
1001	0000	0000	36
1000	0000	0000	38

TABLE VIII

$n = 99, \gamma = 1.5, \Delta t/\Delta x = .25$

ρ	$u/8$	p	k
4990	1252	4986	−28
4984	1254	4977	−26
4975	1257	4963	−24
4961	1261	4943	−22
4942	1267	4914	−20
4917	1275	4876	−18
4883	1285	4827	−16
4841	1298	4765	−14
4790	1314	4691	−12
4730	1333	4604	−10
4660	1355	4504	−8
4582	1380	4393	−6
4496	1408	4272	−4
4403	1438	4144	−2
4306	1470	4011	0
4205	1504	3874	2
4102	1539	3737	4
3999	1575	3602	6
3898	1610	3471	8
3800	1645	3346	10
3708	1679	3229	12
3623	1711	3122	14
3546	1740	3026	16
3480	1766	2943	18
3426	1789	2873	20
3384	1808	2817	22
3357	1822	2772	24
3346	1834	2739	26
3352	1841	2716	28
3378	1847	2700	30
3425	1849	2691	32
3495	1851	2685	34
3586	1850	2680	36
3695	1847	2671	38
3809	1838	2646	40
3898	1817	2583	42
3907	1769	2437	44
3735	1666	2134	46
3273	1461	1613	48
2525	1103	0932	50
1747	0618	0350	52
1253	0211	0075	54
1061	0043	0010	56
1012	0006	0001	58
1002	0000	0000	60
1000	0000	0000	62

$$\textit{Exact solution: } U = \begin{cases} U_i & \text{for } x/t < -0.225, \\ \text{rarefaction wave} & \text{for } -0.225 < x/t < 1.47, \\ U_s & \text{for } 1.47 < x/t < 1.84, \\ U_f & \text{for } 1.84 < x/t. \end{cases}$$

The value of U_s is: $u_s = 1.47$, $p_s = 27.1$, $\rho_s = 50$.

Results of the calculation after 49 time cycles are given in Table VII. There is a rapid transition around $k = 22$, corresponding to shock speed $88/49 = 1.79$, in fairly good agreement with the theoretically calculated shock speed of 1.84.

TABLE IX

$n = 49$, $\gamma = 2$

ρ	$u/8$	p	k
4996	2501	9986	−16
4992	2503	9970	−14
4985	2507	9942	−12
4973	2513	9895	−10
4955	2522	9823	−8
4929	2535	9720	−6
4894	2552	9584	−4
4850	2574	9415	−2
4798	2601	9216	0
4740	2630	8997	2
4678	2661	8768	4
4615	2693	8541	6
4556	2723	8327	8
4501	2751	8138	10
4453	2774	7981	12
4411	2792	7860	14
4372	2805	7773	16
4328	2814	7717	18
4272	2819	7683	20
4193	2822	7667	22
4082	2823	7660	24
3936	2823	7659	26
3762	2823	7657	28
3573	2824	7655	30
3393	2824	7652	32
3241	2824	7654	34
3130	2823	7656	36
3064	2824	7679	38
3004	2805	7573	40
3064	2882	7971	42
1741	1376	2197	44
1007	0007	0002	46
1000	0000	0000	48

Results of the calculation after 99 time cycles are given in Table VIII; the transition here occurs around $k = 46$ which gives a shock speed $184/99 = 1.86$, in good agreement with the theoretically calculated value.

In Table VIII, u and p appear to be fairly constant for awhile behind the shock, the value of u being $(0.184 \pm 0.001) \times 8 = 1.47 \pm 0.01$, the value of p around 27 ± 0.3. These are fairly close to $u_s = 1.47$ and $p_s = 27.1$, in spite of the fact that the value of ρ in this range differs considerably from ρ_s.

(7) *Equation*: Hydrodynamical equations in Eulerian form.

$$\textit{Initial vector}:\ u_i = 2,\quad p_i = 100,\quad \rho_i = 50,$$
$$u_f = 0,\quad p_f = 0,\quad \rho_f = 10,$$
$$\gamma = 2,$$
$$\frac{\Delta t}{\Delta x} = 0.25.$$

$$\textit{Exact solution}:\ U = \begin{cases} U_i & \text{for } x < 0, \\ \text{rarefaction wave} & \text{for } 0 < x/t < 2.26, \\ U_s & \text{for } 2.26 < x/t < 3.40, \\ U_f & \text{for } 3.40 < x/t. \end{cases}$$

The value of $\hat{U}_s$ is: $u_s = 2.26$, $p_s = 76.5$, $\rho_s = 30$.

Results of the calculation after 49, respectively 99 steps are listed in Tables IX and X; u and p appear to be fairly constant behind the shock, the value of u being $8 \times 0.2824 = 2.26$, p equal to 76.5. This is very close to $u_s = 2.26$, $p_s = 76.5$.

Remark: The value of $\Delta t/\Delta x = 0.25$ is *larger* than its maximum permissible value according to the Courant-Fredrichs-Lewy theory at $p = 76.6$, $\rho = 30$. The indicated instability is indeed beginning to show around the shock front.

(8) *Equation*: Hydrodynamic equations in Lagrange mass variables

$$V_t - u_\xi = 0,$$
$$u_t + p_\xi = 0,$$
$$(e + \tfrac{1}{2}u^2)_t + (up)_\xi = 0.$$

V here denotes specific volume; if we introduce V, u and $E = e + \frac{1}{2}u^2$ as new unknowns (volume, momentum and energy per unit mass) the equations read

$$V_t - u_\xi = 0,$$
$$u_t + \left[(\gamma - 1)\frac{E - \frac{1}{2}u^2}{V}\right]_\xi = 0,$$
$$E_t + \left[(\gamma - 1)\frac{uE - \frac{1}{2}u^3}{V}\right]_\xi = 0.$$

TABLE X

$n = 99, \gamma = 2, \Delta t/\Delta x = .25$

ρ	$u/8$	p	k
4997	2501	9989	−24
4995	2502	9982	−22
4992	2503	9970	−20
4988	2505	9952	−18
4981	2509	9927	−16
4972	2513	9891	−14
4960	2519	9844	−12
4945	2527	9783	−10
4926	2536	9708	−8
4903	2548	9617	−6
4875	2562	9511	−4
4844	2577	9391	−2
4810	2595	9259	0
4772	2614	9117	2
4733	2634	8968	4
4692	2655	8816	6
4651	2676	8664	8
4611	2697	8515	10
4572	2717	8373	12
4535	2736	8242	14
4502	2753	8122	16
4472	2769	8017	18
4447	2782	7927	20
4425	2793	7853	22
4408	2802	7794	24
4395	2809	7749	26
4385	2814	7716	28
4377	2817	7692	30
4371	2820	7677	32
4365	2821	7667	34
4358	2822	7661	36
4349	2823	7657	38
4334	2823	7656	40
4312	2823	7655	42
4280	2823	7654	44
4234	2823	7654	46
4172	2823	7654	48
4091	2823	7654	50
3993	2823	7653	52
3878	2823	7653	54
3751	2824	7652	56
3619	2824	7652	58
3489	2824	7651	60
3368	2824	7651	62
3263	2824	7651	64
3178	2824	7650	66

TABLE X—*Continued*

$n = 99, \gamma = 2, \Delta t/\Delta x = .25$

ρ	$u/8$	p	k
3115	2825	7656	68
3066	2820	7628	70
3061	2841	7723	72
2981	2778	7342	74
3175	2949	8228	76
2734	2510	5891	78
3497	3137	9270	80
1993	1515	2139	82
3892	3514	2160	84
1853	1480	2426	86
1035	0045	0020	88
1000	0000	0000	90

Initial vector: $V_i = 1, \quad u_i = 4, \quad p_i = 8,$

$$V_f = 3, \quad u_f = 0, \quad p_f = 0,$$

$$\gamma = 2,$$

$$\frac{\Delta t}{\Delta x} = 0.25.$$

Exact solution: $$U = \begin{cases} U_i & \text{for } \xi/t < 2, \\ U_f & \text{for } 2 < \xi/t. \end{cases}$$

REMARK: The initial position $\xi = 0$ of the separation line of the two states was at $k = 100$. The tabulated results after 52, respectively 104 time cycles are listed in Tables XI and XII; the index k is listed in the leftmost column; the distance of the two consecutive lattice points is $2\Delta\xi$.

There is a rapid transition from the initial to the final state taking place around $k = 113$ for $n = 52$, and around $k = 126$ for $n = 104$, corresponding exactly to a shock speed of 2.

The experimental calculations presented have shown, I believe, the convergence of the method.

A large number of further shock calculations in Lagrange variables varying the initial state U_i and the value of $\Delta t/\Delta x$ were carried out by L. Baumhoff at Los Alamos. These calculations approximated the theoretically expected solutions very accurately with the exception noted further on and indicated that calculations carried out by this numerical scheme have these general features:

(i) *The width of the transition across a shock depends on the magnitude of* $\Delta t/\Delta x$; *it is narrowest if* $\Delta t/\Delta x$ *is taken as large as possible.*

TABLE XI

$n = 52$

k	V	E	u	p
076	1000	1600	4001	7993
077	1002	1602	4010	7959
078	1004	1603	4016	7935
079	1005	1604	4019	7920
080	1005	1604	4022	7911
081	1005	1604	4022	7908
082	1005	1604	4022	7909
083	1005	1604	4021	7913
084	1005	1604	4020	7918
085	1004	1603	4018	7925
086	1004	1603	4016	7932
087	1003	1603	4015	7939
088	1003	1602	4013	7946
089	1003	1602	4011	7953
090	1003	1602	4009	7960
091	1004	1603	4008	7966
092	1006	1605	4007	7971
093	1009	1607	4005	7976
094	1015	1612	4004	7980
095	1023	1618	4003	7984
096	1032	1625	4003	7987
097	1042	1634	4002	7989
098	1052	1642	4001	7992
099	1060	1648	4001	7994
100	1063	1650	4000	7995
101	1062	1650	4000	7996
102	1057	1645	4000	7997
103	1047	1638	4000	7998
104	1036	1629	3999	7998
105	1025	1620	3999	7999
106	1016	1612	3999	7999
107	1008	1606	3999	7999
108	1003	1602	3999	7999
109	1000	1600	3999	7999
110	0999	1599	3999	7999
111	1017	1556	3927	7717
112	1375	0931	2755	4011
113	2234	0190	0880	0680
114	2793	0020	0157	0068
115	2960	0001	0021	0006
116	2993	0000	0002	0000
117	3000	0000	0000	0000

TABLE XII

$n = 104$

k	V	E	u	p
47	1000	1600	4000	8000
56	1005	1604	4022	7912
77	1000	1600	4000	8000
100	1045	1636	4000	8000
114	1000	1600	4000	8000
123	1000	1600	4000	8000
124	1018	1557	3928	7716
125	1376	932	2755	4011
126	2235	191	881	680
127	2793	20	157	61
128	2961	2	21	5
129	2994	0	2	0
130	3000	0	0	0

The values of V, E, u, p vary monotonically between the lattice points $k = 47, 56, 77, 100, 114, 123$.

(ii) *The values of u and p converge faster than the values of ρ and V.*

The statement on the width of the shock is illustrated most strikingly in example 6, Table IX, where $\Delta t/\Delta x$ exceeds its maximum permissible value; here the shock transition takes place across one space interval. In fact, as J. Calkin pointed out, if U_i and U_f can be connected by a single shock, i.e. if $s[U_f - U_i]$ is equal to $F(U_f) - F(U_i)$, s being the shockspeed then, if we take $\Delta x/\Delta t$ to be exactly the shockspeed s, the numerically calculated solution is just $U_k^n = U_i$ for $k \leq sn$, $U_k^n = U_f$ for $k > sn$. Of course in hydrodynamics the character of the function F is such that the shock speed is always subsonic with respect to the state behind it, so that choosing $\Delta x/\Delta t$ to be s is necessarily an unstable mesh ratio. This is in sharp contrast to the linear case where s is the reciprocal of the slope of the characteristic curve involved and so $\Delta x/\Delta t = s$ is precisely the largest permissible mesh ratio.

Inaccuracies in the values of ρ and V are particularly noticeable if there are contact discontinuities present, i.e. lines across which u and p are continuous but not ρ; the reason for this is that contact discontinuities, in contrast to shocks, are very much like discontinuities of solutions of *linear*[7] hyperbolic equations and these, when calculated by the difference scheme described before, spread like $\sqrt{n}$. That is, after n time cycles the width of the contact discontinuity (measured, say, from the point where U is within 1% of U_i to the point where U is within 1% of U_f) is $O(\sqrt{n})$. One may check this by an explicit calculation

[7]In the same sense that solutions with contact discontinuities can be obtained as limits of continuous solutions; for proof of this fact, see a forthcoming paper of the author.

in a simple case, e.g. for the equation $u_t + u_x = 0$. In contrast, the width of shocks remains constant.

Of course whether or not there is any spreading of contact discontinuities depends on the difference scheme used and the magnitude of Δx and Δt. For instance, if we use the difference scheme described before for the equation $u_t + u_x = 0$ and choose $\Delta t/\Delta x = 1$, the solutions of the difference equations just propagate the initial data, unaltered in shape, at unit speed. There are other examples of equations and difference schemes where spreading does not occur. In all these examples the line of discontinuity passes through lattice points, and very likely this is a *necessary* condition on difference schemes for *linear* equations to avoid the spreading of initial discontinuities. This indicates that the spreading of contact discontinuities in difference calculations in *Eulerian* coordinates cannot be prevented unless one uses variable size space intervals which is tantamount to introducing Lagrange coordinates. In Lagrange coordinates a contact discontinuity is a straight line parallel to the t-axis so its path is known in advance. It is indeed possible to write down various schemes which, at least formally, maintain sharp contact discontinuities. The stability, effectiveness and accuracy of such schemes is being investigated by the author and L. Baumhoff, and by J. Calkin and N. Metropolis (who regard them from a somewhat different point of view, see a forthcoming report).

2. *Mixed Initial and Boundary Value Problems*

Let D be a domain bounded by an initial interval (say $-\infty < x \leq x_0, t = 0$) and a boundary curve (say $x = x_0, t \geq 0$) issuing from x_0 into the upper half-plane. In a mixed initial and boundary value problem for a system of equations we prescribe not only the values of all the unknowns (i.e. of all components of U) on the initial interval, but we also prescribe a certain number (say r) of relations between the unknowns along the boundary curve. The problem is to determine the solution in the domain D; this problem has a unique solution if the number r of relations prescribed is equal to the number of those characteristics issuing from x_0 which enter the domain D.

The theory of mixed initial and boundary value problems for *genuine solutions* of linear and nonlinear equations is in a fairly satisfactory state; the corresponding theory for *weak solutions* of nonlinear conservation equations is less well known. Here again one would expect to find, just as in the case of the pure initial value problem, new features which are not present in either the linear theory or the theory of genuine solutions.

Since most hydrodynamic problems arising in physics are mixed problems, there is a great need for effective finite difference schemes to compute their solutions. In this paper I would like to report on a method for handling a particularly simple case: reflection of a gas from a rigid wall (i.e. $u(x_0, t) = 0$). This boundary condition can be taken into account in the finite difference scheme simply by putting the value of u at the first lattice point *beyond* the wall equal to the negative of its value at the last lattice point before the wall, while the values of p, V and E (Lagrange) or p, ρ and E (Euler) are transferred unchanged. I hope to be able to report in the near future on numerical schemes for handling more general boundary conditions.

Calculations were carried out in Lagrange coordinates using the following initial data: $U = U_i$ for all lattice points to the left of $k = 155$, $U = U_f$ for all lattice points between $k = 155$ and $k = 199$; the rigid wall is located at $k = 199$.

U_i and U_f were chosen as in example 8: $u_i = 4$, $V_i = 1$, $p_i = 8$, $u_f = 0$, $V_f = 3$, $p_f = 0$; γ was chosen as 2.

The exact solution is as follows: the shock propagates at the speed 2 until it reaches the wall, then it is reflected and propagated back at the speed of 8; the state of the gas U_r behind the reflected shock is $u_r = 0$, $V_r = 0.5$, $p_r = 40$. The original choice of $\Delta t/\Delta x$ was 0.25, which is the maximum permissible value for $p = 8$ and $V = 1$, and which is much too high at $p = 40$ and $V = 0.5$. Accordingly, $\Delta t/\Delta x$ was changed to 0.07 after 102 time cycles: this ratio is pretty close to the maximum permissible value of $\Delta t/\Delta x$ in the compressed state U_r. Calculations were carried on for an additional 650 time cycles. Taking Δx to be one, this corresponds to a total passage of time of $102 \times 0.25 + 65 \times 0.07 = 71$ units. During this time the shock will have reached the wall and will be reflected 216 units. Since meshpoints are $2\Delta x = 2$ units apart, this would locate the reflected shock at $k = 91$.

TABLE XIII

k	V	u	ϕ
88	0.9989	3.995	8.017
89	.9980	3.992	8.031
90	.9942	3.976	8.094
91	.9792	3.911	8.365
92	.9272	3.653	9.591
93	.7967	2.836	13.96
94	.6187	1.334	25.48
95	.5221	.292	36.41
96	.5039	.071	39.09
97	.5022	.050	39.36
98	.5019	.047	39.40

Table XIII lists the data around $k = 91$ after 752 time cycles. These calculations indicate that the reflected shock is located between $k = 91$ and 96, in pretty good agreement with theory.

The values of u beyond $k = 98$ decrease monotonically to $u = 0$ at $k = 199$; those of p increase monotonically to $p = 39.86$ at $k = 199$. The value of V decreases monotonically to $V = 0.5005$ which is reached at $k = 118$; from then

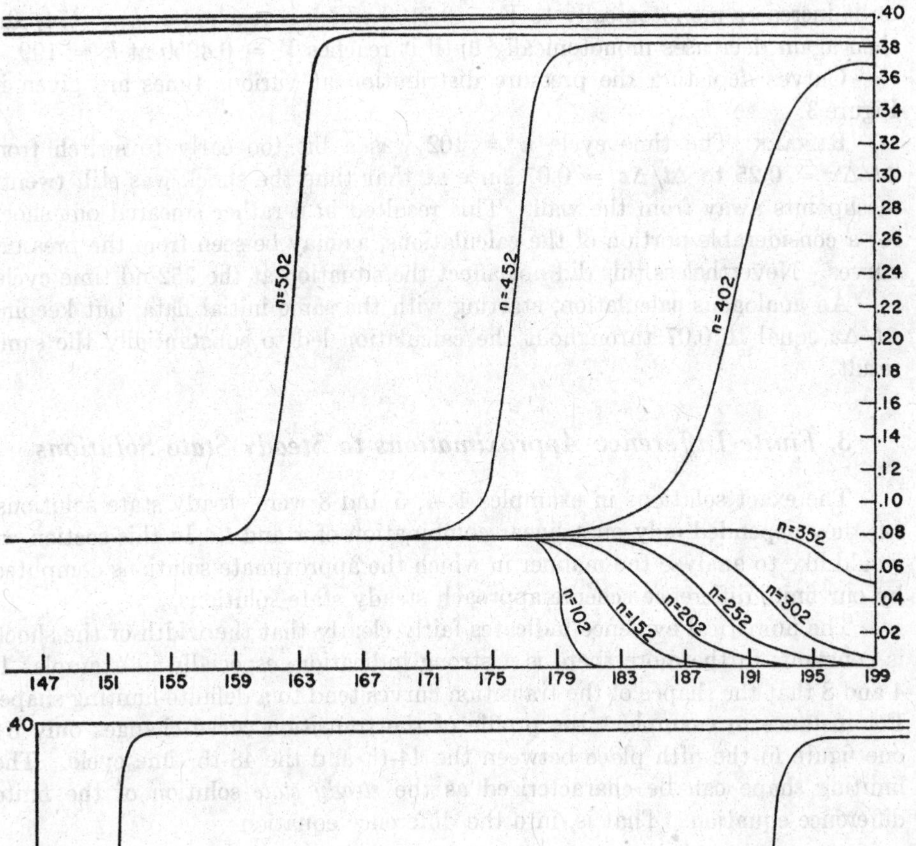
.40
.38
.36
.34
.32
.30
.28
.26
.24
.22
.20
.18
.16
.14
.12
.10
.08
.06
.04
.02
n=502
n=452
n=402
n=352
n=302
n=252
n=202
n=152
n=102
147
151
155
159
163
167
171
175
179
183
187
191
195
199
.40
.30
.20
.10
n=752
n=602
91
95
99
103
107
111
115
119
123
127
131
135
139
143

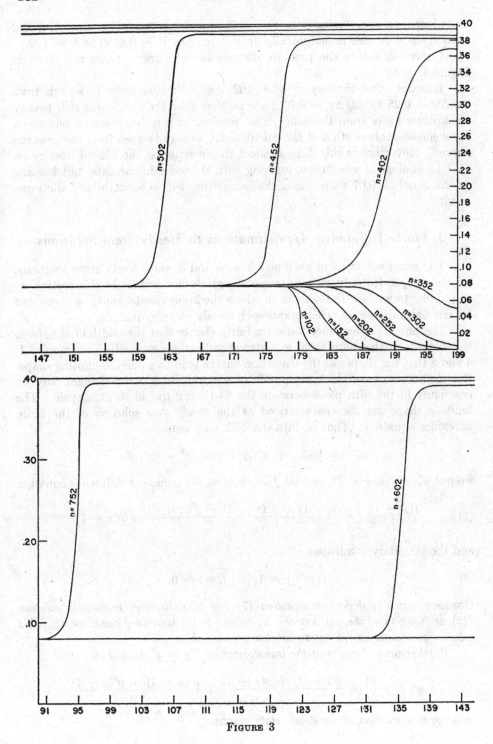

FIGURE 3

on it increases monotonically to $V = 0.5109$ which is reached at $k = 174$, and then again decreases monotonically until it reaches $V = 0.4950$ at $k = 199$.

Curves depicting the pressure distribution at various times are given in Figure 3.

REMARK: The time cycle $n = 102$ was a bit too early to switch from $\Delta t/\Delta x = 0.25$ to $\Delta t/\Delta x = 0.07$ since at that time the shock was still twenty meshpoints away from the wall. This resulted in a rather smeared out shock for a considerable portion of the calculations, as may be seen from the pressure curves. Nevertheless this did not affect the situation at the 752-nd time cycle.

An analogous calculation, starting with the same initial data, but keeping $\Delta t/\Delta x$ equal to 0.07 throughout the calculation led to substantially the same result.

3. *Finite Difference Approximations to Steady State Solutions*

The exact solutions in examples 1, 4, 5 and 8 were steady state solutions, i.e. they depended only on a linear combination of x and t. In this section we would like to analyze the manner in which the approximate solutions computed by our finite difference scheme approach steady state solutions.

The numerical evidence indicates fairly clearly that the width of the shock is constant; furthermore there is a strong indication, especially in examples 1, 4 and 8 that the shapes of the transition curves tend to a definite limiting shape. For instance in example 1 the profile of the transition curve changes only by one figure in the fifth place between the 44-th and the 48-th time cycle. The limiting shape can be characterized as the *steady state* solution of the finite difference equation. That is, into the difference equation

$$u_k^{n+1} = \tfrac{1}{2}(u_{k+1}^n + u_{k-1}^n) + \tfrac{1}{4}(u_{k-1}^n)^2 - (u_{k+1}^n)^2$$

we put $u_k^n = f(k + n/2)$, so that $f(x)$ satisfies the nonlinear difference equation

$$\frac{f(x-1) + f(x+1)}{2} + \frac{f^2(x-1) - f^2(x+1)}{4} = f(x + \tfrac{1}{2}). \tag{7}$$

and the boundary conditions

$$f(-\infty) = 1, \qquad f(\infty) = 0. \tag{8}$$

CONJECTURE: *The difference equation* (7) *has a continuous, monotonic solution* $f(x)$ *as function of the real variable* x, *subject to the boundary condition* (8); *this solution is unique except for an arbitrary phase shift.*

Furthermore: *Iterates of the transformation* $Tg = g'$ *defined by*

$$g'\left(x + \frac{1}{2}\right) = \frac{g(x-1) + g(x+1)}{2} + \frac{g^2(x-1) - g^2(x+1)}{4}$$

converge to a solution of the steady state equation.

TABLE XIV

$n = 44$		$n = 48$	
k	u	k	u
17	1.00000	19	1.00000
19	.99195	21	.99195
21	.71566	23	.71566
23	.17449	25	.17449
25	.01858	27	.01859
27	.00165	29	.00165
29	.00014	31	.00014

This last statement means: If $g_0(x)$ is any function defined over the odd integers and equal to 1 for x large enough negative, 0 for x large enough positive, and if we denote $T^n g_0$ by $g_n(x)$, then $g_n(x)$ tends uniformly to $f(x + \alpha)$, where $f(x)$ is the steady state solution of (7), (8) (made unique, say, by fixing $f(0)$ to be $\frac{1}{2}$). The phase shift α depends of course on the initial distribution $g_0(x)$.

Observe that $g_n(x)$ is defined only at points congruent $n/2$ modulo 2; consequently the only values of $f(x + \alpha)$ that enter this limiting statement are at points congruent 0 or $\frac{1}{2}$ modulo one. The somewhat exceptional situation in this example (and in examples 4 and 8 as well) arose because $\Delta t/\Delta x$ was chosen commensurable with the speed of propagation of the discontinuity.

The calculations in example 1 verify the conjecture. As further check,

TABLE XV

x	g	x	g
−3.0	0.995	−3.5	0.998
−2.5	.982	−3.0	.992
−2.0	.947	−2.5	.972
−1.5	.877	−2.0	.925
−1.0	.768	−1.5	.840
−0.5	.626	−1.0	.716
0.0	.471	−0.5	.566
0.5	.327	0.0	.412
1.0	.211	0.5	.277
1.5	.128	1.0	.174
2.0	.074	1.5	.104
2.5	.042	2.0	.060
3.0	.023	2.5	.034
3.5	.013	3.0	.019
4.0	.007	3.5	.010
		4.0	.006

calculations were carried out for this initial function: $g_0(x) = 1$ for odd negative x less than minus 1, $g_0(-1) = 0.9$, $g_0(x) = 0$ for positive odd x. The results after 44, respectively 48 time cycles, listed in Table XIV, support the conjecture.

Table XV lists g_{45}, g_{46}, g_{47} and g_{48}, both for original and the second choice of g_0.

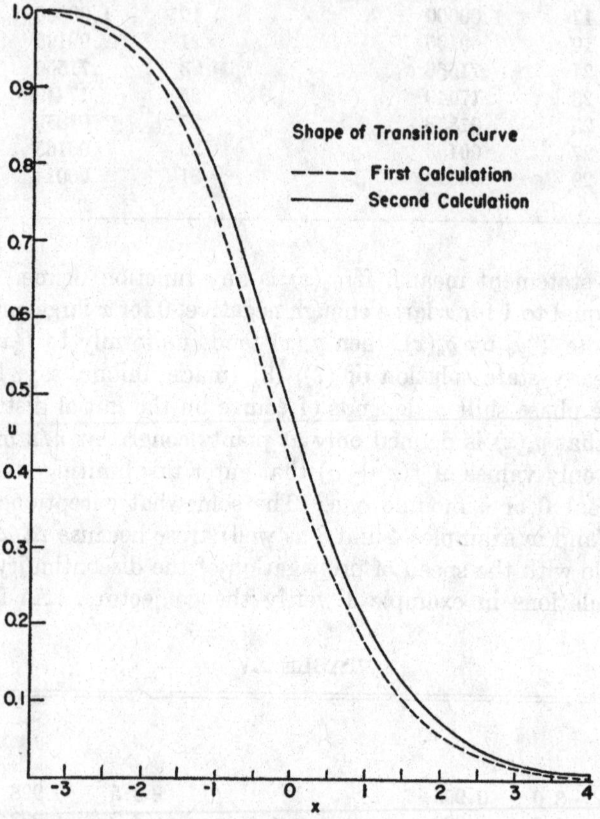

FIGURE 4

Figure 4 shows plots of these values; the two curves do indeed appear to be translates of each other, the phase shift being approximately 0.2.

Similar checks were carried out for examples 4 and 8; in each case the conjecture was verified.

It should be remarked that the nonlinear difference equation (7) has no solution if the boundary values in (8) are switched; i.e., $f(-\infty)$ is required to be zero, $f(\infty)$ one. This expresses the fact that the finite difference method furnishes compression shocks but no rarefaction shocks; I have no rigorous proof of this nonexistence at present. The result, if true, would be an analogue of a known result on viscous flows in steady state (See footnote 4). For completeness I shall present this result for the simplified equation $u_t + uu_x = \lambda u_{xx}$.

Let $u_0(x, t)$ be a steady state solution of this equation; that is, u_0 is a function of $x - ct$ only: $u_0(x, t) = f(x - ct)$. $f(\xi)$ satisfies the nonlinear ordinary differential equation

$$cf' + ff' = \lambda f''.$$

Integrating both sides with respect to ξ gives

$$\lambda f' = cf + \tfrac{1}{2}f^2 + k, \tag{9}$$

k being some constant.

We are interested in those solutions $f(\xi)$ of (9) which exist for all ξ and which approach constant values u_l and u_r as ξ tends to $-\infty$ and $+\infty$ respectively. Clearly if f approaches constant values, there must be two sequences of ξ tending to $+\infty$ and $-\infty$ respectively for which $f'(\xi)$ tends to zero, and so $u_l = f$ and $u_r = f$ are zeros of the quadratic function on the right of (9). But a quadratic function with two real roots is *negative* between its two roots, and so by (9) f' is negative for all ξ; this shows that u_r cannot exceed u_l. Conversely, if u_r is less than u_l, the two states can be connected by a solution of (9); the formula

$$\xi = \frac{2\lambda}{u_l - u_r} \log \frac{u_l - u}{u - u_r}$$

gives the shape of the connecting curve.

We emphasized at the beginning of this paper that the class of weak solutions depends on the form in which the equation is written. I would like to present here an example which shows that if in a conservation equation the exact space derivative is *not* replaced by an exact difference, the limit of the solutions of the difference equations is *not* a weak solution of the original conservation law.

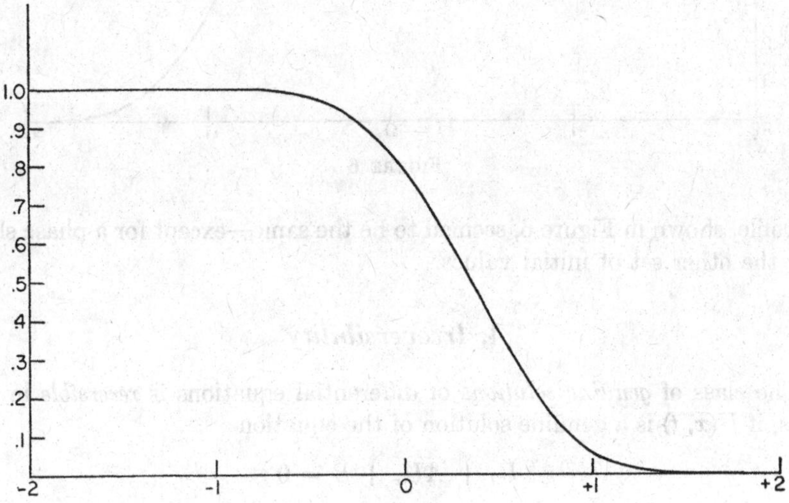

FIGURE 5

The equation in this example is the same as in example 4, $u_t + u^2 u_x = 0$; the following difference scheme was used:

$$u_k^{n+1} = u_k^n - \frac{(u_k^n)^2 + (u_{k-1}^n)^2}{2}(u_k^n - u_{k-1}^n)\frac{\Delta t}{\Delta x};$$

that is, instead of regarding the term $u^2 u_x$ as a perfect derivative, u^2 was regarded as coefficient of u_x. u_k^0 was chosen as 1 for $k < 0$, zero for $k > 0$; $\Delta t/\Delta x$ was chosen as one, and calculations were carried out for 48 time cycles. At all cycles the solution differed from 1 or zero only at two or three space points; the position of the interval of transition seemed to propagate at about the speed of 0.388; this differs appreciably from 0.333, the propagation speed of the exact weak solution. The shape of the transition curves seemed to settle down to a steady state shown in Figure 5.

The calculations were repeated with slightly altered initial data, taking u_1^0 to be 0.5 and u_k^0 for $k \neq 1$ the same as before. The solution again seemed to converge to a steady profile propagating at the speed of 0.388. The shape of

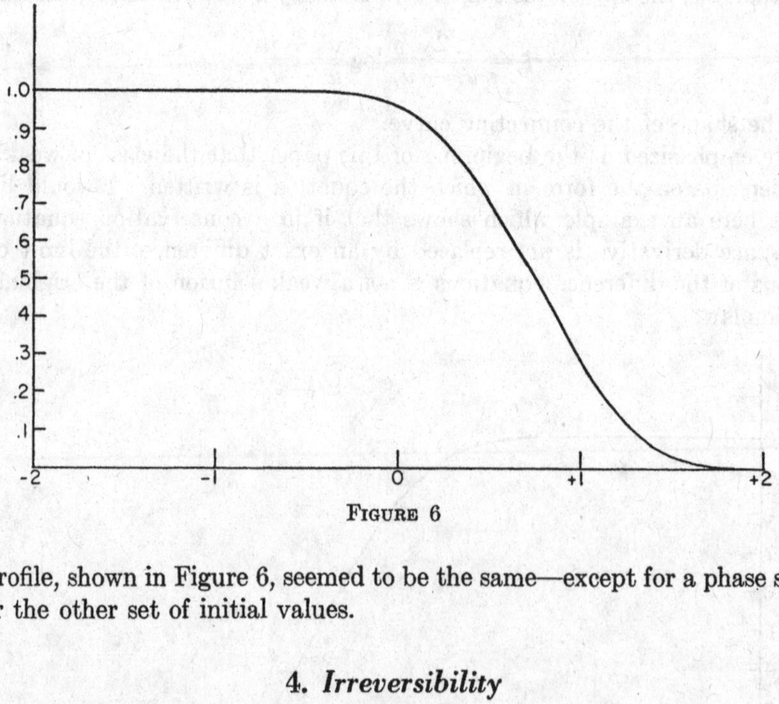

FIGURE 6

the profile, shown in Figure 6, seemed to be the same—except for a phase shift—as for the other set of initial values.

4. *Irreversibility*

The class of *genuine solutions* of differential equations is *reversible* in time; that is, if $U(x, t)$ is a genuine solution of the equation

$$U_t + AU_x + B = 0 \tag{1}$$

(in which we exclude, for sake of simplicity, explicit dependence of the coefficients on time) in the strip $0 < t < T$, then $U'(x, t) = U(x, T - t)$, is, clearly, a genuine solution of

$$U'_t - AU'_x - B = 0.$$

Likewise, if $U(x, t)$ is a *weak solution* of the equation

$$U_t + F_x + B = 0, \tag{2}$$

$U'(x, t) = U(x, T - t)$ is a weak solution of

$$U'_t - F(U')_x - B = 0. \tag{2'}$$

But if $U(x, t)$ is a *physically relevant* solution of (2), U' need not be a physically relevant solution of (2′). This can be summarized thus: *the class of physically relevant weak solutions is irreversible.*

There is nothing paradoxical in this loss of reversibility; it is just another instance of that universal experience that when concepts are generalized, not all properties of the original can be retained. In our case we wanted to generalize the notion of solution of the initial value problem and retain the property of uniqueness and continuous dependence; this turned out to be at the expense of reversibility.

Note that both the viscosity method and the finite difference schemes described in this paper, as well as the von Neumann-Richtmyer scheme, described in [13], distinguish the positive t direstion from the negative one; thus it is not surprising that the class of solutions produced by these methods is irreversible. Conversely, any limiting procedure that can be expected to furnish the physically relevant solutions must be *unsymmetric* in time.

5. *Continuous Dependence of the Solution on the Initial Data*

In the section where the viscosity method and the finite difference scheme were described I asserted that on account of the systematic nature of these processes one would expect the solutions furnished by these methods to depend continuously on the initial data. This is of course merely a plausibility argument, no proof. However, if the number of unknowns is one and the equation is homogeneous in the first derivatives, it is possible to give a rigorous proof of this fact. The most convenient topology for this purpose is a sort of a *weak* topology. More precisely, we introduce the following norm for functions $\varphi(x)$:

$$|\varphi| = \max_{a,b} \left| \int_a^b \varphi(\xi)\, d\xi \right|.$$

We shall refer to this norm as the *weak norm.*

Let u and u' be two solutions of the equation $u_t + f(u)_x = \lambda u_{xx}$ with initial values φ and φ'. We shall show that their deviation, in the sense of the norm introduced, at any positive time t does not exceed twice their deviation at

$t = 0$. This would prove the continuous dependence of solutions of our parabolic equation on the initial data *uniformly* with respect to λ, and thus in the limit $\lambda \to 0$ as well.

Proof: Denote by U and U' integrals of u and u' with respect to x; that is, $U(x, t)$ and $U'(x, t)$ are two functions for which $\partial U/\partial x$ is u, $\partial U'/\partial x$ is u'. Clearly U and U' are determined only modulo a function of t. U and U' satisfy, modulo a function of t, the nonlinear parabolic equation

$$(10) \qquad U_t + f(U_x) = \lambda U_{xx} ;$$

by adding to U and U' a suitable function of t we can achieve that these modified functions satisfy the equation (10) exactly. Imagine such terms already added on; there is still room for an arbitrary additive constant in U and U' which we choose so that $U(0, 0)$ and $U'(0, 0)$ are zero.

Subtract the equations satisfied by U and U' and apply the mean value theorem to the second term on the left; we obtain

$$(U - U')_t + f_u(\bar{u}) \cdot (U - U')_x = \lambda(U - U')_{xx} ,$$

where $\bar{u}$ denotes a value between u and u'. The resulting equation is a linear parabolic equation for $U - U'$; for this the *maximum principle* holds, therefore $| U(x, t) - U'(x, t) |$ is, for t positive,

$$\leq \max_x | U(x, 0) - U' \ (x, 0) | = \max_x \left| \int_0^x u(\xi, 0) - u'(\xi, 0) \, d\xi \right| .$$

This last quantity is clearly $\leq | u(x, 0) - u'(x, 0) | = | \varphi - \varphi' |$. Since $\int_a^b (u(x, t) - u'(x, t)) \, dx$ is equal to $U(b, t) - U'(b, t) - [U(a, t) - U'(a, t)]$, the maximum of this quantity with respect to a and b, $| u(t) - u'(t) |$, does not exceed $2 | \varphi - \varphi' |$.

The above result on solutions of the nonlinear parabolic equation has an analogue for the solutions of the finite difference scheme

$$(11) \qquad u_k^{n+1} = u_{k+1}^n + \tfrac{1}{2}u_{k-1}^n - [f(u_{k+1}^n) - f(u_{k-1}^n)] \frac{\Delta t}{2\Delta x} .$$

If we introduce as new unknown

$$U_k^n = \sum_{j=0}^{k} u_k^n + c^n,$$

c^n being a function of n but not of k, U_k^n will satisfy the equation

$$(12) \qquad U_k^{n+1} = U_{k+1}^n + \tfrac{1}{2}U_{k-1}^n - f(U_{k+1}^n - U_{k-1}^n) \frac{\Delta t}{2\Delta t}$$

modulo a function n; and if the constants c^n are chosen suitably (12) will be satisfied exactly.

Let u_k^n and $\tilde{u}_k^n$ be two solutions of (11) and U_k^n and $\tilde{U}_k^n$ the corresponding solutions of (12). Subtract (12) and ($\widetilde{12}$) and apply the mean value theorem to the second term on the right:

$$(13) \qquad D_k^{n+1} = D_{k+1}^n + \tfrac{1}{2}D_{k-1}^n - f_u(\bar{u})[D_{k+1}^n - D_{k-1}^n] \frac{\Delta t}{2\Delta x} ;$$

D_k^n is an abbreviation for the difference of U_k^n and $\tilde{U}_k^n$. The stability criterion demands that $f_u \ \Delta t/\Delta x$ should not exceed one; assume that this condition is satisfied. Then (13) is a recursion relation for D_k^n with *positive* coefficients whose sum is one, and for these the *maximum principle* applies: D_k^n never exceeds its largest value for $n = 0$. From this result one can deduce that the solutions constructed by the finite difference scheme are continuously dependent on the initial data in the sense of the weak norm.

The result we have just proved about the continuous dependence of the solutions constructed by the viscosity method on the initial data in the *weak* topology has this rather remarkable consequence:

The dependence of u on φ is completely continuous.

More precisely: Denote by $\{\varphi\}$ the collection of measurable functions bounded by some constants and vanishing outside some fixed interval[8], and denote by $\{u\}$ the corresponding weak solutions constructed by the viscosity method. We claim that the set $\{u\}$ is *compact in the L_1 sense.*

REMARK: The same result holds for the solutions constructed by the finite difference scheme.

Proof: Let $[u]$ be an infinite subset of the set of weak solutions considered; we wish to show that $[u]$ has a point of accumulation in the L_1 sense. Let $[\varphi]$ be the set of their initial data. The original collection $\{\varphi\}$ of initial functions is *compact* in the sense of the *weak* norm introduced before; therefore it is possible to select an infinite subsequence φ_i in $[\varphi]$ which converges to a limit φ in the sense of the weak norm. According to the lemma proved before, the sequence of corresponding solutions u_i converges in the sense of the weak norm to the solution u with initial value φ. However, according to a principle stated in one of the earlier sections, a weak limit of weak solutions is a weak solution if and only if it is a strong (i.e. L_1) limit. Therefore we can conclude that the sequence u_i tends to u in the L_1 sense, and this proves the compactness of the set $\{u\}$ in the L_1 set.

For sake of completeness we shall state and prove rigorously the principle about weak limits of weak solutions. We have to assume to start with that our equation $u_t + f(u)_x = 0$ is genuinely nonlinear, meaning that f_u, the coefficient of u_x, genuinely depends on u, i.e. that its derivative with respect to u doesn't vanish: $f_{uu} \neq 0$, say f_{uu} is greater in the relevant range of u than some positive quantity.

LEMMA: If the sequence of functions u_n converges[9] in the *weak sense* to a

[8]These conditions imposed on the set of initial functions can be replaced by others.
[9]The range of the functions $u_n(x, t)$ is assumed to lie in a finite range: $a \leq u \leq b$.

limit u, then $f(u_n)$ converges in the weak sense to $f(u)$, if and only if u_n tends to u in the L_1 sense.

Proof: If $f(u_n)$ converges weakly to $f(u)$, $\int vf(u_n)$ must approach $\int vf(u)$ for every square integrable function v. Take for v any positive function and consider the integral

$$I(u) = \int vf(u)$$

as functional of u. It is an easy consequence of the assumption that f_{uu} is positive, that the second variation of $I(u)$ is positive definite, i.e.

$$I[u_n] - I[u] = \int vf_u(u) \cdot (u_n - u) + R,$$

where R is greater than const. $\int v(u_n - u)^2$. This implies the lower semicontinuity of the functional $I[u]$ in the *strict* sense:

$$\lim\inf I[u_n] \geq I[u]$$

for any sequence u_n weakly convergent to u; the sign of equality holding *if and only if* u is a strong limit of the sequence u_n.

The results derived can be summarized thus:

If the sequence of initial vectors φ_i converges in the weak norm, the corresponding solutions u_i converge in the L_1 sense in every bounded subset of the halfplane $t > 0$.

It follows from this result that for almost all t the functions $u_i(x, t)$ converge in the L_1 sense on every bounded interval of the x-axis. Most likely this is true for all values of t without *any* exception; this fact is expressed concisely by this statement:

The transformation mapping the initial state into the state at time t is completely continuous.

This is a particularly striking manifestation of irreversibility, in sharp contrast with the state of affairs in the theory of linear equations. There the transformations mapping the initial data into their value at time t is invertible.

The completely continuous dependence of the solution on the initial data can also be read off the explicit expression for the solution given in section 1. It would be interesting to know whether something like this is true for systems with more unknowns, in particular the equations of compressible flow.

6. *The Case of More Space Variables*

Much of what was said for the case of one space variable can be generalized if there are more space variables. In particular the same staggered difference scheme can be employed; the space derivatives are still replaced by centered difference quotients, and the time derivative U_t by a forward difference quotient

$$U_k^{n+1} - \overline{U}_k^n/\Delta t,$$

where $\overline{U}_k^n$ is the average of the value of U at the $2m$ neighbors of the lattice point k. The convergence of this scheme under a certain restriction on the ratio of Δt to the space increments has been proved for *linear* symmetric hyperbolic equations by K. O. Friedrichs.

The method can be applied to the Euler form of the flow equations in two (and three) dimensions. The conservation equations are:

$$\rho_t + (u\rho)_x + (v\rho)_y = 0,$$
$$(u\rho)_t + (u^2\rho + p)_x + (uv\rho)_y = 0,$$
$$(v\rho)_t + (uv\rho)_x + (v^2\rho + p)_y = 0,$$
$$E_t + (uE + up)_x + (vE + vp)_y = 0.$$

One would operate with ρ, $u\rho$, $v\rho$, and E as unknown functions.

A large portion of the work reported here was performed for the Los Alamos Scientific Laboratory. The calculations for examples 1, 2, 3, 4 were carried out on IBM calculators The calculations for examples 5, 6 and 7, as well as the calculations of Lester Baumhoff, were carried out on the Los Alamos MANIAC. The calculations for example 8 and for the reflection of a shock from a rigid wall were carried out on the UNIVAC of the Eckert-Mauchly Corporation. My thanks are due to Stewart Schlesinger for coding problems 1, 2, 4 and for programming problems 5, 6, 7, to Paul Stein and Paul Rosenthal for programming and coding the problems for the UNIVAC, to Lois Cook for coding problems 5, 6, 7, and to George Pimbley for coding problem 3. My thanks are also due to members of the staff of the Theoretical Physics Division of the Los Alamos Laboratory for their interest in this work; I had many fruitful conversations with them on finite difference methods.

Bibliography

[1] Becker, R., *Stosswelle und Detonation*, Zeitschrift für Physik, Volume 8, 1922, pp. 321–362.

[2] Courant, R., and Friedrichs, K. O., *Supersonic Flow and Shock Waves*, Interscience Publishers, New York-London, 1948.

[3] Courant, R., and Hilbert, D., *Methoden der mathematischen Physik*, Volume II, Springer, Berlin, 1937; reprinted by Interscience Publishers, New York, 1943.

[4] Courant, R., Isaacson, E., and Rees, M., *On the solution of nonlinear hyperbolic differential equations by finite difference methods*, Communications on Pure and Applied Mathematics, Volume V, 1952, pp. 243–256.

[5] Courant, R., Friedrichs, K. O., and Lewy, H., *Über die partielle Differentialgleichungen der mathematischen Physik*, Mathematische Annalen, Volume 100, 1928, pp. 32–74.

[6] Gilbarg, D., *The existence and limit behaviour of the one-dimensional shock layer*, American Journal of Mathematics, Volume 73, 1951, pp. 256–275.

[7] Grad, H., *The profile of a steady plane shock wave*, Communications on Pure and Applied Mathematics, Volume V, 1952, pp. 257–300.

[8] Hopf, E., *The partial differential equation* $u_t + u u_x = \mu u_{xx}$, Communications on Pure and Applied Mathematics, Volume III, 1950, pp. 201–230.

[9] Keller, J., and Lax, P. D., *Finite difference schemes for hyperbolic systems*, LAMS, 1205, 1950.

[10] Lax, P. D., *On discontinuous initial value problems for nonlinear equations and finite difference schemes*, Los Alamos Scientific Laboratory Report, 1332, 1952.

[11] Lax, P. D., *The initial value problem for nonlinear hyperbolic equations in two independent variables*, Princeton University Annals of Mathematics Studies, No. 33.
[12] von Neumann, J., *Proposal and analysis of a numerical method for the treatment of hydrodynamic shock problems*, National Defense and Research Committee Report AM551, 1944.
[13] von Neumann, J., and Richtmyer, R., *A method for numerical calculation of hydrodynamic shocks*, Journal of Applied Physics, Volume 21, 1950, pp. 232–237.
[14] Thomas, L. H., *Note on Becker's theory of the shock front*, Journal of Chemical Physics, Volume 12, 1944, pp. 449–453.
[15] Weyl, H., *Shock waves in arbitrary fluids*, Communications on Pure and Applied Mathematics, Volume 2, 1949, pp. 103–122.
[16] Germain, P. and Bader, R., *Unicité des écoulements avec chocs dans la mécanique de Burgers*, Office National d'Etudes et de Recherches Aeronautiques, Paris, 1953, pp. 1–13.
[17] Cole, J. D., *On a quasi-linear parabolic equation occurring in aerodynamics*, Quarterly of Applied Mathematics, 1951, pp. 226–236.
[18] Ludford, G. S. S., Polachek, R. J., and Seeger, R., *On unsteady flow of viscous fluids*, Journal of Applied Physics, Volume 24, 1953, pp. 490–495.

COMMUNICATIONS ON PURE AND APPLIED MATHEMATICS, VOL. X, 537–566 (1957)

Hyperbolic Systems of Conservation Laws II*

P. D. LAX

1. Introduction

A conservation law is an equation in divergence form, i.e.

$$u_t + \sum_{j=1}^{3} \frac{\partial}{\partial x_j} f_j = 0.$$

It expresses the fact that the quantity of u contained in any domain G of x-space changes at a rate equal to the flux of the vector-field (f_1, f_2, f_3) into G:

$$\frac{d}{dt}\iiint_G u\,dx = \iint_{BG} f \cdot n\,dS.$$

Many physical laws are conservation laws; the quantities u and f depend on the variables describing the state of a physical system, and on their derivatives. In theories which ignore mechanisms of dissipation such as viscous stresses, heat conduction, ohmic loss, the conservation laws are of first order, i.e., the quantities u and f are functions of the state variables but not of their derivatives. In this paper we shall consider such systems of first order conservation laws in one space variable. The components of u shall be chosen as state variables so that the system is in the form

(1.1) $$u_t + f_x = 0,$$

where u is a vector of n components and $f = f(u)$ a vector valued function of u.

When the differentiation in (1.1) is carried out, a quasilinear system of first order results:

(1.1′) $$u_t + A(u)u_x = 0, \qquad A = \operatorname{grad} f.$$

The system of conservation laws (1.1) is called *hyperbolic* if as a quasilinear system it is hyperbolic, i.e., if the matrix $A = \operatorname{grad} f$ in (1.1′) has real and distinct eigenvalues for all values of the argument u.

The initial value problem consists of determining solutions u of (1.1)

---

*The work for this paper was supported by the Office of Naval Research, United States Navy, under Contract N6ori-201, T. O. No. 1, and by the Los Alamos Scientific Laboratory. It contains an extension of results discussed in "Weak solutions of nonlinear hyperbolic equations and their numerical computation", Comm. Pure Appl. Math., Vol. 7, 1954, pp. 159–193.

from their initial state $u(x, 0) = \phi$ for all future time. In this paper we shall try to develop an adequate theory of the initial value problem for systems of conservation laws.

At any time t_0, the function $u(x, t_0)$ describing the state of the system is required to possess certain properties. Denote by $\{\phi\}$ the set of these "permissible" states.[1] We wish to assign to each state ϕ of this set a solution $u(x, t) = S(t)\phi$ for all time $t \geqq 0$ with the following properties:

(i) $u(x,t) = S(t)\phi$ is a solution of the system of conservation laws (1.1).

(ii) The operators $S(t)$ map the set of permissible.states intot iself.

(iii) The operators $S(t)$ form a one-parameter semigroup,

$$S(t_1+t_2) = S(t_1)S(t_2), \qquad t_1, t_2 \geqq 0, \qquad S(0) = I.$$

(iv) For each t, the operator $S(t)$ is continuous in some topology.

Solutions in the classical sense of the quasilinear hyperbolic system (1.1′) develop singularities (discontinuities) after a finite time, no matter how smooth their initial data, and cannot be continued as regular solutions. They can be continued, however, as solutions in a generalized sense; this kind of generalization is dictated by the integral version of a conservation law which states that the vector field

$$(u, f_1, f_2, f_3)$$

is divergence-free in space-time. Generalized (or weak) solutions are functions u for which the above vector field is divergence-free in a generalized sense.

A precise definition of weak solution is given in Section 2. It turns out (see e.g. [16] for examples) that weak solutions are not determined uniquely by their initial values, therefore, an additional principle is needed for selecting a relevant subclass. This matter is discussed in Sections 5 and 7.

In Section 2 we shall solve the initial value problem for a generalized solution of a *single* conservation law for a single unknown function by an explicit formula. This formula has been derived in [11] by Hopf for a quadratic conservation law and by the author [16] for the general case.

Section 3 contains a brief discussion of the "viscosity method", i.e. of obtaining solutions of systems of conservation laws as limits of solutions of parabolic equations as the coefficient of the dissipative term goes to zero. The results of Hopf [11], Cole [4], Olejnik [23, 24] and Ladyzhenskaya [13] are briefly described.

In Section 4 we prove that a difference scheme proposed by the author in [14] and [16] for solving the initial value problem converges for a particular

[1]These need not form a linear space.

conservation law. This result was announced in [16]. Recently Vvedenskaya [28] succeeded in proving the convergence of this scheme and of related ones for an arbitrary single conservation law.

In Section 5 we state the uniqueness theorems of Germain and Bader [9] and of Olejnik [25], and give a *nonuniqueness* theorem for the *backward* initial value problem.

In Section 6 we discuss the asymptotic behavior of solutions as t tends to infinity. Our results are related to those of Friedrichs [6], Lighthill [19], Hopf [11], Whitham [29] and Keller [12].

Sections 7—9 are about systems of conservation laws. In Section 7 a generalization of the entropy condition is derived and used in developing a theory of shocks. Section 8 contains a theory of simple waves for quasilinear systems. These theories are used in Section 9 to solve the Riemann initial value problem, i.e. the one for which the initial function is piecewise constant.

We show in particular that the equations of compressible flow (time dependent and steady supersonic) are examples of hyperbolic systems of conservation laws. Our studies in Sections 7—9 show that many important properties of such flows are shared by solutions of general systems. In particular, the well-known rule that across a weak shock entropy is constant up to terms of third order in shock strength appears as a special case of a general law. Friedrichs has observed in [7] that the equations of hydromagnetics with infinite conductivity (Lundquist model) furnish another example of systems of conservation laws.

Most of the results of this paper were presented at the 1954 Summer Symposium on Partial Differential Equations in Kansas and are contained in the Proceedings of that Symposium [17].

SINGLE CONSERVATION LAWS

2. An Explicit Formula

DEFINITION. The function $u(x, t)$ is a weak solution of the system of conservation laws (1.1) with initial values ϕ if u and $f(u)$ are integrable functions over every bounded set of the half-plane $t \geqq 0$ and the integral relation

$$\int_0^\infty \int_{-\infty}^\infty \{w_t u + w_x f(u)\}\, dx\, dt + \int_{-\infty}^\infty w(x, 0)\, \phi(x)\, dx = 0 \tag{2.1}$$

is satisfied for all smooth test vectors w which vanish for $|x|+t$ large enough.

This definition expresses the divergence-free character of the vector field (u, f) in the weak sense.

Smooth solutions of (1.1′) are weak solutions and, conversely, if u is a weak solution with continuous first derivatives, then it satisfies the differential equation (1.1′).

LEMMA 2.1. *If u is a piecewise continuous weak solution of* (1.1), then *across the line of discontinuity the jump relation*

$$s[u] = [f] \tag{2.2}$$

holds; here [ ] *denotes the jump across the line of discontinuity, and s is the velocity of propagation*[2] *of the discontinuity.*

Relation (2.2) is a generalization of the Rankine-Hugoniot relation; it expresses the fact that the component of the vector field (u, f) in the direction normal to the line of discontinuity is continuous across the line of discontinuity. It is easy to show that piecewise continuous generalized divergence-free vector fields have this property.

In this section we shall discuss single conservation laws; u and f denote scalar quantities. The conservation law

$$u_t + f_x = 0 \tag{2.3}$$

can be written as a quasilinear equation

$$u_t + a(u)u_x = 0, \tag{2.4}$$

where $a(u)$ denotes $f'(u)$. We shall require that (2.4) be *genuinely nonlinear*, i.e., that a, the coefficient of u_x, should vary with u in the sense that $a'(u)$ is not zero. Since $a = f'$, this means that $f'' \neq 0$, i.e., *the function $f(u)$ is either strictly convex or strictly concave.*

In [15], the author has defined the notion of genuine nonlinearity for quasilinear systems. We shall return to it in Section 8 of this paper.

Given a convex (concave) function $f(u)$ defined for all u we define the conjugate* function $g(s)$ by the relation

$$g(s) = \operatorname*{Max}_{u} (\operatorname{Min}) \{us - f(u)\}.$$

Denote by $u = b(s)$ that value of u where the above maximum (minimum) occurs. It is easy to show that b is the inverse of $a = f'$ and also that b is the derivative of g.

$$b(a(u)) = u, \tag{2.5}$$

$$\frac{d}{ds} g(s) = b(s). \tag{2.6}$$

It is easy to show that $b(s)$ and $g(s)$ are uniquely defined on the range of $a(u)$, that $g(s)$ is convex (concave) there, and that $g(s)$ tends to infinity as s approaches the endpoints of the domain of definition of g.

[2] I.e., if the line of discontinuity is given by $t = t(\sigma)$, $x = x(\sigma)$, then $s = \dot{x}/\dot{t}$.

* *Footnote added in proof:* The theory of conjugate functions (or Lengendre transformation) for functions of one variable has been developed by Mandelbrojt [30]; for functions of a finite number of variables the theory is due to Fenchel [31], and for functions over an infinite-dimensional space, to Friedrichs [32] and Courant-Hilbert (Vol. I, Chapter IV) and, in an abstract setting, to Hörmander [33].

With the aid of these auxiliary functions we shall assign to any bounded measurable initial function $\phi(x)$ a function $u(x, t)$ which will turn out to be a weak solution of our conservation law (2.3) with initial value $\phi(x)$.

First we define $\Phi(y)$ as the integral of ϕ:

$$\Phi(y) = \int_0^y \phi(\eta)\, d\eta.$$

Consider the expression

$$\Phi(y) + tg\left(\frac{x-y}{t}\right) \tag{2.7}$$

which, for fixed x and t, is a continuous function of y. One can easily show that it tends to plus (minus) infinity as y approaches the endpoints of its domain of definition. Therefore, it assumes a finite minimum (maximum) in the interior.

Following Hopf [11] we prove two lemmas:

LEMMA 2.2. *Let y_1 and y_2 be values where the function* (2.7) *assumes its minimum (maximum) for x_1, t and x_2, t, respectively. If $x_1 < x_2$ then $y_1 \leqq y_2$.*

Proof: Take the case when f—and thereby g—is convex. By definition of y_1 and y_2 as minimum points, we have

$$\Phi(y_1) + tg\left(\frac{x_1-y_1}{t}\right) \leqq \Phi(y_2) + tg\left(\frac{x_1-y_2}{t}\right)$$

and

$$\Phi(y_2) + tg\left(\frac{x_2-y_2}{t}\right) \leqq \Phi(y_1) + tg\left(\frac{x_2-y_1}{t}\right).$$

Adding the two inequalities we get

$$g\left(\frac{x_1-y_1}{t}\right) + g\left(\frac{x_2-y_2}{t}\right) \leqq g\left(\frac{x_1-y_2}{t}\right) + g\left(\frac{x_2-y_1}{t}\right).$$

Since g is a convex function, it follows from this inequality that $y_1 \leqq y_2$.

For fixed x, t, denote by $y^+(x, t)$ and $y^-(x, t)$ the largest and smallest value of y for which the function (2.7) assumes its minimum. By definition $y^-(x) \leqq y^+(x)$; on the other hand, we deduce from Lemma 2.2 that $y^+(x_1) \leqq y^-(x_2)$ for $x_1 < x_2$. From these facts it follows that y^- and y^+ can differ only at points of discontinuity. Since y^- and y^+ are nondecreasing functions of x, this can happen only at a denumerable number of points. So we have

LEMMA 2.3. *For given t, with the exception of a denumerable set of values of x, the function* (2.7) *assumes its minimum at a single point which we denote by $y_0(x, t)$.*

With the aid of y_0 we define

$$u(x, t) = b\left(\frac{x-y_0}{t}\right) \tag{2.8}$$

and assert

THEOREM 2.1. *The function u defined by* (2.8) *is a weak solution of the conservation law* (2.3) *with initial value ϕ. This solution of the initial value problem satisfies the properties* (i)—(iv) *stated. in Section* 1.

Proof: We show first that if ϕ is a smooth function, then the function given by (2.8) coincides with the smooth solution of the initial value problem as long as the latter exists. For ϕ continuous the function (2.7) has a continuous first derivative; hence at the point where the minimum (maximum) occurs, the first derivative must vanish:

$$\phi(y_0) - g'\left(\frac{x-y_0}{t}\right) = 0$$

which by (2.6) and (2.7) is the same as

$$\phi(y_0) = b\left(\frac{x-y_0}{t}\right) = u(x, t).$$

Using (2.5) we get

$$\frac{x-y_0}{t} = a(\phi(y_0)).$$

These equations express the fact that the function $u(x, t)$ is constant along straight lines, and that the slope of the line issuing from the point $(y_0, 0)$ is $a(\phi(y_0))$. This is precisely what the differential equation

$$u_t + a(u)u_x = 0$$

asserts.

A systematic derivation of formula (2.8) for differentiable solutions of such a quasilinear equation can be given with the aid of a method of Bellman [1] for expressing solutions of nonlinear equations as suprema of solutions of linear ones.

Next we show that the functions given by formula (2.8) are weak solutions of (2.3).

We can write u as $\lim_{N\to\infty} u_N$, where

$$u_N(x, t) = \frac{\int_{-\infty}^{\infty} b\left(\frac{x-y}{t}\right) \exp\left\{-N\left[\Phi(y) + tg\left(\frac{x-y}{t}\right)\right]\right\} dy}{\int_{-\infty}^{\infty} \exp\left\{-N\left[\Phi(y) + tg\left(\frac{x-y}{t}\right)\right]\right\} dy}. \tag{2.9}$$

Denote the denominator in (2.9) by

$$V_N = \int_{-\infty}^{\infty} \exp\left\{-N\left[\Phi(y) + tg\left(\frac{x-y}{t}\right)\right]\right\} dy. \tag{2.10}$$

In view of relation (2.6) we can write (2.9) as

$$u_N = \frac{1}{N}\frac{(V_N)_x}{V_N} = \left(\frac{1}{N}\log V_N\right)_x = (U_N)_x. \tag{2.9'}$$

Likewise, we can write $f(u)$ as $\lim f_N$, where

$$f_N(x,t) = \frac{\int_{-\infty}^{\infty} f\left[b\left(\frac{x-y}{t}\right)\right]\exp\left\{-N\left[\Phi(y)+tg\left(\frac{x-y}{t}\right)\right]\right\}dy}{V_N}. \tag{2.11}$$

Using the definition of g it is easy to show that

$$f(b(s)) = sb(s)-g(s).$$

Substituting this into (2.11) we get

$$f_N = \frac{1}{N}\frac{(-V_N)_t}{V_N} = -(U_N)_t. \tag{2.11'}$$

From (2.9'), (2.11') we conclude that the vector fields (u_N, f_N) are divergence-free. Therefore, their limit $(u, f(u))$ is divergence-free in the generalized sense. This implies that our $u(x, t)$ is indeed a weak solution. From the relation

$$U_N = \frac{1}{N}\log V_N = \log (V_N)^{1/N}$$

and formula (2.10) for V_N, we can determine the limit of U_N:

$$U(x,t) = \lim_{N\to\infty} U_N = \min_y \left\{\Phi(y)+tg\left(\frac{x-y}{t}\right)\right\}. \tag{2.12}$$

Integrating (4.6') with respect to x and letting N tend to ∞, we obtain in the limit

$$U(x,t) = \int_{-\infty}^{x} u(\xi, t)\, d\xi.$$

Thus we have derived this

COROLLARY TO THEOREM 2.1. *The value of the minimum (maximum) of* (2.7) *is equal to the x-integral of the solution* $u(x, t)$ *defined by formula* (2.8).

We show now that as t tends to zero, $u(x, t)$ tends to ϕ in the weak sense. Denote $\max_x (y_0(x, t)-x)$ by $\delta(t)$. Clearly

$$U(x,t) = \min_{x-\delta(t)<y<x+\delta(t)} \left\{\Phi(y)+tg\left(\frac{x-y}{t}\right)\right\}. \tag{2.12'}$$

Since $g(s)$ is a strictly convex function which tends to ∞ as $|s|$ tends to ∞, it follows that $g(s)/|s|$ also tends to infinity with increasing $|s|$. Since ϕ

was assumed bounded, $\Phi(y)$ is $O(y)$. From these facts, and the form of the function (2.7) it follows easily that $\delta(t)$ tends to zero with t.

Let $\eta(\delta)$ denote the oscillation of $\Phi(y)$ over an interval of length δ; since $\Phi(y)$ is uniformly continuous, $\eta(\delta)$ tends to zero with δ. Denote by m a lower bound for g; we have, from (2.12′), the following lower bound for U:

$$U(x,t) \geqq \Phi(x) - \eta(\delta) + mt.$$

On the other hand, according to the corollary to Theorem 2.1, the value of the function (2.7) at any point, in particular at $y = x$, is an upper bound for U:

$$U(x,t) \leqq \Phi(x) + tg(0).$$

These estimates show that $U(x, t)$ tends to $\Phi(x)$ uniformly as t tends to zero, i.e., that $u(x, t)$ tends to $\phi(x)$ *in the weak sense.*

Next we show that the class of solutions defined in Theorem 2.1 form a semigroup in t.

Proof: Let t_1, t_2 be two positive quantities, x any value. According to the corollary,

$$U(y, t_1) = \min_z \left[U(z, 0) + t_1 g\left(\frac{y-z}{t_1}\right)\right] \tag{2.13}$$

and

$$U(x, t_1+t_2) = \min_z \left[u(z, 0) + (t_1+t_2)\, g\left(\frac{x-z}{t_1+t_2}\right)\right]. \tag{2.14}$$

To exhibit the semigroup property we have to show that $U(x, t_1+t_2)$ can also be obtained as

$$\min_y \left[U(y, t_1) + t_2 g\left(\frac{x-y}{t_2}\right)\right]. \tag{2.15}$$

Substitute (2.13) into (2.15); we get

$$\min_{y,z} \left[U(z, 0) + t_1 g\left(\frac{y-z}{t_1}\right) + t_2 g\left(\frac{x-y}{t_2}\right)\right]. \tag{2.15′}$$

The minimum with respect to the variables y and z can be taken in any order. Evaluating it with respect to y we first get, by differentiating with respect to y,

$$b\left(\frac{y-z}{t_1}\right) - b\left(\frac{x-y}{t_2}\right) = 0.$$

Since b is a monotonic function, we must have

$$\frac{y-z}{t_1} = \frac{x-y}{t_2} = \frac{x-z}{t_1+t_2}.$$

Using this relation to eliminate y from (2.15′) we see that (2.15) is equal to (2.14).

The initial function ϕ can be any function which has a finite integral Φ; in particular, ϕ could be a δ-function or a sum of δ-functions. Such initial configurations represent point sources. Of course, in contrast to the linear theory, we cannot build up more general solutions by superimposing point sources.

Finally we show that the solutions constructed depend continuously on the initial data. In fact, we shall show that the dependence of u on ϕ is *completely continuous*:

THEOREM 2.2. *Let ϕ_n be a sequence of functions which converges weakly to a limit ϕ. Let u_n be the solution assigned to the initial value ϕ_n by formula* (2.8) *and let u be the solution assigned to ϕ. Then the sequence u_n converges to u at all points of continuity of u.*

Proof: Let Φ_n and Φ denote as before the integrals of ϕ_n and ϕ. Let (x, t) be a point where the function (2.7) has a unique minimum y_0. Since ϕ_n converges weakly to ϕ, Φ_n converges uniformly to Φ. Therefore the function

$$\Phi_n(y) + tg\left(\frac{x-y}{t}\right) \tag{2.7$_n$}$$

achieves its minimum at a point y_n which tends to y_0 as n tends to infinity. The rest follows from the explicit formula (2.8).

A different proof of a slightly weaker kind of compactness was given in [16].

The transformation $v = a(u)$ transforms solutions u of the quasilinear equation (2.4) into solutions v of the quadratic equation $v_t + vv_x = 0$. But it is easy to see that it does *not* map *discontinuous* weak solutions of (2.3) into weak solutions of the quadratic conservation law $v_t + \frac{1}{2}(v^2)_x = 0$. Another striking example of this noninvariance of weak solutions under nonlinear transformation occurs in the theory of compressible flows: A discontinuous flow which conserves mass, momentum and energy does not, in contrast to smooth flows, conserve entropy.

An explicit formula similar to (2.8) can be given for solutions with values prescribed along an arbitrary curve in the x, t-plane.

3. The Viscosity Method

For a quadratic conservation law, $f(u) = \frac{1}{2}u^2$, the sequence of approximations u_n used in formula (2.9) of the last section are solutions of the parabolic equation[3]

[3]This equation was first investigated by J. M. Burgers, see e.g. [2].

$$u_t+\tfrac{1}{2}uu_x = \lambda u_{xx}, \qquad \lambda = \frac{1}{2N}.$$

For nonquadratic f it is no longer possible to find such an explicit solution of the parabolic equation

$$u_t+f(u)_x = \lambda u_{xx}.$$

However, Olejnik succeeded [23, 24, 25] in proving that solutions of this parabolic equation with given initial value $u(x, 0) = \phi(x)$ tend to the weak solution of $u_t+f_x = 0$ given in the last section. A simpler proof was found recently [13] by Ladyzhenskaya.

4. The Method of Finite Differences

In [14] and [16] I have proposed a finite difference method for constructing weak solutions, with prescribed initial values, of systems of conservation laws. In addition to presenting numerical evidence, I stated there that I was able to give a rigorous proof of convergence in one case. We present now the details of that proof.

THEOREM 4.1. *Consider the single conservation law* (1.1) *with*

$$f(u) = -\log(a+be^{-u}), \qquad a+b = 1. \tag{4.1}$$

Replace in (1.1) *the time derivative by a forward difference and the space derivative by a left difference*[4]:

$$u(x, t+\Delta) = u(x, t)-\{f(u(x, t))-f(u(x-\Delta, t))\}. \tag{4.2}$$

Let u_Δ be the solution of this difference equation with bounded and measurable initial value $u(x, 0) = \phi(x)$. Then

$$\lim_{\Delta\to 0} u_\Delta = u(x, t) \tag{4.3}$$

exists for fixed t for almost all x, where u is given by (2.8)[5].

Proof: Define

$$U(x, t) = \sum_{k=-\infty}^{0} u_\Delta(x-k\Delta, t);$$

here we assume that $\phi(x)$ — and thereby $u(x, t)$ — is zero for large negative x. Write $x-k\Delta$ for x in (4.2) and sum with respect to k. Taking $f(0) = 0$ — which is true for our choice (4.1) — we get

$$U(x, t+\Delta) = U(x, t)-f(U(x, t)-U(x-\Delta, t)). \tag{4.4}$$

[4]Here $\Delta t = \Delta x = \Delta$. Our scheme differs slightly (in an inessential way) from the one discussed in [16].

[5]An essential feature of the scheme presented here is that the exact derivative f_x in the conservation law is replaced by an exact difference.

We seek a nonlinear transformation

$$U = G(V)$$

which linearizes (4.4). The equation for V is

$$G(V(x, t+\Delta)) = G(V(x, t)) - f[G(V(x, t)) - G(V(x-\Delta, t))] \tag{4.4'}$$

which is linear if the identity

$$G(V) - f[G(V) - G(W)] = G(aV + bW) \tag{4.5}$$

holds for all V and W with some constants a and b. It can be shown that the only function G which satisfies an identity of this kind is

$$U = G(V) = \log V, \tag{4.6}$$

with $f(u)$ as given by (4.1). Equation (4.4') becomes in this case

$$V(x, t+\Delta) = aV(x, t) + bV(x-\Delta, t)$$

which has the well-known solution

$$V(x, t) = \sum_{l=0}^{n} \binom{n}{l} a^{n-l} b^{l} V(x - l\Delta, 0). \tag{4.7}$$

Thus

$$\begin{aligned} u_\Delta(x, t) &= U(x, t) - U(x-\Delta, t) = \log \frac{V(x, t)}{V(x-\Delta, t)} \\ &= \log \frac{\sum_{l=0}^{n} \binom{n}{l} a^{n-l} b^{l} V(x-l\Delta, 0)}{\sum_{l=0}^{n} \binom{n}{l} a^{n-l} b^{l} V(x-(l+1)\Delta, 0)}. \end{aligned} \tag{4.8}$$

Writing $l-1$ for the summation variable in the numerator and abbreviating $x-(l+1)\Delta$ by y, we get

$$u_\Delta(x, t) = \log \frac{\sum_{l=-1}^{n} \frac{t-(x-y)}{x-y} \cdot \frac{b}{a} e^{\Theta/\Delta}}{\sum_{l=0}^{n} e^{\Theta/\Delta}}, \tag{4.8'}$$

where

$$\Theta = \Delta \log \left[\binom{n}{l} a^{n-l} b^{l} \right] + \Delta U(y, 0). \tag{4.9}$$

Here we have made use of relation (4.6)

$$V(y, 0) = e^{U(y,0)}.$$

As Δ tends to 0 (i.e., n tends to ∞), $\Delta U(y, 0)$ tends to $\Phi(y)$ in measure.

We can evaluate the binomial coefficient $\binom{n}{l}$ asymptotically by Sterling's formula obtaining the following expression for (3.12):

$$\Theta = \Phi(y) - t\left[\log\left(1-\frac{x-y}{t}\right) - \frac{x-y}{t}\log\left(\frac{t-x+y}{x-y}\cdot\frac{b}{a}\right)\right] + \text{error}, \tag{4.9'}$$

where the error tends to zero for almost all x. The expression (4.9′) has a unique maximum y_0 in y for all but a denumerable number of x.

By a well-known argument we conclude then that the limit of u_Δ as given by formula (4.8′) is equal to

$$\log\left[\frac{b}{a}\left(\frac{t}{x-y_0}-1\right)\right].$$

This is formula (2.8) of Section 2.

I arrived at formula (2.8) by noting the similarity of the explicit formula for solutions derived by Hopf for $f(u) = \frac{1}{2}u^2$ to the one derived by me in the present case, and generalizing it.

It does not seem possible to solve explicitly the difference equation (4.2) for any choice of $f(u)$ other than (4.1). Vvedenskaya has succeeded recently [28] in proving convergence for any f by another method.

5. Uniqueness and Irreversibility

Weak solutions of conservation laws are not determined uniquely by their initial values. Therefore, some additional principle is needed for preferring some—such as the ones selected in Section 2—to others. In [16] I conjectured that there is just one way of assigning a weak solution u to each ϕ of a permissible class so that the properties (i)–(iv) of Section 1 are satisfied. Even if this proposition were true, it would not yield a practical criterion for picking out a preferred weak solution from many with the same initial value. A different type of criterion is that the preferred weak solution be obtainable by a specific limiting process such as the viscosity method. In order to avoid carrying out the limiting process, it is desirable to give an intrinsic characterization of such weak solutions. It is easy to show that if u is a piecewise continuous weak solution whose discontinuities occur along smooth arcs and which is a limit of solutions of parabolic equations, then the jump $[u]$ from left to right across each discontinuity curve is negative for f convex, positive for f concave. Germain and Bader have shown in [9] that there exists at most one weak solution with prescribed initial value whose jumps satisfy this condition. Their proof, given for the quadratic conservation law, can be extended without alteration to any single conservation law. Another proof has been given by A. Douglis (unpublished).

Olejnik has proved the following uniqueness theorem which completely characterizes weak solutions that are limits of solutions of parabolic equations.

Among all weak solutions of the conservation law (2.3), f convex (concave), there exists exactly one whose x-difference quotients are bounded from above (below):

$$\frac{u(x_1,t)-u(x_2,t)}{x_1-x_2} \leqq K(t) \qquad (\geqq K(t)).$$

The upper (lower) bound may tend to plus (minus) infinity as t tends to zero.

Olejnik's brief and elegant proof is contained in [25].

It is easy to verify that the solutions given in Section 2 have this property.

The condition on the sign of jumps of discontinuous weak solutions derived in the present section will be generalized, on the basis of the theory of mixed initial and boundary value problems, to systems of conservation laws in Section 7.

Theorem 2.2 asserts that the future is a completely continuous function of the past. It follows, therefore, that the past is a discontinuous function of the future. More than that: there exist bounded, piecewise continuous states ψ such that $\psi(x) \neq u(x, t_0)$, $t_0 > 0$, for any u of the class of solutions discussed in Section 2. On the other hand, there exist solutions in this class which are equal for $t \geqq t_0$ but differ for $t < t_0$. Concerning these the following curious result holds:

THEOREM 5.1. *The set of states $\{\phi\}$, for which the corresponding solutions are equal at $t = T$, is convex.*

Proof: Let ϕ_1 and ϕ_2 be two elements of such a set; we wish to show that $s\phi_1+(1-s)\phi_2$, $0 \leqq s \leqq 1$, also belongs to this set. Since the solutions corresponding to ϕ_1 and ϕ_2 were assumed to be equal, it follows from (4.2) that the functions $y_0(x, T)$, corresponding to these two different initial states, are equal for almost all x, i.e., that the two functions

$$\Phi_1(y)+Tg\left(\frac{x-y}{T}\right) \tag{5.1}$$

and

$$\Phi_2(y)+Tg\left(\frac{x-y}{T}\right) \tag{5.2}$$

take on their minimum (maximum) at the same value y for almost all x. If two functions assume their minimum at the same point, so does any linear combination of them with positive weights s and $1-s$. This completes the proof of the theorem.

6. Asymptotic Behavior for Large t

E. Hopf, in [11], discusses the asymptotic behavior for large t of solutions of the special conservation law $u_t + (\frac{1}{2}u^2)_x = 0$. We shall discuss related properties of solutions of an arbitrary conservation law (see also the discussion in Lighthill [19] and the articles of Whitham quoted there).

DEFINITION 6.1. A function $\phi(x)$ has mean value M if

$$\lim_{L\to\infty} \frac{1}{L}\int_a^{a+L} \phi(x)\,dx = M$$

uniformly in a.

THEOREM 6.1. *Let $u(x, t)$ be a bounded weak solution of a system of conservation laws whose initial value has a mean value. Then $u(x, t)$ has the same mean value for all t.*

Proof: It is easy to show that, if u is a weak solution,

$$\int u(x, t)\, w(x)\, dx$$

is independent of t for every smooth test function w. Taking w to be equal to $1/L$ in $a < x < a+L$, zero for $x < a-1$ and $a+L+1 < x$, smooth in between, and letting L tend to infinity we deduce our result.

One can show similarly that if $u(x, t)$ is bounded and integrable at $t = 0$, then $\int u(x, t)dx$ is time independent.

THEOREM 6.2. *Let u be the solution defined in Theorem 4.1 of the single nonlinear conservation law $u_t + f_x = 0$. Assume that the initial value of u has mean value M. Then $u(x, t)$ tends to M uniformly in x as t tends to infinity.*

Proof: The proof is based on the explicit formula, given in Section 4,

$$u(x, t) = b\left(\frac{x-y_0}{t}\right), \tag{6.1}$$

where $y_0 = y_0(x, t)$ is the value of y for which the function

$$\Phi(y) + tg\left(\frac{x-y}{t}\right) \tag{6.2}$$

achieves its minimum (maximum).

Take the case when the mean value M is zero[6], and denote the sound speed in the state $u = 0$ by

$$c = f'(0) = a(0).$$

From the relations (2.5), (2.6) between the functions a, b and g we see that $b(c)$ is zero, and $g(s)$ assumes its minimum (maximum) at $s = c$.

[6]This can be accomplished by introducing $\bar{u} = u - M$ as a new variable.

Introduce the abbreviation

$$y_1 = y_1(x, t) = x - ct. \tag{6.3}$$

We claim that, as t tends to infinity, the quantity

$$\left|\frac{y_1 - y_0}{t}\right| \tag{6.4}$$

tends to zero uniformly in x. We shall prove this by showing that otherwise the function (6.2) is, for large t, smaller at $y = y_1$ than at $y = y_0$, contrary to the definition of y_0.

We form the difference of the values of the function (6.2) at y_0 and y_1:

$$\begin{aligned} &\Phi(y_0) + tg\left(\frac{x-y_0}{t}\right) - \Phi(y_1) - tg\left(\frac{x-y_1}{t}\right) \\ &\qquad = \Phi(y_0) - \Phi(y_1) + t\left\{g\left(c + \frac{y_1 - y_0}{t}\right) - g(c)\right\}. \end{aligned} \tag{6.5}$$

Since $g(s)$ is a strictly convex function which has its minimum at $s = c$, the quotient

$$\frac{g(c+\xi) - g(c)}{|\xi|}$$

is bounded away from zero if $|\xi|$ is, i.e., if to any positive δ there corresponds a positive $\varepsilon = \varepsilon(\delta)$ such that

$$g(c+\xi) - g(c) \geqq \varepsilon|\xi| \qquad \text{for } |\xi| > \delta. \tag{6.6}$$

Since ϕ has mean value zero, there is an $L = L(\varepsilon)$ such that

$$|\Phi(y_0) - \Phi(y_1)| \leqq \varepsilon|y_0 - y_1| \qquad \text{for } |y_0 - y_1| > L. \tag{6.7}$$

Inequalities (6.6), with $\xi = (y_1 - y_0)/t$, and (6.7) show that the quantity (6.5) is positive if $|y_0 - y_1|$ is greater than L and $|y_0 - y_1|/t$ is greater than δ. On the other hand, by the definition of y_0, (6.5) is nonpositive. Therefore $|y_1 - y_0|/t$ is less than δ for t greater than L/δ.

Writing formula (6.1) for u in the form

$$u(x, t) = b\left(c + \frac{y_1 - y_0}{t}\right)$$

and recalling that $b(c)$ is zero, we conclude that $U(x, t)$ tends to zero with $|y_1 - y_0|/t$.

Two interesting special cases of initial functions having mean values are: 1) ϕ is periodic, and 2) ϕ is zero outside of a finite interval. In both these cases one can estimate the rate at which u tends to zero as t increases, and even the asymptotic shape of u.

THEOREM 6.3. *Assume that $\phi(x)$ is periodic with period p, and has mean value zero, and assume that $\Phi(x)$, also periodic, has a unique minimum in each period. Then the maximum of $u(x, t)$ is asymptotically equal to*

$$\frac{\text{const. } p}{t},$$

where the value of the constant does not depend on ϕ. Furthermore the graph of $u(x, t)$ tends to the graph of a saw tooth function.

Proof: We start again by investigating the possible location of y_0, the value of y minimizing the function (6.2). Since the first term $\Phi(y)$ is periodic, y_0 must be within a period p of that value of y which minimizes the second term, i.e. $y = y_1$. In fact we must have

$$y_1 - \frac{p}{2} - \varepsilon \leqq y_0 \leqq y_1 + \frac{p}{2} + \varepsilon, \tag{6.8}$$

where ε tends to zero with t. Since the variation of the second term over this interval is small if t is large, we have

$$y_0 - \xi = \varepsilon \pmod{p}, \tag{6.9}$$

where ξ is the value for which the minimum of $\Phi(x)$ occurs. Clearly, the graph of

$$u(x, t) = b\left(c + \frac{y_1 - y_0}{t}\right)$$

is within ε of the graph of the curve we would obtain if in (6.8) and (6.9) the quantity ε were taken to be zero. Since b is nearly linear within the narrow range defined by (6.8), this latter curve is very nearly a saw tooth function; its maximum occurs at $y_0 = y_1 + p/2$ and its value is

$$b\left(a + \frac{p}{2t}\right) \approx b'(a)\frac{p}{2t}.$$

Observe that the amplitude of the asymptotic shape depends on the *frequency* but not the *amplitude* of the initial configuration. It is clear from our derivation that *the larger* the initial amplitude is, *the sooner* the asymptotic form becomes valid.

The law of decay of periodic disturbances in compressible flows has been investigated by Whitham [29], J. B. Keller [12] and Lighthill [19].

THEOREM 6.4. *Let $\phi(x)$ be an initial function which is zero outside of a finite interval, and denote by $u(x, t)$ the weak solution, given by formulas (6.1), (6.2), of the equation $u_t + f_x = 0$ with initial value ϕ. The asymptotic shape of $u(x, t)$ for large t is*

$$(6.10) \qquad u(x,t) \approx \begin{cases} k\left(\dfrac{x}{t} - c\right) & \text{inside } ct-\alpha\sqrt{t} < x < ct+\beta\sqrt{t} \\ 0 & \text{outside} \end{cases}$$

in the sense that for large t every point of the graph of $u(x, t)$ is within a distance $o(\sqrt{t})$ of the graph of the function (6.10). The constants in (6.10) have the following values:

$$c = f'(0),$$

i.e. the speed of propagation of signals in the state $u = 0$,

$$k = \frac{1}{f''(0)},$$

$$\alpha = \sqrt{\frac{2m_1}{k}}, \qquad \beta = \sqrt{\frac{2m_2}{k}},$$

where

$$(6.11) \qquad m_1 = \max_x \int_x^{-\infty} \phi(\xi)d\xi, \qquad m_2 = \max_x \int_x^{\infty} \phi(\xi)d\xi.$$

(For f concave, the maxima in (6.11) are to be replaced by minima.)

Observe that the maxima in (6.11) occur at the same point x_0, and that $m_2 - m_1 = \int_{-\infty}^{\infty} \phi(\xi)\,d\xi$ is time-invariant.

It follows in particular from Theorem 6.4 that the quantities m_1 and m_2 themselves are time-invariant. This also follows from the explicit formula (4.9) for $U(x, t)$. It would be desirable to find such additional time-invariants of solutions of systems of conservation laws.

On account of the shape of the curve (6.10), the asymptotic form of a finite disturbance is called an N-wave.

Proof: As before, we investigate the solution with the aid of the explicit formulas (6.1), (6.2) by locating asymptotically y_0 as a function of x, t. Denote as before $x-ct$ by y_1; we wish to investigate the position of the minimum of

$$(6.12) \qquad \Phi(y) + tg\left(c + \frac{y_1 - y}{t}\right).$$

As we have shown before, $(y_1-y_0)/t$ tends to zero. Therefore, in the range which is of interest to us, the function g is accurately represented by its Taylor expansion around c:

$$(6.13) \qquad g(c) + \frac{k-\varepsilon}{2}s^2 < g(c+s) \leqq g(c) + \frac{k+\varepsilon}{2}s^2,$$

where ε can be taken arbitrarily small by taking t large enough. It is easy to show on the basis of the estimate (6.12) that

i) in the range $y_1 < -\sqrt{2m_1 t/(k-\varepsilon)}$, the minimum of the function (6.12) is reached at $y = y_1$,

ii) in the range $-\sqrt{2m_1 t/(k+\varepsilon)} < y_1 < \sqrt{2m_2 t/(k+\varepsilon)}$, the minimum of the function (6.12) is reached within ε of the point x_0 where the maxima in (6.11) are assumed,

iii) in the range $\sqrt{2m_2 t/(k-\varepsilon)} < y_1$, the minimum of (6.12) is reached for $y = y_1$.

According to formula (6.1), the solution is equal to

$$u(x, t) = b\left(c + \frac{y_1 - y_0}{t}\right).$$

Since $(y_1 - y_0)/t$ tends to zero, the function b is adequately represented by the first term in its Taylor expansion. This yields, with the above determination of y_0, formula (6.10).

The asymptotic form of the solution of a system of conservation laws whose initial state is zero outside of a finite interval is expected to consist of n distinct N-waves of width of the order $t^{1/2}$, height $t^{-1/2}$, each propagating with one of the characteristic speeds corresponding to the zero state. The propagation of N-waves in compressible flows has been investigated by Chandrasekhar [3], Friedrichs [6] and Whitham [29].

SYSTEMS OF CONSERVATION LAWS

7. The Entropy Condition

As we saw in Section 1, in order to develop a theory in the large for the initial value problem for systems of conservation laws, we need a criterion for selecting the physically relevant weak solutions. In this section we shall give such a criterion.

Suppose the system (1.1)

$$u_t + f_x = 0$$

consists of n scalar conservation laws in n unknowns. We assume that (1.1) is a hyperbolic system, i.e., that the associated quasilinear system (1.1′)

$$u_t + A(u)u_x = 0$$

is hyperbolic. This means that the matrix A has n real, distinct eigenvalues $\lambda_1, \lambda_2, \cdots, \lambda_n$ which are functions of u and which are arranged in increasing order. The corresponding right and left eigenvectors of A are denoted by $r_1, r_2, \cdots, r_n$ and $l_1, l_2, \cdots, l_n$, and are also functions of u. Later on, we shall give a convenient normalization for them.

Throughout this section we shall consider piecewise smooth weak solutions of the system of conservation laws (1.1). These are, we recall from Lemma 2.1, completely characterized by requiring that the Rankine-Hugoniot condition

$$s[u] = [f]$$

hold across the lines of discontinuity. Let P be a point on a line of discontinuity C and let u_l and u_r be the values at P of the solution on the left and right side of the discontinuity C, respectively. Draw, issuing from P in the positive t-direction, those characteristics with respect[7] to the state u_l which stay to the left of C, and those with respect to the state u_r which stay to the right of C.

DEFINITION 7.1. A jump discontinuity in a weak solution is called a shock if the total number of characteristics drawn in this fashion is $n-1$.

The analytical expression for the shock requirement is: *for some index k, $1 \leqq k \leqq n$, the inequalities*

$$\lambda_{k-1}(u_l) < s < \lambda_k(u_l), \tag{7.1}$$
$$\lambda_k(u_r) < s < \lambda_{k+1}(u_r)$$

hold. The characteristic speeds are assumed to be indexed increasingly; *k is the index of the shock*. Thus we see that there are n different kinds of shocks.

We have tacitly assumed in the definition of a shock that the characteristic speeds with respect to the state on the right and to the state on the left are either less or greater than the speed of the propagation of the discontinuity (shock speed). Nevertheless it could happen that the discontinuity C is a characteristic curve with respect to the state on one side. Such a discontinuity is called a *contact discontinuity*, and its nature will be analyzed in detail at the end of Section 8.

Example. The characteristic speeds of the equations of compressible flow are $q-c$, q, $q+c$, where q is the flow velocity and c the sound speed. There are three kinds of discontinuities possible. The second-kind is a contact discontinuity; the first and third are shocks facing the left and the right, respectively.

A shock facing the right is, by (9.1), characterized by the inequalities

$$q_l < s < q_l + c_l,$$
$$q_r + c_r < s.$$

One consequence of these inequalities is that the shock speed s is greater than the flow speed on either side. Since distance is measured positively to the right, it follows that particles cross the shock from the right to the left.

[7]We recall that, for a nonlinear equation, the characteristic speeds λ are functions of u.

Call the state to the right of the shock the front, to the left, the back. In this terminology the rest of the inequalities expresses the following rule: *Shock speed is supersonic with respect to the state in the front, subsonic with respect to the state in the back.*

The same statement holds for shocks facing the right.

It is well known (see e.g. Courant-Friedrichs, Supersonic Flow and Shock Waves) that this property of shocks is entirely equivalent to the requirement that the entropy of particles increase upon crossing the shock.

The motivation for formulating the shock condition in terms of the number of characteristics is furnished by the theory of free boundary value problems, see Goldner [10]. According to that theory, the number of relations to be imposed on the states on the two sides must be equal to the number of characteristics impinging on the discontinuity curve C from either side. Since the Rankine-Hugoniot conditions represent (after elimination of s) $n-1$ conditions, there must be precisely $n-1$ characteristics impinging on C, as required by Definition 7.1.

Given a state u_l, we shall investigate now the set of all states u_r to which u_l can be connected by a k-shock on the right. Such states u_r are subjected to two kinds of conditions: the Rankine-Hugoniot relations, and the inequality embodied in the shock condition. The Rankine-Hugoniot conditions represent $n-1$ relations between u_l and u_r, so that, if u_l is kept fixed, the states u_r (at least the ones which are close enough to u_l) form a one-parameter family of states[8]:

$$u_r = u(\varepsilon), \qquad u(0) = u_l .$$

The shock speed s is also a function of the parameter ε:

$$s = s(\varepsilon).$$

We shall determine now the derivatives of $u(\varepsilon)$, $s(\varepsilon)$ with respect to ε at $\varepsilon = 0$. Differentiating the Rankine-Hugoniot relations once and putting $\varepsilon = 0$ we get

$$s\dot{u} = \dot{f} = \operatorname{grad} f\dot{u} = A\dot{u}.$$

This shows that if $\dot{u}$ is different from zero, $s(0)$ must be one of the eigenvalues of $A = A(u_l)$, say the k-th,

$$s(0) = \lambda_k(u_l), \tag{7.2}$$

and that $\dot{u}$ is parallel to the corresponding right eigenvector,

$$\dot{u}(0) = \alpha r_k(u_l). \tag{7.3}$$

---

[8]It is not difficult to show that the permissible states u_r form a smooth one-parameter family if the condition stated in Definition 7.2 is satisfied.

By changing the parameter ε we can achieve that the constant of proportionality α is one, i.e.,

$$\dot{u}(0) = r_k(u_l). \tag{7.3'}$$

The normalization of r_k will be given below.

From now on we shall omit the subscript k. Differentiating the Rankine-Hugoniot relations once more we get, at $\varepsilon = 0$,

$$s\ddot{u} + 2\dot{s}\dot{u} = A\ddot{u} + \dot{A}\dot{u}.$$

Substituting our previous determination of s and $\dot{u}$ we can write this relation as

$$\lambda\ddot{u} + 2\dot{s}r = A\ddot{u} + \dot{A}r. \tag{7.4}$$

To determine $\dot{s}$ and $\ddot{u}$ we take the relation

$$\lambda r = Ar,$$

valid for all values of u, restrict u to the one-parameter family under consideration, and differentiate it with respect to ε. We then have

$$\lambda\dot{r} + \dot{\lambda}r = A\dot{r} + \dot{A}r. \tag{7.5}$$

Take the scalar product of both (7.4) and (7.5) with the k-th left eigenvector $l = l_k$ of A, and subtract the two relations:

$$\dot{\lambda}(0) = 2\dot{s}(0). \tag{7.6}$$

Subtracting (7.5) from (7.4), we find that $\ddot{u} - \dot{r}$ satisfies the equation

$$\lambda(\ddot{u} - \dot{r}) = A(\ddot{u} - \dot{r})$$

and therefore $\ddot{u} - \dot{r}$ is parallel to r:

$$\ddot{u}(0) = \dot{r} + \beta r. \tag{7.7}$$

By a change of parametrization, we can achieve that the constant β is zero,

$$\ddot{u}(0) = \dot{r} = \dot{u} \cdot \operatorname{grad} r = r \cdot \operatorname{grad} r. \tag{7.7'}$$

DEFINITION 7.2. The k-th characteristic field of a quasilinear system (1.1') is called genuinely nonlinear if grad λ_k and r_k are not orthogonal for any value of u:[9]

$$r \cdot \operatorname{grad} \lambda \neq 0 \qquad \text{for all } u.$$

[9]This condition ensures that the k-th characteristic speed depends truly on the state of the flow which produces "breaking' of waves, i.e. the development of cusps in the characteristic fields and corresponding discontinuities in the solution. I have shown in a previous publication [15], that if there is linear degeneration, i.e., if $r \cdot \operatorname{grad} \lambda$ is identically zero for, say, the k-th characteristic field, then there is no breaking of waves in the k-th mode of propagation. This condition is discussed further at the end of Section 8.

Assume now that the k-th characteristic field is genuinely nonlinear. We normalize the eigenvector r so that

$$r \cdot \operatorname{grad} \lambda \equiv 1. \tag{7.8}$$

With this determination of r we have from (7.3′) and (7.6)

$$\dot{\lambda}(0) = 1, \quad \dot{s}(0) = \tfrac{1}{2}. \tag{7.6′}$$

It is easy to see, in consequence of (7.6′), that the discontinuity line with slope $s(\varepsilon)$ separating the state $u(0)$ on the left from $u(\varepsilon)$ on the right is a k-shock in the sense of our definition *if and only if ε is negative.* We can formulate this result as

THEOREM 7.1. *Any given state u can be connected to a one-parameter family of states $u_r = u(\varepsilon)$, $\varepsilon \leqq 0$, on the right through a k-shock, provided that the k-th family of characteristics is genuinely nonlinear. The parametrization is normalized by* (7.8), (7.3′) *and* (7.7′).

Relation (7.6) can be expressed as

THEOREM 7.2. *The shock speed, up to terms of order ϵ^2, is the arithmetic mean of the sound speeds in the front and the back.*

The discontinuities of solutions given by the explicit formula of Section 4 for single conservation laws are shocks. For, we have shown there that for a convex $f(u)$, the limit from the left is not less than the limit from the right,

$$u_l \geqq u_r .$$

The single characteristic speed $\lambda(u) = f'(u)$ is, by assumption, an increasing function of u; so we have

$$\lambda(u_l) \geqq \lambda(u_r),$$

while from the Rankine-Hugoniot relation

$$s(u_r - u_l) = f(u_r) - f(u_l)$$

and the convexity of f we conclude that the speed s lies between $f'(u_r)$ and $f'(u_l)$,

$$\lambda(u_l) \geqq s \geqq \lambda(u_r).$$

This is precisely the inequality characterizing shocks.

Ideally, the ultimate aim of the theory of shocks is to assign to every initial state a weak solution which exists for all time, and all whose discontinuities are shocks, and to show that this assignment is unique. So far, this has not been achieved, not even in the classical case of compressible flows although many special solutions are known. In Section 9, we construct weak solutions with shocks for very special initial data. Then we show that these solutions are unique, at least within a certain restricted class. Preparatory to this we present in the next section a theory of simple waves.

8. Simple Waves

In this section we shall describe a special class of solutions, called simple waves, of an arbitrary quasilinear hyperbolic system

$$u_t + A(u)u_x = 0. \tag{8.1}$$

It is not necessary for these equations to be conservation laws.

DEFINITION 8.1. A function $v = v(u_1, u_2, \cdots, u_n)$ *is a* k-Riemann invariant of the system (8.1) if it satisfies the condition

$$r_k \cdot \operatorname{grad} v = 0 \tag{8.2}$$

for all values of u, where r_k is the k-th right eigenvector of A.

Condition (8.2) is a single linear homogeneous equation for v as function of u. The classical theory of such equations[10] asserts

THEOREM 8.1. *There exist precisely $n-1$ independent Riemann invariants.*

By independence of functions we mean that their gradients are linearly independent. Since the gradient of a Riemann invariant is perpendicular to r, we see that *the gradients of $n-1$ independent Riemann invariants span the orthogonal complement of r.*

The equations of compressible flow are

$$\varrho_t + q\varrho_x + \varrho q_x = 0,$$

$$q_t + qq_x + \frac{1}{\varrho}p_x = 0,$$

$$S_t + \quad qS_x = 0,$$

where ϱ is the density, S the entropy, p the pressure and q the flow velocity. The pressure is regarded as a function of ϱ and S, and $\sqrt{p_\varrho} = c$ is called the sound velocity.

The characteristic speeds are $q-c$, q and $q+c$; the right eigenvectors are $\begin{pmatrix} \varrho \\ -c \\ 0 \end{pmatrix}$, $\begin{pmatrix} p_s \\ 0 \\ -p_\varrho \end{pmatrix}$ and $\begin{pmatrix} \varrho \\ c \\ 0 \end{pmatrix}$. It is easy to see that the three pairs of Riemann invariants can be chosen as $\{S, q+h\}$, $\{q, p\}$ and $\{S, q-h\}$, where h is a function of ϱ and S, satisfying the relation $h_\varrho = c/\varrho$.

A *constant state* is a domain in the x, t-plane in which a solution is constant.

THEOREM 8.2. *Let G be a domain of the x, t-plane in which a smooth*

[10]See e.g. Courant-Hilbert, Vol. II, Ch. II.

solution u of a quasilinear system is defined. Let C be a smooth arc in G such that u is constant on one side of C. Then, either u is constant also on the other side of C, or C is a characteristic arc.

COROLLARY. *A constant state in the x, t-plane is bounded by characteristics.*

Proof: If C is not characteristic then, according to the classical uniqueness theorem, there is precisely one way of continuing the solution as smooth solution across C. Since u can certainly be continued as a constant, our result follows.

The boundary of a constant state consists, therefore, of characteristics, necessarily straight lines. The solution on the other side of the boundary is characterized by

THEOREM 8.3. *If C is part of the boundary of a constant state and is a characteristic of the k-th field, then all k-Riemann invariants on the other side of C are constant.*

The following version of the proof is due to Friedrichs (see [8]). Let $v_1, v_2, \cdots, v_{n-1}$, denote $n-1$ independent k-Riemann invariants. Take the characteristic form of the equation (8.1),

$$l_j \frac{du}{dj} = 0, \qquad j = 1, 2, \cdots, n, \tag{8.3}$$

where l_j is the j-th left eigenvector of A, and d/dj abbreviates differentiation in the j-th characteristic direction,

$$\frac{d}{dj} = \frac{\partial}{\partial t} + \lambda_j \frac{\partial}{\partial x}.$$

The left and right eigenvectors of a matrix are biorthogonal:

$$l_j \cdot r_k = 0 \quad \text{for } j \neq k.$$

Since the gradients of $n-1$ independent Riemann invariants span the orthogonal complement of r_k, the vectors l_j, $j \neq k$, can be expressed as linear combinations of the gradients of k-Riemann invariants:

$$l_j = \sum_{s=1}^{n-1} b_{js} \operatorname{grad} v_s, \qquad j \neq k.$$

The quantities b_{js} are functions of u. Substituting this expression into (8.3) we get

$$\sum_{s=1}^{n-1} b_{js} \operatorname{grad} v_s \frac{du}{dj} = 0$$

which is the same as

$$\sum_{s=1}^{n-1} b_{js} \frac{dv_s}{dj} = 0, \qquad j \neq k. \tag{8.4}$$

For a given solution u the coefficients b_{js} are given functions, and (8.4) is a linear hyperbolic system of $n-1$ equations for the $n-1$ unknowns $v_1, v_2, \cdots, v_{n-1}$ with characteristic speeds λ_j, $j \neq k$. Since C is no longer a characteristic of this system we conclude by the same uniqueness theorem as was used before that the functions v are constant on the other side of C, as asserted in our theorem.

DEFINITION 8.2. A solution in a domain of the x, t-plane for which all k-Riemann invariants are constant is called a k-simple wave.

According to this terminology Theorem 8.3 asserts that *a solution adjacent to a constant state is a simple wave.*

By making use of the k-th characteristic equation in (8.3) it is easy to show

THEOREM 8.4. *The characteristics of the k-th field in a k-simple wave are straight lines along which the solution is constant.*

We turn to *simple waves centered at the origin*, i.e., which depend on the ratio x/t only. Let $u(x, t)$ be a centered simple wave in the angle $a < x/t < b$, and assume that the corresponding characteristic field is genuinely nonlinear, i.e., that $r \cdot \operatorname{grad} \lambda$ is not zero; normalize r as before so that

$$r \cdot \operatorname{grad} \lambda \equiv 1.$$

Since u is supposed to be centered,

$$u(x, t) = h\left(\frac{x}{t}\right).$$

The function h must be determined so that

$$v_s(h) = c_s, \qquad s = 1, 2, \cdots, n-1, \tag{8.5}$$

the c_s being the assigned constant values of $n-1$ independent Riemann invariants.

According to Theorem 8.4, the lines $x/t = \xi =$ const. are characteristics. Therefore, the function h satisfies the relation

$$\lambda(h(\xi)) = \xi. \tag{8.6}$$

The assumption that $r \cdot \operatorname{grad} \lambda$ is not zero guarantees that (8.5), (8.6) lead to a uniquely determined function h. The relations (8.5) guarantee, as we saw before, that $n-1$ of the characteristic relations (8.3) are satisfied. Equation (8.6) guarantees that the lines $x/t =$ const. are characteristics of the field under consideration. Since along the lines $x/t =$ const. the solution is constant, we see that also the remaining characteristic equation is satisfied.

Denote by u_l and u_r the values of $u(x, t)$ on $x/t = a$ and $x/t = b$. For x/t less than a and greater than b we can define $u(x, t)$ to be equal to u_l and

u_r, respectively. The two constant states, connected through a centered simple wave, are subject to the restriction that they have the same k-Riemann invariants and that $\lambda_k(u_l)$ be less than $\lambda_k(u_r)$. Thus, we have

THEOREM 8.5. *Two states u_l and u_r can be connected from left to right by a centered k-simple wave if and only if they have the same k-Riemann invariants and*

$$\lambda_k(u_l) < \lambda_k(u_r). \tag{8.7}$$

The states u_r form a one-parameter family $u_r = u(\varepsilon)$, $u(0) = u_l$. By assumption, $v(u(\varepsilon))$ is a constant for all Riemann invariants v; differentiating at $\varepsilon = 0$ we get

$$\dot{u} \cdot \text{grad } v = 0.$$

Since the gradients of the Riemann invariants span the orthogonal complement of r, $\dot{u}(0)$ must be proportional to $r(u_l)$. We fix the parametrization so that the constant of proportionality is one:

$$\dot{u}(0) = r(u_l). \tag{8.8}$$

With this choice we have

$$\dot{\lambda} = \dot{u} \cdot \text{grad } \lambda = r \cdot \text{grad } \lambda = 1,$$

i.e., λ is an increasing function of ε at the origin. Therefore, condition (8.7) is satisfied only for positive values of ε.

The two halves of the one-parameter families of states to which a given state u_l can be connected by either a shock or a centered simple wave can be fitted together to form a single one-parameter family. Comparing (7.3′) with (8.8) we see that *the first derivative is continuous across* $\varepsilon = 0$.

Differentiate the relation $v(u(\varepsilon)) =$ const. a second time with respect to ε; we obtain at $\varepsilon = 0$

$$\ddot{u} \cdot \text{grad } v + \dot{u} \cdot (\dot{\text{grad}}\, r) = 0.$$

Substituting the expression (8.8) for $\dot{u}$ we have

$$\ddot{u} \cdot \text{grad } v + r \cdot (\dot{\text{grad}}\, v) = 0. \tag{8.9}$$

Next take the identities

$$r \cdot \text{grad } v = 0,$$

specialize u to $u = u(\varepsilon)$ and differentiate them with respect to ε:

$$\dot{r} \cdot \text{grad } v + r \cdot (\dot{\text{grad}}\, v) = 0. \tag{8.10}$$

Subtract (8.9) from (8.10); we get

$$(\ddot{u} - \dot{r}) \cdot \text{grad } v = 0.$$

Since the gradients of all the Riemann invariants span the orthogonal complement of r, it follows that

$$\ddot{u} = \dot{r} + \beta r.$$

But the parametrization of $u(\varepsilon)$ can be adjusted so that β is zero. Imagine that done; we then have

$$\ddot{u}(0) = \dot{r} = \dot{u} \cdot \operatorname{grad} r = r \cdot \operatorname{grad} r. \tag{8.11}$$

Compare (8.11) with (7.7′); they give identical values for $\ddot{u}$. *Hence our composite one-parameter family has continuous second derivatives even at* $\varepsilon = 0$.

An equivalent expression of this fact is

THEOREM 8.6. *The change in a k-Riemann invariant across a k-shock is of third order in* ε, *the magnitude of the shock.*

This is a well-known useful result for weak shocks in compressible flows. It appears here as a special case of a general law.

We turn now to characteristic fields which are linearly degenerate, i.e., in which $r_k \cdot \operatorname{grad} \lambda_k$ is identically zero. In our terminology, this fact is expressed by saying that λ_k is a k-Riemann invariant.

THEOREM 8.7. *If two nearby states* u_l *and* u_r *have the same k-Riemann invariants with respect to a degenerate field, then the Rankine-Hugoniot conditions are satisfied with* $s = \lambda(u_l) = \lambda(u_r)$.

Such a discontinuity is called a *contact discontinuity.*

Proof: We have to show that

$$s(u_r - u_l) = f(u_r) - f(u_l).$$

Imagine u_r and u_l connected by a differentiable one-parameter family of states $u(\varepsilon)$ for which all Riemann invariants are constant. This is certainly possible if u_r and u_l are nearby states. Since λ_k is one of the Riemann invariants, $\lambda_k(u(\varepsilon)) \equiv s$ for all ε. As before we conclude that $\dot{u}(\varepsilon)$, being orthogonal to the gradients of all Riemann invariants, is proportional to $r_k(u(\varepsilon))$; therefore

$$s\dot{u} = A\dot{u} \quad \text{for all } \varepsilon.$$

Integrating this we obtain the Rankine-Hugoniot relation.

In a previous publication, [15], the following result was stated and its proof briefly sketched.

THEOREM 8.8. *Assume that in the quasilinear system* (8.1) *one of the characteristic fields, say the k-th, is degenerate, i.e.,* $r_k \cdot \operatorname{grad} \lambda_k \equiv 0$. *Let* $u(x, t)$ *be a piecewise smooth solution of* (8.1), *and assume that* u *contains only a contact discontinuity, i.e., that the k-Riemann invariants are continuous across the discontinuity of u. Then, provided that the jump of u is not too large, u can be obtained as the limit of a sequence of smooth solutions.*

9. The Riemann Initial Value Problem

In this section we shall solve the initial value problem for initial states that are piecewise constant, with one jump discontinuity:

$$u(x, 0) = \phi(x) = \begin{cases} u_0 & \text{for } x < 0 \\ u_n & \text{for } 0 < x, \end{cases} \tag{9.1}$$

where u_0, u_n are constant vectors. The determination of the flow in a shock tube after the breaking of a diaphragm at $t = 0$ is such a problem; so is the problem of determining a flow after the head-on collision of two shocks, or the impinging of a shock on a contact discontinuity. Since such problems have been investigated by Riemann, we shall call this class of problems *Riemann's problem.*

As a preliminary observation we note that the solution $u(x, t)$ is a function of the ratio x/t. For, if $u(x, t)$ is a weak solution, with initial value (9.1), the function

$$u_\alpha = u(\alpha x, \alpha t),$$

α any positive constant, is also a weak solution; its discontinuities are shocks and it takes on the same initial value. Since presumably there is a unique solution, we must have $u_\alpha \equiv u$; this is the case if and only if u is a function of x/t only. Such a solution is called *centered at the origin* or, for brevity, centered.

Our solution to the initial value problem (9.1) consists of $n+1$ constant states $u_0, u_1, \cdots, u_n$; the constant states u_{k-1} and u_k are separated by a k-shock or a centered simple wave of the k-th kind, or, if the k-th field is linearly degenerate, by a contact discontinuity. The two end states u_0, u_n being given, we find the intermediate states u_1, u_2, etc. as follows:

According to Section 8, there is a one-parameter family of states u_1 that can be joined to u_0 on the right by waves of the first kind. Similarly, there is a one-parameter family of states u_2 which can be joined to u_1 on the right by a wave of the second kind, etc. Thus we get an n-parameter family of states

$$\begin{aligned} u_n &= u_n(u_0 ; \varepsilon_1, \varepsilon_2, \cdots, \varepsilon_n), \\ u_0 &= u_n(u_0 ; 0, 0, \cdots, 0) \end{aligned} \tag{9.2}$$

that can be joined to u_0; the Jacobian is not zero at the origin since, according to (8.8), the derivative of u_n with respect to the k-th parameter is parallel to the k-th right eigenvector r_k, and these, by assumption, are linearly independent. Therefore by the implicit function theorem, a sufficiently small cube in ε-space is mapped in a one-to-one way onto a neighborhood of u_0. This can be summarized in

THEOREM 9.1. *Every state u_0, has a neighborhood such that, if u_n belongs to this neighborhood, the Riemann initial value problem* (9.1) *has a solution. This solution consists of $n+1$ constant states connected by centered waves. There is exactly one solution of this kind, provided the intermediate states are restricted to lie in a neighborhood of u_0.*

In order to find such a solution when the state u_n is *not* near u_0, i.e. for large values of the parameters ε_i, one must study the n-parameter family (9.2) and verify that u_n is in its range.

Within terms of order ε, the intermediate states $u_1, u_2, \cdots, u_{n-1}$ can be found by decomposing the initial discontinuity as follows:

$$u_n - u_0 = \sum_{k=1}^{n} \varepsilon_k r_k . \tag{9.3}$$

Then

$$u_j = u_0 + \sum_{k=1}^{j} \varepsilon_k r_k + o(\varepsilon). \tag{9.4}$$

The expression (9.4) resembles the solution of Riemann's initial value problem for a linear equation with constant coefficients. In that case, the solution consists of $n+1$ constant states separated by characteristics, and the intermediate states are given by (9.3) and

$$u_j = u_0 + \sum_{k=1}^{j} \varepsilon_k r_k . \tag{9.4'}$$

Bibliography

[1] Bellman, R., *Functional equations in the theory of dynamic programming IV. Positivity and quasi-linearity*, Proc. Nat. Acad. Sci. U.S.A., Vol. 41, 1955, pp. 743–746.

[2] Burgers, J. M., *A mathematical model illustrating the theory of turbulence*, Advances in Applied Mechanics, Vol. 1, 1948, pp. 171–179.

[3] Chandrasekhar, S., *On the Decay of Plane Shock Waves*, Ballistic Research Lab., Aberdeen Proving Ground, Maryland Report No. 423, 1943.

[4] Cole, J. D., *On a quasilinear parabolic equation occurring in aerodynamics*, Quart. Appl. Math., Vol. 9, 1951, pp. 226–236.

[5] Courant, R., and Friedrichs, K. O., *Supersonic Flow and Shock Waves*, Interscience Publishers, Inc., New York, 1948.

[6] Friedrichs, K. O., *Formation and decay of shock waves*, Comm. Pure Appl. Math., Vol. 1, 1948, pp. 211–245.

[7] Friedrichs, K. O., *Wave Motions in Magnetohydrodynamics*, Los Alamos Scientific Lab. Technical Report, 1954.

[8] Friedrichs, K. O., *Nichtlineare Differenzialgleichungen*, Notes of lectures delivered in Göttingen, Summer, 1955.

[9] Germain, P., and Bader, R., *Unicité des écoulements avec chocs dans la mécanique de Burgers*, Office Nationale d'Etudes et de Recherches Aeronautiques, Paris, 1953, pp. 1–13.

[10] Goldner, S., *Existence and Uniqueness Theorems for Two Systems of Non-Linear Hyperbolic Differential Equations for Functions of Two Independent Variables*, Thesis, New York University, 1949.

[11] Hopf, E., *The partial differential equation* $u_t+uu_x=\mu u_{xx}$, Comm. Pure Appl. Math., Vol. 3, 1950, pp. 201–230.
[12] Keller, J. B., *Decay of finite amplitude sound waves*, unpublished note, 1955.
[13] Ladyzhenskaya, A., *On the construction of discontinuous solutions of quasilinear hyperbolic equations as limits to the solutions of the respective parabolic equations when the viscosity coefficient is approaching zero*, Doklady Akad. Nauk SSSR, Vol. 3, 1956, pp. 291–295.
[14] Lax, P. D., *On Discontinuous Initial Value Problems for Nonlinear Equations and Finite Difference Schemes*, L.A.M.S. 1332, Los Alamos, 1952.
[15] Lax, P. D., *The initial value problem for nonlinear hyperbolic equations in two independent variables*, Ann. of Math. Studies 33, Princeton, 1954, pp. 211–229.
[16] Lax, P. D., *Weak solutions of nonlinear hyperbolic equations and their numerical computation*, Comm. Pure Appl. Math., Vol. 7, 1954, pp. 159–193.
[17] Lax, P. D., *Difference Approximation to Solutions of Linear Differential Equations — An Operator Theoretical Approach*, Conference on partial differential equations, Summer, 1954, University of Kansas, Technical Report 14, 1955.
[18] Lax, P. D., and Richtmyer, R. D., *Survey of stability of linear finite difference equations*, Comm. Pure Appl. Math., Vol. 9, 1956, pp. 267–293.
[19] Lighthill, M. J., *Viscosity effects in sound waves of finite amplitude*, Surveys in Mechanics, Taylor Anniversary Volume, Cambridge University Press, 1956.
[19a] Lighthill, M. J., *The energy distribution behind decaying shocks*, Philos. Mag., Ser. 7, Vol. 41, 1950, pp. 1101–1128.
[20] Von Neumann, J., *Proposal and Analysis of a Numerical Method for the Treatment of Hydrodynamical Shock Problems*, National Defense and Research Committee Report AM551, 1944.
[21] Von Neumann, J., and Richtmyer, R. D., *A method for numerical calculation of hydrodynamical shocks*, J. Appl. Phys., Vol. 21, 1950, pp. 232–237.
[22] Olejnik, O. A., *On Cauchy's problem for nonlinear equations in the class of discontinuous functions*, Doklady Akad. Nauk SSSR, Vol. 95, 1954, pp. 451–454.
[23] Olejnik, O. A., *On Cauchy's problem for nonlinear equations in the class of discontinuous functions*, Uspehi Matem. Nauk, Vol. 9, 1954, pp. 231–233.
[24] Olejnik, O. A., *Boundary problems for partial differential equations with small parameter in the highest order and Cauchy's problem for nonlinear equations*, Uspehi Matem. Nauk, Vol. 10, 1955, pp. 229–234.
[25] Olejnik, O. A., *On discontinuous solutions of nonlinear differential equations*, Doklady Akad. Nauk SSSR, Vol. 109, 1956, pp. 1098–1101.
[26] Olejnik, O. A., *Cauchy's problem for nonlinear differential equations of first order with discontinuous initial data*. Trudy Moscow, Mat. Obsc, Vol. 5, 1956, pp. 433–454.
[27] Petrowsky, I. G., *Lectures on Partial Differential Equations*, Moscow, 1953.
[28] Vvedenskaya, N. D., *The difference method solution of Cauchy's problem for a nonlinear equation with discontinuous initial values*, Doklady Akad. Nauk SSSR, Vol. 3, 1956, pp. 517–521.
[29] Whitham, G. B., *The flow pattern of a supersonic projectile*, Comm. Pure Appl. Math., Vol. 5, 1952, pp. 301–348.
[30] Mandelbrojt, S., *Sur les fonctions convexes*, C. R. Acad. Sci. Paris, Vol. 209, 1939, pp. 977–978.
[31] Fenchel, W., *On conjugate convex functions*, Canadian J. Math., Vol. 1, 1949, pp. 73–77.
[32] Friedrichs, K. O., *Ein Verfahren der Variationsrechnung*, Nachr. Akad. Wiss. Göttingen. Math.-Phys. Kl. IIa., 1929, pp. 13–20.
[33] Hörmander, L., *Sur la fonction d'appui des ensembles convexes dans un espace localement convexe*, Ark. Mat., Vol. 3, 1955, pp. 181–187.

Received April, 1957.

COMMUNICATIONS ON PURE AND APPLIED MATHEMATICS, VOL. XIII, 217–237 (1960)

Systems of Conservation Laws*

PETER LAX and BURTON WENDROFF

New York University and Los Alamos Scientific Laboratory

Introduction

In this paper a wide class of difference equations is described for approximating discontinuous time dependent solutions with prescribed initial data of hyperbolic systems of nonlinear conservation laws. Among these schemes we determine the best ones, i.e., those which have the smallest truncation error and in which the discontinuities are confined to a narrow band of 2—3 meshpoints. These schemes are tested for stability and are found to be stable under a mild strengthening of the Courant-Friedrichs-Lewy criterion. Test calculations of one-dimensional flows of compressible fluids with shocks, rarefaction waves and contact discontinuities show excellent agreement with exact solutions. When Lagrange coordinates are used, there is no smearing of interfaces.

The additional terms introduced into the difference scheme for the purpose of keeping the shock transition narrow, when computed specifically for the equations of hydrodynamics, are reminiscent of the artificial viscosity terms introduced by Richtmyer and von Neumann and similar devices considered by other workers in the field. In particular, the methods proposed in this paper have some features similar to the methods proposed and used by Godunov [14].

1. Difference Schemes for Conservation Laws

In this paper we consider systems of conservation laws, i.e., equations of the form

$$u_t = f_x, \tag{1.1}$$

where u is an unknown vector function of x and t with n components, and f a given vector function of u, depending in general nonlinearly on u. When the differentiation on the right side of (1.1) is carried out a quasi-linear system results:

$$u_t = Au_x, \tag{1.1$'$}$$

*This paper represents results obtained under the sponsorship of the U. S. Atomic Energy Commission, Contract W-7405-ENG 36. Reproduction in whole or in part permitted for any purpose of the United States Government.

where $A = A(u)$ is the gradient of f. We require that our system of equations be hyperbolic in the sense that the matrix A has n distinct real eigenvalues $\mu_1, \cdots, \mu_n$ for all values of u. Of course the eigenvalues themselves are functions of u.

The negatives of the quantities μ are the local sound speeds. We shall index them in, say, monotonically increasing order.

The initial value problem is to find a solution of (1.1) with prescribed values at $t = 0$:

$$u(x, 0) = \varphi(x). \tag{1.2}$$

It is well known that no smooth solution will in general exist for all time because of the nonlinearity of (1.1). Instead we have to seek weak solutions of (1.1), (1.2), defined by the requirement that the integral relation

$$\iint (w_t u - w_x f)\,dx\,dt + \int w(x, 0)\varphi(x)\,dx = 0 \tag{1.3}$$

be satisfied for all smooth test vectors w which vanish for $|x|+t$ large enough.

An immediate consequence of the integral relations is that every piecewise continuous weak solution must satisfy the Rankine-Hugoniot relation across a line of discontinuity:

$$s[u]+[f] = 0; \tag{1.4}$$

here the brackets denote the jump across the discontinuity, s the speed of propagation of the discontinuity.

It is well known in the theory of systems of conservation laws (see, e.g., Section 7 of [9]) that in addition one has to impose a so-called entropy condition on discontinuities. This condition can be formulated as follows: *There is an index k such that the shock speed lies between the $(k-1)$-st and the k-th characteristic speed with respect to the state on the left of the shock, and between the k-th and the $(k+1)$-st characteristic speed on the right.*

We shall describe now a fairly large class of difference schemes which can be used to approximate weak solutions with prescribed initial data.

First we choose a vector valued function g of $2l$ vector arguments, related to f by the sole requirement that when all the $2l$ arguments are equal, g reduces to f:

$$g(u, \cdots, u) = f(u). \tag{1.5}$$

Abbreviate the values of $u(x)$ at the lattice points $x+k\Delta x$ and at time t by u_k, $k = -\infty, \cdots, \infty$. We define

$$g(x+\tfrac{1}{2}\Delta x) = g(u_{-l+1}, u_{-l+2}, \cdots, u_l);$$

therefore similarly we have

$$g(x-\tfrac{1}{2}\Delta x) = g(u_{-l}, u_{-l+1}, \cdots, u_{l-1}).$$

We take now the following difference analogue of (1.1):

$$\frac{\Delta u}{\Delta t} = \frac{\Delta g}{\Delta x}, \tag{1.6}$$

where Δu denotes the forward time difference

$$\Delta u = u(x, t+\Delta t) - u(x, t)$$

and Δg the symmetric space difference

$$\Delta g = g(x+\tfrac{1}{2}\Delta x) - g(x-\tfrac{1}{2}\Delta x).$$

From (1.6) we can determine $u(x, t+\Delta t)$:

$$u(x, t+\Delta t) = u(x, t) + \lambda \Delta g, \tag{1.6$'$}$$

where λ abbreviates the quotient $\Delta t/\Delta x$ of the time and space increments. Given the initial values φ of u, we can, using (1.6)′, determine successively the values of u at all times which are integer multiples of Δt. We claim now that as a consequence of (1.5) our difference scheme is *consistent* with the differential equation in the following sense: denote by $v(x, t)$ this solution of the difference equation (which for noninteger multiples t of Δt is defined, for the sake of convenience, as equal to $v(x, t')$, $t' = \Delta t[t/\Delta t]$). This v depends, of course, on Δt and Δx.

THEOREM. *Assume that as Δx and Δt tend to zero, $v(x, t)$ converges boundedly almost everywhere to some function $u(x, t)$. Then $u(x, t)$ is a weak solution of* (1.1) *with initial values φ.*

Proof: Multiply equation (1.6), satisfied by v, by any test vector w, integrate with respect to x and sum over all values of t which are integer multiples of Δt. Apply summation by parts to the left side, and transform the terms involving Δg by replacing the variable of integration x by $x-\frac{1}{2}\Delta x$ and $x+\frac{1}{2}\Delta x$, respectively. We get the identity

$$\begin{aligned} &\sum \int \frac{w(x, t-\Delta t) - w(x, t)}{\Delta t} v(x, t)\,dx\Delta t - \int w(x, 0)\varphi(x)\,dx \\ &\qquad = -\sum \int \frac{w(x+\frac{1}{2}\Delta x) - w(x-\frac{1}{2}\Delta x)}{\Delta x}\, g\, dx\Delta t, \end{aligned} \tag{1.3$'$}$$

where g stands for $g(v_1, \cdots, v_{2l})$, $v_1, \cdots, v_{2l}$ denoting values of v at $2l$ points which are distributed symmetrically around (x, t) and have distance Δx from each other. If v tends boundedly almost everywhere to $u(x, t)$, so do $v_1, \cdots, v_{2l}$, and thus $g(v_1, \cdots, v_{2l})$ tends to $g(u, \cdots, u)$, which, by the consistency requirement (1.5), is $f(u)$. Hence the limit of (1.3)′ is the desired integral relation (1.3).

The following special case is of interest: u is scalar and $f(u)$ has the special form

$$f(u) = \log \sum_{-l+1}^{l} a_k \exp\{ku\},$$

a_k constants whose sum is 1. If g is chosen to be

$$g(u_{-l+1}, \cdots, u_l) = \log \{\sum a_k \exp \sum_{0}^{k} u_j\},$$

then, as shown in [15], the difference equation (1.6)′ can be linearized and solved explicitely. In case that the constants a_k are non-negative the limit of the solutions of the difference equations can be determined explicitly; the value of the limit agrees with an explicit formula, valid for any convex f, given previously by Olejnik and Lax.

2. Minimizing the Truncation Error

Let $u(x, t)$ be an exact smooth solution of equation (1.1). It will then satisfy the difference equation (1.6)′ only approximately; the deviation of the right side from the left side of (1.6)′ is called the truncation error. It is easy to see that if the function g satisfies the consistency condition (1.5), the truncation error is at least $O(\Delta^2)$. In this section we shall determine g so that the truncation error is of as high an order as possible. Specifically we shall consider the case $l = 1$, and show that g can be chosen so that the truncation error is $O(\Delta^3)$.

Expand $u(x, t+\Delta t)$ into a Taylor series up to terms of second order:

$$u(x, t+\Delta t) = u(x, t)+\Delta t u_t+\tfrac{1}{2}(\Delta t)^2 u_{tt}+O(\Delta^3). \tag{2.1}$$

With the help of the differential equation (1.1) which u is supposed to satisfy, we can express the time derivatives of u as space derivatives:

$$\begin{aligned} u_t &= f_x, \\ u_{tt} &= f_{xt} = f_{tx} = (Au_t)_x = (A^2 u_x)_x; \end{aligned} \tag{2.2}$$

what is significant is that all t derivatives are exact x derivatives and therefore can be approximated by exact x differences. Substitute namely (2.2) into (2.1); we get

$$u(x, t+\Delta t) = u+(\Delta t f+\tfrac{1}{2}(\Delta t)^2 A^2 u_x)_x+O(\Delta^3). \tag{2.3}$$

Comparing (2.3) with (1.6)′, we see that the truncation error is $O(\Delta^3)$ if and only if

$$\frac{\Delta g}{\Delta x} = (f+\tfrac{1}{2}\Delta t A^2 u_x)_x+O(\Delta^2).$$

From this we can easily determine the form that g must take.

THEOREM. *The truncation error in the difference scheme* (1.6)′ *is* $O(\Delta^3)$ *if and only if*

$$g(a, b) = \frac{f(a)+f(b)}{2} + \tfrac{1}{2}\lambda A^2 \cdot (b-a) \tag{2.4}$$

plus terms which are $O(|a-b|^2)$ *for* $a-b$ *small.*

Formula (2.4) has a fairly intuitive meaning and can be derived by differencing our differential equation (1.1) as follows: replace time and space derivatives by differences centered at x, $t+\frac{1}{2}\Delta t$. This means that u_t is to be replaced by a forward time difference, and f_x by

$$\frac{f(x+\frac{1}{2}\Delta x, t+\frac{1}{2}\Delta t)-f(x-\frac{1}{2}\Delta x, t+\frac{1}{2}\Delta t)}{\Delta x}.$$

The value of f at $x+\frac{1}{2}\Delta x$, $t+\frac{1}{2}\Delta t$ is evaluated on the basis of the formula

$$f(x\pm\tfrac{1}{2}\Delta x, t+\tfrac{1}{2}\Delta t) = f(x\pm\tfrac{1}{2}\Delta x, t)+\tfrac{1}{2}\Delta t f_t+O(\Delta)^2.$$

We express f_t as $Au_t = A^2u_x$, and approximate u_x by a difference quotient. The value of $f(x\pm\frac{1}{2}\Delta x, t)$ is evaluated as the average of f at x and $x\pm\Delta x$. The resulting formula is precisely (2.4).

The quantity A^2 in (2.4) shall be taken as $\frac{1}{2}\{A^2(a)+A^2(b)\}$ for the sake of symmetry more than anything else; any other choice would make a difference that is quadratic in $a-b$.

Denoting the function (2.4) by g_0, we can write any permissible g in the form

$$g = g_0+\tfrac{1}{2}Q(a, b) \cdot (b-a), \tag{2.5}$$

where $Q(a, b)$ is a matrix which vanishes when its two vector arguments are equal.

Substituting formula (2.5) for g into the difference equation (1.6)′, we get

$$u(x, t+\Delta t) = u(x, t)+\lambda\Delta' f+\tfrac{1}{2}\lambda^2\Delta A^2\Delta u+\tfrac{1}{2}\lambda\Delta Q\Delta u, \tag{2.6}$$

where Δ' denotes the operator $\frac{1}{2}[T(\Delta x)-T(-\Delta x)]$ and Δ the operator $T(\frac{1}{2}\Delta x)-T(-\frac{1}{2}\Delta x)$, $T(s)$ denoting translation of the independent variable by the amount s. We shall call Q the *artificial viscosity*, since it occurs in the difference equations in a way which is similar to the artificial viscosity terms introduced by von Neumann and Richtmyer in [13].

The difference equation (2.6) expresses the value of u at time $t+\Delta t$ as a nonlinear function of u at t; we shall denote this function (operation) by N:

$$u(t+\Delta t) = Nu(t). \tag{2.6$'$}$$

The value of the solution of the difference equation at some later time $k\Delta t$ is obtained from the initial values by application of the k-th power of the operator N.

Our aim is to show that the difference scheme (2.6) is convergent if the size of λ is suitably restricted. Now in the case of linear equations it is well known and easy to show (see, e.g., [10]) that convergence is equivalent to stability defined as the uniform boundedness of all powers N^k of the operator N within some fixed range $p\Delta t \leqq T$. In the nonlinear case von Neumann has made the reasonable assumption (see, e.g., [4]) that the convergence of the scheme would hinge on the stability of the *first variation* of the operator N. The first variation of N is a linear difference operator with variable coefficients; von Neumann has conjectured that such an operation is stable if and only if all the "localized" operators associated with it, i.e., the operators obtained by replacing the variable coefficients by their value at some given point, are stable.[1]

The stability of a difference operator with constant coefficients is easily ascertained by making use of the Fourier transform. In that representation the application of a difference operator with constant coefficients goes over into multiplication by an "amplification matrix". In case the amplification matrix has n linearly independent eigenvectors, the operator is stable if and only if the eigenvalues of the amplification matrix do not exceed $1+O(\Delta t)$ in absolute value.

On the other hand, Courant, Friedrichs and Lewy have observed in their classical paper [2] that a necessary condition for the convergence of a difference scheme is that the rate of propagation of signals in the difference scheme should be at least as large as the true maximum signal speed, i.e.,

$$\frac{\Delta x}{\Delta t} \geqq |\mu|_{\max}, \tag{2.7}$$

where $|\mu|_{\max}$ is the largest eigenvalue of A at any point within the relevant range of values.

Conversely, we shall show:

THEOREM. *If condition* (2.7) *is fulfilled, the difference equation* (2.6) *satisfies von Neumann's condition of stability.*

Proof: The first variation of the operator appearing on the right side of (2.6) is easily computed; its value is

$$I+\lambda A\Delta'+\tfrac{1}{2}\lambda^2A^2\Delta^2+O(\Delta x), \tag{2.8}$$

where Δ' and Δ are defined as before, and $O(\Delta x)$ denotes an operator bounded

[1] In the parabolic case the validity of this conjecture has been established by Fritz John in his important paper [4]. In the hyperbolic case see Lax [8] and [15].

in norm by $O(\Delta x)$, provided that we are perturbing in the neighborhood of a smoothly varying solution, i.e., one where neighboring values differ by $O(\Delta x)$. In this case the influence of the additional artificial viscosity term is $O(\Delta x)$; in Section 3 we shall show how to take into account the effect on stability of the Q term in regions where u does vary rapidly.

To "localize" the operator (2.8) we merely replace the variable matrix A by its value at some point. After making a Fourier transformation, the operator Δ' becomes multiplication by $i \sin \alpha$, and the operator Δ multiplication by $2i \sin \frac{1}{2}\alpha$, so that $\frac{1}{2}\Delta^2$ becomes multiplication by $\cos \alpha - 1$; here α is $\xi \Delta x$, ξ being the dual variable. Hence the amplication matrix of the operator (2.8) is

$$I + i\lambda A \sin \alpha + \lambda^2 A^2(\cos \alpha - 1) + O(\Delta x). \tag{2.9}$$

Denote the eigenvalues of λA by k; then, according to the spectral mapping theorem, the eigenvalues v of the amplification matrix (2.9) are

$$v = 1 + ik \sin \alpha + k^2(\cos \alpha - 1) + O(\Delta x).$$

Since $k = \lambda\mu$ is real, the absolute value of v is given by

$$\begin{aligned} |v|^2 &= (1 - k^2[1 - \cos \xi])^2 + k^2 \sin^2 \xi + O(\Delta x) \\ &= 1 - 2k^2[1 - \cos \xi] + k^4[1 - \cos \xi]^2 + k^2(1 - \cos^2 \xi) + O(\Delta x) \\ &= 1 - (k^2 - k^4)[1 - \cos \xi]^2 + O(\Delta x). \end{aligned}$$

According to our basic stability assumption (2.7) the quantity k does not exceed 1 in absolute value; the above formula for $|v|$ shows that $|v|$ is bounded by $1 + O(\Delta x)$. Thus we have demonstrated that the eigenvalues of the amplification matrix do not exceed $1 + O(\Delta x)$ in absolute value, as required in von Neumann's condition.

3. Artificial Viscosity for Single Conservation Laws

In the last section we have determined the function $g(u, v)$ up to quadratic terms in $u - v$. These undetermined quadratic terms influence neither the order of the truncation error nor the stability of the scheme at points where the solution varies smootly. At points, however, where the solution varies rapidly—across a shock, that is—it is reasonable to expect that the quadratic terms have a controlling influence. In this section we shall show how to choose the additional quadratic term Q in our formula (2.6) so as to assure a fairly narrow shock width.

As observed in the last section, Q enters the difference equation (2.6) like an artificial viscosity. Therefore, in order to insure that Q has a stabilizing influence on the difference equation, it is reasonable to require that:

a) Q is positive.

Another property of Q, already previously noted, is:

b) $Q(a, b) = 0$ when $a = b$,

while on dimensional grounds we must require that:

c) Q has the dimension of A.

We shall discuss first the scalar case, i.e., when the number of components of u (and f) is one; correspondingly the matrix $A = \text{grad } f$ is one-by-one. In this case the restrictions a)—c) show that Q must be a multiple of $|A(a)-A(b)|$:

$$Q(a, b) = \tfrac{1}{2}B|A(a)-A(b)|, \tag{3.1}$$

where B is a dimensionless function (which of course could depend on the value of a and b). With the above choice for Q in the difference equation (3.1) we shall now study the shape of a *steady state shock*, i.e., a time independent solution of (2.6) connecting states $\mu(-\infty)$ and $\mu(\infty)$:

$$\Delta' f + \tfrac{1}{2}\lambda \Delta A^2 \Delta u + \tfrac{1}{2}\Delta Q \Delta u = 0. \tag{3.2}$$

We shall denote the values of u at three successive lattice points by a, b and c. We write,

$$\Delta' f = \tfrac{1}{2}[f(c)-f(b)] + \tfrac{1}{2}[f(b)-f(a)]$$

and make the following approximation:

$$f(c)-f(b) \approx A(b, c)(c-b),$$
$$f(b)-f(a) \approx A(a, b)(b-a),$$

where $A(a, b)$ abbreviates $\tfrac{1}{2}(A(a)+A(b))$. For the simplest nonlinear function f, a quadratic one, these approximations involve no error.

Substituting the above changes into (3.2), we get, after multiplication by 2,

$$\{A+\lambda A^2+Q\}(c-b)+\{A-\lambda A^2-Q\}(b-a) = 0. \tag{3.3}$$

We make now a further approximation by omitting the terms λA^2 in (3.3), under the reasonable assumption that the presence or absence of this term will not alter much the nature of the solution, since this term has the same character as the retained Q terms, and is expected to be much smaller than the Q term in the important region of rapid variation, especially if λ is small.

This assumption was put to the following numerical test: in the difference equation (2.6), A^2 was replaced by WA^2, where $W(a, b)$ is a factor so contrived that its value is near 1 when u and v are close, while its value is near zero when u and v differ greatly.[2] The presence or absence of such a

[2] I.e., W switches off the higher order correction term in the shock zone, where it has no function to fulfill anyway.

factor W made hardly any difference in the shape of the steady state shock profile obtained experimentally.

Making this change, and substituting our choice (3.1) for Q into (3.3), we get

$$\{A(b,c)+\tfrac{1}{2}B|A(c)-A(b)|\}(c-b)+\{A(a,b)-\tfrac{1}{2}B|A(b)-A(a)|\}(b-a)=0.$$

When $B\equiv 1$, this equation can be written in the following simple form:

$$\max\{A(b),A(c)\}(c-b)+\min\{A(a),A(b)\}(b-a)=0. \tag{3.3$'$}$$

Denote by u_L and u_R the two prescribed states at $-\infty$ and $+\infty$ which are supposed to be connected by a solution of (3.3)′. These states have to satisfy the Rankine-Hugoniot condition (1.4) which, for a stationary shock, is

$$f(u_L)=f(u_R). \tag{3.4}$$

In addition we suppose that the entropy condition is satisfied, i.e., that the shock speed is less than $-A(u_L)$, the sound speed to the left of the shock, and greater than $-A(u_R)$, the sound speed to the right:

$$A(u_L)<0<A(u_R). \tag{3.5}$$

It follows from (3.5) that there exists a value u_M such that

$$A(u_M)=0. \tag{3.6}$$

We claim now that the following lattice function is a solution of (3.3)′:

$$u(x)=\begin{cases}u_L & \text{for all negative lattice points,}\\ u_M & \text{for } x=0,\\ u_R & \text{for all positive lattice points.}\end{cases} \tag{3.7}$$

We have to verify that the difference equation (3.3)′ is satisfied for all triplets of three successive lattice points. Since (3.3)′ is trivially satisfied whenever $a=b=c$, only three cases remain to be checked:

I	II	III
$a=u_L$	u_L	u_M
$b=u_L$	u_M	u_R
$c=u_M$	u_R	u_R

In case I, it follows from the first half of inequality (3.5) and from (3.6) that $\max\{A(b),A(c)\}=0$; since $b-a$ is likewise zero, (3.3)′ is satisfied. We can show similarly that (3.3)′ is satisfied in case II.

In this case, the left side of (3.3)′ is

$$A(u_R)[u_R-u_M]+A(u_L)[u_M-u_L].$$

a) Q is positive.

Another property of Q, already previously noted, is:

b) $Q(a, b) = 0$ when $a = b$,

while on dimensional grounds we must require that:

c) Q has the dimension of A.

We shall discuss first the scalar case, i.e., when the number of components of u (and f) is one; correspondingly the matrix $A = \text{grad } f$ is one-by-one. In this case the restrictions a)—c) show that Q must be a multiple of $|A(a)-A(b)|$:

$$Q(a, b) = \tfrac{1}{2}B|A(a)-A(b)|, \tag{3.1}$$

where B is a dimensionless function (which of course could depend on the value of a and b). With the above choice for Q in the difference equation (3.1) we shall now study the shape of a *steady state shock*, i.e., a time independent solution of (2.6) connecting states $\mu(-\infty)$ and $\mu(\infty)$:

$$\Delta' f + \tfrac{1}{2}\lambda\Delta A^2 \Delta u + \tfrac{1}{2}\Delta Q \Delta u = 0. \tag{3.2}$$

We shall denote the values of u at three successive lattice points by a, b and c. We write,

$$\Delta' f = \tfrac{1}{2}[f(c)-f(b)] + \tfrac{1}{2}[f(b)-f(a)]$$

and make the following approximation:

$$f(c)-f(b) \approx A(b, c)(c-b),$$
$$f(b)-f(a) \approx A(a, b)(b-a),$$

where $A(a, b)$ abbreviates $\tfrac{1}{2}(A(a)+A(b))$. For the simplest nonlinear function f, a quadratic one, these approximations involve no error.

Substituting the above changes into (3.2), we get, after multiplication by 2,

$$\{A+\lambda A^2+Q\}(c-b) + \{A-\lambda A^2-Q\}(b-a) = 0. \tag{3.3}$$

We make now a further approximation by omitting the terms λA^2 in (3.3), under the reasonable assumption that the presence or absence of this term will not alter much the nature of the solution, since this term has the same character as the retained Q terms, and is expected to be much smaller than the Q term in the important region of rapid variation, especially if λ is small.

This assumption was put to the following numerical test: in the difference equation (2.6), A^2 was replaced by WA^2, where $W(a, b)$ is a factor so contrived that its value is near 1 when u and v are close, while its value is near zero when u and v differ greatly.[2] The presence or absence of such a

[2] I.e., W switches off the higher order correction term in the shock zone, where it has no function to fulfill anyway.

Section 2, that this factor does not exceed 1 in absolute value for any frequency α if and only if

$$(3.10) \qquad \lambda^2 A^2 + \lambda Q \leqq 1.$$

Denote by μ the largest possible value of $|A|$; it follows from equation (3.1) for Q that then Q will not exceed $\frac{1}{2}BV$, where V is the variation of A. If A does not change sign, $V \leqq \mu$. Substituting the above bounds into (3.10), we get

$$(3.11) \qquad \lambda^2\mu^2 + \tfrac{1}{2}B\lambda\mu \leqq 1,$$

which is equivalent to

$$(3.11)' \qquad \lambda\mu \leqq (1+\tfrac{1}{16}B^2)^{\frac{1}{2}} - \tfrac{1}{4}B.$$

In particular, we get for $B = 1$

$$\lambda\mu \leqq .78.$$

This stability condition is slightly more stringent than the Courant-Friedrichs-Lewy condition (2.7). It bears a strong resemblance to the stability condition derived by von Neumann and Richtmyer in [13] and to the stability criterion derived later by George White of LASL.

If in (3.10) the strict inequality holds, then the amplification factor (3.9) is actually less than 1 in absolute value for all frequencies α except $\alpha = 2\pi n$, n integer. This would lead us to believe that then the steady steat solutions are strongly stable, in the following sense:

Every solution of the difference equation (2.6), *with Q given by* (3.1), *whose initial values tend to u_L and u_R as x tends to $\pm\infty$ approaches a steady state solution with increasing t, provided that the stability condition* (3.11)′ *is satisfied.*

Given an arbitrary initial state, such that the corresponding exact solution of the differential equation (1.1) contains a number of shocks, not necessarily stationary or steady progressing, we would nevertheless expect the corresponding solution of the difference equation (2.6), with the same initial data, to bridge these shocks by transitions similar to the stationary solution we have found before, i.e., we expect the bulk of the transition to be spread over 2—3 mesh widths. This is reasonable, since the rate of variation of the states at the two sides of the shock will in general be much slower than the rapid variation within the shock itself.

Test calculations performed so far have confirmed this expectation. As yet none of these calculations have included very rapidly changing shocks, and hence we do not know at this time how the present method will handle such a situation.

4. Artificial Viscosity for Systems of Conservation Laws

In the case of a system of any number of conservation laws the three properties of Q listed at the beginning of Section 3 no longer suffice to determine Q within a single dimensionless function but still leave a rather bewildering variety of possibilities.

Richtmyer and von Neumann based their choice of the artificial viscosity on a physical analogy, and the various ingenious modifications proposed and tested by Landshoff, Harlow and Longley were likewise partly based on physical analogies. In this section we propose a form for Q which is dictated entirely by mathematical analysis.

Again we study the shape of stationary solutions of (2.6):

$$\Delta' f + \tfrac{1}{2}\lambda \Delta A^2 \Delta + \tfrac{1}{2}\Delta Q \Delta u = 0. \tag{4.1}$$

We denote again by a, b and c three consecutive values of u, and as before make the following approximation:

$$\begin{aligned} 2\Delta' f &= f(c)-f(a) = f(c)-f(b)+f(b)-f(a) \\ &\approx A(b, c)(c-b)+A(a, b)(b-a), \end{aligned} \tag{4.2}$$

where again $A(a, b)$ abbreviates $\frac{1}{2}(A(a)+A(b))$. Substituting this approximation into (4.1), we transform it to the following form:

$$\{A+\lambda A^2+Q\}(c-b)+\{A-\lambda A^2-Q\}(b-a) = 0, \tag{4.3}$$

where A and Q in the first brace are to be evaluated between the points b and c, in the second brace between the points a and b. We wish to reduce the above difference equations for a vector valued function to a scalar equation. Such a reduction is rigorously possible if all the coefficient matrices commute; in this case we can get n difference equations similar to (3.3), where the role of A and Q is played by the eigenvalues of A and Q. This dictates the following choice for Q: $Q(a, b)$ should be a matrix commuting with $A(a, b)$ whose eigenvalues are equal to the absolute values of the differences of the corresponding eigenvalue of $A(a)$ and $A(b)$ times dimensionless factors $B_1, \cdots, B_n$ of the order of magnitude 1.

The requirement that Q should commute with A implies, according to a well known theorem of matrix theory, that Q is a function of A. This function we can take to be a polynomial of degree $n-1$:

$$Q = g_0 I + g_1 A + \cdots + g_{n-1} A^{n-1} \tag{4.4}$$

whose coefficients $g_0, \cdots, g_{n-1}$ are uniquely determined by the proposed choice for the eigenvalues of Q. Finding the coefficients $g_0, \cdots, g_{n-1}$ leads to the Lagrange interpolation problem which is easily solved.

In the next section we shall actually carry out the determination of Q

in the case of equations of motion of a compressible fluid using Lagrange coordinates.

It should be pointed out that even if $Q(a, b)$ is chosen to commute with $A(a, b)$, there is an error involved in replacing (4.3) by scalar equations since the matrices $A(a, b)$ and $A(b, c)$ do *not* commute in general. Still we feel that the above choice for Q comes pretty close to imitating the scalar case.

We close this section by a further observation on the shape of the solutions of the steady state equation (4.1). In analysing this equation in the scalar case we have simplified matters by dropping the term $\frac{1}{2}\lambda A^2$ which was thought to be small compared to the artificial viscosity term in the region of rapid variation. On the other hand, in determining the shape of the tail ends of the transition curve the role of the two terms is reversed. We shall determine now asymptotically the shape of the tail ends. We write

$$a = u+r^k w, \qquad b = u+r^{k+1}w, \qquad c = u+r^{k+2}w,$$

where u is one of the two values u_L or u_R and w some vector independent of k. We substitute this into (4.3), divide by r^k and drop all terms that still contain the factor r^k. In particular the quadratic term Q drops out. We get

$$[\{A+\lambda A^2\}(r^2-r)+\{A-\lambda A^2\}(r-1)]w = 0,$$

where A denotes $A(u)$. This equation has a nontrivial solution if and only if the matrix acting on w has zero as eigenvalue. Now according to the spectral mapping theorem the eigenvalues of that matrix are

$$\{\mu+\lambda\mu^2\}(r^2-r)+\{\mu-\lambda\mu^2\}(r-1),$$

where μ stands for an eigenvalue of A. Setting the above expression equal to zero gives, after eliminating the uninteresting root $r = 1$, the following value for r:

$$r = \frac{\lambda\mu-1}{\lambda\mu+1}.$$

According to the Courant-Friedrichs-Lewy condition (2.7), $|\lambda\mu| \leqq 1$ so that the above expression is always negative. Furthermore, at the left endpoint u_L we must have $|r| > 1$, at the right end u_R we must have $|r| < 1$. This is the case if and only if $\mu(u_L)$ is negative, $\mu(u_R)$ is positive. But according to the entropy condition described at the beginning of Section 1, two states u_L and u_R which can be connected by a stationary shock always possess such eigenvalues.

A similar analysis can be made of the asymptotic shape of the tail of steady, progressing solutions of the difference equation (2.6). We are led to an equation which contains noninteger powers of r and which may have complex solutions.

All this is not very important as a practical consideration, but it does explain a curious phenomenon observed already in calculations performed with the Richtmyer-von Neumann method, and in many calculations using modifications of that method: that the shock transition overshoots and approaches a constant value in an oscillatory fashion. By changing the available parameter, the amount of overshooting could be diminished but never completely eliminated, nor could the oscillatory nature of the approach to a constant value be changed. The present analysis explains this behaviour by the negative value of r.

In contrast, calculations performed by a somewhat crude earlier method proposed by Lax (see [6] and [7]) produced shock transitions which approach their final values monotonically. A similar asymptotic study of the tail end of the steady state solutions for these equations disclosed that all the relevant values of r are positive.

5. Application to Hydrodynamics

In Section 4 we have sketched a method for constructing an artificial viscosity term for use in a numerical scheme for arbitrary systems of conservation laws. In this section we shall carry out the details of this construction in the special case of the equations of compressible flow in Lagrange coordinates.

As dependent variables we shall use specific volume[3], momentum and total energy, denoted by V, v and E. The quantity v is of course velocity, and the total energy E is the sum of internal and kinetic energy:

(5.1) $$E = e+\tfrac{1}{2}v^2.$$

The internal energy e is related to pressure p and specific volume V by the *equation of state*

$$p = p(e, V).$$

The equations of conservation of mass, momentum and energy are

(5.2) $$V_t = v_x, \qquad v_t = -p_x, \qquad E_t = -(vp)_x.$$

Here t is time and x is Lagrange mass variable. Differentiating these equations with respect to time, we obtain

(5.3) $$V_{tt} = v_{tx}, \qquad v_{tt} = -p_{tx}, \qquad E_t = (vp)_{tx}.$$

Using the chain rule, we have

(5.4) $$p_t = p_e e_t + p_V V_t.$$

According to a well known identity in thermodynamics

(5.5) $$pp_e - p_V = C^2,$$

[3] I.e., per unit mass.

where C is the Lagrangian sound speed. We can also write

$$e_t = E_t - vv_t. \tag{5.6}$$

Using (5.4), (5.5), (5.6) and the original differential equations (5.2) to express t derivatives as x derivatives, we can rewrite (5.3) as

$$V_{tt} = -p_{xx}, \qquad v_{tt} = (C^2 v_x)_x, \qquad E_{tt} = (pp_x + C^2 vv_x)_x. \tag{5.7}$$

Recalling the identity valid for arbitrary systems of conservation laws,

$$u_{tt} = (A^2 u_x)_x,$$

we see from (5.7) that in our case

$$A^2 u_x = \begin{pmatrix} -p_x \\ C^2 v_x \\ pp_x + C^2 vv_x \end{pmatrix}. \tag{5.8}$$

Accordingly we shall use the approximation

$$A^2 \Delta u = \begin{pmatrix} -\Delta p \\ C^2 \Delta v \\ p\Delta p + C^2 v \Delta v \end{pmatrix}. \tag{5.9}$$

We turn now to the determination of Q which, according to the recipe given in Section 4, is a quadratic polynomial in $A(a, b)$ whose eigenvalues are constant multiples of the absolute values of the differences of the eigenvalues of $A(a)$ and $A(b)$. Now the eigenvalues of A are $0, \pm C$. We claim that Q is given by

$$Q = \tfrac{1}{2} B \frac{|C(a) - C(b)|}{C^2} A^2;$$

here A and C are to be evaluated at $\frac{1}{2}(a+b)$. Clearly, Q as given by the above formula has the appropriate eigenvalues.

Substituting this form of Q into formulas (2.4) (2.5) for g and making use of (5.9), we have

$$g(a, b) = \begin{pmatrix} \bar{v} - \frac{1}{2}\left\{ B \dfrac{|\Delta C|}{2\bar{C}^2} + \lambda \right\} \Delta p \\ -\bar{p} + \frac{1}{2}\left\{ \dfrac{B|\Delta C|}{2\bar{C}^2} + \lambda \right\} \bar{C}^2 \Delta v, \\ -\overline{vp} + \frac{1}{2}\left\{ B \dfrac{|\Delta C|}{2\bar{C}^2} + \lambda \right\} \{\bar{p}\Delta p + \bar{C}^2 \bar{v} \Delta v\} \end{pmatrix}, \tag{5.10}$$

where the barred quantities are to be evaluated as averages between points a and b.

Two observations about this formula are in order:

1. Consider an initial distribution in which v and p are constant, although V not necessarily; in fact, V may be discontinuous. For such initial values the function g, given by formula (5.10), is a constant; therefore the corresponding solution of the difference equation (5.2) is stationary. This agrees, of course, with the fact that such an initial configuration is — in the absence of diffusion or heat conduction — in hydrodynamic equilibrium and shows that our difference scheme introduces no artificial diffusion or heat conduction.

More generally, we expect that even in nonstationary flows *contact discontinuities are transmitted as sharp discontinuities.*

2. Although our formula (5.10) for g was derived on a purely mathematical basis, it can be given a more intuitive interpretation. First of all, as already observed in Section 2, the second order correction terms can be regarded as merely centering the values of f properly. The additional artificial viscosity term can be given the following interpretation: Define an artificial *velocity* and an artificial *viscous pressure* as

$$v_{\text{art}} = -\frac{B\Delta x^2}{4C^2}|C_x|p_x,$$

$$p_{\text{art}} = -\frac{B\Delta x^2}{4}|C_x|v_x,$$

where the indicated derivatives are to be replaced by centered difference quotients. Clearly, the effect of the Q term in the mass and momentum equations is to augment the value of velocity and pressure — properly centered — by the amounts indicated above. The effect of the Q term on the energy equation can be described similarly if we neglect the difference between $\overline{vp}$ and $\bar{v}\bar{p}$, and if we neglect the product of the artificial velocity and the artificial viscous pressure.

The artificial viscous pressure that has cropped up in this treatment bears a strong resemblance to the one introduced by Rolf Landshoff.

We present the results of two calculations using (1.6) and (5.10) to obtain approximate solutions of (5.2). In the first calculation we initially had two constant states separated by a shock. The exact shock speed is 1. In Table I we show the appearance of the configuration at the 40-th time cycle which, with $\Delta t/\Delta x = .337$, corresponds to $t = 13.5$. At this time the shock which started at $x = 50$ should be at $x = 63.5$. We used 2 for the parameter B, and the second order correction terms were not switched off.

In the second calculation we again initially have two constant states separated by a discontinuity. However, this time the configuration at

40 cycles is a shock moving with speed 1.24, a contact discontinuity at the point of the initial discontinuity, namely $x = 50$, and a rarefaction wave. We used $\Delta t/\Delta x = .377$, $B = 1$. The results of the calculation are listed in Table II.

Figures

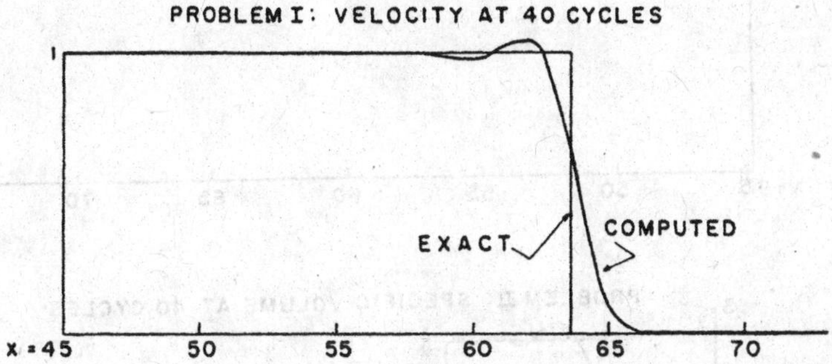

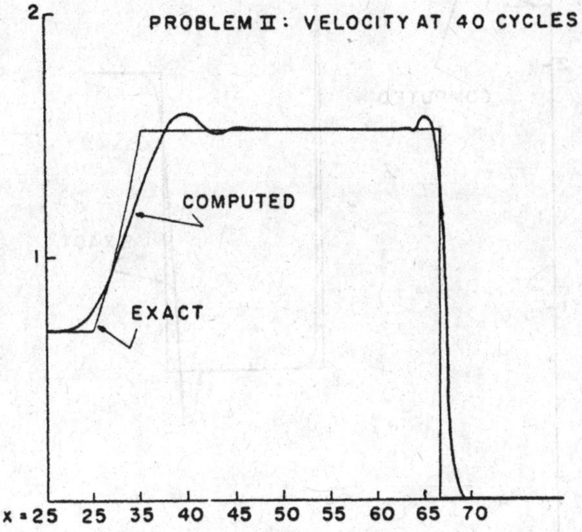

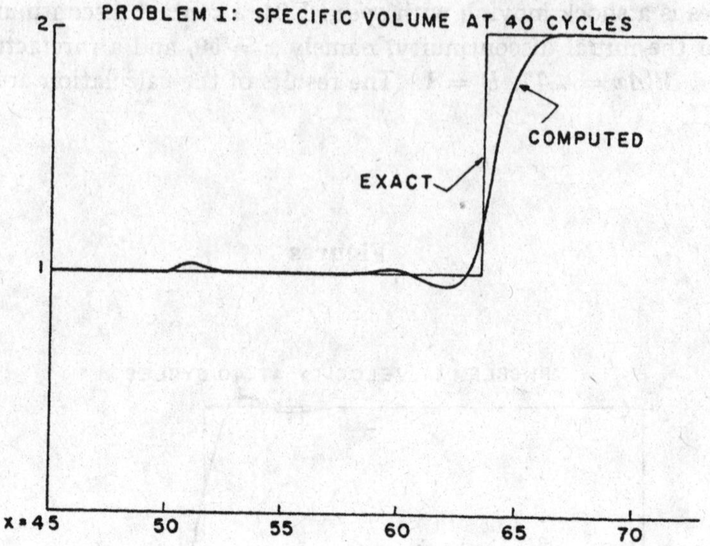
PROBLEM I: SPECIFIC VOLUME AT 40 CYCLES
2
1
COMPUTED
EXACT
X=45
50
55
60
65
70

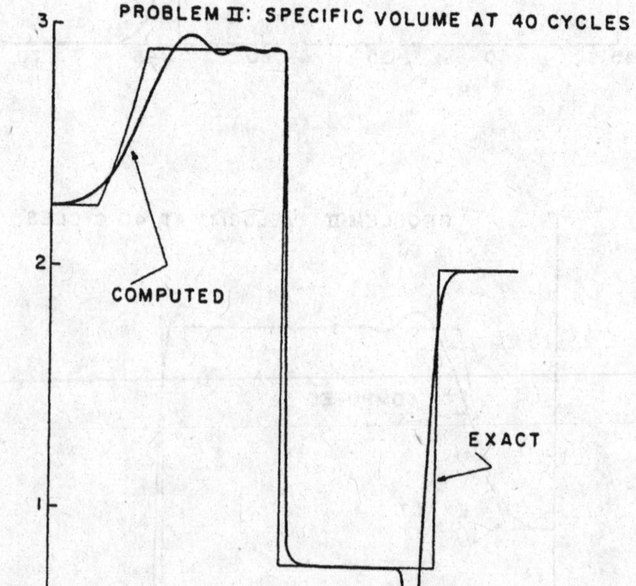
PROBLEM II: SPECIFIC VOLUME AT 40 CYCLES
3
2
1
COMPUTED
EXACT
X=25
30
35
40
45
50
55
60
65
70

Table I

Progressing shock

x	Volume	Velocity	Energy	
44	1.000	1.000	4.429	
45	1.000	1.000	4.429	
46	1.000	1.000	4.429	
47	1.000	1.000	4.429	
48	1.000	1.000	4.429	
49	1.000	1.000	4.428	
50	1.001	1.000	4.434	← Initial Position of Shock
51	1.044	1.000	4.601	
52	1.012	1.000	4.476	
53	1.003	1.001	4.445	
54	1.001	1.001	4.436	
55	1.002	.998	4.426	
56	1.002	.998	4.425	
57	.996	1.006	4.442	
58	.995	1.007	4.444	
59	1.010	.986	4.401	
60	1.014	.980	4.388	
61	.979	1.032	4.496	
62	.965	1.054	4.545	
63	1.103	.864	4.172	← Present Position of Shock
64	1.503	.411	3.384	
65	1.876	.085	2.944	
66	1.983	.011	2.867	
67	1.998	.001	2.858	
68	2.000	.000	2.857	
69	2.000	.000	2.857	
70	2.000	.000	2.857	

Table 2
Progressing Shock, Stationary Contact Discontinuity, and Rarefaction Wave

x	Volume	Velocity	Energy	
24	2.245	.698	20.04	
25	2.246	.699	20.04	$p = 3.528$
26	2.248	.702	20.03	
27	2.253	.709	20.02	
28	2.264	.725	20.00	
29	2.284	.754	19.95	
30	2.316	.800	19.87	
31	2.362	.866	19.78	
32	2.422	.948	19.66	
33	2.495	1.045	19.53	Rarefaction Wave (rows 30–36)
34	2.577	1.150	19.40	
35	2.660	1.259	19.28	
36	2.756	1.366	19.18	
37	2.842	1.463	19.09	
38	2.912	1.541	19.03	Exact Volume = 2.900
39	2.956	1.589	19.00	
40	2.963	1.596	19.00	
41	2.935	1.566	19.02	
42	2.898	1.525	19.05	
43	2.882	1.508	19.06	
44	2.892	1.518	19.05	
45	2.907	1.534	19.04	
46	2.909	1.536	19.04	
47	2.903	1.529	19.05	
48	2.902	1.526	19.06	$p = 2.465$
49	2.907	1.528	19.08	
50	2.887	1.528	18.96	← Initial Discontinuity
51	.825	1.528	6.253	
52	.777	1.528	5.956	
53	.772	1.528	5.930	
54	.774	1.528	5.937	
55	.773	1.528	5.932	Exact Volume = .767
56	.772	1.528	5.924	Exact Velocity = 1.528
57	.771	1.527	5.914	
58	.769	1.528	5.906	
59	.768	1.527	5.900	
60	.767	1.527	5.895	
61	.766	1.528	5.894	
62	.765	1.533	5.903	
63	.766	1.526	5.889	
64	.770	1.519	5.870	
65	.749	1.576	6.011	
66	.754	1.546	5.983	
67	1.204	.850	4.309	← Present Exact Position of Shock
68	1.852	.108	2.979	
69	1.990	.006	2.863	
70	2.000	.000	2.857	$p = .5714$

Bibliography

[1] Courant, R., and Friedrichs, K. O., *Supersonic Flow and Shock Waves*, Interscience Publishers, New York, 1948.

[2] Courant, R., Friedrichs, K. O., and Lewy, H., *Über die Partiellen Differenzengleichungen der Mathematischen Physik*, Math. Ann., Vol. 100, 1928, pp. 32–74.

[3] Harlow, F. H., *Hydrodynamic problems involving large fluid distortions*, J. Assoc. Comp. Mach., Vol. 4, 1957, pp. 137–142.

[4] John, F., *On integration of parabolic equations by difference methods*, Comm. Pure Appl. Math., Vol. 5, 1952, pp. 155–211.

[5] Landshoff, R., *A numerical method for treating fluid flow in the presence of shocks*, Los Alamos Scientific Laboratory Report LA-1930, January, 1955.

[6] Lax, P. D., *On discontinuous initial value problems for nonlinear equations and finite difference schemes*, Los Alamos Scientific Laboratory Report LAMS-1332, December, 1952.

[7] Lax, P. D., *Weak solutions of nonlinear hyperbolic equations and their numerical computation*, Comm. Pure Appl. Math., Vol. 7, 1954, pp. 159–193.

[8] Lax, P. D., *On the scope of the energy method*, Bull. Amer. Math. Soc., Vol. 66, 1960, pp. 32–35.

[9] Lax, P. D., *Hyperbolic systems of conservation laws II*, Comm. Pure Appl. Math., Vol. 10, 1957, pp. 537–566.

[10] Lax, P. D., and Richtmyer, R. D., *Survey of the stability of linear finite difference equations*, Comm. Pure Appl. Math., Vol. 9, 1956, pp. 267–293.

[11] Longley, H. J., *Methods of differencing in Eulerian hydrodynamics*, Los Alamos Scientific Laboratory Report, LAMS-2379.

[12] Von Neumann, J., *Proposal and analysis of a numerical method for the treatment of hydrodynamical shock problems*, National Defense and Research Committee Report AM-551, March, 1944.

[13] Von Neumann, J., and Richtmyer, R. D., *A Method for the numerical calculations of hydrodynamic shocks*, J. Appl. Physics, Vol. 21, 1950, pp. 232–237.

[14] Godunov, S. K., *Difference method for the numerical computation of discontinuous solutions of equations of hydrodynamics*, Mat. Sbornik, N. S., Vol. 47 (89), No. 3, 1959, pp. 271–306.

[15] Lax, P. D., *On difference schemes for solving initial value problems for conservations laws*, Symposium on Questions of Numerical Analysis, Provisional International Computing Center, Rome, 1958, pp. 69–78.

Received August, 1959.

COMMUNICATIONS ON PURE AND APPLIED MATHEMATICS, VOL. XVII, 381–398 (1964)

Difference Schemes for Hyperbolic Equations with High Order of Accuracy*

PETER D. LAX AND BURTON WENDROFF

Introduction

The limitation of the speed and memory of calculating machines places an upper bound on the number of mesh points that may be used in a finite difference calculation. This means that in problems involving many independent variables (and for present-day machines, three is many) the mesh employed is necessarily coarse. Therefore in order to get reasonably accurate final results one must employ highly accurate difference approximations. The purpose of this paper is to set up and analyse such difference schemes for solving the initial value problem for first order symmetric hyperbolic systems of partial differential equations in two space variables.

It is well known that a difference scheme furnishes accurate answers over a reasonably long range of time only if it is stable. The bulk of this paper is devoted to determining the conditions under which the proposed difference schemes are stable.

In Section 1 we give a brief review of the general theory of accuracy and stability of difference schemes. In Section 2 we show that the set of matrices whose field of values belongs to the unit circle forms a stable family. This result is of interest in its own right. In Section 3 we set up some difference schemes of second order accuracy and, with the aid of the criterion described in Section 2, analyse the range of parameters for which these schemes are stable. In Section 4 we give a geometric interpretation of the stable range of the parameters and in this connection we devise a difference scheme with a maximum stable range; this scheme however is accurate only to first order. Section 5 contains some remarks and open questions concerning the effect of the non-constancy of the coefficients of a difference scheme on stability. In Section 6 we show how to set up difference schemes with higher order accuracy for non-linear hyperbolic systems of conservation laws, such as the equations of compressible flow and magnetohydrodynamics.

The difference scheme $(3.3)_S$ discussed in this paper has also been proposed by C. Leith and used by him in meteorological calculations carried out at the

* This paper represents results obtained under the sponsorship of the United States Atomic Energy Commission Contract 0-5741-172. Reproduction in whole or in part is permitted for any purpose of the United States Government.

Livermore National Laboratory. An interesting "two-step" derivation of this difference scheme is given by R. D. Richtmyer in his survey article [16].

In [21] Strang proposed several difference schemes of high order accuracy and analysed their stability. It would be desirable to make a thorough study of the comparative usefulness of all these difference schemes in various situations.

The difference scheme (6.8), (6.9) has been used by S. Burstein in a series of highly successful calculations of shocked flows in a narrowing channel. The calculations were carried out at the A.E.C. Computing Center at N.Y.U., and a portion of them is described in [20].

1. Review of the Notion of Stability and Accuracy

The class of equations under consideration are of the form

(1.1) $$u_t = Au_x + Bu_y ,$$

u a vector function of x, y and t, and A, B symmetric matrices which may depend on x and y; for the sake of convenience we shall not consider explicit dependence on t. On occasion we shall abbreviate the right side of (1.1) by G and write the equation in the form

(1.1)′ $$u_t = Gu ,$$

indicating explicitly only the dependence of u on t. We are interested in the initial value problem, i.e., the problem of finding a solution of (1.1), given the value of $u(0)$.

We shall consider difference approximations to (1.1) of the form

(1.2) $$v(t+h) = S_h v(t) ;$$

here v denotes an approximation to u, h is the time increment and S_h is a difference operator

(1.3) $$S_h = \sum_j c_j T^j,$$

where j is a multi-index (j_x, j_y) and T^j abbreviates $T_x^{j_x} T_y^{j_y}$, where T_x and T_y denote translations by the amounts μh and νh in the x and y directions, respectively, μ and ν being constants independent of h. The coefficient matrices c_j are functions of x, y and of h; they are polynomials in h.

Definition. The difference scheme (1.2), (1.3) approximates the differential equation (1.1) with m-th order accuracy if for all smooth solutions $u(t)$ of (1.1)

(1.4) $$\|u(t+h) - S_h u(t)\| \leqq O(h^{m+1}) ,$$

i.e., if after one time step exact and approximate solutions differ only by $O(h^{m+1})$.

DEFINITION. The difference scheme (1.2) is stable if its solutions are uniformly bounded in a unit time range, i.e., if there exists a constant K such that

$$\|S_h^n\| \leqq K \tag{1.5}$$

for all n,h satisfying $nh \leqq 1$.

In this paper the norm appearing in these definitions will be taken as the L_2 norm.

The following is well known, see [14]:

THEOREM 1. *Let u and v denote solutions of the exact differential equation* (1.1) *and the difference equation* (1.2), *respectively, having the same smooth initial values. Then*

$$\|u(t) - v(t)\| \leqq O(h^m)\,, \qquad t \leqq 1\,, \tag{1.6}$$

for all smooth initial values if the difference scheme is stable and accurate of order m.

Since accuracy and some sort of stability are necessary as well as sufficient in order that the overall error be of the order (1.5), we endeavor to construct difference approximations accurate of order m and also stable. In the present paper we take m to be 2.

There is a simple way of expressing the accuracy of a difference scheme; we shall derive this form first in the case when the coefficients of the differential and difference equations are constant, i.e., independent of x and y. We start with the observation that it suffices to verify the error estimate (1.5) for a dense set of solutions. We choose these solutions as the exponential ones; that is, we prescribe $u(0)$ as

$$u(0) = e^{i(x\xi + y\eta)}\phi\,, \tag{1.7}$$

where ξ, η are arbitrary real numbers and ϕ is an arbitrary vector. The corresponding solution of (1.1) is, as is easy to check,

$$u(t) = e^{it(\xi A + \eta B)}u(0)\,, \tag{1.8}$$

while the corresponding solution of (1.2) is

$$v(h) = C(\mu h\xi, \nu h\eta)u(0)\,, \tag{1.9}$$

where

$$C(\zeta) = \sum_j c_j e^{ij\zeta}\,. \tag{1.10}$$

Here ζ denotes the vector ξ, η.

Comparing (1.8) and (1.9) we see that accuracy of order m means that

$$e^{i(\xi A + \eta B)} = C(\mu\xi, \nu\eta) + O(|\zeta|^{m+1}) \tag{1.11}$$

for ζ near zero.

The function C defined by (1.10) is called the *amplification matrix* of the difference operator (1.2).

We turn now to the question of stability of difference schemes. We shall deal here with the case of constant coefficients; schemes with variable coefficients will be discussed in Section 5.

Denote the Fourier transformation in the space variables by T; then

$$TS_h v = C(h\zeta)\, Tv\,,$$

where ζ denotes the dual variable. Repeated applications of the above identity, gives

$$TS_h^n v = C^n(h\zeta)\, Tv\,.$$

Since Fourier transformation is an isometry in the L_2 norm, the uniform boundedness of $S_h^n v$ is equivalent to the uniform boundedness of their Fourier transforms. The latter is clearly equivalent to the uniform boundedness of the matrices C^n. Thus we have shown:

THEOREM 2. *A difference scheme with constant coefficients is stable if and only if all powers of the associated amplification matrix are bounded, uniformly for all real values of ζ and all powers of the matrix.*

To be able to use the above stability criterion we need to know conditions under which a family of matrices has the property that all powers of its elements are uniformly bounded. As observed by von Neumann, a necessary condition for this is that the eigenvalues of each matrix of the family be not greater than one in absolute value. This condition by itself is not sufficient; there are various additional conditions given in the literature, see [15] and [5] which together with von Neumann's condition guarantee the uniform boundedness of the set $\{C^n\}$. Necessary and sufficient conditions were given by Kreiss [6], [7] and by Buchanan [2]. In the next section we shall give a new sufficient condition and use it in Section 3 to discuss the stability of the difference schemes which are the subject of this paper.

Recently Kreiss has shown that our stability condition is a consequence of a new necessary and sufficient condition formulated by him, see [17], also [18]. An illuminating new proof of the criterion of Kreiss has been given by Morton in [19]; the deduction of our stability criterion from that of Kreiss is given there.

2. A Stability Theorem

THEOREM 3. *Suppose that the field of values of a matrix C lies in the unit disk, i.e., that*

(2.1) $$|(Cu, u)| \leqq 1$$

for all unit vectors u. Then there exists a constant K depending only on the order of C such that

$$|C^n| \leqq K\,, \qquad n = 1, 2 \cdots.$$

Here $|C^n|$ denotes the operator norm of C^n with respect to the Euclidean norm for vectors.

Remark. Since all eigenvalues of C belong to its field of values, (2.1) implies that C satisfies the von Neumann condition.

Proof of Theorem 3: It is well known that every matrix is unitarily equivalent with an upper triangular matrix. Since the field of values of two matrices so related is the same, and since the norms of their powers are the same, we may assume that C is already upper triangular.

We shall treat first the 2×2 case:

$$C = \begin{pmatrix} a & b \\ 0 & c \end{pmatrix}.$$

Then

$$(Cu, u) = a|t|^2 + b\bar{t}z + c\,|z|^2 ,$$

where t, z are the components of u. Choose $|t|^2 = |z|^2 = \frac{1}{2}$, and adjust the argument of t and z so that the middle term has the same argument as the sum of the first and the third. With such a choice

$$|(Cu, u)| = \left|\frac{a+c}{2}\right| + \frac{|b|}{2} .$$

So from (2.1) we have

$$|b| \leqq 2 - |a + c| . \tag{2.2}$$

By the triangle inequality, $|a + c| \geqq 2\,|c| - |a - c|$; this and (2.2) imply that

$$|b| \leqq 2(1 - |c|) + |a - c| . \tag{2.3}$$

Consider now powers of C; it is easy to show recursively that

$$C^n = \begin{pmatrix} a^n & P_n(a, c)b \\ 0 & c^n \end{pmatrix}, \tag{2.4}$$

where

$$P = P_n(a, c) = a^{n-1} + a^{n-2}c + \cdots + c^{n-1} = \frac{a^n - c^n}{a - c} .$$

As remarked before, it follows from (2.1) that the eigenvalues a and c of C do not exceed one in absolute value. Hence using $|a| \leqq 1$, we have from the first form for P the inequality

$$|P| \leqq 1 + |c| + \cdots + |c|^{n-1} \leqq \frac{1}{1 - |c|} . \tag{2.5}$$

From the second form using $|a|^n$, $|c|^n \leqq 1$, we get

$$|P| \leqq \frac{2}{|a - c|} . \tag{2.6}$$

Multiplying (2.5) by $2(1 - |c|)$, (2.6) by $|a - c|$ and adding the two, we get

$$[2(1 - |c|) + |a - c|]\,|P| \leqq 4 . \tag{2.7}$$

Combining (2.3) and (2.7), we get

$$|bP| \leqq 4 \tag{2.8}$$

which shows that the corner element of C^n is at most four.

The above derivation of (2.8) from (2.3) is taken from Buchanan's paper [2].

We shall now prove Theorem 3 for $p \times p$ matrices C inductively on p; this device is borrowed from de Bruijn who has used it in [1].

Let C be an upper triangular $p \times p$ matrix. We write it in block notation as

$$C = \begin{pmatrix} A & \beta \\ 0 & c \end{pmatrix},$$

where A is a $(p-1) \times (p-1)$ upper triangular matrix, β is a column vector with $p-1$ components, and c is a scalar.

Let u be a column vector with p components and length 1. We can write it as

$$u = \begin{pmatrix} tv \\ z \end{pmatrix},$$

where v is a unit vector with $p-1$ components and t, z are complex numbers satisfying

$$|t|^2 + |z|^2 = 1 .$$

With this notation, the field of values of our matrix C may be written in the form

$$\begin{aligned}(Cu, u) &= (Av, v)\,|t|^2 + (\beta, v)\,\bar{t}z + c\,|z|^2 \\ &= a\,|t|^2 + b\bar{t}z + c\,|z|^2 ,\end{aligned}$$

where a and b abbreviate

$$a = (Av, v)\,, \qquad b = (\beta, v)\,. \tag{2.9}$$

Since C satisfies (2.1), it follows that the absolute value of the expression $a\,|t|^2 + b\bar{t}z + c\,|z|^2$ does not exceed 1. Since this expression can be thought of also as the field of values of the 2×2 matrix

$$\begin{pmatrix} a & b \\ 0 & c \end{pmatrix},$$

we conclude that a, b, c satisfy the inequality (2.3), where we now substitute the expression (2.9) for a and b. This yields

$$|(\beta, v)| \leqq 2(1 - |c|) + |((A - cI)v, v)|\,; \tag{2.10}$$

here we wrote $a - c = (Av, v) - c = (Av, v) - c(v, v) = ((A - cI)v, v)$, since v is a unit vector. We also conclude that $|a| \leqq 1$, i.e., that the $(p-1) \times (p-1)$ matrix A satisfies the hypothesis (2.1). Moreover, $|c|$ does not exceed 1.

Our aim is to derive a uniform bound for all powers of C. It is easy to verify recursively that these may be written as

$$C^n = \begin{pmatrix} A^n & P_n(A, c)\beta \\ 0 & c^n \end{pmatrix},$$

where $P_n(A, c)$ is the matrix

$$P = A^{n-1} + cA^{n-2} + \cdots + c^{n-1}I = (A^n - c^nI)(A - cI)^{-1}. \tag{2.11}$$

By the induction hypothesis, there exists a constant K such that

$$|A^n| \leqq K, \qquad n = 1, 2, \cdots. \tag{2.12}$$

Thus in order to find an upper bound for $|C^n|$ it suffices to find one for the norm of the vector $P\beta$.

From the first expression for P in (2.11), using (2.12), we find that

$$|P| \leqq |A^{n-1}| + \cdots + |c^{n-1}| \leqq K(1 + \cdots + |c|^{n-1}) \leqq \frac{K}{1 - |c|}. \tag{2.13}$$

Next we note that, according to a well-known principle, the norm of the vector $P\beta$ can be characterized in terms of inner products:

$$|P\beta| = \sup_{|w|=1} |(P\beta, w)| = \sup_{|w|=1} |(\beta, P^*w)|. \tag{2.14}$$

Thus we can bound $|P\beta|$ from above by finding an upper bound for $(P\beta, w) = (\beta, P^*w)$.

We now choose the unit vector

$$v = \frac{P^*w}{|P^*w|};$$

then (2.10) becomes

$$\frac{|(\beta, P^*w)|}{|P^*w|} \leqq 2(1 - |c|) + \frac{|((A - cI)P^*w, P^*w)|}{|P^*w|^2}. \tag{2.15}$$

The second term on the right can be rewritten as

$$\frac{(P(A - cI)P^*w, w)}{|P^*w|^2};$$

and using the second expression in (2.11) for P we can rewrite this as

$$\frac{((A^n - c^nI)P^*w, w)}{|P^*w|^2}. \tag{2.16}$$

By (2.12) and since $|c| \leqq 1$, the norm of the operator $A^n - c^nI$ is at most $K + 1$; so

$$\|(A^n - c^nI)P^*w\| \leqq (K + 1)\, \|P^*w\|.$$

Therefore, if we estimate the numerator in (2.16) by the Schwarz inequality, then use the above estimate and the fact that w is a unit vector we find that

$$\frac{K + 1}{\|P^*w\|}$$

is an upper bound for (2.16). Substituting this upper bound for the second term on the right in (2.15) we get, after multiplication by $|P^*w|$,

$$|(\beta, P^*w)| \leqq 2(1 - |c|)\, |P^*w| + K + 1 .$$

Since w is a unit vector, $|P^*w| \leqq |P^*|$; the norm of P^* equals that of P, for which we already have the estimate (2.13). Thus we obtain

$$|(\beta, P^*w)| \leqq 3K + 1 \quad \text{for all unit vectors } w.$$

In view of (2.14), the above inequality shows that

$$|P_n\beta| \leqq 3K + 1 ;$$

this is the required uniform bound, and the induction is now completed.

Observe that the value of the uniform bound K derived here depends on the order p of the matrix C, and increases exponentially with p. It would be of some interest to study the dependence on p of the best constant K in Theorem 3.

3. Derivation and Analysis of Difference Schemes of Second Order Accuracy

We shall construct now with the aid of condition (1.11) some difference schemes which are accurate to second order. We take A and B to be constants and assume for simplicity $\mu = \nu = 1$. Expanding in Taylor series near $\xi = \eta = 0$, we have

$$e^{i(\xi A + \eta B)} \equiv I + i(\xi A + \eta B) - \tfrac{1}{2}(\xi A + \eta B)^2 , \tag{3.1}$$

where the symbol $\equiv$ denotes congruence modulo third order terms. Furthermore,

$$\begin{gathered} \xi \equiv \sin \xi , \qquad \eta \equiv \sin \eta , \qquad \xi\eta \equiv \sin \xi \sin \eta , \\ \tfrac{1}{2}\xi^2 \equiv 1 - \cos \xi , \qquad \tfrac{1}{2}\eta^2 \equiv 1 - \cos \eta . \end{gathered} \tag{3.2}$$

Substitute the congruences (3.2) into the right side of (3.1) and denote the resulting function by $C(\xi, \eta)$:

$$\begin{aligned} C = I + i(A \sin \xi + B \sin \eta) - A^2(1 - \cos \xi) \\ - \tfrac{1}{2}(AB + BA) \sin \xi \sin \eta - B^2(1 - \cos \eta) . \end{aligned} \tag{3.3}$$

By the above construction and (1.11), C is the amplification matrix of a difference scheme which is accurate to second order.

Given the amplification matrix C of a difference operator S, we can recover S by replacing $e^{i\xi}$ and $e^{i\eta}$ in C by translations by the amount h in the x and y directions. For C in (3.3) we get

$$S = I + \tfrac{1}{2}AD_{1x} + \tfrac{1}{2}BD_{1y} + \tfrac{1}{2}A^2D_{2x} + \tfrac{1}{8}(AB + BA)D_{1x}D_{1y} + \tfrac{1}{2}B^2D_{2y} , \tag{3.3$_S$}$$

where D_1 and D_2 denote symmetric first and second differences:

$$D_1 = T - T^{-1}, \qquad D_2 = T - 2I + T^{-1}.$$

A more intuitive way of arriving at the difference operator $(3.3)_S$ is to write, by Taylor's theorem, the approximation

$$\begin{aligned} v = u(t+h) &\equiv u + hu_t + \frac{h^2}{2} u_{tt} \\ &= \left(I + hD_t + \frac{h^2}{2} D_t^2\right) u . \end{aligned} \tag{3.4}$$

For solutions of (1.1), time derivatives can be expressed as space derivatives:

$$D_t = AD_x + BD_y, \qquad D_t^2 = A^2D_x^2 + (AB + BA)D_xD_y + B^2D_y^2 . \tag{3.5}$$

Substitute (3.5) into (3.4), and express the first and second space derivatives as symmetric divided first and second differences, respectively. We obtain

$$v = Su,$$

S given by $(3.3)_S$.

(3.3) is not the only nine-point scheme which is accurate to second order. The scheme associated with the amplification matrix

$$C' = C - \frac{A^2 + B^2}{2}(1 - \cos \xi)(1 - \cos \eta), \tag{3.3$'$}$$

C as given by (3.3), has the same accuracy as C, since the added term in C' is of fourth order.

THEOREM 4. *The difference scheme associated with* (3.3) *is stable if*

$$A^2 \leqq \tfrac{1}{8}I, \qquad B^2 \leqq \tfrac{1}{8}I, \tag{3.6}$$

and the scheme associated with (3.3)$'$ *is stable if*

$$A^2 + B^2 \leqq \tfrac{1}{2}I. \tag{3.6$'$}$$

These results are the best possible ones in the sense explained at the end of the proof.

Proof: We shall show that under the conditions stated above both C and C' satisfy the hypothesis of Theorem 3. We start with C; separating it into real and imaginary part we write

$$C = R + iJ.$$

Here

$$J = A \sin \xi + B \sin \eta \tag{3.7}$$

and

$$R = I - K, \tag{3.8}$$

where K abbreviates

$$K = A^2(1 - \cos\xi) + B^2(1 - \cos\eta) + \tfrac{1}{2}(AB + BA)\sin\xi\sin\eta\,. \tag{3.9}$$

It turns out, and this is of importance, that K can be written as a sum of three squares. Using the abbreviations

$$1 - \cos\xi = X\,, \qquad 1 - \cos\eta = Y \tag{3.10}$$

we have the identity

$$K = \tfrac{1}{2}A^2X^2 + \tfrac{1}{2}B^2Y^2 + \tfrac{1}{2}J^2\,. \tag{3.11}$$

The verification of this identity is left to the reader.

Clearly since A and B are symmetric so are R and J.

Our aim is to estimate the quantity (Cu, u) for unit vectors u. We can write

$$(Cu, u) = (Ru, u) + i(Ju, u) = r + ij\,.$$

Since R and J are real and symmetric, r and j are real, and so

$$|(Cu, u)|^2 = r^2 + j^2\,. \tag{3.12}$$

We estimate j by the Schwarz inequality:

$$j^2 = (Ju, u)^2 \leqq |Ju|^2\,. \tag{3.13}$$

By (3.8) we can write

$$r = (Ru, u) = 1 - (Ku, u)\,. \tag{3.14}$$

By (3.11) we can write

$$\begin{aligned}(Ku, u) &= \tfrac{1}{2}\,|Au|^2\,X^2 + \tfrac{1}{2}\,|Bu|^2\,Y^2 + \tfrac{1}{2}\,|Ju|^2 \\ &= \tfrac{1}{2}aX^2 + \tfrac{1}{2}bY^2 + \tfrac{1}{2}\,|Ju|^2\,,\end{aligned} \tag{3.15}$$

where we have used the abbreviations

$$|Au|^2 = a\,, \qquad |Bu|^2 = b\,. \tag{3.16}$$

Squaring (3.14) and using (3.15), we get

$$r^2 = 1 - aX^2 - bY^2 - |Ju|^2 + (Ku, u)^2\,.$$

Adding (3.13) to this we get

$$r^2 + j^2 \leqq 1 - aX^2 - bY^2 + (Ku, u)^2\,. \tag{3.17}$$

Next we turn to estimating (Ku, u); using (3.9) we have

$$(Ku, u) = aX + bY + \mathcal{R}e\,(Au, Bu)\sin\xi\sin\eta\,.$$

Applying the Schwarz inequality to the last term on the right, we get

$$|(Ku, u)| \leqq aX + bY + |Au|\,|Bu|\,|\sin\xi|\,|\sin\eta|\,. \tag{3.18}$$

Estimating the last term in the above by

$$\frac{|Au|^2 \sin^2 \xi + |Bu|^2 \sin^2 \eta}{2}$$

and using the elementary inequalities

$$\frac{\sin^2 \xi}{2} \leqq 1 - \cos \xi \,,$$

we get

$$|(Ku, u)| \leqq 2aX + 2bY \,.$$

Thus

$$(Ku, u)^2 \leqq 8a^2X^2 + 8b^2Y^2 \,; \tag{3.19}$$

substituting this into (3.17), gives

$$r^2 + j^2 \leqq 1 - aX^2(1 - 8a) - aY^2(1 - 8b) \,. \tag{3.20}$$

The expression on the right will not exceed one if a and b are both not greater than $\frac{1}{8}$. According to the definition of a and b, this is the case if condition (3.6) of Theorem 4 is satisfied. Thus combining (3.19) and (3.12), we see that under condition (3.6) the field of values of C lies in the unit disk. According to Theorems 2 and 3 this proves the stability of the associated scheme.

Observe that in the estimates above the Schwarz inequality was used so gently that the sign of equality can hold throughout. In fact if $A = B$, one can easily show, by setting $\xi = \eta$, that condition (3.6) is *necessary* as well; in this sense our result is best possible.

We turn now to the second part of Theorem 4. Following the same line of argument we get, in the place of (3.9) and (3.11),

$$K' = K + \frac{A^2 + B^2}{2} XY.$$

In place of (3.17), we get

$$\begin{aligned} r^2 + j^2 &\leqq 1 - aX^2 - bY^2 - (a + b)XY + (Ku, u)^2 \\ &= 1 - (aX + bY)(X + Y) + (Ku, u)^2 \end{aligned} \tag{3.17)'$$

and in place of (3.18) we get, after replacing $|Au|\,|Bu|$ by $(a + b)/2$,

$$|(Ku, u)| \leqq aX + bY + \frac{a + b}{2} \{|\sin \xi|\,|\sin \eta| + XY\} \,. \tag{3.18)'$$

We estimate the curly brackets by the Schwarz inequality:

$$\begin{aligned} \{|\sin \xi|\,|\sin \eta| + XY\} &\leqq \sqrt{\sin^2 \xi + (1 - \cos \xi)^2}\,\sqrt{\sin^2 \eta + (1 - \cos \eta)^2} \\ &= \sqrt{2(1 - \cos \xi)}\sqrt{2(1 - \cos \eta)} = 2\sqrt{XY} \,. \end{aligned}$$

Substituting this into (3.18)′, gives

$$(3.18)'' \qquad |(Ku, u)| \leqq (a\sqrt{X} + b\sqrt{Y})(\sqrt{X} + \sqrt{Y}).$$

Squaring and using the Schwarz inequality, gives

$$(3.19)' \qquad (Ku, u)^2 \leqq 2(aX + bY)(a + b)(X + Y).$$

Substituting this into (3.17)′, gives

$$(3.20)' \qquad r^2 + j^2 \leqq 1 - (aX + bY)(X + Y)[1 - 2(a + b)].$$

Clearly, the expression on the right will not exceed one if $2(a + b)$ does not exceed one; but this is precisely what is guaranteed by condition (3.6)′. This completes the proof of the second half of Theorem 4.

Again, by setting $A = B$ and $\xi = \eta$, one can easily show that condition (3.6) is also necessary.

So far we have taken $\mu = \nu = 1$; it is easy to show that in general (3.6) and (3.6)′ have to be formulated as follows:

$$\frac{A^2}{\mu^2} \leqq \frac{I}{8}, \qquad \frac{B^2}{\nu^2} \leqq \frac{I}{8} \quad \text{and} \quad \frac{A^2}{\mu^2} + \frac{B^2}{\nu^2} \leqq \frac{I}{2}.$$

Observe that condition (3.6)′ is less restrictive than (3.6), i.e., that the scheme associated with C' is stable in a wider range than that associated with C. This greater stability is the effect of the extra term in C' which introduces in the associated difference operator a corresponding additional term. This additional term is clearly a difference analogue of the operator

$$-h^4 \frac{A^2 + B^2}{2} D_x^2 D_y^2.$$

Such a *higher order negative definite* term is called an artificial viscosity. The effect of such terms has been investigated by several authors, in particular by Kreiss [7] in the linear case, and by von Neumann and Richtmyer [13] and Lax and Wendroff [11] in the non-linear case. The results of this section furnish another illustration of the stabilizing effect of artificial viscosity.

4. The Geometric Meaning of Stability

The difference schemes discussed in the last section are only *conditionally stable*, i.e., they are stable only if the coefficients of the differential equation which they approximate satisfy the inequalities (3.6) and (3.6)′. An intuitive reason why such inequalities are necessary for stability has been given a long time ago by Courant, Friedrichs and Lewy in their classical paper:

Let p be any point and t_0 any time; denote by $D(p, t_0)$ the set of those points on the initial plane $t = 0$ where the values of the initial data influence the value of the solution of the differential equation at p, t_0. Denote by $D_h(p, t_0)$ the analogous set with respect to the difference equation. Then, as Courant, Friedrichs and Lewy have pointed out, if a difference scheme is convergent for all smooth initial data, then for any p and t_0 the set $D(p, t_0)$ must be contained in the set of limit points of $D_h(p, t_0)$ as h tends to zero. Since convergence and stability are equivalent (Theorem 1) this gives a necessary condition for stability.

To use this condition we have to determine the domains of dependence D and D_h. If we deal with the case of constant coefficients, then by reason of homogeneity we may take p to be the origin and t_0 to be one. Taking $\mu = \nu = 1$, for a nine-point scheme the set D_h consists then of all lattice points with mesh width h inside the unit square S. The set of limit points of D_h is therefore the unit square S.

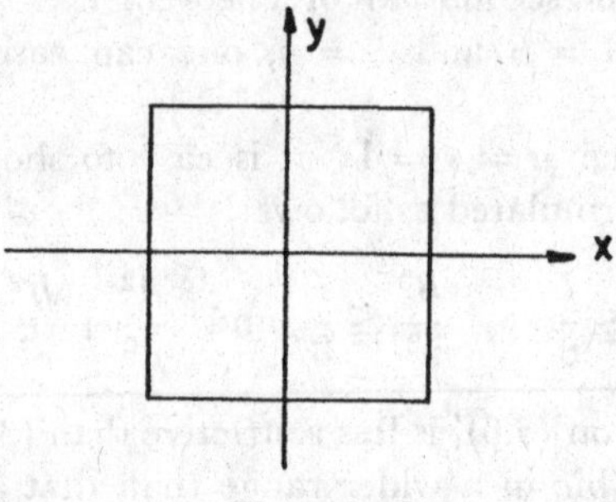

Figure 1

The determination of the domain of dependence D itself is a slightly delicate problem; but the support function of D is easily determined. We recall *that the support function $h_D(\xi, \eta)$ of any closed bounded set in the plane is defined as follows:*

$$h_D(\xi, \eta) = \max_{(x,y) \text{ in } D} (x\xi + y\eta) .$$

A well known result in the theory of hyperbolic equations (see e.g., [9]) asserts:

(4.1) $$h_D(\xi, \eta) = \lambda_{\max}(\xi A + \eta B) ,$$

where $\lambda_{\max}(X)$ denotes the largest eigenvalue of X. (This result is related to the fact that $\lambda_{\max}(\xi A + \eta B)$ is the maximum speed of propagation in the direction (ξ, η).)

It follows from the definition of support function that if one set is contained in a second set, the support function of the first does not exceed that of the second. It is further known from the theory of convex sets that if the second set is convex, then the converse of the above statement holds. Thus we can express the C-F-L condition in this form:

A nine point scheme for equation (1.1) *can be stable only if*

(4.2) $$h_D \leqq h_S$$

for all ξ, η, where S denotes the unit square.

It is easy to show that if (4.2) holds for all (ξ, η) which are perpendicular to a side of S, then it holds for all ξ, η. Thus taking (ξ, η) to be $(\pm 1, 0)$ and $(0, \pm 1)$ we see, using (4.2), that the necessary condition of C-F-L can be stated as follows:

$$-1 \leqq \lambda_{\min}(A) \leqq \lambda_{\max}(A) \leqq 1 ,$$
$$-1 \leqq \lambda_{\min}(B) \leqq \lambda_{\max}(B) \leqq 1 .$$

These inequalities about eigenvalues can be expressed also as matrix inequalities:

$$A^2 \leqq I, \qquad B^2 \leqq I. \tag{4.3}$$

It is something of a curiosity that both (3.6) and (3.6)′ are more stringent than (4.3).

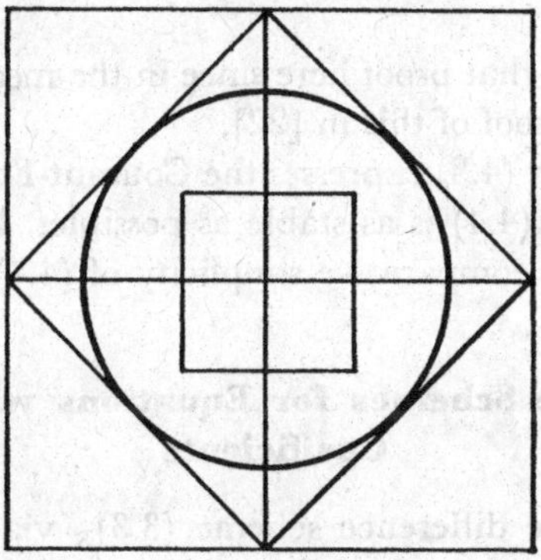

Figure 2

It is easy to give a geometric interpretation of (3.6) and (3.6)′. The former asserts that the domain of dependence D lies inside the small square shown on Figure 2, while (3.6)′ requires D to lie inside the circle shown there. The side of the small square equals the radius of the circle.

We consider next the difference scheme whose associated amplification matrix is

$$G = \tfrac{1}{2}(I + A^2 - B^2) \cos \xi + \tfrac{1}{2}(I - A^2 + B^2) \cos \eta + i(A \sin \xi + B \sin \eta) . \tag{4.4}$$

Since C is a trigonometric polynomial of first degree in ξ and η jointly, it is associated with a *four point* difference scheme based on the four points indicated in Figure 3.

In [14] we have shown using the method of the last section that this difference scheme is stable if the condition

$$\left(\frac{A}{\mu} \pm \frac{B}{\nu}\right)^2 \leqq I \tag{4.5}$$

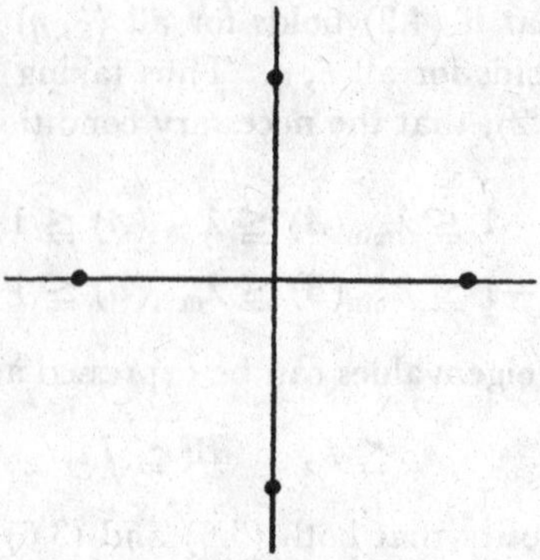

Figure 3

holds. We shall not give that proof here since in the meanwhile Strang has given a very straightforward proof of this in [22].

It is easy to verify that (4.5) expresses the Courant-Friedrichs-Lewy condition. In this sense the scheme (4.4) is as stable as possible. It is not clear however if this greater stability and comparative simplicity of (4.4) compensates for its low accuracy.

5. Difference Schemes for Equations with Variable Coefficients

The derivation of the difference scheme $(3.3)_S$ via (3.4) and (3.5) can be extended without alteration to equations with variable coefficients and yields a difference scheme which is accurate to second order. What remains is to prove its stability. The very general stability criterion of Kreiss contained in [18] is not quite sufficient for this; but in an unpublished note Kreiss has proved the stability of $(3.3)_S$ for variable coefficients provided that Δt is small enough compared to Δx and Δy. This restriction on the size of Δt is more severe than condition (3.6) derived in this paper; presumably it is due to an imperfection in the proof (to be corrected soon) rather than to the true state of affairs. At least nothing like it has shown up in the calculations carried out by Burstein, who on the contrary found stability for values of Δt *larger* than those permitted by (3.6). This must be due to the non-commutativity of the matrices A and B which occur in the equations of compressible flow.

6. Systems of Conservation Laws

In this section we shall adapt the difference schemes described in Section 3 to the construction of approximate weak solutions[1] of systems of conservation

[1] See e.g., [8] for a discussion of the theory of weak solutions of systems of conservation laws.

laws, i.e., of equations of the form

$$u_t = f_x + g_y , \tag{6.1}$$

u being a vector of unknown functions, f and g non-linear vector valued functions of u. Carrying out the differentiation on the right, brings (6.1) into the form

$$u_t = Au_x + Bu_y , \tag{6.1$'$}$$

where

$$A = \operatorname{grad} f , \qquad B = \operatorname{grad} g . \tag{6.2}$$

We assume that the matrices A and B can be made symmetric by the same similarity transformation; this guarantees that (6.1)′ is hyperbolic.

It was shown in [11], for systems of conservation laws in one space variable —and the proof carries over to several variables—that if we approximate (6.1) by a difference equation in conservation form, then the strong limit of the approximate solution is a weak solution of the conservation law. By conservation form (see a fuller discussion in [11]) we mean that

$$v(t + h) = v(t) + D_x^h F + D_y^h G , \tag{6.3}$$

where D_x^h and D_y^h denote the centered difference operators

$$(D_x^h u)(x) = u(x + \tfrac{1}{2}h) - u(x - \tfrac{1}{2}h) , \tag{6.4}$$

similar expressions holding in y, and where F and G denote functions of the values of $T^{\frac{1}{2}+j}u$, j ranging over some finite set such that if all the arguments $T^{\frac{1}{2}+j}u$ are set equal, F reduces to f, G to g. We have shown in [11] how to construct, in the case of one space variable, difference equations in conservation form which approximate a given system of conservation laws with second order accuracy. Here we extend this to two space variables.

Following the method described in Section 3 we write

$$u(t + h) \equiv u + hu_t + \tfrac{1}{2}h^2 u_{tt} \tag{6.5}$$

modulo terms of third order in h. Differentiating (6.1) with respect to t, we get using (6.2)

$$\begin{aligned} u_{tt} &\equiv f_{tx} + g_{ty} = (Au_t)_x + (Bu_t)_y \\ &= (Af_x + Ag_y)_x + (Bf_x + Bg_y)_y . \end{aligned}$$

Substituting this and (6.1) into (6.5), gives

$$\begin{aligned} u(t + h) = u &+ [hf + \tfrac{1}{2}h^2(Af_x + Ag_y)]_x \\ &+ [hg + \tfrac{1}{2}h^2(Bf_x + Bg_y)]_y . \end{aligned} \tag{6.6}$$

The important point about the above formula is that the right side is in conservation form and so can be approximated by a difference expression of the same kind. Using the abbreviation

$$M_x^h u = \tfrac{1}{2}[u(x + \tfrac{1}{2}h) + u(x - \tfrac{1}{2}h)]\,, \tag{6.7}$$

we approximate the right side of (6.6) by $S_h u$ defined as follows:

$$\begin{aligned} S_h u = u + D_x^h M_x^h f + D_y^h M_y^h g + \tfrac{1}{2} D_x^h A D_x^h f \\ + \tfrac{1}{2} D_y^h B D_y^h g + \tfrac{1}{2} D_x^h M_x^h A D_y^h M_y^h g + \tfrac{1}{2} D_y^h M_y^h B D_x^h M_x^h f\,. \end{aligned} \tag{6.8}$$

Clearly with S_h defined by (6.8),

$$v(t + h) = S_h v \tag{6.9}$$

is of the general form (6.3).

It is easy to see that (6.9) is accurate to second order, and that the amplification matrix associated with the linearized form of (6.9) is given by (3.3). This indicates that (6.9) is stable, at least away from regions where u is discontinuous.

Define S_h' as

$$S_h' = S_h - \tfrac{1}{2} D_x^h (D_y^h)^2 A D_x^h f - \tfrac{1}{2} D_y^h (D_x^h)^2 B D_y^h g\,; \tag{6.8$'$}$$

the amplification matrix associated with the linearized form of

$$v(t + h) = S_h' v \tag{6.9$'$}$$

is (3.3)′. (6.9) and (6.9)′ are our proposed difference schemes.

The main use of difference equations of the form (6.9) and (6.9)′ is to calculate approximations to discontinuous solutions of (6.1). In such calculations it is important to keep discontinuities in the approximate solutions fairly sharp. The method of artificial viscosity [13] was developed to accomplish this, and in [11] we have devised a way of introducing artificial viscosity for arbitrary systems of conservation laws in one space variable. We hope in a future publication to do the same in the case of two space variables.

Bibliography

[1] de Bruijn, *Inequalities between quadratic and Hermitian forms*, to appear.

[2] Buchanan, M., *A necessary and sufficient condition for stability of difference schemes for second order initial value problems*, Adelphi College, Dept. Grad. Math., Res. Rep. No. 104, 1962.

[3] Hahn, S. G., *Stability criteria for difference schemes*, Comm. Pure Appl. Math., Vol. 11, 1958, pp. 243–257.

[4] John, F., *On the integration of parabolic equations by difference methods*, Comm. Pure Appl. Math., Vol. 5, 1952, pp. 155–211.

[5] Kato, T., *Estimation of iterated matrices, with application to the von Neumann condition*, Numer. Math., Vol. 2, 1960, pp. 22–29.

[6] Kreiss, H. O., *Über die approximative Lösung von linearen partiellen Differentialgleichungen mit Hilfe von Differenzengleichungen*, Trans. Roy. Inst. Tech. Stockholm, No. 128, 1958.

[7] Kreiss, H. O., *Über die Lösung des Cauchyproblems*, Acta Math., Vol. 101, 1959, pp. 180–199.

[8] Lax, P. D., *Hyperbolic systems of conservations laws*, Comm. Pure Appl. Math., Vol. 10, 1957, pp. 537–566.

[9] Lax, P. D., *Differential equations, difference equations and matrix theory*, Comm. Pure Appl. Math., Vol. 11, 1958, pp. 175–194.

[10] Lax, P. D., *On the stability of difference approximations to solutions of hyperbolic equations with variable coefficients*, Comm. Pure Appl. Math., Vol. 14, 1961, pp. 497–520.

[11] Lax, P. D., and Wendroff, B., *Systems of conservation laws*, Comm. Pure Appl. Math., Vol. 13, 1960, pp. 217–237.

[12] Lax, P. D., and Wendroff, B., *On the stability of difference schemes with variable coefficients*, Comm. Pure Appl. Math., Vol. 15, 1962, pp. 363–371.

[13] von Neumann, J., and Richtmyer, R. D., *A method for the numerical calculation of hydrodynamical shocks*, J. Appl. Phys., Vol. 21, 1950, pp. 232–237.

[14] Lax, P. D., and Wendroff, B., *Difference schemes with high order of accuracy for solving hyperbolic equations*, New York Univ., Courant Inst. Math. Sci., Res. Rep. No. NYO-9759, 1962.

[15] Richtmyer, R. D., *Difference Methods for Initial Value Problems*, Tracts in Pure and Applied Math., Vol. 4, Interscience Publishers, New York, 1957.

[16] Richtmyer, R. D., *A survey of difference methods for non-steady fluid dynamics*, NCAR Tech. Notes 63-2, 1962.

[17] Kreiss, H. O., *Über die Stabilitätdefinition für Differenzengleichungen die partielle Differentialgleichungen approximieren*, BIT, Vol. 2, 1962, pp. 153–181.

[18] Kreiss, H. O., *On difference approximations of the dissipative type for hyperbolic differential equations* Comm. Pure Appl. Math., this issue.

[19] Morton, K. W., *On a matrix theorem due to H. O. Kreiss*, Comm. Pure Appl. Math., this issue.

[20] Burstein, S. Z., *Numerical calculations of multidimensional shocked flows*, New York Univ., Courant Inst. Math. Sci., Res. Rep. NYO-10,433, 1963.

[21] Strang, G., *Accurate partial difference methods I: linear Cauchy problems*, Arch. Rational Mech. Anal., Vol. 12, 1963, pp. 392–402.

[22] Strang, G., *Accurate partial difference methods II: non-linear problems*, to appear.

Received February, 1964.

Reprinted from:
CONTRIBUTIONS TO NONLINEAR FUNCTIONAL ANALYSIS

Shock Waves and Entropy

PETER LAX

Introduction: We study systems of the first order partial differential equations in conservation form:

$$(1)\qquad \partial_t u^j + \partial_x f^j = 0\ , \quad j = 1,\dots,m,\ f^j = f^j(u^1,\dots,u^m).$$

In many cases all smooth solutions of (1) satisfy an additional conservation law

$$(2)\qquad \partial_t U + \partial_x F = 0\ , \quad F = F(u)\ ,$$

where U is a convex function of u. We study weak solutions of (1) which satisfy in addition the "entropy" inequality

$$(3)\qquad \partial_t U + \partial_x F \le 0\ .$$

We show that all weak solutions of (1) which are limits of solutions of modifications of (1) by the introduction of various kinds of dissipation satisfy the entropy inequality (3). We show that for weak solutions which contain discontinuities of moderate strength, (3) is equivalent to the usual shock condition involving the number of characteristics impinging on the shock. Finally we study all possible entropy conditions of form (3) which can be associated to a given hyperbolic system of two conservation laws.

1. We consider systems of first order nonlinear partial differential equations in conservation form:

(1.1) $$\partial_t u^j + \partial_x f^j = 0 , \quad j-1,\ldots,m ,$$

where ∂_t and ∂_x denote partial differentiation with respect to t and x, and where each f^j is a function of $u = \{u^1,\ldots,u^m\}$, in general nonlinear. For simplicity we take the number of space variables to be 1. Carrying out the differentiations in (1.1) we get the equations

(1.2) $$\partial_t u^j + \sum f^j_\ell \, \partial_x u^\ell = 0$$

where the subscript ℓ denotes partial differentiation with respect to u^ℓ.

If u is a solution of (1.1) which is zero for $|x|$ large integrating (1.1) we obtain that

$$\int \partial_t u^j dx = 0$$

which means that the quantities

(1.3) $$\int u^j dx , \quad j = 1,\ldots,m$$

are conserved, i.e. independent of t.

Let U be some function of $u^1,\ldots,u^m$; when does U satisfy a conservation law, i.e. an equation of form

(1.4) $$\partial_t U + \partial_x F = 0 \; ?$$

(1.4) can be written in the form

(1.5) $$\sum U_j f^j_\ell = F_\ell \quad \text{for} \quad \ell = 1,\ldots,m.$$

We suppose now that (1.5) has a solution, so that (1.1) implies an additional conservation law (1.4); and we suppose furthermore that the new conserved quantity U is a <u>convex</u> function of $u^1,\ldots,u^m$.

We observe that if the system (1.2) is symmetric, i.e.

$$f^j_\ell = f^\ell_j$$

then

$$U = \sum u_j^2$$

satisfies a conservation law, with

$$F = \sum u^j f^j - g \ ,$$

where g is a function satisfying

$$g_\ell = f^\ell \ .$$

As observed in [3], if the convex function U satisfies a conservation law then multiplying (1.2) by $U_{j,\ell}$ puts the system into symmetric hyperbolic form. For the purposes of section 2 we shall require a little more, that (1.2) be strictly hyperbolic. This means that the matrix

(1.8) $$f' = f^j_\ell$$

has real and distinct eigenvalues. We denote these eigenvalues arranged in increasing order, by $c_1, c_2, \ldots, c_m$. They are called sound speeds, and are functions of u.

It is well known that the initial value problem is properly posed for symmetric hyperbolic systems, i.e. we may prescribe the values of u as arbitrary smooth functions at $t = 0$. Solutions are uniquely determined by their initial data but, since the governing equations are nonlinear, in general they exist only for a limited time range. For solutions which exist for all $t > 0$ we have to turn to weak solutions, which may be discontinuous solutions or slightly worse; these satisfy the conservation laws (1.1) only in the weak, or integral sense, i.e.

(1.9) $$\iint_{t \geq 0} [u^j \partial_t \phi + f^j \partial_x \phi] dx dt + \int u(x,0)\, \phi(x,0) dx = 0$$

for all smooth test functions ϕ with bounded support in $t \geq 0$. As is well known, for piecewise continuous solutions (1.9) is equivalent to the Rankine-Hugoniot jump conditions

$$s[u^j] - [f^j] = 0 \tag{1.10}$$

where s is the speed with which the discontinuity is propagating, and $[u]$ denotes the difference between the values of u on the two sides of the discontinuity.

It is equally well known, see e.g. [8] for some simple examples, that weak solutions of conservation laws are not uniquely determined by their initial values. To pick out the physically relevant solutions among the many, some additional physical principle has to be introduced. This additional principle usually identifies the relevant solutions as limits of solutions of equations with some dissipation. Specifically, we consider the equations with artifical viscosity:

$$\partial u_t^j + \partial_x f^j = \varepsilon\, \partial_x^2 u^j, \quad \varepsilon > 0 . \tag{1.11}$$

Suppose that a sequence $u(x,t;\varepsilon)$ of solutions of (1.11) tends to a limit $u(x,t)$ boundedly, almost everywhere; then $\varepsilon\, \partial_x^2\, u(\varepsilon)$ tends to 0 in the topology for distributions, so that the limit u satisfies (1.1) in the weak sense.

We show next how to characterize such limit solutions directly, with the aid of the function U:

Multiplying (1.11) by U_j and summing we get

$$\partial_t U + \partial_x F = \varepsilon \sum U_j\, \partial_x^2 u^j . \tag{1.12}$$

Using the identity

$$\partial_x^2 U = \sum U_j \partial_x^2 u^j + \sum U_{jk} \partial_x u^j \partial_x u^k$$

and the convexity of U we deduce that

$$\partial_x^2 U \geq \sum U_j \partial_x^2 u^j .$$

Since ε is > 0, using this to estimate the right side of 91.12) we get

$$\partial_t U + \partial_x F \leq \varepsilon \, \partial_x^2 F .$$

Letting $\varepsilon \to 0$ the right side tends to 0 in the topology of distributions and we deduce

<u>Theorem 1.1:</u> Let (1.1) be a system of conservation laws which implies an additional conservation law (1.4) where U is a strictly convex function. Then every weak solution of (1.1) which is the limit, boundedly a.e., of solutions of the viscous equation (1.11) satisfies the inequality

(1.13) $$\partial_t U + \partial_x F \leq 0 .$$

<u>Remark A:</u> Suppose u satisfies (1.13) and has compact support in x. Integrating (1.13) with respect to x gives

$$\int \partial_t U dx \leq 0$$

which implies that

(1.14) $$\int U dx$$

is a <u>decreasing</u> function of t.

<u>Remark B:</u> Suppose that u is a piecewise continuous weak solution of (1.1); then it is easy to deduce either from (1.13) or (1.14) that at a point of discontinuity

(1.15) $$s[U] - [F] \leq 0 ,$$

where s is the velocity with which the discontinuity propagates, and $[U]$, $[F]$ denote the jumps $U_{left} - U_{right}$ and $F_{left} - F_{right}$, respectively.

For compressible fluid flow (1.14) corresponds to the increase of total negative entropy, and (1.15) states that the classical entropy of particles upon crossing a shock increases. For this reason we shall call (1.13) and (1.15) entropy conditions.

The addition of a viscous term as in equation (1.11) is only one of many ways of introducing a slight amount of artificial dissipation into the system (1.1). Another way is to discretize the differential equations; one of the standard ways of doing this is to replace the operator ∂_t and ∂_x by the following difference operators. Denote by $T(h)$ translation in t by the amount h, and by $S(k)$ translation in x by the amount k. Define

(1.16)
$$D_t = \frac{1}{\Delta t}\{T(\Delta t) - \frac{S(\Delta x) + S(-\Delta x)}{2}\},$$
$$D_x = \frac{S(\Delta x) - S(-\Delta x)}{2\Delta x}.$$

We consider now the difference equation

(1.17)
$$D_t u_s + D_x f^j = 0$$

and study limits of solutions of (1.17) as $\Delta t, \Delta x \to 0$ while the ratio $\frac{\Delta t}{\Delta x} = \lambda$ remains constant. We assume that $u(x,t)$ is the limit, boundedly and almost everywhere, of solutions $u(x,t,\Delta t,\Delta x) = u(\Delta)$ of (1.17); it follows that $D_t u(\Delta)$ and $D_x f(u(\Delta))$ tend, in the topology of distributions, to $\partial_t u$ and $\partial_x f(u)$, so that it follows that the limit function u satisfies the system of conservation laws (1.1). We shall show now, under an additional restriction on λ, that such a limit u satisfies the entropy inequality (1.13). It suffices to show that every solution of (1.17) satisfies the inequality

(1.18)
$$D_t U + D_x F \leq 0,$$

for the limit of (1.18) in the topology of distributions is (1.13). We introduce the following vector notation:

$$u(x,t+\Delta t) = u,\ u(x-\Delta x,t) = v,\ u(x+\Delta x,t) = w.$$

We regard u, v and w as column vectors. Then

$$D_t u = \frac{1}{\Delta t}\{u - \frac{v+w}{2}\}, \quad D_x f = \frac{1}{2\,\Delta x}\{f(w)-f(v)\};$$

substituting this into (1.17) and solving for u we get, with the notation

$$\frac{\Delta t}{\Delta x} = \lambda$$

(1.19) $$u = \frac{v+w}{2} + \frac{\lambda}{2}[f(v) - f(w)] .$$

Inequality (1.18) asserts that

(1.20) $$U(u) \leq \frac{U(v) + U(w)}{2} + \frac{\lambda}{2}[F(v) - F(w)] .$$

We deform v continuously into w ; set

$$v(s) = sv + (1-s)w .$$

Since $v(0) = w$, both sides of (1.20) equal $U(w)$ for $s = 0$. So we can write the difference of the right and the left side in (1.20) as the integral of the difference of their derivatives with respect to s . This difference is

(1.21) $$\frac{1}{2}U'(v)(v-w) + \frac{\lambda}{2}F'(v)(v-w) - U'(u)\frac{du}{ds}$$

where U' and F' are the gradients of U and F , regarded as row vectors. From (1.19) we get

(1.22) $$\frac{du}{ds} = \frac{v-w}{2} + \frac{\lambda}{2}f'(v)(v-w);$$

f' , the gradient of the vector quantity f , is a matrix.

In this notation we can write identity (1.6) as follows

(1.23) $$U'f' = F' .$$

Substituting (l. 23) and (l. 22) into (l. 21) we get

$$(1.24) \qquad \frac{1}{2}[U'(v) - U'(u)][I + \lambda f'(v)](v-w) \ .$$

Next we set

$$w(r) = rv(s) + (1-r)w = rsv + (1-rs)w \ .$$

Since $w(1) = v(s)$, $w(0) = w$, we have

$$U'(v) - U'(u) \int_0^1 \frac{d}{dr} U'(u)dr \ .$$

We write

$$\frac{d}{dr} U'(u) = \frac{du^t}{dr} U'' ,$$

where U'' is the matrix of second derivatives of U and u^t the transpose of u. From (l. 19) we get

$$\frac{du}{dr} = \frac{s}{2}[v-w-\lambda f'(w)(v-w)] \ .$$

Substituting these relations back into (l. 25) we see that the difference between the right and left side of (l. 20) is the double integral from 0 to 1 with respect to s and r of

$$(1.25) \qquad \frac{s}{4}[[I-\lambda f'(w)](v-w]^t \cdot U'' [I+\lambda f'(v)](v-w) \ .$$

By assumption U'' is positive definite; this implies that <u>for λ small enough</u> (l. 25) <u>is positive</u>. The precise restriction on λ is as follows:

Denote by m and M the minimum and maximum eigenvalue of U'' in that portion of u-space in which we are operating, and denote by c the norm of f' there. Denote $v - w$ by z. The the following is a lower bound for the expression to the right of $s/4$ in (l. 25):

$$m\|z\|^2 - M(2c\lambda+c^2\lambda^2)\|z\|^2 \ .$$

Clearly this is positive for $z \neq 0$ if

(1. 26) $$c\lambda \leq \sqrt{1+m/M} - 1 .$$

Thus we have proved

Theorem 1. 2: Suppose all differentiable solutions of the system of conservation laws (1. 1) satisfy an additional conservation law (1. 4), where U is a strictly convex function of u. Then all weak solutions of (1. 1) which are the limits, boundedly a. e., of solutions of the difference equation (1. 17) satisfy the entropy inequality (1. 13), provided that condition (1. 26) is fulfilled.

Remark: In the case of f' symmetric, the norm c of f' equals the absolute value c_{max} of the largest eigenvalue of f'. In this case $U = \sum u_j^2$ is a conserved quantity; for this U, $U'' = 2I$, so $m/M = 1$ and condition (1. 26) becomes

$$c_{max} \frac{\Delta t}{\Delta x} \leq \sqrt{2} - 1 = .414 .$$

This is slightly more stringent than the Courant-Friedrichs-Lewy necessary condition for convergence,

$$c_{max} \frac{\Delta t}{\Delta x} \leq 1 .$$

In the nonsymmetric case (1. 26) is a still more stringent version of the C-F-L criterion, since then the norm c of f' is $> c_{max}$, and the condition number m/M of U'' is < 1. We remark that Theorem 1. 1 was also proved by Kružkov at the end of [14], and he has suggested condition (1. 13) as a generalized entropy condition.

2. In this section we assume that (1. 1) is strictly hyperbolic, i. e. that the matrix $f' = (f^j)$ has distinct real eigenvalues $c_1, \ldots, c_m$, indexed in increasing order. The c_j are the characteristic speeds; they are functions of u. We also assume that the system is genuinely nonlinear in the sense of [8].

At a point of discontinuity of a solution we shall denote the value of u on the left, respectively right side of the discontinuity as follows:

$$u_{left} = v , \quad u_{right} = w . \tag{2.1}$$

A point of discontinuity is called a <u>k-shock</u> if

a) The Rankine-Hugoniot relation

$$s[v-w] = f(v) - f(w) \tag{2.2}$$

holds

b) There are exactly k-1 of the characteristic speeds $c_j(v) < s$ and exactly (m-k) speeds $c_j(w) > s$:

$$c_{k-1}(v) < s < c_k(v) ,$$
$$(2.3)$$
$$c_k(w) < s < c_{k+1}(w) .$$

We shall call (2.3) the <u>shock condition.</u>

<u>Theorem 2.1:</u> Suppose that the system of conservation laws (1.1) is strictly hyperbolic, and that there is a strictly convex function U of u which satisfies the additional conservation law (1.4). Let u be a weak solution of (1.1) which has a discontinuity propagating with speed s , and suppose that the values of v and w on the left and right sides of the discontinuity are close. Then the shock condition (2.3) is satisfied if and only if the strict entropy condition (1.15):

$$s[U(v) - U(w)] - F(v) + F(w) < 0 \tag{2.4}$$

is satisfies.

<u>Proof:</u> It was shown in [8] that all states w near v which satisfy the R-H condition (2.2) form m one-parameter families $w_k(r)$, $s_k(r)$ where $w_k(0) = v$. If the parametrization is so taken that

$$(2.5) \qquad \left.\frac{ds_k}{dr}\right|_{r=0} > 0$$

then those w_k which correspond to $r < 0$ satisfy the shock condition (2.3).

To prove (2.4) we shall substitute for w one of these families $w_j(r)$ and expand the left side of (2.4) in powers of r; the crux of the argument is to show that the lowest power r^p which is different from 0 is odd, and that the coefficient of r^p is positive.

Let's denote differentiation with respect to r by a dot $\cdot$ and, as before, the gradient with respect to u by prine $'$. The crucial exponent p turns out to be 3, so we have to calculate the first 3 derivatives of the left side of (2.4) at $r = 0$. Differentiating (2.2) we get

$$(2.6) \qquad \dot{s}[v-w] - s\dot{w} = -\dot{f}(w).$$

The derivative of the left side of (2.4) in r is

$$(2.7) \qquad \dot{s}[U(v) - U(w)] - s\dot{U} + \dot{F}.$$

Using relation (1.6), $F' = U'f'$, we can write

$$\dot{F} = F'\dot{w} = U'f'\dot{w} = U'\dot{f}.$$

Substituting for $\dot{f}$ from (2.6) we get

$$\dot{F} = \dot{s}U'[w-v] + sU'\dot{w}.$$

Substituting this into (2.7) and noting that $\dot{U} = U'\dot{w}$ we get the following expression:

$$\dot{s}[U(v) - U(w)] + \dot{s}U'[w-v].$$

Differentiating once more we get

$$\ddot{s}[U(v)-U(w)] + \ddot{s}U'[w-v] + \dot{s}\dot{U}'[w-v].$$

Since $w(0) = v$, this is clearly zero at $r = 0$.

Differentiating once more, and setting $r = 0$ we get, after eliminating, those terms which are zero when $w = v$,

$$\dot{s}\, U' \dot{\dot{w}}\,.$$

The remaining term can be written as

$$\dot{s}\, \dot{w} + U'' \dot{w}\,; \tag{2.8}$$

since according to (2.5) the parametrization is so chosen that $\dot{s}$ is positive, and since U'' is positive because of the strict convexity of U, it follows that (2.8) is positive. This proves inequality (2.4) of theorem 2.1.

A noncalculational proof can be given using the following result of Foy, [2]:

If two nearby states v and w can be connected through a shock, then they can be connected through a viscous profile, i.e. a steady progressing solution of (1.11) of the form

$$u(x,t,\varepsilon) = w\left(\frac{x-st}{\varepsilon}\right), \quad w(-\infty) = v, \quad w(\infty) = w\,. \tag{2.9}$$

Substituting this form of u into (1.11) gives for the function w the ordinary differential equation

$$-s\dot{w} + \dot{f} = \dot{w}\,. \tag{2.10}$$

Clearly, the discontinuous solution

$$u(x,t) = \begin{cases} v \text{ for } x < st \\ w \text{ for } st < x \end{cases} \tag{2.11}$$

is the weak limit of $w\left(\frac{x-st}{\varepsilon}\right)$ as $\varepsilon \to 0$. Therefore according to theorem 1.1, the solution (2.11) satisfies the entropy condition.

Recently Conley and Smoller, [1], have shown that, for a fairly general class of systems of two conservation laws, any two states v and w which can be connected through a shock can also be connected through a viscous profine. It follows from the above argument that for such systems the restriction that v and w be close can be removed from theorem 2.1.

In his important paper [4] Glimm constructs solutions of systems of conservation laws as the limit of approximate solutions. These are piecewise continuous weak solutions in each strip $k\Delta t < t < (k+1)\Delta t$, and all their discontinuities are shocks; in addition the oscillation of these solutions is small. If there is a convex function which satisfies an additional conservation law, it follows from theorem 2.1 that the entropy condition

(2.12) $$\partial_t U + \partial_x F \leq 0$$

is satisfied by each approximate solution in each strip. Let ϕ be a smooth, positive test function with compact support. Multiply (2.12) by ϕ, integrate over each strip; integrating by parts with respect to t over each strip and summing over all strips we get

(2.13) $$\sum_{k=1}^{\infty} \int \phi(x, k\Delta)[U(x, k\Delta+) - U(x, k\Delta-)]dx + \iint (-(\partial_t \phi)U + \phi\partial_x F)dxdt \leq 0 .$$

Lemma (5.1) in Glimm's paper shows that the sum in (2.13) tends to zero for a suitably selected subsequence. This leaves us in the limit with

$$\iint (-\phi_t U + \phi F_x)dxdt \leq 0$$

for all positive test function ϕ supported in $t > 0$. Integrating by parts with respect to t we get that

$$\iint \phi[\partial_t U + \partial_x F]dxdt \leq 0$$

for all such ϕ. Clearly this implies (2.12). Thus we have proved

Theorem 2.2: Suppose that the system of conservation laws (1.1) is strictly hyperbolic, and that there is a convex function U of u which satisfies the additional conservation law (1.4). Then all weak solutions of (1.1) constructed by Glimm's method satisfy the entropy inequality (2.12).

3. What systems admit an additional conservation law where the additional conserved quantity U is a convex function of the original ones? We saw that symmetric systems do, and so does the system consisting of the laws of conservation of mass, momentum and energy for a compressible gas. A systematic search for additional conservation laws was carried out by Rozdestvenskii, [12]; in this section we record some observations on the existence and utility of additional convex conservation laws.

We start with a single conservation law:

$$\partial_t u + \partial_x f = 0 . \tag{3.1}$$

In this case we may choose for U any convex function; F is then determined by integrating the compatibility relation (1.6):

$$U'f' = F' . \tag{3.2}$$

The entropy condition,

$$\partial_t U + \partial_x F \leq 0 , \tag{3.3}$$

was derived for smooth U only; by passing to the limit in the topology of distributions we deduce (3.3) for any convex U, smooth or not.

Every convex function lies in the convex cone generated by the functions $U(u) = |u-z|$, z some constant, and by the linear functions. In [7], Krushkov takes (3.3) for all U of this form to be the definition of the relevant class of weak solutions of the analogue of (3.1) for n space variables; he proves existence of such solutions with arbitrary initial data, and announces their uniqueness.

We present now some known consequences of the entropy inequality (3.3); the first was found independently by Krushkov and by Hopf, [6]:

Suppose that u is a piecewise continuous weak solution of (3.1) which satisfies (3.3); let's denote the speed of propagation of a discontinuity by s, and denote by v and w the values of u on the left, respectively right side of the discontinuity. According to (1.15), (3.3) implies that

$$s[U(v) - U(w)] - F(v) + F(w) < 0 . \tag{3.4}$$

Suppose that $w < v$; let z be any number between w and v, and set

$$U(u) = \begin{cases} 0 & \text{for } u < z \\ u - z & \text{for } z < u \end{cases} \tag{3.5}$$

It follows from (3.2) that then

$$F(u) = \begin{cases} 0 & \text{for } u < z \\ f(u) - f(z) & \text{for } z < u . \end{cases} \tag{3.6}$$

Substituting these into (3.4), and using the jump relation

$$s = \frac{f(v) - f(w)}{v - w} \tag{3.7}$$

we get a relation which, after rearrangement, becomes

$$f(z) \leq \frac{v-z}{v-w} f(v) + \frac{z-w}{v-w} f(v) \quad \text{for } w \leq z \leq v . \tag{3.8$_+$}$$

The geometrical meaning of (3.8) is that the graph of f over the interval $[w, z]$ lies below the chord connecting $(w, f(w))$ with $(v, f(v))$; for $w > v$ the opposite inequality $(3.8)_-$ obtains; these inequalities are Oleinik's celebrated condition (E), see [10].

We remark that in [11] B. Quinn has shown that $(3.8)_+$ <u>is necessary and sufficient for</u> L_1 <u>contraction,</u> more precisely:

If a pair of piecewise continuous solutions u_1 and u_2 both satisfy $(3.8)_\pm$, then

$$\int |u_1(x,t) - u_2(x,t)|\, dx \tag{3.9}$$

is a decreasing function of t; conversely, if (3.9) is a decreasing function of t for a certain piecewise continuous u_1 and every continuous u_2, then u_1 satisfies $(3.8)_\pm$.

We shall derive now another consequence of (3.3):

Let u be a weak solution of (3.1) which satisfies (3.3), and which is 0 at $x = \pm\infty$. Then as observed in (1.14), for U convex, and $U(0) = 0$,

$$\int U(u(x,t_2)dx \le \int U(u(x,t_1)dx \quad \text{for } t_1 < t_2. \tag{3.10}$$

Choose for U the convex function

$$U(u) = |M - u| - M \tag{3.11}$$

where

$$M = \text{ess.}\sup_x u(x,t_1)\,. \tag{3.12}$$

Since $u(x,t_1) \le M$, a.e., it follows that

$$U(u(x,t_1)) = -u(x,t_1) \text{ a.e.}; \tag{3.13}$$

on the other hand, since $|M - u| \ge M - u$,

$$U(u(x,t_2)) \ge -u(x,t_2)\,. \tag{3.14}$$

Substituting (3.13) and (3.14) into (3.10) we get that

$$-\int u(x,t_2)dx \le -\int u(x,t_1)dx\,. \tag{3.15}$$

On the other hand it follows from (3.1) that

$$\int u(x,t)dt$$

is independent of t; therefore in (3.15) the sign of equality holds. But this can be if and only if in (3.14) equality holds a.e.; this is the case if and only if

$$u(x,t_2) \leq M \quad \text{for almost all } x .$$

In view of the definition of (3.12) of M this result can be expressed as follows:

$$\operatorname*{Ess\,sup}_x u(x,t)$$

is a decreasing function of t for every weak solution u which satisfies (3.3) for all convex U. Similarly,

$$\operatorname*{Ess\,inf}_x u(x,t)$$

is an increasing function of t.

We turn now to pairs of conservation laws:

$$(3.16) \qquad \begin{aligned} \partial_t u^1 + \partial_x f^1 &= 0 \\ \partial_t u^2 + \partial_x f^2 &= 0 , \end{aligned}$$

these can be written in the form

$$(3.17) \qquad \partial_t u + f' \partial_x u = 0 .$$

The compatibility relation (1.6) is

$$(3.18) \qquad F' = U' f' ,$$

a pair of first order equations for the two functions F and U. These equations are linear; it is easy to show that if the nonlinear system (3.17) is strictly hyperbolic, so is† (3.18). For suppose that the matrix f' has 2 distinct real eigenvalues; let r be a right eigenvector of f':

(3.19) $$f'r = cr.$$

Multiplying (3.18) by r on the right gives

(3.20) $$F'r = U'f'r = cU'r,$$

a linear combination in which both F and U are differentiated in the direction r. The existence of two such directions shows that (3.18) is hyperbolic, and its characteristic directions are those of the right eigenvectors of f'.

We can eliminate F from (3.18) by differentiating the first equation with respect to u^2, the second with respect to u^1, and subtracting one from the other. We get a homogeneous 2nd order equation for U of the form

(3.21) $$SU = a_{11}U_{11} + a_{12}U_{12} + a_{22}U_{22} = 0$$

where S is the second order operator with coefficients

(3.22) $$a_{11} = -f^1_2,\ a_{12} = f^1_1 - f^2_2,\ a_{22} = f^2_1.$$

Equation (3.21), being derived from a hyperbolic first order system, is itself hyperbolic, which means that the quadratic form

(3.23) $$a_{11}\xi^2 + a_{12}\xi\eta + a_{22}\eta^2$$

is indefinite.

We turn now to the question: does equation (3.21) have convex solutions? It is pretty easy to see that it does in the small, on the basis of this

---

†and similarly, as shown by Loewner, if (3.17) is elliptic, so is (3.18).

Lemma 3.1: Let a_{ij} be a symmetric $n \times n$ matrix which is indefinite; then there exists a positive definite matrix U_{ij} such that

$$\sum a_{ij} U_{ij} = 0 . \tag{3.24}$$

Proof: Since a_{ij} is indefinite, there exist vectors $\{\xi_i\} = \xi$ such that

$$\sum a_{ij} \xi_i \xi_j = 0 . \tag{3.25}$$

The set of these separates the set of those vectors where the quadratic form $\xi a \xi$ is positive from the set where the form is negative. It follows that the set of ξ satisfying (3.25) spans the whole space; denote by $\xi^1, \ldots, \xi^n$ a spanning set. Now define U_{ij} by

$$U_{ij} = \sum_k \xi_i^k \xi_j^k ;$$

since the ξ^k span the whole space, U_{ij} is positive definite; on the other hand it follows from (3.25) that condition (3.24) holds; this completes the proof of the lemma.

Applying the lemma to the quadratic form (3.23) at some point v we conclude that there exists a positive definite U_{ij} which satisfies (3.21) at v. By solving an appropriate Cauchy problem we can construct a solution U whose second derivatives at v equal U_{ij}; this solution will be convex near v. Thus we have proved

Theorem 3.2: A homogeneous second order hyperbolic equation has a convex solution in the neighborhood of every point.

We show now that the compatibility equation (3.18) has solutions with U convex in any domain G where a certain inequality, see (3.39), is satisfied. We do not claim that this condition is necessary.

We shall construct a one-parameter family of such solutions; we start with approximate solutions of the form

(3.26) $$U_{approx} = e^{k\phi}V, \quad F_{approx} = e^{k\phi}H,$$

where

(3.27) $$V = \sum_0^N V^j/k^j, \quad H = \sum_0^N H^j/k^j;$$

ϕ, V^j and H^j are independent of k. Solutions of this sort, with $i\phi$ in place of ϕ, were constructed in [9]. For this reason we only sketch the details.

Substituting (3.26) in (3.18) gives, after division by ek^{ϕ}, the equation

$$k\phi'Vf' + V'f' = k\phi'H + H'.$$

We substitute (3.27) into the above equation; equating coefficients of various powers of k we get

(3.28) $$V^o\phi'f' = H^o\phi',$$

and

(3.29) $$V^j\phi'f' + V^{(j-1)'} = H^h\phi' + H^{(j-1)'}.$$

Equation (3.28) asserts that ϕ' is a left eigenvector of f':

(3.30) $$\phi'f' = c\phi'$$

with

(3.31) $$cV^0 = H^0.$$

Such a function ϕ, called a <u>phase function,</u> is easily constructed since a left eigenvector is characterized by orthogonality to the right eigenvector r corresponding to the other eigenvalue:

(3.32) $$\phi'r = 0$$

where

(3.33) $$f'r = sr\ , \quad s \neq c\ .$$

Substituting (3.30) into (3.29) gives

$(3.34)^j$ $$(cV^j - H^j)\phi' = H^{(j-1)'} - V^{(j-1)'}f'.$$

The first step in solving equations (3.34) is to multiply $(3.34)^j$ by r; using (3.33) we get

$$0 = (H^{j-1} - sV^{j-1})'r\ .$$

Using (3.31) we get

$$(c-s)\,V^{o}{}'r + c'rV^{o} = 0\ .$$

This is a first order equation for V^o which can be solved once we prescribe the value of V^o on a noncharacteristic initial curve. Notice that if we prescribe positive values for V^o initially, V^o is positive everywhere.

Having determined V^o and H^o, equation $(3.34)^j$ gives one linear relation between V^1 and H^1; proceeding recursively we can determine all V^j and H^j.

The functions U_{approx} and F_{approx} of form (3.26), (3.27) constructed in this fashion satisfy the approximate equation

$$U'_{approx}f' = F'_{approx} + e^{k\phi}R_n/k^N\ .$$

We shall now construct another solution

(3.35) $$U'_N f' = F'_N + e^{k\phi}R_N/k^N$$

of this same equation such that

(3.36) $$U_N\ , F_N = e^{k\phi}O(1/k^N).$$

Then

$$U = U_{approx} - U_N, \; F = F_{approx} - F_N$$

are exact solutions of $U'f' = F'$, of which the leading term is (3. 26).

To construct a solution of (3. 35) which satisfies (3. 36) in some domain G of the u-plane we assign initial values zero for U_N, F_N along a non-characteristic curve C with these properties:

i) C doesn't intersect G .

ii) G is contained in the domain of determinacy of C .

In what follows we assume that ϕ has no ciritcal points, i. e. $\phi' \neq 0$ everywhere. Then ϕ is monotonic along every curve whose tangent is not parallel to r appearing in (3. 32); in particular ϕ is monotonic along characteristics corresponding to the other eigenvector of f'. Our last condition is

iii) Along these other characteristics ϕ increases in the direction from C toward G .

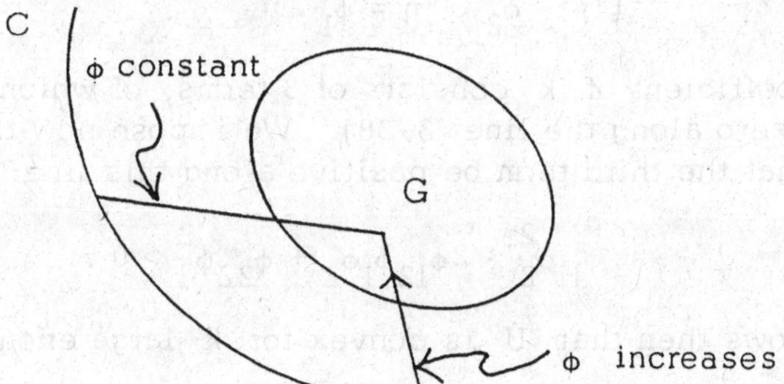

Condition iii) determines the side of G on which C lies.

It is easy to show, using standard estimates in the maximum norm for the hyperbolic equation (3. 35), that if the initial values of U_N and F_N are chosen to be zero on C , then $U_N \leq O(e^{k\phi}/k^N)$ in G as $k \to +\infty$. This completes the construction of exact solutions.

When is the function U just constructed convex? The answer can be read off from the first 2 leading terms in the asymptotic expression for the quadratic form of U'':

$$e^{-k\phi}\{U_{11}\xi^2 + 2U_{12}\xi\eta + U_{22}\eta^2\} = k^2\{\phi_1^2\xi^2 + 2\phi_1\phi_2\xi\eta + \phi_2^2\eta^2\}V^o +$$

(3.37)

$$+ k\{\phi_1^2\xi^2 + 2\phi_1\phi_2\xi\eta + \phi_2^2\eta^2\}V^1$$

$$+ 2k\{\phi_1 V_1^o \xi^2 + (\phi_2 V_1^o + \phi_1 V_2^o \xi\eta + \phi_2 V_2^o \eta^2\}$$

$$+ k\{\phi_{11}\xi^2 + 2\phi_{12}\xi\eta + \phi_{22}\eta^2\}V^o .$$

The coefficient of k^2 is

$$(\phi_1\xi + \phi_2\eta)^2 V^o ,$$

a positive quantity except along

$$\xi = -\phi_2 , \quad \eta = \phi_1 . \tag{3.38}$$

The coefficient of k consists of 3 terms, of which the first 2 are zero along the line (3.38). We impose now the condition that the third term be positive along this line:

$$\phi_{11}\phi_2^2 - 2\phi_{12}\phi_1\phi_2 + \phi_{22}\phi_1^2 > 0 . \tag{3.39}$$

It follows then that U is convex for k large enough, positive.

Let u be any differentiable solution of (3.17):

$$\partial_t u + f'\partial_x u = 0 .$$

Multiplying by ϕ' and using (3.30) we get

$$\partial_t\phi + c\partial_x\phi = 0 ;$$

this equation asserts that ϕ is constant along one of the characteristics of the nonlinear equation (3.17). Such a function is called a Riemann invariant; thus the phase functions of the linear compatibility equation (3.18) are the Rieman invariants of the nonlinear equation (3.17).

Any function $p(\phi)$ of a Riemann invariant is another Riemann invariant. Denoting $p(\phi)$ by ψ and differentiating with respect to ϕ by a dot we have

$$\begin{aligned} &\psi_{11}\xi^2 + 2\psi_{12}\xi\eta + \psi_{22}\eta^2 \\ (3.40) \qquad &= \ddot{p}\,\{\phi_1^2\xi^2 + 2\phi_1\phi_2\xi\eta + \phi_2^2\eta^2\} \\ &+ \dot{p}\,\{\phi_{11}\xi^2 + 2\phi_{12}\xi\eta + \phi_{22}\eta^2\}\,. \end{aligned}$$

We deduce from this that if ϕ satisfies (3.38), so does every increasing function ψ of ϕ. This shows that property (3.39), except for the sign, does not depend on the particular choice for ϕ.

It is easy to decide when (3.39) can be satisfied. Denote by dot differentiation in the direction

$$(3.41) \qquad \dot{u} = r\,,$$

where r is the right eigenvector appearing in (3.32). Differentiating (3.30) in the above direction and multiplying by r we get, using (3.32) and (3.33) that

$$(c-s)\,\dot{\phi}'\,r = \phi'\dot{f}'r\,.$$

In view of (3.41), $\dot{\phi}'r$ is the left side of (3.39); therefore (3.39) can be satisfied if and only if

$$(3.42) \qquad \phi'\dot{f}'r \neq 0\,,$$

a condition on the derivatives and right and left eigenvectors of f'.

We shall now use the solutions (3.26) constructed to prove

Theorem 3.3: Let $u(x,t)$ be a weak solution of the conservation laws (3.17), defined for $t \geq 0$, which satisfies the entropy condition

(3.43) $$U_t + F_x \leq 0$$

for all convex solutions U of (3.18). Let ϕ be a Riemann invariant which satisfies (3.39) in a domain G which contains all values of $u(x,t)$. Then

(3.44) $$\operatorname*{Max}_x \phi(u(x,t))$$

is a decreasing function of t.

Actually we shall prove a sharper theorem of which Theorem 3.3 is a corollary:

Theorem 3.4: Denote by c_{min} and c_{max} the minimum and maximum of $c(u)$ in G, where c is the eigenvalue in (3.30). Then at any point (x,t), $t > 0$,

(3.45) $$\varphi(u(x,t)) \leq \operatorname*{Sup}_{a \leq y \leq b} \phi(u(y,0))$$

where

(3.46) $$a = x - tc_{max}, \quad b = x - tc_{min}.$$

Proof: Integrate (3.43) over the triangle T shown below,

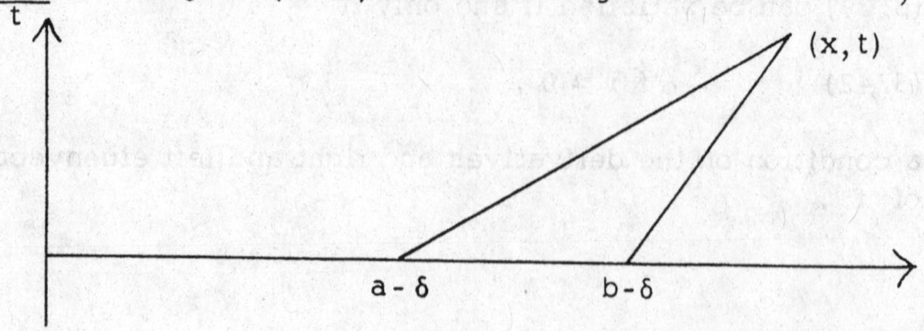

where δ is some positive quantity. Using the divergence theorem we get

$$(3.47)\qquad \int_{a-\delta}^{(x,t)} + \int_{(x,t)}^{b+\delta} [Un_t + Fn_x]ds \leq \int_{a-\delta}^{b+\delta} U(y,0)dy\ ,$$

where (n_x, n_t) is the outward normal to T. Substituting the special solutions with leading term (3.26), the leading term on the left in (3.47), can be written, using (3.31), as

$$\text{const} \int e^{k\phi}[c_{max} - c + \frac{\delta}{t}]\, V^o ds$$

$$+ \text{const} \int e^{k\phi}[c - c_{min} + \frac{\delta}{t}]V^o ds\ ;$$

since V^o and δ are >0, this is bounded from below by

$$\text{const} \int_{a-\delta}^{(x,t)} \int_{(x,t)}^{b+\delta} e^{k\phi} ds\ .$$

On the other hand the right side of (3.47) is bounded from above by

$$\text{const} \int_{a-\delta}^{b+\delta} e^{k\phi} dy\ .$$

The kth root of the former tends, as $k \to \infty$ to $\text{Max}\ \phi(u)$ on the segments connecting $a - \delta$, (x,t) and $b + \delta$, while the kth root of the latter tends to $\text{Max}\ \phi(u)$ along $(a-\delta, b+\delta)$. Therefore inequality (3.47) implies, if we take the kth root, let $k \to \infty$ and $\delta \to 0$, that (3.45) holds.

We have shown in Theorems 1.1 and 2.2 that weak solutions which are the limits of solutions of the viscous equation, or of Glimm's scheme, satisfy all entropy inequalities (3.43). It follows therefore that the estimates for the Riemann invariants asserted in Theorems 3.3 and 3.4 hold for such weak solutions. This conclusion is not new; we indicate the relation of condition (3.39) to known results.

We start with the observation that if ϕ satisfies (3.39) and if the function p is chosen so that $\ddot{p}$ is very much larger than $\dot{p}$, then the first term on the right in (3.40) is larger than the second except near these values where $\phi_1\xi + \phi_2\eta = 0$. For such values the second term is, by (3.39) positive, so that the right side of (3.40) is positive. But that means that $\delta = p(\phi)$ is <u>convex</u>. Thus (3.39) <u>implies the existence of a convex</u>† <u>Riemann invariant</u> ψ.

Let ψ be a convex Riemann invariant; let $u(x, t, \varepsilon)$ be solutions of the viscous equation

$$(3.47) \qquad \partial_t u + f'\partial_x u = \varepsilon\partial_x^2 u\,, \quad \varepsilon > 0\,.$$

Multiply this equation by ψ'; using relation (3.30): $\psi' f' = c\psi'$ we get

$$\partial_t\psi + c\partial_x\psi = \varepsilon\psi'\partial_x^2 u\,.$$

Using the identity

$$\partial_x^2\psi = \psi'\partial_x^2 + \partial_x u^t\psi''\partial_x u$$

and the convexity of ψ we conclude that

$$\partial_t\psi + c\partial_x\psi \le \varepsilon\partial_x^2\psi\,.$$

The maximum principle holds for solutions of such a differential inequality and tells us that

$$\operatorname*{Max}_x \psi(u(x,t;\varepsilon))$$

is a decreasing function of t. But then the same is true of their a.e. limits as $\varepsilon \to 0$; this proves Theorem 3.3 for this class of solutions.

We turn now to Glimm's scheme; in Theorem 2.1 we have shown that if U is a convex function which satisfies an additional conservation law, then U decreases across

---

† Smoller and Johnson in [13] have shown that condition (3.42) implies that the curves ϕ = const. are convex. This implies the existence of a convex $\psi = p(\phi)$.

shocks; a similar result holds for convex Riemann invariants:

<u>Lemma 3.5:</u> If the Riemann invariant ϕ satisfies (3.39), then ϕ decreases across a shock of the family opposite to ϕ.

<u>Remark:</u> This decrease was stipulated in Glimm-Lax, [5], precisely for the purpose of proving that the Riemann invariant is a decreasing function.

<u>Sketch of proof:</u> Consider all states w close to a given state v which can be connected on the right to v through a shock of a fixed kind, i.e. which satisfy the jump relations

$$s[w-v] + f(v) - f(w) = 0 \tag{3.48}$$

and the shock inequality (2.3). We saw in Section 2 that these states w form a one parameter family $w(p)$, $p \leq 0$, under the normalization $w(0) = v$, $s(0) > 0$, where dot denotes differentiation with respect to the parameter.

Differentiating (3.48) gives

$$\dot{s}(w-v) + s\dot{w} - f'\dot{w} = 0 . \tag{3.49}$$

Multiply (3.49) by ϕ' on the left; using (3.30) and that $\phi'\dot{w} = \dot{\phi}$ we get

$$\phi'\dot{s}(w-v) = (c-s)\dot{\phi}. \tag{3.50}$$

Since $w(0) = v$, (3.50) implies that $\dot{\phi}(0) = 0$. Differentiate (3.50) with respect to p; using $\dot{\phi}(0) = 0$ we deduce $\ddot{\phi}(0) = 0$; differentiating once more we get

$$\dddot{\phi}(0) = \frac{\dot{s}}{c-s}\dot{\phi}'\dot{w} . \tag{3.51}$$

By (3.39), $\dot{\phi}'\dot{w} > 0$ and by choice of normalization $\dot{s} > 0$; furthermore we are restricted to negative values of the parameter. So it follows from (3.51) that for w near v,

(3.52) $$\operatorname{sgn}(\phi(v)-\phi(w)) = \operatorname{sgn}(c-s).$$

Consider now a flow with a single shock going with speed s greater than the sound speed c of the opposite family:

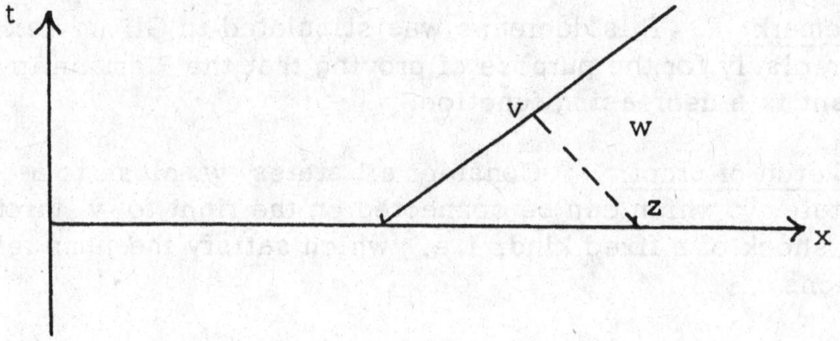

It follows from (3.52) that

$$\phi(v) < \phi(w) ;$$

on the other hand $\phi(w) = \phi(z)$, where z is the value of u at that point on the initial line which can be connected to the shock by a characteristic of the opposite family. So the value v of ϕ along the other side of the shock is < the value of ϕ at some point of the initial line. The same conclusion holds for shocks of the other family; this proves Lemma 3.5.

In Glimm's difference scheme the initial interval is divided into subintervals and the initial function is approximated by one which is constant in each subinterval; this problem is solved exactly in a time interval taken so short that the waves issuing from the point of discontinuity do not interact. At the end of that time interval the solution is approximated by a piecewise constant function obtained by setting u in each subinterval equal to its value at a randomly chosen point in that subinterval. It follows from Lemma 3.5 that for any Riemann invariant which satisfies (3.39) each approximate solution u_Δ satisfies the conclusion of

Theorem 3.3. But then so does their limit.

It is not hard to show that for almost all choices of the random points, the approximate solutions u_Δ satisfy the conclusions of Theorem 3.4. Therefore so do their limits, for almost all random choices.

Another difference scheme, devised by Godunov, starts similarly by solving exactly a piecewise constant initial value problem for a short time, but the conversion into piecewise constant data at the end of that time interval is accomplished differently: in each subinterval u is set equal to its average over that subinterval. If ψ is a convex Riemann invariant, this process decreases the maximum value of $\psi(u)$; so these approximate solutions satisfy the conclusion of Theorem 3.3. But then so does their limit.

It would be useful to determine all convex solutions of the second order equation (3.21), so that one can study weak solutions which satisfy the entropy condition with respect to all additional convex conservation laws. The most important question is: are such solutions uniquely determined by their initial data? Another interesting task is to derive from these entropy conditions an analogue for systems of Oleinik's condition E.

REFERENCES

1. Conley, C. C. and Smoller, J., Shock waves as limits of standing wave solutions of higher order equations, Comm. Pure Appl. Math., Vol. 24, No. 2, 1971.

2. Foy, R. L., Steady state solutions of hyperbolic systems of conservation laws with viscosity terms, Comm. Pure Appl. Math., Vol. 17, 1964, pp. 177-188.

3. Friedrichs, K. O. and Lax, P. D., to appear in Proc. Nat. Acad. Sci.

4. Glimm, J., Solutions in the large for nonlinear hyperbolic systems of equations, Comm. Pure Appl. Math., Vol. 18, pp. 697-715 (1965).

5. Glimm, J., and Lax, P. D., Decay of solutions of systems of nonlinear hyperbolic conservation laws, Memoirs of the Amer. Math. Soc., No. 101, 1970.

6. Hopf, E., On the right weak solution of the Cauchy problem for quasilinear equation of first order, J. of Math. and Mech., Vol. 19, 1969, pp. 483-487.

7. Kružkov, S. N., Results on the character of continuity of solutions of parabolic equations and some of their applications, Math. Zametky, Vol. 6, 1969, pp. 97-108.

8. Lax, P. D., Hyperbolic systems of conservation laws, II., Comm. Pure Appl. Math., Vol. 10, pp. 537-566 (1957).

9. Lax, P. D., Asymptotic solutions of oscillatory initial value problems, Duke Math. J., Vol. 24, 1957, pp. 627-646.

10. Oleinik, O. A., On the Uniqueness of the generalized solution of the Cauchy problem for a non-linear system of equations occurring in mechanics, Usp. Mat. Nauk. Vol. 78, pp. 169-176 (1957).

11. Quinn, B., Solutions with shocks, an example of an L_1-contractive semigroup, Comm. Pure Appl. Math., Vol. 24, No. 1, 1971.

12. Rozdestvenskii, B., Discontinuous solutions of hyperbolic systems of quasi-linear equations, Ups. Mat. Nauk. Vol. 15, pp. 59-117 (1960). English translation in Russ. Math. Surv. Vol. 15, pp. 55-111 (1960).

13. Smoller, J. and Johnson, J. L., Global solutions for an extended class of hyperbolic systems of conservation laws, Arch. For Rat. Mech. and Anal., Vol. 32, 1969, pp. 169-189.

14. Kružkov, S. N., First order quasilinear equations in several independent variables, Math. USSR Sbornik Vol. 10 (1970), No. 2.

The work presented in this paper is supported by the AEC Computing and Applied Mathematics Center, Courant Institute of Mathematical Sciences, New York University, under Contract AT(30-1)-1480 with the U. S. Atomic Energy Commission.

Courant Institute of Mathematical Sciences
AEC Computing and Applied Mathematics Center
New York University
New York, New York

Received May 6, 1971

Reprinted from
Proc. Nat. Acad. Sci. USA
Vol. 68, No. 8, pp. 1686–1688, August 1971

Systems of Conservation Equations with a Convex Extension

K. O. FRIEDRICHS AND P. D. LAX

Courant Institute of Mathematical Sciences, New York University, New York, N.Y. 10012

Contributed by P. D. Lax, May 18, 1971

ABSTRACT We discuss first-order systems of nonlinear conservation laws which have as a consequence an additional conservation law. We show that if the additional conserved quantity is a convex function of the original ones, the original system can be put into symmetric hyperbolic form. Next we derive an entropy inequality, which has also been suggested by I. Kružhkov, for discontinuous solutions of the given system of conservation laws.

Systems of first-order nonlinear partial differential equations are conservation equations if they can be written in the form

$$\partial_t u^j + \partial_k f^{j,k} = 0, \qquad j = 1,\dots,m. \tag{1}$$

Here ∂_t and ∂_k stand for partial differentiation with respect to the time t and the space coordinates x^k, summation with respect to k is implied. The $f^{j,k}$ are functions of the quantities $\{u^l,\dots,u^m\} = u$. If the quantities u^j, as functions of the x^k and t, satisfy Eq. 1, the integrals

$$\iint \dots u^j(t,x)dx^1dx^2\dots \text{ are constant in } t. \tag{2}$$

Therefore the u^j will be referred to as the conserved quantities.

The question is, when is a new conservation law,

$$\partial_t U + \partial_k F^k = 0, \tag{3}$$

with U and F^k functions of u, a consequence of the old ones?

Denoting partial differentiation with respect to u^l by a subscript l, and writing (1) and (3) as

$$\partial_t u^j + f_l^{j,k}\partial_k u^l = 0, \quad \text{and} \tag{1$'$}$$

$$U_j\partial_t u^j + F_l^k\partial_k u^l = 0, \tag{3$'$}$$

we see that (3)′ follows from (1)′ if and only if

$$U_j f_l^{j,k} = F_l^k. \tag{4}$$

We assume that (4) is satisfied; differentiating this relation with respect to u^h we get

$$U_{j,h} f_l^{j,k} + U_j f_{l,h}^{j,k} = F_{l,h}^k.$$

The second term on the left and the right side are symmetric in the subscripts l and h. Therefore, so is the first term,

$$U_{j,h} f_l^{j,k} = U_{j,l} f_h^{j,k}. \tag{5}$$

Multiplying (1)′ by $U_{j,h}$ and summing with respect to j we get

$$U_{j,h}\partial_t u^j + U_{j,h} f_l^{j,k}\partial_k u^l = 0. \tag{6}$$

This system is equivalent to (1) if the matrix $(U_{j,h})$ is nonsingular. The noteworthy feature of Eq. 6 is that it is *symmetric;* i.e., the matrices which multiply $\partial_t u$ and $\partial_k u$ are symmetric, by (5).

If the coefficient matrix of $\partial_t u$ is *positive definite,* a symmetric system is called *symmetric hyperbolic.* The importance of this concept is that *Cauchy's initial value problem* for symmetric hyperbolic systems *is well posed;* that is, the values of their solution on a surface t = const., or on any other space-like surface may be prescribed as arbitrary smooth functions of x. A unique solution of the equation will then exist at least in a neighborhood of the initial surface.

The system (6) is evidently symmetric hyperbolic if the matrix $U_{j,h}$ is positive definite, i.e., if the matrix of second derivatives of U with respect to the u^j is positive definite; in other words, if U is *convex* as function of the u_k.

Thus we have shown: *If a system of conservation laws* (1) *implies a new conservation law* (3) *such that the new conserved quantity U is a convex function of the original conserved quantities u^j, the initial value problem for Eq.* 1 *is well posed.*

We observe that in case Eq. 1′ is already symmetric,

$$f_l^{j,k} = f_j^{l,k},$$

we can derive from it a new conservation law with

$$U = \tfrac{1}{2}\sum_l (u^l)^2, \qquad F^k = u^j f^{j,k} - g^k,$$

where g^k satisfies $g_l^k = f^{l,k}$. Clearly U is convex.

We remark that the conclusion derived above is also valid for nonhomogeneous conservation laws, i.e., equations of the form

$$\partial_t u^j + \partial_k f^{j,k} = h^j,$$

where the h^j are functions of $u^1,\dots,u^m$.

Symmetric hyperbolic systems are not the only ones for which the initial value problem is properly posed; for example, this problem is also well posed for all strictly hyperbolic systems, i.e., systems with real and distinct characteristics. The requirement of distinctness of the characteristics, an unnatural one for mathematical physics, is not needed in the symmetric hyperbolic case.

Most basic equations of mathematical physics can be written as systems of conservation laws (1) that have a convex extension (3). This is, for example, the case of the equations of Maxwellian electromagnetism, of elasticity, of the dynamics of compressible fluids in Eulerian form, and of magneto-fluid-dynamics, both nonrelativistic and relativistic.

The original conserved quantities in these equations may be the densities of mass and momentum, but may also be other quantities. The role of new conserved quantity will frequently be played by the energy density; but possibly also by the density of the negative entropy. Circumstances under which this is so will be indicated at the end of this note. Details about convexly extensible systems of conservation laws in mathematical physics will be described in a future publication, in which also the relationship of the Lagrangian formalism with the present one will be discussed.

In handling concrete problems it would be awkward if one forced oneself to take the original conserved quantities u^j as unknowns. If the u^j as well as the $f^{j,k}$ are considered functions of other unknowns $v = \{v^h\}$, one will postulate the existence of multipliers $\lambda^j(v)$—in place of U_j—such that

$$\lambda^j(\partial_t u^j + \partial_k f^{j,k}) \equiv \partial_t U + \partial_k F^k.$$

Then the matrix $(\partial\lambda^j/\partial v^k)(\partial u^j/\partial v^l)$—in place of $U_{j,h}$—will serve as the symmetrizer and the convexity condition is replaced by the condition that the matrix

$$(\partial^2 U/\partial v^h \partial v^l - \lambda^j \partial^2 u^j/\partial v^h \partial v^l) \text{ is positive definite.} \quad (\#)$$

This remark is helpful if one wants to take the derived conserved quantity U as an original one and the negative of one of the original ones, $-u^m$ say, as the new derived conserved quantity. Since the new multipliers are $\{\lambda^1/\lambda^m, \ldots, \lambda^{m-1}/\lambda^m, -1/\lambda^m\}$, the new convexity condition is that the matrix

$$\left(-\partial^2 u^m/\partial v^h \partial v^l - (\lambda^m)^{-1} \sum_{j=1}^{m-1} \lambda^j \partial^2 u^j/\partial v^h \partial v^l - (\lambda^m)^{-1} \partial^2 U/\partial v^h \partial v^l\right)$$

should be positive definite. This is the same as condition (#) if the factor λ^m is positive. Thus, the quantities U and u^m may be switched if $\lambda^m > 0$. In fluid dynamics one may take U and u^m as the densities of energy and entropy; then λ^m is the temperature and hence positive. The switch between energy and negative entropy is therefore permitted.

We must return to our statement that the initial problem of our conservation Eq. 1 have a unique solution in a neighborhood of the initial surface. The latter qualification is necessary inasmuch as, in general, continuous solutions of nonlinear equations exist only for a limited time interval. Such solutions can, nevertheless, be continued for all future time as discontinuous solutions which satisfy the conservation laws in the integrated sense (2). Across a discontinuity surface these integrated laws imply the jump relations

$$[u^j]n_t + [f^{j,k}]n_k = 0, \qquad (7)$$

in which $n = (n_t, n_k)$ is the normal to the discontinuity surface in (t,x) − space and $[f]$ denotes the jump in f across the discontinuity surface in the direction of n. It is well known that discontinuous solutions which satisfy the conservation laws in the form (7), in addition to (1), are not determined uniquely by their initial values. The physically relevant ones are singled out as the limits of solutions of modified equations that incorporate dissipative forces. We shall indicate how to single out the physically relevant solutions directly, without such a limit process, provided the equations imply an additional convex conservation law.

The modified equations* will be taken in the form

$$u_t^j + \partial_k f^{j,k} = \epsilon \Delta u^j, \qquad \epsilon > 0, \qquad (8)$$

in which $\Delta = \partial_k \partial_k$ is the spatial Laplace operator. The term $\epsilon \Delta u^j$ may be regarded as representing viscous forces. Multiplying Eq. 8 by U_j and summing with respect to j we get

$$\partial_t U + \partial_k F^k = \epsilon U_j \Delta u^j. \qquad (9)$$

From the identity

$$\Delta U = U_j \Delta u^j + U_{j,l} \partial_k u^j \partial_k u^l$$

and the convexity of U, we deduce that

$$\Delta U \geq U_j \Delta u^j.$$

Since ϵ is positive, it follows that the solutions of (9) satisfy the inequality

$$U_t + \partial_k F^k \leq \epsilon \Delta U. \qquad (10)$$

We now let ϵ tend to zero and suppose that a sequence of solutions $u^j(x,t,\epsilon)$ of (8) tends, as $\epsilon \to 0$, boundedly a.e. to a limit function $u^j(x,t)$. Since then $\epsilon \Delta u^j$ tends to 0 in the sense of distributions, the limit function satisfies the conservation Eq. 1. At the same time letting $\epsilon \to 0$ in Eq. 10 gives

$$U_t + \partial_k F^k \leq 0. \qquad (11)$$

From this inequality we draw two conclusions concerning the quantities U and F^k as functions of the limit functions $U_t(x,t)$: A) The integral

$$\int U dx \qquad (12)$$

is a decreasing function of t.

B) Across a surface at which U is discontinuous the

* It has been shown in [2] that also other dissipative limiting procedures lead to inequality (11).

inequality

$$[U]n_t + [F^k]n_k < 0 \tag{13}$$

holds.

Inequality (13) together with the jump condition (7), in addition to Eq. 1, presumably determine the solution of the initial value problem uniquely for all time. Under appropriate circumstances this can be proved to be true. For a more detailed discussion, see a later publication [1].

If the derived conserved quantity U is taken as the energy density, the quantity F^k is the energy flux across the discontinuous surface. Eq. 13 then implies that the energy decreases across such a surface. This may be interpreted by saying that energy is lost by dissipation.

Actually, energy is not lost but is converted into internal energy. In some processes, such as the flow of compressible fluids, one must keep track of this conversion; in these cases the total energy must be taken as one of the original conserved quantities. Negative entropy is then a derived conserved quantity; it is indeed a convex function of the other conserved quantities, as follows from the remark after formula (#). Now, inequality (13) stipulates that the entropy increases when the fluid crosses a shock. This entropy inequality has also been suggested by Kružhkov [2].

The work presented in this paper is supported by the Air Force Office of Scientific Research, Contract SF-49(638)-1719 and the U.S. Atomic Energy Commission, Contract AT(30-1)-1480. Reproduction in whole or in part is permitted for any purpose of the U.S. Government.

1. Lax, P. D., "A concept of entropy", ed. E. Zarantonello, *Proc. Wisconsin Conference* (1971), to appear.
2. Kružhkov, S. N., "First order quasilinear equations in several independent variables", *Math. USSR Sbornik* (1970), Vol. 10, No. 2.

Computational Fluid Dynamics JOURNAL vol.5. no.2 July 1996 (pp.133–156)
Special Issue dedicated to Professor Antony Jameson for His 60th Birthday

POSITIVE SCHEMES FOR SOLVING MULTI-DIMENSIONAL HYPERBOLIC SYSTEMS OF CONSERVATION LAWS

Xu-Dong LIU † Peter D. LAX ‡

Abstract

This paper introduces a new positivity principle for numerical schemes for solving hyperbolic systems of conservation laws in many space variables. A family of positive schemes has been designed that are second order accurate in the space variables. Second order accuracy in time is achieved by a second order accurate energy conserving Runge-Kutta method due to Shu, [17] and [18].

Our positive schemes have a very simple structure using characteristic decomposition, treating each space variable separately. Only about 150 (180) FORTRAN 77 lines are needed for solving the two(three)-dimensional Euler equations. Positive schemes can be constructed for all hyperbolic conservation laws where Roe averages are known, and where the eigenvectors and eigenvalues of the Roe matrix can be evaluated explicitly.

Excellent numerical results have been obtained. Comparison with successful high order accurate schemes shows that the positive schemes are competitive with the best schemes available today.

1. INTRODUCTION

Over the last decade a great deal of effort has been devoted to designing total variation diminishing (TVD) numerical schemes for solving hyperbolic conservation laws, see Harten's basic paper [4]. Strictly speaking, TVD schemes exist only for scalar conservation laws, and for linear hyperbolic systems in one space variable. Hyperbolic systems of conservation laws in one space variable can be treated in a formal manner. No TVD schemes can exist for non-scalar hyperbolic systems in more than one space variable, linear or non-linear. The possibility of focusing shows that total variation norms are not bounded. The only functional known to be bounded for solutions of linear hyperbolic equations is energy. This is particularly easy to

† Courant Institute of Mathematical Sciences, 251 Mercer street, New York, NY 10012, xliu@cims.nyu.edu.
The author was supported by National Science Foundation grants DMS-9112654 and DMS-9406467, and Air Force grant AFOSR F49620-94-1-0132
‡ Courant Institute of Mathematical Sciences, 251 Mercer street, New York, NY 10012. The author was supported by Applied Mathematical Sciences Program of the U.S. Department of Energy under contract DE-FG02-88ER25053

show, as Friedrichs has done in [3], for symmetric hyperbolic linear systems

$$U_t + \sum_{s=1}^{d} A_s U_{x_s} = 0, \tag{1}$$

A_s real symmetric matrices, whose dependence on x is Lipschitz continuous. It is easy to show in this case that the L^2 norm of a solution is of bounded growth

$$\|U(t)\| \leq e^{ct}\|U(0)\|, \tag{2}$$

where c is related to the Lipschitz constant. Friedrichs [3] has shown that solutions of such equations can be approximated by solutions of difference equations of form

$$U_J^{m+1} = \sum_K C_K U_{J+K}^m, \tag{3}$$

J a lattice point, U_J^m an approximation to the value of the solution $U(x,t)$ of (1) at the point $x = (j_1\Delta x_1, \cdots, j_d\Delta x_d)$ at time $t = m\Delta t$. The coefficient matrices C_K are required to have the following properties:

i) C_K is symmetric,
ii) C_K is nonnegative,
iii) $\sum_K C_K = I$, where I is the identity matrix,
iv) $C_K = 0$ except for a finite set of K,
v) C_K depends Lipschitz continuously on x.

We show now that the l^2 norm of solutions of difference schemes satisfying these properties has bounded growth:

$$\|U^{m+1}\| \leq (1 + const\Delta)\,\|U^m\|, \tag{4}$$

where the discrete l^2 norm is defined as

$$\|U\|^2 = \sum |U_J|^2.$$

The value of the constant in (4) depends on the Lipschitz constant.

Proof: Take the scalar product of (3) with U_J^{m+1}

$$|U_J^{m+1}|^2 = \sum (U_J^{m+1}, C_K U_{K+J}^m). \tag{5}$$

Since C_K is symmetric and nonnegative, $(U, C_K V)$ can be regarded as an inner product, to which the Schwarz inequality applies; combined with the inequality between arithmetic and geometric mean we get

$$(U, C_K V) \leq \sqrt{(U, C_K U)}\sqrt{(V, C_K V)} \leq \frac{1}{2}(U, C_K U) + \frac{1}{2}(V, C_K V). \tag{6}$$

Using this on the right side of (5) gives

$$|U_J^{m+1}|^2 \leq \frac{1}{2}\sum (U_J^{m+1}, C_K U_J^{m+1}) + \frac{1}{2}\sum (U_{J+K}^m, C_K U_{J+K}^m). \tag{7}$$

Carrying out the summation with respect to K, using condition iii) and multiplying by 2 gives

$$|U_J^{m+1}|^2 \leq \sum \left(U_{J+K}^m, C_K U_{J+K}^m\right). \tag{8}$$

Now sum with respect to J, and introduce $K + J = N$ as new index of summation:

$$\sum_J |U_J^{m+1}|^2 \leq \sum_{N,K} (U_N^m, C_K(N-K)U_N^m). \tag{9}$$

Replace on the right $C_K(N-K)$ by $C_K(N)$; using Lipschitz continuity, and the fact that K ranges over a finite stencil, we get that the right side of (9) is

$$\leq \sum_{N,K} (U_N^m, C_K(N)U_N^m) + O(\Delta)\sum |U_N^m|^2,$$

which, using iii), is equal to

$$\sum |U_N^m|^2 (1 + O(\Delta)).$$

Setting this into (9) gives (4). Q.e.d..

In the usual variables the Euler equations are not symmetric but symmetrizable, i.e. can be written in the form

$$B_0 U_t + \sum B_s U_{x_s} = 0, \tag{10}$$

where all the B_s are symmetric and B_0 is positive. Multiplying (10) by B_0^{-1} gives

$$U_t + \sum A_s U_{x_s} = 0, \tag{11}$$

where

$$A_s = B_0^{-1} B_s. \tag{12}$$

Since B_0 is positive, it has a symmetric square root S:

$$B_0 = S^2, \qquad S^T = S. \tag{13}$$

Multiplying (12) by S on the left, S^{-1} on the right gives

$$S A_s S^{-1} = S^{-1} B_s S^{-1}. \tag{14}$$

The matrix on the right is symmetric; therefore so is the matrix on the left. In words: the matrices A_s can be symmetrized by the same similarity transformation.

In section 2.we show how to construct difference schemes that can be put in form (3), where the $C_K(J)$ have the following two properties:

a) $\sum_K C_K(J) = I$,

b) C_K is of the form,

$$C_K(J) = \sum A_s(J \pm \frac{1}{2}e_s),$$

where the matrix $\tilde{A}_s(J \pm \frac{1}{2}e_s)$ commutes with $A_s(J \pm \frac{1}{2}e_s)$, and has positive eigenvalues; here e_s is the unit vector in the x_s direction. It follows from a) and b) that

$$S\tilde{A}_s S^{-1},$$

is a symmetric and nonnegative matrix, where $S = S(J \pm \frac{1}{2}e_s)$.

Assume now that the matrices B_s depend Lipschitz continuously on x; then so does $S = B_0^{1/2}$ and C_K. We define D_K by

$$D_K(J) = S(J)C_K(J)S^{-1}(J). \tag{15}$$

It follows that the $D_K(J)$ have the following properties:
i) D_K differs by $O(\Delta)$ from a symmetric matrix,
ii) The symmetric part of D_K has eigenvalues that are $\geq -O(\Delta)$,
iii) $\sum_K D_K = I$,
iv) $D_K = 0$ except for a finite number of K,
v) D_K depends Lipschitz continuously on x.

Under these conditions we can deduce an analogue of inequality (4). We introduce $V_J = S(J)U_J$ as new variable and multiply (3) by $S(J)$; we obtain

$$V_J^{m+1} = S(J)U_J^{m+1} = \sum S(J)C_K U_{J+K}^m = \sum S(J)C_K S^{-1}(J)V_{J+K}^m = \sum D_K(J)V_{J+K}^m.$$

We can now proceed as before and deduce that

$$\|V^{m+1}\| \leq (1 + O(\Delta))\|V^m\|.$$

In terms of U, this can be expressed as

$$\|U^{m+1}\|_S^2 \leq (1 + O(\Delta))\|U^m\|_S^2, \tag{16}$$

where $\|U\|_S^2$ is defined as

$$\|U\|_S^2 = \sum(V_J, V_J) = \sum(SU_J, SU_J) = \sum(U_J, S^2U_J) = \sum(U_J, B_0U_J).$$

If B_0 depends on t as well, as it does in our case, then also the norm $\|U\|_S$ depends on t. Since B_0 is assumed to depend Lipschitz continuously on t, it follows that

$$\|U\|_{S(t_{m+1})} \leq (1 + O(\Delta))\|U\|_{S(t_m)}.$$

Combined with (16) we deduce recursively that

$$\|U\|_{S(t_m)} \leq (1 + mO(\Delta))\|U\|_{S(0)}.$$

Q.e.d..

The difference schemes studied in this paper are nonlinear; when we write them in linear form (3), the coefficient matrices C_K depend on the solution being computed. In the applications of interest these solutions contain shocks and contact discontinuities; therefore the coefficient matrices $C_K(J)$ not only fail to be Lipschitz continuous, they are not even continuous. So our analysis of the boundedness of the l^2 norm of the solution is not applicable. A possible way of salvaging our argument is to note that in inequality (6) the left side is substantially smaller than the right side, unless the vector U and V are nearly equal. For unless the vectors U and V are nearly proportional, the Schwarz inequality is a strict inequality. Similarly, unless (U, CU) and (V, CV) are nearly equal, their geometric mean is substantially less than their arithmetic mean. This shows that at a discontinuity the left side of (8) is substantially less than the right side. We don't at this moment see how to show that this gain is enough to counterbalance what we may lose when we replace on the right in inequality (9) the matrix $C_K(N-K)$ by $C_K(N)$, but at least we have found a plausible reason why our scheme is as stable as it appears to be in numerical experiments.

In this paper we construct finite difference approximation in conservation form to symmetric and symmetrizable hyperbolic systems of conservation laws that are second order accurate and yet can be written in positive form (3). This is achieved by combining a second order accurate scheme with an upwind scheme, using an appropriate flux limiter, see LeVeque [13] for a general discussion.

In verifying positivity we use the fact that the sum of positive matrices is positive, and that the product of commuting positive matrices is positive. We use a combination of several such schemes. One of them, which we call the least dissipative, is, in one space dimension, similar to one proposed by Helen Yee in [23]. In [9], Jameson discusses this scheme in the scalar case; he observes that it has the local extremum diminishing (LED) property. Fluxes in the several coordinate directions are handled independently of each other. Second order accuracy in time is achieved by using an energy preserving second order time step.

Our positive schemes have the following advantages.

Great simplicity and very low cost. No Riemann solvers or MUSCL-type reconstruction is needed. By using dimension-by-dimension split and the method of lines, cost for each time step is proportional to dimension. Thanks to Roe averages only simple algebraic operations are involved, programs are short and run fast. For solving the two-dimensional Euler equations, the positive schemes need less than 150 lines in FORTRAN 77 (excluding input/output, initial, boundary condition subroutines). For three-dimension, the positive schemes need less than 180 lines. Positive schemes are at least as fast as MUSCL schemes in multi-dimension. One-dimensional positive schemes are 40% faster than MUSCL schemes.

Wide range of applicability. Positive schemes can be applied to all hyperbolic conservation laws where Roe averages are known, and where the eigenvalues and eigenvectors of the Roe matrix can be evaluated explicitly.

The disadvantage of our positive schemes is that they are limited to first order accuracy at extrema, although second order elsewhere. ENO schemes [7], [18], [14] and [10] have no such limitation.

High resolution numerical experiments show that positive schemes are capable of resolving multi-dimensional hyperbolic systems of conservation laws. The numerical results are comparable to the numerical results using ENO [7] [18] and PPM [22].

The organization of this paper is as follows. Section 2.1 describes the notion of a positive scheme for solving symmetric nonlinear equations. In section 2.2 we construct two positive schemes, both second order accurate. The scheme used in practice is an appropriate combination of them. Another family of one-dimensional positive schemes based on Lax-Wendroff are constructed in the section. The Runge-Kutta method to achieve second order accuracy in time is described in section 2.2 In section 2.3we explain how to combine the fluxes in all space direction. In section 2.4 we explain why the scheme constructed in section 2.3 work also for symmetrizable systems.

The numerical experiments are described in section 3. In section 3.1several one-dimensional problems are solved, a Riemann problem, an interaction of two waves, and low density and internal energy Riemann problem. In section 3.2 we present a calculation of a plane shock diffracted by a wedge; the flow is rather complicated, containing double Mach reflection. The second example is flow in a wind tunnel obstructed by a step; this flow, too, is rather complicated, containing a strong rarefaction wave, a Mach reflection and many simple reflections. Most details are well resolved by our positive scheme.

We thank Tony Jameson for his valuable comments, and Stanley Osher and Chi-Wang Shu for their valuable comments on ENO. We thank Bjorn Sjogreen for calling our attention to references [23] and [9]. We also thank Smadar Karni for helpful discussions.

2. POSITIVITY PRINCIPLE AND POSITIVE SCHEMES

2.1 Positivity Principle

We consider multi-dimensional hyperbolic systems of conservation laws

$$U_t + \sum_{s=1}^{d} F_s(U)_{x_s} = 0, \tag{17}$$

where $U = (u_1, \cdots, u_n)^T \in \mathbf{R^n}$ and $x = (x_1, x_2, \cdots, x_d) \in \mathbf{R^d}$. We assume that all Jacobian matrices $A_s = \nabla F_s$ are symmetric or simultaneously symmetrizable with the same similarity transformation. We construct a uniform Cartesian grid $\{\Omega_J\}$ in $\mathbf{R^d}$, where $J = (j_1, j_2, \cdots, j_d)$ is a lattice point in which all j_s are integers. In this uniform grid we denote cell-averages as $U_J = \frac{1}{|\Omega|} \int_{\Omega_J} U(x,t)\, dx$, where $|\Omega|$ is the volume of the cell Ω_J.

Conservative schemes are of the form

$$U_J^* = U_J - \sum_{s=1}^{d} \frac{\Delta t}{\Delta x_s} [F_{J+1/2e_s} - F_{J-1/2e_s}] \tag{18}$$

where Δt is the time step and Δx_s is the spatial step in the x_s dimension.

We call the schemes (18) **positive**, if we can write the right side as

$$U_J^* = \sum_K C_K U_{J+K}, \tag{19}$$

so that the coefficient matrices C_K, which themselves depend on all the U_{J+K} that occur in (18), have the following properties

(i) Each C_K is symmetric positive definite i.e. $C_K \geq 0$;

(ii) $\sum_K C_K = I$.

Note that since the right side of (18) is a non-linear function of the U_{J+K}, there are many ways of writing it in form (19).

In section 2.4we shall extend the notion of positive scheme to symmetrizable systems.

2.2 Positive Scheme in One-Dimension

We consider one-dimensional hyperbolic systems of conservation laws

$$U_t + F(U)_x = 0. \tag{20}$$

We consider schemes in conservation form, i.e.

$$U_j^* = U_j - \frac{\Delta t}{\Delta x}[F_{j+1/2} - F_{j-1/2}]. \tag{21}$$

The numerical flux F is obtained by mixing a second order accurate scheme with numerical flux F^{acc}, with another dissipative scheme with numerical flux F^{diss}. We combine them using the flux limiting philosophy [1] and [8] of having a numerical flux of the form

$$F = F^{diss} + L(F^{acc} - F^{diss}), \tag{22}$$

where the flux limiter L is near the identity when the flow is smooth, and near zero otherwise. For the accurate scheme we take not Lax-Wendroff but central differencing as in [23] and [9]. For the dissipative scheme we have two candidates of upwind type; one is the least dissipative, the other more dissipative. Different flux limiters are chosen in the two cases. This way we come up with two distinct positive schemes; an appropriate combination of the two positive schemes is again positive. Note that although centered differencing leads to an unconditionally unstable scheme, the combined scheme is stable.

The schemes are second order in the smooth regions with respect to space discretization. To achieve second order accuracy in time we employ the second order energy preserving Runge-Kutta of Shu [17] and [18].

Both dissipative schemes use the Roe average $\hat{U} = \hat{U}(U, V)$, which has the property

$$F(U) - F(V) = A(\hat{U})(U - V), \tag{23}$$

where $A = \nabla F$. Roe [16] has shown how to construct $\hat{U}$ for the Euler flux in all dimensions. We write $F(U_{j+1}) - F(U_j) = A_{j+1/2}(U_{j+1} - U_j)$ and then drop the subscript $j + 1/2$ to avoid clutter. The spectral resolution of A is $A = R\Lambda R^{-1}$, Λ diagonal. The entries λ^i of Λ are the eigenvalues of A, the columns of R the right eigenvectors, the rows of R^{-1} the left eigenvectors.

For the least dissipative scheme we define the "absolute value" of A as $|\mathring{A}| = R\,|\Lambda|\,R^{-1}$ where $|\Lambda| = diag\,(|\lambda^i|)$. We construct the numerical flux $F_{j+1/2}$ in (21) as a mixture of a 2nd order accurate centered difference flux

$$F^c_{j+1/2} = \frac{F(U_j) + F(U_{j+1})}{2} \tag{24}$$

and a first order accurate upwind flux

$$F^{up}_{j+1/2} = \frac{F(U_j) + F(U_{j+1})}{2} - \frac{1}{2}\,|\mathring{A}|\,(U_{j+1} - U_j). \tag{25}$$

The mixture is accomplished with the aid of a limiter L^0:

$$F^0_{j+1/2} = \frac{F(U_j) + F(U_{j+1})}{2} - \frac{1}{2} \mid \mathring{A} \mid (U_{j+1} - U_j) + \frac{1}{2} L^0 \mid \mathring{A} \mid (U_{j+1} - U_j) \tag{26}$$

where L^0 is a limiter i.e. matrix that is near I in a smooth region of the solution, and near 0 otherwise. This makes $F^0_{j+1/2}$ the 2nd order accurate central difference flux in the smooth regions and the least dissipative flux otherwise.

We choose L^0 to commute with A, specially we take

$$L^0 = R\Phi^0 R^{-1}, \quad \text{where} \quad \Phi^0 = diag\left(\phi^0(\theta^i)\right) \quad \text{is a diagonal matrix.}$$

We take $\phi^0(\theta)$ as some limiter function such that, see Sweby [20],

$$0 \le \phi^0(\theta), \frac{\phi^0(\theta)}{\theta} \le 2, \quad \phi^0(1) = 1.$$

Here $\phi^0(\theta^i)$ could be different for each i.

We define θ^i in terms of the i-th component of differences of the characteristic variables:

$$\theta^i = \frac{l^i(U_{j'+1} - U_{j'})}{l^i(U_{j+1} - U_j)} \qquad 1 \le i \le n, \qquad \text{where} \qquad j' = \begin{cases} j-1 & \text{if} \quad \lambda^i \ge 0 \\ j+1 & \text{otherwise.} \end{cases} \tag{27}$$

The l^i are the left eigenvectors of A associated with λ^i, This completes the construction of the least dissipative scheme.

Our choice in (27) is somewhat different from Helen Yee's scheme in [23], where she define θ^i as

$$\theta^i = \frac{l^i_{j'+1/2}(U_{j'+1} - U_{j'})}{l^i_{j+1/2}(U_{j+1} - U_j)}.$$

We exploit the precise form of (27) in proving positivity.

We now prove that scheme (26) is positive. We define

$$\mathring{A}^+ = R\Lambda^+ R^{-1}, \qquad \mathring{A}^- = R\Lambda^- R^{-1}, \tag{28}$$

where two diagonal matrices are

$$\Lambda^+ = \begin{pmatrix} \max(\lambda^1, 0) & & \\ & \ddots & \\ & & \max(\lambda^n, 0) \end{pmatrix}, \qquad \Lambda^- = \begin{pmatrix} \min(\lambda^1, 0) & & \\ & \ddots & \\ & & \min(\lambda^n, 0) \end{pmatrix}.$$

Obviously

$$\begin{aligned} A &= \mathring{A}^+ + \mathring{A}^-, \qquad & \mid \mathring{A} \mid &= \mathring{A}^+ - \mathring{A}^-, \\ \mathring{A}^+ &\ge 0 & \mathring{A}^- &\le 0. \end{aligned} \tag{29}$$

By definition (27)

$$\theta^i = \frac{l^i(U_j - U_{j-1})}{l^i(U_{j+1} - U_j)} \quad \text{when} \quad \lambda^i \geq 0. \tag{30}$$

It follows from (30) that for $\lambda^i \geq 0$,

$$e_i R^{-1}(U_{j+1} - U_j) = l^i(U_{j+1} - U_j) = \frac{1}{\theta^i} l^i(U_j - U_{j-1}) = \frac{1}{\theta^i} e_i R^{-1}(U_j - U_{j-1}), \tag{31}$$

where e_i is the i-th unit row vector and $l^i = e_i R^{-1}$. Define the matrix $\tilde{\Phi}^0$ as

$$\tilde{\Phi}^0 = \begin{pmatrix} \frac{\phi^0(\theta^1)}{\theta^1} & & \\ & \ddots & \\ & & \frac{\phi^0(\theta^n)}{\theta^n} \end{pmatrix};$$

we obtain from (31) that for $\lambda^i \geq 0$

$$e_i \Phi^0 R^{-1}(U_{j+1} - U_j) = e_i \tilde{\Phi}^0 R^{-1}(U_j - U_{j-1}),$$

or equivalently

$$\max(\lambda^i, 0) e_i \Phi^0 R^{-1}(U_{j+1} - U_j) = \max(\lambda^i, 0) e_i \tilde{\Phi}^0 R^{-1}(U_j - U_{j-1}).$$

Using (28) we can write above equation in matrix form:

$$\Lambda^+ \Phi^0 R^{-1}(U_{j+1} - U_j) = \Lambda^+ \tilde{\Phi}^0 R^{-1}(U_j - U_{j-1}).$$

Multiply by R on the left and obtain the crucial equation

$$\mathring{A}^+ R \Phi^0 R^{-1}(U_{j+1} - U_j) = \mathring{A}^+ R \tilde{\Phi}^0 R^{-1}(U_j - U_{j-1}). \tag{32}$$

Here and below we denote

$$\tilde{L}^0 = R \tilde{\Phi}^0 R^{-1}.$$

Therefore, from (26), (29) and (32), and since L^0 commutes with $| \mathring{A} |$,

$$\begin{aligned} F^0_{j+1/2} &= F^{up}_{j+1/2} + \tfrac{1}{2} | \mathring{A} | L^0(U_{j+1} - U_j) \\ &= F^{up}_{j+1/2} + \tfrac{1}{2}\mathring{A}^+ L^0(U_{j+1} - U_j) - \tfrac{1}{2}\mathring{A}^- L^0(U_{j+1} - U_j) \\ &= F^{up}_{j+1/2} + \tfrac{1}{2}\mathring{A}^+ \tilde{L}^0(U_j - U_{j-1}) - \tfrac{1}{2}\mathring{A}^- L^0(U_{j+1} - U_j). \end{aligned} \tag{33}$$

We put the subscript $j + 1/2$ back in (33):

$$F^0_{j+1/2} = F^{up}_{j+1/2} + \frac{1}{2}\mathring{A}^+_{j+1/2}\tilde{L}^0_{j+1/2}(U_j - U_{j-1}) - \frac{1}{2}\mathring{A}^-_{j+1/2} L^0_{j+1/2}(U_{j+1} - U_j). \tag{34}$$

Similarly we obtain the flux $F_{j-1/2}$ as

$$F^0_{j-1/2} = F^{up}_{j-1/2} + \frac{1}{2}\mathring{A}^+_{j-1/2}L^0_{j-1/2}(U_j - U_{j-1}) - \frac{1}{2}\mathring{A}^-_{j-1/2}\tilde{L}^0_{j-1/2}(U_{j+1} - U_j). \tag{35}$$

Using (23) and (25) we can write

$$F^{up}_{j+1/2} - F^{up}_{j-1/2} = \frac{1}{2}(A_{j+1/2} - |\mathring{A}_{j+1/2}|)(U_{j+1} - U_j) + \frac{1}{2}(A_{j-1/2} + |\mathring{A}_{j-1/2}|)(U_j - U_{j-1}). \tag{36}$$

We obtain from (21,34-36)

$$\begin{aligned} U^*_j = U_j - \tfrac{\Delta t}{2\Delta x}[\ &(A_{j+1/2} - |\mathring{A}_{j+1/2}|)(U_{j+1} - U_j) + (A_{j-1/2} + |\mathring{A}_{j-1/2}|)(U_j - U_{j-1}) \\ &+\mathring{A}^+_{j+1/2}\tilde{L}^0_{j+1/2}(U_j - U_{j-1}) - \mathring{A}^-_{j+1/2}L^0_{j+1/2}(U_{j+1} - U_j) \\ &-\mathring{A}^+_{j-1/2}L^0_{j-1/2}(U_j - U_{j-1}) + \mathring{A}^-_{j-1/2}\tilde{L}^0_{j-1/2}(U_{j+1} - U_j)\,]. \end{aligned} \tag{37}$$

We claim that (37) is the desired positive expression of U^*_j as

$$U^*_j = C_{-1}U_{j-1} + C_0U_j + C_1U_{j+1}.$$

We denote $A_{j\pm1/2}$ as $A_{\pm1/2}$. ¿From (37) we have the coefficient matrix C_1

$$C_1 = -\frac{\Delta t}{2\Delta x}\left((A_{1/2} - |\mathring{A}_{1/2}|) - \mathring{A}^-_{1/2}L^0_{1/2} + \mathring{A}^-_{-1/2}\tilde{L}^0_{-1/2}\right). \tag{38}$$

Using (29) we can rewrite C_1 as

$$C_1 = -\frac{\Delta t}{2\Delta x}\left(\mathring{A}^-_{1/2}(2I - L^0_{1/2}) + \mathring{A}^-_{-1/2}\tilde{L}^0_{-1/2}\right) \tag{39}$$

Since ϕ^0 was chosen so that $0 \le \phi^0(\theta), \frac{\phi^0(\theta)}{\theta} \le 2$, it follows that $0 \le \tilde{L}^0,\ L^0 \le 2I$. By construction $\mathring{A}^- \le 0$. Since the sum of positive matrices is positive, and the product of commuting positive matrices is positive, it follows that $C_1 \ge 0$.

Similarly the coefficient matrix $C_{-1} \ge 0$.

¿From (37), the coefficient matrix C_0 is,

$$\begin{aligned} C_0 = I - \frac{\Delta t}{2\Delta x}\Big(&-(A_{1/2} - |\mathring{A}_{1/2}|) + (A_{-1/2} + |\mathring{A}_{-1/2}|) \\ &+\mathring{A}^+_{1/2}\tilde{L}^0_{1/2} + \mathring{A}^-_{1/2}L^0_{1/2} - \mathring{A}^+_{-1/2}L^0_{-1/2} - \mathring{A}^-_{-1/2}\tilde{L}^0_{-1/2}\Big). \end{aligned} \tag{40}$$

Using (29) we can rewrite C_0 as

$$\begin{aligned} C_0 &= I - \frac{\Delta t}{2\Delta x}\left(-2\mathring{A}^-_{1/2} + \mathring{A}^+_{1/2}\tilde{L}^0_{1/2} + 2\mathring{A}^+_{-1/2} - \mathring{A}^-_{-1/2}\tilde{L}^0_{-1/2} + \mathring{A}^-_{1/2}L^0_{1/2} - \mathring{A}^+_{-1/2}L^0_{-1/2}\right) \\ &\ge I - \frac{\Delta t}{2\Delta x}[-2\mathring{A}^-_{1/2} + \mathring{A}^+_{1/2}\tilde{L}^0_{1/2}] - \frac{\Delta t}{2\Delta x}[2\mathring{A}^+_{-1/2} - \mathring{A}^-_{-1/2}\tilde{L}^0_{-1/2}]. \end{aligned} \tag{41}$$

The matrices in the brackets are positive. Under the CFL condition

$$\frac{\Delta t}{\Delta x} \max_{1\leq i\leq n,U} \mid \lambda^i \mid \leq \frac{1}{2} \tag{42}$$

each is $\leq \frac{1}{2}I$; therefore the coefficient matrix $C_0 \geq 0$.

We can read off from (37) that the sum of the three coefficients is the identity matrix I. Therefore the least dissipative scheme (21, 26) is positive.

This scheme is very close to the ones described in [23] from point of view of TVD and in [9] from point of view of local extremum diminishing. The scheme gives sharp shocks but admits entropy violating solutions, because it has zero dissipation in the field of zero eigenvalue. To overcome this difficulty, Harten [4] suggested an entropy fix increasing the magnitude of the eigenvalue up to certain artificial amount.

In the following we shall construct schemes that are more dissipative than the least dissipative positive scheme including those with an entropy fix. Entropy fix destroys the positivity principle for most limiter functions; by choosing the minmod limiter, we obtain dissipative schemes that are positive. However these schemes are smeary because of the minmod limiter and the added dissipation. We then combine the least dissipative positive scheme and the more dissipative positive schemes in an appropriate way to have the best features of both. These combined schemes again are positive.

The more dissipative schemes are constructed as follows. We define a class of "absolute values" of A as $\mid \dot{A} \mid = R diag(\mu^i) R^{-1}$, where the diagonal matrix $diag(\mu^i) \geq \mid \Lambda \mid$; for the least dissipative scheme equality holds. We construct the flux $F_{j+1/2}$ in (21) as

$$F^1_{j+1/2} = \frac{F(U_{j+1}) + F(U_j)}{2} - \frac{1}{2} \mid \dot{A} \mid (U_{j+1} - U_j) + \frac{1}{2} L^1 \mid \dot{A} \mid (U_{j+1} - U_j), \tag{43}$$

where the limiter $L^1 = R\, diag\,(\phi^1(\theta^i))\, R^{-1}$. Note that L^1 commutes with A. Here $\phi^1(\theta)$ is the minmod limiter function

$$\phi^1(\theta) = \begin{cases} 0 & \theta \leq 0, \\ \theta & 0 < \theta \leq 1, \\ 1 & \theta > 1. \end{cases}$$

This has the properties $0 \leq \phi^1(\theta), \frac{\phi^1(\theta)}{\theta} \leq 1, \phi^1(1) = 1$. Every θ^i is defined as in (27). We shall prove below that also the class of more dissipative schemes is positive under an appropriate CFL condition.

We define

$$\dot{A}^+ = R diag\left(sgn_+(\lambda^i)\mu^i\right) R^{-1} \qquad \dot{A}^- = R diag\left(sgn_-(\lambda^i)\mu^i\right) R^{-1}, \tag{44}$$

where

$$sgn_+(a) = \begin{cases} 1 & a \geq 0 \\ 0 & a < 0 \end{cases}, \quad sgn_-(a) = \begin{cases} 0 & a \geq 0 \\ -1 & a < 0 \end{cases}$$

Note that $\dot{A}^+$ and $\dot{A}^-$ commute with A. Clearly

$$\begin{aligned} |\dot{A}| &= \dot{A}^+ - \dot{A}^-, \\ \dot{A}^+ \geq \mathring{A}^+, &\qquad \dot{A}^- \leq \mathring{A}^-. \end{aligned} \tag{45}$$

However $A \neq \dot{A}^+ + \dot{A}^-$ in general; this is why ϕ^1 is restricted to the minmod. Following the same analysis as before, the coefficient matrix C_1 of U_{j+1} is

$$C_1 = -\frac{\Delta t}{2\Delta x}\left((A_{1/2} - |\dot{A}_{1/2}|) - \dot{A}^-_{1/2}L^1_{1/2} + \dot{A}^-_{-1/2}\tilde{L}^1_{-1/2}\right). \tag{46}$$

Using (29,45), we can rewrite C_1 as

$$C_1 = -\frac{\Delta t}{2\Delta x}\left((\mathring{A}^+_{1/2} - \dot{A}^+_{1/2}) + \mathring{A}^-_{1/2} + \dot{A}^-_{1/2}(I - L^1_{1/2}) + \dot{A}^-_{-1/2}\tilde{L}^1_{-1/2}\right). \tag{47}$$

We claim that all four terms in the parentheses on the right are negative. Since ϕ^1 is the minmod limiter function, $0 \leq \phi^1(\theta)$, $\frac{\phi^1(\theta)}{\theta} \leq 1$; it follows that $0 \leq \tilde{L}^1$, $L^1 \leq I$. By construction $\mathring{A}^- \leq 0, \dot{A}^- \leq 0$, and $\mathring{A}^+_{1/2} - \dot{A}^+_{1/2} \leq 0$. Since all matrices with the same subscript commute, this proves that $C_1 \geq 0$.

Similarly the coefficient matrix $C_{-1} \geq 0$.

The coefficient matrix C_0 is

$$\begin{aligned} C_0 = I - \tfrac{\Delta t}{2\Delta x}\big(&-(A_{1/2} - |\dot{A}_{1/2}|) + (A_{-1/2} + |\dot{A}_{-1/2}|) \\ &+\dot{A}^+_{1/2}\tilde{L}^1_{1/2} + \dot{A}^-_{1/2}L^1_{1/2} - \dot{A}^+_{-1/2}L^1_{-1/2} - \dot{A}^-_{-1/2}\tilde{L}^1_{-1/2}\big) \end{aligned} \tag{48}$$

Using (29,45), we can rewrite C_0 as

$$\begin{aligned} C_0 = I - \tfrac{\Delta t}{2\Delta x}\big(&\dot{A}^+_{1/2}(I + \tilde{L}^1_{1/2}) - \mathring{A}^-_{1/2} - \dot{A}^-_{1/2} + \mathring{A}^+_{-1/2} + \dot{A}^+_{-1/2} - \dot{A}^-_{-1/2}(I + \tilde{L}^1_{-1/2}) \\ &-\mathring{A}^+_{1/2} + \dot{A}^-_{1/2}L^1_{1/2} + \mathring{A}^-_{-1/2} - \dot{A}^+_{-1/2}L^1_{-1/2}\big) \\ \geq I - \tfrac{\Delta t}{2\Delta x}&[\dot{A}^+_{1/2}(I + \tilde{L}^1_{1/2}) - \mathring{A}^-_{1/2} - \dot{A}^-_{1/2}] - \tfrac{\Delta t}{2\Delta x}[\mathring{A}^+_{-1/2} + \dot{A}^+_{-1/2} - \dot{A}^-_{-1/2}(I + \tilde{L}^1_{-1/2})]. \end{aligned} \tag{49}$$

As before the matrices in bracket are positive. Under the following CFL condition

$$\frac{\Delta t}{\Delta x}\max_{1\leq i\leq n, U}\mu^i \leq \frac{1}{2}, \tag{50}$$

each is $\leq \frac{1}{2}I$, therefore the coefficient matrix $C_0 \geq 0$. As before the sum of three coefficient matrices is the identity. Therefore the more dissipative schemes (21,43) are positive.

We can construct now a family of schemes by combining the least dissipative positive scheme and a more dissipative one:

$$F^{\alpha,\beta}_{j+1/2} = \frac{F(U_j) + F(U_{j+1})}{2} - \frac{1}{2}\left(\alpha\,|\mathring{A}_{1/2}|\,(I - L^0_{1/2}) + \beta\,|\dot{A}_{1/2}|\,(I - L^1_{1/2})\right)(U_{j+1} - U_j) \tag{51}$$

where α and β are constant satisfying $0 \leq \alpha \leq 1$ and $\alpha + \beta \geq 1$. The combined scheme (21,51) is a positive scheme under the following CFL condition

$$\frac{\Delta t}{\Delta x}\left(\alpha \max_{1\leq i\leq n,U} \mid \lambda^i \mid +\beta \max_{1\leq i\leq n,U} \mu^i\right) \leq \frac{1}{2}. \tag{52}$$

We show that as follows; from (21,51) we obtain

$$\begin{aligned}
U_j^* = U_j - \tfrac{\Delta t}{2\Delta x}[&A_{1/2}(U_{j+1}-U_j) + A_{-1/2}(U_j - U_{j-1}) \\
&-\alpha\left(\mid \mathring{A}_{1/2} \mid (U_{j+1}-U_j) - \mathring{A}^+_{1/2}\tilde{L}^0_{1/2}(U_j - U_{j-1}) + \mathring{A}^-_{1/2}L^0_{1/2}(U_{j+1}-U_j)\right) \\
&-\beta\left(\mid \dot{A}_{1/2} \mid (U_{j+1}-U_j) - \dot{A}^+_{1/2}\tilde{L}^1_{1/2}(U_j - U_{j-1}) + \dot{A}^-_{1/2}L^1_{1/2}(U_{j+1}-U_j)\right) \\
&+\alpha\left(\mid \mathring{A}_{-1/2} \mid (U_j - U_{j-1}) - \mathring{A}^+_{-1/2}L^0_{-1/2}(U_j - U_{j-1}) + \mathring{A}^-_{-1/2}\tilde{L}^0_{-1/2}(U_{j+1}-U_j)\right) \\
&+\beta\left(\mid \dot{A}_{-1/2} \mid (U_j - U_{j-1}) - \dot{A}^+_{-1/2}L^1_{-1/2}(U_j - U_{j-1}) + \dot{A}^-_{-1/2}\tilde{L}^1_{-1/2}(U_{j+1}-U_j)\right)].
\end{aligned} \tag{53}$$

Hence we obtain with the help of $\mathring{A}^+ \leq \dot{A}^+$ and $\alpha + \beta \geq 1$

$$\begin{aligned}
C_1 &= -\tfrac{\Delta t}{2\Delta x}A_{1/2} - \tfrac{\Delta t}{2\Delta x}\alpha\left(- \mid \mathring{A}_{1/2} \mid - \mathring{A}^-_{1/2}L^0_{1/2} + \mathring{A}^-_{-1/2}\tilde{L}^0_{-1/2}\right) \\
&\quad -\tfrac{\Delta t}{2\Delta x}\beta\left(- \mid \dot{A}_{1/2} \mid - \dot{A}^-_{1/2}L^1_{1/2} + \dot{A}^-_{-1/2}\tilde{L}^1_{-1/2}\right) \\
&= -\tfrac{\Delta t}{2\Delta x}\alpha\left(\mathring{A}^-_{1/2}(2I - L^0_{1/2}) + \mathring{A}^-_{-1/2}\tilde{L}^0_{-1/2}\right) \\
&\quad -\tfrac{\Delta t}{2\Delta x}\left((1-\alpha)\mathring{A}^-_{1/2} + \left((1-\alpha)\mathring{A}^+_{1/2} - \beta\dot{A}^+_{1/2}\right) + \beta\dot{A}^-_{1/2}(I - L^1_{1/2}) + \beta\dot{A}^-_{-1/2}\tilde{L}^1_{-1/2}\right) \\
&\geq 0.
\end{aligned}$$

Similarly the coefficient matrix $C_{-1} \geq 0$.

$$\begin{aligned}
C_0 &= I - \tfrac{\Delta t}{2\Delta x}\left(-A_{1/2} + A_{-1/2}\right) \\
&\quad -\tfrac{\Delta t}{2\Delta x}\alpha\left(\mid \mathring{A}_{1/2} \mid + \mid \mathring{A}_{-1/2} \mid + \mathring{A}^+_{1/2}\tilde{L}^0_{1/2} - \mathring{A}^-_{-1/2}\tilde{L}^0_{-1/2} + \mathring{A}^-_{1/2}L^0_{1/2} - \mathring{A}^+_{-1/2}L^0_{-1/2}\right) \\
&\quad -\tfrac{\Delta t}{2\Delta x}\beta\left(\mid \dot{A}_{1/2} \mid + \mid \dot{A}_{-1/2} \mid + \dot{A}^+_{1/2}\tilde{L}^1_{1/2} - \dot{A}^-_{-1/2}\tilde{L}^1_{-1/2} + \dot{A}^-_{1/2}L^1_{1/2} - \dot{A}^+_{-1/2}L^1_{-1/2}\right) \\
&\geq I - \tfrac{\Delta t}{2\Delta x}\alpha\left(-A_{1/2} + A_{-1/2} + \mid \mathring{A}_{1/2} \mid + \mid \mathring{A}_{-1/2} \mid + \mathring{A}^+_{1/2}\tilde{L}^0_{1/2} - \mathring{A}^-_{-1/2}\tilde{L}^0_{-1/2}\right) \\
&\quad -\tfrac{\Delta t}{2\Delta x}\left(-(1-\alpha)A_{1/2} + (1-\alpha)A_{-1/2} + \beta \mid \dot{A}_{1/2} \mid + \beta \mid \dot{A}_{-1/2} \mid + \beta\dot{A}^+_{1/2}\tilde{L}^1_{1/2} - \beta\dot{A}^-_{-1/2}\tilde{L}^1_{-1/2}\right) \\
&\geq I - \tfrac{\Delta t}{2\Delta x}\alpha[-2\mathring{A}^-_{1/2} + \mathring{A}^+_{1/2}\tilde{L}^0_{1/2}] - \tfrac{\Delta t}{2\Delta x}\alpha[2\mathring{A}^+_{-1/2} - \mathring{A}^-_{-1/2}\tilde{L}^0_{-1/2}] \\
&\quad -\tfrac{\Delta t}{2\Delta x}[-(1-\alpha)\mathring{A}^-_{1/2} - \beta\dot{A}^-_{1/2} + \beta\dot{A}^+_{1/2} + \beta\dot{A}^+_{1/2}\tilde{L}^1_{1/2}] \\
&\quad -\tfrac{\Delta t}{2\Delta x}[(1-\alpha)\mathring{A}^+_{-1/2} + \beta\dot{A}^+_{-1/2} - \beta\dot{A}^-_{-1/2} - \beta\dot{A}^-_{-1/2}\tilde{L}^1_{-1/2}].
\end{aligned}$$

The matrices in the brackets are positive, therefore $C_0 \geq 0$ under the CFL condition (52). As before we can read off from (53) that $C_1 + C_0 + C_{-1} = I$. The proof is completed.

To achieve second order accuracy in time we use the following second order accurate energy preserving Runge-Kutta method, due to Shu:

$$U_j^* = U_j^m - \tfrac{\Delta t}{\Delta x}[F_{j+1/2}^{\alpha,\beta} - F_{j-1/2}^{\alpha,\beta}],$$
$$U_j^{**} = U_j^* - \tfrac{\Delta t}{\Delta x}[F_{j+1/2}^{\alpha,\beta,*} - F_{j-1/2}^{\alpha,\beta,*}],$$
$$U_j^{m+1} = \tfrac{1}{2}U_j^m + \tfrac{1}{2}U_j^{**}.$$

Here F^* abbreviates the positive numerical flux evaluated at U^*. If we assume that the positive schemes (51) does not increase the l^2 norm, it follows that neither does Shu's Runge-Kutta method:

$$\|U^{**}\| \leq \|U^*\| \leq \|U^m\|,$$

and so

$$\|U^{m+1}\| \leq \frac{1}{2}\|U^m\| + \frac{1}{2}\|U^{**}\| \leq \|U^m\|.$$

Remark 1: An essential difference of our schemes from others is that we combine schemes using two different limiters. This makes a difference for systems. One of our scheme is the least dissipative positive scheme; the other contains more dissipation. We remark that we can construct positive schemes with as much dissipation as needed.

Remark 2: The 2nd order accurate numerical flux $F_{j+1/2}^{\alpha,\beta}$ (51) can be expressed in a simple compact form as

$$F_{j+1/2}^{\alpha,\beta} = \frac{F(U_j) + F(U_{j+1})}{2} - \frac{1}{2}R\left(\alpha \mid \Lambda \mid (I - \Phi^0) + \beta diag(\mu^i)(I - \Phi^1)\right) R^{-1}(U_{j+1} - U_j). \quad (54)$$

The leading cost of computing one flux in the above compact form is $3n^2$ scalar multiplications, where n is the order of the system.

Remark 3: For one-dimensional systems, we can mix Lax-Wendroff with upwind and obtain the second family of positive schemes as follows

$$F_{j+1/2}^{\alpha,\beta} = \frac{F(U_j) + F(U_{j+1})}{2} - \frac{1}{2}\left(\alpha \mid A \mid (I - L^0) + \beta \mid A \mid (I - L^1) + \frac{\Delta t}{\Delta x}A^2\right)(U_{j+1} - U_j). \quad (55)$$

The numerical flux (55) is more dissipative than (54) by the amount of $\frac{\Delta t}{\Delta x}A^2(U_{j+1} - U_j)$. The family of schemes (21,55) are positive under the following CFL condition

$$\frac{\Delta t}{\Delta x}\left(\alpha \max_{1\leq i\leq n,U} \mid \lambda^i \mid + \beta \max_{1\leq i\leq n,U} \mu^i + \frac{1}{2}\frac{\Delta t}{\Delta x} \max_{1\leq i\leq n,U} \mid \lambda^i \mid^2\right) \leq \frac{1}{2}. \quad (56)$$

It follows that the condition below is sufficient for positivity:

$$\frac{\Delta t}{\Delta x}\left(\alpha \max_{1\leq i\leq n,U} \mid \lambda^i \mid + \beta \max_{1\leq i\leq n,U} \mu^i\right) \leq \sqrt{2} - 1. \quad (57)$$

Note that the family of positive schemes (21,55) are second order accurate both in space and time. No Runge-Kutta is needed, which saves $1 - \frac{1}{4(\sqrt{2}-1)} \approx 40\%$ *of computations.*

Remark 4: For one-dimensional systems, we can also mix Lax-Wendroff with upwind in another way and obtain a positive scheme as follows

$$F^{\alpha,\beta}_{j+1/2} = \frac{F(U_j)+F(U_{j+1})}{2}$$
$$-\tfrac{1}{2}\left(\alpha \mid \mathring{A} \mid (I-L^0) + \beta \mid \dot{A} \mid (I-L^1) + \tfrac{\Delta t}{\Delta x}A^2(\alpha L^0 + (1-\alpha)L^1)\right)(U_{j+1}-U_j). \tag{58}$$

This scheme is positive under the following CFL condition

$$\frac{\Delta t}{\Delta x}\left(\alpha \max_{1\le i\le n,U} \mid \lambda^i \mid + \beta \max_{1\le i\le n,U} \mu^i + \frac{\alpha+1}{2}\frac{\Delta t}{\Delta x} \max_{1\le i\le n,U} \mid \lambda^i \mid^2\right) \le \frac{1}{2}. \tag{59}$$

The proof is similar and is omitted here.

2.3 Positive Schemes for Multi-dimensional Systems of Conservation Laws

We consider multi-dimensional hyperbolic systems of conservation laws (17). In each dimension x_s, $s = 1, \cdots, d$, we construct a family of fluxes exactly as in one-dimension (51), for $0 \le \alpha_s \le 1$ and $\alpha_s + \beta_s \ge 1$,

$$F^{\alpha_s,\beta_s}_{J+1/2e_s} = \frac{F_s(U_{J+e_s}) + F_s(U_J)}{2} - \frac{1}{2}\alpha_s \mid \mathring{A} \mid (I-L^0)(U_{J+e_s}-U_J) - \frac{1}{2}\beta_s \mid \dot{A} \mid (I-L^1)(U_{J+e_s}-U_J). \tag{60}$$

Here α_s, β_s could be different for each dimension.

The family of schemes (18, 60) are positive under the following CFL condition

$$\sum_{s=1}^{d}\frac{\Delta t}{\Delta x_s}\left(\alpha_s \max_{1\le i\le n,U} \mid \lambda^{s,i} \mid + \beta_s \max_{1\le i\le n,U} \mu^{s,i}\right) \le \frac{1}{2}. \tag{61}$$

We use the same 2nd order accurate energy preserving Runge-Kutta of Shu to achieve 2nd order accuracy in time: for $m = 0, 1, \cdots$,

$$U^*_J = U^m_J - \sum_{s=1}^{d} \tfrac{\Delta t}{\Delta x_s}[F^{\alpha_s,\beta_s}_{J+1/2e_s} - F^{\alpha_s,\beta_s}_{J-1/2e_s}],$$
$$U^{**}_J = U^*_J - \sum_{s=1}^{d} \tfrac{\Delta t}{\Delta x_s}[F^{\alpha_s,\beta_s,*}_{J+1/2e_s} - F^{\alpha_s,\beta_s,*}_{J-1/2e_s}],$$
$$U^{m+1}_J = \tfrac{1}{2}U^m_J + \tfrac{1}{2}U^{**}_J.$$

2.4 Symmetrizable Systems

A system of conservation laws (17) is symmetrizable if the Jacobian of the flux vectors

$$A_s = \nabla F_s,$$

can be symmetrized by the same similarity transformation:

$$SA_sS^{-1}.$$

The matrix S depends on U. It is well known that the equations of compressible flow, and of MHD, are symmetrizable. As explained as in the introduction, our schemes (51) (55) and their multi-dimensional analogue (60) can also be made symmetric and positive by the same similarity transformation, modulo terms that can be expressed as spatial differences of U. Therefore the positivity philosophy applies to symmetrizable systems as well.

3. NUMERICAL EXPERIMENTS

In the applications described in this section no special problem-dependent techniques were used and no post-processing were used.

In all of our numerical experiments we choose $|\dot{A}| = \max_{1\leq i\leq n, U} |\lambda^i| I$ for the more dissipative scheme. The least dissipative positive scheme $\alpha = 1, \beta = 0$ gives us sharp shocks but admits entropy violating solutions, which are not physically relevant. The more dissipative scheme $\alpha = 0, \beta = 1$ gives us smeary shocks but seems to converge to physically relevant solutions. The compromise $\alpha = 0.9, \beta = 0.1$ seems to combine the best features of both; these are the schemes used in experiments described here unless otherwise stated.

3.1 One-dimensional Euler Equations of Gas Dynamics

In this subsection we apply our schemes (21,51) and (21,55) to the Euler equation of gas dynamics for a polytropic gas.

$$
\begin{aligned}
U_t + F(U)_x &= 0 \\
U &= (\rho, m, E)^T \\
F(U) &= (m, \rho q^2 + P, q(E+P))^T \\
P &= (\gamma - 1)(E - \tfrac{1}{2}\rho q^2) \\
m &= \rho q
\end{aligned}
$$

In the following computation $\gamma = 1.4$. For details of the Jacobian $F'(U)$, its eigenvalues, eigenvectors, etc., see [7].

We used the scheme (21,51) in example 1 and the scheme (21,55) in examples 2 and 3.

Example 1. We consider the following Riemann problems:

$$
U_0(x) = \begin{cases} U_l & x < 0 \\ U_r & x > 0. \end{cases}
$$

Two sets of initial date are used. One is proposed by Sod in [19]:

$$
(\rho_l, q_l, P_l) = (1, 0, 1); \quad (\rho_r, q_r, P_r) = (0.125, 0, 0.1).
$$

The other is used by Lax [12]:

$$
(\rho_l, q_l, P_l) = (0.445, 0.698, 3.528); \quad (\rho_r, q_r, P_r) = (0.5, 0, 0.571).
$$

The numerical results, showed in Figure 1a, are obtained by using the positive scheme (21,51) based on centered differencing with 100 points as in [7] and [18]. We used the superbee limiter function

$$\phi^0(\theta) = \begin{cases} 0, & \theta \le 0 \\ 2\theta & 0 < \theta \le 1/2 \\ 1 & 1/2 < \theta \le 1 \\ \theta & 1 < \theta \le 2 \\ 2 & \theta > 2 \end{cases}$$

for ϕ^0. We observe that the positive scheme does much better than the 4-th order ENO scheme at corners of rarefactions (discontinuities in derivatives) and better at contact discontinuity; but is smearer at shocks. Hence our numerical results are comparable to the 4-th order ENO scheme [7].

In the numerical results shown in Figure 1b we used Van Leer's limiter function

$$\phi^0(\theta) = \frac{\theta + |\theta|}{1 + |\theta|}.$$

The positive scheme does better than the flux based 3-rd order ENO scheme [18] at corners of rarefactions (discontinuities in derivatives) and comparable at contact discontinuity and shocks in Figure 1b. Hence our numerical results are comparable the flux based 3-rd order ENO scheme without sharpening [18]. With sharpening, ENO resolves the contact discontinuity much better. At extrema in smooth regions of the solutions, ENO schemes have high order accuracy; our positive schemes have only first order accuracy.

Example 2. Interacting Blast Wave [22] with initial data,

$$U_0(x) = \begin{cases} U_l & 0 \le x < 0.1, \\ U_m & 0.1 \le x < 0.9, \\ U_r & 0.9 \le x < 1. \end{cases}$$

$$(\rho_l, m_l, E_l) = (1, 0, 1000); \quad (\rho_m, m_m, E_m) = (1, 0, 0.01); \quad (\rho_r, m_r, E_r) = (1, 0, 100).$$

Reflecting boundary conditions are applied at both ends. The numerical results were obtained by scheme (21,55) based on Lax-Wendroff. In the calculations displayed in Figure 2a, we used Van Leer's limiter for ϕ^0. The circled lines are numerical solution with 200 points, and the solid lines are numerical solution with 1200 points, same as the resolutions in [22]. The positive scheme captures important structures on the fine grid, and the results on both grids are comparable to MUSCL scheme in [22]. The cost of the positive scheme is about 60% cost of MUSCL scheme. In Figure 2b, we show the numerical result of the positive scheme with the superbee limiter for ϕ^0; the solid lines are the same as in Figure 2a. The resolution is improved dramatically; this shows that the numerical resolution of this problem depends very much on the limiter.

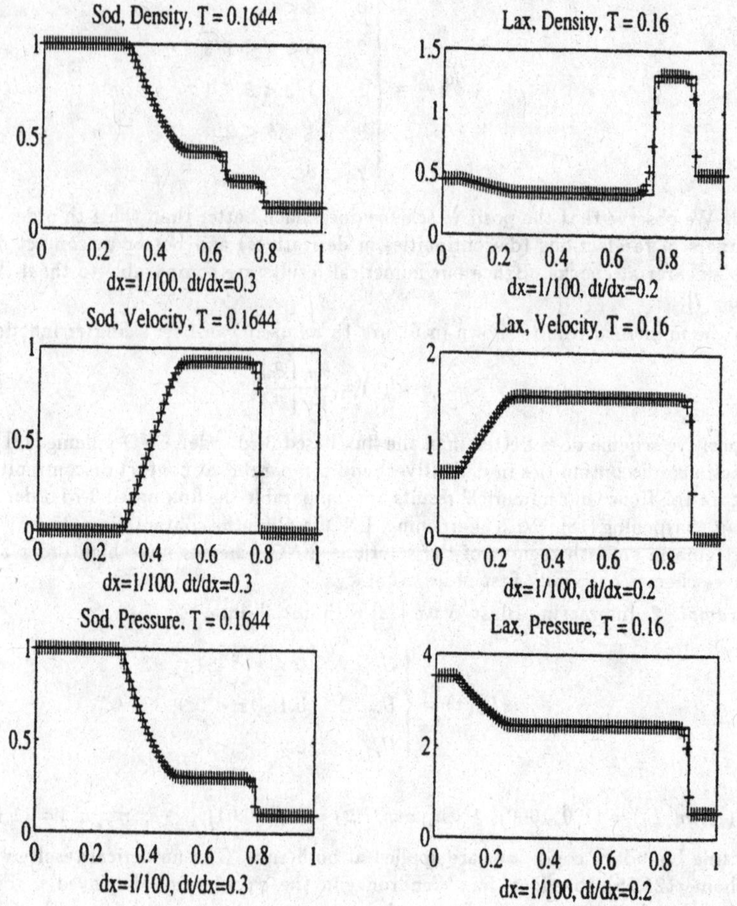

Fig.1a:

Example 3. Low density and internal energy Riemann problem [2] with initial data,

$$U_0(x) = \begin{cases} U_l & 0 \leq x < 0.5, \\ U_r & 0.5 \leq x \leq 1. \end{cases}$$

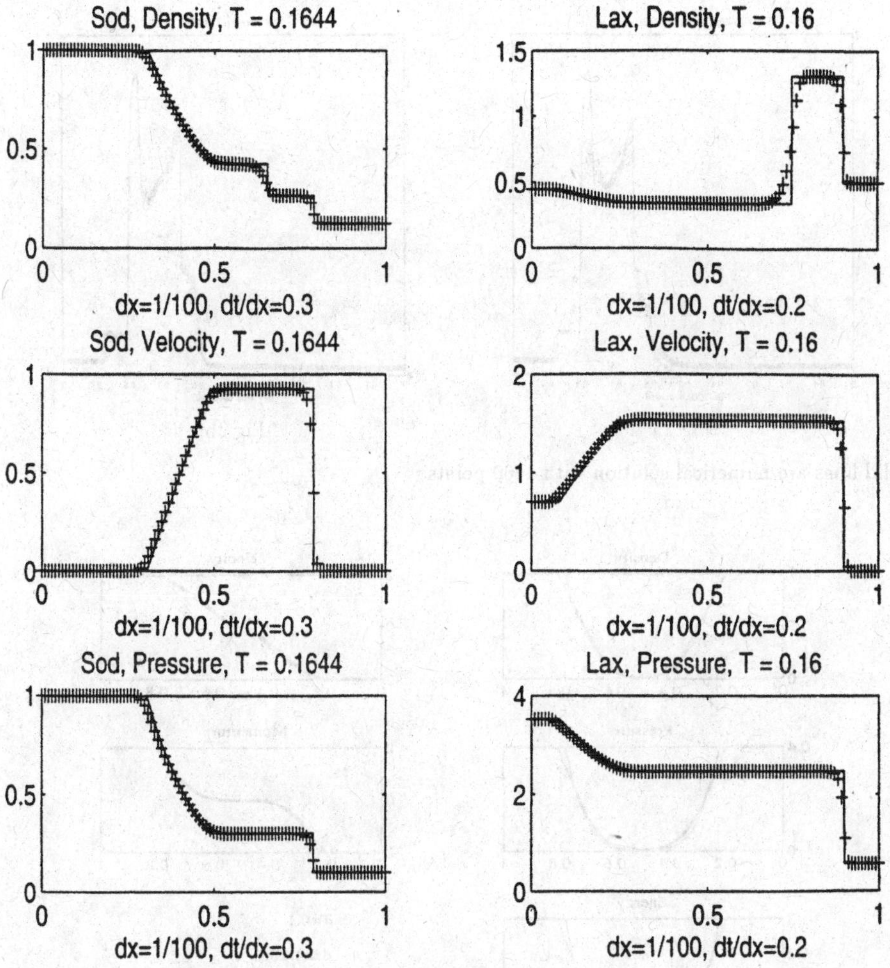

Fig. 1b

$$(\rho_l, m_l, E_l) = (1, -2, 3); \quad (\rho_r, m_r, E_r) = (1, 2, 3).$$

It has been observed that schemes which are based on a linearized Riemann solver (e.g., Roe's scheme) may lead to an unphysical negative density or internal energy during the computational process [2]. The positive scheme (21, 55) with $\alpha = 0.9, \beta = 0.1$ does suffer from negative internal energy. However after we add more dissipation by choosing $\alpha = 1, \beta = 4$, this positive scheme does succeed. In Figure 3, the dotted lines are numerical solution with 200 points, and the

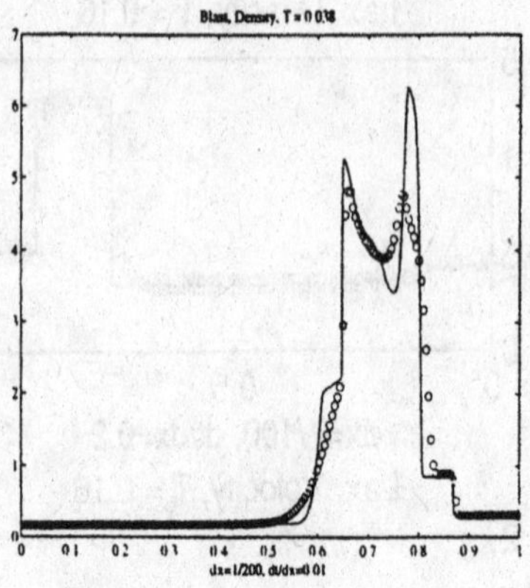

Fig.2a:

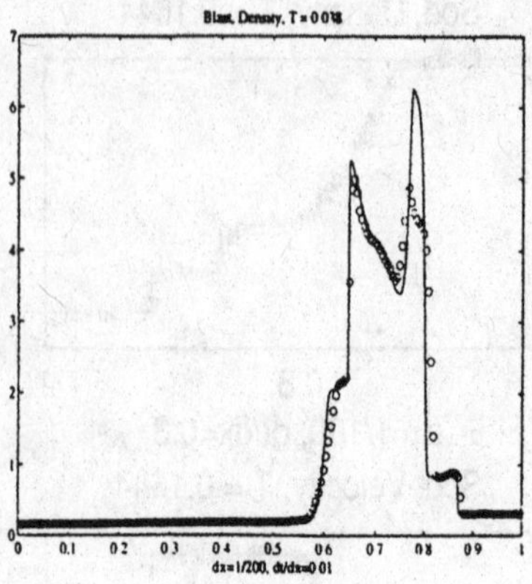

Fig.2b:

solid lines are numerical solution with 1600 points.

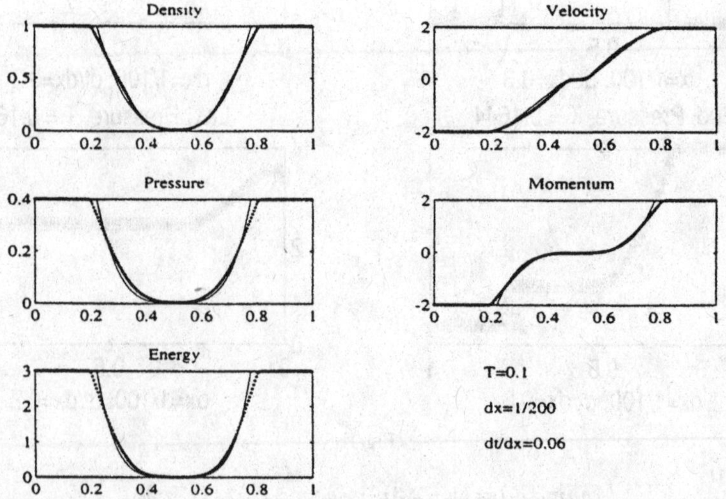

Fig.3:

3.2 Multi-dimensional Euler equations of Gas Dynamics

We approximate solutions of the two-dimensional Euler equations of Gas Dynamics,

$$U_t + F_1(U)_x + F_2(U)_y = 0$$
$$U = (\rho, m, n, E)^T$$
$$F_1(U) = (m, \rho u^2 + P, \rho uv, u(E+P))^T$$
$$F_2(U) = (n, \rho uv, \rho v^2 + P, v(E+P))^T$$
$$P = (\gamma - 1)(E - \tfrac{1}{2}\rho(u^2+v^2))$$
$$m = \rho u \qquad n = \rho v.$$

For details of the Jacobian $F_1'(U)$ and $F_2'(U)$, their eigenvalues, eigenvectors, etc., see [18].

Example 4: Double Mach Reflection. A planar shock is incident on an oblique wedge at a 60° angle. The test problem involves a Mach 10 shock in air, $\gamma = 1.4$. The undisturbed air ahead of the shock has a density of 1.4 and a pressure of 1. We use the boundary conditions described in [22]. Van Leer's limiter function is used for ϕ^0. The flow at time $t = 0.2$, computed by our positive scheme, is plotted in Figure 4 with $\Delta x = \Delta y = 1/120$, $\Delta t = \frac{5}{3} \times 10^{-4}$. In each plot 30 equally spaced contours are shown. There is no visible oscillation behind the strong shock. There are three difficulties in computing this flow mentioned in [22]. The first difficulty is the rather weak second Mach shock; it dies out entirely by the time it reaches the contact discontinuity from the first Mach reflection. Figure 4 shows that the second Mach shock is perfectly captured. The second difficulty is the jet, formed when the flow of the denser fluid is deflected by a pressure gradient built up in the region where the first contact discontinuity approaches the reflecting wall. Figure 4 shows that the jet is extremely well captured. The third difficulty is caused by the region bounded by the second Mach shock, the curved reflected shock, and the reflecting wall. The double Mach reflection contains both steady and unsteady structures. The curved reflected shock is moving rapidly at its right end and is not moving at all at its left end; this causes oscillations for many difference schemes, including PPM, MUSCL and Jin and Xin's ingenious relaxing scheme [11]. For the positive scheme there is no oscillation at all; thus the positive scheme overcomes extremely well all three numerical difficulties.

Example 5: A Mach 3 Wind Tunnel with a Step. This problem has been a useful test for schemes for many years. The tunnel is 3 length units long and 1 length unit wide, with a step which is 0.2 length units high and 0.6 length units away from the left end of the tunnel. The state behind the incoming shock is density 1.4, pressure 1.0, and velocity 3 from left to right. These are used as boundary condition at the left; at the right all horizontal gradients are assumed to vanish. Along the walls of the tunnel and the obstacle reflecting conditions are applied in the perpendicular direction. The corner of the step is the center of a rarefaction fan and hence is a singular point of the flow. At the corner we reflect the solution using a "wall" which is 45° both to the top and the side of the step. No problem dependent boundary conditions are used as in PPM. Van Leer's limiter function is used for ϕ^0. The density and pressure contours in the tunnel at time 4 are displayed in Figure 5 with $\Delta x = \Delta y = 1/80$, $dt = \frac{3}{16} \times 10^{-2}$. The flow at time 4 is still unsteady.

The general position and shape of the shocks are accurate. The shocks are well captured. There is no numerical noise. The contact discontinuities are resolved and are only slightly

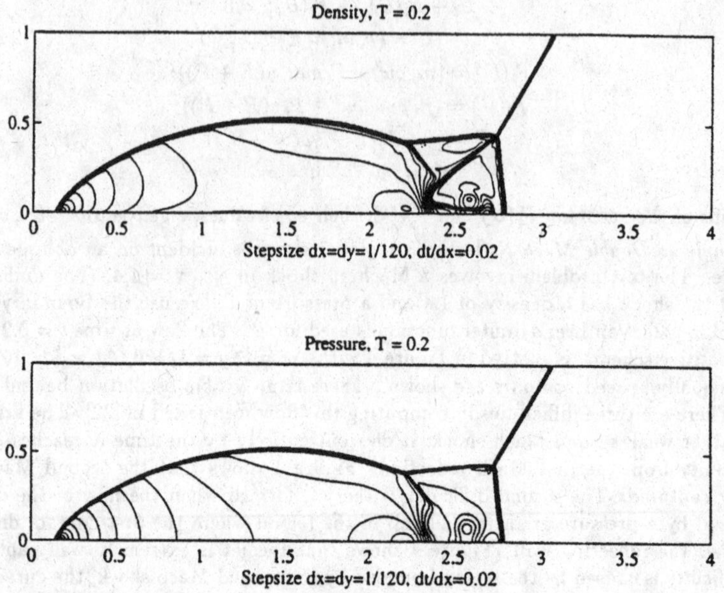

Fig.4:

broader than ones of PPM and MUSCL. The weak oblique shock is resolved.

REFERENCES

[1] J. P. Boris and D. L. Book, "Flux Corrected Transport I, SHASTA, A Fluid Transport Algorithm that Works," *J. Comput. Phys.*, 11(1973), pp. 38-69.

[2] B. Einfeldt, C.D. Munz, P.L. Roe, and B. Sjogreen, "On Godunov-Type Methods near Low Densities," *Journal of Computational Physics*, Vol. 92, 1991, pp. 273-295.

[3] K. O. Friedrichs, "Symmetric Hyperbolic Linear Differential Equations," *Commun. Pure Appl. Math.* 7, (1954), pp. 345-392.

[4] A. Harten, "On a Class of High Resolution Total-Variation-Stable Finite-Difference Schemes," *SIAM J. Numer. Anal.*, Vol. 21, No. 1, 1984, pp. 1-23.

[5] A. Harten, "High Resolution Schemes for Hyperbolic Conservation Laws," *J. Comput. Phys.*, Vol 49, 1983, pp.357-393.

[6] A. Harten, "On the Symmetric Form of Systems of Conservation Lwas with Entropy," *Journal of Computational Physics*, V49, pp. 151-164, 1983.

[7] A. Harten, B. Engquist, S. Osher and S. Chakravarthy, "Uniformly High Order Accurate Essentially Non-Oscillatory Schemes III," *Journal of Computational Physics*, V71, pp.

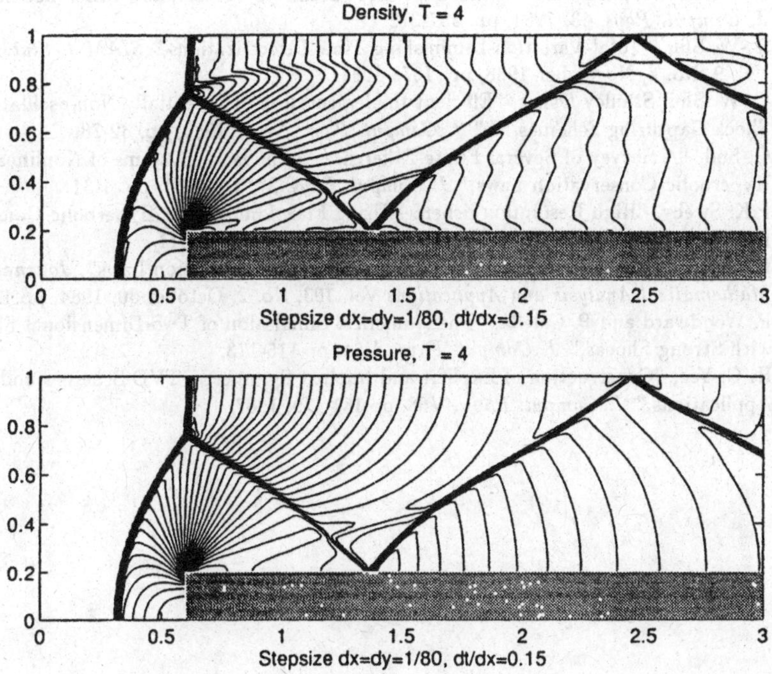

Fig.5:

231-303, 1987; also ICASE Report No. 86-22, April 1986.

[8] A. Harten and G. Zwas, "Self-Adjusting Hybrid Schemes for Shock Computations," *Journal of Computational Physics*, V9, pp.568-583, 1972.

[9] A. Jameson, "Artificial Diffusion, Upwind Biasing, Limiters and Their Effect on Accuracy and Multigrid Convergence in Transonic and Hypersonic Flows," *AIAA-93-3359.*

[10] G.-S. Jiang and C.W. Shu, "Efficient Implementation of Weighted ENO Schemes," in preparing.

[11] S. Jin and Z. Xin, "The Relaxing Schemes for Systems of Conservation Laws in Arbitrary Space Dimensions," to appear in CPAM.

[12] P. Lax. "Weak Solutions of Nonlinear Hyperbolic Equations and Their Numerical Computation," *Commun. Pure Appl. Math.* 7, (1954), pp. 159-193.

[13] R. LeVeque, "Numerical Methods for Conservation Laws," Lectures in Mathematics, Birkhäuser Verlag.

[14] X.-D. Liu, S. Osher and T. Chan, "Weighted Essentially Non-oscillatory Schemes," *J. Comput. Phys.* 115, 1994, pp. 200-212.

[15] C. Morawetz, "Potential Theory for Regular and Mach Reflection of a Shock at a Wedge," *Commun. Pure Appl. Math.* Vol XLVII, (1994), pp. 593-624.

[16] P.L. Roe, "Approximate Riemann Solvers, Parameter Vectors, and Difference Schemes," *J. Comput. Phys.* 43, 1981, pp. 357-372.

[17] C.-W. Shu, "Total-Variation-Diminishing Time Discretizations," *SIAM J. Stat. Comput.* Vol. 9, No. 6, November 1988, pp. 1073-1084.

[18] C.-W. Shu, Stanley Osher, "Efficient Implementation of Essentially Non-oscillatory Shock-Capturing Schemes, II," *J. Comput. Phys.*, V83 , 1989, pp. 32-78.

[19] G. Sod, "A Survey of Several Finite Difference Methods for Systems of Nonlinear Hyperbolic Conservation Laws," *J. Comput. Phys.* 27, 1 (1978), pp. 1-31.

[20] P.K. Sweby, "High Resolution Schemes Using Flux Limiters for Hyperbolic Conservation Laws" *SIAM J. Numer. Anal.*, Vol 21, No. 5, 1984, pp.995-1011.

[21] E. Tadmor, "Skew-Selfadjoint Form for Systems of Conservation Laws", *Journal of Mathematical Analysis and Applications* Vol. 103, No. 2, October 30, 1984, pp.428-442.

[22] P. Woodward and P. Colella, "The Numerical Simulation of Two-Dimensional Fluid Flow with Strong Shocks," *J. Comput. Phys.*, V54, pp.115-173.

[23] H. C. Yee, "Construction of Explicit and Implicit Symmetric TVD Schemes and Their Applications," *J. Comput. Phys.*, V68, pp.151-179, 1987.

Commentary on Part III

7

Peter Lax is one of the pioneers in developing numerical methods for time-dependent nonlinear PDEs in concert with their mathematical theory. This paper has major contributions to each topic.

On the computational side, Lax introduces the subsequently named *Lax–Friedrichs scheme* for practical numerical solutions of nonlinear hyperbolic equations. Peter's teacher, Friedrichs, also used the same difference scheme for theoretical purposes for linear hyperbolic equations. In this paper, Lax writes down exact solutions for the Riemann problem with piecewise constant initial data for gas dynamics and uses them to test the accuracy of his proposed numerical procedure. He also points out the poor accuracy near contact discontinuities of the numerical procedure; over twenty years later, Peter's brilliant Ph.D. student, Ami Harten, developed the "artificial compression method" in his Ph.D. thesis to cure the undesirable numerical diffusion of contact discontinuities. The Lax–Friedrich scheme is one of the first explicit conservation form numerical methods for nonlinear hyperbolic equations. From a contemporary computational viewpoint, the Lax–Friedrichs scheme has been replaced by a suite of methods with less numerical diffusion; the interested reader can consult the book by Randall Leveque *Finite Volume Methods for Hyperbolic Problems*, Cambridge University Press, 2002. Recently, Tadmor and his collaborators have found the Lax–Friedrichs scheme to be very useful as a simple nonlinear flux limiter in combination with basic higher-order central difference schemes in diverse applications.

On the theoretical side, in Section 5, Lax presents the first study of solutions of hyperbolic conservation laws in the weak topology. He shows for scalar convex conservation laws that there is a subtle inherent compactness in the space of solutions. In particular, bounded weakly convergent sequences of solutions automatically have subsequences that converge strongly. This result anticipates by over twenty years the novel use of the weak topology and compensated compactness methods developed by Tartar and Diperna in the late 1970s and 1980s for hyperbolic systems of conservation laws. See the book by C. Dafermos *Hyperbolic Conservation Laws in Continuum Physics*, Springer-Berlin, 2000, for extensive references.

A. Majda

18

This is one of the landmark papers in the theory of nonlinear hyperbolic equations. Sections 7, 8, and 9, present the first systematic mathematical discussion of solutions of general systems of nonlinear hyperbolic conservation laws in a single space dimension with general piecewise constant "Riemann" initial data. Lax presents a beautiful nonlinear generalization involving shock and rarefaction waves of the progressing wave solution of linear hyperbolic systems with discontinuous data. In the 1960s, Glimm used Lax's solution of the Riemann problem in an ingenious fashion as building blocks to construct solutions of hyperbolic conservation laws with general small initial data.

In Section 2 of this same paper, Lax presents an explicit solution formula for the solution of scalar convex hyperbolic equations, generalizing earlier work of E. Hopf and developed independently by Oleinik. In Section 4, he discusses the explicit solution of a finite difference scheme for a special nonlinearity by consciously mimicking the solution procedure from Section 2. Section 5 contains Lax's pioneering use of the explicit solution to describe the general long-time behavior of scalar convex conservation laws with arbitrary initial data in both all of space and periodic settings. Glimm and Lax generalized these results to systems of conservation laws in a single space variable in a very influential memoir of the American Mathematical Society from 1969. Dafermos, Diperna, and T.P. Liu, among others, have presented significant generalizations of this work. Subsequently, P.L. Lions, Crandall, and Souganidis have all made substantial use of the Lax–Oleinik formula in their penetrating studies of Hamilton–Jacobi equations. See the book by C. Dafermos *Hyperbolic Conservation Laws in Continuum Physics*, Springer-Berlin, 2000, for extensive references.

A. Majda

23, 34

These two papers describe and analyze the second-order-accurate Lax–Wendroff scheme for hyperbolic systems of conservation laws, in one and two space dimensions. We show in two space dimensions that if a CFL-type condition is satisfied, then the numerical range of the amplification matrix of the linearized scheme does not exceed 1 in absolute value; this is sufficient for stability.

P.D. Lax

58

Although appearing as a paper in conference proceedings, this is a remarkably influential work both for the mathematical theory and also the numerical computation of nonlinear hyperbolic conservation laws. The main idea involves the systematic mathematical generalization of the important physical fact that (minus) the entropy is an additional convex conserved quantity for smooth solutions of compressible gas dynamics. In a related paper, Lax and Friedrichs show that systems of equations with an additional convex conserved quantity are a systematic nonlinear generalization of Friedrichs' theory of linear symmetric hyperbolic systems. In the present paper, Lax formulates an "entropy condition" to guarantee uniqueness of weak solutions for such systems with an additional convex conservation law. He also asks the question whether finite difference schemes have solutions that converge to this entropy solution and checks whether such conditions are satisfied for the Lax–Friedrichs scheme under a suitable hypothesis. Finally, he uses linear geometric optics approximations in an ingenious fashion to show that pairs of conservation laws locally have many additional convex conserved quantities.

The paper blends mathematical theory and numerical analysis in a nonobvious fashion, and the results from the paper briefly mentioned above have established important research trends over several decades in both numerical computation and mathematical theory. On the numerical side, this paper pioneered the central issue of designing difference schemes that satisfy a discrete form of the physical entropy condition. This was continued by Lax and his two Ph.D. students Harten and Hyman in an important paper from 1976 and followed in significant work by Osher, Majda, Engquist, and Tadmor, among others, through the 1970s and 1980s for a wide variety of new numerical methods.

On the theoretical side, in the 1980s Diperna used Lax's "geometric optics constructions" of additional convex conservation laws for pairs of equations together with ingenious arguments utilizing compensated compactness to study the existence of solutions in a single space dimension by Tartar's approach through the weak topology. More recently, in the late 1990s Bressan used the basic ideas in this paper among many others to solve the important uniqueness problem for characterizing the vanishing viscosity limit in a single space dimension. For extensive references or subsequent computational and theoretical developments inspired by this paper the interested reader can consult the textbooks by R. Leveque, *Finite Volume Methods for Hyperbolic Problems*, Cambridge University Press (2002); D. Serre, *Systems of Conservations Laws*, Vols. I & II, Cambridge University Press (1999 and 2000); A. Bressan, *Hyperbolic Systems of Conservation Laws: The Cauchy Problem*, Oxford University Press (2000).

A. Majda

137

For linear symmetric hyperbolic equations of first order, Friedrichs has devised positive schemes, that is, schemes of the form

$$U^{m+1}(x) = \sum C_J u^n(x - J\Delta),$$

where the coefficients C_J are symmetric positive matrices. When the dependence of C_J on x is Lipschitz continuous, such a scheme is stable. On the other hand, it is not hard to show that except in trivial cases, such a positive scheme is only of first-order accuracy.

First-order systems of conservation laws that have a convex entropy are symmetric hyperbolic, and one can devise positive schemes for them. These schemes can be combined with higher-order schemes with the aid of a switch that turns off the higher-order scheme at those points where the solution is not smooth. We show that such a hybrid scheme has positive coefficients, although these coefficients are discontinuous. Numerical experiments show that these schemes are stable, and that they produce the accurate, highly resolved approximations to initial value problems that give rise to very complicated wave patterns, such as the diffraction of a strong wave by a wedge, propagation of a shock wave in a partly occluded channel, and Riemann initial value problems.

References

[1] Kurganov, A.; Tadmor, E. New high resolution central schemes for nonlinear conservation laws and convection diffusion equations, *J. of Comp. Physics* **160** (2000), 214–282.

[2] Liu, Z.D.; Lax, P.D. Solution of two-dimensional Riemann problems of gas dynamics by positive schemes, *SIAM J. of Sci. Comp.* **19** (1998), 319–340.

[3] Tadmor, E. Entropy stability theory for difference approximations of nonlinear conservation laws and related time dependent problems, to appear in Acta Numerica, 2005.

P.D. Lax

PART IV

INTEGRABLE SYSTEMS

COMMUNICATIONS ON PURE AND APPLIED MATHEMATICS, VOL. XXI, 467–490 (1968)

Integrals of Nonlinear Equations of Evolution and Solitary Waves*

PETER D. LAX

Abstract

In Section 1 we present a general principle for associating nonlinear equations of evolutions with linear operators so that the eigenvalues of the linear operator are integrals of the nonlinear equation. A striking instance of such a procedure is the discovery by Gardner, Miura and Kruskal that the eigenvalues of the Schrödinger operator are integrals of the Korteweg-de Vries equation.

In Section 2 we prove the simplest case of a conjecture of Kruskal and Zabusky concerning the existence of double wave solutions of the Korteweg-de Vries equation, i.e., of solutions which for $|t|$ large behave as the superposition of two solitary waves travelling at different speeds. The main tool used is the first of a remarkable series of integrals discovered by Kruskal and Zabusky.

§1. In this paper we study the equation

$$u_t + uu_x + u_{xxx} = 0 \tag{1.1}$$

introduced by Korteweg and de Vries in their approximate theory of water waves, [3]; we shall refer to it as the KdV equation. Subsequently the KdV equation was found to be relevant for the description of hydromagnetic waves, [2], and in the description of acoustic waves in an anharmonic crystal, [8]. Equation (1.1) is a special instance of a *nonlinear evolution equation* of the form

$$u_t = K(u) \, . \tag{1.2}$$

We shall study C^∞ solutions of (1.1) defined for all x in $(-\infty, \infty)$, which tend to zero as $x \to \pm\infty$, together with all their x derivatives. It is easy to show that such solutions are *uniquely determined by their initial values*. Let v be another solution of (1.1):

$$v_t + vv_x + v_{xxx} = 0 \, . \tag*{$(1.1)_v$}$$

Subtracting this from (1.1) and denoting $u - v$ by w, we obtain the linear equation

$$w_t + uw_x + wv_x + w_{xxx} = 0$$

* This research represents results obtained at the Courant Institute, New York University, under the sponsorship of the Atomic Energy Commission, contract AT(30-1)-1480. Reproduction in whole or in part is permitted for any purpose of the United States Government.

for w. Multiplying by w and integrating with respect to x over $(-\infty, \infty)$, we obtain, after integration by parts, the relation

$$\frac{d}{dt}\frac{1}{2}\int w^2\,dx + \int (v_x - \tfrac{1}{2}u_x)w^2\,dx = 0\,. \tag{1.3}$$

Here we have used the fact that w and w_{xx} tend to zero as $x \to \pm\infty$. Denoting $\frac{1}{2}\int w^2\,dx$ by $E(t)$ and $\max|2v_x - u_x|$ by m we obtain from (1.3) the inequality

$$\frac{d}{dt}E(t) \leqq mE(t)\,.$$

This differential inequality yields

$$E(t) \leqq E(0)e^{mt} \tag{1.4}$$

which implies, in particular, that if $E(0)$ is zero then so is $E(t)$—and thereby w—for all t. Furthermore, Sjöberg has shown in [6] that (1.1) has a solution with arbitrarily prescribed initial value $f(x)$ provided that f is smooth enough and tends to zero with sufficient rapidity as $|x|$ tends to infinity.

Equation (1.1) has travelling wave solutions, i.e., solutions of the form $u(x, t) = s(x - ct)$, c being the speed of the wave. To see this we substitute into (1.1):

$$-cs_x + ss_x + s_{xxx} = 0\,. \tag{1.5}$$

Integrating with respect to x and imposing the boundary condition that s and its derivatives vanish at $x = \pm\infty$, we get

$$-cs + \tfrac{1}{2}s^2 + s_{xx} = 0\,. \tag{1.6}$$

Multiplication by $2s_x$ and one more integration gives

$$-cs^2 + \tfrac{1}{3}s^3 + s_x^2 = 0\,. \tag{1.7}$$

From this relation and the assumption $s_x(0) = 0$, s can be determined explicitly:

$$s(x) = 3c\,\mathrm{sech}^2\,\tfrac{1}{2}x\sqrt{c}\,. \tag{1.8}$$

Thus we see that *for every positive speed c,* (1.5) *has a solution vanishing at $x = \pm\infty$ uniquely determined except for a shift which can be so chosen that the maximum of s occurs at* $x = 0$. *We denote this normalized s, explicitly described by* (1.8), *by $s(x, c)$; $s(x, c)$ is symmetric in x, decays exponentially as $|x| \to \infty$, and $s(0, c) = 3c$.*

On account of its shape, s is called a *solitary wave.*

For linear equations it often happens that all solutions are superpositions of special solutions; e.g., all solutions of linear equations with constant coefficients are superpositions of exponential solutions. For nonlinear equations one cannot in general form new solutions out of old ones and hence special families of solutions are not expected to play any special role in the description of all solutions. It was therefore very surprising when Kruskal and Zabusky observed, by analyzing numerically computed solutions, that all solutions of the KdV equation have hidden in them solitary waves. A precise formulation of this observation is as follows:

Let u be any solution of (1.1) *which is defined for all x and t and which vanishes at* $x = \pm\infty$. *Then there exist a discrete set of positive numbers* $c_1, \cdots, c_N$*—called the eigenspeeds of u—and sets of phase shifts* $\theta_j^{\pm}$ *such that*

$$\lim_{t\to\pm\infty} u(x + ct, t) = \begin{cases} s(x - \theta_j^{\pm}, c_j) & \text{if } c = c_j, \\ 0 & \text{if } c \neq c_j. \end{cases} \tag{1.9}$$

It seems likely that the method described in [1] can be used to prove this conjecture.

Note. Consider any equation of evolution (1.2) which does not involve x and t explicitly, i.e., whose set of solutions is invariant under translation with respect to x and t. Suppose that $u(x, t)$ is a solution of such an equation, and that for a certain value of c

$$\lim_{t\to+\infty} u(x + ct, t)$$

exists uniformly on compact sets in x-space. Clearly the limit is a travelling wave solution; one could then define the eigenspeeds of a solution u as those values of c for which the above limit exists and is different from zero. It is far from obvious whether—as is the case for the KdV equation—the eigenspeeds which appear in the limit as $t \to \infty$ and $t \to -\infty$ are the *same.*

The eigenspeeds c_j are unequivocally determined by the solution u under consideration and thus can be regarded as *functionals* of solutions. It is clear from the definition that if u is translated by the amounts a and b in the x and t direction, the eigenspeeds remain the same while the phase shifts change by the amount $a - cb$.

We saw earlier that solutions are uniquely determined by their initial values; therefore, the eigenspeeds also can be regarded as functionals of the initial values. It follows from translation invariance that the eigenspeeds are *invariant functionals* (also called *integrals*); that is, if f and f' denote the value of u at two different times, then $c_j(f) = c_j(f')$. It follows similarly that the difference $\theta_j^+ - \theta_j^-$ is an integral.

Since the number of eigenspeeds appears to be unbounded, this analysis shows that if solutions of the KdV equation behave, for large t, as indicated in (1.9), then the KdV equation has an infinity of integrals. Indeed, Kruskal, Gardner and Miura [5] succeeded in constructing explicitly an infinite sequence of integrals; the simplest of these will be described and used in Section 2. We present now a general method for constructing an infinite set of integrals for equations of evolution (1.2).

Let $\mathscr{B}$ be some space of functions chosen so that, for each t, $u(t)$ lies in $\mathscr{B}$. Suppose that to each function u in $\mathscr{B}$ we can associate a selfadjoint operator

$L = L_u$ over some Hilbert space,

$$u \to L_u \,, \tag{1.10}$$

with the following property: if u changes with t subject to the equation

$$u_t = K(u) \,,$$

the operators $L(t)$, which also change with t, remain unitarily equivalent. If this is the case, then the *eigenvalues of* L_u constitute a set of integrals for the equation under consideration.

The unitary equivalence of the operators $L(t)$ means that there is a one-parameter family of unitary operators $U(t)$ such that

$$U(t)^{-1}L(t)U(t) \tag{1.11}$$

is independent of t. This fact can be expressed by setting the t derivative of (1.11) equal to zero:

$$-U^{-1}U_tU^{-1}LU + U^{-1}L_tU + U^{-1}LU_t = 0 \,. \tag{1.12}$$

A one-parameter family of unitary operators satisfies a differential equation of the form

$$U_t = BU \,, \tag{1.13}$$

where $B(t)$ is an antisymmetric operator. Conversely, every solution of (1.13) with $B^* = -B$ is a one-parameter family of unitary operators. Substituting (1.13) into (1.12) we get, after multiplication by U on the left and by U^{-1} on the right,

$$-BL + L_t + LB = 0$$

which is the same as

$$L_t = BL - LB = [B, L] \,. \tag{1.14}$$

If u satisfies the equation $u_t = K(u)$, then L_t can be expressed in terms of u, and all that remains to verify is that equation (1.14) has an antisymmetric solution B.

The drawback of this method is that it requires one to guess correctly the relation (1.10) between the function u and the operator L. Now Gardner, Kruskal and Miura have made the remarkable discovery, [1], that the eigenvalues of the Schrödinger operator

$$L = D^2 + \tfrac{1}{6}u \tag{1.15}$$

are invariant if u varies according to the KdV equation. We shall presently verify this fact with the aid of the linear operator equation (1.14); more generally, we shall use equation (1.14) to find a class of differential equations under which the operators (1.15) are unitarily equivalent for all t.

With the choice (1.15) the operator L_t reduces to multiplication by $\frac{1}{6}u_t$ so that according to (1.14) we have to find an antisymmetric operator B whose commutator with L is multiplication. An obvious choice is

$$B_0 = D \,. \tag{1.16$_0$}$$

Indeed, an easy calculation gives

$$[B_0, L] = \tfrac{1}{6}u_x .$$

This shows that if u varies according to the equation

$$u_t = u_x , \tag{1.17}$$

the operators (1.15) are unitarily equivalent. Alas, this is a trivial fact since changing u according to (1.17) amounts to replacing the potential u by a translate of u, which obviously results in an equivalent operator. For a less trivial result we try a third order antisymmetric operator

$$B_1 = D^3 + bD + Db , \tag{1.16$_1$}$$

the coefficient b to be chosen. A brief calculation yields the following value for the commutator:

$$[B_1, D] = \tfrac{1}{2}u_x D^2 + \tfrac{1}{2}u_{xx} D + \tfrac{1}{6}u_{xxx} - 4b_x D^2 - 4b_{xx} D - b_{xxx} + \tfrac{1}{3}bu_x .$$

Clearly, to eliminate all but the zero order terms we have to choose

$$b = \tfrac{1}{8}u .$$

With this choice, $[B_1, L]$ is multiplication by

$$\tfrac{1}{24}[u_{xxx} + uu_x] .$$

Setting $B = 24B_1$, we verify that

$$[B, D] = K(u) ,$$

where $K(u) = u_t$ is the KdV equation (1.1)!

Clearly this process can be generalized; we could choose B_q as a skew symmetric differential operator of any odd order $2q + 1$:

$$B_q = D^{2q+1} + \sum_1^q (b_j D^{2j-1} + D^{2j-1} b_j) .$$

Since $[B_q, L]$ is symmetric, the requirement that it be of degree zero imposes q conditions; these uniquely determine the q coefficients b_j, and the zero order term of $[B_q, L] = K_q$ determines a higher order KdV equation

$$u_t = K_q(u) ,$$

which shares with the KdV equation the property that the eigenvalues of the Schrödinger equation with u as potential are its integrals.

Gardner has discovered[1] an interesting relation between the higher order KdV equations and the explicit sequence of invariants mentioned earlier; this relation will be described in Section 2.

[1] Personal communication.

The process presented above can be generalized in a straightforward way to yield the following theorem.

THEOREM 1.1. *Suppose that L is a selfadjoint operator depending on u in the following fashion:*

$$L_u = L_0 + M_u ,$$

where L_0 is independent of u and M depends linearly on u. Suppose that there exists an antisymmetric operator $B = B_u$ such that

$$[B, L_u] = M_{K(u)} .$$

Then the eigenvalues of L_u are integrals of

$$u_t = K(u) .$$

As example of this procedure we take u to be a $p \times p$ symmetric matrix variable and take L to be the matrix operator $L = D^2 + \frac{1}{6}u$. If we choose B to be a third order matrix operator we obtain the *matrix* KdV *equation*

$$u_t + \tfrac{1}{2}(uu_x + u_x u) + u_{xxx} = 0 .$$

Other choices for the operator L_u should lead to other classes of equations.

Having shown that the eigenvalues $\lambda_1(u), \cdots, \lambda_N(u)$ of the Schrödinger operator (1.15) are integrals for the KdV equation, one asks how these integrals are related to the eigenspeeds $c_1(u), \cdots, c_N(u)$ which appear in the asymptotic description (1.9). Gardner and Kruskal have found the answer:

$$c_j(u) = 4\lambda_j(u) . \tag{1.18}$$

We give here a derivation of this result which makes use of a general relation involving integrals of nonlinear equations.

This relation is an extension of the following well-known fact about *quadratic* integrals Q for *linear* equations: the *bilinear* functional $Q(u, v)$ is independent of t for any *pair* of solutions u and v. This result can be deduced from the invariance of Q for the solutions $u + v$ and $u - v$. The corresponding result for integrals of a nonlinear equation

$$u_t = K(u) \tag{1.19}$$

is derived by considering one-parameter families u_ε of solutions; these can be constructed e.g., by making the initial value of $u_\varepsilon(t)$ a function of ε:

$$u_\varepsilon(0) = u_0 + \varepsilon f .$$

We assume that the nonlinear operator K depends differentiably on u, i.e., that

$$\frac{d}{d\varepsilon} K(u + \varepsilon v) \big|_{\varepsilon=0} = V(u)v \tag{1.20}$$

exists and is a *linear* function of v. We call the linear operator $V(u)$ the *variation* of K. Differentiating the equation (1.19) with respect to ε, we obtain the *variational equation*

$$v_t = V(u)v \tag{1.21}$$

for the quantity

$$v = \left.\frac{du_\varepsilon}{d\varepsilon}\right|_{\varepsilon=0} . \tag{1.22}$$

Let $I(u)$ be an integral of the equation under consideration; we assume that $I(u)$ is differentiable in the Frechet sense, i.e., that

$$\frac{d}{d\varepsilon} I(u + \varepsilon v)\Big|_{\varepsilon=0}$$

exists and is a *linear* functional of v. This linear functional can be represented as

$$\frac{d}{d\varepsilon} I(u + \varepsilon v) = (G(u), v) , \tag{1.23}$$

where (,) is some convenient bilinear functional; $G(u)$ is called the *gradient* of I.

Let $u_\varepsilon(t)$ be the one-parameter family of solutions considered before; then

$$I(u_\varepsilon(t))$$

is independent of t, for every value of ε. Differentiating with respect to ε, we see that $(G(u), v)$ is independent of t. We formulate this result as

LEMMA 1.1. *Let $u(t)$ be any solution of the equation* (1.19), *$v(t)$ any solution of the corresponding linear variational equation* (1.21). *Let I be any integral for* (1.19), *G its gradient; then*

$$(G(u(t)), v(t)) \tag{1.24}$$

is independent of t.

We suppose now that equation (1.19) is translation invariant and that it has a solitary wave solution $s(x - ct)$. We also assume that the bilinear functional (,) is symmetric as well as translation invariant.

Let v be any solution of (1.21); according to (1.24),

$$(G(s(x - ct)), v(x, t))$$

is independent of t. Using the translation independence of (,) and introducing the abbreviation

$$v(x + ct, t) = w(x, t) , \tag{1.25}$$

we see that

$$(G(s(x)), w(x, t))$$

is equal to the above quantity and therefore also independent of t. Differentiating with respect to t, we obtain that

(1.26) $$(G(s), w_t) = 0 .$$

From (1.25),

$$w_t = cv_x + v_t ;$$

using equation (1.21) and the fact that V commutes with translation, we get

$$w_t = [cD + V(s)]w ,$$

where D denotes $\partial/\partial x$. Substituting this into (1.26) we have

(1.27) $$(G(s), [cD + V(s)]w) = 0 .$$

Denote the adjoint of V by V^*. Since (,) is translation invariant, $D^* = -D$; hence we see from (1.27) that

$$([-cD + V^*(s)]G(s), w) = 0 .$$

The value of w at any particular time, say $t = 0$, can be prescribed arbitrarily; therefore, it follows from the above relation that in fact

(1.28) $$[-cD + V^*(s)]G(s) = 0 .$$

Next we make our final assumption about the differential equation (1.19); we assume namely that it is *energy preserving*, i.e., that for a solution u of (1.19)

$$(u(t), u(t))$$

is independent of t. Differentiating with respect to t and using (1.19), we see that this amounts to requiring that, for all u,

$$0 = 2(u, u_t) = 2(u, K(u)) .$$

The initial value of u is arbitrary. Putting u_ε in place of u, we see after differentiating with respect to ε that

$$(v, K(u)) + (u, V(u)v) = 0 .$$

Using the adjoint of V this can be written as

$$(v, K(u) + V^*(u)u) = 0 .$$

Since v is arbitrary, this implies that

(1.29) $$K(u) + V^*(u)u = 0 .$$

We turn now to the equation for solitary waves:

$$0 = cs_x + K(s) .$$

Expressing K from (1.29) we can rewrite this as

$$[cD - V^*(s)]s = 0 ;$$

i.e., the solitary wave s belongs to the nullspace of the linear operator $cD - V^*(s)$.

We can state the following lemma.

LEMMA 1.2. *Suppose the equation of evolution* (1.19) *has the following properties:*

(a) $K(u)$ *depends differentiably on* u; *denote its variation by* $V(u)$.

(b) (1.19) *is translation invariant and preserves a positive translation invariant quadratic functional, which we call energy.*

(c) (1.19) *has a solitary wave solution* s.

(d) *The only function annihilated by* $cD - V^*(s)$ *and vanishing at* $\pm\infty$ *is a multiple of* s.

Let $I(u)$ *be an integral for* (1.19) *such that*

(e) $I(u)$ *is differentiable,*

and

(f) $G(s)$ *vanishes at* $\pm\infty$, *where* $G(u)$ *is the gradient of* $I(u)$ *with respect to energy.*

Then

$$G(s) = \kappa s, \tag{1.30}$$

where κ *depends on* I *and* c, *i.e., every solitary wave is an eigenfunction of the gradient of an integral.*

It is easy to verify that the KdV equation is energy preserving with respect to the L_2 scalar product. In that case, $K(u) = -uu_x - u_{xxx}$ so that

$$V(u)v = -(uv)_x - v_{xxx},$$

and thus

$$V^*(s) = uD + D^3.$$

It is not hard to show that every function annihilated by

$$(C - s)D - D^3$$

and vanishing at $\pm\infty$ is a constant multiple of s.

We apply now the foregoing to the integral $I(u) = \lambda(u)$, where λ is an eigenvalue of $L = D^2 + \frac{1}{6}u$:

$$Lw = \lambda w. \tag{1.31}$$

To compute the gradient of λ we replace u by $u + \varepsilon v$ and differentiate with respect to ε. Denoting $d/d\varepsilon$ by a dot and using the fact that $\dot{L} = \frac{1}{6}\dot{u} = \frac{1}{6}v$, we get

$$L\dot{w} + \tfrac{1}{6}vw = \dot{\lambda}w + \lambda\dot{w}.$$

We take the L_2 scalar product with w; using the fact that L is symmetric and that (1.31) is satisfied, we get rid of $\dot{w}$ and obtain

$$(vw, w) = \dot{\lambda}(w, w).$$

Assuming that w is normalized so that $(w, w) = 1$, we get

$$\dot{\lambda} = (\tfrac{1}{6}vw, w) = \int \tfrac{1}{6}vw^2\,dx = (\tfrac{1}{6}w^2, v)\,.$$

Since $\dot{\lambda} = (d/d\varepsilon)I = (G(u), v)$, this shows that the gradient G of λ is given by

(1.32) $$G(u) = \tfrac{1}{6}w^2\,.$$

Since eigenfunctions w vanish at $\pm\infty$, the hypothesis preceding (1.30) is fulfilled; therefore we conclude from (1.30) that $G(s)$ is a constant multiple of s:

$$G(s) = \tfrac{1}{6}w^2 = \kappa s\,,$$

i.e., that the eigenfunction w of

$$L = D^2 + \tfrac{1}{6}s$$

is

(1.33) $$w = \text{const. } s^{1/2}\,.$$

This relation is easily verified by an explicit calculation. Taking the constant to be 1 we get

$$w_x = \tfrac{1}{2}s_x s^{-1/2}\,,$$

$$w_{xx} = \tfrac{1}{2}s_{xx}s^{-1/2} - \tfrac{1}{4}s_x^2 s^{-3/2}\,.$$

Using relations (1.6) and (1.7) for the solitary wave we have

$$w_{xx} = \tfrac{1}{2}(cs - \tfrac{1}{2}s^2)s^{-1/2} + \tfrac{1}{4}(\tfrac{1}{3}s^3 - cs^2)\,.$$

Substituting this into the eigenvalue equation

$$Lw = w_{xx} + \tfrac{1}{6}sw\,,$$

we obtain after a brief calculation

(1.34) $$Lw = Ls^{1/2} = \tfrac{1}{4}cs^{1/2}\,.$$

This proves that

$$c(s) = 4\lambda(s)\,.$$

Let u be any solution of the KdV equation which contains a solitary wave travelling with speed c, i.e., such that, given any positive ε and X, there exists a T such that

(1.35) $$|u(x + cT, T) - s(x - \theta)| < \varepsilon$$

for all $|x| < X$. We claim that then the operator $L_T = D^2 + u(T)$ has $\frac{1}{4}c$ as an approximate eigenvalue and

$$w_T(x) = s^{1/2}(x - cT - \theta)$$

as approximate eigenfunction, in the sense that

(1.36) $$\|L_T w - \tfrac{1}{4}cw\| \leq \delta\|w\|\,,$$

where δ tends to zero as $\varepsilon \to 0$ and $X \to \infty$. To see this we use the fact that, according to (1.35), in the interval

$$cT - X < x < cT + X \tag{1.37}$$

u differs by ε from $s(x - cT - \theta)$; therefore, using (1.34) we conclude that in the interval (1.37)

$$|L_T w_T - \tfrac{1}{4}cw_T| < \varepsilon w_T\,. \tag{1.38}$$

On the other hand, it follows from formula (1.8) that, outside of the interval (1.37), w_T and its second derivative are bounded by $\exp\{-\text{const.}\,|X - x|\}$. Denoting by M the supremum[2] $u(x, t)$, it follows that

$$|L_T w_T - \tfrac{1}{4}cw_T| < (\text{const.} + M)e^{-\text{const.}|X-x|}, \tag{1.38$'$}$$

outside the interval (1.37). Combining (1.38) and (1.38)′ we deduce (1.36).

According to spectral theory, inequality (1.36) implies that $\frac{1}{4}c$ lies within δ of a point of the spectrum of L_T; since we have seen earlier that the spectrum of L_T is independent of T, it follows that $\frac{1}{4}c$ is an eigenvalue of L.

This proves one half of (1.18); the second half—that if λ is an eigenvalue of L_u, then 4λ is an eigenspeed of u—has not yet been demonstrated.

§2. In [9] Kruskal and Zabusky have studied the interaction of solitary waves; in particular they posed the following problem:

Let d be a solution of KdV which for t large negative represents two solitary waves travelling with speed c_1, respectively c_2, approaching each other; what is the asymptotic behaviour of $d(x, t)$ for large positive values of t?

They solved this problem by computing d; for $-T$ large negative they set

$$d(x, -T) = s(x, c_1) + s(x - X, c_2)\,, \tag{2.1}$$

where the separation distance X was chosen so large that the two solitary waves overlap only by a negligible amount. Since solitary waves die down exponentially, even a moderate value of X accomplishes this. The speeds were chosen so that $c_1 > c_2$, and X was taken to be positive so that at $t = -T$ the faster wave lies to the left of the slower one.

If the equation governing the motion were linear, the solitary waves would not interact at all and so after the elapse of $2X/(c_1 - c_2)$ time the relative position of the two would be merely interchanged. Numerical calculation of the solution of the KdV equation with initial values (2.1) showed that for $S > 2X/(c_1 - c_2)$

$$d(x, -T + S) = s(x - c_1S - \theta_1, c_1) + s(x - X - c_2S - \theta_0, c_2)\,,$$

except for deviations that could be accounted for by truncation error, i.e., the same as would be obtained in a linear case except for phase shifts θ_1 and θ_2.

The actual process of interaction, i.e., the behaviour of $d(x, t)$ around the time $-T + X/(c_1 - c_2)$ is far from being a mere superposition. In fact, Kruskal and

[2] In Section 2 we shall demonstrate the uniform boundedness of solutions of the KdV equation.

Zabusky observed that in cases when $c_1 \gg c_2$, i.e., when the first wave was much higher (and therefore faster) than the second one, the big wave swallows up the small one during the interaction, and reemits it later. In cases where c_1 and c_2 were comparable, the two waves interact as follows: As soon as the big wave comes reasonably close to the smaller one in front of it, the big wave begins to shrink and the smaller one begins to grow, until the two waves interchange their roles; thereafter they separate.

In this section we present a rigorous proof of the fact that the KdV equation indeed has solutions which behave like two solitary waves approaching, interacting, and then separating; furthermore we give a precise estimate for the ratio of speeds which lead to the two different kinds of interaction described above. It turns out that there is yet a third manner, intermediate between the other two, in which the interaction can take place.

We shall call these special solutions *double waves*, and denote them by $d(x, t; c_1, c_2)$. Note that for fixed c_1 and c_2 there is a two-parameter family of double waves, the parameters being the phases of each solitary wave at $t = -\infty$.

In [8], Kruskal and Zabusky derived an ordinary differential equation with respect to x which $d(x, t; c_1 c_2)$ satisfies for each fixed t. The argument in [8] is formal; here we present an (almost) rigorous derivation.

Our starting point is Lemma 1.1, the constancy of the functional (1.24) for pairs of solutions u, v of the nonlinear equation and of the variational equation, respectively. Taking u to be d we see that, for any solution v of (1.21),

$$(G(d), v) , \tag{2.2}$$

where G is the gradient of some integral, is independent of t. By definition, for t large negative

$$d(x, t; c_1, c_2) = s(x - c_1 t - \theta_1, c_1) + s(x - c_2 t - \theta_2, c_2) + \text{error}(t) . \tag{2.3}$$

The error term in (2.3) is due to the interaction of the tails of the solitary waves into which the double wave d decomposes. Since the tails of solitary waves decay exponentially, and since the separation of the two solitary waves is proportional to t, we expect *the error term in* (2.3) *to decay exponentially in t*. Suppose that the gradient G is a local operator; it follows from (2.3) that

$$G(d) = G(s_1) + G(s_2) + \text{error} , \tag{2.4}$$

where the error in (2.4) also tends to zero exponentially as $t \to -\infty$.

Suppose that G satisfies

$$G(s_1) = 0 , \quad G(s_2) = 0 , \tag{2.5}$$

then it follows from (2.4) that $V(d)$ *tends to zero exponentially as* $t \to -\infty$.

We turn now to the functions v; since these satisfy a linear equation, it can be shown, by an argument similar to the one which led to inequality (1.4), that any v increases at most exponentially. In fact, since exponentially increasing solutions of the variational equation usually indicate an instability, and since on the other hand numerical evidence indicates that double waves are stable, it is reasonable to expect that *all solutions v of the variational equation grow at a rate slower than exponential.*

We return now to the functional (2.2); we have shown that as $t \to -\infty$ the first factor $G(d)$ tends to zero exponentially, and that the second factor v tends to infinity slower than exponentially. It follows then that $(G(d), v)$ tends to zero as $t \to -\infty$; on the other hand, according to (2.2), this functional is independent of t. Therefore it follows that $(G(d), v)$ is zero for all t.

At any particular time t_0 the initial values of v may be prescribed arbitrarily; therefore it follows that

$$G(d) = 0 \tag{2.6}$$

at time t_0, that is for any time.

There remains the task of constructing an integral I whose gradient G is local and satisfies (2.5), i.e., annihilates both solitary waves s_1 and s_2. Here we rely on Lemma 1.2, relation (1.30), according to which solitary waves are eigenfunctions of the gradient of *every* integral:

$$G(s_1) = \kappa(G)s_1\,, \qquad G(s_2) = \kappa(G)s_2\,.$$

It follows that, given three independent integrals, an appropriate linear combination of them will have a gradient which annihilates any two given solitary waves.

We turn now to the task of finding three independent integrals whose gradients are local operators. We have already noted in Section 1 that the energy

$$I_1(u) = \int \tfrac{1}{2}u^2\,dx = \tfrac{1}{2}(u, u) \tag{2.7$_1$}$$

is an integral for the KdV equation. Another integral was found by Whitham [7]:

$$I_2(u) = \int (\tfrac{1}{3}u^3 - u_x^2)\,dx\,. \tag{2.7$_2$}$$

A third one was discovered by Kruskal and Zabusky:

$$I_3(u) = \int (\tfrac{1}{4}u^4 - 3uu_x^2 + \tfrac{9}{5}u_{xx}^2)\,dx\,. \tag{2.7$_3$}$$

Subsequently, Kruskal and Zabusky found two more explicit integrals and Miura four more; an infinite sequence [3] of such integrals was constructed in [5].

[3] Denote by I_n the n-th integral of this sequence and by G_n its gradient. The relation discovered by Gardner between I_n and the n-th generalized KdV operator K_n described in Section 1 is

$$K_n = DG_n\,.$$

For the discussion of the double wave we need only the first three integrals. The gradients of these are

$$G_1(u) = u\,, \tag{2.8$_1$}$$

$$G_2(u) = u^2 + 2u_{xx}\,, \tag{2.8$_2$}$$

$$G_3(u) = u^3 + 3u_x^2 + 6uu_{xx} + \tfrac{18}{5}u_{xxxx}\,. \tag{2.8$_3$}$$

Note that these are indeed local operators. The proof that I_1, I_2, I_3 are integrals of KdV consists in verifying that the product of $G_n(u)$ with $K(u) = uu_x + u_{xxx}$ is a perfect x derivative. Indeed an explicit calculation gives

$$G_1(u)K(u) = (uu_{xx} - \tfrac{1}{2}u_x^2 + \tfrac{1}{3}u^3)_x = H_1(u)_x\,, \tag{2.9$_1$}$$

$$G_2(u)K(u) = (u_{xx}^2 + u^2u_{xx} + \tfrac{1}{4}u^4)_x = H_2(u)_x\,, \tag{2.9$_2$}$$

$$G_3(u)K(u) = (\tfrac{9}{5}u_{xxx}^2 + \tfrac{18}{5}u_{xxx}u_xu + \tfrac{6}{5}u_{xx}^2u - \tfrac{3}{5}u_{xx}u_x^2 + \tfrac{3}{2}u_x^2u + u_{xx}u^3 + \tfrac{1}{5}u^5)_x = H_3(u)_x\,. \tag{2.9$_3$}$$

We know that solitary waves are eigenfunctions of G_n; a calculation gives the eigenvalues $\kappa(G_n)$ as follows:

$$G_1(s) = s\,, \tag{2.10$_1$}$$

$$G_2(s) = 2cs\,, \tag{2.10$_2$}$$

$$G_3(s) = \tfrac{18}{5}c^2s\,. \tag{2.10$_3$}$$

We form now the linear combination

$$I = I_3 + AI_2 + BI_1 \tag{2.11}$$

whose gradient is

$$G = G_3 + AG_2 + BG_1\,. \tag{2.12}$$

In view of (2.10),

$$G(s) = (\tfrac{18}{5}c^2 + 2Ac + B)s\,.$$

Thus G annihilates s_1 and s_2 if c_1 and c_2 satisfy the equation

$$S(c) = \tfrac{18}{5}c^2 + 2Ac + B = 0\,. \tag{2.13}$$

In view of the relation between coefficients and roots of a quadratic equation this means that

$$A = -\tfrac{9}{5}(c_1 + c_2)\,, \tag{2.14$_A$}$$

$$B = \tfrac{18}{5}c_1c_2\,. \tag{2.14$_B$}$$

In view of (2.6) we conclude: Let d be a double wave with speeds c_1 and c_2, and define the constants A, B by (2.14); then for each t, d satisfies

$$G(d) = G_3(d) + AG_2(d) + BG_1(d) = 0\,. \tag{2.15}$$

This is a nonlinear ordinary differential equation of fourth order[4]; therefore its solutions form a 4-parameter family of functions. On the other hand, the double waves form a 2-parameter family. We shall accordingly deduce from (2.15) a second order equation satisfied by all double waves. To do this we make use of the earlier observation (see (2.9)) that the invariance of $I(u)$ for solutions of KdV is equivalent with the fact that $G(u)K(u)$ is a perfect x derivative. Multiplying (2.15) by $K(u)$ and using (2.9) we see that

$$(H_3 + AH_2 + BH_1)_x = 0 .$$

Integrating this and using the fact that d and all its derivatives are zero at $x = \pm\infty$, we deduce

$$H_3(d) + AH_2(d) + BH_1(d) = 0 . \tag{2.16}$$

Next we make use of the *translation invariance* of the integrals I under consideration, i.e., of the fact that $I(u_\varepsilon)$ is independent of ε, where u_ε is the translate of u by ε in the x direction. Differentiating with respect to ε, we get

$$0 = \frac{d}{d\varepsilon} I(u_\varepsilon)\big|_{\varepsilon=0} = \int G(u)u_x \, dx .$$

This implies that $G(u)u_x$ is a perfect x derivative. Indeed, an explicit calculation gives

$$G_1(u)u_x = (\tfrac{1}{2}u^2)_x = J_1(u)_x , \tag{2.17$_1$}$$

$$G_2(u)u_x = (u_x^2 + \tfrac{1}{3}u^3)_x = J_2(u)_x , \tag{2.17$_2$}$$

$$G_3(u)u_x = (\tfrac{18}{5}u_x u_{xxx} - \tfrac{9}{5}u_{xx}^2 + 3uu_x^2 + \tfrac{1}{4}u^4)_x . \tag{2.17$_3$}$$

Thus multiplying (2.15) by u_x we obtain, after integration and use of the fact that d and its derivatives vanish at ∞,

$$J_3(d) + AJ_2(d) + BJ_1(d) = 0 . \tag{2.18}$$

Both (2.16) and (2.18) are third order differential equations. Expressing d_{xxx} from (2.18), substituting into (2.16) and multiplying by d_x^2, we get an equation of second order and fourth degree in d_{xx} which we write symbolically as

$$Q(d_{xx}, d_x, d) = 0 . \tag{2.19}$$

From this equation d_{xx} can be expressed as a 4-valued function of d and d_x; since x does not appear explicitly in these equations, this second order equation is equivalent to a first order autonomous system of equations for d and d_x. By studying carefully the geometry of all four branches of the corresponding vector-field one can show that (2.19) indeed has solutions which tend to zero as $x \to \pm\infty$ and has the shape of a double wave, i.e., has two maxima and one minimum.

[4] This equation appears in [8].

I shall not present the details because

(i) I did not carry them out completely,
(ii) the formulas are horribly complicated,
(iii) an explicit formula for double waves (indeed N-tuple waves) was derived recently in [1].

We show now how to use equation (2.19) to study the time history of double waves. For this purpose we consider the *maximum value* of $d(x, t)$ with respect to x, or rather the relative maxima of d as functions of time. Denote by $m = m(t)$ the value of a relative maximum of $d(x, t)$; at the point y where the relative maximum occurs

$$d_x = 0 \,. \tag{2.20}$$

It follows from (2.20) by the implicit function theorem that if $d_{xx} < 0$ at y, then y is a differentiable function of t. Hence $m = d(y, t)$ also is differentiable and satisfies

$$m_t = d_x y_t + d_t = d_t \,.$$

Since d satisfies the KdV equation, we have, in view of (2.20),

$$m_t = -K(d) = -d_{xxx} \,. \tag{2.21}$$

We proceed now to determine d_{xxx} at a local maximum. At a point where $d_x = 0$, equation (2.16) simplifies considerably; using formulas (2.9) and denoting the value of d by m, we get

$$\tfrac{9}{5}d_{xxx}^2 + \tfrac{6}{5}d_{xx}^2 m + d_{xx}m^3 + \tfrac{1}{5}m^5 + A(d_{xx}^2 + m^2 d_{xx} + \tfrac{1}{4}m^4) + B(md_{xx} + \tfrac{1}{3}m^3) = 0. \tag{2.22}$$

Similarly, at a point where $d_x = 0$, equation (2.18) becomes

$$d_{xx}^2 = \tfrac{5}{9}(\tfrac{1}{4}m^4 + A\tfrac{1}{3}m^3 + B\tfrac{1}{2}m^2) = P(m) \,. \tag{2.23}$$

From (2.23) we deduce that

$$d_{xx} = -P^{1/2}(m) \,, \tag{2.24}$$

the negative sign being chosen since the second derivative is nonpositive at a local maximum point. Substituting (2.24) into (2.22) we get

$$d_{xxx}^2 = R(m) \,, \tag{2.25}$$

where

$$R(m) = -a(m) + b(m)P^{1/2}(m) \,, \tag{2.26}$$

with

$$a(m) = \tfrac{2}{3}mP(m) + \tfrac{1}{9}m^5 + \tfrac{5}{9}AP(m) + \tfrac{5}{36}Am^4 + \tfrac{5}{27}Bm^3 \tag{2.27$_a$}$$

and

$$b(m) = \tfrac{5}{9}m^3 + \tfrac{5}{9}Am^2 + \tfrac{5}{9}Bm \,. \tag{2.27$_b$}$$

Combining (2.21) and (2.25) we have

$$m_t = \pm R(m)^{1/2} ; \tag{2.28}$$

the sign to be taken as positive when m increases, negative when m decreases. Thus m as function of t is governed by equation (2.28).

To study the behaviour of m we have to know something about the function $R(m)$. First of all we investigate the sign of $P(m)$; since $P(m)$ is $\frac{5}{9}m^2$ times the quadratic polynomial

$$\tfrac{1}{4}m^2 + \tfrac{1}{3}Am + \tfrac{1}{2}B , \tag{2.29}$$

it will be positive if the discriminant of (2.29) is:

$$\text{discr} = \tfrac{1}{2}B - \tfrac{1}{9}A^2 .$$

Using formulas (2.14) for A and B we have

$$\text{discr} = \tfrac{9}{25}(3c_1c_2 - c_1^2 - c_2^2) .$$

This quadratic form is positive if and only if

$$\frac{c_1}{c_2} \leqq \frac{3 + \sqrt{5}}{2} = 2.62 . \tag{2.30}$$

Thus we have proved

LEMMA 2.1. *If c_1 and c_2 satisfy* (2.30), $P(m)$ *is positive for all real values of* m. *If* (2.30) *is violated,* $P(m)$ *is negative in the interval* (n_1, n_2) ,

$$n_{1,2} = \tfrac{2}{5}[m_1 + m_2 \pm (m_1^2 + m_2^2 - 3m_1 m_2)^{1/2}] . \tag{2.31}$$

It is easy to verify that the interval (n_1, n_2) lies inside (m_1, m_2).

We turn now to $R(m)$; a lengthy calculation, the gist of which is described in [10], yields the following

LEMMA 2.2. $R(m)$ *has a double zero at* m_1 *and at* m_2 *and* d^2R/dm^2 *is positive at these points.*

LEMMA 2.3. (a) *In the range*

$$\frac{c_1}{c_2} < \frac{3 + \sqrt{5}}{2} ,$$

the function $R(m)$ *is positive in* (m_1, m_2).

(b) *In the range*

$$\frac{3 + \sqrt{5}}{2} < \frac{c_1}{c_2} < 3 ,$$

R *is positive in* (m_1, n_1) *and in* (m_2, n_2), *where* n_1, n_2 *are defined by* (2.31).

(c) *In the range*

$$3 < \frac{c_1}{c_2},$$

R is positive on (m_2, n_2) *and on* $(m_1, m_1 - m_2)$, *and* $R(m_1 - m_2) = 0$.

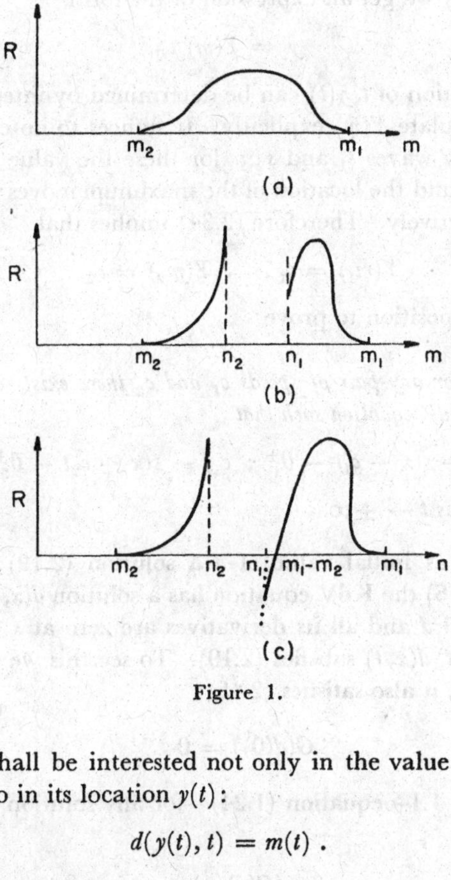

Figure 1.

Eventually we shall be interested not only in the value of the maximum as function of t but also in its location $y(t)$:

$$d(y(t), t) = m(t) .$$

The criterion for a maximum is

$$d_x(y, t) = 0 ;$$

differentiating this with respect to t we get

$$d_{xx} y_t + d_{xt} = 0 ,$$

or

$$y_t = -\frac{d_{xt}}{d_{xx}}. \tag{2.32}$$

Differentiating the KdV equation we have

$$d_{tx} = -d_x^2 - dd_{xx} - d_{xxxx} . \tag{2.33}$$

Using equation (2.15) we can express d_{xxxx} as function of d, d_x, d_{xx}, and d_{xxx}. Furthermore, at a local maximum point we have $d_x = 0$, and d_{xx}, d_{xxx} can be expressed as in (2.24) and (2.25) as functions of m. Substituting these expressions into (2.33) and (2.32) we get an expression of the form

$$y_t = Y(m) . \tag{2.34}$$

If we know m as function of t, $y(t)$ can be determined by integration.

We shall not calculate $Y(m)$ explicitly. It suffices to note that the discussion applies to the solitary waves s_1 and s_2; for these the value of the maximum is m_1, respectively m_2, and the location of the maximum moves with speed $c_1 = \frac{1}{3}m_1$ and $c_2 = \frac{1}{3}m_2$, respectively. Therefore (2.34) implies that

$$Y(m_1) = c_1 , \qquad Y(m_2) = c_2 . \tag{2.35}$$

We are now in a position to prove

THEOREM 2.1. *For any pair of speeds c_1 and c_2 there exists a double wave, i.e., a solution $d(x, t)$ of the KdV equation such that*

$$d(x, t) - s(x - c_1 t - \theta_1^{\pm} ; c_1) - s(x - c_2 t - \theta_2^{\pm} ; c_2) \tag{2.36}$$

tends to zero uniformly as $t \to \pm\infty$.

Proof: We take as initial value of d a solution (2.19); according to the existence theorem in [6] the KdV equation has a solution $d(x, t)$ for all time with these initial data, and d and all its derivatives are zero at $x = \pm\infty$. We claim that for all values of t, $d(x, t)$ satisfies (2.19). To see this we note first that since $d(x, 0)$ satisfies (2.19), it also satisfies (2.15):

$$G(d(0)) = 0 . \tag{2.37}$$

According to Lemma 1.1—equation (1.24)—for any solution v of the variational equation,

$$(G(d), v) \tag{2.38}$$

is independent of t. Relation (2.37) shows that, at $t = 0$, (2.38) is zero; therefore (2.38) is zero for all time. Since the value of v can be prescribed arbitrarily at any particular time, it follows that $G(d)$ is zero at each time. From this, and from the fact that d and its derivatives are zero at $x = \pm\infty$, we deduce as before that d satisfies equation (2.19) at each t.

We turn now to case (a) of Lemma 2.3 and claim:

For any time t, $d(x, t)$ has exactly two local maxima.

Proof: In case (a) the number of local maxima is independent of t, since at the time of the creation or disappearance of a local maximum $d_{xx} = 0$; on the other hand, according to (2.23), $d^2_{xx} = P(m)$ at a local maximum, and in case (a), $P(m)$ is positive for all m.

The time history of each local maximum is governed by equation (2.33). In case (a), R is positive in (m_1, m_2) and has a double zero at m_1 and at m_2. It follows from this that each solution of (2.33) whose values [5] lie in (m_1, m_2) goes from m_1 to m_2 as t goes from $+\infty$ to $-\infty$, or in the other direction, depending on the sign in (2.33). Furthermore, the approach of $m(t)$ to m_1 or m_2 as $t \to \pm\infty$ is exponential. It follows from this that the function $Y(m)$ occurring in (2.34), (2.35) tends to c_1, respectively c_2, exponentially as $t \to \pm\infty$. Integrating (2.34) we conclude therefore that

$$(2.39) \qquad y(t) - c_{1,2}t$$

tends to a limiting value as $t \to \pm\infty$.

Next we show that, as $m(t) \to m_1$ or m_2, the shape of the curve $d(x, t)$ around the maximum point $y(t)$ tends to the shape of a solitary wave, i.e., that

$$(2.40) \qquad \lim_{t \to \pm\infty} d(x - y(t), t) = s_{1,2}(x),$$

uniformly on bounded x intervals. To prove this we merely note that for each fixed t, $s_1(x)$, $s_2(x)$ and $d(x, t)$ all satisfy the fourth order equation (2.15). Furthermore, it follows from relations (2.20), (2.24), (2.35) and (2.25) that the Cauchy data of d at $x = y$ for a fourth order ordinary differential equation, i.e., the values of d and of its first three x derivatives tend as $t \to \pm\infty$ to the Cauchy data of $s_1(x)$, respectively $s_2(x)$, at $x = 0$. Relation (2.40) follows then from the continuous dependence of solutions of ordinary differential equations on their Cauchy data.

From (2.40) we can deduce our assertion about the number of maxima; for, assume that there were more than two of them. Then at least two would tend to the same limit, say m_1, as $t \to \infty$. It follows from (2.39) that the separation of these two locations of maxima tends to a constant; but this is incompatible with relation (2.39) which says that d looks like s_1 centered around *either* location.

Likewise it is impossible for d to have only one maximum; since then, $d(x - y(t), t)$ would be monotonically decreasing in x on either side of $y(t)$ and so would tend *uniformly* to one solitary wave as $t \to -\infty$, the other as $t \to +\infty$. We claim that, for all t,

$$(2.41) \qquad \int_{|x-y(t)|>X} d^2(x, t)\, dx < \varepsilon(X),$$

[5] It is not hard to show that the values of any solution of (2.19) which is zero at $x = \pm\infty$ lie in (m_1, m_2).

where $\varepsilon(X)$ tends to zero as $X \to \infty$. If this were so, we could deduce that

$$\lim_{t\to\pm\infty} \int d^2(x - y(t), t)\, dx = \begin{cases} \int s_1^2(x)\, dx\,, \\ \int s_2^2(x)\, dx\,. \end{cases} \tag{2.42}$$

But this is a contradiction. For, on the one hand, for $c_1 \neq c_2$,

$$\int s_1^2\, dx \neq \int s_2^2\, dx\,,$$

on the other hand, $\int d^2\, dx$ is invariant under both x and t translation, and so the integral on the left side of (2.42) is independent of t. But then the limits as $t \to +\infty$ and $-\infty$ cannot have different values.

To prove (2.41) we merely note that $\int d(x, t)\, dx = I_0(d)$ is also an integral for the KdV equation, from which (2.41) follows with

$$\varepsilon(X) = s(X)\,,$$

because of the monotonicity of d on either side of $y(t)$.

Having shown that $d(x, t)$ has exactly two maxima, it follows easily by the arguments already presented that, as $t \to \pm\infty$, $d(x, t)$ tends uniformly to the superposition of two solitary waves, each travelling at its own speed. This completes the proof of Theorem 2.1 in case (a).

The analysis presented above shows that the time history of the two maxima is as follows: as t goes from $-\infty$ to $+\infty$ the height of the larger solitary wave decreases from m_1 to m_2, while the height of the smaller one increases from m_2 to m_1. Thus in this case the two solitary waves interchange their roles without passing through each other, as observed by Kruskal and Zabusky in their calculations.

We turn now to case (b); here $R(m)$ is not defined inside (n_1, n_2), which means (see equation (2.23)) that no relative maximum of $d(x, t)$ can lie in that interval. We assert that *the absolute maximum lies above that interval at all times.* For, suppose on the contrary that at some time the absolute maximum is less than n_2; then, since the absolute maximum is a continuous function of t, and since on the other hand the value of the absolute maximum cannot cross (n_1, n_2), it follows that $d(x, t)$ does not exceed n_2 for any value of x and t. Let $m(t)$ denote a local maximum at time t. As we saw earlier, $m(t)$ satisfies the differential equation (2.33): $m_t = \pm R^{1/2}(m)$. In case (b), those solutions of this differential equation whose values lie in (m_2, n_2) behave as follows: depending on the sign in (2.33), $m(t)$ tends with increasing (decreasing) t to n_2:

$$\lim_{t\to t_0} m(t) = n_2\,.$$

Denote as before the location of the maximum by $y(t)$, and set

$$\lim_{t\to t_0} y(t) = y_0\,.$$

Relations (2.20) and (2.24) imply that

$$d_x(y_0, t_0) = 0 , \qquad d_{xx}(y_0, t_0) = 0 ,$$

while relation (2.25) and the fact that $R(n_2) > 0$ imply that

$$d_{xxx}(y_0, t_0) > 0 .$$

This means that the function $d(x, t_0)$ does not have a local—much less global—maximum at $x = y_0$. Since the value of d at $x = y_0$ is n_2, it follows that the global maximum of $d(x, t_0)$ exceeds n_2; this contradicts our previous assumption, and so proves the assertion.

Denote by $M(t)$ the absolute maximum of $d(x, t)$; denote by $m(t)$ that local maximum which at, say, $t = 0$ equals $M(0)$ and which satisfies the differential equation (2.33) for local maxima. The function $m(t)$ is either increasing or decreasing. Suppose it is increasing, then, since all local maxima satisfy (2.33), it follows that $m(t)$ remains the absolute maximum for $t > 0$; it further follows from $(2.33)_+$ that $m(t)$ tends to m_1 as t tends to $+\infty$, and that for some finite time t_0

$$\lim_{t \to t_0} m(t) = n_1 .$$

When m reaches the value n_1, it ceases to be a local maximum, so that

$$M(t_0) > m(t_0) .$$

Denote by t_1 the infimum of those values of t_1 for which $m(t)$ is the absolute maximum; clearly $t_0 < t < 0$. At time t_1 there must be at least two points where the absolute maximum is assumed. Denote by $n(t)$ that local maximum which is equal to $m(t)$ at $t = t_1$ but is a decreasing function of t; $n(t)$ satisfies equation $(2.33)_-$, and it follows that, as $t \to -\infty$, $n(t)$ tends to m_1.

Denote by $y(t)$ the location at time t of the absolute maximum. As before we deduce that

$$\lim_{|t| \to \infty} d(x + y(t), t) = s_1(x) ,$$

uniformly on bounded x intervals.

The analysis presented in the previous case, when applied to this situation, shows that for $|t|$ large there can be at most one additional local maximum; suppose it exists and is located at $z(t)$. Then it follows as above that

$$\lim_{|t| \to \infty} d(x + z(t), t) = s_2(x) ,$$

uniformly on bounded x intervals.

A careful examination of equation (2.33) satisfied by d shows that d does indeed have another local maximum for $|t|$ large; this proves the theorem in case (b).

Case (c) can be analyzed in a similar fashion. The main difference is that the function $R(m)$ vanishes at $m = m_1 - m_2$; it is easy to show that this causes solutions of $m_t = \pm R^{1/2}(m)$ in the range $m_1 - m_2 \leqq m(t) < m_1$ to behave in the following fashion:

There is a value t_0 such that $m(t_0) = m_1 - m_2$, and m satisfies

$$m_t = \begin{cases} -R^{1/2}(m) & \text{for} \quad t < t_0 , \\ R^{1/2}(m) & \text{for} \quad t_0 < t . \end{cases}$$

In particular $m(t)$ tends to m_1 as $|t|$ tends to ∞.

As in case (b), one can show that for $|t|$ large there is exactly one absolute and one local maximum, and that the asymptotic relation (2.36) holds. This completes the proof of Theorem 2.1.

The time history of the local maxima is as follows:

As t goes from $-\infty$ to ∞, the amplitude of the smaller solitary wave increases until it reaches the value n_2, at which point the local maximum disappears. In the meanwhile the amplitude of the larger solitary wave decreases steadily; in case (c), this amplitude reaches its minimum value $m_1 - m_2$ at some time T and some point X. The double wave is symmetric with respect to this occurrence, i.e.,

$$d(X - x, T - t) = d(x, t) . \tag{2.43}$$

In case (b), the amplitude of the larger solitary wave decreases until it reaches the value n_1, at which point the local maximum disappears. Before this happens however another local maximum is created which starts increasing. Denote by T the time when the two maxima are equal, and denote by X the midpoint between the two local maxima; the double wave satisfies the symmetry relation (2.43).

Speaking qualitatively we might say that in cases (b) and (c) the big wave first absorbs, then re-emits the small wave, and that in case (b) the absorption of the small wave raises a secondary peak on the big wave.

Up to a certain point one can give a similar analysis of N-tuple waves, that is, solutions of the KdV equation which as $t \to \pm\infty$ split apart into a superposition of n-solitary waves with speeds $c_1, c_2, \cdots, c_N$. Using the first N of the sequence of integrals constructed in [5] one can derive an ordinary differential equation of order $2N$ with respect to x. An elegant argument of Gardner (personal communication) shows that the generalized KdV operators K_n, $n = 1, \cdots, N$, are integrating factors for this differential operator. In this fashion we can obtain by elimination a differential equation of order N for the N-tuple wave, but the resulting equation is too complicated to yield any useful information. Happily, these N-tuple waves have been described explicitly in [1].

Acknowledgement. I acknowledge with pleasure a number of stimulating conversations with Martin Kruskal and Norman Zabusky, and an enlightening correspondence with Clifford Gardner.

Bibliography

[1] Gardner, C. S., Greene, J. M., Kruskal, M. D., and Miura, R. M., *A method for solving the Korteweg-de Vries equation*, Phys. Rev. Letters, Vol. 19, 1967, pp. 1095–1097.

[2] Gardner, C. S., and Morikawa, G. K., *Similarity in the asymptotic behavior of collision-free hydromagnetic waves and water waves*, New York Univ., Courant Inst. Math. Sci., Res. Rep. NYO-9082, 1960.

[3] Korteweg, D. J., and de Vries, G., *On the change of form of long waves advancing in a rectangular channel and on a new type of long stationary wave*, Philos. Mag., Vol. 39, 1895, pp. 422–443.

[4] Miura, R. M., *Korteweg-de Vries equation and generalizations. I. A remarkable explicit nonlinear transformation*, J. Math. Phys., Vol. 9, 1968, pp. 1202–1204.

[5] Miura, R. M., Gardner, C. S., and Kruskal, M. D., *Korteweg-de Vries equation and generalizations. II. Existence of conservation laws and constants of motion*, J. Math. Phys., Vol. 9, 1968, pp. 1204–1209.

[6] Sjöberg, A., *On the Korteweg-de Vries equation*, Uppsala Univ., Dept. of Computer Sci., Report, 1967.

[7] Whitman, G. B., *Nonlinear dispersive waves*, Proc. Roy. Soc., Ser. A., Vol. 283, 1965, pp. 238–261.

[8] Zabusky, N. J., *A synergetic approach to problems of nonlinear dispersive wave propagation and interaction, Nonlinear Partial Differential Equations*, Academic Press, New York, 1967.

[9] Zabusky, N. J., and Kruskal, M. D., *Interaction of solutions in a collisionless plasma and the recurrence of initial states*, Phys. Rev. Letters, Vol. 15, 1965, pp. 240–243.

[10] Peter D. Lax, *Integrals of nonlinear equations of evolution and solitary waves*, New York Univ., Courant Inst. Math. Sciences, Report NYO-1480-87, Jan. 1968.

Received February, 1968.

COMMUNICATIONS ON PURE AND APPLIED MATHEMATICS, VOL. XXVIII, 141–188 (1975)

Periodic Solutions of the KdV Equation*

PETER D. LAX

Dedicated to Arne Beurling

Abstract

In this paper we construct a large family of special solutions of the KdV equation which are periodic in x and almost periodic in t. These solutions lie on N-dimensional tori; very likely they are dense among all solutions.

The special solutions are characterized variationally; they minimize $F_N(u)$, subject to the constraints $F_j(u) = A_j$, $j = -1, \cdots, N-1$; here F_j denote the remarkable sequence of conserved functionals discovered by Kruskal and Zabusky. The above minimum problem was originally suggested by them. In exploring the manifold of solutions of this minimum problem we make essential use of Gardner's discovery that these functionals are in involution with respect to a suitable Poisson bracket.

Gardner, Greene, Kruskal and Miura have shown that the eigenvalues of the Schrödinger operator are conserved functionals if the potential is a function of t and satisfies the KdV equation. In Section 6 a new set of conserved quantities is constructed which serve as a link between the eigenvalues of the Schrödinger operator and the F_j.

Another result in Section 6 is a slight sharpening of an earlier result of the author and J. Moser: for the special solutions constructed above all but $2N+1$ eigenvalues of the Schrödinger operator are double.

The simplest class of special solutions, $N = 1$, are cnoidal waves. In an appendix, M. Hyman describes the results of computing numerically the next simplest case, $N = 2$. These calculations show that the shape of these solutions recurs exactly after a finite time, in a shifted position. The theory verifies this fact.

1. Introduction

The equation

$$u_t + uu_x + u_{xxx} = 0 \tag{1.1}$$

is called after its discoverers the Korteweg–de Vries equation (see [12]), abbreviated as KdV; here u is a real-valued scalar variable, t is time and x a single spatial variable. The initial value problem is to relate solutions of (1.1) to their initial values, i.e., their values at $t = 0$. Recently, the initial value problem has been studied extensively for sufficiently differentiable solutions which tend to zero rapidly as $|x|$ tends to ∞. It is easy to show (see

---

* Results obtained at the Courant Institute of Mathematical Sciences, New York University, under Contract AT(11-1)-3077 with the U.S. Atomic Energy Commission. Reproduction in whole or in part is permitted for any purpose of the United States Government.

e.g. [13], p. 467) that solutions in this class are uniquely determined by their initial values. It has been further shown (see e.g. [20] or [15]) that the initial values of u may be prescribed arbitrarily in this class.

In 1965, Kruskal and Zabusky [25] devised a numerical scheme which they used to construct and study solutions of the KdV equation. They observed that after the elapse of some time each solution contained a number of waves, each of which propagates with a positive speed and without altering its shape. They identified these waves as the *solitary waves* already discovered by Korteweg and deVries, i.e., waves which are zero at $x = \pm\infty$ and which propagate with constant speed and unaltered shape. The culmination of these investigations was [7], in which the initial value problem for KdV was reduced to the inverse problem in scattering theory for potentials vanishing at $x = \pm\infty$. This led to a procedure for constructing solutions of the KdV equation in terms of their initial data which involved solving the Gelfand–Levitan equation. The resulting expression turned out to be the perfect tool for studying the long-time behavior of solutions (see [1], [21], and [8]).

The initial value problem for solutions which are periodic in x turns out to be more difficult than for solutions which vanish at $x = \pm\infty$. Here too the initial steps were taken by Kruskal and Zabusky, who observed in their numerical calculations that solutions with sinusoidal initial values develop for a while into a rather complicated wave pattern, but that at certain times the waves recombine into a reasonable semblance of the initial shape; this led them to suspect that these *solutions of* KdV *are almost periodic in time.*

In this paper we construct a large class of solutions of KdV which are periodic in x and almost periodic in t. I believe that the initial values of these special solutions are dense among all periodic initial values, and I further believe that all spatial periodic solutions of KdV are almost periodic in time.

The last section contains new information on the relation of the KdV equation to the Schrödinger equation.

In their recent study, Henry McKean and Pierre van Moerbeke [17] have constructed explicit formulas for the special class of solutions constructed in this paper, and for the KdV flow on this class of solutions. Their formulas involve hyperelliptic functions.

I learned from Prof. Fomenko at the Vancouver International Congress of Mathematicians that S. Novikov and Dubrovin have done similar work on this subject (see [19]).

Acknowledgement. I acknowledge with pleasure many stimulating conversations with Jürgen Moser.

This paper is dedicated to Arne Beurling, in recognition of his influence in the development of inverse problems of spectral theory.

2. Conserved Functionals of Equations of Evolution

Consider a nonlinear evolution equation

$$u_t = K(u)\,. \tag{2.1}$$

We assume that solutions of (2.1) exist for all t, positive and negative, are uniquely determined by their initial values, and that their initial values can be prescribed arbitrarily within a certain set in a linear space. Solutions of (2.1) induce a *flow* on this set. Let $F(u)$ be a *functional* which is *conserved* under the flow (2.1), i.e., one for which $F(u(t))$ is independent of t when $u(t)$ is a solution of (2.1). If F is a conserved functional, the sets $F(u)=c$ are invariant sets for the flow. These sets have codimension one and so are rather large; we show now how to construct with the aid of F much smaller invariant sets.

THEOREM 2.1. *Let F be a conserved functional of* (2.1) *which is Frechet differentiable. Then the set of stationary points of F forms an invariant set for the flow* (2.1).

Remark 1. The result is intuitively clear for stationary points which are local extremals, i.e., which are maxima or minima, since the flow carries a local maximum into a local maximum.

Remark 2. Take the very special case of linear, unitary flows in a Hilbert space, i.e., one governed by an equation of the form

$$u_t = Au\,. \tag{2.1)'$$

A skew selfadjoint. In this case we can construct invariant *quadratic* functionals; they are of the form

$$F(u) = (u, Bu)\,, \tag{2.2}$$

where B is a linear operator which *commutes* with A. The stationary points of (2.2) are the null vectors of B; so in this case Theorem 2.1 says that the null space of a linear operator B which commutes with A is an invariant subspace for solutions of (2.1)′. This is well known, of course.

Proof of Theorem 2.1: Frechet differentiability means that for every u and v the derivative

$$\frac{d}{d\varepsilon} F(u+\varepsilon v)$$

exists at $\varepsilon=0$ and is a linear functional of v. We assume that the linear space in which solutions lie is equipped with an *inner product* (,). Then linear functionals can be expressed as an inner product; in particular,

$$\frac{d}{d\varepsilon}F(u+\varepsilon v)\Big|_{\varepsilon=0}=(G_F(u),v)\,. \tag{2.3}$$

$G_F(u)$ is called the *gradient* of F at u; G_F is a nonlinear operator defined on the space. The stationary points of F form the *null set* of G_F. We want to show that if $u(t)$ is a solution of (2.1) whose initial value belongs to the null set of G_F, then all values $u(t)$ belong to the null set. To see this we imbed the solution $u(t)$ in a differentiable one-parameter family of solutions $u(t,\varepsilon)$ which for $\varepsilon=0$ reduces to $u(t)$. For example, we may define $u(t,\varepsilon)$ as that solution of (2.1) whose value at $t=T$ is

$$u(T,\varepsilon)=u(T)+\varepsilon w\,, \tag{2.4}$$

where T is an arbitrary but fixed time and w an arbitrary but fixed vector. We assume that there is a dense set of vectors w for which the dependence on ε of u thus defined is differentiable.

Since F is conserved, $F(u(t,\varepsilon))$ is independent of t, for every fixed ε. Then the same is true of its ε-derivative at $\varepsilon=0$. Using the definition of the gradient given in (2.3), we can write this derivative as

$$(G_F(u(t)),v(t))\,, \tag{2.5}$$

wnere

$$v(t)=\frac{d}{d\varepsilon}u(t,\varepsilon)\Big|_{\varepsilon=0}\,. \tag{2.6}$$

By assumption, $u(0)$ belongs to the null set of G_F; therefore, (2.5) is zero at $t=0$. Since (2.5) is independent of t, it is zero for all t. In particular, it is zero at $t=T$:

$$(G_F(u(T)),v(T))=0\,. \tag{2.7}$$

Comparing (2.6) and (2.4) we see that

$$v(T)=w\,,$$

so that (2.7) says that $G_F(u(T))$ is orthogonal to w. Since w is an arbitrary element, it follows that $G_F(u(T))=0$, and since T is an arbitrary time, it follows that $G_F(u(t))=0$ for all t. This completes the proof of Theorem 2.1.

We derive now a further result along these lines which will be useful in Section 5.

THEOREM 2.2. *Let F be a differentiable conserved functional for the evolution equation* (2.1). *Denote by $M(u)$ the derivative of the nonlinear operator $K(u)$ appearing on the right in* (2.1):

$$Mv = \frac{d}{d\varepsilon} K(u+\varepsilon v)\Big|_{\varepsilon=0}\,; \tag{2.8}$$

M is a linear operator. Let $u = u(t)$ be a solution of (2.1); *then $G_F = G_F(u(t))$ satisfies*

$$\left(\frac{\partial}{\partial t} + M^*\right) G_F = 0\,. \tag{2.9}$$

where M^* is the adjoint of M with respect to the inner product used to define the gradient.

Proof: Let $u = u(t, \varepsilon)$ be the one-parameter family of solutions of (2.1) introduced earlier. Differentiating (2.1) with respect to ε at $\varepsilon = 0$, we get, using v as defined by (2.6) and M defined by (2.8), the following linear equation for v:

$$v_t = Mv\,. \tag{2.10}$$

Since (2.5) is independent of t, its t derivative is zero:

$$\left(\frac{\partial}{\partial t} G_F,\, v\right) + \left(G_F,\, \frac{\partial}{\partial t} v\right) = 0\,.$$

Substituting Mv for v_t and applying the adjointness relation we get the relation

$$\left(\left(\frac{\partial}{\partial t} + M^*\right) G_F,\, v\right) = 0$$

for all t. Since at any particular value of t the values of v can be prescribed arbitrarily. it follows that equation (2.9) is satisfied as asserted in Theorem 2.2.

THEOREM 2.3. *Denote by $G_F'(u)$ the second derivation of F; $G_F'(u)$ is an operator defined by*

$$\frac{d}{d\eta} G_F(u+\eta w)\Big|_{\eta=0} = G_F'(u) w\,. \tag{2.11}$$

Assertion: G'_F *is a symmetric operator with respect to the parentheses* (,) *used to define* G_F.

Proof: Substituting the definition (2.3) of G_F into (2.11) we get

$$\frac{\partial}{\partial\eta}\frac{\partial}{\partial\varepsilon}F(u+v+\eta v)\Big|_{\substack{\varepsilon=0\\ \eta=0}}=(G'_F(u)w,\,v)\,.$$

The symmetry of $G'_F(u)$ follows from the equality of mixed partial derivatives $\partial^2/\partial\eta\,\partial\varepsilon$ and $\partial^2/\partial\varepsilon\,\partial\eta$.

3. Conserved Functionals for the KdV Equation

The usefulness of the results contained in Section 2 hinges on the existence of plenty of conserved functionals whose gradient has a nontrivial but not too large null set. The KdV equation (1.1) is rich in such functionals; three of them are classical:

$$F_{-1}(u)=\int u\,dx\,,$$

(3.1) $$F_0(u)=\int \tfrac{1}{2}u^2\,dx\,.$$

$$F_1(u)=\int(-\tfrac{1}{3}u^3+u_x^2)\,dx\,.$$

Kruskal and Zabusky made the remarkable discovery that there are further invariant functionals of which the first is

(3.1)$_2$ $$F_2(u)=\int(\tfrac{1}{4}u^4-3uu_x^2+\tfrac{9}{5}u_{xx}^2)\,dx\,.$$

Eventually, Gardner, Kruskal and Miura showed in [18] that these four are merely the first of an infinite sequence of conserved functionals; the n-th functional has the form

(3.1)$_n$ $$F_n(u)=\int P_n\,dx\,,$$

where P_n is a polynomial in u and its derivatives up to order n.

The following characterization of the polynomials P_n is given in [18]: *Each term in* P_n *has weight* $n+2$, *where the weight of a term is defined as the sum of the weights of its factors, and the weight of* $u^{(k)}=\partial^k u$ *is defined as* $1+\frac{1}{2}k$.

It follows in particular that P_n is at most quadratic in $u^{(n)}$:

$$P_n = a_n u^{(n)^2} + b_n u^{(n)} + c_n \,, \tag{3.2}$$

where a_n is a constant, and b_n and c_n are polynomials in $u^{(j)}$, $j<n$, of weight $1+\frac{1}{2}n$ and $2+n$. It was shown in [18] that a_n is non-zero; we choose it to be positive.

It further follows that, for $n>1$, P_n is quadratic in $u^{(n)}$ and $u^{(n-1)}$:

$$P_n = au^{(n)^2} + bu^{(n)}u^{(n-1)} + cu^{(n-1)^2} + du^{(n)} + eu^{(n-1)} + f, \tag{3.2$'$}$$

a as before and $b, \cdots, f$ polynomials in $u^{(j)}$, $j<n-1$.

THEOREM 3.1. (a) *Let* u *be a smooth periodic function,* n *some non-negative integer; the quantities*

$$\max\{|u(x)|, |u^{(1)}(x)|, \cdots, |u^{(n-1)}(x)|\}, \qquad \int u^{(n)^2}\, dx \tag{3.3}$$

can be bounded in terms of $F_0(u), F_1(u), \cdots, F_n(u)$.

(b) *The functional*

$$F_n(u) \geqq \tfrac{1}{2}a_n \int u^{(n)^2} - f_{n-1} \,, \tag{3.4}$$

where the quantity f_{n-1} *depends on* $F_0(u), F_1(u), \cdots, F_{n-1}(u)$.

Proof: The case $n=0$ is trivial; to prove the result for $n=1$ we introduce the abbreviations

$$M = \max |u(x)| \,, \qquad S = \int u_x^2 \,. \tag{3.5}$$

By the mean value theorem, there is a point x_0 where

$$u^2(x_0) = \frac{1}{p}\int u^2\, dx = \frac{F_0(u)}{p} \,,$$

p the period. For any x,

$$u(x) = u(x_0) + \int_{x_0}^{x} u_x\, dx \,;$$

using the Schwarz inequality, we get

$$u^2(x) \leqq 2u^2(x_0) + 2p \int u_x^2 \, dx ,$$

which implies that

$$M^2 \leqq \frac{2F_0}{p} + 2pS . \tag{3.6}$$

By definition $(2.8)_1$ of F_1,

$$S = \int u_x^2 \, dx = F_1 + \int \tfrac{1}{3} u^3 \, dx .$$

Using the definition $(2.8)_0$ of F_0 we get the inequality

$$S \leqq F_1 + \frac{MF_0}{3} . \tag{3.7}$$

Substituting this into (3.6) we have

$$M^2 \leqq \frac{2F_0}{p} + 2pF_1 + \frac{2pMF_0}{3} ;$$

from this we easily estimate M in terms of F_0, F_1, and p. From (3.7) we get a similar estimate for S. This completes the proof of part (a) for $n = 1$. For $n > 1$ we proceed by induction. Using the quadratic character (3.2)′ of P_n and the positivity of the coefficient a we can easily estimate $\int u^{(n)^2} dx$ in terms of $F_n(u)$ and quantities already estimated by the induction assumption; $\max |u^{(n-1)}|$ can then be estimated in terms of $\int u^{(n)^2} dx$ and F_0. This completes the inductive proof of part (a).

To prove part (b) for $n > 1$ we use formula (3.2)′ for P_n; since the coefficients $b, \cdots, f$ depend on $u^{(j)}$, $j < n-1$, it follows from part (a) that they, as well as $\int u^{(n-1)^2} dx$, are bounded.

Using the Schwarz inequality one can show that the sum of terms beyond the first is less than $\frac{1}{2} a \int u^{(n)^2} dx + \text{const}$; this implies inequality (3.4).

The case $n = 1$ can be handled as in part (a).

We consider now the following extremum problem, originally posed by Kruskal and Zabusky:

Minimize $F_N(u)$, *given the values of* $F_{-1}(u), \cdots, F_{N-1}(u)$:

$$F_j(u) = A_j , \qquad j = -1, \cdots, N-1 . \tag{3.8}$$

The constants A_j have to be so chosen that the constraints (3.8) can be satisfied by some admissible function, and furthermore so that A_j is *not* an extremal value of $F_j(u)$ when the other constraints are imposed. Analytically this means that for any function u satisfying the constraints (3.8), the gradients $G_j(u)$, $j = -1, \cdots, N-1$, are *linearly independent*. We call such constraints *admissible*.

THEOREM 3.2. (a) *Suppose the constraints* (3.8) *are admissible; then there is a periodic function* u *which minimizes* $F_N(u)$ *among all functions satisfying the constraints.*

(b) *Every minimizing function* u *satisfies an Euler equation of the form*

$$G(u) = G_N(u) + \sum_{-1}^{N-1} a_j G_j(u) = 0 , \tag{3.9}$$

where the a_j *are some numbers, and* G_j *abbreviates* G_{F_j}, *the gradient of* F_j *with respect to the* L_2 *scalar product.*

Proof: It follows from inequality (3.4) that $F_N(u)$ is bounded from below for all u satisfying the constraints. Let u_n be a minimizing sequence of functions satisfying the constraints. It follows from part (a) of Theorem 3.1 that $u^{(j)}(x)$ are bounded for $j<N-1$ uniformly for all n. From inequality (3.4) we know that also the sequence $\int |u_n^{(N)}|^2\, dx$ is bounded; therefore so is $u^{(N-1)}(x)$.

We can select a subsequence for which $u_n^{(N)}$ converges in the weak topology in L_2 to some limit $u^{(N)}$; since the L_2 norm is lower semicontinuous in the weak topology,

$$\int u^{(N)2}\, dx \leqq \liminf_{n\to\infty} \int u_n^{(N)2}\, dx .$$

It follows from the Rellich compactness theorem that for a further subsequence, $u_n^{(j)}(x)$ converges uniformly to some limit $u^{(j)}$ for all $j<N$. Clearly, $u^{(j)}$ is the j-th derivative of $u^{(0)}$ in the sense of distributions; equally clearly, $u^{(0)} = u$ satisfies the constraints, and minimizes F_N. This completes the proof of part (a) of Theorem 3.2.

To prove part (b) we consider one-parameter families of deformations of u of the form

$$u(\varepsilon) = u + \varepsilon v + \sum_{-1}^{N-1} a_k v_k . \tag{3.10}$$

where v is an arbitrary function and the v_k are N fixed functions so chosen that the matrix

$$(G_j(u), v_k) \tag{3.11}$$

is nonsingular. It is possible to choose such v_k since for admissible constraints the functions $G_j(u)$, $j = -1, \cdots, N-1$, are linearly independent. The coefficients $a_k = a_k(\varepsilon)$ are so chosen that the constraints

$$F_j(u + \varepsilon v + \sum a_k v_k) = A_j , \tag{3.12}$$

$j = -1, \cdots, N-1$. are satisfied. According to the implicit function theorem, we can choose such a_k for every sufficiently small ε, so that $a_k(0) = 0$ and a_k depends differentiably on ε. Differentiating (3.12) at $\varepsilon = 0$, we get

$$(G_j(u), v + \sum \dot{a}_k(0) v_k) = 0 .$$

Since (3.11) is nonsingular, this determines $\dot{a}_k(0)$ uniquely. In particular, if v satisfies

$$(G_j(u), v) = 0 , \qquad j = -1, \cdots, N-1 , \tag{3.13}$$

then

$$\dot{a}_k(0) = 0 \tag{3.14}$$

for all k.

Since $u = u(0)$ minimizes $F_N(u)$, it follows that

$$\frac{d}{d\varepsilon} F_N(u(\varepsilon))|_{\varepsilon=0} = 0 ;$$

this can be expressed as

$$(G_N(u), v + \sum \dot{a}_k(0) v_k) = 0 .$$

In particular, if v satisfies (3.13), all $\dot{a}_k(0) = 0$, so that then

$$(G_N(u), v) = 0 . \tag{3.15}$$

This can happen only when $G_N(u)$ is a linear combination of $G_j(u)$, $j=-1,\cdots,N-1$, as asserted in (3.9), completing the proof of Theorem 3.2.

Equation (3.9) states that u is a stationary point of

$$F(u)=F_N(u)+\sum a_j F_j(u)\,. \tag{3.16}$$

F, being the linear combination of conserved functionals, is itself conserved; we conclude therefore from Theorem 2.1:

THEOREM 3.3. *Solutions of* (3.9) *form an invariant set for the* KdV *flow.*

According to $(3.1)_n$, F_n is the integral of a polynomial of u and its derivatives up to order n. It follows that G_n is a differential operator of order $2n$. Based on formulas (3.1) we can easily compute the gradients of the first four conserved functionals:

$$\begin{aligned} G_{-1}(u)&=1\,,\\ G_0(u)&=u\,,\\ G_1(u)&=-2u_{xx}-u^2\,,\\ G_2(u)&=\tfrac{18}{5}u_{xxxx}+6uu_{xx}+6u_x^2+u^3\,. \end{aligned} \tag{3.17}$$

For $N=0$, equation (3.9) is

$$u+a_{-1}=0\,, \tag{3.18_0}$$

whose solutions are constant, a very uninteresting case. For $N=1$, we have

$$-2u_{xx}-u^2+a_0u+a_{-1}=0\,. \tag{3.18_1}$$

Multiplying this by u_x we get, after integration,

$$u_x^2+\tfrac{1}{3}u^3-a_0\tfrac{1}{2}u^2-a_{-1}u=\text{const}\,. \tag{3.19}$$

Solutions of this equation are *elliptic functions.*

THEOREM 3.4. *If $u(x)$ is a solution of* $(3.18)_1$, *then*

$$u(x,t)=u(x-a_0t) \tag{3.20}$$

is a solution of the KdV *equation.*

Proof: Differentiating $(3.18)_1$ with respect to x we get, after dividing by -2,

$$u_{xxx}+uu_x-a_0u_x=0\,. \tag{3.21}$$

On the other hand, substituting (3.20) into the KdV equation (1.1), we have

$$u_t+uu_x+u_{xxx}=-a_0u_x+uu_x+u_{xxx}$$

which according to (3.21) is zero.

Solutions of the form (3.20) travel with constant speed a_0, without altering their shape. The existence of such solutions, and the differential equation (3.21) governing them, was discovered by Korteweg and deVries; they called these solutions *cnoidal waves.*

In an appendix, M. Hyman presents a numerical study of the case $N=2$. Solutions of equation (3.9) and the KdV flow on this manifold of solutions are described.

In Section 5 we present a theoretical study of solutions of (3.9) for general N; we shall show that the set of solutions constitutes an N-dimensional torus, and that the KdV flow on this torus is quasi-periodic.

4. Hamiltonian Mechanics

In this section we give a brief review of some topics in classical Hamiltonian mechanics which will be used in Section 5. The Hamiltonian form of the equations of motion is

$$\frac{d}{dt}q_n=\frac{\partial H}{\partial p_n},\qquad \frac{d}{dt}p_n=-\frac{\partial H}{\partial q_n},\qquad n=1,\cdots,N. \tag{4.1$_H$}$$

where H, the Hamiltonian, is some function of the $2N$ variables p_n, q_n, $n=1,\cdots,N$. The *Poisson bracket* of any pair of functions F and K of the variables p, q is defined as follows:

$$[F,K]=\sum\frac{\partial F}{\partial q_n}\frac{\partial K}{\partial p_n}-\frac{\partial F}{\partial p_n}\frac{\partial K}{\partial q_n}\,. \tag{4.2}$$

A straightforward calculation shows that

(a) the Poisson bracket is a bilinear, alternating function of F and K,

(b) the Jacobi identity

$$[[F,H],K]+[[H,K],F]+[[K,F],H]=0 \tag{4.3}$$

is satisfied.

When p and q change according to $(4.1)_H$, the rate of change of any function F of p and q equals the Poisson bracket of F and H:

$$\frac{dF}{dt}=\sum\frac{\partial F}{\partial q}\frac{dq}{dt}+\frac{\partial F}{\partial p}\frac{dp}{dt} \tag{4.4}$$

$$=\sum\frac{\partial F}{\partial q}\frac{\partial H}{\partial p}-\frac{\partial F}{\partial p}\frac{\partial H}{\partial q}=[F,H]$$

It follows from this that F *is constant along the trajectories of the flow* $(4.1)_H$ *if and only if* $[F,H]=0$.

More generally, let v be a function of p, q and t. Relation (4.4) implies that

$$\frac{dv}{dt}=\frac{\partial v}{\partial t}+\frac{\partial v}{\partial q}\frac{dq}{dt}+\frac{\partial v}{\partial p}\frac{dp}{dt}$$

$$=\frac{\partial v}{\partial t}+[v,H].$$

Thus we conclude that v *is constant along trajectories of the flow* $(4.1)_H$ *if and only if* v *satisfies*

$$\frac{\partial v}{\partial t}+[v,H]=0. \tag{4.5_H}$$

For fixed H, $[v,H]$ is a first-order partial differential operator acting on v, called the *Liouville operator* and denoted by L_H:

$$L_Hv=[v,H]. \tag{4.6}$$

Equation (4.5) is called the *Liouville equation.* Its solution can be written symbolically as

$$v(t)=e^{-tL_H}v(0); \tag{4.7_H}$$

e^{-tL_H} is the *solution operator* for the Liouville equation $(4.5)_H$. Let L_H and L_K be two Liouville operators. It is an immediate consequence of the Jacobi identity (4.3) that

$$L_KL_H-L_HL_K=L_{[H,K]}. \tag{4.8}$$

In particular, *if* $[H,K]=0$, *the operators* L_K *and* L_H *commute.* It follows then that also their exponentials, $\exp\{-tL_H\}$ and $\exp\{-tL_K\}$, the solution operators of $(4.5)_H$ and $(4.5)_K$, commute. Since solutions of these equations are characterized as functions of p, q, t which are constant along trajectories of $(4.1)_H$ and $(4.1)_K$, respectively, we have established

THEOREM 4.1. *If the Poisson bracket of K and H is zero, the Hamiltonian flows $(4.1)_H$ and $(4.1)_K$ commute.*

5. The KdV Equation as a Hamiltonian System

In [6], Clifford Gardner has given a Hamiltonian formulation of the KdV equation in the periodic case. In [3], Faddeev and Zacharov have given a Hamiltonian formulation of the KdV equation on the whole real axis; they have shown, using the inverse method of Gardner, Green, Kruskal and Miura, that this Hamiltonian system is completely integrable.

In this section we show how to use Gardner's results to study the manifold of periodic solutions of (5.1) and the KdV flow on this manifold.

Gardner defines a Poisson bracket for the class of C^∞ functionals F, H of smooth periodic functions u as follows:

$$[F, H] = (G_F, \partial G_H)\,, \tag{5.1}$$

where the parenthesis (,) is the L_2 scalar product, G_F and G_H the gradients of F and H with respect to (,), and $\partial = d/dx$.

THEOREM 5.1. (a) *The Poisson bracket defined by* (5.1) *is a bilinear and alternating function of F and H.*

(b) *The Jacobi identity* (4.3) *holds.*

Proof: Bilinearity is obvious, and the alternative character follows from the antisymmetry of the operator ∂.

To prove the Jacobi identity we first compute the gradient of $[F, H]$:

$$\begin{aligned} G_{[F,H]} &= \frac{d}{d\varepsilon}[F(u+\varepsilon v), H(u+\varepsilon v)]|_{\varepsilon=0} \\ &= \frac{d}{d\varepsilon}(G_F(u+\varepsilon v), \partial G_H(u+\varepsilon v)) \\ &= (G'_F(u)v, \partial G_H(u)) + (G_F(u), \partial G'_H(u)v)\,. \end{aligned} \tag{5.2}$$

According to Theorem 2.3, G'_F and G'_H are symmetric; using this and the antisymmetry of ∂ we can rewrite (5.2) as

$$(v, G'_F \partial G_H) - (G'_H \partial G_F, v).$$

This shows that

$$G_{[F,H]} = G'_F \partial G_H - G'_H \partial G_F . \tag{5.3}$$

Using (5.3) we can write

$$\begin{aligned}
&[[F, H], K]+[[H, K], F]+[[K, F], H] \\
&\qquad = (G'_F \partial G_H - G'_H \partial G_F, \partial G_K) + (G'_H \partial G_K - G'_K \partial G_H, \partial G_F) \\
&\qquad\quad + (G'_K \partial G_F - G'_F \partial G_K, \partial G_H) .
\end{aligned}$$

Using once more the symmetry of the operators G'_F, G'_H, G'_K we conclude that the above sum is zero. This proves the Jacobi identity.

We turn now to the KdV equation (1.1). Using formula (3.17) for the gradient G_1 of the conserved functional F_1 we can write the KdV equation in the following form:

$$u_t = \tfrac{1}{2} \partial G_1 .$$

Let F be any functional; by definition of gradient,

$$\frac{dF}{dt} = (G_F, u_t) . \tag{5.4}$$

Using (5.4) and the definition (5.1) of the Poisson bracket we can write this as

$$\frac{dF}{dt} = (G_F, \tfrac{1}{2} \partial G_1) = [F, \tfrac{1}{2} F_1] . \tag{5.5}$$

Comparing this with (4.4) we recognize *the* KdV *equation as a Hamiltonian flow for the Poisson bracket* (5.1) *and with Hamiltonian* $H = \frac{1}{2} G_1$.

It follows from (5.5) that F is conserved for the KdV flow if and only if $[F, F_1] = 0$; since the functionals F_n defined by (2.8) are conserved, we conclude that

$$[F_n, F_1] = 0 , \qquad n = -1, 0, \cdots . \tag{5.6}$$

Gardner [6] has proved the following remarkable result:

THEOREM 5.2 (Gardner):

$$[F_n, F_m] = 0 \quad \textit{for} \quad \textit{all} \quad n, m . \tag{5.7}$$

Remark 1. The proof of this result first appeared in [6]; I learned this result from Gardner in the Spring of 1967, and remarked on its possible use at the end of [13].

Remark 2. The case $m=-1$ is trivial. For $m=0$, (5.7) expresses the invariance of the functionals F_n under translation with respect to x. For $m=1$, (5.7) expresses the fact that the F_n are conserved by the KdV flow.

Remark 3. In Section 6 we shall present the proof of a related result from which Theorem 5.2 follows.

We define the m-th generalized KdV *equation as*

$$u_t = \partial G_m(u)\,. \tag{5.8$_m$}$$

THEOREM 5.3. (a) *All functionals* F_n *are conserved for each equation* (5.8)$_m$.

(b) *The flows defined by equations* (5.8)$_m$ *commute with each other.*

Proof: Part (a) follows from Theorem 5.2, and part (b) from Theorems 4.1 and 5.2.

Remark. In view of Theorem 3.1 (a), u and its derivatives are bounded in terms of the functionals $F_n(u)$. Therefore we can derive, using part (a) of Theorem 5.3, *a priori* bounds for solutions of the generalized KdV equations (5.8)$_m$ in terms of their initial data. It can be shown that the initial value problem is properly posed for these equations.

We shall denote by $S_m(t)$ the *solution operator* of the m-th KdV equation; that is, $S_m(t)$ *relates the initial values of solutions of* (5.8)$_m$ *to their values at time* t. Part (b) of Theorem 5.3 states that *the operators* $S_m(t)$ *commute.*

THEOREM 5.4. *Let* u_0 *be a solution of equation* (3.9). *Then so is*

$$\prod_0^{N-1} S_m(t_m)u_0 = u(t_0, \cdots, t_{N-1}) \tag{5.9}$$

for every value of the parameters $t_0, \cdots, t_{N-1}$.

Proof: Equation (3.9) is

$$G(u)=0\,,$$

where G is the gradient of F:

$$F = F_N + \sum_{-1}^{N-1} a_j F_j \,.$$

According to part (a) of Theorem 5.3, F is a conserved quantity for the m-th generalized KdV flow; therefore it follows from Theorem 2.1 that $S_m(t)$ maps a solution of (3.9) into a solution of (3.9). This proves Theorem 5.4.

DEFINITION. We denote the n-parameter family of functions defined by (5.9) by S.

Since (3.9) is an ordinary differential equation of order $2N$, its solutions are uniquely determined by the Cauchy data $\{u(x_0), u^{(1)}(x_0), \cdots, u^{(2N-1)}(x_0)\}$ at an arbitrary point x_0. It is convenient to regard the set S as imbedded via its Cauchy data in $\mathbb{R}^{2N}$.

THEOREM 5.5. *The set S defined by* (5.9) *is a bounded subset of* $\mathbb{R}^{2N}$.

Proof: Since by Theorem 5.3 each F_n is a conserved functional for each flow $(5.8)_m$, $F_n(u) = F_n(u_0)$ for all u, $n = 0, 1, \cdots, 2N$. It follows then from part (a) of Theorem 3.1 that the Cauchy data of u are bounded uniformly for all u in S; this proves Theorem 5.5.

THEOREM 5.6. *Suppose u_0 is a solution of the variational problem discussed in Theorem* 3.2; *then S is an N-dimensional manifold immersed in* $\mathbb{R}^{2N}$.

Proof: It was assumed in Theorem 3.2 that the constraints (3.8) are admissible. This implies, as we observed just before stating Theorem 3.2, that the gradients $G_j(u)$, $j = -1, \cdots, N-1$, form a linearly independent set of functions for every u satisfying the constraints. Since by (3.17), $G_{-1}(u) = 1$, the functions ∂G_j, $j = 0, \cdots, N-1$, are linearly independent. According to (5.8), $\partial G_j(u)$ is the direction in which the j-th generalized KdV flow starts.

Let $u = u(t_0, \cdots, t_{N-1})$ be any point of S. As we have shown above, $F_n(u) = F_n(u_0)$ for all n; consequently u satisfies the same constraints as u_0. It follows from the above that the directions $\partial G_j(u)$, $j = 0, \cdots, N-1$, are linearly independent. Since, by part (b) of Theorem 5.3, the flows S_j commute,

$$u(t_0 + \varepsilon_0, \cdots, t_{N-1} + \varepsilon_{N-1}) = S_0(\varepsilon_0), \cdots, S_{N-1}(\varepsilon_{N-1})u \,.$$

The linear independence of $\partial G_j(u)$, $j = 0, \cdots, N-1$, implies that for $\varepsilon_0, \cdots, \varepsilon_{N-1}$ small enough these points form an open subset of a smoothly imbedded N-dimensional manifold. This completes the proof of Theorem 5.6.

Remark 1. The tangent space of S at u is spanned by $\partial G_0(u), \cdots, \partial G_{N-1}(u)$. We observe that, by continuity, $F_n(u) = F_n(u_0)$ not only at all points u of S but also for all points u of $\bar{S}$. Therefore we deduce:

COROLLARY 5.7. *Let u be a point of $\bar{S}$, the closure of S; then*

$$S_0(\varepsilon_0) \cdots S_{N-1}(\varepsilon_{N-1})u, \qquad |\varepsilon_0| < \varepsilon,$$

forms an open subset of a smoothly imbedded N-dimensional manifold.

The next 8 lemmas lead up to showing that S is an algebraic variety. The first one is a calculus lemma:

LEMMA 5.8. *Suppose that Q is a polynomial in $u^{(0)}, u^{(1)}, \cdots, u^{(j)}$ such that, for every periodic function u of period p,*

$$\int_0^p Q(u)\, dx = 0. \tag{5.10}$$

Then there exists a polynomial J in $u^{(0)}, u^{(1)}, \cdots, u^{(j-1)}$ such that

$$Q = \partial J. \tag{5.11}$$

Proof: Let y be an arbitrary point, $0 < y < p$. Let u be a C^∞ function such that

$$u(0) = u^{(1)}(0) = \cdots = u^{(j-1)}(0) = 0. \tag{5.12}$$

Define the function v as follows:

$$v(x) = \begin{cases} u(x) & \text{for} \quad 0 \leqq x \leqq y, \\ q(x) & \text{for} \quad y \leqq x \leqq y+\varepsilon, \\ 0 & \text{for} \quad y+\varepsilon \leqq x \leqq p. \end{cases} \tag{5.13}$$

Here $q(x)$ is defined as that polynomial of degree $2j-1$ which satisfies the $2j$ interpolation conditions

$$q(y) = u(y), \quad q^{(1)}(y) = u^{(1)}(y), \quad \cdots, \quad q^{j-1}(y) = u^{(j-1)}(y). \tag{5.14$_1$}$$

and

$$q(y+\varepsilon)=q^{(1)}(y+\varepsilon)=\cdots=q^{(j-1)}(y+\varepsilon)=0\,. \tag{5.14$_2$}$$

It follows from $(5.14)_1$ that v and its derivatives of order less than j are continuous, and it follows from $(5.14)_2$ and (5.12) that they are periodic. The j-th derivative of v may have jump discontinuities at $x=y$ and $x=0$. Identity (5.10) is true for such a function:

$$\int_0^p Q(v)\,dx=0\,.$$

Using the definition of v we get by (5.13)

$$\int_0^y Q(u)\,dx=-\int_y^{y+\varepsilon} Q(q)\,dx\,. \tag{5.15}$$

The right side is a polynomial in the coefficients of q and ε; q in turn is completely determined by the data $(5.14)_1$. On the other hand, the left side is independent of ε. So we conclude from (5.15) that

$$\int_0^y Q(u)\,dx=J$$

for all u satisfying (5.12), where J is a polynomial in $u^{(0)}(y), u^{(1)}(y), \cdots, u^{(j-1)}(y)$. Differentiating this relation with respect to y we obtain (5.11); this completes the proof of Lemma 5.8.

Remark. J is uniquely determined if its constant term is taken to be zero.

According to Theorem 5.3,

$$[F_m, F_n]=\int G_m\,\partial G_n\,dx=0\,. \tag{5.16}$$

Both G_m and G_n are polynomials in u. Applying Lemma 5.8 to

$$H=G_m\,\partial G_n$$

we have

LEMMA 5.9.

$$G_m\,\partial G_n=\partial J_{mn}\,, \tag{5.17}$$

where J_{mn} *is a polynomial of* $u^{(0)}, u^{(1)}, \cdots, u^{(l)}$,

$$l=\begin{cases}2n & \text{if} \quad m\leqq n,\\ 2m-1 & \text{if} \quad m>n.\end{cases} \tag{5.17$'$}$$

Remark 1. The following relations are obviously true:

$$J_{mn}+J_{nm}=G_m G_n .$$

Remark 2. The leading term in J_{mn} is $u^{(2m)}u^{(2n)}$ for $m\leqq n$, and $u^{(2m-1)}u^{(2n+1)}$ for $m>n$.

The second derivative $G'_m(u)$ of F_m is a linear differential operator whose coefficients depend on u and its derivatives. According to Lemma 2.3, G'_m is symmetric; this means that, for any periodic u and any pair of periodic functions v and w,

$$\int_0^p (vG'_m(u)w - wG'_m(u)v)\,dx = 0 .$$

This implies that the integrand is a perfect derivative and proves

LEMMA 5.10.

$$vG_m w - wG'_w v = \partial C_m(v, w)\,, \tag{5.18}$$

where C_m *is a bilinear differential operator, whose coefficients depend on* u *and derivatives of* u.

LEMMA 5.11. *Suppose* u *satisfies the generalized* KdV *equation* $(5.8)_m$; *then* J_{jk} *satisfies*

$$\frac{\partial}{\partial t}J_{jk}+C_m[\partial G_j, \partial G_k]-G_j G'_m\,\partial G_k=0 . \tag{5.19}$$

Proof: We show first that the expression on the left in (5.19) is independent of x. The x derivative of the first term, $\partial\frac{\partial}{\partial t}J_{jk}$, can be expressed with the aid of (5.17) as

$$\left(\frac{\partial}{\partial t}G_j\right)(\partial G_k)+G_j\,\partial\frac{\partial}{\partial t}G_j\,. \tag{5.20$_1$}$$

The x derivative of the second term can be expressed from (5.18) with $v=\partial G_j$ and $w=\partial G_k$ as

$$(5.20)_2 \qquad (\partial G_j)G_m\,\partial G_k-(\partial G_k)G_m\,\partial G_j\,.$$

The x derivative of the third term is

$$(5.20)_3 \qquad -(\partial G_j)G'_m\,\partial G_k-G_j\,\partial G'_m\,\partial G_k\,.$$

Adding these three expressions we get

$$(5.21) \qquad (G_j\,\partial)\left[\frac{\partial}{\partial t}-G'_m\,\partial\right]G_k+(\partial G_k)\left[\frac{\partial}{\partial t}-G'_m\,\partial\right]G_j\,.$$

Next we turn to Theorem 2.2 and take for K the generalized KdV operator

$$K=\partial G_m\,.$$

The derivative M of K is then

$$M=\partial G'_m\,.$$

Using the antisymmetry of ∂ and the symmetry of G'_m we have

$$(5.22) \qquad M^*=-G'_m\,\partial\,.$$

For the conserved functional F occurring in Theorem 2.2 we take F_k and F_j; setting G_k and G_j for G_F in (2.9), and using the value (5.22) of M^*, we obtain

$$\left(\frac{\partial}{\partial t}-G'_m\,\partial\right)G_k=0$$

and

$$\left(\frac{\partial}{\partial t}-G'_m\,\partial\right)G_j=0\,.$$

Using these relations in (5.21) we conclude that (5.21) is zero; since (5.2) is the x-derivative of (5.19), we see that (5.19) is independent of x.

Relation (5.19) is a polynomial in u and its derivatives with respect to x and t. If u satisfies the generalized KdV equation $(5.8)_m$, the t derivatives can be expressed in terms of x derivatives, so that (5.19) becomes a polynomial in the x derivatives of u alone. The initial values of u are arbitrary, so this polynomial is independent of x for arbitrary u. But this can be true only if all non-constant terms in the polynomial are zero. On the other hand, the constant term in this polynomial is zero since the ingredients of (5.19), J_{kj}, G_k, and G_j, all have zero constant terms. This proves that (5.19) is zero, and completes the demonstration of Lemma 5.11.

Suppose u is a solution of (3.9):

$$G(u)=\sum_{-1}^{N} a_j G_j = 0\,, \qquad a_N = 1\,. \tag{5.23}$$

We multiply this by ∂G_k; using the definition of J_{kj} and relation (5.17), we can write

$$\partial \sum a_j J_{jk} = 0\,.$$

Introducing the abbreviation

$$J_k = \sum a_j J_{jk}\,, \tag{5.24}$$

we deduce from the result above the following:

LEMMA 5.12. *If u satisfies $G(u)=0$, $J_k(u)$ is independent of x.*

Suppose $u=u(x,t)$ satisfies the generalized KdV equation $(5.8)_m$; then according to Lemma 5.11 the relations (5.19) hold. Multiplying (5.19) by a_j and summing, we get a relation which, using the definition J_k in (5.24) and (5.23) of G, can be written as follows:

$$\frac{\partial}{\partial t} J_k + C_m[\partial G, \partial G_k] - G G'_m\, \partial G_k = 0\,. \tag{5.25}$$

LEMMA 5.13: *$J_k(u)$ has the same value for all u in S.*

Proof: According to Theorem 5.4, $G(u)=0$ for every u in S. By definition of S, every point of S can be reached from u_0 by generalized KdV flows. Along such a flow, (5.25) holds. Since $G(u)=0$ on S, (5.25) becomes $\partial J_k/\partial t = 0$; since according to Lemma 5.12, $J_k(u)$ is independent of x for every u on S, the conclusion of Lemma 5.13 follows.

We nave noted in Remark 2 following Lemma 5.9 that for $m>n$ the leading term of J_{mn} is $u^{(2m-1)}u^{(2n+1)}$; hence the leading term of J_k is $u^{(2N-1)}u^{(2k+1)}$. From this one easily deduces

LEMMA 5.14. *The polynomials* J_k, $k=0,1,\cdots,N-1$, *in the* $2N$ *variables* $u^{(0)},\cdots,u^{(2N-1)}$ *are algebraically independent.*

Denote by c_k the constants $J_k(u_0)$, and denote by V the set in $\mathbb{R}^{2N}$ satisfying

$$J_k=c_k\,, \qquad k=0,\cdots,N-1\,. \tag{5.26}$$

According to Lemma 5.14, V is an N-dimensional algebraic variety; according to Lemma 5.13, V contains S. We shall show next that S is a component of V; to see this we look at $\bar{S}$, the closure of S.

LEMMA 5.15. (a) $\bar{S}$ *is invariant under the generalized* KdV *flows.*
(b) $\bar{S}\subset V$.
(c) $\bar{S}$ *is compact.*

Denote as before the solution operator of the m-th KdV flow by S_m. Let u be some point of $\bar{S}$; by definition,

$$u=\lim u_n\,, \qquad u_n\in S\,.$$

Since $S_m(t)$ is a continuous operator, $S_m(t)u_n$ converges to $S_m(t)u$; this proves part (a) of the lemma.

Part (b) follows from the continuity of the functions J_k. Finally, according to Theorem 5.5, S is bounded, proving part (c).

According to Corollary 5.7, for any point u of $\bar{S}$, the set $S(u,\varepsilon)$ defined by

$$S(u,\varepsilon)=\prod S_j(t_j)u\,, \qquad |t_j|<\varepsilon\,, \tag{5.27}$$

is an open subset of a smoothly imbedded N-dimensional manifold. We claim that $S(u,\varepsilon)$ contains a neighborhood of u in $\bar{S}$. For suppose this is not the case; then there would be a sequence of elements u_n of $\bar{S}$ such that

$$\text{(i)}\quad u_n\to u\,, \tag{5.28}$$
$$\text{(ii)}\quad u_n\notin S(u,\varepsilon)\,.$$

According to part (a) of Lemma 5.15 the set $S(u_n,\varepsilon)$ belongs to $\bar{S}$. We claim that $S(u_n,\tfrac{1}{2}\varepsilon)$ and $S(u,\tfrac{1}{2}\varepsilon)$ do not intersect. Suppose they did:

$$\prod S_j(t_j)u=\prod S_j(t_j')u_n\,, \qquad |t_j|<\tfrac{1}{2}\varepsilon,\ |t_j'|<\tfrac{1}{2}\varepsilon\,.$$

Since the S_j commute, this implies

$$\prod S_j(t_j - t_j')u = u_n$$

which would make u_n belong to $S(u, \varepsilon)$, contrary to $(5.28)_{ii}$.

We construct now a subsequence of u_n, again denoted by u_n, so that u_{n+1} does not belong to any $S(u_j, \varepsilon)$, $j \leqq n$; since each $S(u_j, \varepsilon)$ has a positive distance from u, this will be the case if u_{n+1} is close enough to u. Arguing as above we see that the sets $S(u_n, \frac{1}{2}\varepsilon)$ are pairwise disjoint. By part (a) of Lemma 5.15 these sets belong to $\bar{S}$, and so by part (b) of Lemma 5.15 they belong to V. But an algebraic variety V cannot have such a strudel-like structure with infinitely many leaves; so we conclude that $S(u, \varepsilon)$ contains a neighborhood of u in $\bar{S}$.

By definition of $\bar{S}$, every neighborhood of u in $\bar{S}$ contains points of S; thus some points of $S(u, \varepsilon)$ belong to S, which means that they are of the form $\prod S_j(r_j)u_0$. But then, again using the commutativity of the S_j, u itself is of this form. This shows that $\bar{S} = S$. Combining Lemma 5.15, Theorem 5.6 and the definition of S we can state

THEOREM 5.16. *S is a compact, connected, open and closed subset of the algebraic variety V: every point of S is a regular point.*

Next we turn to showing that S is an N-dimensional torus; we need

LEMMA 5.17. *There is a number T such that*

$$S \subset S(u_0, T). \tag{5.29}$$

Proof: By part (c) of Lemma 5.15, $\bar{S}$ is compact. Therfore, there are a finite number of points, $u_1, \cdots, u_m$ such that $\{S(u_j, \varepsilon)\}$ cover S. Each pair of points in one neighborhood can be connected to the other by applying $\prod S_k(t_k)$, $|t_j| < 2\varepsilon$; since each point can be linked to the base point u_0 by a chain of at most m links, (5.29) follows with $T = 2m\varepsilon$.

Lemma 5.17 implies that there must exist among the operators S_m relations of the form

$$\prod_0^{N-1} S_m(t_m) = I. \tag{5.30}$$

Now consider the set of all N-tuples $(t_0, t_1, \cdots, t_{N-1})$ for which (5.30) holds. These form a *module* $\mathcal{M}$ in $\mathbb{R}^N$ over the integers. It follows from Theorem 5.6 that (5.30) does not hold for $|t_m| < \varepsilon$ if ε is small enough except for $t_m \equiv 0$; this shows that the module $\mathcal{M}$ is *discrete*. According to linear algebra, a discrete module is a *lattice*, i.e., can be represented as

$$\sum_{j=1}^{K} n_j \omega_j ; \tag{5.31}$$

as the n_j range over all integers, each point of $\mathcal{M}$ has exactly one representation of form (5.31). K is the *dimension* of $\mathcal{M}$, where $K \leqq N$.

By Lemma 5.15, $S = \mathbb{R}^N \pmod{\mathcal{M}}$ is compact; according to linear algebra this implies that the lattice $\mathcal{M}$ is N-dimensional. In this case the quotient $\mathbb{R}^N/\mathcal{M}$ is an N-dimensional torus. This proves

THEOREM 5.18.

$$\prod S_m(t_m) : \mathbb{R}^N/\mathcal{M} \to \mathbb{R}^{2N} \tag{5.32}$$

is a one-to-one mapping of the torus $\mathbb{R}^N/\mathcal{M}$ *onto* S. *In particular,* S *is a torus; the* KdV *flow, given by* $S_1(t_1)$ *on* $\mathbb{R}^N/\mathcal{M}$, *is quasi-periodic.*

The case $N=1$ is trivial from the point of view of Theorem 5.18; since, according to (3.17), $\partial G_0 = \partial$, $S_0(t)$ is translation across the x-axis. The module $\mathcal{M}$ is generated by a single relation which says that functions in S are periodic; the KdV flow also is translation along the x-axis.

The case $N=2$ is more interesting; here $\mathcal{M}$ has two generators, at least one of which, call it $\omega = (t_0, t_1)$, has second component $t_1 \neq 0$. From (5.30) we see that on S

$$S_1(t_1) = S_0(-t_0) . \tag{5.33}$$

Since S_0 is translation, (5.33) says that if an initial function belonging to S is subjected to the KdV flow, it resumes its initial shape after time t_1, translated by the amount t_0. The calculations of M. Hyman, presented in the appendix bear this out strikingly.

Actually, the calculations were carried out first and were helpful in pointing the theory in the right direction. In particular, Lemma 5.13 was suggested by numerical evidence.

6. The Spectrum of the Schrödinger Operator

Let L denote the Schrödinger operator

$$L = \partial^2 + v , \tag{6.1}$$

$v(x)$ being some periodic potential, acting on periodic functions. L has as discrete spectrum $\{\lambda_j\}$, $j=1, 2, \cdots$. Whereas the eigenvalues are completely determined by the potential v, the converse is not true: operators L with different potentials v may very well have the same spectrum. In fact, Gardner, Kruskal and Miura have shown in [7] that if $u(x, t)$ is a solution of the KdV equation, then, for

$$v(x, t)=\tfrac{1}{6}u(x, t)\,, \tag{6.2}$$

the spectrum of the operator $L=L(t)$ defined by (6.1) is independent of t. In [13], I gave a new derivation of this fact and indicated how to construct an infinite sequence of differential equations whose solutions have the same property; Gardner has identified these equations as the generalized KdV equations discussed in Section 5. We start by reproducing very briefly the derivation of these results.

Let $L(t)$ be a one-parameter family of selfadjoint operators in Hilbert space. Clearly the spectrum of $L(t)$ is independent of t if the operators $L(t)$ are unitarily equivalent to each other, i.e., if there is a one-parameter family of unitary operators $U(t)$ such that, for all t,

$$U^*(t)L(t)U(t)=L(0)\,. \tag{6.3}$$

Assume that both $L(t)$ and $U(t)$ depend differentiably on t. Differentiating with respect to t the relation

$$U(t)U^*(t)=I\,,$$

we see that

$$U_tU^*=B \tag{6.4}$$

is *antisymmetric:*

$$B^*=-B\,. \tag{6.5}$$

From (6.4) and the relation $U^*U=I$ we deduce that

$$U_t=BU\,. \tag{6.6}$$

Conversely, given $B(t)$ as an antisymmetric operator-valued function of t, we can construct an operator $U(t)$ satisfying (6.6) by solving the initial value problem

$$h_t(t)=B(t)h(t)\,, \qquad h(t_0)=h_0\,, \tag{6.6$'$}$$

and setting $U(t)h(0)=h(t)$. Clearly the operator $U(t)$ defined in this way is isometric; if the initial value problem (6.6)′ can be solved for a dense set of h_0 for every t_0, then both $U(t)$ and its inverse are densely defined, so that the closure of $U(t)$ is unitary. Differentiate (6.3) with respect to t; using (6.6) and (6.5) we get

(6.7) $$L_t = BL - LB\,.$$

This condition in turn implies the unitary equivalence (6.3); so we have proved

THEOREM 6.1. *Let $L(t)$ be a one-parameter family of selfadjoint operators, $B(t)$ a one-parameter family of anti-selfadjoint operators. Suppose that* (6.7) *holds, and that* (6.6)′ *can be solved for a dense set of h_0; then the operators $L(t)$ are unitarily equivalent.*

We apply this theorem to L given by (6.1) where v depends on t. In this case, $L_t = v_t$, and (6.7) becomes

(6.7)′ $$v_t = BL - LB\,.$$

In [13] it is described how to satisfy this equation by differential operators B_m, with real coefficients and leading coefficient 1:

(6.8)$_m$ $$B_m = \partial^{2m+1} + \sum_{0}^{m-1} b_j^{(m)}\partial^{2j+1} + \partial^{2j+1} b_j^{(m)}\,.$$

Clearly, B_m is antisymmetric. Relation (6.7)′ requires the commutator of B_m and L to be multiplicative. Now the commutator of B_m and L is a differential operator of order $2m$; furthermore, this operator is symmetric. Therefore the requirement that this operator be actually of order zero imposes m conditions, which can be satisfied by appropriate choice for the m coefficients $b_j^{(m)}$ in (6.8)$_m$.

Explicit calculation for the cases $m=0$ gives

$$B_0 L - LB_0 = u_x\,;$$

so (6.7)$'_0$ is

(6.9)$_0$ $$v_t = v_x\,.$$

For $m=1$ we obtain, with $b = b_0'$,

$$B_1 L - LB_1 = \partial(3v_x - 4b_x)\,\partial + v_{xxx} - b_{xxx} + 2bv_x\,.$$

Setting $b=\frac{1}{3}v$, and $v=\frac{1}{6}u$ we get in $(6.7)_1'$:

$$(6.9)_1 \qquad 4u_t=uu_x+u_{xxx}.$$

This is, except for an inessential change of sign and scale in t, the KdV equation, and of course $(6.9)_0$ is the zero-th KdV equation. We introduce the notation

$$(6.10)_m \qquad B_mL-LB_m=K_m.$$

As already reported in [13], Gardner has observed that K_m is related to the m-th conserved functional of KdV:

THEOREM 6.2 (Gardner). *Let L be the operator*

$$(6.11) \qquad L=\partial^2+\tfrac{1}{6}u$$

and B_m the operator of form $(6.8)_m$ so chosen that $(6.10)_m$ is satisfied. Then

$$(6.12)_m \qquad K_m=c_m\,\partial G_m,$$

G_m being the gradient of the m-th conserved functional F_m, c_m some constant.

Thus equation

$$(6.9)_m \qquad u_t=K_m$$

is, except for a change of scale in time, the m-th generalized KdV equation $(5.8)_m$.

Using Theorem 6.1 and relation $(6.10)_m$ we see that if u satisfies $(6.9)_m$, the operators $L(t)$ are unitarily equivalent. The eigenvalues λ_j of L are functionals of the potential u appearing in L. We recognize this dependence of λ_j on u by writing

$$(6.13) \qquad \lambda_j(u).$$

The invariance of the spectrum of $L(t)$ when u varies subject to $(6.9)_m$ can be expressed as follows:

THEOREM 6.3. *The eigenvalues (6.13) are conserved functionals under all generalized KdV flows.*

For $m=0$, the KdV flow is just translation along the x-axis; for $m=1$, this result was discovered by Gardner, Green, Kruskal and Miura (see [7]).

We have now two different kinds of conserved functionals for the generalized KdV equations: the functionals $(3.1)_n$ described in Section 3 and the eigenvalues (6.13) of the operator (6.11). Already Kruskal and Zabusky have observed that these functionals are not independent: the functionals F_m defined in $(3.1)_m$ can be related to the asymptotic behavior of the eigenvalues for large n. An elegant derivation of these relations is given by McKean and van Moerbeke in [17].

We turn now to the functional λ_j:

THEOREM 6.4. *The functionals* $\lambda_j(w)$ *are in involution; i.e.,*

$$[\lambda_j, \lambda_k]=0 \tag{6.14}$$

for $j\neq k$, *where* [,] *is the Poisson bracket* (5.1).

Since the functionals F_n are functions of the λ_j, Theorem 6.4 gives another proof of Gardner's result contained in Theorem 5.2.

For the proof we need the following lemmas:

LEMMA 6.5. *Let* λ *be a simple eigenvalue of* L, *and denote by* w *the eigenfunction*

$$Lw=\lambda w \tag{6.15}$$

normalized by $(w, w)=1$. *The gradient of* λ *is*

$$G_\lambda=\tfrac{1}{6}w^2 . \tag{6.16}$$

Proof: We recall that the gradient of λ is defined by

$$\frac{d}{d\varepsilon}\lambda(u+\varepsilon v)|_{\varepsilon=0}=(G_\lambda(u), v), \tag{6.17}$$

where (,) denotes the L_2 scalar product. Replace u in (6.15) by $u+\varepsilon v$, differentiate with respect to ε and set $\varepsilon=0$. Recalling the definition (6.11) of L we get

$$L\frac{dw}{d\varepsilon}+\tfrac{1}{6}vw=\lambda\frac{dw}{d\varepsilon}+\left(\frac{d\lambda}{d\varepsilon}\right)w .$$

Taking the scalar product with w, using the symmetry of L, the eigenvalue equation (6.15) and $(w, w)=1$, we obtain

$$\frac{1}{6}\int vw^2=\frac{d\lambda}{d\varepsilon} .$$

Comparing this with (6.17) we deduce (6.16).

Remark. If λ is a double eigenvalue at u, it cannot be defined as a single-valued functional for nearby u. However, $\lambda_1+\lambda_2$ is an unequivocally defined functional. It is not hard to show that

$$G_{\lambda_1+\lambda_2}=w_1^2+w_2^2, \tag{6.16$'$}$$

where w_1, w_2 is any orthonormal pair of eigenfunctions.

LEMMA 6.6. *Suppose* w *satisfies the differential equation* (6.15); *then* w^2 *satisfies the following third order differential equation:*

$$Hw^2=4\lambda\,\partial w^2, \tag{6.18}$$

where H *is the linear differential operator*

$$H=\partial^3+\tfrac{2}{3}u\,\partial+\tfrac{1}{3}u_x. \tag{6.19}$$

Remark. This lemma is well known.

Proof: We have

$$\partial^1 w^2=2ww_x,$$

$$\partial^2 w^2=2ww_{xx}+2w_x^2.$$

$$\partial^3 w^2=2ww_{xxx}+6w_xw_{xx}.$$

Substituting this into (6.19), we get

$$Hw^2=2ww_{xxx}+6w_xw_{xx}+\tfrac{4}{3}uww_x+\tfrac{1}{3}u_xw^2. \tag{6.20}$$

According to (6.15) and the definition of L in (6.11),

$$w_{xx}+\tfrac{1}{6}uw=\lambda w. \tag{6.21}$$

Differentiating with respect to x we have

$$w_{xxx}+\tfrac{1}{6}uw_x+\tfrac{1}{6}u_xw=\lambda w_x. \tag{6.21$'$}$$

Multiply (6.21) by $6w_x$ and (6.21)′ by $2w$ and add; using (6.20) we obtain (6.18).

Observe that *the operator* H *defined by* (6.19) *is antisymmetric:*

$$H^* = -H.$$

We are now ready to evaluate the Poisson bracket (6.14). By definition (5.1) of the bracket,

$$[\lambda_j, \lambda_k] = (G_{\lambda_j}, \partial G_{\lambda_k})$$

which by (6.16) equals

$$\tfrac{1}{36}(w_j^2, \partial w_k^2). \tag{6.22}$$

Using $(6.18)_k$, we can rewrite this as

$$\frac{1}{36\times 4\lambda_k}(w_j^2, Hw_k^2).$$

Since H is antisymmetric, this equals

$$\frac{-1}{36\times 4\lambda_k}(Hw_j^2, w_k^2).$$

Using $(6.18)_j$ and then the antisymmetry of ∂ we get

$$\frac{-\lambda_j}{36\times\lambda_k}(\partial w_j^2, w_k^2) = \frac{\lambda_j}{36\lambda_k}(w_j^2, \partial w_k^2).$$

Comparing this with (6.22) we see that, for $\lambda_j \neq \lambda_k$,

$$(w_j^2. \partial w_k^2) = 0.$$

This completes the proof of Theorem 6.4.

The generalized KdV flows are not the only ones which leave the spectrum of L invariant and which commute with each other. In Theorem 6.10 we present a whole one-parameter family of such flows. First we need some auxiliary results.

Let α be any real number, and consider the equation

$$(L-\alpha)w = 0, \tag{6.23}$$

where L is given by (6.11). The Wronskian of any two solutions w_1 and w_2 of (6.23),

$$w_1\,\partial w_2 - w_2\,\partial w_1\,, \tag{6.24}$$

is independent of x. Hence, according to Floquet's classical theory, (6.23) has two particular solutions w_1 and w_2 which satisfy

$$w_1(x+p)=\kappa w_1(x)\,, \tag*{(6.25)$_1$}$$

$$w_2(x+p)=\kappa^{-1} w_2(x)\,. \tag*{(6.25)$_2$}$$

The parameter κ, called the *Floquet exponent*, is uniquely determined by the potential u and the parameter α which occurs in equation (6.23). If α lies in a stability interval, κ is complex of modulus 1. If α is an eigenvalue of L operating on functions with period p, $\kappa=1$; if α is an eigenvalue of L operating on functions of period $2p$, $\kappa=-1$. In these cases, and these cases only, w_1 and w_2 can coincide.

For fixed α, κ is a functional of u; we shall compute now the gradient of $\kappa(u)$. We normalize w_1, say, by $(w_1, w_1)=1$. Set $u+\varepsilon v$ in place of u in (6.23):

$$\partial^2 w + (\tfrac{1}{6}(u+\varepsilon v)-\alpha)w=0\,,$$

and differentiate with respect to ε; we get at $\varepsilon=0$

$$(L-\alpha)\frac{dw_1}{d\varepsilon}+\tfrac{1}{6}vw_1=0\,. \tag{6.26}$$

Differentiating (6.25)$_1$ with respect to ε gives

$$\frac{dw_1}{d\varepsilon}(x+p)=\kappa\frac{dw_1}{d\varepsilon}(x)+\frac{d\kappa}{d\varepsilon}\,w_1(x)\,. \tag{6.27}$$

We multiply (6.26) by w_2 and integrate by parts:

$$\begin{aligned} 0&=\int_0^p w_2(L-\alpha)\frac{dw_1}{d\varepsilon}+\tfrac{1}{6}vw_1w_2\,dx \\ &=\int (L-\alpha)w_2\frac{dw_1}{d\varepsilon}+\tfrac{1}{6}vw_1w_2\,dx + w_2\,\partial\frac{dw_1}{d\varepsilon}-\partial w_2\frac{dw_1}{d\varepsilon}\bigg|_0^p\,. \end{aligned} \tag{6.28}$$

Since w_2 satisfies (6.23), $(L-\alpha)w_2=0$; the boundary terms in (6.28) can be evaluated with the aid of the relations (6.27) and $(6.25)_2$ and we get

$$W\frac{1}{\kappa}\frac{d\kappa}{d\varepsilon}=\frac{1}{6}\int vw_1w_2\,dx\,, \tag{6.29}$$

where W is the Wronskian (6.24).

We recall the definition of the gradient:

$$\left.\frac{d\kappa(u+\varepsilon v)}{d\varepsilon}\right|_{\varepsilon=0}=(G_\kappa(u),v)\,,$$

and conclude from (6.29) that the following lemma holds.

LEMMA 6.7. *Let* κ *be the Floquet exponent of equation* (6.23) *as defined by relation* (6.25). *The gradient of* κ *with respect to* u *is*

$$G_\kappa=\tfrac{1}{6}\kappa w_1w_2=\tfrac{1}{6}\kappa n\,, \tag{6.30}$$

where w_1, w_2 *are solutions of* (6.23) *satisfying* (6.25) *and so normalized that their Wronskian equals* 1. *The symbol* n *is an abbreviation:*

$$n=w_1w_2\,. \tag{6.31}$$

Remark. Although w_1 and w_2 separately are not periodic, it follows easily from (6.25) that their product n is periodic.

LEMMA 6.8. *The product* n *of any two solutions of* (6.23) *satisfies the equation*

$$Hn=4\alpha\,\partial n\,, \tag{6.32}$$

where H *is the linear operator* (6.19).

Proof: This lemma is a corollary of Lemma 6.6, according to which squares of solutions of (6.23) satisfy (6.32). For, the product $n=w_1w_2$ can be written as a linear combination of $(w_1+w_2)^2$, w_1^2 and w_2^2.

THEOREM 6.9. (a) *The functionals* $\kappa(u,\alpha)$ *are in involution, i.e.,*

$$[\kappa(\alpha),\kappa(\beta)]=0 \tag{6.33}$$

for $\alpha\neq\beta$, *neither* α *nor* β *being equal to an eigenvalue of* L.

(b) *The functionals* $\kappa(u,\alpha)$ *and* $\lambda_j(u)$ *are in involution, i.e.,*

(6.33)′ $$[\kappa(\alpha), \lambda_j]=0\,.$$

Here [.] *is the Poisson bracket defined by* (5.1).

Proof: By definition of the Poisson bracket in (5.1),

$$[\kappa(\alpha), \kappa(\beta)]=(G_{\kappa(\alpha)}, \partial G_{\kappa(\beta)})$$

and

$$[\kappa, \lambda]=(G_\kappa, \partial G_\lambda^{\cdot})\,.$$

Using (6.30) and (6.16) we can write the above expressions as

$$\tfrac{1}{36}\kappa(\alpha)\kappa(\beta)(n(\alpha), \partial n(\beta))$$

and

$$\tfrac{1}{36}\kappa(n, \partial w^2)\,.$$

According to Lemma 6.6.

$$Hw^2=4\lambda\,\partial w^2\,,$$

and according to Lemma 6.8.

$$Hn=4\alpha\,\partial n\,.$$

Using these relations, and the antisymmetry of ∂ and H, we can complete the proof of Theorem 6.9 along the lines of the proof of Theorem 6.4.

For the next step we need the following observation, easily deduced using the standard tools of the theory of differential equations:

w_1 *and* w_2, *measured in the* C^1 *norm, depend Lipschitz continuously on u, measured in the* C^1 *norm.* From this it follows by the Banach space version of the Picard existence theorem that the differential equation

(6.34) $$u_t=\partial n(\alpha, u)$$

has a unique solution, local in time, for arbitrary continuous initial function u_0.

THEOREM 6.10. (a) *The flows (6.34) corresponding to different values of* α *commute.*

(b) *The spectrum of* L *does not change under the flows (6.34).*

(c) *For* C^∞ *initial data, the functionals* $F_m(u)$ *introduced in Section 3 are conserved under the flows (6.34).*

(d) *For* C^1 *initial data, the flows (6.34) exist for all time.*

Proof: Parts (a) and (b) follow from parts (a) and (b) of Theorem 6.9, using Theorem 4.1 of Hamiltonian theory. Part (c) follows from the fact mentioned earlier that for C^∞ potentials the functionals $F_m(u)$ are completely determined by the spectrum of L.

To prove (d) we use Theorem 3.1, according to which the maximum of $u(x)$ can be estimated in terms of $F_0(u)$ and $F_1(u)$. According to part (c) above, $F_0(u(t))$ and $F_1(u(t))$ are independent of t; therefore, we have an estimate for $|u(x,t)|$ which is uniform for all t. This implies that the Lipschitz constant occurring in the dependence of w_1 and w_2 on u is uniform; under these conditions, equation (6.34) has a solution for all t.

It is easy to show, on the basis of perturbation theory, that n defined in Lemma 6.8 depends analytically on α:

$$n=\sum_0^\infty n_j\alpha^j . \tag{6.35}$$

Substituting this into (6.32) gives the following relation among the n_j:

$$Hn_j=4\partial n_{j-1}; \tag{6.36$_j$}$$

here we make the convention that $n_{-1}=1$.

These equations can be easily solved recursively; since the operator H has a one-dimensional null space spanned by n_0, its range is orthogonal to n_0. It is easy to verify that the right sides of $(6.36)_j$ are orthogonal to n_0. Making use of the antisymmetry of ∂, then of equation $(6.36)_1$, then of the antisymmetry of H, and finally of equation $(6.36)_{j-1}$, we can write

$$\begin{aligned}(4\partial n_{j-1}, n_0)&=-(n_{j-1}, 4\partial n_0)\\ &=-(n_{j-1}, Hn_1)=(Hn_{j-1}, n_1)=(4\partial n_{j-2}, n_1) .\end{aligned}$$

Repeating this a number of times we obtain an expression of one of two forms:

$$(\partial n_k, n_k) \qquad \text{or} \qquad (Hn_k, n_k) ;$$

since both ∂ and H are antisymmetric and since H and n_k are real. the above scalar products are zero.

Since by Lemma 6.7. $n(\alpha)$ is the gradient of $6 \log \kappa(\alpha)$. it follows that n_j is the gradient of $(6/j!)(d/d\alpha)^j \log \kappa$. Therefore the flow

$$(6.37)_j \qquad u_t = \partial n_j(u)\,, \qquad j = 0, 1, \cdots,$$

is a Hamiltonian flow.

THEOREM 6.11. (a) *The flows* $(6.37)_j$ *commute with each other.*
(b) *The flows* $(6.37)_j$ *commute with the generalized* KdV *flows.*
(c) *The spectrum of* L *is invariant under the flows* $(6.37)_j$.
(d) *For* C^1 *initial data, the flows* $(6.37)_j$ *exist for all* t.

Proof: Denote the flow governed by the differential equation (6.34) by $S(\alpha, t)$. According to part (a) of Theorem 6.10,

$$S(\alpha, t)S(\beta, r) = S(\beta, r)S(\alpha, t)\,;$$

differentiating this j times with respect to α. and k times with respect to β gives part (a) of Theorem 6.11.

Part (c) of Theorem 6.11 can be deduced by a similar argument from part (b) of Theorem 6.10. Since, as mentioned earlier, the functionals F_m are determined by the spectrum of L. it follows that $F_m(u)$ are conserved along the flows $(6.37)_j$. This implies that F_m and n_j are in involution, which yields part (b) of Theorem 6.11. Part (d) follows by the argument used to prove part (d) of Theorem 6.10.

According to a result of A. Lenard (see equation (5.11) of [8]) the gradients G_m of the functionals F_m satisfy the recursion formula

$$(6.38)_m \qquad HG_m = 4\partial G_{m+1}\,.$$

(One has to replace u by $-6u$ in Lenard's formula to account for a different form of L used in [8].) This is exactly the recursion relation (6.36) satisfied by the functionals n_j, except for a change in sign in the index, showing that the sequence n_j can be regarded as a continuation of G_m to negative indices through the relation

$$(6.39) \qquad G_{-m} = n_{m-2}\,, \qquad m = 2, \cdots.$$

Therefore. the sequence of conserved functionals F_m too can be extended to negative values of the index

$$F_{-m} = \frac{6}{(m-2)!}\left(\frac{d}{d\alpha}\right)^{m-2} \log \kappa(\alpha)\bigg|_{\alpha=0}. \tag{6.40}$$

We conclude these observations with some remarks on $\log \kappa$ as function of complex values of α:

$\log \kappa$ is a 2-valued function of α, with branchpoints of order 2 at the simple eigenvalues of L for functions of period p and $2p$. Furthermore, the growth of κ is exponential:

$$|\log \kappa(\alpha)| < \text{const.} + p\,|\alpha|^{1/2}. \tag{6.41}$$

Inequality (6.41) can be deduced from standard estimates for the theory of ordinary differential equations.

In his paper [23], Peter Ungar employs the function $\log \kappa$ to give a simple proof of the following theorem of Borg [2]:

If all but one of the eigenvalues of $L = \partial^2 + v$, *acting on functions with period* $2p$, *are double, then* v *is constant.*

In his proof, Ungar eliminates the branchpoint at the lowest eigenvalue λ_0 by introducing the function

$$f(z) = \log \kappa(\lambda_0 + z^2).$$

It follows from (6.41) that $f(z)$ has a simple pole at ∞. If λ_0 is the only simple eigenvalue, $f(z)$ is a linear function.

If there are in addition to λ_0 a finite number $2N$ of simple eigenvalues, then $f(z)$ is a double-valued analytic function with branch points of order 2 at the simple eigenvalues, and a simple pole at ∞; the theory of such functions plays an important role in the work of McKean and van Moerbeke.

The last 3 results of this section deal with the spectrum of L when u is one of the special potentials satisfying an equation of the form (3.9).

The next theorem is a slight sharpening of one due to J. Moser and the author given in [14]:

THEOREM 6.12. *Let* u *be a solution of* (3.9):

$$G(u) = \sum_{-1}^{N} a_i G_i(u) = 0, \qquad a_N = 1. \tag{6.42}$$

Then all but $2N+1$ *eigenvalues of the operator* L *given by* (6.11) *are double.*

Proof: Multiply equation $(6.10)_j$ by a_j/c_j and sum. We introduce the abbreviation

$$\sum \frac{a_j}{c_j} B_j = B ; \tag{6.43}$$

using (6.12), we can write the resulting relation as

$$BL - LB = \partial \sum a_j G_j . \tag{6.44}$$

If u satisfies (6.42),

$$BL - LB = 0 , \tag{6.45}$$

i.e., B commutes with L. Now suppose that λ is a simple eigenvalue of L. with the corresponding eigenvector w. It follows from (6.45) that

$$LBw = BLw = \lambda Bw ,$$

i.e., that Bw is also an eigenfunction of L. Since λ is assumed simple,

$$Bw = \kappa w . \tag{6.46}$$

B, being a linear combination of antisymmetric operators, is itself antisymmetric; so its spectrum is pure imaginary. Therefore, κ in (6.39), which is real, is zero:

$$Bw = 0 . \tag{6.47}$$

Thus all eigenfunctions with simple eigenvalues satisfy the *same* equation (6.47). Now eigenfunctions corresponding to distinct eigenvalues are linearly independent; on the other hand, B being a linear differential operator of order $2N+1$ cannot have more than $2N+1$ linearly independent solutions. This proves Theorem 6.12.

Remark. I suspect, but cannot prove, that the $2N+1$ simple eigenvalues are the lowest ones. Since the lowest eigenvalue is always simple, this is the case for $N=0$. In [14], an argument was given to show that for $N=1$ the lowest 3 eigenvalues are indeed simple. The argument is based on the observation that the eigenfunction w, in addition to satisfying the equation (6.47) of order $2N+1$, also satisfies the second-order equation (6.21). This equation can be used to express all derivatives of w of any order as a linear combination of w and w_x. Using this in (6.47) yields a first order equation for w of the form

$$aw_x + bw = 0\,, \tag{6.48}$$

where $a = a(\lambda, x)$ is a polynomial in λ of degree N whose leading coefficient is 1.

All but the lowest eigenfunctions w have zeros; in fact the $(2m-1)$-st and $2m$-th have $2m$ zeros. Let x_0 be such a zero: $w(x_0) = 0$. Since w satisfies $Lw = \lambda w$, and is not equal to 0, $w_x(x_0) \neq 0$; it follows then from (6.48) that

$$a(\lambda, x_0) = 0\,. \tag{6.49}$$

Since a is a polynomial in λ, we can get from this relation an upper bound for the simple eigenvalues λ in terms of estimates for u and its derivatives up to order $2N-1$.

We return now to Theorem 6.12. We have seen in Section 3, equation (3.19), that solutions of (3.9) are elliptic functions; it follows therefore from Theorem 6.12 that if the potential u occurring in L of the form (6.11) is such an elliptic function, then all but the first 3 eigenvalues of L are double. This fact is known (see Magnus and Winkler [16]). In a remarkable paper [11], H. Hochstadt has raised and answered the converse proposition: i.e., he has shown that if all but the first 3 eigenvalues of L are double, then the potential u is an elliptic function. Furthermore, he has shown that if L has only a finite number of simple eigenvalues, then u is C^∞.

In [14], after proving that if u satisfies (3.9), then L has only a finite number of simple eigenvalues, I raised the question whether the converse might be true. This has been answered affirmatively by W. Goldberg [9], using the method of Hochstadt. Another proof has been given by H. Flaschka [4]. Here is yet another proof of this fact by McKean and van Moerbeke, based on the following observation:

If L_0 has only $2N+1$ simple eigenvalues, the set of potentials u for which L has the same spectrum as L_0 forms an N-parameter family.

The proof of this relies on Borg's uniqueness theorem in [2].

THEOREM 6.13. *Suppose L_0 has only $2N+1$ simple eigenvalues; then u_0 satisfies an equation of the form* (6.42).

Proof: According to the theorem of Hochstadt quoted above, such a u_0 is C^∞; therefore u_0 belongs to the domain of each $S_m(t)$, the solution operator of the m-th generalized KdV flow. It follows from Theorem 6.3 that if we take u to be

$$u = u(t_0, \cdots, t_N) = \prod_0^N S_m(t_m) u_0 , \tag{6.50}$$

then L is unitarily equivalent with L_0. According to the observation quoted above, the set of such u forms an N-parameter family. The parameters are continuous functions of the $N+1$ parameters appearing in (6.50); hence the parameters appearing in (6.50) cannot all be independent. This is so if and only if the directions in which S_m start are linearly dependent; these directions are $G_m(u_0)$, and therefore there is a linear relation among the $G_m(u_0)$:

$$\sum_0^N a_m G_m(u_0) = 0 .$$

This is an equation of the form (6.42). We note that a_N cannot be zero; otherwise, according to Theorem 6.12, the operator L_0 would have only $2J+1$ simple eigenvalues, where J is the index of the highest nonzero a_m. This completes the proof of Theorem 6.13.

Theorem 6.12 shows that if u satisfies equation (6.42), the eigenfunctions of L corresponding to simple eigenvalues satisfy a differential equation (6.47). It is remarkable that also the squares of eigenfunctions satisfy an equation of the same kind.

THEOREM 6.14. *Suppose* u *is a solution of* (6.42):

$$G(u) = 0 .$$

L *denotes the operator* (6.11).

(a) *Let* w *be an eigenfunction of* L *corresponding to a simple eigenvalue. Then*

$$G'(u)\, \partial w^2 = 0 . \tag{6.51$_1$}$$

(b) *Let* w_1 *and* w_2 *be a pair of orthonormal eigenfunctions of* L *corresponding to a double eigenvalue. Then*

$$G'(u)\, \partial (w_1^2 + w_2^2) = 0 . \tag{6.51$_2$}$$

Proof: We appeal to Theorem 2.2 about time dependent solutions of nonlinear equations of the form

$$u_t = K(u)\,;$$

according to Theorem 2.2, if $F(u)$ is a conserved quantity for this equation, its gradient $G_F(u)$ satisfies the linear equation

$$\left(\frac{\partial}{\partial t}+M^*\right)G_F=0\,, \tag{6.52}$$

where M^* is the adjoint of M, the derivative of κ defined in (2.8). In our present situation where u satisfies $G(u)=0$, we regard u as time independent solution of the evolution equation

$$u_t=\partial G(u)\,. \tag{6.53}$$

For this equation, $K=\partial G(u)$ whose derivative is $M=\partial G(u)$. Since by Theorem 2.3 G' is selfadjoint,

$$M^*=-G'\,\partial\,.$$

So it follows from (6.52) that if F is a conserved functional,

$$G'(u)\,\partial G_F=0\,. \tag{6.54}$$

According to Theorem 6.3, the eigenvalues of L are conserved functionals for (6.53); according to Lemma 6.5, if λ is differentiable, $G_\lambda=\frac{1}{6}w^2$. Setting this into (6.54) we obtain $(6.51)_1$. According to the remark following Lemma 6.5, for a double eigenvalue, $\lambda_1+\lambda_2$ is differentiable, and

$$G_{\lambda_1+\lambda_2}=\tfrac{1}{6}(w_1^2+w_2^2)\,.$$

Setting this into (6.47) we obtain $(6.51)_2$; this completes the proof of Theorem 6.14.

For G of the form (6.42), G' is an operator of order $2N+1$; therefore its null space is at most $(2N+1)$-dimensional. It follows therefore that there is a linear relation between $w_0^2, \cdots, w_{2N}^2$, the squares of the simple eigenfunctions, and $w_{(1)}^2+w_{(2)}^2$ for any double eigenfunction. I suspect that the latter can be expressed as a linear combination of the former. This is certainly so in the trivial case $N=0$, when $w_0^2=1$, and $w_{(1)}=\sin nx$, $w_{(2)}=\cos nx$.

Appendix

James M. Hyman

In this appendix we describe how the construction of special solutions of the KdV equation which minimize F_2 subject to the constraint $F_i = A_i$ was implemented numerically. We saw in Section 3 that a minimizing function satisfies the Euler equation

$$G_2 + \sum_{-1}^{1} a_i G_i = 0 . \tag{A.1}$$

A solution of (A.1) is an extremal for

$$F = F_2 + \sum_{-1}^{1} a_i F_i , \tag{A.2}$$

and presumably can be obtained by minimizing that functional. This is indeed what we did. We chose the constants a_{-1}, a_0, a_1, then discretized the functional (A.2) by specifying u at N equidistant points and expressed the first and second derivatives of u in (A.2) by difference quotients. The resulting function of N variables was minimized using A. Jameson's version[1] of the Fletcher-Powell-Davidov algorithm [5]. The resulting discretized function is a somewhat crude approximation to the function, u, we are looking for. To get u more accurate we proceeded as follows:

We are looking for a periodic solution of (A.1); since this is a fourth order equation, its solutions are parametrized by their four Cauchy data at, say, $x=0$. Periodicity requires the Cauchy data at $x=p$ to be equal to the Cauchy data at $x=0$. We sought to satisfy this requirement by a sequence of approximations constructed by "shooting". The Cauchy data of u_{n+1} at $x=0$ were chosen by applying the rule of false position to $u_n(p)-u_n(0)$, $u'_n(p)-u'_n(0)$, $u_{n-1}(p)-u_{n-1}(0)$ and $u'_{n-1}(p)-u'_{n-1}(0)$. Then $u_{n+1}(p)$ and $u'_{n+1}(p)$ were computed numerically. For this purpose we used the ODE package developed by A. Hindmarsh [10].

The following observations were helpful:

(a) Instead of matching all four Cauchy data for the fourth order equation (A.1), it sufficed to match only two, since according to Lemma 5.12 of Section 5 there exist two polynomials J_0 and J_1 of the Cauchy data which are independent of x.

(b) The initial guesses for the Cauchy data come from the approximate

---

[1] My thanks are due to A. Jameson for acquainting me with his FPD package.

solution obtained by the variational procedure. Without such a good initial guess we were unable to construct a periodic solution of (A.1).

The periodic solution of (A.1) constructed above was then used as initial value data. We solved numerically the initial value problem for the KdV equation using Fred Tappert's[2] method [22] and code. The accuracy of the solution was monitored by checking the constancy of the functionals F_0, F_1, F_2, J_0 and J_1, and the extent to which the solution satisfied the ODE (A.1). The solution constructed by Tappert's method passed these tests of accuracy reasonably well. Solutions constructed using earlier methods of Kruskal and Zabusky [25] and Vliegenthart [24] were less accurate and were not used in this study. The finest vindication of Tappert's KdV solver was that after a finite elapse of time the solution resumed its initial shape, in a shifted position. During the intervening time the shape of the solution underwent considerable gyrations.

It was proved rigorously at the end of Section 5 that the shape of the solution recurs exactly after a finite time. Here is another proof, based on an idea of Lax in [13]:

According to Lemma 5.13 a solution $u(x, t)$ of KdV whose initial values satisfy equation (A.1) also satisfies

$$J_0 = c_0, \qquad J_1 = c_1, \tag{A.3}$$

where c_0 and c_1 are constants and J_0 and J_1 are polynomials in u and its x derivatives up to order 3. J_0 and J_1 can be calculated explicitly from formulas (3.17), (5.17) and (5.24).

Equations (A.3) can be solved to express u_{xx} and u_{xxx} in terms of u and u_x. These expressions become particularly simple when $u_x = 0$; they are of the form

$$u_{xx} = P^{1/2}(u), \tag{A.4}$$

and

$$u_{xxx} = R^{1/2}(u), \tag{A.5}$$

where P is a polynomial in u and R is a polynomial in u and $P^{1/2}$.

Let $m = m(t)$ denote the value of the maximum of $u = u(x, t)$ with respect to x. At the point $y = y(t)$ where the maximum is achieved,

$$u_x(y(t), t) = 0. \tag{A.6}$$

Therefore formulas (A.4) and (A.5) are valid at this point. Note that since at a maximum $u_{xx} < 0$, the square root in (A.4) has to be taken as the negative root.

---

[2] My thanks are due to Fred Tappert for acquainting me with his KdV solver.

Differentiating

$$m(t) = u(y(t), t)\,.$$

we get

$$m_t = u_x y_t + u_t\,.$$

From the KdV equation we have

$$u_t = -uu_x - u_{xxx}\,.$$

Combining these two and using (A.6) we obtain

$$m_t = -u_{xxx}\,.$$

Using (A.5) we deduce that

(A.7) $$m_t = -R^{1/2}(m)\,.$$

Solutions of (A.7) behave as follows:

If at $t=0$ the value of u_{xxx} is greater than 0 at $y(0)$, then $m(t)$ decreases until it reaches the nearest zero of $R(m)$. After that u_{xxx} changes sign and $m(t)$ starts increasing until it reaches the next zero of $R(m)$. After that u_{xxx} changes sign again and $m(t)$ starts decreasing until it reaches the value $m(0)$. At this time T, and at the point of $y(T)$, the Cauchy data u, u_x, u_{xx} and u_{xxx} have the same value as at $t=0$ and at $y(0)$. Since both $u(x,0)$ and $u(x,T)$ are solutions of (A.1), it follows that

(A.8) $$u(x, T) = u(x-L, 0)\,,$$

where

(A.9) $$L = y(T) - y(0)\,.$$

Since equation (A.7) can be integrated by quadrature, T can be expressed as

(A.10) $$T = 2\int_a^b \frac{dm}{R^{1/2}(m)}\,,$$

where

$$R(a) = R(b) = 0\,.$$

Specifically, a and b are those zeros of $R(m)$ between which the maximum of $u(x,0)$ is located.

The value of $L(T)$ can also be determined by quadrature. Differentiating (A.6) with respect to t gives

$$u_{xx} y_t + u_{xt} = 0\,,$$

so that

$$y_t = -\frac{u_{xt}}{u_{xx}}\,. \tag{A.11}$$

Differentiating KdV with respect to x gives

$$u_{tx} = -u_x^2 - uu_{xx} - u_{xxxx}\,.$$

Substituting this into (A.11) we get, using $u_x = 0$,

$$y_t = u + \frac{u_{xxxx}}{u_{xx}}\,. \tag{A.12}$$

Using (A.1) we can express u_{xxxx} in terms of derivatives of u of lower order. The values of these at $x = y(t)$ have already been expressed in terms of m; so from (A.12) we can get a relation of the form

$$y_t = Y(m)\,, \tag{A.13}$$

where $Y(m)$ is an explicitly computable function of m. Integrating (A.13) gives, using (A.7),

$$L = y(T) - y(0) = \int_0^T y_t\,dt = \int_0^t Y(m)\,dt$$

$$= \int Y(m)\frac{dm}{dt}\,dt = \int Y(m)R^{1/2}(m)\,dm\,.$$

Hence, altogether,

$$L = 2\int_a^b Y(m)R^{1/2}(m)\,dm\,. \tag{A.14}$$

We conclude by presenting some numerical data:

Figures 1 and 2 give the graphs of functions minimizing $F(u)$ of the form

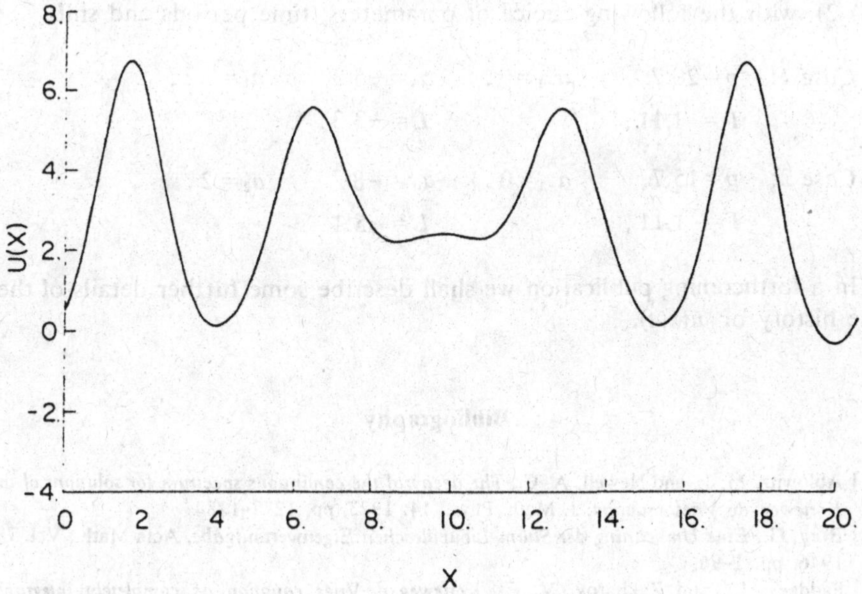

Figure 1. A periodic solution to equation (A.1) with $a_{-1}=0$, $a_0=-5$, $a_1=-2$.

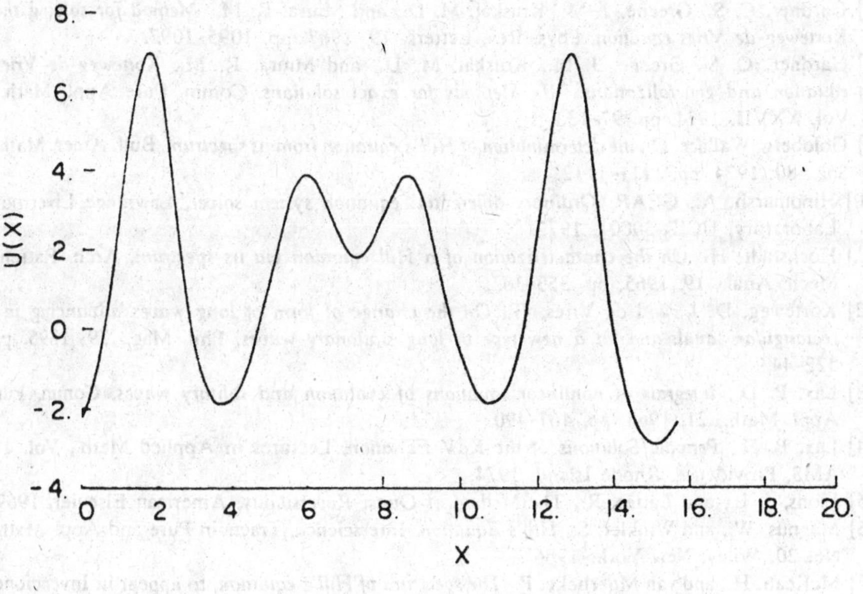

Figure 2. A periodic solution to equation (A.2) with $a_{-1}=0$, $a_0=-8$, $a_2=2$.

(A.2), with the following choice of parameters, time periods and shift:

Case 1: $p=20.7$, $a_{-1}=0$, $a_0=-5$, $a_1=-2$,
$T=1.11$, $L=-3.3$,

Case 2: $p=15.7$, $a_{-1}=0$, $a_0=-8$, $a_2=2$,
$T=1.11$, $L=-5.1$.

In a forthcoming publication we shall describe some further details of the time history of $u(x,t)$.

Bibliography

[1] Ablowitz, M. J., and Newell, A. C., *The decay of the continuous spectrum for solutions of the Korteweg-de Vries equation*, J. Math. Phys., 14, 1973, pp. 1277–1284.

[2] Borg, G., *Eine Umkehrung der Sturm-Liouvilleschen Eigenwertaufgabe*, Acta Math., Vol. 78, 1946, pp. 1–96.

[3] Faddeev, L., and Zakharov, V. E., *Korteweg-de Vries equation as completely integrable Hamiltonian system*, Funktsional. Anal. i Prilozhen. 5, 1971, pp. 18–27. (In Russian.)

[4] Flaschka, H., *Integrability of the Toda lattice*, Theoretical Physics, Vol. 51, 1974, p. 703.

[5] Fletcher, R., and Powell, M. J. D., *A rapidly convergent descent method for minimization*, Comput. J., Vol. 6, 1963/64, pp. 163–168.

[6] Gardner, C. S., *Korteweg-de Vries equation and generalizations. IV. The Korteweg-de Vries equation as a Hamiltonian system*, J. Mathematical Phys. 12, 1971, pp. 1548–1551.

[7] Gardner, C. S., Greene, J. M., Kruskal, M. D., and Miura, R. M., *Method for solving the Korteweg-de Vries equation*, Phys. Rev. Letters, 19, 1967, pp. 1095–1097.

[8] Gardner, C. S., Greene, J. M., Kruskal, M. D., and Miura, R. M., *Korteweg-de Vries equation and generalizations. VI. Methods for exact solutions*, Comm. Pure Appl. Math., Vol. XXVII, 1974, pp. 97–133.

[9] Goldberg, Wallace, *On the determination of Hill's equation from its spectrum*, Bull. Amer. Math. Soc., 80, 1974, pp. 1111–1112.

[10] Hindmarsh, A., *GEAR: Ordinary differential equation system solver*, Lawrence Livermore Laboratory, UCID-30001, 1972.

[11] Hochstadt, H., *On the characterization of a Hill equation via its spectrum*, Arch. Rational Mech. Anal., 19, 1965, pp. 353–362.

[12] Korteweg, D. J., and de Vries, G., *On the change of form of long waves advancing in a rectangular canal, and on a new type of long stationary waves*, Phil. Mag., 39, 1895, pp. 422–443.

[13] Lax, P. D., *Integrals of nonlinear equations of evolution and solitary waves*, Comm. Pure Appl. Math., 21, 1968, pp. 467–490.

[14] Lax, P. D., *Periodic Solutions of the KdV Equation*, Lectures in Applied Math., Vol. 15, AMS, Providence, Rhode Island, 1974.

[15] Lions, J. L., and Lattès, R., *The Method of Quasi-Reversibility*, American Elsevier, 1969.

[16] Magnus, W., and Winkler, S., *Hill's Equation*, Interscience, Tracts in Pure and Appl. Math., No. 20, Wiley, New York, 1966.

[17] McKean, H., and van Moerbeke, P., *The spectrum of Hill's equation*, to appear in Invenciones Mat.

[18] Miura, R., Gardner, C. S., and Kruskal, M. D., *Korteweg-de Vries equation and generalizations. II. Existence of conservation laws and constants of motion*, J. Math. Phys., 9, 1968, pp. 1204–1209.

[19] Novikov, S. P., *The periodic problem for the Korteweg-de Vries equation. I.* Funk. Anal. Prilozh., 8, No. 3, 1974, pp. 54–66.

[20] Sjoberg, A., *On the Korteweg-de Vries equation*, J. Math. Anal. Appl., 29, 1970, pp. 569–579.

[21] Tanaka, S., *Korteweg-de Vries equation: Asymptotic behavior of solutions*, to appear.

[22] Tappert, F., *Numerical Solution of the KdV Equation and its Generalizations by the Split-Step Fourier Method*, Lectures Appl. Math., Vol. 15, AMS, Providence, Rhode Island, 1974.

[23] Ungar, P., *Stable Hill equation*, Comm. Pure Appl. Math., Vol. XIV, 1961, pp. 707–710.

[24] Vliegenthart, A. C., *On finite difference methods for the Korteweg-de Vries equation*, J. of Engineering Math., Vol. 5, 1971, pp. 137–155.

[25] Zabusky, N. J., and Kruskal, M. D., *Interaction of "solitons" in a collisionless plasma and the recurrence of initial states*, Phys. Rev. Letters, 15, 1965, pp. 240–243.

Received November, 1974.

SIAM REVIEW
Vol. 18, No. 3, July 1976

ALMOST PERIODIC SOLUTIONS OF THE KdV EQUATION*

Dedicated to Joachim Weyl, on the occasion of his 60th birthday, and in recognition of his role in nurturing applied mathematics. Through his influence on science policy and his eloquent advocacy of applications, he has encouraged many young mathematicians in the United States to choose secular rather than monastic mathematics.

PETER D. LAX†

Abstract. In this talk we discuss the almost periodic behavior in time of space periodic solutions of the KdV equation

$$u_t + uu_x + u_{xxx} = 0.$$

We present a new proof, based on a recursion relation of Lenart, for the existence of an infinite sequence of conserved functionals $F_n(u)$ of form $\int P_n(u)\,dx$, P_n a polynomial in u and its derivatives; the existence of such functionals is due to Kruskal, Zabusky, Miura and Gardner. We review and extend the following result of the speaker: the functions u minimizing $F_{N+1}(u)$ subject to the constraints $F_j(u) = A_j$, $j = 0, \cdots, N$, form N-dimensional tori which are invariant under the KdV flow. The extension consists of showing that for certain ranges of the constraining parameters A_j the functional $F_{N+1}(u)$ has minimax stationary points; these too form invariant N-tori. The Hamiltonian structure of the KdV equation, discovered by Gardner and also by Faddeev and Zakharov, which is used in these studies, is described briefly. In an Appendix, M. Hyman describes numerical studies of the stability of some invariant 2-tori for the KdV flow; the numerical evidence points to stability.

1. Introduction. A recent series of investigations of nonlinear wave motion, commencing with Kruskal and Zabusky's paper [29], have led to the unexpected discovery that an astonishingly large number of important differential equations of mathematical physics are completely integrable Hamiltonian systems. Included among these are the Korteweg–de Vries (KdV) and Boussinesque equations for waves in shallow water, the equations governing self-induced transparency, and self-focusing and self-modulating waves in optics, the vibrations of the Toda lattice, the motion of particles under an inverse square potential, and some others.

These equations have been studied under two kinds of boundary conditions:

(i) Solutions are required to be periodic in space.

(ii) Solutions propagate in free space but are required to vanish at ∞.

We shall call (i) the compact, (ii) the noncompact case. It turns out that solutions behave quite differently in the two cases. In terms of Hamiltonian theory the difference can be explained in the following way:

A Hamiltonian system

$$\frac{dq_j}{dt} = H_{p_j}, \qquad \frac{dp_j}{dt} = -H_{q_j} \tag{1.1}$$

* Received by the editors April 30, 1975. Presented by invitation at the symposium on Nonlinear Waves I, supported in part by the Office of Naval Research, Department of the Navy, at the 1974 Fall Meeting of Society for Industrial and Applied Mathematics held at Alexandria, Virginia, October 23–25, 1974.

† Department of Mathematics, Courant Institute of Mathematical Sciences, New York University, New York, New York 10012. This work was supported by the Energy and Research Development Administration under Contract AT(11-1)-3077.

is completely integrable if there is a canonical transformation introducing new conjugate variables $\bar{p}_j$, $\bar{q}_j$ and a new Hamiltonian $\bar{H}$ so that $\bar{H}$ is a function of $\bar{p}$ alone and is independent of $\bar{q}$. The Hamiltonian system in these new variables is

$$\frac{d\bar{p}_j}{dt} = -H_{\bar{q}_j} = 0$$

which implies that each $\bar{p}_j$ is constant; therefore

$$\frac{d\bar{q}_j}{dt} = H_{\bar{p}_j} = \text{const.},$$

so that

$$\bar{q}_j(t) = \bar{q}_j(0) + tH_{\bar{p}_j}. \tag{1.2}$$

The difference between the compact and noncompact case is this: in the compact case the $\bar{q}_j$ are angle variables, i.e., the original variables are periodic functions of the $\bar{q}_j$, whereas in the noncompact case there is no such periodicity. When we express the original coordinates q_j and p_j in terms of $\bar{q}_j$ and $\bar{p}_j$, we see that in the compact case every flow is a function of periodic motions; since the periods are, in general, incommensurable, we see that flows in the compact case are *almost periodic*. In the noncompact case, on the other hand, the time dependence in (1.2) represents a genuine linear growth which describes the manner in which particles or waves tend to infinity.

Starting with the work of Gardner, Greene, Kruskal and Miura [10], Faddeev and Zakharov [6] have shown that the KdV equation

$$u_t + uu_x + u_{xxx} = 0 \tag{1.3}$$

constitutes, for solutions defined on the whole real axis and zero at $x = \pm\infty$, a completely integrable Hamiltonian system, whose action and angle variables are simply related to the so-called *scattering data* of the associated Schrödinger operator, see (3.22). For solutions which are periodic with respect to x no such formulas are known; nevertheless it is strongly suspected that in this case too we are dealing with a completely integrable Hamiltonian system; the following items are evidence for this:

1. numerical calculations by Kruskal and Zabusky which indicate that solutions of KdV which are periodic in x are almost periodic in t, [29];
2. the construction of infinitely many conserved quantities by Gardner, Kruskal and Miura;
3. Gardner's observation that these functionals are in involution;
4. the existence of compact submanifolds of arbitrary finite dimension which are invariant under the KdV equation.

In this talk we review these facts:

In § 2 we describe a relation between conserved functionals and invariant submanifolds of flows. In § 3 we give a new proof of the existence of infinitely many conserved functionals, and show that they are in involution. In § 4 we display the Hamiltonian structure of the KdV equation; in § 5 we describe the construction with the aid of a minimum problem of invariant submanifolds which are N-dimensional tori, and on which solutions of KdV are almost periodic in t. In an Appendix, M. Hyman describes a calculation of some invariant 2-dimensional tori, and demonstrates by means of another calculation the remarkable stability of these manifolds.

At the end of § 5 we give an indication how invariant tori might be constructed with the aid of a minimax problem.

2. Equations of evolution, invariant submanifolds and conserved functionals. We consider equations of evolution of the form

(2.1) $$\frac{d}{dt}u = K(u),$$

K an operator, in general nonlinear, mapping a linear space into itself. We assume that the initial value problem is properly posed, i.e., that solutions of (2.1) are uniquely determined by their values at $t = 0$, that the initial value of u can be prescribed arbitrarily, and that solutions exist for all t. The mapping of initial data of solutions of (2.1) into data at time t can be thought of as a *flow*.

Let $F(u)$ be a *functional*, i.e., a numerically-valued function, in general nonlinear, defined on the underlying linear space. F is called differentiable if the directional derivative

$$\lim_{\varepsilon \to 0} \frac{F(u+\varepsilon v) - F(u)}{\varepsilon}$$

exists for all u and v and is a linear functional of v. We assume that our linear space is equipped with a *scalar product* $(\cdot, \cdot)$; since linear functionals can be expressed as scalar products, we can write

(2.2) $$\frac{d}{d\varepsilon}F(u+\varepsilon v)\Big|_{\varepsilon=0} = (G_F(u), v).$$

$G_F(u)$ is called the *gradient* of F at u with respect to the specified scalar product. G_F is a nonlinear operator.

Let F be a functional for which $F(u(t))$ is independent of t for all solutions of (2.1). Such an F is called a *conserved functional* of the flow.

Differentiating, using the definition of gradient and equation (2.1) gives

(2.3) $$\frac{d}{dt}F(u(t)) = (G_F(u), u_t) = (G_F(u), K(u));$$

we deduce the following theorems from this.

THEOREM 2.1. *$F(u)$ is an invariant functional of* (2.1) *if for all u,*

$$(G_F(u), K(u)) = 0.$$

THEOREM 2.2. *Let F be a differentiable conserved functional of the flow* (2.1). *Then solutions of*

(2.4) $$G_F(u) = 0$$

form an invariant manifold of the flow (2.1).

A formal proof of Theorem 2.2 is given in [18]; here is a simple intuitive argument:

Solutions of (2.4) are stationary points of $F(u)$, i.e., points u_0 such that for every smooth curve $u(\varepsilon)$ issuing from $u_0 = u(0)$,

$$\frac{d}{d\varepsilon} F(u(\varepsilon))\Big|_{\varepsilon=0} = 0. \tag{2.5}$$

Since the flow carries smooth curves into smooth curves, and conserves the values of F, (2.5) is true at all points along the trajectory issuing from u_o.

3. Conserved functionals of the KdV equation. The usefulness of Theorem 2.2 depends on the existence of many conserved functionals whose gradients have tractable null sets. The KdV equation is rich in such functionals; three of them are classical:

$$\begin{aligned} F_0(u) &= \int 3u\,dx, \\ F_1(u) &= \int \tfrac{1}{2}u^2\,dx, \\ F_2(u) &= \int (\tfrac{1}{6}u^3 - \tfrac{1}{2}u_x^2)\,dx. \end{aligned} \tag{3.1}$$

The gradients of these functionals are

$$\begin{aligned} G_0 &= 3, \\ G_1 &= u, \\ G_2 &= \tfrac{1}{2}u^2 + u_{xx}. \end{aligned} \tag{3.2}$$

To prove that the functionals (3.1) are conserved for KdV we have to verify condition (2.3) of Theorem 2.1, with G_F given by (3.2) and $K(u) = -u_{xxx} - uu_x$.

Kruskal and Zabusky made the remarkable discovery that there are further conserved functionals, of which

$$F_3(u) = \int (\tfrac{5}{72}u^4 - \tfrac{5}{6}uu_x^2 + \tfrac{1}{2}u_{xx}^2)\,dx \tag{3.1$_3$}$$

is the first example. Eventually Gardner, Kruskal and Miura proved, see [11], that these four are the beginning of an infinite sequence of conserved functionals F_n of the form

$$F_n(u) = \int P_n\,dx, \tag{3.1$_n$}$$

P_n a polynomial in u and its derivatives up to order $n-1$. We give now a new proof of the existence of these functionals F_n, based on a remarkable recursion formula for their gradients discovered by Andrew Lenart some years ago, first published in [11]:

$$HG_n = \partial G_{n+1}, \tag{3.3}$$

where H is the third order operator

$$H=\partial^3+\tfrac{2}{3}u\partial+\tfrac{1}{3}u_x, \qquad \partial=d/dx; \tag{3.4}$$

note that H is antisymmetric:

$$H^*=-H. \tag{3.5}$$

It is easy to verify by a calculation that (3.3) holds for $n=0$, 1 and 2. Next we show the following theorem.

THEOREM 3.1. *There exists a sequence G_n of polynomials in u and its derivatives up to order $2n-2$ which satisfy* (3.3); *G_n is uniquely determined if we set the constant term equal to zero.*

Proof. We use the following simple calculus lemma: Suppose that Q is a polynomial in derivatives of u up to order j, such that for every periodic function u of period p

$$\int_0^p Q(u)\,dx=0.$$

Then there exists a polynomial G in derivatives of u up to order $j-1$ such that

$$Q=\partial J.$$

We assume that G_j has been constructed for all $j\leqq n$; to construct G_{n+1} we have to solve (3.3). According to the above calculus lemma, we have to show that for all u,

$$\int HG_n\,dx=0.$$

Since according to $(3.2)_0$, $G_0=3$, we can rewrite this equation as

$$(HG_n, G_0)=0. \tag{3.6}$$

Using repeatedly the antisymmetry of H and ∂ and relation (3.3) we can write the following sequence of identities:

$$\begin{aligned}(HG_n, G_0)&=-(G_n, HG_0)=-(G_n, \partial G_1)\\ &=(\partial G_n, G_1)=(HG_{n-1}, G_1)=\cdots\\ &=\begin{cases}(HG_{n/2}, G_{n/2}) & \text{if } n \text{ is even,}\\ (\partial G_{(n+1)/2}, G_{(n+1)/2}) & \text{if } n \text{ is odd.}\end{cases}\end{aligned}$$

Because H and ∂ are antisymmetric, both of these expressions are zero; this proves that the compatibility relation (3.6) is fulfilled and completes the proof of Theorem 3.1.

The argument presented above can be used, with trivial alterations, to prove the following result of Gardner [9] which plays an important role in the Hamiltonian theory of the KdV equation.

THEOREM 3.2. *For all m and n and all u,*

$$(G_m, \partial G_n)=0. \tag{3.7}$$

Next we show the following theorem.

THEOREM 3.3. *G_n is the gradient of a functional of form* $(3.1)_n$.

Proof. Just as in finite-dimensional spaces, gradients G are characterized by the *symmetry* of their derivatives. Set

$$\frac{d}{d\varepsilon} G(u+\varepsilon v)\bigg|_{\varepsilon=0} = N(u)v. \tag{3.8}$$

Suppose G is the gradient of F; then

$$\frac{d^2}{d\varepsilon\, d\eta} F(u+\varepsilon v+\eta w)\bigg|_{\varepsilon=\eta=0}$$

is equal to

$$(N(u)w, u) \quad \text{or} \quad (N(u)v, w)$$

depending on whether we differentiate first with respect to ε or η. Since mixed partials are equal, the symmetry of $N(u)$ follows.

To prove the converse, take a smooth path $u(\varepsilon)$ connecting u_0 and u_1; denote $(d/d\varepsilon)u$ by u_ε. Integrating

$$\frac{d}{d\varepsilon} F(u(\varepsilon)) = (G(u), u_\varepsilon)$$

with respect to ε we get

$$F(u_1) - F(u_0) = \int (G(u), u_\varepsilon)\, d\varepsilon. \tag{3.9}$$

Given G whose derivative is symmetric we define F by formula (3.9); to verify that G is the gradient of F so defined we have to show that definition (3.9) is independent of the path connecting u_0 and u_1. To verify this independence we consider one parameter families of curves $u = u(\varepsilon, \eta)$ with common endpoints u_0 and u_1. The derivative of the right side with respect to η is

$$\int (G(u), u_{\varepsilon\eta})\, d\varepsilon + \int (N(u)u_\eta, u_\varepsilon)\, d\varepsilon.$$

We integrate the first term by parts; there are no boundary terms since $u_\eta = 0$ at the endpoints, and so we get

$$-\int (N(u)u_\varepsilon, u_\eta)\, d\varepsilon + \int (N(u)u_\eta, u_\varepsilon)\, d\varepsilon.$$

Clearly this quantity is zero if N is a symmetric operator.

Next we compute the derivative of G_n; let $u(\varepsilon)$ be a smooth curve, and denote differentiation with respect to ε by prime. Differentiating (3.3) with respect to ε we get

$$H'G_n + HG_n' = \partial G_{n+1}'. \tag{3.10}$$

Using the definition (3.4) of H we have

$$H'G_n = \tfrac{2}{3}u'G_{n_x} + \tfrac{1}{3}u_x'G_n$$

which can be rewritten as $K_n u'$, where K_n is the operator

$$K_n = \tfrac{1}{3}G_n\partial + \tfrac{2}{3}G_{n_x}. \tag{3.11}$$

Denote the derivative of G_n by N_n; by definition of derivative

$$G'_n = N_n u', \qquad G'_{n+1} = N_{n+1} u';$$

substituting this and the previous relation into (3.10) we get

$$K_n + HN_n = \partial N_{n+1}. \tag{3.12}$$

LEMMA 3.4.

$$K_{n-1}H - K_n\partial \tag{3.13}$$

is self-adjoint.

Proof. An explicit calculation using the definition (3.4) of H and $(3.11)_{n-1}$ of K_{n-1} gives the following formula:

$$3K_{n-1}H = G_{n-1}\partial^4 + 2G_{n-1_x}\partial^3 + \tfrac{2}{3}G_{n-1}u\partial^2 + (G_{n-1}u_x + \tfrac{4}{3}G_{n-1_x}u)\partial + \cdots.$$

A slightly more tedious calculation gives

$$3K_{n-1}H - 3(K_{n-1}H)^* = a_{n-1}\partial + b_{n-1}, \tag{3.14}$$

where

$$a_{n-1} = 2G_{n-1_{xxx}} + \tfrac{4}{3}G_{n-1_x} + \tfrac{2}{3}G_{n-1}u_x$$

and

$$b_{n-1} = G_{n-1_{xxxx}} + \tfrac{2}{3}G_{n-1_{xx}}u + G_{n-1_x}u_x + \tfrac{1}{2}G_{n-1}u_{XX}.$$

We observe, using (3.4), that

$$a_{n-1} = 2HG_{n-1}, \qquad b_{n-1} = \partial HG_{n-1}.$$

Using relation $(3.3)_{n-1}$ we get

$$a_{n-1} = 2G_{n_x}, \quad b_{n-1} = G_{n_{xx}}.$$

Substituting these into (3.14) we get

$$K_{n-1}H - (K_{n-1}H)^* = \tfrac{2}{3}G_{n_x} + \tfrac{1}{3}G_{n_{xx}}. \tag{3.15}$$

On the other hand, a straightforward calculation gives

$$K_n\partial - (K_n\partial)^* = \tfrac{2}{3}G_{n_x} + \tfrac{1}{3}G_{n_{xx}}. \tag{3.16}$$

Subtracting (3.16) from (3.15) gives Lemma 3.4.

Now take equations $(3.12)_{n-1}$ and $(3.12)_n$:

$$K_{n-1} + HN_{n-1} = \partial N_n,$$

$$K_n + HN_n = \partial N_{n+1}.$$

We multiply the first equation by H on the right, the second by ∂ on the right, and subtract the first from the second; we get

$$\partial N_{n+1}\partial = K_n\partial - K_{n-1}H + \partial N_n H + HN_n\partial - HN_{n-1}H. \tag{3.17}$$

We assume as induction hypothesis that N_n and N_{n-1} are symmetric; since H is antisymmetric, it follows that the last term on the right is symmetric, and that the sum of the 3rd and 4th terms is symmetric. According to Lemma 3.4 the sum of the first two terms is symmetric; so we conclude from (3.17) that $\partial N_{n+1}\partial$ is symmetric.

From the symmetry of $\partial N_{n+1}\partial$ we conclude that N_{n+1} is symmetric on the subspace consisting of periodic functions whose mean value is zero. For such functions f_0 and g_0 can be written as derivatives of other periodic functions f and g; thus

$$(f_0, Ng_0) = (\partial f, N\,\partial g) = -(f, \partial N\,\partial g) = -(\partial N\,\partial f, g)$$
$$= (N\,\partial f, \partial g) = (Nf_0, g_0).$$

We claim that N_{n+1} is symmetric over the whole space; to show this we define F_{n+1} by formula (3.9) with $G = G_{n+1}$, taking $u_0 = 0$ and choosing as path of integration the straight line segments connecting $u_0 = 0$ to u. Since G_{n+1} is a polynomial in u and its derivatives, F_{n+1} is of the form

$$F_{n+1}(u) = \int P_{n+1}(u)\,dx, \tag{3.18}$$

P_{n+1} a polynomial in u and its derivatives. We claim that the gradient of F_{n+1} is G_{n+1}; denote the gradient of F_{n+1} by $\tilde{G}_{n+1}$. Since F_{n+1} is of form $(3.1)_{n+1}$, $\tilde{G}_{n+1}$ is a polynomial in u and its derivatives. N_{n+1} is symmetric on the subspace of functions with mean value zero, it follows that for any u and v in that subspace

$$\frac{d}{d\varepsilon}F_{n+1}(u+\varepsilon v)\Big|_{\varepsilon=0} = (G_{n+1}(u), v).$$

Subtracting this from the definition of the gradient

$$\frac{d}{d\varepsilon}F_{n+1}(u+\varepsilon v) = (\tilde{G}_{n+1}(u), v),$$

we get that

$$(G_{n+1}(u) - \tilde{G}_{n+1}(u), v) = 0$$

for all v with mean value 0. This implies that

$$G_{n+1}(u) - \tilde{G}_{n+1}(u) = \text{const.}$$

We claim that this implies that $G_{n+1} = \tilde{G}_{n+1}$; for both G_{n+1} and $\tilde{G}_{n+1}$ are polynomials in u and its derivatives, without constant term. If they were not identical, one could easily construct a function of mean value zero such that $G_{n+1}(u) - \tilde{G}_{n+1}(u)$ is not constant. This completes the proof that G_{n+1} is a gradient; this implies that N_{n+1} is a symmetric operator, and the inductive step for the proof of Theorem 3.3 is complete.

THEOREM 3.5. *The functionals F_m are conserved for solutions of the* KdV *equation.*

Proof. Using formula $(3.2)_2$ for G_2 we see that the KdV equation can be rewritten in the form

$$u_t + \partial G_2(u) = 0, \tag{3.19}$$

i.e., $K(u) = -\partial G_2(u)$. Using Theorem 2.1 we conclude that F is a conserved functional of KdV if and only if

$$(G_F, \partial G_2) = 0. \tag{3.20}$$

According to relation (3.7) of Theorem 3.2, with $n=2$, the functionals F_m all satisfy this condition; therefore all F_m are conserved. This completes the proof of Theorem 3.5.

The same argument, when combined with the full force of Theorem 3.2 yields a more general result. We define the nth generalized KdV equation to be

$$(3.21)_n \qquad u_t + \partial G_n(u) = 0.$$

The proof given above also serves to prove the following theorem.

THEOREM 3.6. *Each F_m is a conserved functional for all generalized* KdV *equations* (3.21).

We turn now to another class of conserved functionals. Gardner, Kruskal and Miura, see [11], have shown that the eigenvalues of the Schrödinger operator

$$(3.22) \qquad L = \partial^2 + u/6$$

are conserved functionals of the KdV equation. The author has shown that they are conserved functionals of all generalized KdV equations. Another proof of this has been given by Lenart, see [11]; here we present yet another proof.

Suppose λ is a simple eigenvalue of L:

$$(3.23) \qquad Lw = \lambda w.$$

Then λ is a differentiable functional of the potential u occurring in L; the gradient of $\lambda(u)$ is easily computed by considering one parameter families $u(\varepsilon)$ and differentiating (3.22) with respect to ε. We get

$$Lw' + \frac{u'}{6} w = \lambda w' + \lambda' w,$$

where prime denotes derivative with respect to ε. Multiply this equation by u and integrate; using the symmetry of L and equation (3.23) we can eliminate w' and end up with this expression for λ':

$$(3.24) \qquad \lambda' = \tfrac{1}{6} \int u' w^2 \, dx = \left(\frac{w^2}{6}, u' \right);$$

here we have assumed that w is normalized so that $(w, w) = 1$. By definition of the gradient G_λ,

$$\lambda' = (G_\lambda, u').$$

Comparing this with (3.24) we conclude that

$$(3.25) \qquad G_\lambda = \tfrac{1}{6} w^2.$$

THEOREM 3.7. *Each λ is a conserved functional for all generalized* KdV *equations.*

Proof. As we saw earlier, for any functional F and any solution $u(t)$ of $(3.21)_m$,

$$\frac{dF(u(t))}{dt} = (G_F, u_t) = (G_F, \partial G_m),$$

so that F is conserved if and only if for all u,

$$(G_F, \partial G_m) = 0. \tag{3.26}$$

Applying this to $F = \lambda$ and using (3.25) to express the gradient of λ, we get

$$(w^2, \partial G_m) = 0 \tag{3.27}$$

as condition of invariance of λ. Next we make use of the following obscure but well-known lemma.

LEMMA 3.8. *Suppose w is an eigenfunction of L, satisfying* (3.23). *Then w^2 satisfies the differential equation*

$$Hw^2 = 4\lambda\, \partial w^2, \tag{3.28}$$

where H is defined by (3.4).

Remark. This relation can be verified by a simple calculation.

Irrelevant remark. The jth powers of the eigenfunctions satisfy a $(j+1)$st order equation.

By using the antisymmetry of H and ∂, equation (3.28) and the recursion relation (3.3) we get the following string of identities:

$$\begin{aligned}(w^2, \partial G_m) &= (w^2, HG_{m-1}) = -(Hw^2, G_{m-1})\\ &= -4\lambda(\partial w^2, G_{m-1}) = 4\lambda(w^2, \partial G_{m-1}) = \cdots\\ &= (4\lambda)^m (w^2, \partial G_0) = 0.\end{aligned}$$

In the last step we used the fact that by $(3.2)_0$, G_0 is a constant. This completes the proof of (3.27) and thus of Theorem 3.7.

Remark. In case of a double root we choose for the functional $F(u) = \lambda_1 + \lambda_2$, whose gradient is $w_1^2 + w_2^2$, where w_1, w_2 is any pair of orthonormal eigenvectors. The rest of the proof proceeds as before.

We return now to the Lenart recursion relation

$$HG_n = \partial G_{n+1} \tag{3.29}$$

which can be solved, starting with $G_0 = 1$, for all positive integers. These recursion relations can also be solved for negative integers n. Since $\partial G_0 = 0$, we have for $n = -1$,

$$HG_{-1} = 0.$$

We have shown in § 6 of [18] that this equation has nontrivial periodic solutions; here we offer a different, topological proof for this fact;

The operator H is antisymmetric; therefore its eigenvalues are purely imaginary. Since H is real, the eigenvalues are located symmetrically around the origin. Now consider a one-parameter family of functions $u(\varepsilon)$ entering H. The spectrum of $H(\varepsilon)$ is symmetric around the origin; therefore the multiplicity of 0 as eigenvalue changes by an even number. When $u = 0$, $H = \partial^3$; this operator has 0 as eigenvalue of multiplicity 1; so it follows from the previous argument that for any other u, H has 0 as eigenvalue with multiplicity 1 or 3.

Remark. This intuitive argument can easily be made rigorous, but it is hardly worthwhile to do so since the proof in [18] is perfectly straightforward.

Having shown that a nontrivial G_{-1} exists we can show that equation (3.3) has a solution G_n for $n=-2, -3, \cdots$. The compatibility relation in this case is that the right side, ∂G_{n+1}, be orthogonal to the nullspace of H. Since that nullspace is spanned by G_{-1} the condition is

$$(\partial G_{n+1}, G_{-1})=0.$$

This can be verified in the same manner as (3.7) was for Theorem 3.2. The only difference is that for n negative the G_n are no longer polynomials in u and their derivatives.

One could show that, if properly normalized, G_n is a gradient. However this is unnecessary since we have given in [18] an explicit formula for the functionals F_n, $n=-1, -2, \cdots$, whose gradients G_n satisfy the recursion relation (3.3). They are expressible in terms of the *Floquet exponent* of the operator L, defined as follows:

For any real α, the equation

$$(L-\alpha)w=0$$

has two distinguished solutions $w_\pm$ satisfying

$$w_\pm(x+p)=K^{\pm 1}w_\pm(x).$$

$K=K(\alpha)$ is called the Floquet exponent; it is real in the so-called instability intervals and of modulus one in the stability intervals. Of course K is a functional of u as well as a function of α.

In § 6 of [18] we have shown the following theorem.

THEOREM 3.9. *The gradients of the functionals*

$$F_n=\frac{1}{(-n-1)!}\left(\frac{d}{d\alpha}\right)^{-n-1}\log K(\alpha, u)\bigg|_{\alpha=0}$$

satisfy relations (3.3) *for* $n=-1, -2, \cdots$.

4. Hamiltonian formalism. The Hamiltonian form of equations of motion is

$$(4.1)_H \qquad \frac{d}{dt}q_j=\frac{\partial H}{\partial p_j}, \qquad \frac{d}{dt}p_j=-\frac{\partial H}{\partial q_j}, \qquad j=1, \cdots, N;$$

the Hamiltonian H is a function of the $2N$ variables p_j, q_j, $j=1, \cdots, N$. Such equations can be cast in another form with the aid of the notion of the *Poisson bracket*, defined for any pair of functions F, K of the $2N$ variables as follows:

$$(4.2) \qquad [F, K]=\sum_j \frac{\partial(F, K)}{\partial(q_j, p_j)}.$$

The important properties of the Poisson bracket are:

(a) $[F, K]$ is a bilinear, alternating function of F and K.

(b) The Jacobi identity

$$(4.3) \qquad [[F, H], K]+[[H, K], F]+[[K, F], H]=0.$$

In terms of the Poisson bracket the Hamiltonian equations can be expressed as follows:

Let $p(t), q(t)$ be a solution of (4.1), F any function of p, q; then

$$\frac{d}{dt}F(p(t), q(t)) = [F, H]. \tag{4.4}$$

This formulation implies the following theorem.

THEOREM 4.1. *F is a conserved function for all solutions of* $(4.1)_H$ *if and only if*

$$[F, H] = 0. \tag{4.5}$$

It follows from this and the Jacobi identity (4.3) that if F and K are a pair of conserved functions, so is their Poisson bracket. Two functions whose Poisson bracket is zero are said to be *in involution.*

Two further results of Hamiltonian mechanics are the following:

THEOREM 4.2. *Suppose H and K are in involution, i.e.,* $[H, K]=0$. *Then the Hamiltonian flows generated by H and K, respectively, commute. That is, we denote by $S_H(t)$ and $S_K(t)$ the operator which links initial position to position at time t of a point (p, q) under the Hamiltonian flows* $(4.1)_H$ *and* $(4.1)_K$, *respectively; then*

$$S_H(t)S_K(r) = S_K(r)S_H(t). \tag{4.6}$$

THEOREM 4.3. *Suppose there exist N independent functions $F_1, \cdots, F_N$ in involution which are conserved under the Hamiltonian flow* $(4.1)_H$. *Then* $(4.1)_H$ *is completely integrable.*

5. Hamiltonian structure of KdV and invariant manifolds. C. Gardner in [9] and Faddeev and Zakharov in [6] have independently introduced a Hamiltonian structure for the KdV equation. Here we follow Gardner's line of development.

Gardner introduces the Poisson bracket

$$[F, H] = (G_F, \partial G_H), \tag{5.1}$$

where as before G_F, G_H denote the gradients of F and H with respect to the L_2 scalar product $(\cdot, \cdot)$ and $\partial = d/dx$.

THEOREM 5.1. *The bracket defined by* (5.1) *is*

(a) *bilinear,*

(b) *alternating,*

(c) *satisfies the Jacobi identity.*

We sketch the proof given in [18]. Part (a) is obvious, and part (b) follows by integration by parts. To prove (c) we compute the gradient of $[F, H]$. Using the symmetry of the derivative of G_F and G_H, we can easily show that

$$G_{[F,H]} = N_F\,\partial G_H - N_H\,\partial G_F, \tag{5.2}$$

where N_F and N_H are the second derivatives of F, H. The Jacobi identity (4.3) follows from this if we use once more the symmetry N_F, N_H and N_K.

According to formula (2.3), if $u(t)$ satisfies equation (2.1) and F is any functional,

$$dF/dt = (G_F, K).$$

According to (3.19), for the KdV equation,

$$K = -\partial G_2.$$

Therefore we can write, using (5.1)

$$dF/dt = (G_F, -\partial G_2) = [F, -F_2].$$

Comparing this with (4.4) we see that the KdV equation is of Hamiltonian form, with $H = -F_2$.

Again using (5.1) we can rewrite equations (3.7) as

$$[F_m, F_n] = 0,$$

and equation (3.27) as

$$[\lambda, F_m] = 0.$$

One can show analogously, see [18], that

$$[\lambda, \mu] = 0$$

for any pair of eigenvalues of L. These relations can be expressed by saying *the functionals F_n and λ are in involution.*

Since we don't know an infinite-dimensional analogue of Theorem 4.3, we cannot use this plethora of conserved functions in involution to conclude directly that KdV is integrable. However Theorem 4.2 is true in infinite-dimensional space, and can be used to construct invariant submanifolds of the KdV flow; we outline briefly how:

We start with a variational problem originally suggested by Kruskal and Zabusky:

Given the values

(5.3) $$F_i(u) = A_i, \qquad i = 0, \cdots, n-1,$$

find u which minimizes $(-1)^{N-1} F_N(u)$. The constants A_j have to be so chosen that the constraint (5.3) is satisfied by some function, and so that A_j is not a stationary value of $F_j(u)$ when the other constraints are imposed. This is equivalent to saying that for any function u satisfying the constraints, the gradients

$$G_0(u), G_1(u), \cdots, G_{N-1}(u)$$

are linearly independent. We shall call such constraints *admissible.*

The following theorem was proved in [18].

THEOREM 5.2. *For admissible constraints* (5.3) *the functional* $(-1)^{N-1} F_N(u)$ *is minimized by some function u_0, and every minimizing function satisfies an Euler equation of the form*

(5.4) $$G(u_0) = 0,$$

where

(5.5) $$G(u) = G_N(u) + \sum_0^{N-1} a_j G_j(u).$$

The function G in (5.5) is the gradient of

(5.6) $$F(u) = F_N(u) + \sum_0^{N-1} a_j F_j(u).$$

As we have shown previously,

$$[F_j, F_k]=0,$$

from which we deduce that

$$[F, F_k]=0.$$

This implies that F is a conserved functional for all generalized KdV flows

$$u_t=\partial G_k(u). \tag{5.7$_k$}$$

Denote by $S_k(t)$ the solution operator for equation (5.7)$_k$. It follows from Theorem 2.2 that if u_0 satisfies equation (5.4), then so does every function of the form

$$u=\prod_1^{N-1} S_k(t_k)u_0, \qquad -\infty<t_k<\infty. \tag{5.8}$$

Denote the set (5.8) by $S=S(A_0, \cdots, A_{N-1})$; the following result was proved in [18].

THEOREM 5.3. *S is a compact* $(N-1)$*-dimensional manifold.*

The operators $S_k(t)$ map S onto itself, and they commute. Denote by Ω the collection of those vectors $(\omega_1, \cdots, \omega_{N-1})$ for which

$$\prod S_k(\omega_k)=I.$$

Ω is the module of periods. It follows from the definition of S that

$$S=\mathbb{R}^M/\Omega.$$

Since S is $(N-1)$-dimensional and compact, it follows that Ω is a lattice (i.e., discrete) and $(N-1)$-dimensional; from this we conclude the following theorem.

THEOREM 5.4. *S is an* $(N-1)$*-dimensional torus, and each* $S_k(t)$ *is almost periodic on S.*

Every point on $S(A_0, \cdots, A_{N-1})$ minimizes $F_N(u)$ subject to the constraints (5.3). There may be points not on S which minimize $F_N(u)$; but it follows from the above analysis that the minimizing set is a union of disjoint $(N-1)$-dimensional tori.

The case $N=1$ is trivial and the case $N=2$ is classical, going back to Korteweg–de Vries [15].

Using formulas (3.2) we get the following expression for equation (5.4), $N=2$:

$$u_{xx}+\tfrac{1}{2}u^2+a_1u+a_0=0.$$

Multiplying this by $2u_x$ we obtain an equation of the form

$$u_x^2=Q(u),$$

where Q is a cubic polynomial. From this we can express x as function of u by an elliptic integral, which shows that u_0 is an elliptic function of x. It can be shown that in this case the minimizing set is a single circle formed by the translates of an elliptic function u_0. It is easy to verify that the function

$$u_0(x+a_1t)$$

is a solution of the KdV equation. This traveling wave has been called "cnoidal wave" by Korteweg and de Vries.

Benjamin [2] and Bona [4] have proved the stability of simple cnoidal waves; i.e., they have shown that if $u_1(s)$ is sufficiently near $u_0(x)$ in an appropriate metric, then for any value of t, $S_2(t)u_1$, the solution of KdV with initial value u_1, lies near $u_0(x+\theta)$ for some θ.

We surmise that all solutions $S_2(t)u_0$ constructed in this section are stable in the above sense, as long as u_0 is an absolute—or even just local—minimum of $F_N(u)$ among all u satisfying constraints (5.3). The numerical experiments carried out by M. Hyman and described in an Appendix to this paper certainly very strongly suggest this.

Next we sketch a simple argument which shows that F_N has stationary points on the constrained set (5.3) which are not minima. We use Morse theory, according to which such stationary points exist if the homology of the set (5.3) is not trivial.

THEOREM 5.5. *For suitably chosen constants $A_0, \cdots, A_{N-1}$ the constrained set* (5.3) *is not simply connected.*

Proof. We shall handle the case $N=3$. A_0 and A_1 are taken as arbitrary, and we denote by M the minimum of $-F_N$ on the set (5.3), $N=2$. As remarked earlier, in this case the minimizing set is a single circle consisting of all translates of an elliptic function $u_0(x)$.

Let n be an index $\neq 0$ for which the nth Fourier coefficient of u_0 is $\neq 0$:

$$\int u_o(x)\, e^{-inx}\, dx \neq 0. \tag{5.9}$$

Consider the set of functions u which satisfy the constraints (5.3), $N=2$ and the additional constraint

$$\int u(x)\, e^{-inx}\, dx = 0. \tag{5.9'}$$

Since this constraint excludes the solutions of the previous minimum problem, it follows that the minimum value M_1 of $-F_2(u)$ subject to this new constraint exceeds the old minimum M of $-F_2(u)$:

$$M < M_1.$$

We choose now a function u_1 different from u_0 but so close to it that

$$-F_2(u_1) < M_1. \tag{5.10}$$

We claim that the circle

$$\theta \to u_1(x+\theta)$$

cannot be deformed to a point on the set of those u which satisfy

$$F_j(u) = F_j(u_1), \qquad j = 0, 1, 2. \tag{5.11}$$

To see this we introduce the projection P onto the nth Fourier coefficient. If $u_1(x, \theta, s)$, $0 \leqq s \leqq 1$, were a deformation of $u_1(x+\theta)$ to a point, then $Pu_1(x, \theta, s)$ would be a deformation of $a_n e^{-in\theta}$ to a point in the complex plane. Since $a_n e^{-in\theta}$, $n \neq 0$, winds around the origin, such a deformation would have to cross the origin; but then for some value of θ and s the nth Fourier coefficient of the function $u = u(x+\theta, s)$ is zero. This implies by (5.9) that

$$-F_2(u) \geqq M_1,$$

which when combined with (5.11) contradicts (5.10).

Having proved the existence of a curve in the set (5.11) which is not homotopic zero we consider all curves C in the same homotopy class, and determine that one for which the maximum of $F_3(u)$ on C is as small as possible. I surmise that this minimax problem has a solution; such a solution satisfies an equation of the form (5.4), (5.5). As in the case of the minimum problem, the solutions of the minimax problem form a 2-dimensional torus.

I suspect the solutions of equations of the form (5.4), (5.5), with a_j and N arbitrary, are dense among all C^∞ periodic functions.

The author and Jurgen Moser have shown, see [17] and [18], that if u satisfies an equation of the form (5.4), then all but a finite number of eigenvalues of the Schrödinger operator L defined by (3.22) are double. Using this connection, and a method of Hochstadt [14], McKean and van Moerbeke were able to use the inverse method in spectral theory to study periodic solutions of (5.4). They have given a new proof of Theorem 5.4, and were able to express solutions of (5.4) as hyperelliptic functions. There is hope that their approach can be used to settle the question of integrability of the KdV equation in the class of periodic functions.

Very recently McKean and Trubowitz succeeded in showing that all solutions of KdV which are periodic in x are indeed almost periodic in t.

APPENDIX

JAMES M. HYMAN

In this Appendix we describe how the construction of special solutions of the KdV equation which minimize F_2 subject to the constraint $F_j = A_j$ can be implemented numerically. We saw in § 3 that a minimizing function satisfies the Euler equation

$$G_3 + \sum_0^2 a_j G_j = 0. \tag{A.1}$$

A solution of (A.1) is an extremal for

$$F = F_3 + \sum_0^2 a_j F_j, \tag{A.2}$$

and presumably can be obtained by minimizing that functional. This is indeed what we did. We chose the constants a_{-1}, a_0, a_1, then discretized the functional (A.2) by specifying u at N equidistant points and expressed the first and second derivatives of u in (A.2) by difference quotients. The resulting function of N variables was minimized using A. Jameson's version[1] of the Fletcher–Powell–Davidon algorithm [8]. The resulting discretized function is a somewhat crude

[1] My thanks are due to A. Jameson for acquainting me with his FPD package.

approximation to the function u we are looking for. To make u more accurate we proceeded as follows:

We are looking for a periodic solution of (A.1); since this is a fourth order equation, its solutions are parametrized by their four Cauchy data at, say, $x=0$. Periodicity requires the Cauchy data at $x=p$ be equal to the Cauchy data at $x=0$. We sought to satisfy this requirement by a sequence of approximations constructed by "shooting." The Cauchy data of u_{n+1} at $x=0$ was chosen by applying the rule of false position to $u_n(p)-u_n(0)$, $u'_n(p)-u'_n(0)$, $u_{n-1}(p)-u_{n-1}(0)$ and $u'_{n-1}(p)-u'_{n-1}(0)$. $u_{n+1}(p)$ and $u'_{n+1}(p)$ were then computed numerically. For this purpose we used the ODE package developed by A. Hindmarsh [13].

The following observations were helpful:

(a) Instead of matching all four Cauchy data for the fourth order equation (A.1), it sufficed to match only two, since according to § 5 periodic solutions form a two-parameter family, and therefore two of the matching conditions must be consequences of the other two.

(b) The initial guesses for the Cauchy data come from the approximate solution obtained by the variational procedure. Without such a good initial guess we were unable to construct a periodic solution of (A.1).

The periodic solution of (A.1) constructed above was then used as initial value data. We solved numerically the initial value problem for the KdV equation using Fred Tappert's[2] method [26] and code. The accuracy of the solution was monitored by checking the constancy of the functionals F_0, F_1, F_2, and the extent

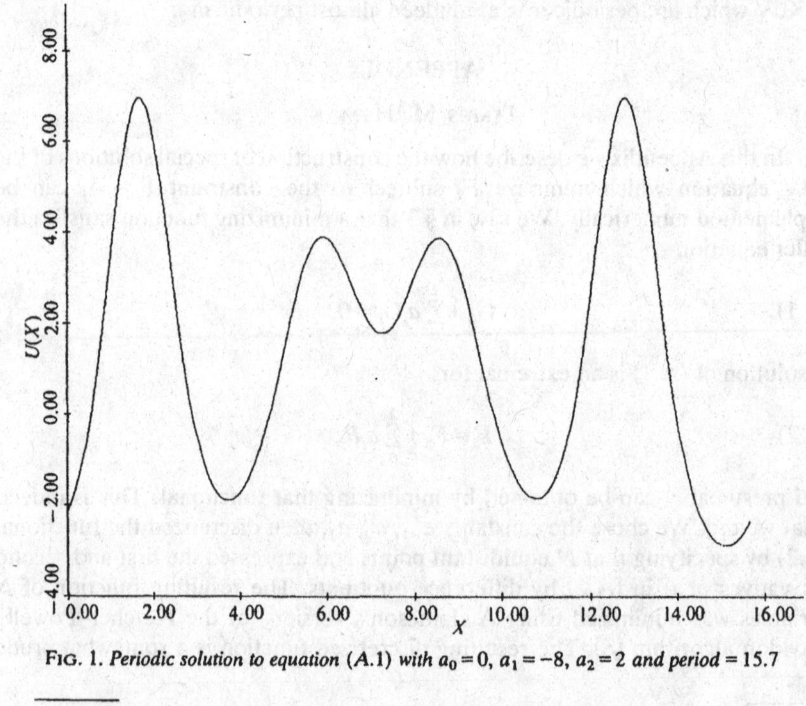

FIG. 1. *Periodic solution to equation (A.1) with $a_0=0$, $a_1=-8$, $a_2=2$ and period* $=15.7$

[2] My thanks are due to Fred Tappert for acquainting me with his KdV solver.

to which the solution satisfied the ODE (A.1). The solution constructed by Tappert's method passed these tests of accuracy reasonably well. Solutions constructed using earlier methods of Kruskal and Zabusky [29] and Vliegenthart [28] were less accurate and were not used in this study. The finest vindication of Tappert's KdV solver was that after a finite elapse of time the solution resumed its initial shape, in a shifted position. During the intervening time the shape of the solution underwent considerable gyrations.

In Fig. 1 we present a periodic solution of equation (A.1), with $a_0=0$, $a_1=-8$, $a_2=2$ and period = 15.7. Figures 2 to 5 show the value at times $t=.28$, .56, .84, and 1.12 of the solution to the KdV equation with initial values shown in Fig. 1. Note that the function shown in Fig. 5 has the same shape as the initial function in Fig. 1, except for a shift by the amount −5.1.

Figures 6 to 10 show the time history of a solution of KdV where the initial value was obtained by superimposing a random disturbance[3] on the initial function shown in Fig. 1. Note that the disturbance is not magnified during the flow, and that the averages of these disturbed signals are very close at all times shown to the undisturbed signals pictured in Figs. 2–5. This calculation demonstrates convincingly the great stability of the KdV flow pictured here. It also demonstrates the ability to Tappert's KdV solver to deal accurately with solutions containing high frequency disturbances.

Figure 11 shows a double cnoidal wave over two periods. The solution of KdV with this initial value propagates with speed $c=.3$ without altering its shape. Figure 12 shows a function obtained from Fig. 11 by raising the first peak by 20% and lowering the second by 20%. Figure 13 shows the solution at time $t=0.52$ using the function in Fig. 12 as initial data for the KdV equation. At time $t=1.04$ the solution returns to its original shape in Fig. 12 while translating with speed $c\approx .3$. This and other calculations indicate that the double cnoidal wave is stable.

In [3], Benjamin presents an elegant argument to show that the second variation of $-F_2$ under constraint of F_0 and F_1 is indefinite for a double cnoidal wave, and therefore a double cnoidal wave is not even a local minimum of $-F_2$ under the constraints; he raises the question whether this implies instability of the double cnoidal wave. The numerical study reported above and others unreported here indicate stability. We remark that this is not surprising from the point of view of Hamiltonian theory.

[3] We learned recently that in 1965 Zabusky tested the stability of solutions of the KdV equation by imposing random disturbances on their initial data; he too observed that solutions were remarkably stable under such perturbations.

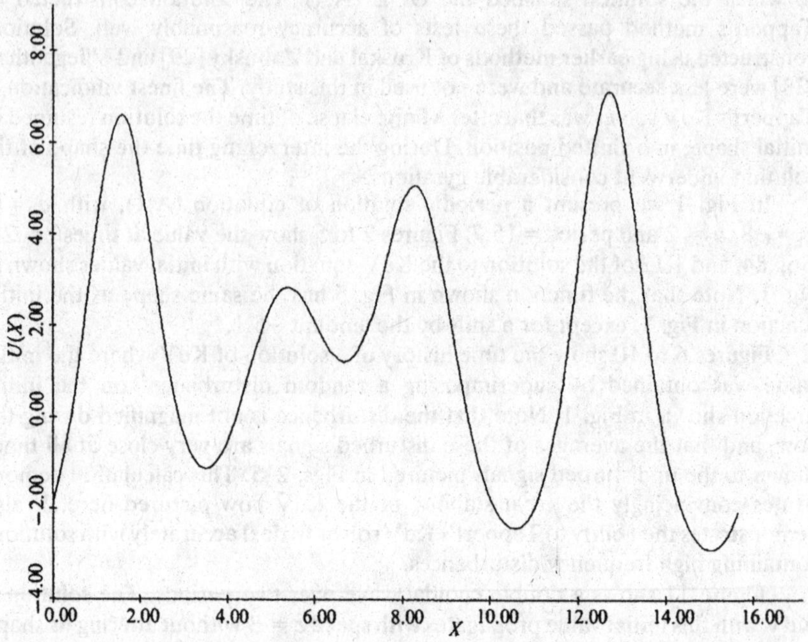

FIG. 2. *The solution of the KdV equation at time t = .28 with initial value shown in Fig.* 1

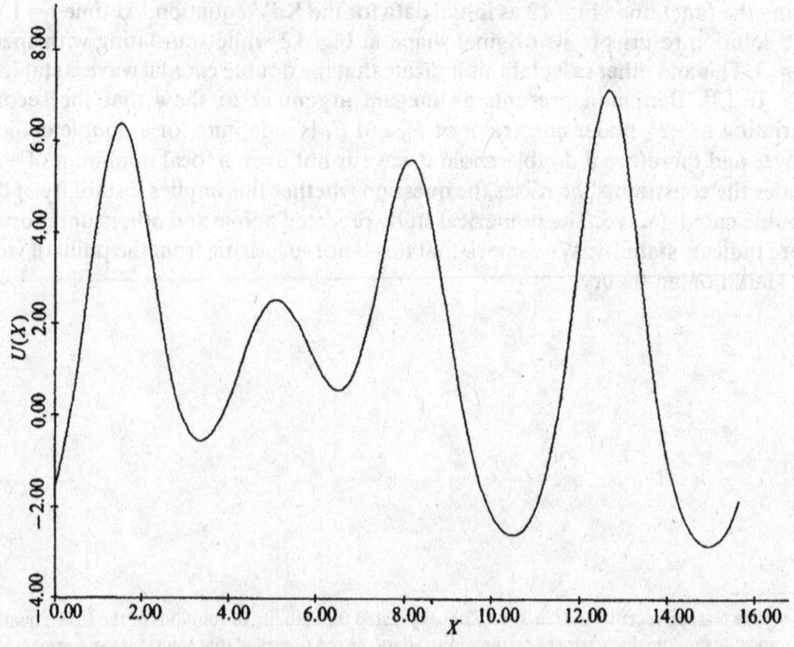

FIG. 3. *The solution of the KdV equation at time t = .56 with initial value shown in Fig.* 1

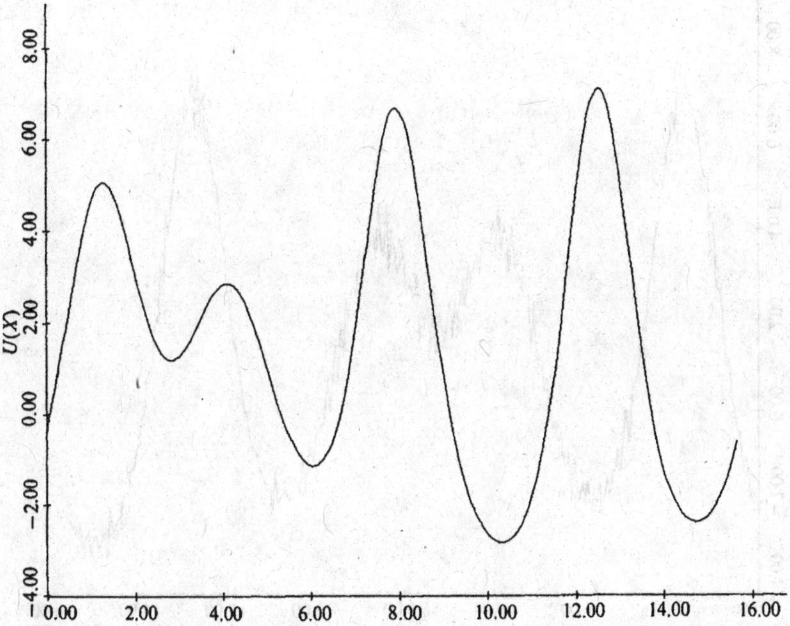

FIG. 4. *The solution of the KdV equation at time $t = .84$ with initial value shown in Fig. 1*

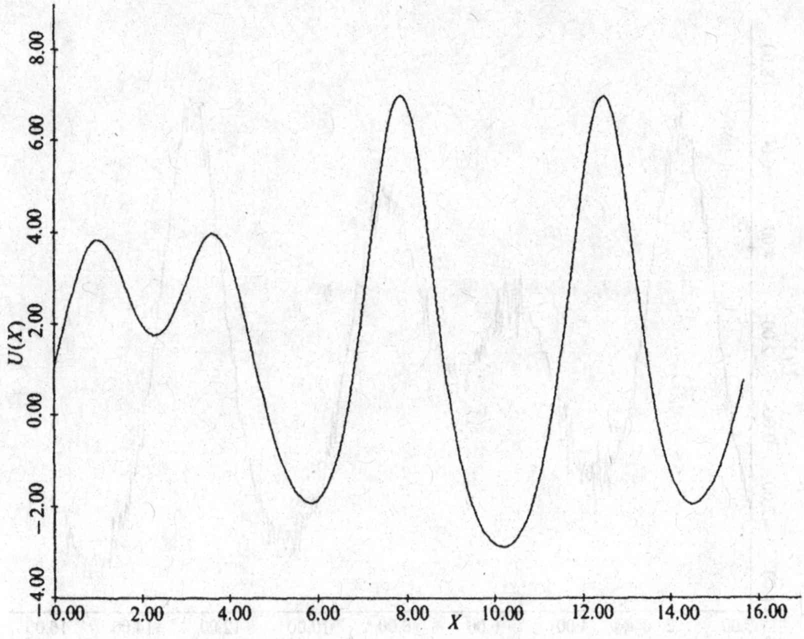

FIG. 5. *The solution of the KdV equation at time $t = 1.12$ with the initial value shown in Fig. 1*

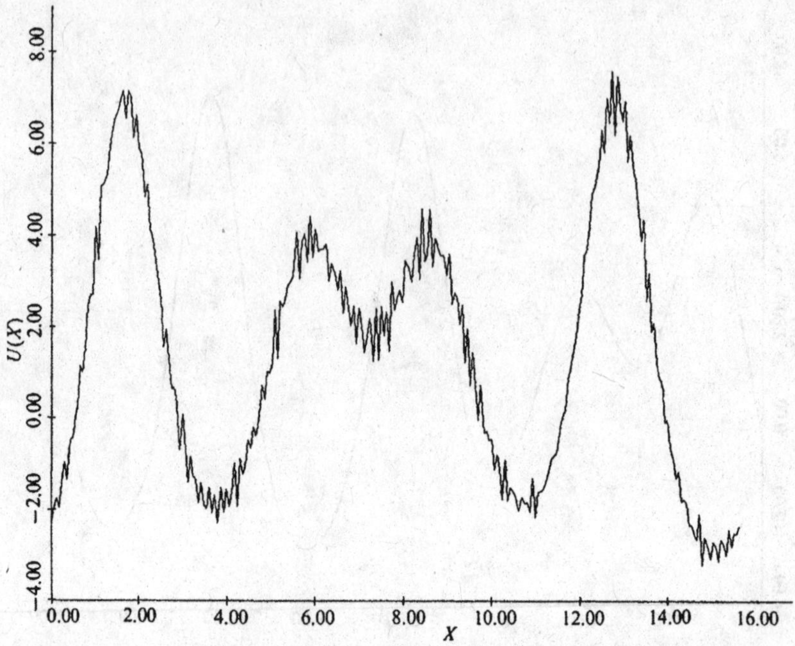

FIG. 6. *A random disturbance is superimposed on the solution shown in Fig.* 1

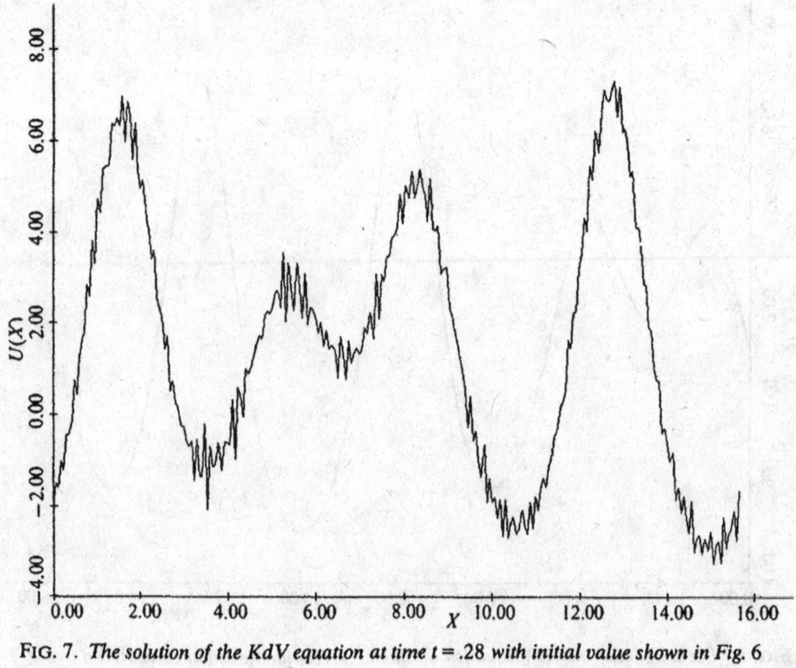

FIG. 7. *The solution of the KdV equation at time* $t = .28$ *with initial value shown in Fig.* 6

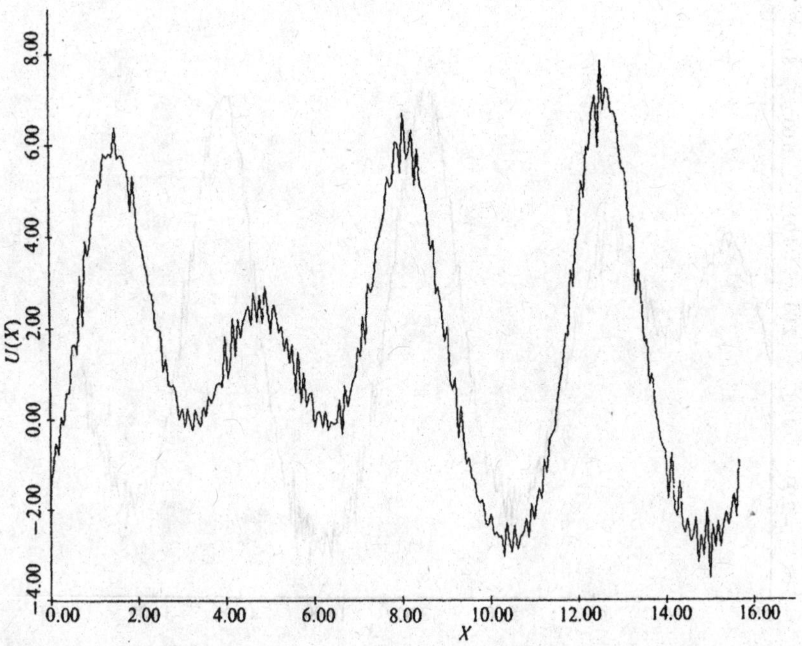

FIG. 8. *The solution of the KdV equation at time $t = .56$ with initial value shown in Fig. 6*

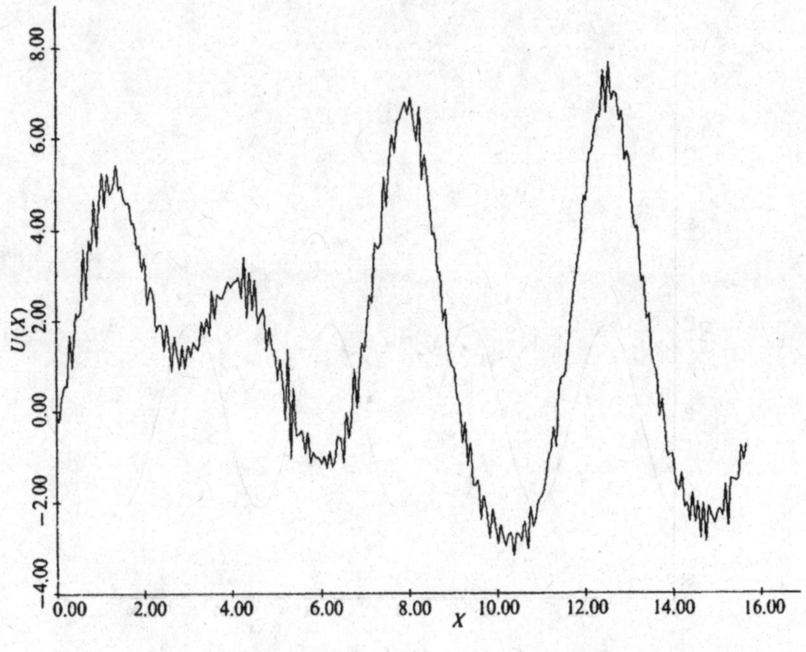

FIG. 9. *The solution of the KdV equation at time $t = .84$ with initial value shown in Fig. 6*

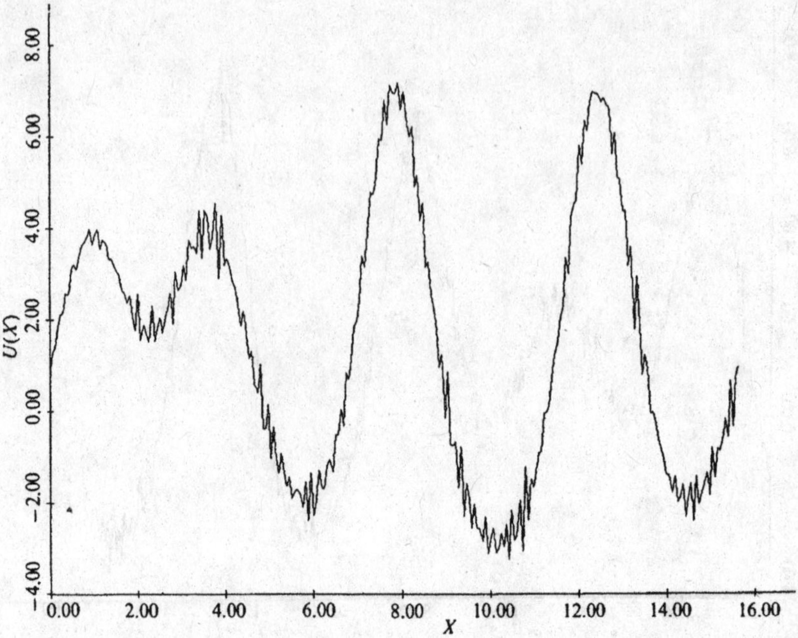

FIG. 10. *The solution of the KdV equation at time $t = 1.12$ with the initial value shown in Fig. 6*

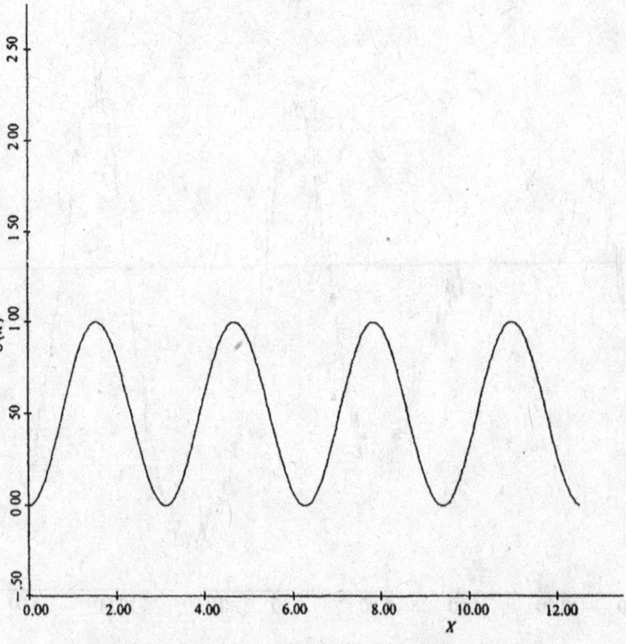

FIG. 11. *A double cnoidal wave over two periods*

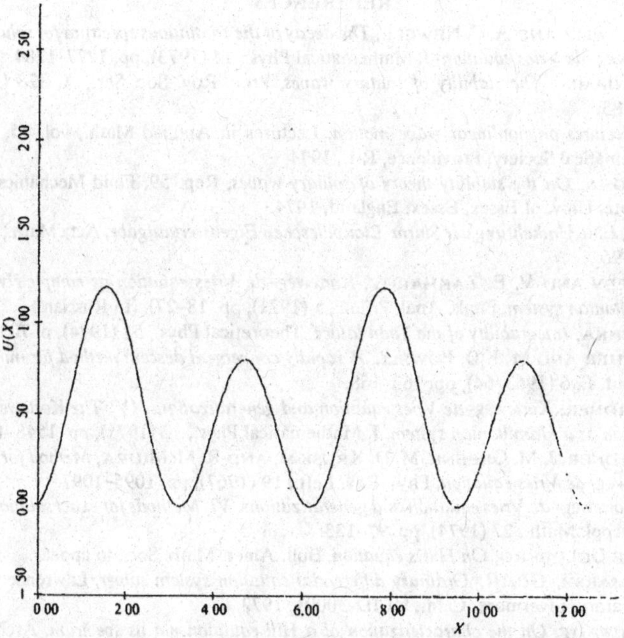

FIG. 12. *The first peak of the cnoidal wave in Fig. 11 was raised by 20% and the second was lowered by 20%*

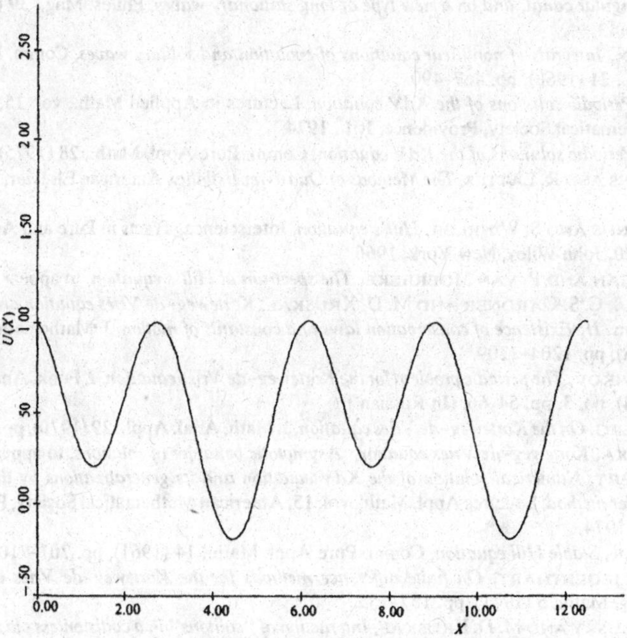

FIG. 13. *The solution to the KdV equation at $t = .52$ using the function shown in Fig. 12 as initial data. At $t = 1.04$ the solution returns to approximately the shape in Fig. 12*

REFERENCES

[1] M. J. ABLOWITZ AND A. C. NEWELL, *The decay of the continuous spectrum for solutions of the Korteweg–de Vries equation*, J. Mathematical Phys., 14 (1973), pp. 1277–1284.

[2] T. B. BENJAMIN, *The stability of solitary waves*, Proc. Roy. Soc. Ser., A, 328 (1972), pp. 153–183.

[3] ———, *Lectures on nonlinear wave motion*, Lectures in Applied Math., vol. 15, American Mathematical Society, Providence, R.I., 1974.

[4] JERRY BONA, *On the stability theory of solitary waves*, Rep. 59, Fluid Mechanics Research Institute, Univ. of Essex, Essex, England, 1974.

[5] G. BORG, *Eine Umkehrung der Sturm-Liouvilleschen Eigenwertaufgabe*, Acta Math., 78 (1946), pp. 1–96.

[6] L. FADDEEV AND V. E. ZAKHAROV, *Korteweg–de Vries equation as completely integrable Hamiltonian system*, Funk. Anal. Priloz., 5 (1971), pp. 18–27). (In Russian.)

[7] H. FLASCHKA, *Integrability of the Toda lattice*, Theoretical Phys., 51 (1974), p. 703.

[8] R. FLETCHER AND M. J. D. POWELL, *A rapidly convergent descent method for minimization*, Comput. J., 6 (1963/64), pp. 163–168.

[9] C. S. GARDNER, *Korteweg–de Vries equation and generalizations. IV: The Korteweg–de Vries equation as a Hamiltonian system*, J. Mathematical Phys., 12 (1971), pp. 1548–1551.

[10] C. S. GARDNER, J. M. GREENE, M. D. KRUSKAL AND R. M. MIURA, *Method for solving the Korteweg–de Vries equation*, Phys. Rev. Lett., 19 (1967), pp. 1095–1097.

[11] ———, *Korteweg–de Vries equation and generalizations. VI. Methods for exact solutions*, Comm. Pure Appl. Math., 27 (1974), pp. 97–133.

[12] WALLACE GOLDBERG, *On Hill's equation*, Bull. Amer. Math. Soc., to appear.

[13] A. HINDMARSH, *GEAR: Ordinary differential equation system solver*, Lawrence Livermore Laboratory, Livermore, Calif., UCID-30001, 1972.

[14] H. HOCHSTADT, *On the characterization of a Hill equation via its spectrum*, Arch. Rational Mech. Anal., 19 (1965), pp. 353–362.

[15] D. J. KOORTEWEG AND G. DE VRIES, *On the change of form of long waves advancing in a rectangular canal, and on a new type of long stationary waves*, Philos. Mag., 39 (1895), pp. 422–443.

[16] P. D. LAX, *Integrals of nonlinear equations of evolution and solitary waves*, Comm. Pure Appl. Math., 21 (1968), pp. 467–490.

[17] ———, *Periodic solutions of the KdV equation*, Lectures in Applied Math., vol. 15, American Mathematical Society, Providence, R.I., 1974.

[18] ———, *Periodic solutions of the KdV equation*, Comm. Pure Appl. Math., 28 (1975).

[19] J. L. LIONS AND R. LATTES, *The Methods of Quasi-reversibility*, American Elsevier, New York, 1969.

[20] W. MAGNUS AND S. WINKLER, *Hill's equation*, Interscience, Tracts in Pure and Appl. Math., No. 20, John Wiley, New York, 1966.

[21] H. MCKEAN AND P. VAN MOERBEKE, *The spectrum of Hill's equation*, to appear.

[22] R. MIURA, C. S. GARDNER AND M. D. KRUSKAL, *Korteweg–de Vries equation and generalizations. II: Existence of conservation laws and constants of motion*, J. Mathematical Phys., 9 (1968), pp. 1204–1209.

[23] S. P. NOVIKOV, *The periodic problem for the Korteweg–de Vries equation. I*, Funk. Anal. Priloz, 8 (1974), no. 3, pp. 54–66. (In Russian.)

[24] A. SJOBERG, *On the Korteweg–de Vries equation*, J. Math. Anal. Appl., 29 (1970), pp. 569–579.

[25] S. TANAKA, *Korteweg–de Vries equation: Asymptotic behavior of solutions*, to appear.

[26] F. TAPPERT, *Numerical solution of the KdV equation and its generalizations by the split-step Fourier method*, Lectures Appl. Math., vol. 15, American Mathematical Society, Providence, R.I., 1974.

[27] P. UNGAR, *Stable Hill equation*, Comm. Pure Appl. Math., 14 (1961), pp. 707–710.

[28] A. C. VLIEGENTHART, *On finite difference methods for the Korteweg–de Vries equation*, J. Engrg. Math., 5 (1971), pp. 137–155.

[29] N. J. ZABUSKY AND M. D. KRUSKAL, *Interaction of "solitons" in a collisionless plasma and the recurrence of initial states*, Phys. Rev. Letters, 15 (1965), pp. 240–243.

The Small Dispersion Limit of the Korteweg–deVries Equation. I

PETER D. LAX
Courant Institute

AND

C. DAVID LEVERMORE
Lawrence Livermore Laboratory

To Harold Grad and Joe Keller, bis Zweihundertvierzig

Abstract

In Part I the scattering transform method is used to study the weak limit of solutions to the initial value problem for the Korteweg–deVries (KdV) equation as the dispersion tends to zero. In that limit the associated Schrödinger operator becomes semiclassical, so the exact scattering data is replaced by its corresponding WKB expressions. Only nonpositive initial data are considered; in that case the limiting reflection coefficient vanishes. The explicit solution of Kay and Moses for the reflectionless inverse transform is then analyzed, and the weak limit, valid for all time, is characterized by a quadratic minimum problem with constraints. This minimum problem is reduced to a Riemann–Hilbert problem in function theory.

In Parts II and III we use function theoretical methods to solve the Riemann–Hilbert problem in terms of solutions to an auxiliary initial value problem.

The weak limit satisfies the KdV equation with the dispersive term dropped until its derivatives become infinite. Up to that time the weak limit is a strong L^2-limit. At later times the weak limit is locally described by Whitham's averaged equations or, more generally, by the equations found by Flaschka et al. using multiphase averaging. For large times, behavior of the weak limit is studied directly from the minimum problem.

0. The Zero Dispersion Limit for the KdV Equation

Introduction

In this study we analyze the behavior of solutions $u(x, t; \varepsilon)$ of

$$u_t - 6uu_x + \varepsilon^2 u_{xxx} = 0 \tag{1}$$

as $\varepsilon \to 0$ while the initial values are fixed:

$$u(x, 0; \varepsilon) = u(x). \tag{2}$$

It is known from computer studies that for t greater than a critical time, independent of ε, dependent only on the initial data, $u(x, t; \varepsilon)$ becomes oscillatory

Communications on Pure and Applied Mathematics, Vol. XXXVI, 253–290 (1983)
 CCC 0010-3640/83/030253-38$04.80

as $\varepsilon \to 0$. The wave length of these oscillations is $O(\varepsilon)$, their amplitude independent of ε. This shows that

$$\lim_{\varepsilon \to 0} u(x, t; \varepsilon) \tag{3}$$

exists only in the weak sense.

Our strategy for studying the limit (3) is similar to one employed by Hopf [5] and Cole [1] to study a similar but simpler singular perturbation, Burgers' equation:

$$u_t + uu_x = \varepsilon u_{xx}, \qquad \varepsilon > 0. \tag{4}$$

Hopf and Cole introduced a new variable V, defined by

$$U_x = u, \qquad U = -2\varepsilon \log V \tag{5}$$

which linearizes equation (4):

$$V_t = \varepsilon V_{xx}.$$

Solutions of the linear heat equation can be explicitly expressed by an integral formula. This formula for V can be used to express u:

$$u(x, t; \varepsilon) = \int \frac{x-y}{t} \exp\left(-\frac{D}{\varepsilon}\right) dy \Big/ \int \exp\left(-\frac{D}{\varepsilon}\right) dy, \tag{6}$$

where

$$D = U(y) + \frac{(x-y)^2}{2t}, \qquad U(y) = \int_{-\infty}^{y} u(z)\, dz, \tag{7}$$

and $u(z)$ is the initial value. Letting $\varepsilon \to 0$ in (6) we get

$$\bar{u}(x, t) = \lim_{\varepsilon \to 0} u(x, t; \varepsilon) = \frac{x - y^*}{t}, \tag{8}$$

where $y^*(x, t)$ is that value of y which minimizes $D(x, y, t)$. The function y^* is defined for almost all x, t.

The key step in the method of Hopf and Cole sketched above is the transformation (5) that linearizes (4). Gardner, Green, Kruskal, and Miura have discovered (see [3]) that the KdV equation can be linearized by the so-called scattering transformation. In this study we use this linearization to trace the dependence of solutions on ε and to determine their limit as $\varepsilon \to 0$. The limit in this case is, for t large enough, only a weak limit. The initial functions we consider are assumed to tend to zero fast enough as $|x| \to \infty$, and to be nonpositive. The latter condition is used essentially in the analysis; it has been shown in the dissertation of Venakides [13] how to remove this restriction.

The organization of the paper is as follows:

In Section 1, we solve the direct scattering problem for given initial data, assumed for simplicity to have a single local minimum. The direct problem is solved asymptotically; we then replace the exact initial data $u(x)$ by an approximate one, $u(x;\varepsilon)$, whose exact scattering data are equal to the approximate data of $u(x)$. As shown subsequently, $\lim_{\varepsilon\to 0} u(x;\varepsilon)=u(x)$ in the $L_2(R)$ sense.

In Section 2, we use the Kay–Moses explicit solution of the reflectionless inverse problem, and carry out the limit $\varepsilon\to 0$. We show that $\lim_{\varepsilon\to 0} u(x,t;\varepsilon)=\bar{u}(x,t)$ exists in the sense of weak convergence in $L_2(R)$ with respect to x, and that the weak limit $\bar{u}$ can be described as

$$\bar{u}=\partial_{xx}Q^*. \tag{9}$$

The function $Q^*(x,t)$ is determined by solving a quadratic programming problem, i.e.,

$$Q^*(x,t)=\min_{0\leqq\psi\leqq\phi} Q(\psi;x,t). \tag{10}$$

Here $Q(\psi;x,t)$ is a quadratic functional of ψ, which depends linearly on the parameters x, t. The function ϕ is determined by the initial data:

$$\phi(\eta)=\mathcal{R}e\int\frac{\eta}{(-u(y)-\eta^2)^{1/2}}\,dy. \tag{11}$$

We also show that $u^2(x,t;\varepsilon)$ tends, in an appropriate weak sense, to a limit.

In Section 3, we show that Q is continuous in a weak sequential topology, and that the space of admissible functions is compact in that topology. We further show that Q is a strictly convex function; since the admissible functions form a convex set, this implies not only that the minimum of Q is taken on at a unique function, but that this function is the only one which satisfies variational conditions.

The variational conditions are then converted to a Riemann–Hilbert problem, i.e., to the problem of determining an analytic function of class H^p in the upper half-plane whose real and imaginary parts are prescribed on alternate intervals of the real axis.

In Section 4 of Part II, we solve the Riemann–Hilbert problem for all times t that do not exceed t_b, called the break-time, defined as the time beyond which the initial value problem

$$u_t-6uu_x=0,\qquad u(x,0)=u(x) \tag{12}$$

has no solution; equation (12) is obtained by setting $\varepsilon=0$ in the KdV equation (1). We show that, for $t<t_b$, $u(x,t;\varepsilon)$ tends to the solution of (10), and that this convergence is strong convergence in $L_2(R)$ with respect to x.

In Section 5, we solve the Riemann–Hilbert problem for any x and t and verify that the solution so obtained satisfies the variational conditions and therefore solves the minimum problem. The solution $\bar{u}$ determined from (9) has the form suggested by Whitham [14] and more generally by Flaschka, Forest, and McLaughlin [2].

In Section 6, we determine the asymptotic behavior of $\bar{u}(x, t)$ for large t. For $0<x/t<4m^2$, where

$$m=-\min u(x),$$

we have

$$\bar{u}(x,t)\simeq-\frac{1}{2\pi t}\phi\left(\frac{x}{4t}\right)^{1/2};$$

here ϕ is the function defined in (11). In Sections 7 and 8 we extend the class of admissible data.

1. Asymptotic Analysis of the Direct Scattering Problem

In a remarkable paper [3], Gardner, Greene, Kruskal, and Miura (GGKM) have shown how to solve the initial value problem for the KdV equation

(1.1) $$u_t-6uu_x+\varepsilon^2 u_{xxx}=0$$

for initial data $u(x)$ which, as $|x|\to\infty$, tend to zero so fast that

(1.2) $$\int(1+x^2)|u(x)|\,dx<\infty.$$

Their method associates to each solution of KdV a one-parameter family of Schroedinger operators $\mathscr{L}(t)$:

(1.3) $$\mathscr{L}(t)=-\varepsilon^2\partial_x^2+u(x,t).$$

Recall that for a Schroedinger operator $\mathscr{L}$ whose potential u satisfies (1.2) one can define *scattering data*; these consist of

(i) the reflection coefficient $R(k)$,

(ii) the eigenvalues $-\eta_n^2$, $n=1,\cdots,N$, in the point spectrum of $\mathscr{L}$ ($\eta_n>0$ by convention),

(iii) the norming constants c_n associated with the eigenfunctions f_n.

We recall the definition of the norming constants: Let f be the eigenfunction of $\mathscr{L}$ with eigenvalue $-\eta^2$,

(1.4) $$\mathscr{L}f=-\varepsilon^2 f_{xx}+uf=-\eta^2 f,$$

normalized by

(1.4)′ $$\int_{-\infty}^{\infty}f^2\,dx=1.$$

It is easy to show using condition (1.2) that, as $|x|\to\infty$, $f(x)$ decays exponentially. More precisely, there is a constant c such that

$$f(x)\cong c\exp\{-\eta x/\varepsilon\} \quad \text{as} \quad x\to+\infty. \tag{1.5}$$

This constant c, chosen to be positive, is the norming constant.

The operation relating the potential u of $\mathscr{L}$ to the scattering data $\{R(k), \eta_n, c_n\}$ is called the *scattering transform*. It can be inverted with the aid of the Gelfand–Levitan–Marchenko (GLM) equation (cf. [4]). GGKM have shown that if the potential u varies with a parameter t so that the KdV equation (1.1) is satisfied, then the scattering data vary in a particularly simple manner with t:

$$\begin{aligned} &\text{(i)} \quad R(k,t)=R(k)\exp\{4ik^3t/\varepsilon\},\\ &\text{(ii)} \quad \eta_n \text{ is independent of } t,\\ &\text{(iii)} \quad c_n(t)=c_n\exp\{4\eta^3t/\varepsilon\}. \end{aligned} \tag{1.6}$$

Formulas (1.6) have been derived from those given by GGKM, who treat the case $\varepsilon=1$, by the simple rescaling

$$u(x,t)=v(x/\varepsilon, t/\varepsilon). \tag{1.7}$$

A similar rescaling has been applied at the end of this section to the formulas of Kay and Moses for the inverse scattering transform.

The GGKM solution of the initial value problem for KdV is to take the scattering transform of the initial data, then apply (1.6) to determine the scattering data at time t, and then invert the scattering transform.

In this paper we shall study initial data which satisfy two further conditions in addition to (1.2):

$$u(x)\leqq 0, \tag{1.8}$$

$$u(x) \text{ is } C^1 \text{ and has only a finite number of critical points.} \tag{1.8'}$$

Condition (1.2) is there to make the machinery of scattering theory work smoothly. Formula (1.8) is a genuine restriction that allows us to neglect reflection. Condition (1.8)′ is there for technical reasons. In fact in this study we shall assume that $u(x)$ has a single critical point; we denote by x_0 its location. The changes necessary to handle the more general case (1.8)′ are not hard to make.

We study now the asymptotic behavior of the scattering data as $\varepsilon\to 0$. The distribution of the eigenvalues is governed by

THEOREM 1.1 (Weyl's law). *Let*

$$\mathscr{L}=-\varepsilon^2\partial_x+u$$

be a Schroedinger operator whose potential satisfies (1.2). *Denote by* $N(\varepsilon)$ *the number of eigenvalues of* $\mathcal{L}$; *then*

$$N(\varepsilon) \cong \frac{1}{\varepsilon\pi} \mathcal{R}e \int (-u(y))^{1/2}\, dy. \tag{1.9}$$

More generally, the number $N(\varepsilon, \eta)$ *of eigenvalues less than* $-\eta^2$ *is*

$$N(\varepsilon, \eta) \cong \frac{1}{\varepsilon\pi} \mathcal{R}e \int (-u(y) - \eta^2)^{1/2}\, dy. \tag{1.10}$$

Suppose u satisfies (1.8) and has a single minimum. Then for $-\eta^2 > \min u$ there are two functions $x_+(y)$ and $x_-(\eta)$, defined by

$$u(x_-) = u(x_+) = -\eta^2, \qquad x_- < x_+. \tag{1.11}$$

Clearly,

$$u(x) < -\eta^2 \quad \text{for} \quad x_- < x < x_+,$$

so that (1.10) can be written as

$$N(\varepsilon, \eta) \approx \frac{1}{\pi\varepsilon} \Phi(\eta), \tag{1.12}$$

where

$$\Phi(\eta) = \int_{x_-(\eta)}^{x_+(\eta)} (-u(y) - \eta^2)^{1/2}\, dy \tag{1.13}$$

(see Figure 1).

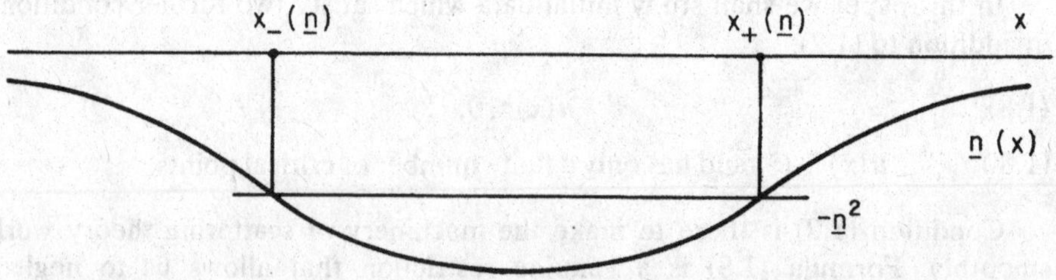

Figure 1

We shall from now on assume that

$$u(x_0) = \min u(x) = -1. \tag{1.13$'$}$$

Then the domain of the functions x_-, x_+ and Φ is $0 \leqq \eta \leqq 1$.

We turn next to the asymptotic determination of the norming constants. We shall use a crude WKB method, i.e., we represent the eigenfunction f in (1.4) as

$$f(x) = c \exp\{-\theta(x)/\varepsilon\}. \tag{1.14}$$

Substituting this into (1.4), we obtain

$$-\theta_x^2+\varepsilon\theta_{xx}+u+\eta^2=0.$$

We now set $\varepsilon=0$ in this relation and define θ as the solution of

$$\theta_x=(\eta^2+u)^{1/2}$$

which yields

$$\theta(\eta,x)=\eta x+\int_x^\infty \eta-(\eta^2+u(y))^{1/2}\,dy. \tag{1.14)$'$}$$

We shall show below, see inequality (1.26), that

$$\int(-u(y))^{1/2}\,dy<\infty.$$

From this it follows easily that the integral (1.14)′ converges, and that, as $x\to\infty$,

$$\theta(\eta,x)=\eta x+O(1). \tag{1.15}$$

As x decreases from $+\infty$, θ is real and decreases until x reaches $x_+(\eta)$. We denote this minimum value of θ by

$$\theta_+(\eta)=\eta x_+(\eta)+\int_{x_+(\eta)}^\infty \eta-(\eta^2+u(y))^{1/2}\,dy. \tag{1.16}$$

We now define

$$c(\eta)=\exp\{\theta_+(\eta)/\varepsilon\} \tag{1.17}$$

and let f be given by formula (1.14), with θ and c defined by (1.14)′ and (1.17). We see that f satisfies the eigenvalue equation (1.4) crudely, and the normalization condition (1.4)′ just as crudely. It follows from (1.15) that f satisfies the asymptotic relation (1.5).

If we apply the WKB method, crude or refined, to calculate the reflection coefficient, we get $R(k)\equiv 0$.

DEFINITION. We define the modified initial function $u(x;\varepsilon)$ as the function whose scattering data are:

(i) $R(k)\equiv 0$,

(ii) η_n given by

$$\Phi(\eta_n)=(n-\tfrac{1}{2})\varepsilon\pi, \qquad n=1,\cdots,N(\varepsilon), \tag{1.18}$$

$$N(\varepsilon)=\left[\frac{1}{\varepsilon\pi}\Phi(0)\right],$$

(iii) $c_n=c(\eta_n)$, $\qquad n=1,\cdots,N(\varepsilon)$,

where [ ] is the "integer part" function, $\Phi(\eta)$ is the function defined by (1.13) and $c(\eta)$ is the function defined by (1.17) and (1.16). Definition $(1.18)_{\text{ii}}$ of η_n is the WKB result and is consistent with Theorem 1.1.

THEOREM 1.2. *The function $u(x;\varepsilon)$, defined above, approximates the prescribed initial data u in the following sense:*

$$L^2-\lim_{\varepsilon\to 0} u(x;\varepsilon)=u(x).$$

The proof of this theorem will be presented in Section 4.

Combining (1.18) and (1.6) gives the scattering data of $u(x, t; \varepsilon)$, the solution of KdV with initial data $u(x;\varepsilon)$:

$$c_n(t)=\exp\{(4\eta^3t+\theta_+(\eta))/\varepsilon\}, \qquad \eta=\eta_n. \tag{1.19}$$

On account of $(1.18)_{\text{i}}$, the functions $u(x, t; \varepsilon)$ are reflectionless potentials. For these, Kay and Moses have given a simple solution of the GLM equation which leads to the following explicit expression of the potential u in terms of its scattering data:

$$u(x,t;\varepsilon)=\partial_x^2 W(x,t;\varepsilon), \tag{1.20}$$

where

$$W(x,t;\varepsilon)=-2\varepsilon^2\log\det(I+G(x,t;\varepsilon)). \tag{1.20$'$}$$

Here G is the matrix

$$G=\varepsilon\left(\frac{\exp\{-(\eta_n x+\eta_m x)/\varepsilon\}}{\eta_n+\eta_m}c_nc_m\right). \tag{1.21}$$

Using (1.19) we can rewrite this as

$$G=\varepsilon\left(\frac{g_ng_m}{\eta_n+\eta_m}\right), \tag{1.21$'$}$$

where

$$g_n(x,t;\varepsilon)=\exp\{-a(\eta_n,x,t)/\varepsilon\}. \tag{1.22}$$

Here

$$a(\eta,x,t)=\eta x-4\eta^3t-\theta_+(\eta), \tag{1.23}$$

where $\theta_+(\eta)$ is defined in (1.16).

We define now some further auxiliary functions that will be of use in Section 2, and prove a few of their properties:

$$\theta_-(\eta)=\eta x_-(\eta)-\int_{-\infty}^{x_-(\eta)}\eta-(\eta^2+u(y))^{1/2}\,dy. \tag{1.24}$$

We denote the derivative of Φ, defined in (1.13), by $-\phi$:

$$\phi(\eta)=-\frac{d}{d\eta}\Phi(\eta)=\int_{x_-(\eta)}^{x_+(\eta)}\frac{\eta}{(-u(y)-\eta^2)^{1/2}}\,dy. \tag{1.25}$$

LEMMA 1.3. *ϕ is a non-negative function which belongs to $L^1(0,1)$.*

Proof: This follows from the fact that Φ is a decreasing function of η which is bounded. The first assertion is obvious; to verify the second we state

$$\begin{aligned}\Phi(0)&=\int_{-\infty}^{\infty}(-u(y))^{1/2}\,dy=\int(-u(y))^{1/2}(1+y^2)^{1/2}(1+y^2)^{-1/2}\,dy\\&\leqq\left[\int|u(y)|(1+y^2)\,dy\int\frac{1}{1+y^2}\,dy\right]^{1/2}<\infty.\end{aligned} \tag{1.26}$$

Here we used the Schwarz inequality, and assumption (1.2).

LEMMA 1.4. *The functions θ_+ and θ_-, defined by* (1.16) *and* (1.24), *respectively, are continuous and satisfy*

$$\begin{aligned}&\theta_+(0)=\theta_-(0)=0,\\&\theta_-(\eta)<\eta x_0<\theta_+(\eta)\quad\text{for}\quad 0<\eta<1.\end{aligned} \tag{1.27}$$

Proof: It follows directly from the definitions of θ_+ and θ_- that they are continuous for $\eta\neq 0$. To show continuity, and (1.27), at $\eta=0$ we argue as follows:

By assumption, $|u(x)|$ is a decreasing function of x for x large enough. Thus, for large x,

$$2\int_{x/2}^{x}(-u(y))^{1/2}\,dy>x(-u(x))^{1/2}. \tag{1.28}$$

Since, according to (1.26),

$$\int_{-\infty}^{\infty}(-u(y))^{1/2}\,dy<\infty,$$

it follows from (1.28) that

$$\lim_{x\to\infty}x(-u(x))^{1/2}=0. \tag{1.29}$$

By (1.11), $-u(x_+)=\eta^2$; from this and (1.29) we conclude that

$$\lim_{\eta\to 0}x_+(\eta)\eta=0. \tag{1.29$'$}$$

This proves that the first term in the definition (1.16) of $\theta_+(\eta)$ tends to 0 as $\eta\to 0$.

We show now that the same is true of the second term. Introducing μ as new variable of integration by

$$u(y) = -\mu^2, \qquad y = x_+(\mu),$$

we have, after integrating by parts,

$$\int_{x_+(\eta)}^{\infty} \eta - (\eta^2 + u(y))^{1/2}\, dy = -\int_0^{\eta} \eta - (\eta^2 - \mu^2)^{1/2} \frac{dx_+(\mu)}{d\mu}\, d\mu$$

$$= \int_0^{\eta} \frac{\mu x_+(\mu)}{(\eta^2 - \mu^2)^{1/2}}\, d\mu - \eta x_+(\eta) \leqq \text{const.} \max_{\mu \leqq \eta} \mu x_+(\mu).$$

It follows from (1.29)′ that the above tends to 0 as $\eta \to 0$. This completes the proof of Lemma 1.4.

2. Asymptotic Analysis of the Inverse Scattering Problem

In this section we carry out an asymptotic analysis as $\varepsilon \to 0$ of $u(x, t; \varepsilon)$, given by formula (1.20):

$$u(x, t; \varepsilon) = \partial_x^2 W(x, t; \varepsilon), \tag{2.1}$$

where W is defined by (1.20)′:

$$W(x, t; \varepsilon) = -2\varepsilon^2 \log \det (I + G(x, t; \varepsilon)) \tag{2.2}$$

and $G(x, t; \varepsilon)$ is given by formula (1.21).

We expand $\det (I + G)$ by grouping together all terms which only contain factors from G with indices from a set S. The resulting formula is

$$\det (I + G) = \sum_S \det G_S, \tag{2.3}$$

where G_S is the principal minor of G obtained by striking out all rows and columns whose index lies outside S. In the sum (2.3), S ranges over all subsets of the N indices and we take $\det G_S = 1$ when S is the null set.

We can factor G, given by (1.21)′, as

$$G = \varepsilon D\left(\frac{1}{\eta_n + \eta_m}\right) D,$$

where D is the diagonal matrix with entries $g_1, \cdots, g_N$. Since the principal minors G_S of G have the same form as G, a similar expression holds for them. The matrix in the middle is of Cauchy type; its determinant is given by

$$\prod_{S \times' S} |\eta_n - \eta_m| \Big/ \prod_{S \times S} (\eta_n + \eta_m),$$

where $S\times' S$ means $S\times S$ minus the diagonal. Since $\det D=\prod_S g_n$, we conclude that

$$\det G_S=(\tfrac{1}{2}\varepsilon)^{|S|}\prod_S \frac{g_n^2}{\eta_n}\prod_{S\times' S}\left|\frac{\eta_n-\eta_m}{\eta_n+\eta_m}\right|.$$

Using (1.22) we can write this as

$$\det G_S=(\tfrac{1}{2}\varepsilon)^{|S|}\exp\{-Q_S/2\varepsilon^2\}, \tag{2.4}$$

where

$$Q_S(x,t;\varepsilon)=4\varepsilon\sum_S a(\eta_n,x,t)-\varepsilon^2\sum_{S\times' S}\log\left(\frac{\eta_n-\eta_m}{\eta_n+\eta_m}\right)^2+2\varepsilon^2\sum_S\log\eta_n. \tag{2.5}$$

Using (2.4) to express (2.3), we get

$$\det(I+G)=\sum_{\text{all } S}(\tfrac{1}{2}\varepsilon)^{|S|}\exp(-Q_S/2\varepsilon^2). \tag{2.6}$$

We shall show that for small ε this sum is dominated by its largest term. Define

$$Q^*(x,t;\varepsilon)=\min_S\{Q_S(x,t;\varepsilon)\}. \tag{2.7}$$

Since all terms in (2.6) are positive, and since there are $2^{N(\varepsilon)}$ of them, we get the following two-sided inequality:

$$(\tfrac{1}{2}\varepsilon)^N\exp\{-Q^*/2\varepsilon^2\}\leqq\det(I+G)\leqq 2^N\exp\{-Q^*/2\varepsilon\}.$$

Taking the logarithm and multiplying by $2\varepsilon^2$ gives

$$2\varepsilon^2 N\log(\tfrac{1}{2}\varepsilon)-Q^*\leqq 2\varepsilon^2\log\det(I+G)\leqq 2\varepsilon^2 N\log 2-Q^*.$$

The center term is just $-W$ as defined by (2.2) so we obtain the estimate

$$|W(x,t;\varepsilon)-Q^*(x,t;\varepsilon)|\leqq 2\varepsilon^2 N(\varepsilon)\log(2/\varepsilon).$$

Since according to (1.9), $N(\varepsilon)=O(\varepsilon^{-1})$, we have proved

THEOREM 2.1.

$$\lim_{\varepsilon\to 0}[W(x,t;\varepsilon)-Q^*(x,t;\varepsilon)]=0, \tag{2.8}$$

uniformly in x and t.

We shall show that the limit of $Q^*(x,t;\varepsilon)$ exists and can be characterized, just as each $Q^*(x,t;\varepsilon)$, as a solution of a minimum problem. To find this minimum problem we rewrite $Q_S(x,t;\varepsilon)$ given by (2.5) as a Stieltjes integral.

For each S we define $\psi_S(\eta,\varepsilon)$ as the distribution

$$\psi_S(\eta,\varepsilon)=\varepsilon\pi\sum_S\delta(\eta-\eta_n) \tag{2.9}$$

and rewrite (2.5) as

$$
\begin{aligned}
Q_S(x,t;\varepsilon) = & \frac{4}{\pi}\int_0^1 a(\eta,x,t)\psi_S(\eta,\varepsilon)\,d\eta \\
& -\frac{1}{\pi^2}\int_0^1\int_0^1 \log\left(\frac{\eta-\mu}{\eta+\mu}\right)^2 \psi_S(\mu,\varepsilon)\times'\psi_S(\eta,\varepsilon)\,d\mu\,d\eta \\
& +\frac{2\varepsilon}{\pi}\int_0^1 \log\eta\,\psi_S(\eta,\varepsilon)\,d\eta;
\end{aligned} \tag{2.10}
$$

here $\psi_S(\mu,\varepsilon)\times'\psi_S(\eta,\varepsilon)$ is the product of the distributions minus the diagonal terms,

$$\psi_S(\mu,\varepsilon)\times'\psi_S(\eta,\varepsilon) = \varepsilon^2\pi^2 \sum_{S\times' S} \delta(\mu-\eta_m)\delta(\eta-\eta_n). \tag{2.11}$$

We define $\phi(\eta,\varepsilon)$ by

$$\phi(\eta,\varepsilon) = \varepsilon\pi \sum_{n=1}^{N(\varepsilon)} \delta(\eta-\eta_n). \tag{2.12}$$

Clearly,

$$0 \leqq \psi_S(\eta,\varepsilon)\,d\eta \leqq \phi(\eta,\varepsilon)\,d\eta \tag{2.13}$$

as measures, for every S.

Since the η_n satisfy Weyl's law (Theorem 1.1), it follows that

$$
\begin{aligned}
& \lim_{\varepsilon\to 0} \phi(\eta,\varepsilon)\,d\eta = \phi(\eta)\,d\eta, \\
& \lim_{\varepsilon\to 0} \phi(\mu,\varepsilon)\times'\phi(\eta,\varepsilon)\,d\mu\,d\eta = \phi(\mu)\phi(\eta)\,d\mu\,d\eta,
\end{aligned} \tag{2.14}
$$

in the sense of weak sequential convergence of measures, where $\phi(\eta)$ is defined by (1.25). These considerations suggest a minimum problem satisfied by the limit $Q^*(x,t;\varepsilon)$:

THEOREM 2.2.

$$\lim_{\varepsilon\to 0} Q^*(x,t;\varepsilon) = Q^*(x,t) \tag{2.15}$$

uniformly on compact subsets of x, t, where

$$Q^*(x,t) = \min\{Q(\psi;x,t)\colon \psi \text{ in } A\}; \tag{2.16}$$

the admissible set A consisting of all Lebesgue measurable functions ψ on [0, 1],

$$0 \leqq \psi(\eta) \leqq \phi(\eta), \tag{2.17}$$

and $Q(\psi; x, t)$ *denoting the quadratic form*

$$Q(\psi; x, t) = \frac{4}{\pi}\int_0^1 a(\eta, x, t)\psi(\eta)\, d\eta \tag{2.18}$$
$$-\frac{1}{\pi^2}\int_0^1\int_0^1 \log\left(\frac{\eta-\mu}{\eta+\mu}\right)^2 \psi(\mu)\psi(\eta)\, d\mu\, d\eta.$$

The proof is based on a series of lemmas and theorems.

Let A_ε denote the set of distributions $\psi_s(\eta, \varepsilon)$ defined by (2.9) for all possible sets of indices s and $\varepsilon > 0$,

$$A_\varepsilon = \{\psi_S(\eta, \varepsilon) : S, \varepsilon\}. \tag{2.19}$$

To the elements ψ of both the admissible sets A_ε and A we assign positive measures $\psi\, d\eta$ on $[0, 1]$. If $\psi_S \in A_\varepsilon$, then from (2.13), (2.12), and $(1.18)_{ii}$ we have

$$\int_0^1 \psi_S(\eta, \varepsilon)\, d\eta \leqq \int_0^1 \phi(\eta, \varepsilon)\, d\eta = \varepsilon\pi N(\varepsilon) \leqq \int_{-\infty}^{\infty} (-u(y))^{1/2}\, dy, \tag{2.20}$$

while if $\psi \in A$, then from (2.17) and (1.25)

$$\int_0^1 \psi(\eta)\, d\eta \leqq \int_0^1 \phi(\eta)\, d\eta = \int_{-\infty}^{\infty} (-u(y))^{1/2}\, dy. \tag{2.21}$$

Thus we conclude from Lemma 1.3:

LEMMA 2.3. *The total variations of the measures* $\psi\, d\eta$, ψ *in* A_ε *and* A *are uniformly bounded.*

DEFINITION. A sequence of distributions ψ_k is said to w-converge to ψ if the corresponding measures $\psi_k(\eta)\, d\eta$ converge to $\psi(\eta)\, d\eta$ in the sense of weak sequential convergence of measures, that is if

$$\lim_{k\to\infty} (\chi, \psi_k) = (\chi, \psi) \tag{2.22}$$

for every continuous function χ. Here the parentheses denote the usual duality on $[0, 1]$:

$$(\chi, \psi) = \int_0^1 \chi(\eta)\psi(\eta)\, d\eta. \tag{2.23}$$

We shall abbreviate (2.22) as w-convergence and write

$$\text{w-}\lim_{k\to\infty} \psi_k = \psi.$$

The next lemma characterizes the admissible set A as w-limits of the distribution in A_ε.

LEMMA 2.4. (a) *Let ψ_k denote a sequence of elements in A_ε, $\psi_k(\eta)=\psi_{S_k}(\eta,\varepsilon_k)$, such that, as $\varepsilon_k\to 0$,*

$$\text{w-}\lim_{k\to\infty}\psi_k(\eta)=\psi(\eta); \tag{2.24}$$

then ψ belongs to A.

(b) *Conversely, for each ψ in A we can choose $S(\varepsilon)$ so that*

$$\text{w-}\lim_{\varepsilon\to 0}\psi_{S(\varepsilon)}(\eta,\varepsilon)=\psi(\eta). \tag{2.25}$$

(c)

$$\text{w-}\lim_{\varepsilon\to 0}\phi(\eta,\varepsilon)=\phi(\eta). \tag{2.26}$$

Proof: By (2.13), the ψ_k satisfy

$$0\leqq\psi_k(\eta)\,d\eta\leqq\phi(\eta,\varepsilon_k)\,d\eta.$$

Since inequalities between measures are preserved for their weak limits, it follows from (2.24), and (2.26) that

$$0\leqq\psi(\eta)\,d\eta\leqq\phi(\eta)\,d\eta.$$

From this we conclude by the Radon–Nikodym theorem that ψ is a measurable function and that (2.17) is satisfied; thus $\psi\in A$.

Part (b) is straightforward and part (c) is contained in (2.14).

The main step in proving Theorem 2.2 is in relating $Q_S(x,t;\varepsilon)$ as given in (2.10) to $Q(\psi;x,t)$ as given by (2.18).

THEOREM 2.5. *Denote by ψ_k a sequence of elements in A_ε, $\psi_k(\eta)=\psi_{S_k}(\eta,\varepsilon_k)$, such that, as $\varepsilon_k\to 0$,*

$$\text{w-}\lim_{k\to\infty}\psi_k(\eta)=\psi(\eta);$$

then

$$\lim_{k\to 0}Q_{S_k}(x,t;\varepsilon_k)=Q(\psi;x,t) \tag{2.27}$$

for every x and t.

The mild technical difficulties in the proof arise in handling the second and third terms of the right-hand side of (2.10). This will be accomplished with the aid of lemmas, and the introduction of the operator L:

$$L\psi(\eta)=\frac{1}{2\pi}\int_0^1\log\left(\frac{\eta-\mu}{\eta+\mu}\right)^2\psi(\mu)\,d\mu. \tag{2.28}$$

LEMMA 2.6. *$L\phi$ is a continuous function on $[0, \infty)$ which tends to zero as $\eta \to \infty$. In the interval $[0, 1]$,*

$$-L\phi(\eta) = \theta_+(\eta) - \theta_-(\eta). \tag{2.29}$$

Proof: Since it was shown in Lemma 1.4 that $\theta_\pm(\eta)$ are continuous finctions, the continuity of $L\phi$ in $[0, 1]$ follows from (2.29). The continuity of $L\phi$ on $[1, \infty)$ and behavior at infinity can be seen directly from formula (2.28).

Formula (2.29) will be derived at the end of Section 3.

COROLLARY 2.7. (a) *For every ψ in the admissible set A, $L\psi$ is continuous and*

$$L\phi(\eta) \leqq L\psi(\eta) \leqq 0 \quad \text{for} \quad \eta \geqq 0. \tag{2.29$'$}$$

(b) *$L\psi$ is an odd function of η.*

Proof: (b) is obvious from (2.28); thus we restrict our attention to $\eta \geqq 0$. Inequality (2.29)′ of (a) follows from the fact that, by inequality (2.17), admissible functions are non-negative, and that the kernel of the integral operator L is negative.

To show the continuity of $L\psi$ at η, $\eta \geqq 0$, we take any sequence $\{\eta_k\}$ converging to η with $\eta_k \geqq 0$. The integrands of $L\psi(\eta_k)$ appearing in (2.28) converge almost everywhere to the integrand of $L\psi(\eta)$ and are bounded below by the integrands of $L\phi(\eta_k)$. Since, by the continuity of $L\phi$, we know that the $L\phi(\eta_k)$ converge to $L\phi(\eta)$, we conclude by a dominated convergence argument that the $L\psi(\eta_k)$ converge to $L\psi(\eta)$. The continuity of $L\psi$ follows, completing the proof.

We use the operator L defined by (2.28) in conjunction with the duality notation of (2.23) to rewrite the functional $Q(\psi; x, t)$ of (2.18) as

$$Q(\psi; x, t) = \frac{4}{\pi}[(a(\eta, x, t), \psi) - \tfrac{1}{2}(L\psi, \psi)]; \tag{2.30}$$

note that, by Lemma 1.4, $a(\eta, x, t)$ given by (1.23) is continuous. In particular this shows that $Q(\psi, x, t)$ is finite for all ψ in the admissible set A.

Preparatory to proving Theorem 2.5, we break up the functionals $Q(\psi)$ and Q_S into two parts depending on a parameter B, which eventually tends to infinity:

$$Q(\psi) = Q^B(\psi) + Q_B(\psi) \tag{2.31}$$

and

$$Q_S = Q_S^B + Q_{B,S}. \tag{2.32}$$

Setting

$$l(\eta, \mu) = -\frac{1}{2\pi} \log\left(\frac{\eta - \mu}{\eta + \mu}\right)^2$$

and

(2.33) $$l^B(\eta,\mu)=\min\{l(\eta,\mu),B\},$$

we define

(2.34) $$Q^B(\psi)=\frac{4}{\pi}\int_0^1 a(\eta)\psi(\eta)\,d\eta+\frac{2}{\pi}\int_0^1\int_0^1 l^B(\eta,\mu)\psi(\eta)\psi(\mu)\,d\eta\,d\mu,$$

(2.35) $$Q_S^B=\frac{4}{\pi}\int_0^1 a(\eta)\psi_S(\eta)\,d\eta+\frac{2}{\pi}\int_0^1\int_0^1 l^B(\eta,\mu)\psi_S(\eta)\psi_S(\mu)\,d\eta\,d\mu.$$

LEMMA 2.8. (a) *If the sequence* $\psi_k=\psi_{S_k}(\eta,\varepsilon_k)$ *satisfies*

$$\text{w-}\lim_{k\to\infty}\psi_k(\eta)=\psi,$$

then

$$\lim_{k\to\infty}Q_{S_k}^B(\varepsilon_k)=Q^B(\psi).$$

(b) *For all* ψ *in* A,

(2.36) $$|Q_B(\psi)|<\delta(B),$$

where

(2.37) $$\lim_{B\to 0}\delta(B)=0.$$

(c) *For all* $\psi(\varepsilon)$ *in* A_ε,

$$|Q_{S,B}|<\delta(B)+\gamma(\varepsilon),$$

where

$$\lim_{\varepsilon\to 0}\gamma(\varepsilon)=0.$$

Proof: Since both functions $a(\eta)$ and $l^B(\eta,\mu)$ are continuous, part (a) is a direct consequence of the definition of weak convergence (2.22).

To prove (2.36) and (2.37) we write

(2.38) $$Q_B(\psi)=\frac{2}{\pi}\iint_{l>B}(l(\eta,\mu)-B)\psi(\eta)\psi(\mu)\,d\eta\,d\mu.$$

Since, for admissible ψ,

$$0\leqq\psi\leqq\phi,$$

it follows that

$$|Q_B(\psi)|\leqq Q_B(\phi).$$

This proves (2.36) with $\delta(B)=Q_B(\phi)$; (2.37) follows from the observation that $Q(\phi)<\infty$.

For the proof of (c) we refer to [9]; we remark that it relies on the choice (1.18) of the η_j to show separately that

$$\varepsilon^2 \sum_{l>B}' l(\eta_n, \eta_m) - B < \delta(B),$$

and that

$$\varepsilon^2 \sum_1^N \log \eta_j < \gamma(\varepsilon).$$

In showing that $\gamma(\varepsilon)\to 0$ as $\varepsilon\to 0$ one has to use the fact that, for a potential u that satisfies (1.2),

$$\int (-u(x))^{1/2} \log(-u(x))\, dx < \infty.$$

Theorem 2.5 is an immediate consequence of Lemma 2.8.

The pointwise limit (2.69) of Theorem 2.5 is uniform on compact subsets of the x, t-space. This follows from

LEMMA 2.9. *The family* $\{Q_S(x, t; \varepsilon)\colon S, \varepsilon\}$ *of linear functions of* x *and* t *defined in* (2.10) *is equicontinuous and equibounded on compact subsets of the* x, t-*space.*

Proof: Since the functions $Q_S(x, t; \varepsilon)$ are linear, it suffices to show that the coefficients of x, t and the quantity $Q_S(0, 0; \varepsilon)$ are uniformly bounded. The coefficients of x and t are uniformly bounded since

$$\varepsilon \sum_S \eta_n \leqq \varepsilon N,$$

$$\varepsilon \sum_S \eta_n^3 \leqq \varepsilon N,$$

respectively, and since, by (1.9), εN is bounded. A uniform bound for $Q_S(0, 0; \varepsilon)$ will be demonstrated by considering the three terms of the right-hand side of (2.10) separately. According to Lemma 1.4, $\theta_+(\eta)$ is continuous and therefore bounded, say $\theta_+(\theta)\leqq M$. The first term of $Q_S(0, 0; \varepsilon)$ is then uniformly bounded since

$$\varepsilon \sum_S \theta_+(\eta_n) \leqq \varepsilon NM$$

and, by (1.9), εN is bounded. The uniform bounds of the second and third terms of $Q_S(0, 0; \varepsilon)$ follow directly from Lemma 2.8.

We now turn to Theorem 2.2.

Proof: Formula (2.15) of Theorem 2.2 will follow from two facts proved below:

$$\limsup_{\varepsilon\to 0} Q^*(x,t;\varepsilon) \leqq \inf\{Q(\psi;x,t)\colon \psi \text{ in } A\} \tag{2.39}$$

and

$$\limsup_{\varepsilon\to 0} Q^*(x,t;\varepsilon) = Q(\psi^*;x,t), \tag{2.40}$$

for some ψ^* in the admissible set A. Comparing (2.39) and (2.40) we see that

$$Q(\psi^*;x,t) = \inf\{Q(\psi;x,t)\colon \psi \text{ in } A\}. \tag{2.41}$$

Thus the minimum asserted in the definition (2.16) of $Q^*(x,t)$ is attained,

$$Q^*(x,t) = Q(\psi^*;x,t), \tag{2.42}$$

and (2.15) follows.

To prove (2.39), let ψ be any element of the admissible set A. By Lemma 2.4, there exists $S(\varepsilon)$ such that

$$\operatorname*{w-lim}_{\varepsilon\to 0} \psi_{S(\varepsilon)}(\eta,\varepsilon) = \psi(\eta). \tag{2.43}$$

Let $\{\varepsilon_k\}$ be a sequence such that $\varepsilon_k \to 0$ and

$$\lim_{k\to\infty} Q^*(x,t;\varepsilon_k) = \limsup_{\varepsilon\to 0} Q^*(x,t;\varepsilon). \tag{2.44}$$

By definition (2.7) and Lemma 2.9, the $Q^*(x,t;\varepsilon)$ are the minima over subfamilies of an equibounded family of functions and so are themselves equibounded, thus insuring that the limits of (2.44) are finite.

By (2.7),

$$Q^*(x,t;\varepsilon_k) \leqq Q_{s_k}(x,t;\varepsilon_k).$$

Using (2.43) and applying Theorem 2.5 we see that the right side above tends to $Q(\psi;x,t)$. Using (2.44) on the left we obtain in the limit

$$\limsup_{\varepsilon\to 0} Q^*(x,t;\varepsilon) \leqq Q(\psi;x,t), \tag{2.45}$$

for all ψ in A. Since the left-hand side of (2.45) is independent of ψ, we take the infimum of the right, obtaining (2.39).

To prove (2.40), let $S(\varepsilon)$ be such that the minimum of (2.7) is attained,

$$Q_{S(\varepsilon)}(x,t;\varepsilon) = Q^*(x,t;\varepsilon). \tag{2.46}$$

Let $\{\varepsilon_k\}$ be a sequence such that $\varepsilon_k \to 0$ and

$$\lim_{k\to\infty} Q^*(x, t; \varepsilon_k) = \liminf_{\varepsilon\to 0} Q^*(x, t; \varepsilon). \tag{2.47}$$

Since by Lemma 2.3 the total variations of the measures $\psi_{S(\varepsilon_k)}(\eta, \varepsilon)\, d\eta$ are uniformly bounded, we apply the Helly selection theorem and extract a w-convergent subsequence. Passing to this subsequence we know by Lemma 2.4 that the w-limit ψ^* lies in the admissible set A. Applying Theorem 2.5 to this sequence and using (2.46) we obtain

$$Q(\psi^*; x, t) = \lim_{k\to\infty} Q^*(x, t; \varepsilon_k), \tag{2.48}$$

which, along with (2.48), proves (2.40).

The limit in (2.15) is uniform on compact subsets of x and t since, by definition (2.7) and Lemma 2.9, the $Q^*(x, t; \varepsilon)$ are minima over subfamilies of an equicontinuous family of functions and so are themselves equicontinuous. The result then follows from the Arzela–Ascoli theorem and the proof of Theorem 2.2 is complete.

Combining the limit (2.15) of Theorem 2.2 with (2.8) of Theorem 2.1 we conclude that

$$\lim_{\varepsilon\to 0} W(x, t; \varepsilon) = Q^*(x, t) \tag{2.49}$$

uniformly on compact subsets of x and t.

Let us denote convergence in the sense of distributions of functions of x and t by d-lim. Since derivatives of a uniformly convergent sequence of functions converge in the distribution sense we conclude from (2.1) and (2.49)

THEOREM 2.10. *Let $u(x, t; \varepsilon)$ be the solution of the* KdV *equation* (1.5) *with initial data $u(x; \varepsilon)$ given by* (1.18); *then*

$$\text{d-}\lim_{\varepsilon\to 0} u(x, t; \varepsilon) = \bar{u}(x, t) \tag{2.50}$$

and

$$\bar{u}(x, t) = \partial_{xx} Q^*(x, t), \tag{2.51}$$

where $Q^*(x, t)$ is defined by (2.16).

Remark. Actually we have proved more; we have shown that $u(x, t; \varepsilon)$ tends to $\bar{u}(x, t)$ in the x-distribution sense. That is, for any C_0^∞ function w of x,

$$\lim_{\varepsilon\to 0} (u(t; \varepsilon), w) = (\bar{u}(t), w); \tag{2.52}$$

furthermore, the limit is uniform over compact subsets of t. We shall further strengthen this result in Theorem 2.12.

Exploiting the concept of convergence in the distribution sense we shall now deduce as a corollary of Theorem 2.10 the following results:

THEOREM 2.11. *Let $u(x, t; \varepsilon)$ be as in Theorem 2.10, then the d-limit*

(2.53) $$\text{d-}\lim_{\varepsilon\to 0} u^2(x, t; \varepsilon) = \overline{u^2}(x, t)$$

exists, and

(2.54) $$\overline{u^2}(x, t) = \tfrac{1}{3}\partial_{xt} Q^*(x, t).$$

Similarly,

(2.55) $$\text{d-}\lim_{\varepsilon\to 0} [u^3(x, t; \varepsilon) + \tfrac{3}{4}\varepsilon^2 u_x^2(x, t; \varepsilon)] = \overline{u^3}(x, t)$$

exists, and

(2.56) $$\overline{u^3}(x, t) = \tfrac{1}{12}\partial_{tt} Q^*(x, t).$$

Proof: We rewrite the KdV equation (1.5) in conservation form

$$u_t - (3u^2)_x + \varepsilon^2 u_{xxx} = 0.$$

Substituting (2.1) into the first term and integrating with respect to x we see that $u(x, t; \varepsilon)$ satisfies

$$\partial_{xt} W - 3u^2 + \varepsilon^2 u_{xx} = 0.$$

Here we ascertain that the constant of integration is zero by letting $x \to \infty$. Solving for u^2 we obtain

(2.57) $$u^2 = \tfrac{1}{3}\partial_{xt} W + \tfrac{1}{3}\varepsilon^2 u_{xx}.$$

The first term on the right-hand side of (2.57) has, by (2.49), the d-limit $\frac{1}{3}\partial_{xt} Q^*$ while, by (2.50), the d-limit of the last term in (2.57) is zero. This shows that (2.53) and (2.54) hold.

To prove (2.55) and (2.56) we multiply the KdV equation (1.5) by $2u$ and write the result in conservation form:

$$(u^2)_t - (4u^3)_x + \varepsilon^2((u^2)_{xx} - 3u_x^2)_x = 0.$$

Substituting (2.57) into the first term and integrating with respect to x we see that $u(x, t; \varepsilon)$ satisfies

$$\tfrac{1}{3}\partial_{tt} W + \tfrac{1}{3}\varepsilon^2 u_{xt} - 4u^3 + \varepsilon^2 (u^2)_{xx} - 3\varepsilon^2 u_x^2 = 0.$$

Here, as before, we ascertain that the constant of integration is zero by letting $x \to \infty$. Solving for $u^3 + \frac{3}{4}\varepsilon^2 u_x^2$ we obtain

(2.58) $$u^3 + \tfrac{3}{4}\varepsilon^2 u_x^2 = \tfrac{1}{12}\partial_{tt} W + \tfrac{1}{12}\varepsilon^2 u_{xt} + \tfrac{1}{4}\varepsilon^2 (u^2)_{xx}.$$

The first term on the right-hand side of (2.58) has, by (2.49), the d-limit $\frac{1}{12}\partial_{tt}Q^*$, while the d-limits of the last two terms in (2.58) are, by (2.50) and (2.53), zero. This shows that (2.55) and (2.56) hold, completing the proof.

Theorem 2.10 can be strengthened:

THEOREM 2.12. *Let $u(x, t; \varepsilon)$ be as in Theorem 2.10. As functions of x, $u(\cdot, t; \varepsilon)$ converge to $\bar{u}(\cdot, t)$ weakly in $L^2(R)$ as $\varepsilon \to 0$ uniformly over compact subsets of t.*

The proof requires

LEMMA 2.13.

$$\lim_{\varepsilon\to 0} \int_{-\infty}^{\infty} u^2(x, t; \varepsilon)\, dx = \int_{-\infty}^{\infty} u^2(x)\, dx \tag{2.59}$$

uniformly in t.

Remark. It is not surprising that (2.59) is independent of t, since $\int u^2\, dx$ is conserved for solutions of KdV.

Proof: Making use of the explicit form of $W(x, t; \varepsilon)$, (2.2)–(2.5) and the fact that u_x vanishes as $|x| \to \infty$, we integrate (2.57):

$$\int_{-\infty}^{\infty} u^2(x, t; \varepsilon)\, dx = \tfrac{1}{3}\partial_t W\Big|_{-\infty}^{\infty}$$

$$= -\frac{16}{3}\varepsilon \left.\frac{\sum_{\text{all } S} (\sum_S \eta_n^3) \det G_S}{\sum_{\text{all } S} \det G_S}\right|_{-\infty}^{\infty} \tag{2.60}$$

$$= \tfrac{16}{3}\varepsilon \sum_{n=1}^{N} \eta_n^3.$$

Here we have taken limits of a weighted mean of $\sum_S \eta_n^3$, where the weights, $\det G_S$, are decreasing exponential functions of x, given by (2.4). At the upper limit the dominant term occurs when S is the null set while at the lower limit the dominant term occurs when S consists of all indices.

It follows from (2.26) of Lemma 2.4 and the definition (2.22) of w-convergence that

$$\lim_{\varepsilon\to 0} \varepsilon\pi \sum_{n=1}^{N} \eta_n^3 = (\eta^3, \phi). \tag{2.61}$$

Using the definition (1.25) of $\phi(\eta)$, interchanging the order of integration and performing the η integral, we obtain

$$\begin{aligned}(\eta^3, \phi) &= \int_0^1 \int_{x_-(\eta)}^{x_+(\eta)} \frac{\eta^4}{(-u(y)-\eta^2)^{1/2}}\,dy\,d\eta \\ &= \int_{-\infty}^{\infty} \int_0^{(-u(y))^{1/2}} \frac{\eta^4}{(-u(y)-\eta^2)^{1/2}}\,d\eta\,dy \qquad (2.62)\\ &= \tfrac{3}{16}\pi \int_{-\infty}^{\infty} u^2(y)\,dy.\end{aligned}$$

Combining (2.60)–(2.62), we obtain (2.59) and prove the lemma.

Proof of Theorem 2.12: By Lemma 2.13 we can find an $M>0$ such that, for all t and ε,

$$\|u(\cdot, t; \varepsilon)\|_{L^2} = \left(\int_{-\infty}^{\infty} u^2(x, t; \varepsilon)\,dx\right)^{1/2} \leqq M. \qquad (2.63)$$

Given any $v \in L^2(R)$ and $\delta>0$, we can find a $C_0^\infty(R)$ function w such that

$$\|v - w\|_{L^2} \leqq \delta. \qquad (2.64)$$

We write

$$(u(t; \varepsilon), v) = (u(t; \varepsilon), w) + (u(t; \varepsilon), v - w). \qquad (2.65)$$

According to (2.52), the first term on the right converges, uniformly for bounded t, to

$$(\bar{u}(t), w).$$

The second term is by the Schwarz inequality and the estimates (2.63) and (2.64) less than $M\delta$. It follows from this that the left side of (2.65) tends to a limit for every v. This proves that $u(x, t; \varepsilon)$ has a weak limit as $\varepsilon \to 0$, uniformly for bounded t. In view of (2.52), this weak limit can be identified with $\bar{u}$. This shows that $\bar{u}(x, t)$ is in L_2 with respect to x, and proves Theorem 2.12.

We end this section by remarking that one can show

$$\lim_{\varepsilon \to 0} \int_{-\infty}^{\infty} u(x, t; \varepsilon)\,dx = \int_{-\infty}^{\infty} u(x)\,dx. \qquad (2.66)$$

3. Analysis of the Minimum Problem

Theorem 2.10 gives an explicit formula, (2.51), for $\bar{u}(x, t)$, the limit in the distribution sense of solutions to the KdV equation as the coefficient of dispersion tends to zero. The formula involves the function $Q^*(x, t)$ characterized by a minimum problem, (2.16), where the quantity $Q(\psi; x, t)$ to be minimized, (2.30), is a quadratic function of ψ and the functions ψ of the admissible set A are subject to two linear inequalities, (2.17), a so-called quadratic programming problem. In this section we begin to use this formula to study a variety of properties of the limit $\bar{u}(x, t)$ and the related limits $\overline{u^2}(x, t)$ and $\overline{u^3}(x, t)$ given by formulas (2.54) and (2.56) of Theorem 2.11.

We start with some easy observations:

THEOREM 3.1. *As a function of x and t, Q^* is*
(a) *less than or equal to* 0,
(b) *continuous,*
(c) *concave,*
(d) *increasing in x,*
(e) *decreasing in t.*

Proof: Since $0 \in A$, it follows from (2.16) that $Q^*(x, t) \leqq Q(0; x, t) = 0$; this proves (a). Using the definition (1.23) of $a(\eta, x, t)$ in the formula (2.30) for $Q(\psi; x, t)$ one sees that

$$Q(\psi; x, t) = \frac{4}{\pi}(\eta, \psi)x - \frac{16}{\pi}(\eta^3, \psi)t - \frac{4}{\pi}(\theta_+, \psi) - \frac{2}{\pi}(L\psi, \psi). \tag{3.1}$$

Since the admissible ψ satisfy $0 \leqq \psi \leqq \phi$ and ϕ belongs to $L^1[0, 1]$, it follows easily from (3.1) that $\{Q(\psi; x, t): \psi \in A\}$ is an equicontinuous family of functions of x and t. It follows that Q^*, the infimum of an equicontinuous family of functions, is itself continuous.

Each function $Q(\psi; x, t)$, being linear in x, t, has properties (c), (d) and (e). It follows that so does Q^*, their infimum.

We draw now some conclusions from Theorem 3.1. From the concavity of Q^* we deduce that the matrix of second derivatives of Q^* is negative, in the distribution sense. Using expressions (2.51), (2.54), and (2.56) for these second derivatives yields

COROLLARY 3.2.

$$\xi^2\bar{u} + 6\xi\tau\overline{u^2} + 12\tau^2\overline{u^3} \leqq 0, \tag{3.2}$$

for all real ξ, τ. In particular,

$$\bar{u} \leqq 0 \quad \textit{and} \quad \overline{u^3} \leqq 0. \tag{3.3}$$

We remark that there is no maximum principle like (3.3) for solutions of the KdV equation. That is, if the initial values of a solution $u(x, t)$ of the KdV equation satisfy $u(x, 0) \leqq 0$, one cannot conclude that $u(x, t) \leqq 0$. This may be seen by noting that if $u(x_0, 0) = 0$, the value of the third x-derivative of u at x_0 could well be negative. It follows then from the KdV equation that $u_t(x_0, 0) > 0$, and so $u(x_0, t) > 0$ for small $t > 0$. Assumption (1.8) is that $u(x) \leqq 0$; thus once we show $\bar{u}(x, 0) = u(x)$, then (3.3) can be interpreted as a kind of maximum principle for the zero dispersion limit of solutions to KdV.

To derive further information, some properties of the integral operator L, defined by (2.28), are needed. The first is a strengthening of Corollary 2.7 which asserted that, for every ψ in the admissible set A, $L\psi(\eta)$ is a continuous function on R that vanishes at infinity.

THEOREM 3.3. *The set of functions* $\{L\psi: \psi \in A\}$ *is an equicontinuous subset of* $C_0(R)$.

Proof: We break up the operator L into two parts depending on a parameter B which will tend to infinity:

$$L = L^B + L_B, \tag{3.4}$$

where

$$L^B\psi(\eta) = -\int_0^1 l^B(\eta, \mu)\psi(\mu)\, d\mu \tag{3.5}$$

and l^B is defined by (2.33).

As B tends to infinity, $L^B\phi(\eta)$ will tend to $L\phi(\eta)$, monotonically for all $\eta \in R$. Since these functions are in $C_0^\infty(R)$, it follows from Dini's theorem that the convergence is uniform in η. That is, if we define

$$\delta(B) = \max |L_B\phi|, \tag{3.6}$$

then

$$\lim_{B\to 0} \delta(B) = 0. \tag{3.7}$$

Let ψ be any admissible function; since admissibility means that

$$0 \leqq \psi(\eta) \leqq \phi(\eta),$$

and since the kernel of L_B is of one sign, it follows by (3.6) that, for any η,

$$|L_B\psi| \leqq |L_B\phi| \leqq \delta(B). \tag{3.8}$$

The kernel of L^B, defined in (2.33), is uniformly continuous. It follows that L^B maps any set bounded in the L^1 norm into an equicontinuous set. Since the admissible functions are L^1-bounded, it follows that they are mapped by L^B into an equicontinuous set. Since, by (3.8), $L^B\psi$ differs from $L\psi$ at most by $\delta(B)$, and since $\delta(B) \to 0$ as $B \to \infty$, Theorem 3.3 follows.

Recalling the definition of w-convergence (2.22) for sequences in A, we use the previous result to prove the continuous dependence of L and Q on ψ:

THEOREM 3.4. *If ψ_k is a w-convergent sequence of elements in A:*

$$\text{w-lim}\ \psi_k = \psi,$$

then

$$\lim_{k\to\infty} L\psi_k(\eta) = L\psi(\eta) \tag{3.9}$$

uniformly on R, and

$$\lim_{k\to\infty} Q(\psi_k; x, t) = Q(\psi; x, t), \tag{3.10}$$

uniformly on compact subsets of x and t.

Proof: By Theorem 3.3, the functions $L\psi_k$ are equicontinuous, so by the Arzela–Ascoli theorem it is enough to show the convergence of (3.9) pointwise. Since the kernel of L^B is continuous, it follows from the definition of w-convergence that, for each η,

$$\lim L^B\psi_k(\eta) = L^B\psi(\eta).$$

Since, by Theorem 3.3, $L^B\psi_k$ differs at most by $\delta(B)$ from $L\psi_k$, (3.9) follows.

Since the family of $Q(\psi; x, t)$ is equicontinuous in x and t, it suffices to show the pointwise convergence of (3.10). We consider separately the two terms on the right side of formula (2.30) for $Q(\psi; x, t)$. According to Lemma 1.4, $a(\eta, x, t)$ given by (1.22) is a continuous function of η and thus, by the definition (2.22) of w-convergence,

$$\lim_{k\to\infty} (a, \psi_k) = (a, \psi) \tag{3.11}$$

for every x and t.

Next we consider the identity

$$(L\psi_k, \psi_k) - (L\psi, \psi) = (L\psi_k - L\psi, \psi_k) + (L\psi, \psi_k - \psi). \tag{3.12}$$

The first term on the right is bounded in absolute value by

$$\|L\psi_k - L\psi\|_{L^\infty}\|\psi_k\|_{L^1}.$$

Formula (3.9) shows that the first factor tends to zero; since ψ is admissible, the second factor is bounded by $\|\phi\|_{L^1}$. Thus the first term on the right of (3.12) vanishes as $k \to \infty$. Corollary 2.7 asserts that $L\psi$ is continuous; hence the second term on the right of (3.12) tends to zero as $k \to \infty$ by the w-convergence of ψ_k to ψ. This proves that

$$\lim_{k\to\infty} (L\psi_k, \psi_k) = (L\psi, \psi). \tag{3.13}$$

Combining (3.11) and (3.13) according to formula (2.30) for $Q(\psi; x, t)$ we obtain (3.10) as asserted in the theorem.

An immediate consequence of (3.10) in Theorem 3.4 is

COROLLARY 3.5. *The functional $Q(\psi; x, t)$ is a continuous function over $A \times R \times R$, where the admissible set A is given the weak sequential topology.*

Proof: The result follows from (3.10) and the fact that $Q(\psi; x, t)$ is jointly continuous in x and t.

THEOREM 3.6. *The set of admissible functions A, defined by* (2.17), *is compact in the weak sequential topology; that is, every sequence in A contains a subsequence w-convergent to an element of A.*

Proof: By Lemma 2.3, the total variations of the measures associated with A are uniformly bounded. It follows then from the Helly selection principle that every sequence of associated measures contains a subsequence $\{\psi_k(\eta)\, d\eta\}$ convergent to a measure $d\sigma$ in the sense that

$$\lim_{k\to\infty} \int \chi(\eta)\psi_k(\eta)\, d\eta = \int \chi(\eta)\, d\sigma \quad .$$

for every continuous χ. We conclude from (2.17) that, for all non-negative χ,

$$0 \leqq \int \chi(\eta)\, d\sigma \leqq \int \chi(\eta)\phi(\eta)\, d\eta.$$

This implies that the measure $d\sigma$ satisfies

$$0 \leqq d\sigma \leqq \phi(\eta)\, d\eta$$

from which we conclude by the Radon–Nikodym theorem that $d\sigma = \psi(\eta)\, d\eta$, where ψ satisfies (2.17). Thus ψ_k is w-convergent to $\psi \in A$ and the proof of Theorem 3.6 is complete.

Combining the compactness of A asserted by Theorem 3.6 with the continuity of Q stated in Corollary 3.5, it will follow from a general topological principle that, for each x and t, $Q(\psi; x, t)$ takes on a minimum value over A. This gives a direct argument that the minimum problem (2.16) of Theorem 2.2 has a solution.

To prove uniqueness of the solution we need another property of the operator L.

THEOREM 3.7. *The integral operator is negative definite over all ψ in $L^1[0, 1]$ for which $L\psi$ is in $L^\infty(R)$. This means*

$$-(L\psi, \psi) > 0 \tag{3.14}$$

for all $\psi \neq 0$ in that class.

Proof: It is convenient to extend the functions ψ, originally defined on the interval $[0, 1]$, to all R by first setting $\psi(\eta) = 0$ for $\eta > 1$ and second by making

ψ odd setting $\psi(-\eta) = -\psi(\eta)$. This allows us to rewrite the operator L, defined by (2.28), as a convolution operator,

$$L\psi(\eta) = \frac{1}{2\pi}\int_{-\infty}^{\infty} \log(\eta-\mu)^2\psi(\mu)\,d\mu = -l * \psi(\eta), \tag{3.15}$$

where

$$l(\eta) = -\frac{1}{2\pi}\log \eta^2. \tag{3.16}$$

Since $L\psi(\eta)$ is odd, we have

$$-(L\psi, \psi) = \tfrac{1}{2}\langle l * \psi, \psi\rangle, \tag{3.17}$$

where the brackets on the right denote the usual duality on R,

$$\langle \chi, \psi\rangle = \int_{-\infty}^{\infty} \chi(\eta)\psi(\eta)\,d\eta.$$

The definiteness of convolution operators is the subject of theorems going back to Carathéodory, Bochner, and L. Schwartz. By (3.17), Theorem 3.7 follows from

LEMMA 3.8. *If $\psi \in L^1(R)$ is odd with compact support and $l * \psi \in L^\infty(R)$, then*

$$\langle l * \psi, \psi\rangle = \int_{-\infty}^{\infty} \frac{1}{|s|}|\hat{\psi}(s)|^2\,ds, \tag{3.18}$$

where $\hat{\psi}(s)$ is the Fourier transform given by

$$\hat{\psi}(s) = \frac{1}{(2\pi)^{1/2}}\int_{-\infty}^{\infty} e^{-i\eta s}\psi(\eta)\,d\eta. \tag{3.19}$$

Proof: Since l is a tempered distribution, for all ψ in the Schwartz class S the Fourier transform of the convolution $l * \psi$ is $(2\pi)^{1/2}\hat{l}\hat{\psi}$. By the Parseval relation,

$$\langle l * \psi, \psi\rangle = (2\pi)^{1/2}\langle \hat{l}\hat{\psi}, \bar{\hat{\psi}}\rangle = (2\pi)^{1/2}\langle \hat{l}, |\hat{\psi}|^2\rangle. \tag{3.20}$$

The Fourier transform of l in the sense of the theory of distributions is

$$\hat{l}(s) = \frac{1}{(2\pi)^{1/2}}\,\text{F.P.}\,\frac{1}{|s|},$$

where the F.P. indicates the integral is a Hadamard finite part. This means that, for every χ in S,

$$\langle \hat{l}, \chi\rangle = \frac{1}{(2\pi)^{1/2}}\lim_{\varepsilon\to 0}\left[\int_{|s|>\varepsilon}\frac{1}{|s|}\chi(s)\,ds + 2\chi(0)\log\varepsilon\right].$$

In particular if $\chi(0)=0$, then

$$\langle \hat{l}, \chi\rangle = \frac{1}{(2\pi)^{1/2}}\int_{-\infty}^{\infty}\frac{1}{|s|}\chi(s)\,ds. \tag{3.21}$$

We apply this now to $\chi(s)=|\hat{\psi}(s)|^2$; taking ψ to be an odd function, (3.19) shows that

$$\hat{\psi}(0)=\frac{1}{(2\pi)^{1/2}}\int_{-\infty}^{\infty}\psi(\eta)\,d\eta=0$$

and allows us to employ (3.21) in conjunction with (3.20) to obtain

$$\langle l*\psi, \psi\rangle = \int_{-\infty}^{\infty}\frac{1}{|s|}|\hat{\psi}(s)|^2\,ds \tag{3.22}$$

for all odd ψ in S.

Now let ψ be any odd compactly supported member of $L^1(R)$ for which $l*\psi$ is in $L^\infty(R)$. We approximate ψ by a sequence ψ_n of functions in S defined as follows:

$$\psi_n = j_n * \psi, \tag{3.23}$$

where the mollifier j_n is given by

$$\begin{aligned} j_n(\eta) &= nj(n\eta),\\ j(\eta) &= \frac{1}{(2\pi)^{1/2}}\exp\{-\tfrac{1}{2}\eta^2\}. \end{aligned} \tag{3.24}$$

Since ψ is odd with compact support and the j_n are even elements of S, each ψ_n is an odd member of S and therefore satisfies (3.22). By classical properties of mollifiers we know that since ψ is in $L^1(R)$,

$$\lim_{n\to\infty}\|\psi_n-\psi\|_{L^1}=0, \tag{3.25}$$

and since $l*\psi$ is in $L^\infty(R)$,

$$\|l*\psi_n\|_{L^\infty}=\|j_n*l*\psi\|_{L^\infty}\leqq\|l*\psi\|_{L^\infty}. \tag{3.26}$$

Using (3.26) we get the estimate

$$\begin{aligned} |\langle l*\psi_n,\psi_n\rangle-\langle l*\psi,\psi\rangle| &= |\langle l*\psi_n+l*\psi,\psi_n-\psi\rangle|\\ &\leqq 2\|l*\psi\|_{L^\infty}\|\psi_n-\psi\|_{L^1} \end{aligned}$$

which by (3.25) implies

$$\lim_{n\to\infty}\langle l*\psi_n,\psi_n\rangle=\langle l*\psi,\psi\rangle. \tag{3.27}$$

Since (3.22) is satisfied for ψ_n, we deduce from (3.27) that

$$\lim_{n\to\infty}\int_{-\infty}^{\infty}\frac{1}{|s|}|g\hat{\psi}_n(s)|^2\,ds=\langle l*\psi,\psi\rangle. \tag{3.28}$$

However from (3.24) we know that

$$\hat{j}_n(s)=\frac{1}{(2\pi)^{1/2}}\exp\{-s^2/2n^2\}$$

which along with (3.23) gives

$$\begin{aligned}\hat{\psi}_n(s)&=(2\pi)^{1/2}\hat{j}_n(s)\hat{\psi}(s)\\&=\exp\{-s^2/2n^2\}\hat{\psi}(s);\end{aligned}$$

hence the integral on the left-hand side of (3.28) converges to the integral in (3.18), completing the proof of both the lemma and Theorem 3.7.

An important consequence of the negative definiteness of L is:

THEOREM 3.9. *For each given x and t, the functional $Q(\psi; x, t)$ assumes its minimum at exactly one element, denoted $\psi^*(x, t)$, of the admissible set A.*

Proof: Every ψ in A satisfies $0\leqq\psi\leqq\phi$ and ϕ is in $L^1[0, 1]$ so that we know that ψ is also in $L^1[0, 1]$. By Corollary 2.7, $L\psi$ is an element of $C_0(R)$ and therefore also of $L^\infty(R)$. We can then apply Theorem 3.7 to the difference of functions in A to prove that $Q(\psi; x, t)$ is strictly convex over the set A. Since A is a convex set, $Q(\psi)$ assumes its minimum at exactly one admissible ψ.

THEOREM 3.10. (a) *$\psi^*(x, t)$ depends continuously on x, t in the sense of w-convergence.*

(b) *$L\psi^*(x, t)$ depends continuously on x, t in the maximum norm.*

Proof: Suppose $(x_n, t_n)\to(x, t)$. Since, according to Theorem 3.6, A is compact, the sequence $\psi^*(x_n, t_n)$ has a cluster point ψ_* in A; a subsequence of $\psi^*(x_n, t_n)$ w-converges to ψ_*. Since, by Corollary 3.5, $Q(\psi; x, t)$ is a continuous function of ψ, and since, by definition, $Q^*(x_n, t_n)=Q(\psi^*(x_n, t_n), x_n, t_n)$, we conclude that over this subsequence

$$\lim_{n\to\infty}Q^*(x_n, t_n)=\lim_{n\to\infty}Q(\psi^*(x_n, t_n); x_n, t_n)=Q(\psi_*; x, t). \tag{3.29}$$

On the other hand, since by part (b) of Theorem 3.1, Q^* is a continuous function of x and t,

$$\lim_{n\to\infty}Q^*(x_n, t_n)=Q^*(x, t).$$

Comparing this with (3.29), we see $Q^*(x, t)=Q(\psi_*; x, t)$ which means that ψ_* solves the minimum problem at (x, t) and thus, by the uniqueness of Theorem

3.9, $\psi_* = \psi^*(x, t)$. Since $\psi^*(x, t)$ is therefore the only cluster point of $\psi^*(x_n, t_n)$ and A is compact, we conclude

$$\text{w-}\lim_{n\to\infty} \psi^*(x_n, t_n) = \psi^*(x, t);$$

this proves the continuity of ψ^*. The continuity of $L\psi^*$ follows since, by Theorem 3.4, L is sequentially continuous from A into $C_0(R)$ and the proof is complete.

The argument given in Theorem 3.1 for the continuity of $Q^*(x, t)$ shows that Q^* is even Lipschitz continuous in x and t. But now we can show that more is true.

THEOREM 3.11. *$Q^*(x, t)$ is a continuously differentiable function of x and t with*

$$\partial_x Q^*(x, t) = \frac{4}{\pi}(\eta, \psi^*(x, t)),$$

(3.30)

$$\partial_t Q^*(x, t) = -\frac{16}{\pi}(\eta^3, \psi^*(x, t)).$$

Proof: We deduce from formula (3.1) that

$$Q(\psi; x, t) - Q(\psi; y, s) = \frac{4}{\pi}(\eta, \psi)(x-y) - \frac{16}{\pi}(\eta^3, \psi)(t-s) \tag{3.31}$$

for all $\psi \in A$. Since ψ^* minimizes Q, we have the two-sided inequality

$$\begin{aligned} Q(\psi^*(x, t); x, t) - Q(\psi^*(x, t); y, s) &\leqq Q^*(x, t) - Q^*(y, s) \\ &\leqq Q(\psi^*(y, s); x, t) - Q(\psi^*(y, s); y, s) \end{aligned}$$

which when combined with (3.31) gives

$$\begin{aligned} &\frac{4}{\pi}(\eta, \psi^*(x, t))(x-y) - \frac{16}{\pi}(\eta^3, \psi^*(x, t))(t-s) \\ &\leqq Q^*(x, t) - Q^*(y, s) \leqq \frac{4}{\pi}(\eta, \psi^*(y, s))(x-y) - \frac{16}{\pi}(\eta^3, \psi^*(y, s))(t-s). \end{aligned}$$

Applying the continuity of ψ^* in the w-topology, proved in Theorem 3.10, to both sides of the above two-sided inequality yields the formulas (3.30) and shows that Q^* is continuously differentiable, proving the theorem.

The solution of a minimum problem satisfies variational conditions. For a minimum problem with constraints, such as the one we are dealing with, the variational conditions take the form (suppressing x and t)

$$\frac{d}{d\varepsilon} Q(\psi^* + \varepsilon\chi)|_{\varepsilon=0} \geqq 0 \tag{3.32}$$

for all functions χ such that $\psi^* + \chi$ is admissible. Since

$$Q(\psi + \varepsilon\chi) = Q(\psi) + \varepsilon\frac{4}{\pi}(a - L\psi, \chi) - \varepsilon^2\frac{2}{\pi}(L\chi, \chi),$$

we can restate (3.22) as

$$(a - L\psi^*, \chi) \geqq 0. \tag{3.32$'$}$$

DEFINITION. We say that ψ in A satisfies the variational condition for minimizing $Q(\psi)$ in A if

$$\frac{d}{d\varepsilon}Q(\psi + \varepsilon x)|_{\varepsilon=0} = (a - L\psi, \chi) \geqq 0 \tag{3.33}$$

for all χ such that

$$\psi + \chi \quad \text{in} \quad A. \tag{3.33$'$}$$

Suppose ψ satisfies (3.33); set

$$\chi_+ = \begin{cases} -\psi & \text{where} \quad a - L\psi > 0, \\ 0 & \text{otherwise,} \end{cases}$$

$$\chi_- = \begin{cases} \phi - \psi & \text{where} \quad a - L\psi < 0, \\ 0 & \text{otherwise.} \end{cases}$$

Clearly, χ_+ and χ_- satisfy (3.33)$'$. By construction, $(a - L\psi, \chi_\pm) \leqq 0$; this is the opposite of (3.33), which implies that

$$(a - L\psi, \chi_\pm) = 0.$$

This yields

$$\psi = \begin{cases} 0 & \text{where} \quad a - L\psi > 0, \\ \phi & \text{where} \quad a - L\psi < 0. \end{cases} \tag{3.34}$$

It is easy to verify that, conversely, if ψ satisfies (3.34), then (3.33) holds for all χ satisfying (3.33)$'$.

THEOREM 3.12. *If ψ in A satisfies the variational condition, then $\psi = \psi^*$.*

Proof: Since $\chi = \psi^* - \psi$ satisfies (3.33)$'$, ψ satisfies (3.33) with $\chi = \psi^* - \psi$,

$$(a - L\psi, \psi^* - \psi) \geqq 0.$$

Since the minimizing ψ^* also satisfies the variational condition, and $\chi = \psi - \psi^*$ satisfies (3.33)$'$, it satisfies (3.32)$'$:

$$(a - L\psi^*, \psi - \psi^*) \geqq 0.$$

We add these inequalities and obtain

$$(L(\psi^*-\psi), \psi^*-\psi) \geqq 0.$$

Using the negative definiteness of L we conclude that $\psi^*-\psi=0$.

The importance of Theorem 3.12 lies in this: in order to show that ψ^* solves the minimum problem (2.16) it suffices to verify that it satisfies the variational conditions (3.34). We shall exploit this idea in Sections 4 and 5 of Part II.

An easy direct consequence of the variational conditions (3.34) that we shall need in later sections is

THEOREM 3.13. *For all x and t,*

$$\psi^*(x,t) = \begin{cases} 0 & \text{where} \quad \eta x - 4\eta^3 t - \theta_+(\eta) > 0, \\ \phi & \text{where} \quad \eta x - 4\eta^3 t - \theta_-(\eta) < 0. \end{cases} \tag{3.35}$$

Proof: By (2.29) of Corollary 2.7,

$$a \leqq a - L\psi^* \leqq a - L\phi.$$

Since, by Lemma 2.6, $-L\phi(\eta) = \theta_+(\eta) - \theta_-(\eta)$ and, by definition (1.22), $a(\eta, x, t) = \eta x - 4\eta^3 t - \theta_+(\eta)$, we see the inequality can be rewritten as

$$\eta x - 4\eta^3 t - \theta_+(\eta) \leqq a - L\psi^* \leqq \eta x - 4\eta^3 t - \theta_-(\eta).$$

Clearly, (3.35) follows from this inequality and the variational conditions (3.34); and the theorem is proved.

Motivated by the variational conditions, we partition $[0,1] \times R \times R$ into three disjoint subsets:

$$\begin{aligned} I^0 &= \{(\eta, x, t) : a - L\psi^* = 0\}, \\ I^+ &= \{(\eta, x, t) : a - L\psi^* > 0\}, \\ I^- &= \{(\eta, x, t) : a - L\psi^* < 0\}. \end{aligned} \tag{3.36}$$

From Theorem 3.10 we see that $a - L\psi^*$ is continuous on $[0,1] \times R \times R$ and so the sets I^+ and I^- are open while I^0 is closed and separates I^+ and I^-. We denote the (x, t) slices of I^0, I^+, and I^- by $I^0(x,t)$, $I^+(x,t)$ and $I^-(x,t)$, respectively, where, for instance

$$I^0(x,t) = \{\eta : (\eta, x, t) \text{ in } I^0\}. \tag{3.37}$$

Given the sets $I^0(x,t)$, $I^+(x,t)$ and $I^-(x,t)$, one can determine $\psi^*(x,t)$ by

THEOREM 3.14. *If $\psi \in L^1[0,1]$ and $L\psi \in L^\infty(R)$ such that ψ satisfies*

$$L\psi = a \quad \text{on} \quad I^0(x,t) \tag{3.38}$$

and

(3.39)
$$\psi = 0 \quad on \quad I^+(x,t),$$
$$\psi = \phi \quad on \quad I^-(x,t),$$

then $\psi = \psi^*(x,t)$.

Proof: That $\psi^*(x,t)$ satisfies (3.38) and (3.39) is clear from the construction (3.36) of the sets I^0, I^+, and I^-, and from the variational conditions (3.34). Now if ψ satisfies (3.38) and (3.39), then the difference $\psi_0 = \psi - \psi^*(x,t)$ satisfies the homogeneous version of (3.38) and (3.39):

(3.40)
$$L\psi_0 = 0 \quad \text{on} \quad I^0(x,t),$$
$$\psi_0 = 0 \quad \text{off} \quad I^0(x,t).$$

This implies that the product $\psi_0 \cdot L\psi_0 = 0$ for all η, so that

$$(L\psi_0, \psi_0) = 0.$$

But, according to Theorem 3.7, L is negative definite and we conclude $\psi_0 = 0$, proving the theorem.

We now establish the last major property of the operator L. We extend the functions ψ defined on $[0, 1]$ to all of R by setting $\psi(\eta) = 0$ for $\eta > 1$ and taking ψ to be odd: $\psi(-\eta) = -\psi(\eta)$. The set of all functions obtained by extending elements of $L^p[0, 1]$, $p \geqq 1$, in this fashion we denote by $L_0^p(R)$. Clearly, $L_0^p(R)$ is just the set of all odd $L^p(R)$ functions supported in $[-1, 1]$.

THEOREM 3.15. *If* $\psi \in L_0^p(R)$ *for some* p, $1 < p < \infty$, *then* $L\psi \in C_0(R)$ *and*

(3.41)
$$L\psi(\eta) = \int_0^\eta H\psi(\tau)\, d\tau,$$

where H *is the Hilbert transform*:

(3.42)
$$H\psi(\eta) = \frac{1}{\pi} \text{P.V.} \int_{-\infty}^{\infty} \frac{1}{\eta - \mu} \psi(\mu)\, d\mu.$$

Proof: Since the Hilbert transform takes L^p functions into L^p functions for $1 < p < \infty$ such that

$$H\psi(\tau) = L^p - \lim_{\varepsilon \to 0} \frac{1}{\pi} \int_{-\infty}^{\infty} \frac{\tau - \mu}{(\tau - \mu)^2 + \varepsilon^2} \psi(\mu)\, d\mu,$$

we conclude that

$$\int_0^\eta H\psi(\tau)\, d\tau = \lim_{\varepsilon \to 0} \int_0^\eta \frac{1}{\pi} \int_{-\infty}^{\infty} \frac{\tau - \mu}{(\tau - \mu)^2 + \varepsilon^2} \psi(\mu)\, d\mu\, d\tau.$$

The order of integration can then be exchanged by Fubini's theorem and the τ integration can be carried out to obtain

$$
\begin{aligned}
\int_0^\eta H\psi(\tau)\,d\tau &= \lim_{\varepsilon\to 0}\int_{-\infty}^{\infty}\frac{1}{2\pi}\log\frac{(\eta-\mu)^2+\varepsilon^2}{\mu^2+\varepsilon^2}\psi(\mu)\,d\mu \\
&= \lim_{\varepsilon\to 0}\frac{1}{2\pi}\int_0^{\infty}\log\frac{(\eta-\mu)^2+\varepsilon^2}{(\eta+\mu)^2+\varepsilon^2}\psi(\mu)\,d\mu.
\end{aligned}
\tag{3.43}
$$

Here the oddness of ψ was used in the last step. Since, for fixed η,

$$
L^q-\lim_{\varepsilon\to 0}\frac{1}{2\pi}\log\frac{(\eta-\mu)^2+\varepsilon^2}{(\eta+\mu)^2+\varepsilon^2}=\frac{1}{2\pi}\log\left(\frac{\eta-\mu}{\eta+\mu}\right)^2,
$$

where $1/p+1/q=1$, we see from (3.43) that

$$
\int_0^\eta H\psi(\tau)\,d\tau=\frac{1}{2\pi}\int_0^1\log\left(\frac{\eta-\mu}{\eta+\mu}\right)^2\psi(\mu)\,d\mu \tag{3.44}
$$

which, by the definition of L, (2.28), proves (3.41).

The continuity of $L\psi$ follows directly from formula (3.41). It can be seen from the definition of L that the kernel $(1/2\pi)\log((\eta-\mu)/(\eta+\mu))^2$ tends to zero uniformly over the support of ψ as $|\eta|$ tends to infinity. Thus $L\psi$ vanishes at infinity and the theorem is established.

Recall now the relation of the Hilbert transform to analytic functions: Let H^p, $1\leqq p<\infty$, denote the space of all functions $f(\zeta)$ satisfying:

(i) $f(\zeta)$ is analytic in the upper half-plane $\mathcal{I}m(\zeta)>0$;

(ii) there exists a constant $C>0$ such that

$$
\left(\int_{-\infty}^{\infty}|f(\eta+i\tau)|^p\,d\eta\right)^{1/p}<C \quad \text{for all} \quad \tau>0.
$$

Such functions have boundary values in $L^p(R)$ in the sense that

$$
L^p\text{-}\lim_{\tau\to 0}f(\eta+i\tau)=f(\eta)
$$

exists.

The following result is classical (cf. [7]).

THEOREM 3.16. *If $\psi\in L^p$ for some p, $1<p<\infty$, and ψ is real, then ψ and $H\psi$ are the real and imaginary parts of the boundary values of a function in H^p. That is,*

$$
\begin{aligned}
\psi(\eta)&=\mathcal{R}e\,(f(\eta)),\\
H\psi(\eta)&=\mathcal{I}m\,(f(\eta)),
\end{aligned}
\tag{3.45}
$$

for some f in H^p.

Using this result, the solution of equations (3.38), (3.39) can be reduced to a Riemann–Hilbert problem in potential theory.

Rather than determining $\psi^*(x, t)$ directly, we shall determine the partial derivatives, $\partial_x\psi^*(x, t)$ and $\partial_t\psi^*(x, t)$. We shall later make assumptions that imply that ψ^* depends differentiably on x and t in the w-topology over the η-variable with derivatives in $L^p[0, 1]$ for some p, $1<p<2$. This means that for every χ in $C[0, 1]$ the function $(\chi, \psi^*(x, t))$ is a differentiable function of x and t, and there exist functions $\psi_x^*(x, t)$ and $\psi_t^*(x, t)$ in $L^p[0, 1]$ such that

$$\begin{aligned} \partial_x(\chi, \psi^*) &= (\chi, \psi_x^*), \\ \partial_t(\chi, \psi^*) &= (\chi, \psi_t^*). \end{aligned} \tag{3.46}$$

Whenever the support of χ lies inside the open sets $I^+(x, t)$ or $I^-(x, t)$ we see by (3.39) that the left-hand side of (3.46) vanishes. Therefore,

$$\psi_x^* = \psi_t^* = 0 \quad \text{off} \quad I^0(x, t). \tag{3.47}$$

For any continuous function χ, $L\chi$ is also continuous and so, since L is a symmetric operator, we deduce that $(\chi, L\psi^*(x, t)) = (L\chi, \psi^*(x, t))$ is a differentiable function of x and t. By (3.46),

$$\begin{aligned} \partial_x(\chi, L\psi^*) &= (L\chi, \psi_x^*) = (\chi, L\psi_x^*), \\ \partial_t(\chi, L\psi^*) &= (L\chi, \psi_t^*) = (\chi, L\psi_t^*). \end{aligned} \tag{3.48}$$

Let I_0 be the interior of the closed set I^0 given by (3.36) in $[0, 1]\times R\times R$ and let $I_0(x, t)$ denote its (x, t) slice defined as in (3.37). Choose the support of χ in (3.48) to lie inside the open set $I_0(x, t)$; then, by (3.38), $L\psi^* = a$ on the support of χ:

$$(\chi, L\psi^*) = (\chi, a).$$

Using the definition (1.23) of a, we deduce that

$$\partial_x(\chi, L\psi^*) = (\chi, \eta), \qquad \partial_t(\chi, L^*\psi) = -4(\chi, \eta^3). \tag{3.49}$$

Combining (3.48) and (3.49) we find

$$L\psi_x^* = \eta, \quad L\psi_t^* = -4\eta^3 \quad \text{on} \quad I_0(x, t). \tag{3.50}$$

By Theorem 3.15, both sides of (3.49) are continuous functions of η, so the equality may be extended to $\bar{I}_0(x, t)$, the closure of $I_0(x, t)$ in $[0, 1]$.

Our later assumptions will imply that $I_0(x, t)$ and $I^0(x, t)$ differ by at most a finite number of points. Then $\bar{I}_0(x, t)$ and $I^0(x, t)$ will have a common interior, denoted by $I(x, t)$, on which (3.50) holds. Since $I^0(x, t)$ and $I(x, t)$ differ by a set of measure zero, we can replace $I^0(x, t)$ by $I(x, t)$ in (3.47). We summarize:

ψ_x^* is in $L^p[0, 1]$ for some p, $1<p<\infty$, and satisfies

$$\begin{aligned} L\psi_x^*(x, t) &= \eta \quad \text{on} \quad I(x, t), \\ \psi_x^*(x, t) &= 0 \quad \text{off} \quad I(x, t). \end{aligned} \tag{3.51}$$

Similarly, ψ_t^* is in $L^p[0, 1]$ for some p, $1<p<\infty$, and satisfies

$$(3.52)\qquad \begin{aligned} L\psi_t^* &= -4\eta^3 \quad \text{on} \quad I(x,t), \\ \psi_t^* &= 0 \quad \text{off} \quad I(x,t). \end{aligned}$$

Remark. Arguing as in Theorem 3.14, it follows that ψ_x^* and ψ_t^* are the unique L^p solution of (3.51) and (3.52), respectively.

We call (3.51) and (3.52) the differentiated variational conditions. Note that the inhomogeneous terms on the right (i.e., η and $-4\eta^3$) do not depend on the initial data $u(x)$.

If we extend $I(x, t)$ to the whole real line, R, by reflection about zero, then (3.51) and (3.52) hold for the extended functions. We differentiate the top equations of (3.51) and (3.52) with respect to η on the open set $I(x, t)$; using (3.41) we obtain

$$(3.53)\qquad H\psi_x^* = 1 \quad \text{on} \quad I(x,t),$$

and

$$(3.54)\qquad H\psi_t^* = -12\eta^2 \quad \text{on} \quad I(x,t).$$

Denote by f the function in H^p whose real part is ψ_x^*:

$$(3.55)\qquad \mathcal{R}e(f) = \psi_x^*.$$

We claim that

$$(3.56)\qquad \begin{aligned} \mathcal{I}m(f) &= 1 \quad \text{on} \quad I(x,t), \\ \mathcal{R}e(f) &= 0 \quad \text{off} \quad I(x,t). \end{aligned}$$

The first relation is obtained by combining (3.53) and Theorem 3.16; the second relation is obtained from (3.51).

Similarly, if g denotes the H^p function whose real part is ψ_t^*:

$$(3.57)\qquad \mathcal{R}e(g) = \psi_t^*,$$

then

$$(3.58)\qquad \begin{aligned} \mathcal{I}m(g) &= -12\eta^2 \quad \text{on} \quad I(x,t), \\ \mathcal{R}e(g) &= 0 \quad \text{off} \quad I(x,t). \end{aligned}$$

We shall exploit this fact in the next two sections to construct explicit solutions $\psi_x^*(x, t)$ and $\psi_t^*(x, t)$ of the differentiated variational conditions. We shall then verify that the function ψ^* satisfies the original variational condition.

We next show how to express $\bar{u}$, $\overline{u^2}$, and $\overline{u^3}$ in terms of ψ_x^* and ψ_t^*:

THEOREM 3.17.

$$\bar{u}(x,t)=\frac{4}{\pi}(\eta,\psi_x^*(x,t)),$$

$$\overline{u^2}(x,t)=\frac{4}{3\pi}(\eta,\psi_t^*(x,t))=-\frac{16}{3\pi}(\eta^3,\psi_x^*(x,t)), \tag{3.59}$$

$$\overline{u^3}(x,t)=-\frac{4}{3\pi}(\eta^3,\psi_t^*(x,t)).$$

Proof: We use formula (3.30) of Theorem 3.11:

$$\partial_x Q^*(x,t)=\frac{4}{\pi}(\eta,\psi^*(x,t)),$$

$$\partial_t Q^*(x,t)=-\frac{16}{\pi}(\eta^3,\psi^*(x,t)).$$

Using formula (3.46) to differentiate these with respect to x and t, we get

$$\partial_{xx}Q^*(x,t)=\frac{4}{\pi}(\eta,\psi_x^*(x,t)),$$

$$\partial_{xt}Q^*(x,t)=\frac{4}{\pi}(\eta,\psi_t^*(x,t))=-\frac{16}{\pi}(\eta^3,\psi_x^*(x,t)), \tag{3.60}$$

$$\partial_{tt}Q^*(x,t)=-\frac{16}{\pi}(\eta^3,\psi_t^*(x,t)).$$

If we compare (3.60) with definitions (2.51), (2.54), and (2.56) for $\bar{u}$, $\overline{u^2}$ and $\overline{u^3}$ we conclude that (3.59) holds, proving the theorem.

We conclude this section by proving (2.29). Using the definition (1.25) of ϕ, the fact that L and $d/d\eta$ commute, and relation (3.41), we have

$$L\phi=-L\frac{d}{d\eta}\Phi=-\frac{d}{d\eta}L\Phi=H\Phi.$$

According to Theorem 3.16, $L\phi$ is the imaginary part of the function of class H^p whose real part is Φ. But that function is

$$f(\eta)=\int_{-\infty}^{\infty}[(-u(y)-\eta^2)^{1/2}-i\eta]\,dy,$$

as may be seen from (1.13).

Using (1.16) and (1.24) we see that

$$\mathscr{I}m\,(f)=\theta_-(\eta)-\theta_+(\eta).$$

This proves (2.29).

Acknowledgment. The research of the first author was supported by the U.S. Department of Energy, Office of basic Energy Sciences, Applied Mathematical Sciences Research program under Contract DE-AC02-76ER03077, that of the second by the John and Fannie Hertz Foundation.

Bibliography

[1] Cole, J. D., *On a quasi-linear parabolic equation occurring in aerodynamics*, Quart. Appl. Math. 9, 1951, pp. 225–236.

[2] Flaschka, H., Forest, M. G., and McLaughlin, D. W., *Multiphase averaging and the inverse spectral solution of the Korteweg–de Vries equation*, Comm. Pure Appl. Math. 33, 1980, pp. 739–784.

[3] Gardner, C. S., Greene, J. M., Kruskal, M. D., and Miura, R. M., *Method for solving the Korteweg–de Vries equation*, Phys. Rev. Lett. 19, 1967, pp. 1095–1097.

[4] Gelfand, I. M., and Levitan, B. M., *On the determination of a differential equation by its spectral function*, Izv. Akad. Nauk SSR Ser. Mat. 15, 1951, pp. 309–360. (In Russian.)

[5] Hopf, E., *The partial differential equation* $u_t + uu_x = \mu u_{xx}$, Comm. Pure Appl. Math. 3, 1950, pp. 201–230.

[6] Kay, I., and Moses, H. E., *Reflectionless transmission through dielectrics and scattering potentials*, J. Appl. Phys. 27, 1956, pp. 1503–1508.

[7] Koosis, P., *Introduction to* H^p *spaces*, LMS Lecture Notes 40, 1980.

[8] Lax, P. D., and Levermore, C. D., *The zero dispersion limit for KdV equation*, Proc. Nat. Acad. Sci. USA, 1979, pp. 3602–3606.

[9] Levermore, C. D., *The small dispersion limit of the Korteweg–de Vries equation*, Dissertation, FGAS, 1982, New York University.

[10] Miura, R. M., and Kruskal, M. D., *Application of a nonlinear WKB method to the Korteweg–de Vries equation*, SIAM J. Appl. Math. 26, 1974, pp. 376–395.

[11] Muskhelishvili, N. I., *Singular Integral Equations*, P. Noordhoff, Groningen, 1953.

[12] Tanaka, S., *On the N-tuple wave solutions of the Korteweg–de Vries equation*, Res. Inst. Math. Sci., Kyoto 8, 1973, pp. 419–427.

[13] Venakides, S., *The zero dispersion limit of the Korteweg–de Vries equation*, Dissertation, FGAS, 1982, New York University.

[14] Whitham, G. B., *Linear and Nonlinear Waves*, Wiley-Interscience, New York, 1974.

Received June, 1982.

The Small Dispersion Limit of the Korteweg–de Vries Equation. II*

PETER D. LAX
Courant Institute
AND
C. DAVID LEVERMORE
Lawrence Livermore Laboratory

Abstract

In Part I* we have shown, see Theorem 2.10, that as the coefficient of u_{xxx} tends to zero, the solution of the initial value problem for the KdV equation tends to a limit $\bar{u}$ in the distribution sense. We have expressed $\bar{u}$ by formula (3.59), where ψ_x^* is the partial derivative with respect to x of the function ψ^* defined in Theorem 3.9 as the solution of the variational problem formulated in (2.16), (2.17). ψ^* is uniquely characterized by the variational condition (3.34); its partial derivatives satisfy (3.51) and (3.52), where I is the set I^0 defined in (3.36). In Section 4 we show that for $t<t_b$, I consists of a single interval, and the $\bar{u}$ satisfies $\bar{u}_t - 6\bar{u}\bar{u}_x = 0$; here t_b is the largest time interval in which (12) has a continuous solution. In Section 5 we show that when I consists of a finite number of intervals, $\bar{u}$ can be described by Whitham's averaged equation or by the multiphased averaged equations of Flaschka, Forest, and McLaughlin.

Equation numbers refer to Part I.

4. The Solution Until Break-time

The variational problem (2.16)–(2.18) contains x and t as parameters. We first investigate the case $t=0$. We make the assumption, justified *a posteriori*, that for $t<t_b$ the set I consists of a single interval

$$I=(-\beta,\beta), \tag{4.1}$$

where β is a differentiable function of x.

We claim that the solution of the Riemann–Hilbert problem posed in (3.56) is

$$f(\zeta)=i-\frac{\zeta}{(\beta^2-\zeta^2)^{1/2}}, \tag{4.2}$$

---

* Part I of this series appeared in Comm. Pure Appl. Math. 36, 1983, pp. xxx–xxx.

Communications on Pure and Applied Mathematics, Vol. XXXVI 571–593 (1983)
 CCC 0010-3640/83/050571-23$03.30

where the sign of the square root is chosen so that

$$i(\beta^2-\eta^2)^{1/2}>0 \quad \text{for} \quad \eta>\beta. \tag{4.3}$$

Clearly $f(\zeta)$ is analytic in the upper half-plane. On account of (4.3),

$$(\beta^2-\zeta^2)^{1/2} \sim -i\zeta \quad \text{for} \quad |\zeta| \text{ large}$$

which, by (4.2), shows that

$$f(\zeta)=O\left(\frac{1}{|\zeta|^2}\right). \tag{4.4}$$

Since $f(\zeta)$ has singularities of the form $(\zeta-\beta)^{-1/2}$ on the real axis, we see from (4.4) that for $1 \leqq p<2$ there exists a constant $C>0$ such that

$$\left(\int_{-\infty}^{\infty}|f(\eta+i\tau)|^p\right)^{1/p}<C$$

for all $\tau>0$. This shows that f belongs to H^p for $1 \leqq p<2$.

For η real, it follows from (4.2) and (4.3) that

$$\mathcal{R}e\,(f)=\begin{cases}-\dfrac{\eta}{(\beta^2-\eta^2)^{1/2}} & \text{on } I,\\ 0 & \text{off } I,\end{cases} \tag{4.5}$$

and

$$\mathcal{I}m\,(f)=\begin{cases}1 & \text{on } I,\\ 1-\dfrac{\eta}{(\eta^2-\beta^2)^{1/2}} & \text{off } I,\end{cases} \tag{4.6}$$

where the square root in (4.6) has the same sign as η. These relations show that conditions (3.56) are satisfied, Re (f) is odd with support in I, and Im (f) is even. We set (4.5) into (3.55):

$$\psi_x^*=\begin{cases}-\dfrac{\eta}{(\beta^2-\eta^2)^{1/2}} & \text{for } |\eta|<\beta,\\ 0 & \text{for } |\eta| \geqq \beta.\end{cases} \tag{4.7}$$

Now using formula (3.41) of Theorem 3.15, with

$$H\psi_x^*=\mathcal{I}m\,(f)$$

given by (4.6), we have

$$L\psi_x^*=\begin{cases}\eta & \text{for } 0 \leqq \eta \leqq \beta,\\ \eta-\displaystyle\int_\beta^\eta \frac{\mu}{(\mu^2-\beta^2)^{1/2}}\,d\mu & \text{for } \eta>\beta.\end{cases} \tag{4.8}$$

Carrying out the integral in (4.8) we obtain

$$L\psi_x^* = \begin{cases} \eta & \text{for } 0 \leqq \eta \leqq \beta, \\ \eta - (\eta^2 - \beta^2)^{1/2} & \text{for } \eta > \beta. \end{cases} \tag{4.9}$$

The value of β is so far undetermined; for that we turn to formula (3.59),

$$\bar{u} = \frac{4}{\pi}(\eta, \psi_x^*). \tag{4.10}$$

We expect that

$$\bar{u}(x, 0) = u(x). \tag{4.11}$$

Combining (4.10) with (4.11) and expressing ψ_x^* in (4.10) by (4.7) yields

$$u(x) = \frac{4}{\pi}(\eta, \psi_x^*) = -\frac{4}{\pi}\int_0^\beta \frac{\eta^2}{(\beta^2 - \eta^2)^{1/2}}\, d\eta = -\beta^2; \tag{4.12}$$

thus we set $\beta(x) = (-u(x))^{1/2}$. Since $-u(x) > \eta^2$ between $x_-(\eta)$ and $x_+(\eta)$ and $-u(x) < \eta^2$ outside, see Figure 1, formula (4.7) may be rewritten as

$$\psi_x^*(x, 0) = \begin{cases} -\dfrac{\eta}{(-u(x) - \eta^2)^{1/2}} & \text{for } x_-(\eta) < x < x_+(\eta), \\ 0 & \text{for } x \leqq x_-(\eta) \text{ or } x \geqq x_+(\eta). \end{cases} \tag{4.13}$$

This completes the determination of $\psi_x^*(x, 0)$.

To determine $\psi^*(x, 0)$ itself we integrate $\psi_x^*(x, 0)$ as given by (4.13). Theorem 3.13 states that $\psi^*(x, 0) = 0$, where $\eta x - \theta_+(\eta) > 0$; in particular, it follows that, for fixed η, $\psi^*(x, 0) = 0$ for x large enough. Thus

$$\psi^*(x, 0) = -\int_x^\infty \psi_x^*(y, 0)\, dy,$$

i.e.,

$$\psi^*(x, 0) = \begin{cases} 0 & \text{for } x \geqq x_+(\eta), \\ \displaystyle\int_x^{x_+(\eta)} \frac{\eta}{(-u(y) - \eta^2)^{1/2}}\, dy & \text{for } x_-(\eta) < x < x_+(\eta), \\ \phi(\eta) & \text{for } x \leqq x_-(\eta). \end{cases} \tag{4.14}$$

In (4.14) we have used the definition (1.25) of $\phi(\eta)$:

$$\phi(\eta) = \int_{x_-(\eta)}^{x_+(\eta)} \frac{\eta}{(-u(y) - \eta^2)^{1/2}}\, d\eta.$$

Clearly, $0 \leqq \psi^*(x, 0) \leqq \phi$ so $\psi^*(x, 0)$ is in the admissible set A.

We shall now show that $\psi^*(x, 0)$, given by (4.14), satisfies the variational conditions (3.34), thus justifying our assumption (4.1). Using (4.13) we may rewrite formula (4.9):

$$(4.15)\quad L\psi_x^*(x,0)=\begin{cases}\eta & \text{for}\quad x_-(\eta)\leqq x\leqq x_+(\eta),\\ \eta-(\eta^2+u(x))^{1/2} & \text{for}\quad x<x_-(\eta)\quad\text{or}\quad x>x_+(\eta).\end{cases}$$

It is easy to see from Theorem 3.13 that

$$\operatorname*{w-lim}_{x\to\infty}\psi^*(x,0)=0,$$

hence we see by Theorem 3.4 that

$$\lim_{x\to\infty}L\psi^*(x,0)=0.$$

We then obtain, by integrating (4.15),

$$-L\psi^*(x,0)=\int_x^\infty L\psi_x^*(y,0)\,dy$$

$$(4.16)\qquad =\begin{cases}\displaystyle\int_x^\infty \eta-(\eta^2+u(y))^{1/2}\,dy & \text{for } x>x_+(\eta),\\ -\eta x+\theta_+(\eta) & \text{for } x_-(\eta)\leqq x\leqq x_+(\eta),\\ \displaystyle -\eta x+\theta_+(\eta)-\int_x^{x_-(\eta)}(\eta^2+u(y))^{1/2}\,dy & \text{for } x\leqq x_-(\eta).\end{cases}$$

In (4.16) we have used the definition (1.16) of $\theta_+(\eta)$:

$$\theta_+(\eta)=\eta x_+(\eta)+\int_{x_+(\eta)}^\infty \eta-(\eta^2+u(y))^{1/2}\,dy.$$

Since $a(\eta,x,0)=\eta x-\theta_+(\eta)$, we see from (4.16) that

$$(4.17)\qquad a-L\psi^*=\begin{cases}\displaystyle\int_{x_+(\eta)}^x(\eta^2+u(y))^{1/2}\,dy & \text{for } x>x_+(\eta),\\ 0 & \text{for } x_-(\eta)\leqq x\leqq x_+(\eta),\\ \displaystyle -\int_x^{x_-(\eta)}(\eta^2+u(y))^{1/2}\,dy & \text{for } x<x_-(\eta),\end{cases}$$

which clearly shows that $\psi^*(x,0)$ given by (4.14) satisfies the variational conditions (3.34).

According to Theorem 3.12, the only admissible function that satisfies the variational conditions is ψ^*. Thus we have proved

THEOREM 4.1. *$\psi^*(x,0)$ given by* (4.14) *solves the variational problem* (2.16)–(2.18) *at* $t=0$.

Formula (4.14) indicates that $\psi^*(x, 0)$ does indeed depend differentiably on x and $\psi^*_x(x, 0)$ is given by (4.13). As shown in Theorem 3.17 the differentiability of ψ^* implies that (4.10) holds; comparing this with (4.13) we conclude from Theorem 2.12:

THEOREM 4.2.

(4.18) $$\bar{u}(x, 0) = \text{weak } L^2 - \lim_{\varepsilon \to 0} u(x, 0; \varepsilon) = u(x).$$

This result was expected (see (4.11)) and provides justification for the replacement of the exact scattering data by the asymptotic data. In Theorem 4.5 we shall sharpen Theorem 4.2.

We now begin our investigation of non-zero values of t. If in the KdV equation one sets $\varepsilon = 0$,

(4.19) $$u_t - 6uu_x = 0$$

results. In the introduction we have pointed out that there is a break-time, t_b, such that equation (4.19) has a smooth solution for $t < t_b$ but not beyond. Its value can be explicitly determined from the initial data $u(x)$:

$$t_b = [6 \max_x u_x(x)]^{-1}.$$

Denote by $u(x, t)$ the solution of (4.19) which takes on the prescribed initial values $u(x)$.

THEOREM 4.3. *For* $0 \leqq t < t_b$,

(4.20) $$\bar{u}(x, t) = \text{weak } L^2 - \lim_{\varepsilon \to 0} u(x, t; \varepsilon) = u(x, t)$$

uniformly in t.

Proof: We shall reduce this to Theorem 4.1. As $u(x, t)$ evolves according to (4.19) it will, until break-time, remain of the same class as the initial data $u(x)$. We shall consider now how the asymptotic scattering data corresponding to $u(x, t)$ change with t. Define $x_+(\eta, t)$ and $x_-(\eta, t)$ as the two roots of

(4.21) $$u(x_\pm, t) + \eta^2 = 0$$

with $x_-(\eta, t) < x_+(\eta, t)$ for η in (0, 1). By replacing $u(x)$ by $u(x, t)$ and $x_\pm(\eta)$ by $x_\pm(\eta, t)$ in formulas (1.12) and (1.16) we define

(4.22) $$\Phi(\eta, t) = \int_{x_-(\eta,t)}^{x_+(\eta,t)} (-u(y, t) - \eta^2)^{1/2}\, dy$$

and

(4.23) $$\theta_+(\eta, t) = \eta x_+(\eta, t) + \int_{x_+(\eta,t)}^{\infty} (\eta^2 + u(y, t))^{1/2}\, dy.$$

LEMMA 4.4. *For* $0 \leqq t < t_b$,

$$x_\pm(\eta, t) = x_\pm(\eta) + 6\eta^2 t, \tag{4.24}$$

$$\Phi(\eta, t) = \Phi(\eta), \tag{4.25}$$

$$\theta_+(\eta, t) = \eta_+(\eta) + 4\eta^3 t. \tag{4.26}$$

Proof: Differentiating (4.21) with respect to t and using (4.19) and (4.21) we obtain

$$\partial_t x_\pm = -\frac{u_t}{u_x} = -6u = 6\eta^2.$$

Formula (4.24) now follows upon integrating this and using the fact that $x_\pm(\eta, 0) = x_\pm(\eta)$. Differentiating (4.22) with respect to t and using (4.19) we obtain

$$\begin{aligned} \partial_t \Phi(\eta, t) &= -\frac{1}{2}\int_{x_-}^{x_+} \frac{u_t}{(-u-\eta^2)^{1/2}}\,dy \\ &= -3\int_{x_-}^{x_+} \frac{u}{(-u-\eta^2)^{1/2}}\,u_x\,dy \\ &= (2u - 4\eta^2)(-u-\eta^2)^{1/2}\Big|_{x_-}^{x_+} = 0, \end{aligned}$$

by (4.21). This proves (4.25). Performing the same procedure on (4.23) we get

$$\begin{aligned} \partial_t \theta_+(\eta, t) &= -\frac{1}{2}\int_{x_+}^{\infty} \frac{u_t}{(\eta^2+u)^{1/2}}\,dy \\ &= -3\int_{x_+}^{\infty} \frac{u}{(\eta^2+u)^{1/2}}\,u_x\,dy \\ &= (4\eta^2 - 2u)(\eta^2+u)^{1/2}\Big|_{x_+}^{\infty} = 4\eta^3, \end{aligned}$$

which after integrating yields (4.26) and establishes Lemma 4.4.

From (4.25) of Lemma 4.4 we see that

$$\phi(\eta) = -\partial_\eta \Phi(\eta, t)$$

and, using (4.26) in the definition (1.23) of a, that

$$a(\eta, x, t) = \eta x - \theta_+(\eta, t).$$

Thus the variational problem (2.16)–(2.18) for $0 \leqq t < t_b$ is the same as the variational problem for $t = 0$ except that $u(x)$ has been replaced by $u(x, t)$. By Theorem 4.1, its solution is given by formula (4.14) with the same replacements, and Theorem 4.3 follows from Theorem 4.2.

The weak limit in formula (4.20) is in fact a strong limit.

THEOREM 4.5. *For* $0 \leqq t_0 < t_b$,

$$\bar{u}(x,t) = L^2 - \lim_{\varepsilon \to \infty} u(x,t;\varepsilon) = u(x,t) \tag{4.27}$$

uniformly in t.

REMARK. In particular this proves Theorem 1.2.

Proof: Lemma 2.13 gives

$$\lim_{\varepsilon \to 0} \int_{-\infty}^{\infty} u^2(x,t,\varepsilon) = \int_{-\infty}^{\infty} u^2(x)\,dx$$

uniformly in t. Since $\int u^2\,dx$ is a conserved quantity for equation (4.19),

$$\lim_{\varepsilon \to \infty} \int_{-\infty}^{\infty} u^2(x,t;\varepsilon)\,dx = \int_{-\infty}^{\infty} u^2(x,t)\,dx \tag{4.28}$$

uniformly in t for $0 \leqq t < t_b$.

By Theorem 4.3,

$$\lim_{\varepsilon \to 0} \int_{-\infty}^{\infty} w(x)u(x,t;\varepsilon)\,dx = \int_{-\infty}^{\infty} w(x)u(x,t)\,dx, \tag{4.29}$$

uniformly in t for $0 \leqq t < t_b$ and over w in any compact subset of $L^2(R)$. But it is easily seen that for $0 \leqq t \leqq t_b$, $u(x,t)$ is a continuous function of t into $L^2(R)$ of the x variable. Thus the set $\{u(x,t): 0 \leqq t \leqq t_b\}$ is a compact subset of $L^2(R)$; setting $w(x) = u(x,t)$ in (4.29) we conclude that

$$\lim_{\varepsilon \to 0} \int_{-\infty}^{\infty} u(x,t)u(x,t;\varepsilon)\,dx = \int_{-\infty}^{\infty} u^2(x,t)\,dx \tag{4.30}$$

uniformly in t for $0 \leqq t < t_b$.

Consider the identity

$$\int_{-\infty}^{\infty} (u(x,t;\varepsilon) - u(x,t))^2\,dx = \int_{-\infty}^{\infty} u^2(x,t;\varepsilon)\,dx - 2\int_{-\infty}^{\infty} u(x,t)u(x,t;\varepsilon)\,dx + \int_{-\infty}^{\infty} u^2(x,t)\,dx.$$

By (4.28) and (4.30) the right-hand side tends to zero uniformly over $0 \leqq t < t_b$ as ε tends to zero, proving the theorem.

In the proof of Theorem 4.3 we relied on the knowledge that $\bar{u}(x,t) = u(x,t)$ satisfies (4.19) for $0 \leqq t < t_b$. We show now how (4.19) can be deduced directly from the variational conditions.

Assume that the set I consists of a single interval, see (4.1), where β is a differentiable function of x and t. This assumption will be justified *a posteriori*.

We claim that the solution of the Riemann–Hilbert problem posed in (3.58) of Theorem 3.18 is

$$g(\zeta) = -12\zeta^2 i + \frac{12\zeta^3 - 6\beta^2\zeta}{(\beta^2-\zeta^2)^{1/2}}, \tag{4.31}$$

where the sign of the square root is again chosen to satisfy (4.3). The same arguments that applied to $f(\zeta)$ given by (4.2) can now be applied to $g(\zeta)$ to show that g belongs to H^p for $1 \leqq p < 2$, and for η real

$$\mathcal{R}e\,(g) = \begin{cases} \dfrac{12\eta^3 - 6\beta^2\eta}{(\beta^2-\eta^2)^{1/2}} & \text{on } I, \\ 0 & \text{off } I, \end{cases} \tag{4.32}$$

and

$$\mathcal{I}m\,(g) = \begin{cases} -12\eta^2 & \text{on } I, \\ -12\eta^2 + \dfrac{12\eta^3 - 6\beta^2\eta}{(\eta^2-\beta^2)^{1/2}} & \text{off } I. \end{cases} \tag{4.33}$$

As before, the square root in (4.33) has the same sign as η. These relations show that conditions (3.58) are satisfied, $\mathcal{R}e\,(g)$ is odd, with support in I, and $\mathcal{I}m\,(g)$ is even.

We set (4.32) into (3.57) to obtain

$$\psi_t^* = \begin{cases} \dfrac{12\eta^3 - 6\beta^2\eta}{(\beta^2-\eta^2)^{1/2}} & \text{for } |\eta| < \beta, \\ 0 & \text{for } |\eta| \geqq \beta. \end{cases} \tag{4.34}$$

Using formula (3.41) of Theorem 3.15 with

$$H\psi_t^* = \mathcal{I}m(g)$$

given by (4.33) we have

$$L\psi_t^* = \begin{cases} -4\eta^3 & \text{for } 0 \leqq \eta \leqq \beta, \\ -4\eta^3 + \displaystyle\int_\beta^\eta \frac{12\mu^3 - 6\beta^2\mu}{(\mu^2-\beta^2)^{1/2}}\,d\mu & \text{for } \eta > \beta. \end{cases} \tag{4.35}$$

Carrying out the integral in (4.35) we obtain

$$L\psi_t^* = \begin{cases} -4\eta^3 & \text{for } 0 \leqq \eta \leqq \beta, \\ -4\eta^3 + (4\eta^2 + 2\beta^2)(\eta^2-\beta^2)^{1/2} & \text{for } \eta > \beta. \end{cases} \tag{4.36}$$

Now ψ_x^* given by (4.7) and ψ_t^* given by (4.34) must satisfy the compatibility condition $\partial_t\psi_x^* = \partial_x\psi_t^*$. This implies

$$\begin{aligned} 0 &= \partial_t\left(\frac{\eta}{(\beta^2-\eta^2)^{1/2}}\right) + \partial_x\left(\frac{12\eta^3 - 6\beta^2\eta}{(\beta^2-\eta^2)^{1/2}}\right) \\ &= -\frac{\eta\beta}{(\beta^2-\eta^2)^{3/2}}(\beta_t + 6\beta^2\beta_x), \end{aligned} \tag{4.37}$$

from which we conclude that

(4.38) $$\beta_t + 6\beta^2\beta_x = 0.$$

Expressing ψ_x^* in formula (4.10) by (4.7), we obtain (see 4.12)

(4.39) $$\bar{u}(x, t) = \frac{4}{\pi}(\eta, \psi_x^*) = -\beta^2(x, t).$$

Multiplying (4.38) by -2β and using (4.39) we deduce

(4.40) $$\bar{u}_t - 6\bar{u}\bar{u}_x = 0;$$

this is relation (4.19), valid for $t \leqq t_b$.

The above procedure will be generalized in the next section to obtain the time dependence of solutions of the variational problem for $t > t_b$.

We close this section with an observation. Set ψ_x^* given by (4.7) and ψ_t^* given by (4.34) into (3.59) of Theorem 3.17:

$$\overline{u^2}(x, t) = -\frac{16}{3\pi}(\eta^3, \psi_x^*) = \frac{4}{3\pi}\int_0^\beta \frac{4}{(\beta^2 - \eta^2)^{1/2}}\, d\eta,$$

$$\overline{u^3}(x, t) = -\frac{4}{3\pi}(\eta^3, \psi_t^*) = -\frac{4}{3\pi}\int_0^\beta \frac{12\eta^6 - 6\beta^2\eta^4}{(\beta^2 - \eta^2)^{1/2}}\, d\eta.$$

A brief calculation shows that

(4.41) $$\begin{aligned} \overline{u^2}(x, t) &= \beta^4(x, t) = \bar{u}^2(x, t), \\ \overline{u^3}(k, t) &= -\beta^6(x, t) = \bar{u}^3(x, t). \end{aligned}$$

From the definitions (2.51), (2.54) and (2.56) for $\bar{u}$, $\overline{u^2}$ and $\overline{u^3}$ we must have

(4.42) $$\begin{aligned} &\bar{u}_t - 3\overline{u^2_x} = 0 \\ &\overline{u^2_t} - 4\overline{u^3_x} = 0. \end{aligned}$$

Using (4.41) to express $\overline{u^2}$ and $\overline{u^3}$ in terms of $\bar{u}$, we see that each of (4.42) will hold if and only if (4.40) holds, showing again the validity of (4.40).

5. The Solution Beyond Break-time

In this section we describe the solution of the variational problem (2.16)–(2.18) when t is larger than the break-time, t_b. We make the assumption that the set I consists of a finite union of disjoint intervals, say

(5.1) $$I = (\beta_{2n}, \beta_{2n-1}) \cdots (\beta_2, \beta_1),$$

where

(5.2) $$\beta_{2n} < \beta_{2n-1} < \cdots < \beta_2 < \beta_1.$$

Since I is symmetric about $\eta = 0$, we see that

$$\beta_{2n-k+1} = -\beta_k$$

and that the set I is completely determined by its set of positive end points

$$B = \{\beta_k\}_{k=1}^{n}. \tag{5.3}$$

We shall work as if n were odd since, as we shall later see, this will be the case except when the initial function $u(x)$ takes on the value zero; in that case we make n odd by setting $\beta_n = 0$.

As first step we shall construct all possible solutions of the Riemann–Hilbert problems (3.56) and (3.58) that ψ_x^* and ψ_t^* satisfy. The solution will be expressed in terms of a function $R(\zeta)$ defined by

$$R^2(\zeta) = \prod_{k=1}^{n} (\beta_k^2 - \zeta^2); \tag{5.4}$$

we choose its sign so that

$$iR(\eta) > 0 \quad \text{for} \quad \eta > \beta_1. \tag{5.5}$$

LEMMA 5.1. *All functions of the form*

$$f(\zeta) = i - \frac{P(\zeta)}{R(\zeta)}, \tag{5.6}$$

$$g(\zeta) = -12\zeta^2 i + \frac{Q(\zeta)}{R(\zeta)} \tag{5.7}$$

are solutions of the Riemann–Hilbert problems (3.56) *and* (3.58), *respectively, where $P(\zeta)$ and $Q(\zeta)$ are odd polynomials with real coefficients*

$$P(\zeta) = \zeta^n + a_2\zeta^{n-2} + \cdots + \alpha_{n-1}\zeta, \tag{5.8}$$

$$Q(\zeta) = 12\zeta^{n+2} - 6(\textstyle\sum \beta_k^2)\zeta^n + \gamma_2\zeta^{n-2} + \cdots + \gamma_{n-1}\zeta. \tag{5.9}$$

Remark. It is not hard to show that these are all solutions (cf. [9]).

Proof: Clearly $f(\zeta)$ and $g(\zeta)$ are analytic in the upper half-plane. By (5.4), (5.5),

$$iR(\zeta) \sim \zeta^n - \tfrac{1}{2}\Big(\sum_{k=1}^{n} \beta_k^2\Big)\zeta^{n-2} + O(|\zeta|^{n-4}) \tag{5.10}$$

for $|\zeta|$ large, which, along with (5.6)–(5.9), shows that

$$f(\zeta) = O\Big(\frac{1}{|\zeta|^2}\Big), \qquad g(\zeta) = O\Big(\frac{1}{|\zeta|^2}\Big). \tag{5.11}$$

Since $f(\zeta)$ and $g(\zeta)$ have singularities of the form $(\zeta-\beta)^{1/2}$ on the real axis, we see from (5.11) that f and g belong to H^p for $1 \leqq p < 2$.

For η real, the function $R(\eta)$ defined by (5.4) is real and even on I, imaginary and odd off I, where I is defined by (5.1). Since $P(\eta)$ and $Q(\eta)$ are real and odd for η real, it follows from (5.6) and (5.7) that

$$\mathcal{R}e\,(f) = \begin{cases} -\dfrac{P(\eta)}{R(\eta)}. & \text{on } I, \\ 0 & \text{off } I, \end{cases} \tag{5.12}$$

and

$$\mathcal{I}m\,(f) = \begin{cases} 1 & \text{on } I, \\ 1+i\dfrac{P(\eta)}{R(\eta)} & \text{off } I, \end{cases} \tag{5.13}$$

while

$$\mathcal{R}e\,(g) = \begin{cases} \dfrac{Q(\eta)}{R(\eta)} & \text{on } I, \\ 0 & \text{off } I, \end{cases} \tag{5.14}$$

and

$$\mathcal{I}m\,(g) = \begin{cases} -12\eta^2 & \text{on } I, \\ -12\eta^2 - i\dfrac{Q(\eta)}{R(\eta)} & \text{off } I. \end{cases} \tag{5.15}$$

These relations show that $f(\zeta)$ and $g(\zeta)$ satisfy conditions (3.56) and (3.58) as asserted in Lemma 5.1.

Using (3.55) and (3.57) we set

$$\psi_x^* = -\mathcal{R}e\left(\frac{P(\eta)}{R(\eta)}\right), \tag{5.16}$$

$$\psi_t^* = \mathcal{R}e\left(\frac{Q(\eta)}{R(\eta)}\right). \tag{5.17}$$

We now show how to choose the coefficients α_j and γ_j so that the differentiated variational conditions, (3.51) and (3.52), are satisfied. Using formula (3.41) with

$$H\psi_x^* = \mathcal{I}m\,(f)$$

given by (5.13) and with

$$H\psi_t^* = \mathcal{I}m\,(g)$$

given by (5.15), we obtain

$$L\psi_x^* = \eta - \int_0^\eta \mathcal{I}m\left(\frac{P(\mu)}{R(\mu)}\right) d\mu \tag{5.18}$$

and

$$L\psi_t = -4\eta^3 + \int_0^\eta \mathscr{I}m\left(\frac{Q(\mu)}{R(\mu)}\right) d\mu. \tag{5.19}$$

If the differentiated variational conditions (3.51) and (3.52) are to be satisfied, the integrals on the right-hand sides of (5.18) and (5.19) must vanish on I. Since the integrands are even functions which vanish on I, we see that it is both necessary and sufficient to require

$$\int_{\beta_{2k+1}}^{\beta_{2k}} \frac{P(\eta)}{R(\eta)} d\eta = 0, \tag{5.20}$$

$$\int_{\beta_{2k+1}}^{\beta_{2k}} \frac{Q(\eta)}{R(\eta)} d\eta = 0, \tag{5.21}$$

for $k = 1, 2, \cdots, \frac{1}{2}(n-1)$. Recalling the form of P and Q in (5.8) and (5.9), we see that (5.20) is a system of $\frac{1}{2}(n-1)$ inhomogeneous linear equations for the $\frac{1}{2}(n-1)$ unknowns $\alpha_2, \alpha_4, \cdots, \alpha_{n-1}$ and (5.21) is a system of $\frac{1}{2}(n-1)$ inhomogeneous linear equations for the $\frac{1}{2}(n-1)$ unknowns $\gamma_2, \gamma_4, \cdots, \gamma_{n-1}$.

LEMMA 5.2. *The systems of equations* (5.20) *and* (5.21) *have unique solutions.*

First Proof: If this were not the case, then according to the alternative principle of linear algebra, the common corresponding homogeneous system would have a nontrivial solution. But then there would be a nontrivial odd polynomial $N(\eta)$ with real coefficients and of degree at most $n-2$ that satisfies

$$\int_{\beta_{2k+1}}^{\beta_{2k}} \frac{N(\eta)}{R(\eta)} d\eta = 0 \tag{5.22}$$

for $k = 1, 2, \cdots, \frac{1}{2}(n-1)$. This implies that

$$\psi_0 = \mathscr{R}e\left(\frac{N(\eta)}{R(\eta)}\right)$$

satisfies

$$L\psi_0 = 0 \quad \text{on} \quad I,$$

$$\psi_0 = 0 \quad \text{off} \quad I.$$

According to the uniqueness argument given for (3.40), such a ψ_0 is zero, and thus $N(\eta)$ is zero, a contradiction.

Second Proof: If this were not the case, then let $N(\eta)$ satisfy (5.22) as above. But this contradicts the following lemma.

LEMMA 5.3. *Let* $N(\eta)$ *be any nontrivial odd polynomial with real coefficients that satisfies* (5.22). *Then* $N(\eta)$ *must have at least one root in each of the* $\frac{1}{2}(n-1)$ *intervals* $(\beta_{2k+1}, \beta_{2k})$ *and the degree of* $N(\eta)$ *must be at least* n.

Proof: Since $R(\eta)$ does not change sign in $(\beta_{2k+1}, \beta_{2k})$, then $N(\eta)$ must do so if (5.22) is to hold. Thus $N(\eta)$ has at least one positive root in each of the $\frac{1}{2}(n-1)$ intervals $(\beta_{2k+1}, \beta_{2k})$. Since N is odd, it has an equal number of negative roots and a root at zero, altogether a minimum of n roots, proving the lemma.

According to (5.20), the polynomial $P(\eta)$ satisfies hypothesis (5.22) of Lemma 5.3. Since P is of degree n, we conclude

COROLLARY 5.4. *$P(\eta)$ has a zero inside each of the intervals $(\beta_{2k+1}, \beta_{2k})$, $k=1,\cdots,n-1$, and at $\eta=0$, and nowhere else.*

According to Lemma 5.2, we can solve (5.20) and (5.21) uniquely for α_j and γ_j, $j=2,4,\cdots,n-1$, as functions of the points β_k in B, see (5.3). Observe that the coefficients and the inhomogeneous terms appearing in the linear systems (5.20) and (5.21) are complete hyperelliptic integrals. Thus the α_j and γ_j are ratios of determinants whose entries are such integrals.

We have now proved

THEOREM 5.5 *Given the set B, the function ψ_x^*, as determined by* (5.16) *and* (5.20), *is the solution of the differentiated variational conditions* (3.51), *similarly ψ_t^*, as determined by* (5.17) *and* (5.21), *is the solution of the differentiated variational condition* (3.52).

The functions ψ_x^* and ψ_t^* satisfy the compatibility condition $\partial_t\psi_x^* = \partial_x\psi_t^*$; or equivalently,

$$\partial t\frac{P(\zeta)}{R(\zeta)} + \partial_x\frac{Q(\zeta)}{R(\zeta)} = 0. \tag{5.23}$$

We shall see what conditions this imposes on the set B.

Rewrite the left side of (5.23) as

$$\partial_t\frac{P(\zeta)}{R(\zeta)} + \partial_x\frac{Q(\zeta)}{R(\zeta)} = \frac{R^2(\partial_t P + \partial_x Q) - \frac{1}{2}(P\partial_t R^2 + Q\partial_x R^2)}{R^3}. \tag{5.24}$$

We claim that the numerator on the right is an odd polynomial of degree at most $3n-2$. Since from the definitions (5.4), (5.8) and (5.9) we see that, for large $|\zeta|$,

$$R^2(\zeta) = (-\zeta^2)^n + \left(\sum_{k=1}^n \beta_k^2\right)(-\zeta^2)^{n-1} + O(|\zeta|^{2n-4}),$$

$$P(\zeta) = \zeta^n + O(|\zeta|^{n-2}),$$

$$Q(\zeta) = 12\zeta^{n+2} - 6\left(\sum_{k=1}^n \beta_k^2\right)\zeta^n + O(|\zeta|^{n-2}),$$

it follows that

$$R^2(\partial_t P + \partial_x Q) = (-\zeta^2)^n\partial_x\left(-6\sum_{k=1}^n \beta_k^2\right)\zeta^n + O(|\zeta|^{3n-2}) \tag{5.25}$$

and

$$\tfrac{1}{2}(P\partial_t R^2 + Q\partial_x R^2) = \tfrac{1}{2}(12\zeta^{n+2})\partial_x\Big(\sum_{k=1}^{n}\beta_k^2\Big)(-\zeta^2)^{n-1} + O(|\zeta|^{3n-2}). \tag{5.26}$$

Subtracting (5.26) from (5.25) we find the leading terms on the right cancel, demonstrating the claim.

Since $R^2 = 0$ at $\zeta = \beta_k$, the numerator on the right of (5.24) vanishes at $\zeta = \beta_k$ if and only if its second term vanishes at $\zeta = \beta_k$. That second term may be written as

$$\tfrac{1}{2}(P\partial_t R^2 + Q\partial_x R^2) = \sum_{k=1}^{n}\Big[\beta_k(P(\zeta)\partial_t\beta_k + Q(\zeta)\partial_x\beta_k)\prod_{j\neq k}(\beta_j^2 - \zeta^2)\Big].$$

Clearly this vanishes at $\zeta = \beta_k$ if and only if β_k satisfies

$$P(\beta_k)\partial_t\beta_k + Q(\beta_k)\partial_x\beta_k = 0. \tag{5.27}$$

This time evolution of the β_k allows us to factor $R^2(\zeta)$ from the numerator of (5.24) and write

$$\partial_t\frac{P(\zeta)}{R(\zeta)} + \partial_x\frac{Q(\zeta)}{R(\zeta)} = \frac{N(\zeta)}{R(\zeta)}, \tag{5.28}$$

where $N(\zeta)$ is the odd polynomial of degree at most $n-2$ given by

$$N(\zeta) = \partial_t P(\zeta) + \partial_x Q(\zeta) + \sum_{k=1}^{n}\Big[\frac{P(\zeta)-P(\beta_k)}{\zeta^2-\beta_k^2}\beta_k\partial_t\beta_k + \frac{Q(\zeta)-Q(\beta_k)}{\zeta^2-\beta_k^2}\beta_k\partial_x\beta_k\Big].$$

In fact, we shall show that $N(\zeta) = 0$.

For ε sufficiently small we employ the fundamental theorem of calculus to obtain

$$\begin{aligned}\partial_t\int_{\beta_{2k+1}+\varepsilon}^{\beta_{2k}-\varepsilon}\frac{P(\eta)}{R(\eta)}\,d\eta + \partial_x\int_{\beta_{2k+1}+\varepsilon}^{\beta_{2k}-\varepsilon}\frac{Q(\eta)}{R(\eta)}\,d\eta &= \int_{\beta_{2k+1}+\varepsilon}^{\beta_{2k}-\varepsilon}\partial_t\frac{P(\eta)}{R(\eta)} + \partial_x\frac{Q(\eta)}{R(\eta)}\,d\eta \\ &\quad + \frac{P(\beta_{2k}-\varepsilon)\partial_t\beta_{2k} + Q(\beta_{2k}-\varepsilon)\partial_x\beta_{2k}}{R(\beta_{2k}-\varepsilon)} \\ &\quad - \frac{P(\beta_{2k+1}+\varepsilon)\partial_t\beta_{2k+1} + Q(\beta_{2k+1}+\varepsilon)\partial_x\beta_{2k+1}}{R(\beta_{2k+1}+\varepsilon)}\end{aligned} \tag{5.29}$$

for $k = 1, 2, \cdots, \frac{1}{2}(n-1)$. As ε goes to zero, the denominators of the second and third terms on the right of (5.29) will, by (5.4), vanish like $\varepsilon^{1/2}$, but, by (5.27), the numerators of the same terms are $O(\varepsilon)$; thus the limit of those terms as $\varepsilon \to 0$ is zero. Using (5.28) in the first term, we then obtain

$$\text{(5.30)} \quad \lim_{\varepsilon\to 0}\left[\partial_t \int_{\beta_{2k+1}+\varepsilon}^{\beta_{2k}-\varepsilon} \frac{P(\eta)}{R(\eta)}\,d\eta + \partial_x \int_{\beta_{2k+1}+\varepsilon}^{\beta_{2k}-\varepsilon} \frac{Q(\eta)}{R(\eta)}\,d\eta\right] = \int_{\beta_{2k+1}}^{\beta_{2k}} \frac{N(\eta)}{R(\eta)}\,d\eta$$

uniformly on compact subsets of x and t. On the other hand, by (5.20) and (5.21),

$$\text{(5.31)} \quad \begin{aligned} &\lim_{\varepsilon\to 0} \int_{\beta_{2k+1}+\varepsilon}^{\beta_{2k}-\varepsilon} \frac{P(\eta)}{R(\eta)}\,d\eta = 0,\\ &\lim_{\varepsilon\to 0} \int_{\beta_{2k+1}+\varepsilon}^{\beta_{2k}-\varepsilon} \frac{Q(\eta)}{R(\eta)}\,d\eta = 0, \end{aligned}$$

uniformly on compact subsets of x and t. Since the divergence is a closed operator, we conclude from (5.30) and (5.31) that

$$\int_{\beta_{2k+1}}^{\beta_{2k}} \frac{N(\eta)}{R(\eta)}\,d\eta = 0.$$

Using Lemma 5.3, we then conclude that $N(\eta) = 0$. Thus we have shown

THEOREM 5.6. *If the β_k in B satisfy* (5.27), *then ψ_x^* and ψ_t^* given by* (5.16) *and* (5.17) *are compatible.*

We deduce from Corollary 5.4 that

$$P(\beta_k) \neq 0, \qquad k = 1, \cdots, n.$$

Dividing (5.27) by $P(\beta_k)$ gives

$$\text{(5.32)} \quad \partial_t\beta + V(\beta; B)\partial_x\beta \quad \text{for} \quad \beta \quad \text{in} \quad B,$$

where

$$V(\eta; B) = \frac{Q(\eta)}{P(\eta)}$$

indicates the dependence of V on the set B through the dependence of α_j and γ_j on B.

Formula (5.32) is a coupled system of n partial differential equations where each β in B is a Riemann invariant with corresponding characteristic velocity $V(\beta; B)$. The set B evolves according to (5.32) so long as the solutions remain regular and n does not change. We have not yet said how the set B, and its cardinality, change when a singularity is encountered, nor have we addressed how $B(x, t)$ is related to the initial data $u(x)$. We shall deal with these issues by introducing a pair of functions $x_+(\eta, t)$ and $x_-(\eta, t)$, $0 \leqq \eta \leqq 1$, already defined in Section 4 for $t < t_b$ but here extended to all t; we shall construct the set B in terms of these functions.

We start by defining the notion of crossing:

DEFINITION. A pair of functions $x_+(\cdot)$, $x_-(\cdot)$ cross the value x at η if the union of the images of any neighborhood of η under $x_+(\cdot)$ and $x_-(\cdot)$ is a neighborhood of x. Such an η is called a crossing point of the pair at x.

Note that if $0<\eta<1$ and $x_\pm(\eta)=x$, then η is a crossing point at x unless $x_\pm(\cdot)$ has a local extremum at η.

For a given pair of functions $x_+(\eta)$, $x_-(\eta)$ and a given value x we define the set $B(x)$ by

$$B(x)=\{\eta: \eta \text{ is a crossing point of } x_+(\cdot), x_-(\cdot) \text{ at } x\}.$$

Note that if $u(x)$ is our initial function and $x_\pm(\eta)$ are defined by (1.11), then $B(x)$ consists of the single point $\beta(x)$ defined by (4.13):

$$\beta(x)=(-u(x))^{1/2}.$$

Suppose the functions $x_\pm=x_\pm(\cdot, t)$ depend on the parameter t; then so does the set B defined above:

$$B(x,t)=\{\eta; \eta \text{ is a crossing point of } x_+(\cdot,t), x_-(\cdot,t) \text{ at } x\}. \tag{5.33}$$

We formulate the following initial value problem for the functions $x_+(\eta,t)$, $x_-(\eta,t)$:

$$\begin{aligned} \frac{dx_+}{dt}&=V(\eta; B(x_+,t)), \qquad x_+(\eta,0)=x_+(\eta),\\ \frac{dx_-}{dt}&=V(\eta; B(x_-,t)), \qquad x_-(\eta,0)=x_-(\eta), \end{aligned} \tag{5.34}$$

where $x_+(\eta)$ and $x_-(\eta)$ are defined by (1.11). Note that the set $B(x,t)$ depends on the functions x_+, x_-. As long as that dependence is Lipschitz continuous in a suitable norm for x_+, x_-, (5.34) has a unique solution. We assume that (5.34) has unique solutions such that:

(a) As functions of η and t, x_+ and x_- are C^1 in $(0,1)\times R$ and continuous in $(0,1]\times R$.
(b) The limits $\lim_{\eta\to 0} x_\pm(\eta,t)$ exist, possibly as $\pm\infty$. If finite we call them the boundary values of $x_\pm$ at $\eta=0$, denote them by $x_\pm(0,t)$ and assume they are C^1 in t.
(c) The number of critical points of $x_+(\cdot,t)$ and $x_-(\cdot,t)$ is finite for all t.
(d) If for some η, $x_+(\eta)=x_-(\eta)$, then $x_+(\eta,t)=x_-(\eta,t)$ for all t.

Note that (c) implies that the cardinality of $B(x,t)$ is bounded. Since $x_+(1)=x_-(1)$, it follows that $x_+(1,t)=x_-(1,t)$; we denote this common value by $x_0(t)$. From (d) and the fact that $x_-(\eta)<x_+(\eta)$ for $0<\eta<1$ we conclude that $x_-(\eta,t)<x_+(\eta,t)$ for all t as well.

Note that, according to the definition of crossing, $\eta = 1$ is a crossing point at $x = x_0(t)$ if, for η near 1,

$$x_-(\eta, t) < x_0(t) < x_+(\eta, t).$$

LEMMA 5.7. *Suppose that $\bar{x}$ is not a critical value of $x_+(\cdot, \bar{t})$ or $x_-(\cdot, \bar{t})$, nor a boundary value of $x_\pm(\cdot, \bar{t})$ (i.e., $\bar{x} \neq x_\pm(0, \bar{t})$ or $\bar{x} \neq x_0(t)$), then, for all x, t in a neighborhood of $\bar{x}$, $\bar{t}$,*

(a) *the cardinality of $B(x, t)$ is constant,*
(b) *$B(x, t)$ is a differentiable function of x, t.*

Proof: It follows from our assumptions that we can find a neighborhood of $\bar{x}$, $\bar{t}$ such that $\partial_\eta x_\pm$ are bounded away from zero and x is bounded away from the boundary values of $x_\pm(\cdot, t)$. By the implicit function theorem, there exists a (maybe smaller) neighborhood N of $\bar{x}$, $\bar{t}$ such that every solution of $x = x_\pm(\eta, t)$ is given by $\eta = \beta(x, t)$, where $\beta(\cdot, \cdot)$ is a C^1 function in N. The conclusions of Lemma 5.7 readily follow.

Now construct $I(x, t)$ from $B(x, t)$ according to (5.1).

LEMMA 5.8. *For any x, t and any η in* $(0, 1)$,

$$x_-(\eta, t) < x < x_+(\eta, t) \Rightarrow \eta \in I(x, t), \tag{5.35}$$

$$x < x_-(\eta, t) \quad or \quad x > x_+(\eta, t) \Rightarrow \eta \notin I(x, t). \tag{5.36}$$

Proof: By (5.1) if η is not in $B(x, t)$, then η is in $I(x, t)$ if and only if the number of elements of $B(x, t)$ in $[\eta, 1]$ is odd. If $x \neq x_0(t)$, then by (5.33) the number of elements of $B(x, t)$ in $[\eta, 1]$ is equal to the number of times that $x_+(\mu, t)$ or $x_-(\mu, t)$ crosses the value x for μ in $[\eta, 1]$. Consider the case $x_-(\eta, t) < x < x_+(\eta, t)$; if $x < x_0(t) = x_\pm(1, t)$, then $x_+(\mu, t)$ must cross the value x at an even number of μ in $[\eta, 1]$ while $x_-(\mu, t)$ must cross the value x at an odd number of μ in $[\eta, 1]$. Thus, the total number of crossings being odd, we conclude $\eta \in I(x, t)$. The argument is essentially the same for all the cases, provided $x \neq x_0(t)$. In that case one must consider the behavior of x_+ and x_- near $\mu = 1$ but it is just as easy.

Now we use formulas (5.16) and (5.17) to construct $\psi_x^*(x, t)$ and $\psi_t^*(x, t)$ from $B(x, t)$.

LEMMA 5.9. *If $\bar{x}$ is not a critical or boundary value of $x_+(\cdot, \bar{t})$ or $x_-(\cdot, \bar{t})$, then, in a neighborhood of $\bar{x}$, $\bar{t}$,*

(a) *$\psi_x^*(x, t)$ and $\psi_t^*(x, t)$ are differentiable functions of x, t,*
(b) *the compatibility condition*

$$\partial_t \psi_x^* = \partial_x \psi_t^*$$

is satisfied.

Proof: The coefficients α_j and γ_j of P and Q are functions of $\beta_1, \cdots, \beta_n$, expressible as rational functions of complete hyperelliptic integrals. Therefore they are analytic functions of $\beta_1, \cdots, \beta_n$ except if two β_k are equal. It follows from Lemma 5.7 that this does not happen under the hypothesis of Lemma 5.9. It then follows from Lemma 5.7 that the β_k are differentiable functions of x, t. Part (a) is a consequence of these two observations.

To prove part (b), we differentiate with respect to x, and then t, the relation

$$x = x_\pm(\beta, t), \qquad \beta = \beta(x, t);$$

we get

$$1 = \partial_\eta x_\pm \partial_x \beta, \qquad 0 = \partial_\eta x_\pm \partial_t \beta + \partial_t x_\pm.$$

Using equation (5.34) to express $\partial_t x_\pm$ as V we obtain,

$$\partial_t \beta = -\partial_t x_\pm / \partial_\eta x_\pm = -V \partial_x \beta.$$

This is equation (5.32), equivalent to (5.27). We appeal now to Theorem 5.6 to conclude that (b) holds.

LEMMA 5.10. *For* $\eta > 0$,

$$\mathcal{Re}\left(\frac{P(\eta)}{R(\eta)}\right) \geqq 0, \tag{5.37}$$

$$\int_0^\eta \mathcal{Im}\left(\frac{P(\mu)}{R(\mu)}\right) d\mu \geqq 0. \tag{5.38}$$

Proof: By (5.12), $\mathcal{Re}\,(P(\eta)/R(\eta))$ is non-zero only on the set I. By Corollary 5.4, the sign of $P(\eta)$ alternates on the intervals that make up I, and by (5.8) it is positive in the rightmost interval. On the other hand, by (5.4) and (5.5), $R(\eta)$ has the same property, so (5.37) follows.

Condition (5.20) ensures that $\int_0^\eta \mathcal{Im}\,(P(\mu)/R(\mu))\, d\mu$ is zero on the set I. By Corollary 5.4, we see that $P(\mu)$ alternatively crosses from negative to positive and positive to negative on the intervals $(\beta_{2k+1}, \beta_{2k})$ in such a way that it is positive on (β_1, ∞). Since, by (5.4) and (5.5), the sign of $iR(\mu)$ alternates on those intervals and is positive on (β_1, ∞), we conclude that $\mathcal{Im}\,(P(\mu)/R(\mu))$ crosses from positive to negative on the intervals $(\beta_{2k+1}, \beta_{2k})$ and is positive on (β_1, ∞). This fact implies (5.38), completing the proof of the lemma.

To construct the solution of the variational problem (2.16)–(2.18), $\psi^*(x, t)$, we integrate $\psi_x^*(x, t)$ as given by (5.16). Theorem 3.13 requires that $\psi^*(x, t) = 0$, where $\eta x - 4\eta^3 t - \theta_+(\eta) > 0$; in particular, it follows that, for fixed η and t, $\psi^*(x, t) = 0$ for x large enough. Thus we set

$$\psi^*(x, t) = -\int_x^\infty \psi_x^*(y, t)\, dy \equiv \int_x^\infty \mathcal{Re}\left(\frac{P(\eta)}{Q(\eta)}\right) dy. \tag{5.39}$$

Lemma 5.8 shows that, for fixed η and t, the integrand is zero for x large. From Lemma 5.9 we get that $\partial_t\psi^*(x,t)$ as calculated from (5.39) equals $\psi_t^*(x,t)$ as given by (5.17). Inequality (5.37) of Lemma 5.10 implies that ψ^* is a positive and decreasing function of x. In the x, t-region, where $x<x_-(\eta,t)$, we have by Lemma 5.8 and (5.16), (5.17) that

$$\psi^* = \int_{x_-(\eta,t)}^{x_+(\eta,t)} \frac{P(\eta)}{R(\eta)}\,dy$$

and

$$\partial_x\psi^* = \partial_t\psi^* = 0.$$

Since by (4.14), $\psi^*(x,0)=\phi$ for $x<x_-(\eta)$ we conclude that

$$\int_{x_-(\eta,t)}^{x_+(\eta,t)} \frac{P(\eta)}{R(\eta)}\,dy = \phi(\eta). \tag{5.40}$$

Finally, since we have seen above that $\psi^*(x,t)$ given by (5.39) is a positive, decreasing function of x, it follows from (5.40) that $0\leqq\psi^*\leqq\phi$, i.e., that ψ^* belongs to the admissible set A.

The main result of this section is

THEOREM 5.11. *The solution of the variational problem* (2.16)–(2.18) *is given by*

$$\psi^*(x,t)=\begin{cases} 0 & \text{for} \quad x\geqq x_+(\eta,t),\\ \displaystyle\int_x^{x_+(\eta,t)} \frac{P(\eta)}{R(\eta)}\,dy & \text{for} \quad x_-(\eta,t)<x<x_+(\eta,t),\\ \phi(\eta) & \text{for} \quad x\leqq x_-(\eta,t). \end{cases} \tag{5.41}$$

Proof: We shall show that $\psi^*(x,t)$ given above satisfies the variational conditions (3.34). Introducing

$$\theta_+(\eta,t)=\eta x_+(\eta,t)-L\psi^*(x_+(\eta,t),t), \tag{5.42}$$

we integrate (5.18) with respect to x from $x_+(\eta,t)$ to x to obtain

$$L\psi^*(x,t)=\begin{cases} \eta x-\theta_+(\eta,t)-\displaystyle\int_{x_+(\eta,t)}^{x}\int_0^{\eta} \mathscr{I}m\left(\frac{P(\mu)}{R(\mu)}\right)d\mu\,dy & \text{for} \quad x>x_+(\eta,t),\\ \eta x-\theta_+(\eta,t) \quad \text{for} \quad x_-(\eta,t)\leqq x\leqq x_+(\eta,t), & \\ \eta x-\theta_+(\eta,t)+\displaystyle\int_{x}^{x_-(\eta,t)}\int_0^{\eta} \mathscr{I}m\left(\frac{P(\mu)}{R(\mu)}\right)d\mu\,dy & \text{for} \quad x<x_-(\eta,t). \end{cases} \tag{5.43}$$

Differentiating (5.42) and using (3.51), (3.52), we have

$$\partial_t\theta_+=\eta\,\partial_t x_+-L\psi_x^*(x_+,t)\partial_t x_+-L\psi_t^*(x_+,t)=4\eta^3. \tag{5.44}$$

We know from (4.16) that

$$\theta_+(\eta, 0) = \eta x_+(\eta, 0) - L\psi^*(x_+(\eta, 0), 0)$$
$$= \eta x_+(\eta) - L\psi^*(x_+(\eta), 0) = \theta_+(\eta).$$

Therefore, integrating (5.44) we see that

$$\theta_+(\eta, t) = 4\eta^3 t + \theta_+(\eta) \tag{5.45}$$

and, using the definition (1.23) of a, that

$$a(\eta, x, t) = \eta x - \theta_+(\eta, t). \tag{5.46}$$

Thus, using (5.46), formula (5.43) becomes

$$a - L\psi^* = \begin{cases} \displaystyle\int_{x_+(\eta,t)}^{x} \int_0^{\eta} \mathscr{I}m\left(\frac{P(\mu)}{R(\mu)}\right) d\eta\, dy & \text{for} \quad x > x_+(\eta, t), \\ 0 & \text{for} \quad x_-(\eta, t) \leqq x \leqq x_+, \\ \displaystyle -\int_{x}^{x_-(\eta,t)} \int_0^{\eta} \mathscr{I}m\left(\frac{P(\mu)}{R(\mu)}\right) d\mu\, dy & \text{for} \quad x < x_-(\eta, t). \end{cases} \tag{5.47}$$

Using (5.38) of Lemma 5.10 we conclude from this that $\psi^*(x, t)$ given by (5.41) satisfies the variational conditions (3.34). This proves Theorem 5.11.

We may obtain explicit formulas for $\bar{u}(x, t)$, $\overline{u^2}(x, t)$ and $\overline{u^3}\ (x, t)$ by substituting ψ_x^* given by (5.16) and ψ_t^* given by (5.17) into (3.59) of Theorem 3.17:

$$\bar{u} = \frac{4}{\pi}(\eta, \psi_x^*) = -\frac{2}{\pi}\int_I \eta \frac{P(\eta)}{R(\eta)}\, d\eta, \tag{5.48}$$

$$\overline{u^2} = -\frac{16}{3\pi}(\eta^3, \psi_x^*) = \frac{8}{3\pi}\int_I \eta^3 \frac{P(\eta)}{R(\eta)}\, d\eta \tag{5.49}$$

$$= \frac{4}{3\pi}(\eta, \psi_t^*) = \frac{2}{3\pi}\int_I \eta \frac{Q(\eta)}{R(\eta)}\, d\eta, \tag{5.50}$$

$$\overline{u^3} = -\frac{4}{3\pi}(\eta^3, \psi_t^*) = -\frac{2}{3\pi}\int_I \eta^3 \frac{Q(\eta)}{R(\eta)}\, d\eta. \tag{5.51}$$

These integrals can be evaluated by contour integration. Slit the plane along the intervals of the set I given by (5.1); the integral along each interval is then equal to half the integral of a clockwise contour around its slit. By Cauchy's integral theorem we can deform the contour to a large clockwise circle around the origin containing I. Then, letting the radius of this circle go to infinity we can evaluate the integral by computing its residue at infinity.

For large $|\zeta|$ we have, by (5.4),

$$R^2(\zeta) = \prod_{k=1}^{n} (\beta_k^2 - \zeta^2)$$

$$\doteq (-\zeta^2)^n \prod_{k=1}^{n} \left(1 - \frac{\beta_k^2}{\zeta^2}\right) \tag{5.52}$$

$$= (-\zeta^2)^n \left[1 - \left(\sum_k \beta_k^2\right)\frac{1}{\zeta^2} + \left(\sum_{ij} \beta_i^2\beta_j^2\right)\frac{1}{\zeta^4} - \left(\sum_{ijk} \beta_i^2\beta_j^2\beta_k^2\right)\frac{1}{\zeta^6} + O\left(\frac{1}{|\zeta|^8}\right)\right],$$

where the sums over more than one index are understood never to repeat an index value in any term. Since

$$(1+\omega)^{-1/2} = 1 - \tfrac{1}{2}\omega + \tfrac{3}{8}\omega^2 - \tfrac{5}{16}\omega^3 + O(\omega^4),$$

we see from (5.52) and (5.5) that

$$\begin{aligned}\frac{1}{R(\zeta)} = \frac{i}{\zeta^n}\Bigg[1 &+ \frac{1}{2}\left(\sum_k \beta_k^2\right)\frac{1}{\zeta^2} + \left(\frac{3}{8}\left(\sum_k \beta_k^2\right)^2 - \frac{1}{2}\sum_{ij}\beta_i^2\beta_j^2\right)\frac{1}{\zeta^4} \\ &+ \left(\frac{5}{16}\left(\sum_k \beta_k^2\right)^3 - \frac{3}{4}\left(\sum_k \beta_k^2\right)\left(\sum_{ij}\beta_i^2\beta_j^2\right) + \frac{1}{2}\sum_{ijk}\beta_i^2\beta_j^2\beta_k^2\right)\frac{1}{\zeta^6} + O\left(\frac{1}{|\zeta|^8}\right)\Bigg].\end{aligned} \tag{5.53}$$

From (5.8) and (5.9) we have

$$P(\zeta) = \zeta^n\left[1 + \frac{\alpha_2}{\zeta^2} + \frac{\alpha_4}{\zeta^4} + O\left(\frac{1}{|\zeta|^6}\right)\right],$$

$$Q(\zeta) = \zeta^{n+2}\left[12 - 6\left(\sum_k \beta_k^2\right)\frac{1}{\zeta^2} + \frac{\gamma_2}{\zeta^4} + \frac{\gamma_4}{\zeta^6} + O\left(\frac{1}{|\zeta|^8}\right)\right].$$

Combining this with (5.53) gives

$$\begin{aligned}\frac{P(\zeta)}{R(\zeta)} = i\Bigg[1 &+ \left(\alpha_2 + \frac{1}{2}\left(\sum_k \beta_k^2\right)\right)\frac{1}{\zeta^2} \\ &+ \left(\alpha_4 + \tfrac{1}{2}\left(\sum_k \beta_k^2\right)\alpha_2 + \frac{3}{8}\left(\sum_k \beta_k^2\right)^2 - \frac{1}{2}\sum_{ij}\beta_i^2\beta_j^2\right)\frac{1}{\zeta^4} + O\left(\frac{1}{|\zeta|^6}\right)\Bigg],\end{aligned} \tag{5.54}$$

$$\begin{aligned}\frac{Q(\zeta)}{R(\zeta)} = i\Bigg[12\zeta^2 &+ \left(\gamma_2 + \frac{3}{2}\left(\sum_k \beta_k^2\right)^2 - 6\sum_{ij}\beta_i^2\beta_j^2\right)\frac{1}{\zeta^2} \\ &+ \Bigg(\gamma_4 + \frac{1}{2}\left(\sum_k \beta_k^2\right)\gamma_2 + \frac{3}{2}\left(\sum_k \beta_k^2\right)^3 \\ &- 6\left(\sum_k \beta_k^2\right)\left(\sum_{ij}\beta_i^2\beta_j^2\right) + 6\sum_{ijk}\beta_i^2\beta_j^2\beta_k^2\Bigg)\frac{1}{\zeta^4} + O\left(\frac{1}{|\zeta|^6}\right)\Bigg].\end{aligned} \tag{5.55}$$

Using (5.54) and (5.55) to find the residues, we compute from (5.48)–(5.51):

$$\bar{u} = \frac{1}{\pi}\oint \zeta \frac{P(\zeta)}{R(\zeta)}\,d\zeta = -\Big(\sum_k \beta_k^2\Big) - 2\alpha_2, \tag{5.56}$$

$$\overline{u^2} = -\frac{4}{3\pi}\oint \zeta^3 \frac{P(\zeta)}{R(\zeta)}\,d\zeta = \Big(\sum_k \beta_k^2\Big)^2 - \frac{4}{3}\Big(\sum_{ij} \beta_i^2\beta_j^2\Big) + \frac{4}{3}\Big(\sum_k \beta_k^2\Big)\alpha_2 + \frac{8}{3}\alpha_4, \tag{5.57}$$

$$\overline{u^2} = -\frac{1}{3\pi}\oint \zeta \frac{Q(\zeta)}{R(\zeta)}\,d\zeta = \Big(\sum_k \beta_k^2\Big)^2 - 4\Big(\sum_{ij} \beta_i^2\beta_j^2\Big) + \frac{2}{3}\gamma_2, \tag{5.58}$$

$$\begin{aligned}\overline{u^3} = \frac{1}{3\pi}\oint \zeta^3 \frac{Q(\zeta)}{R(\zeta)}\,d\zeta &= -\Big(\sum_k \beta_k^2\Big)^3 + 4\Big(\sum \beta_k^2\Big)\Big(\sum_{ij} \beta_i^2\beta_j^2\Big)\\ &\quad - 4\Big(\sum_{ijk} \beta_i^2\beta_j^2\beta_k^2\Big) - \frac{1}{3}\Big(\sum_k \beta_k^2\Big)\gamma_2 - \frac{2}{3}\gamma_4.\end{aligned} \tag{5.59}$$

Note that, together, (5.57) and (5.58) yield an algebraic identity:

$$\gamma_2 = 4\sum_{ij} \beta_i^2\beta_j^2 + 2\Big(\sum_k \beta_k^2\Big)\alpha_2 + 4\alpha_4.$$

Now we consider the region of x and t where n, the cardinality of $B(x, t)$, is equal to one. In that case formulas (5.56)–(5.59) become

$$\bar{u} = -\beta^2, \qquad \overline{u^2} = \beta^4, \qquad \overline{u^3} = -\beta^6 \tag{5.60}$$

which are (4.39) and (4.41).

Acknowledgment. The research of the first author was supported by the U.S. Department of Energy, Office of Basic Energy Sciences, Applied Mathematical Sciences Research Program under Contract DE-AC02-76ER03077, that of the second by the John and Fannie Hertz Foundation.

Bibliography

[1] Cole, J. D., *On a quasi-linear parabolic equation occurring in aerodynamics*, Quart. Appl. Math. 9, 1951, pp. 225–236.

[2] Flaschka, H., Forest, M. G., and McLaughlin, D. W., *Multiphase averaging and the inverse spectral solution of the Korteweg-de Vries equation*, Comm. Pure Appl. Math. 33, 1980, pp. 739–784.

[3] Gardner, C. S., Greene, J. M., Kruskal, M. D., and Miura, R. M., *Method for solving the Korteweg–de Vries equation*, Phys. Rev. Lett. 19, 1967, pp. 1095–1097.

[4] Gelfand, I. M., and Levitan, B. M., *On the determination of a differential equation by its spectral function*, Izv. Akad. Nauk SSR Ser. Mat. 15, 1951, pp. 309–360. (In Russian.)

[5] Hopf, E., *The partial differential equation* $u_1 + uu_x = \mu_{xx}$, Comm. Pure Appl. Math. 3, 1950, pp. 201–230.

[6] Kay, I., and Moses, H. E., *Reflectionless transmission through dielectrics and scattering potentials*, J. Appl. Phys. 27, 1956, pp. 1503–1508.

[7] Koosis, P., *Introduction to H^p spaces*, LMS Lecture Notes 40, 1980.

[8] Lax, P. D., and Levermore, C. D., *The zero dispersion limit for KdV equation*, Proc. Nat. Acad. Sci. USA, 1979, pp. 3602–3606.

[9] Levermore, C. D., *The small dispersion limit of the Korteweg–de Vries equation*, Dissertation, FGAS, 1982, New York University.

[10] Miura, R. M., and Kruskal, M. D., *Application of a non linear WKB method to the Korteweg–de Vries equation*, SIAM J. Appl. Math. 26, 1974, pp. 376–395.

[11] Muskhelishvili, N. I., *Singular Integral Equations*, 1953, P. Noordhoff, Groningen.

[12] Tanaka, S., *On the N-tuple wave solutions of the Korteweg–de Vries equation*, Res. Inst. Math. Sci., Kyoto 8, 1973, pp. 419–427.

[13] Venakides, S., *The zero dispersion limit of the Korteweg–de Vries equation*, Dissertation, FGAS, 1982, New York University.

[14] Whitham, G. B., *Linear and Nonlinear Waves*, Wiley-Interscience, New York, 1974.

Received June, 1982.

The Small Dispersion Limit of the Korteweg–deVries Equation. III

PETER D. LAX
Courant Institute

AND

C. DAVID LEVERMORE
Lawrence Livermore Laboratory

Abstract

In Parts I and II we have derived explicit formulas for the distribution limit $\bar{u}$ of the solution of the *KdV* equation as the coefficient of u_{xxx} tends to zero. This formula contains n parameters $\beta_1, \cdots, \beta_n$ whose values, as well as whose number, depends on x and t. In Section 4 we have shown that for $t < t_b$, $n = 1$, and the value of β, was determined. In Section 5 we have shown that the parameters β_j satisfy a nonlinear system of partial differential equations.

In Part III, Section 6 we show that for t large, $n = 3$, and we determine the asymptotic behavior of β_1, β_2, β_3, and of $\bar{u}$ and $\overline{u^2}$, for t large. The explicit formulas show that $\bar{u}$ and $\overline{u^2}$ are $O(t^{-1})$ and $O(t^{-2})$ respectively (see formulas (6.2) and (6.24)).

In Section 7 we study initial data whose value tends to zero as $x \to +\infty$, and to -1 as $x \to -\infty$. If one accepts some plausible guesses about the behavior of solutions with such initial data, we derive an explicit formula for the solution and determine the large scale asymptotic behavior of the solution:

$$\bar{u}(x,t)\begin{cases} \cong -1 & \text{for } x < -6t \\ = s(x/t) & \text{for } -6t < x < 4t \\ \sim 0 & \text{for } 4t < x. \end{cases}$$

The function $s(\xi)$ is expressible in terms of complete elliptic integrals; a similar formula is derived for $\overline{u^2}$.

In Section 8 we indicate how to extend the treatment of this series of papers to multihumped (but still negative) initial data.

6. Asymptotic Behavior for Large Time

The main result of this section concerns the values of $\bar{u}(x,t)$ as $t \to \infty$:

THEOREM 6.1. *Let δ be any positive quantity.*
(a) *For $x/t < -\delta$,*

$$\bar{u}(x,t) = O(t^{-2}). \tag{6.1}$$

Communications on Pure and Applied Mathematics, Vol. XXXVI 809–830 (1983)

CCC 0010–3640/83/060809-21 $03.10

(b) *For* $\delta < x/t < 4-\delta$,

$$\bar{u}(x,t) = -\frac{1}{2\pi t}\phi(\sigma(x,t)) + o(t^{-1}), \tag{6.2}$$

where

$$\sigma(x,t) = \tfrac{1}{2}(x/t)^{1/2}. \tag{6.2$'$}$$

(c) *For* $x/t > 4+\delta$,

$$\bar{u}(x,t) = O(t^{-2}). \tag{6.3}$$

The proof is based on the variational characterization of $\bar{u}$. We use condition (3.25):

$$\psi^*(\eta, x, t) = \begin{cases} 0 & \text{where } \eta x - 4\eta^3 t - \theta_+(\eta) > 0, \\ \phi(\eta) & \text{where } \eta x - 4\eta^3 t - \theta_-(\eta) < 0. \end{cases} \tag{6.4}$$

According to Lemma 1.4, $\theta_+(\eta)$ and $\theta_-(\eta)$ are continuous functions that vanish at $\eta = 0$. From this and (6.4) we deduce

LEMMA 6.2. (a) *For* $x/t < -\delta$,

$$\psi^*(\eta) = \phi(\eta) \textit{ for } O(t^{-1}) < \eta < 1. \tag{6.5}$$

(b) *For* $\delta < x/t < 4-\delta$,

$$\psi^*(\eta) = \begin{cases} 0 & \textit{for } O(t^{-1}) < \eta < \sigma(x,t) - O(t^{-1}) \\ \phi(\eta) & \textit{for } \sigma(x,t) + O(t^{-1}) < \eta < 1. \end{cases} \tag{6.6}$$

(c) *For* $x/t > 4+\delta$,

$$\psi^*(\eta) = 0 \textit{ for } O(t^{-1}) < \eta \leqq 1. \tag{6.7}$$

We show now that parts (a) and (c) of Theorem 6.1 follow from parts (a) and (c) of Lemma 6.2. It follows from (6.5) and (6.7) that in both cases the parameters β_j introduced in Section 5 lie in the range $(0, O(t^{-1}))$. We assume that there is in fact only a single β.

According to formula (5.60),

$$\bar{u} = -\beta_1^2;$$

since $\beta_1 \leqq O(t^{-1})$, (6.1) and (6.3) follow.

We turn now to the interesting case (b); here it follows from (6.6) that the parameters β_j are contained in the intervals

$$[0, O(t^{-1})] \quad \text{or} \quad [\sigma(x,t) - O(t^{-1}), \sigma(x,t) + O(t^{-1})]$$

with at least two β's in the second interval. We assume that in fact there are exactly two, β_1 and β_2, in the second interval, and one, β_3, in the first. Thus

(6.6) can be rewritten as

$$\psi^*(\eta)\begin{cases} =0 & \text{for } \eta \text{ in } [\beta_3,\beta_2],\\ =\phi(\eta) & \text{for } \eta \text{ in } [\beta_1,1],\\ \text{is in } (0,\phi(\eta)) & \text{for } \eta \text{ in } (0,\beta_3) \text{ or } (\beta_2,\beta_1). \end{cases} \tag{6.8}$$

Here

$$|\beta_j-\sigma(x,t)|\leqq O(t^{-1}), \qquad j=1,2, \tag{6.9}$$

and

$$\beta_3=O(t^{-1}). \tag{6.9$'$}$$

The next lemma nails down more precisely the distance between β_1 and β_2. We assume that ϕ is continuous in (0, 1) and that θ_+ is differentiable.

LEMMA 6.3.

$$\lim_{t\to\infty}\frac{\log(\beta_1-\beta_2)}{x}\phi(\sigma(x,t))=-2\pi. \tag{6.10}$$

Proof: According to the variational condition (3.34),

$$L\psi^*=a \tag{6.11}$$

at every point η where $0<\psi^*(\eta)<\phi(\eta)$. In particular, according to (6.8), (6.11) holds for η in (β_2,β_1), and so by continuity also at $\eta=\beta_i$, $i=1,2$. We now set $\eta=\beta_i$ in (6.11); using the definition (2.20) of L we get

$$a(\beta_i)=(L\psi^*)(\beta_i)=\frac{1}{2\pi}\int_0^1\log\left(\frac{\beta_i-\mu}{\beta_i+\mu}\right)^2\psi^*(\mu)\,d\mu, \qquad i=1,2. \tag{6.11$'$}$$

According to (6.8), $\psi^*(\mu)=0$ for $\beta_3\leqq\mu\leqq\beta_2$ and $\psi^*(\mu)=\phi(\mu)$ for $\beta_1\leqq\mu\leqq 1$. Using this information, we obtain the following:

$$\begin{aligned} a(\beta_1)-a(\beta_2)&=(L\psi^*)(\beta_1)-(L\psi^*)(\beta_2)\\ &=\frac{1}{2\pi}\int_0^{\beta_3}\log\left(\frac{(\beta_1-\mu)(\beta_2+\mu)}{(\beta_2-\mu)(\beta_1+\mu)}\right)^2\psi^*(\mu)\,d\mu\\ &\quad+\frac{1}{2\pi}\int_{\beta_2}^1\log\left(\frac{\beta_2+\mu}{\beta_1+\mu}\right)^2\psi^*(\mu)\,d\mu\\ &\quad+\frac{1}{2\pi}\int_{\beta_2}^{\beta_1}\log\left(\frac{\beta_1-\mu}{\beta_2-\mu}\right)^2\psi^*(\mu)\,d\mu\\ &\quad+\frac{1}{2\pi}\int_{\beta_1}^1\log\left(\frac{\beta_1-\mu}{\beta_2-\mu}\right)^2\phi(\mu)\,d\mu. \end{aligned} \tag{6.12}$$

We claim that the first three terms on the right are $O(\beta_1-\beta_2)$. This is obvious for the first two terms; in the third term we introduce as new variable of integration

$$\lambda=\frac{\mu-\beta_2}{\beta_1-\beta_2}.$$

We get

$$(\beta_1-\beta_2)\int_0^1 \log\left(\frac{1-\lambda}{\lambda}\right)\psi^*\,d\lambda.$$

Since $\psi^*\leqq\phi\leqq$const., the above quantity is indeed $O(\beta_1-\beta_2)$. Thus we can rewrite (6.12) as

$$a(\beta_1)-a(\beta_2)=\frac{1}{2\pi}\int_{\beta_1}^1 \log\left(\frac{\beta_1-\mu}{\beta_2-\mu}\right)^2\phi(\mu)\,d\mu+O(\beta_1-\beta_2). \tag{6.13}$$

We further split the integral on the right into two

$$\int_{\beta_1}^1=\int_{\beta_1}^{\beta_1+\gamma}+\int_{\beta_1+\gamma}^1,$$

γ a small quantity. For fixed γ, the second integral is $O(\beta_1-\beta_2)$; in the first integral we write

$$\phi(\mu)=\phi(\beta_1)+\varepsilon(\mu);$$

$\varepsilon(\mu)$ is small in $[\beta_1,\beta_1+\gamma]$ when γ is small. We introduce

$$\kappa=\mu-\beta_1$$

as a new variable of integration; this allows us to rewrite the right side of (6.13) as

$$\begin{aligned}&\frac{1}{\pi}\int_0^\gamma[\phi(\beta_1)+\varepsilon]\log\frac{\kappa}{\kappa+(\beta_1-\beta_2)}\,d\kappa+O(\beta_1-\beta_2)\\&=\frac{1}{\pi}[\phi(\beta_1)+\varepsilon](\beta_1-\beta_2)\log(\beta_1-\beta_2)+O(\beta_1-\beta_2).\end{aligned} \tag{6.13$'$}$$

We turn now to the left side of (6.13) and apply the mean value theorem; we get

$$a(\beta_1)-a(\beta_2)=\overline{a_\eta}(\beta_1-\beta_2), \tag{6.14}$$

where

$$\overline{a_\eta}=a_\eta(\bar\beta),\qquad \beta_2<\bar\beta<\beta_1. \tag{6.15}$$

It follows from (6.9) that

$$\bar\beta=\sigma(x,t)+O(t^{-1}). \tag{6.16}$$

We employ definition (1.23) of $a(\eta, x, t)$ to write

$$a_\eta = x - 12t\eta^2 - d\theta_+/d\eta.$$

Using this and (6.16) in (6.15) we get

$$\overline{a_\eta} = x - 12tx/4t + O(1) = -2x + O(1).$$

Inserting this into (6.14) we obtain

$$a(\beta_1) - a(\beta_2) = -2x(\beta_1 - \beta_2) + O(\beta_1 - \beta_2).$$

This is equal to (6.13)′; dividing by $(\beta_1 - \beta_2)$ one has

$$-2x + O(1) = \frac{1}{\pi}(\phi(\beta_1) + \varepsilon)\log(\beta_1 - \beta_2) + O(1).$$

Using (6.9) once more we note that

$$\phi(\beta_1) = \phi(\sigma(x, t)) + \varepsilon(t),$$

where $\varepsilon(t) \to 0$ as $t \to \infty$; relation (6.10) of Lemma 6.3 follows.

We make use of formula (5.56) for the limit solution:

$$\bar{u} = -(\beta_1^2 + \beta_2^2 + \beta_3^2) - 2\alpha_2; \tag{6.17}$$

α_2 is a function of $\beta_1, \beta_2, \beta_3$ defined by (5.20). With $n = 3$,

$$P(z) = z^3 + \alpha_2 z$$

so that (5.20) becomes

$$\int_{\beta_3}^{\beta_2} \frac{\eta^3 + \alpha_2\eta}{R(\eta)}\, d\eta = 0.$$

Thus

$$\alpha_2 = -J^{(3)}/J^{(1)}, \tag{6.18}$$

where

$$J^{(j)} = \int_{\beta_3}^{\beta_2} \frac{\eta^j}{R(\eta)}\, d\eta, \qquad j = 1, 3, \tag{6.19}$$

and

$$R^2(\eta) = (\eta^2 - \beta_3^2)(\beta_2^2 - \eta^2)(\beta_1^2 - \eta^2). \tag{6.20}$$

According to (6.9), (6.9)′, as $t \to 0$, $\beta_1 - \beta_2 \to 0$ and $\beta_3 \to 0$. We shall now determine the asymptotic behavior of α_2 under these circumstances.

LEMMA 6.4 *As $\beta_1 - \beta_2$ and $\beta_3 \to 0$,*
(a)

$$\beta_1^2 J^{(1)} - J^{(3)} - \beta_1 \to 0, \tag{6.21}$$

(b)

$$J^{(1)} \cong \frac{1}{2\beta_1} |\log(\beta_1 - \beta_2)|. \tag{6.21$'$}$$

Proof: (a) Using (6.19) and (6.20) we get

$$\begin{aligned} \beta_1^2 J^{(1)} - J^{(3)} &= \int_{\beta_3}^{\beta_2} \frac{\beta_1^2 \eta - \eta^3}{R(\eta)} d\eta \\ &= \int_{\beta_3}^{\beta_2} \frac{\eta}{(\eta^2 - \beta_3^2)^{1/2}} \left(\frac{\beta_1^2 - \eta^2}{\beta_2^2 - \eta^2} \right)^{1/2} d\eta. \end{aligned} \tag{6.22}$$

The integrand on the right in (6.22) tends to 1 uniformly on every closed subset of (β_3, β_2), and is dominated over that interval by const./$[(\eta - \beta_3)(\beta_2 - \eta)]^{1/2}$, an integrable function. Therefore, by the principle of dominated convergence, as $\beta_3 \to 0$ and $\beta_2 - \beta_1 \to 0$, (6.22) approximates β_1, as asserted in (6.21).

To prove (6.21)′ we separate those factors of R which are small at the end point β_1:

$$R(\eta) = [(\eta^2 - \beta_3^2)(\beta_2 + \eta)(\beta_1 + \eta)]^{1/2}[(\beta_1 - \eta)(\beta_2 - \eta)]^{1/2}.$$

The value of the first factor at $\eta = \beta_2$ is

$$2\beta_1^2 + O(\beta_1 - \beta_2) + O(\beta_3).$$

From this we deduce easily that

$$J^{(1)} \cong \frac{1}{2\beta_1} \int_{\beta_3}^{\beta_2} \frac{d_\eta}{[(\beta_1 - \eta)(\beta_2 - \eta)]^{1/2}}; \tag{6.22$'$}$$

elementary calculus shows that the integral on the right in (6.22)′ is $\cong$ $-\log(\beta_1 - \beta_2)$; this completes the proof of part (b) of Lemma 6.4.

Using (6.21), (6.21)′ in (6.18) gives

$$\begin{aligned} \alpha_2 = -\frac{J^{(3)}}{J^{(1)}} &= \frac{\beta_1^2 J^{(1)} - J^{(3)}}{J^{(1)}} - \beta_1^2 \\ &= -\beta_1^2 + \frac{2\beta_1^2}{\log(\beta_1 - \beta_2)} + o\left(\frac{1}{\log(\beta_1 - \beta_2)} \right). \end{aligned}$$

Inserting this into (6.17) and making use of (6.10) we get

$$\bar{u}(x, t) = \beta_1^2 - \beta_2^2 - \beta_3^2 - \frac{2\beta_1^2}{\pi x} \phi(\sigma(x, t)) + o(t^{-1}). \tag{6.23}$$

It follows from (6.10) that

$$|\beta_1 - \beta_2| \leq O(e^{-c/t}).$$

Using this to estimate the first two terms in (6.23), (6.9)′ to estimate the third term and (6.9) to estimate the fourth term we obtain (6.2). This completes the proof of Theorem 6.1.

We conclude with a number of observations:

(i) A similar argument, based on formula (5.57) for $\overline{u^2}$, yields

(6.24) $$\overline{u^2}(x,t) = \frac{x}{6\pi t^2}\phi(\sigma(x,t)) + o(t^{-1}).$$

The first relation of (4.42) is

(6.25) $$\bar{u}_t = 3\overline{u_x^2};$$

the leading terms in the asymptotic descriptions (6.2) and (6.24) are consistent with this relation.

(ii) Relation (6.6) suggests the following crude approximation to $\psi^*(\eta, x, t)$ in case (b):

$$\psi^*(\eta, x, t) = H(\eta - \sigma(x,t))\phi(\eta),$$

where

$$H(\mu) = 0 \quad \text{for} \quad \mu < 0, \qquad H(\mu) = 1 \quad \text{for} \quad \mu > 0.$$

Then, from (6.2)′ we have

$$\psi_x^*(\eta, x, t) = -\delta(\eta - \sigma(x,t))\frac{1}{4(xt)^{1/2}}\phi(\eta).$$

Using this expression in formula (3.59) for $\bar{u}$ gives

$$\bar{u}(x,t) = \frac{4}{\pi}(\eta, \psi_x^*) = -\frac{4}{\pi}\sigma(x,t)\frac{\phi(\sigma(x,t))}{4(xt)^{1/2}}$$
$$= -\frac{1}{2\pi t}\phi(\sigma(x,t)),$$

in agreement with the leading term in (6.2).

(iii) Next we show that (6.2) can be derived from the zero dispersion limit of Tanaka's description of the large t behavior of solutions of KdV. Tanaka shows that, for large t and for $0 < x < 4t$, every solution is a wave train, i.e., approximately a superposition of solitons

(6.26) $$u(x, t; \varepsilon) \cong \sum_1^N s(x - 4\eta_n^2 t - \delta_n, \eta_n);$$

here

(6.26)′ $$s(x, \eta) = -2\eta^2 \operatorname{sech}^2(\eta x/\varepsilon).$$

The width of such a soliton is

$$\text{const. } \varepsilon/\eta. \tag{6.27}$$

It follows from the precise form $(1.18)_{\text{ii}}$ of Weyl's law that

$$\eta_{n+1}-\eta_n \cong \frac{\pi\varepsilon}{\phi(\bar\eta_n)}, \qquad \bar\eta_n \quad \text{in} \quad [\eta_n, \eta_{n+1}]. \tag{6.28}$$

Peaks of the wave train (6.26) are located at

$$4\eta_n^2 t+\delta_n. \tag{6.29}$$

In view of (6.28) we see that for t large the peaks are separated by the distance

$$\cong 8\eta_n t(\eta_{n+1}-\eta_n)+\delta_{n+1}-\delta_n \cong \frac{8\pi\eta_n t\varepsilon}{\phi(\eta_n)}. \tag{6.29$'$}$$

Comparing this to (6.27) we conclude that for large t the individual solitons in the wave train are well separated.

The wave number η of the soliton that peaks at x at time t is, by (6.29) and (6.2)$'$,

$$\eta_n \cong \sigma(x,t)=(x/4t)^{1/2}, \tag{6.29$''$}$$

provided that t is large and $0<x/t<4$. Setting this in (6.29)$'$ we conclude that for t large the density of solitons at x is

$$\cong \phi(\sigma(x,t))/4\pi\varepsilon\,(xt)^{1/2}. \tag{6.30}$$

It follows from (6.26)$'$ that the area under a soliton is $4\eta\varepsilon$; at the wave number (6.29)$''$ this is

$$2(x/t)^{1/2}\varepsilon. \tag{6.30$'$}$$

The asymptotic area density in the wave train (6.26) is the product of (6.30) and (6.30)$'$:

$$(2\pi t)^{-1}\phi(\sigma(x,t)).$$

Since the asymptotic area density is the weak limit $\bar u$, we obtain yet another derivation of (6.2).

(iv) In deriving (6.2) we have assumed that the function ϕ is continuous. When the potential has several local minima, the function ϕ has discontinuities; in these cases, Theorem 6.1 remains true as long as $\sigma(x,t)$ is bounded away from the points of discontinuity of ϕ.

7. Initial Data with $u(-\infty) \neq u(\infty)$

In this section we study solutions of

$$u_t - 6uu_x + \varepsilon^2 u_{xxx} = 0 \tag{7.1}$$

whose initial value u satisfies

$$\lim_{x\to-\infty} u(x) = -1, \qquad \lim_{x\to\infty} u(x) = 0, \tag{7.2}$$

and is an increasing function of x.

The solution of such an initial value problem can be obtained as the limit of solutions whose initial data u_n satisfy the conditions laid down in Section 1:

$$\lim_{n\to\infty} u_n(x) = u(x), \tag{7.3}$$

uniformly for $x > x_0$, x_0 any number. Denoting by $u_n(x, t; \varepsilon)$ the solution of (7.1) with initial data $u_n(x)$, it can be shown that

$$\lim_{n\to\infty} u_n(x, t; \varepsilon) = u(x, t; \varepsilon). \tag{7.3$'$}$$

Although for the solutions of KdV equations signals propagate with infinite speed, the dependence of the solution at fixed x, t on the values of the initial data at y diminishes to zero as y tends to ∞, uniformly in ε. Therefore we conjecture that the limits $n \to \infty$ and $\varepsilon \to 0$ can be interchanged:

CONJECTURE 7.1.

$$\text{d-}\lim_{\varepsilon\to 0} u(x, t; \varepsilon) = \lim_{n\to\infty} \text{d-}\lim_{\varepsilon\to 0} u_n(x, t; \varepsilon). \tag{7.4}$$

The inner limit on the right in (7.4) is characterized in Theorem 2.10 in terms of the solution of a minimum problem (2.16):

$$\bar{u}_n = \text{d-}\lim u_n(x, t; \varepsilon) = \partial_x^2 Q_n^*, \tag{7.5}$$

where

$$Q_n^* = \min_{0 \leqq \psi \leqq \phi_n} Q_n(\psi; x, t), \tag{7.5$'$}$$

and

$$Q_n(\psi) = \frac{4}{\pi}(a_n, \psi) - \frac{2}{\pi}(L\psi, \psi). \tag{7.6}$$

By (1.23),

$$a_n = x\eta - 4t\eta^3 - \theta_+^{(n)}(\eta), \tag{7.7}$$

by (1.25),

$$\phi_n(\eta)=\int_{x_-^{(n)}(\eta)}^{x_+^{(n)}(\eta)}\frac{\eta}{|u_n(y)+\eta^2|^{1/2}}\,dy, \tag{7.8}$$

and by (1.16),

$$\theta^{(n)}(\eta)=\eta x_+^{(n)}(\eta)+\int_{x_+^{(n)}(\eta)}^{\infty}(\eta-(u_n(y)+\eta^2)^{1/2})\,dy; \tag{7.9}$$

$x_\pm^{(n)}(\eta)$ are defined by (1.11):

$$u_n(x_\pm^{(n)}(\eta))=-\eta^2. \tag{7.10}$$

Since, by (7.3), u_n tend to u, and u satisfies (7.2), it follows that

$$\lim_{n\to\infty} x_-^{(n)}(\eta)=-\infty,\qquad \lim_{n\to\infty} x_+^{(n)}(\eta)=x_+(\eta). \tag{7.11}$$

From this we deduce easily that

$$\lim_{n\to\infty}\theta_+^{(n)}(\eta)=\theta_+(\eta),$$

$$\lim_{n\to\infty} a_n(x,t,\eta)=a(x,t,\eta),$$

$$\lim_{n\to\infty}\phi_n(\eta)=\infty,$$

uniformly in η. It follows that, for fixed ψ in L^1,

$$\lim_{n\to\infty} Q_n(\psi;x,t)=Q(\psi,x,t). \tag{7.11$'$}$$

We surmise that the limit of the minima Q_n^* defined in (7.5)′ can be characterized as the minimum of (7.11)′:

CONJECTURE 7.2.

$$\lim_{n\to\infty} Q_n^*=Q^*, \tag{7.12}$$

where

$$Q^*=\min_{0\leq\psi} Q(\psi), \tag{7.13}$$

$$Q(\psi)=\frac{4}{\pi}(a,\psi)-\frac{2}{\pi}(L\psi,\psi). \tag{7.13$'$}$$

Combining Conjectures 7.1 and 7.2 we deduce

CONJECTURE 7.3.

(7.14)
$$\operatorname*{d-lim}_{\varepsilon\to 0} u(x,t;\varepsilon) = \partial_x^2 Q^*,$$

where Q^* is defined by (7.13).

It is far from obvious that the minimum problem (7.13) has a solution. Nevertheless it follows from the positive definiteness of Q that if ψ^* satisfies the variational conditions for the minimum problem, then $Q(\psi)$ achieves its minimum at ψ^*, and only at ψ^*. The variational conditions (3.34) in the present case are:

(7.15)
$$\psi^* = 0, \quad \text{where} \quad a - L\psi^* > 0,$$

(7.15)′
$$a - L\psi^* \geqq 0 \quad \text{for all} \quad \eta.$$

We turn now to the special initial function

(7.16)
$$u(x) = \begin{cases} -1 & \text{for} \quad x < 0, \\ 0 & \text{for} \quad x > 0. \end{cases}$$

In this case, $x_+(\eta) \equiv 0$, $\theta_+(\eta) \equiv 0$, and

(7.17)
$$a = x\eta - 4t\eta^3.$$

Equation (7.13) can be solved explicitly:

THEOREM 7.4. *For a given by* (7.17), *the solution ψ^* of the variational problem* (7.13) *is as follows*:

(a) *For*

(7.18)
$$x < -6t,$$

(7.18)′
$$\psi^* = \frac{12t\eta^3 - (6t + x)\eta}{(1-\eta^2)^{1/2}}.$$

(b) *For*

(7.19)
$$-6t < x < 4t,$$

(7.19)′
$$\psi^* = \begin{cases} 0 & \textit{for} \quad \eta < \beta, \\ S/R_0 & \textit{for} \quad \beta < \eta < 1, \end{cases}$$

where

(7.20)
$$R_0^2(\eta) = (1-\eta^2)(\eta^2 - \beta^2),$$

(7.21)
$$S(\eta, x, t) = 12tQ - xP,$$

(7.21)′
$$Q(\eta) = \eta^4 + \gamma_1\eta^2 + \gamma_2, \qquad P(\eta) = \eta^2 + \alpha.$$

Here

$$\alpha = -I^{(2)}/I^{(0)}, \tag{7.22}$$

$$\gamma_1 = -\tfrac{1}{2}(1+\beta^2), \tag{7.23}$$

$$\gamma_2 = -(\gamma_1 I^{(2)} + I^{(4)})/I^{(0)}. \tag{7.24}$$

$I^{(j)}$ *denotes the complete elliptic integral*

$$I^{(j)} = \int_0^\beta \frac{\eta^j}{R(\eta)}\, d\eta \tag{7.25}$$

$$R^2(\eta) = (1-\eta^2)(\beta^2-\eta^2). \tag{7.25'}$$

The parameter β *is related to* x/t *through*

$$S(\beta, x, t) = 0. \tag{7.26}$$

(c) *For*

$$4t < x, \tag{7.27}$$

$$\psi^* \equiv 0. \tag{7.27'}$$

Proof: We verify (c) first. Clearly ψ^* is non-negative and so admissible. We show now that the variational conditions are satisfied. Since $\psi^* = 0$, $L\psi^* = 0$; combining this with (7.17) we obtain

$$a - L\psi^* = x\eta - 4t\eta^3. \tag{7.28}$$

Clearly, if (7.27) holds, (7.28) is greater than or equal to 0 in the range $0 \leqq \eta \leqq 1$; thus (7.15) and (7.15)′ are satisfied.

Next we take case (a).

As defined by (7.18)′ $\psi^* = \mathcal{R}e\, g$ on $0 \leqq \eta \leqq 1$, $g = i(x - 12t\eta^2) + (12t\eta^3 - (6t+x)\eta)/(1-\eta^2)^{1/2}$. Using (3.41) we deduce that $L\psi^* = x\eta - 4\eta^3$ on $0 \leqq \eta \leqq 1$, so, by (7.17), $L\psi^* = a$.

This shows that the variational conditions are satisfied. It further follows from (7.18) that ψ^* as defined by (7.18)′ is greater than or equal to 0 for $0 \leqq \eta \leqq 1$ and therefore is admissible.

We turn now to case (b). Fix β in [0, 1], and define x/t through the relation (7.26).

We can write ψ^*, given by (7.19)′, as

$$\psi^* = x\psi_1^* - 12t\psi_2^*, \tag{7.29}$$

where

$$\psi_1^* = \begin{cases} 0 & \text{for} \quad \eta < \beta, \\ P/R_0,\ P = \eta^2 + \alpha & \text{for} \quad \beta < \eta < 1, \end{cases}$$

(7.30)

$$\psi_2^* = \begin{cases} 0 & \text{for} \quad \eta < \beta, \\ Q/R_0,\ Q = \eta^4 + \gamma_1 \eta^2 + \gamma_2 & \text{for} \quad \beta < \eta < 1; \end{cases}$$

see formulas (5.16), (5.17), $n = 2$.

We see that ψ_1^* and ψ_2^* are real parts on the real axis of the analytic functions (see (5.16), (5.17), $n = 2$)

$$f_1(\eta) = \frac{P(\eta)}{R_0(\eta)} - i,$$

(7.31)

$$f_2(\eta) = \frac{Q(\eta)}{R_0(\eta)} - i\eta^2.$$

Clearly, f_1 is $O(\eta^{-1})$ in the upper half-plane; using (7.23) we see that so is f_2.

From (7.29), (7.31) and (7.21) we get

$$\psi^* = \mathscr{R}e\,(f), \tag{7.32}$$

$$f(\eta) = \frac{S(\eta)}{R_0(\eta)} + i(x - 12t\eta^2). \tag{7.32$'$}$$

Using relation (3.41) between the operator L and the Hilbert transform H we deduce from (7.32), (7.32)′ that

$$L\psi^* = \int_0^\eta H\psi^*\, d\mu = \int_0^\eta \mathscr{I}m\,(f)\, d\mu$$

$$= \begin{cases} -\displaystyle\int_0^\eta \frac{S(\mu)}{R(\mu)}\, d\mu + x\eta - 4\eta^3 t, & \eta < \beta, \\ -\displaystyle\int_0^\beta \frac{S(\mu)}{R(\mu)}\, d\mu + x\eta - 4\eta^3 t, & \eta \geqq \beta, \end{cases} \tag{7.33}$$

where R, defined by (7.25)′, is positive on $(0, \beta)$.

Relations (7.22) and (7.24) guarantee that

$$\int_0^\beta \frac{S(\mu)}{R(\mu)}\, d\mu = 0. \tag{7.34}$$

Combining this with (7.33) and using (7.17) we obtain

$$L\psi^* = \begin{cases} \displaystyle\int_0^\eta \frac{S}{R}\, d\mu + a & \text{for} \quad \eta < \beta, \\ a & \text{for} \quad \eta > \beta. \end{cases} \tag{7.35}$$

We are now ready to verify the variational conditions. Since $R(\mu)$ is of one sign in $(0, \beta)$, (7.34) implies that $S(\mu)$ has a zero in $(0, \beta)$. According to (7.26), $S(\beta) = 0$; since S is a quadratic polynomial in μ^2, it can have no more than these two roots on the positive axis and furthermore these two roots are simple; we draw now some consequences:

Since the leading term in S is $12t\mu^4$, $S(\mu) > 0$ for μ large. It follows therefore that $S(\mu) > 0$ for $\mu > \beta$; then, since S has a simple root at β, $S(\mu) < 0$ for $\beta < \mu$ and near β. Moreover, since S has exactly one simple root in $(0, \beta)$, $S(\mu) > 0$ for μ small, see Figure 1.

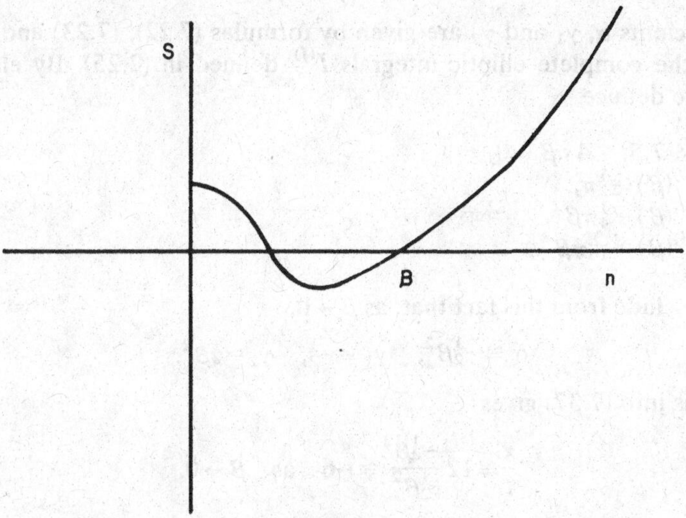

Figure 1

Since R is positive in $(0, \beta)$, it follows that, for η small and greater than 0,

$$\int_0^\eta \frac{S(\mu)}{R(\mu)} d\mu > 0. \tag{7.36}$$

We claim that (7.36) holds for all $\eta < \beta$; for clearly, it holds for $\eta < \eta_0$, where η_0 is the root of S in $(0, \beta)$. Since S is negative on (η_0, β), we deduce from (7.34) that, for η between η_0 and β,

$$\int_0^\eta \frac{S}{R} d\mu > \int_0^\beta \frac{S}{R} d\mu = 0.$$

This proves our contention.

Combining (7.33), (7.34) and (7.36) we have

$$a - L\psi^* \begin{cases} >0 & \text{for} \quad \eta < \beta, \\ =0 & \text{for} \quad \beta < \eta. \end{cases}$$

In view of (7.19)′ this shows that the variational conditions (7.15), (7.15)′ are satisfied.

We note finally that, since $S(\mu)>0$ for $\mu>\beta$, it follows from (7.19)′ that $\psi^*\geqq 0$ and therefore admissible.

To complete the proof of part (b), we have to show that as β varies from 0 to 1, x/t varies between -6 and 4. From (7.21) and (7.26) we can express x/t as a function of β:

$$\frac{x}{t}=12\frac{\beta^4+\gamma_1\beta^2+\gamma_2}{\beta^2+\alpha}. \tag{7.37}$$

The coefficients α, γ_1 and γ_2 are given by formulas (7.22), (7.23) and (7.24) in terms of the complete elliptic integrals $I^{(j)}$ defined in (7.25). By elementary calculus we deduce

LEMMA 7.5. *As* $\beta\to 0$,
(i) $I^{(0)}(\beta)\cong\frac{1}{2}\pi$,
(ii) $I^{(2)}(\beta)\cong\frac{1}{4}\pi\beta^2$,
(iii) $I^{(4)}(\beta)\cong\frac{3}{16}\pi\beta^4$.

We conclude from this fact that, as $\beta\to 0$,

$$\alpha\cong-\tfrac{1}{2}\beta^2,\quad \gamma_1\cong-\tfrac{1}{2},\quad \gamma_2\cong\tfrac{1}{4}\beta^2.$$

Setting this into (7.37) gives

$$\frac{x}{t}\cong 12\frac{-\frac{1}{4}\beta^2}{\frac{1}{2}\beta^2}\cong-6 \quad\text{as}\quad \beta\to 0. \tag{7.38}$$

Using intermediate calculus we obtain

LEMMA 7.6. *As* $\beta\to 1$,
(i) $I^0(\beta)\cong\frac{1}{2}|\log(1-\beta)|$,
(ii) $I^{(0)}(\beta)-I^{(2)}(\beta)\cong 1$,
(iii) $I^{(2)}(\beta)-I^{(4)}(\beta)\cong\frac{1}{3}$.

We deduce from this that

$$\alpha\cong-1+I_0^{-1},\quad \gamma_1\cong-1,\quad \gamma_2\cong\tfrac{1}{3}I_0^{-1}.$$

Setting this into (7.37) gives

$$\frac{x}{t}\cong 12\frac{\frac{1}{3}I_0^{-1}}{I_0^{-1}}=4 \quad\text{as}\quad \beta\to 1. \tag{7.38′}$$

Since x/t as defined by (7.37) is a continuous function of β, it follows that, as β varies from 0 to 1, x/t varies between -6 and 4. Since according to Theorem 3.12 for each x, t there can be at most one function ψ^* that satisfies the variational condition, it follows that the relation of β to x/t is one-to-one. This completes the proof of Theorem 7.4.

According to formulas (5.56), (5.57),

$$\bar{u} = -1 - \beta^2 - 2\alpha,$$
$$\overline{u^2} = (1+\beta^2)^2 - \tfrac{4}{3}\beta^2 + \tfrac{4}{3}(1+\beta^2)\alpha.$$

Setting the definitions (7.22)–(7.24) into these formulas yields

$$\bar{u} = -1 - \beta^2 + 2I^{(2)}(\beta)/I^{(0)}(\beta) \tag{7.39}$$

and

$$\overline{u^2} = 1 + \beta^4 + \tfrac{2}{3}\beta^2 - \tfrac{4}{3}(1+\beta^2)I^{(2)}(\beta)/I^{(0)}(\beta). \tag{7.40}$$

THEOREM 7.7. (a) *For* $x < -6t$,

$$\bar{u}(x, t) = 1, \qquad \overline{u^2}(x, t) = 1.$$

(b) *For* $-6t < x < 4t$, $\bar{u}$ *and* $\overline{u^2}$ *are given by* (7.39) *and* (7.40), *where* β *is a function of* x/t *defined by* (7.37) *and* (7.22)–(7.24).

(c) *For* $4t < x$,

$$\bar{u}(x, t) = 0, \qquad \overline{u^2}(x, t) = 0.$$

Note that weak limits satisfy

$$0 \leqq \bar{u}^2 \leqq \overline{u^2}. \tag{7.41}$$

Using formulas (7.39), (7.40) we get amusing inequalities for complete elliptic integrals from (7.41).

We close this section by presenting numerical values of $\bar{u}(x, 1)$ and $\overline{u^2}(x, 1)$ in the interval of interest, $-6 \leqq x \leqq 4$, obtained by evaluating numerically the complete elliptic integrals occurring in formulas (7.37), (7.39), and (7.40), using approximations given in Abramowitz and Stegun. We thank John Scheuermann for carrying out the calculations.

X	$\bar{u}$	$\overline{u^2}$
−6.0	−1.0	1.0
−5.0	−0.9984	1.0029
−4.0	−0.9931	1.0106
−3.0	−0.9837	1.0215
−2.0	−0.9691	1.0334
−1.0	−0.9481	1.0437
0	−0.9188	1.0484
1.0	−0.8777	1.0411
2.0	−0.8183	1.0107
3.0	−0.7225	0.9293
3.9933	−0.2514	0.3353
4.0	0	0

The calculations indicate that $\bar{u}$ increases monotonically from -1 to 0 as x increases from -6 to 4. On the other hand, $\overline{u^2}$ increases as x goes from -6 to 0, and then decreases as x goes from 0 to 4. Two aspects of this behavior can be deduced from our theory:

Since $\bar{u}(x, t)$ is a function of x/t, it follows that $\bar{u}(0, t) = \text{const.}$; therefore, $\bar{u}_t(0, t) = 0$. Setting this into the first relation in (4.42) we conclude that

$$\overline{u_x^2}(0, t) = 0,$$

as indicated by the calculations.

The calculations also indicate that $\overline{u^2}$ attains values greater than 1; this may be seen as follows:

Given two numbers, a, b, define $I(t)$ by

$$I(t) = \int_{at}^{bt} \overline{u^2}(x, t)\, dx. \tag{7.42}$$

Then

$$\frac{dI}{dt} = -a\overline{u^2}(at, t) + b\overline{u^2}(bt, t) + \int_{at}^{bt} \overline{u_t^2}\, dx.$$

Using the second equation in (4.42) we get

$$\frac{dI}{dt} = -a\overline{u^2}(at, t) - 4\overline{u^3}(at, t) + b\overline{u^2}(bt, t) + 4\overline{u^3}(bt, t). \tag{7.43}$$

If we choose $a < -6$ and $b > 4$, then by Theorem 7.7

$$\overline{u^2}(at, t) = 1, \qquad \overline{u^2}(bt, t) = 0,$$
$$\overline{u^3}(at, t) = -1, \qquad \overline{u^3}(bt, t) = 0.$$

Setting this into (7.43) we get

$$\frac{dI}{dt} = -a + 4,$$

so

$$I = (-a + 4)t.$$

Letting a tend to -6, b to 4 we obtain

$$\int_{-6t}^{4t} \overline{u^2}(x, t)\, dx = 10t. \tag{7.44}$$

Since the length of the interval of integration is $10t$, and since formula (7.40) shows the integrand to be continuous, it follows from (7.44) that $\overline{u^2}(x, t)$ is greater than 1 somewhere.

It follows from formulas (7.39) and (7.40), and elementary facts about the integrals entering these formulas, that the modulus of continuity of $\bar{u}(x, 1)$ and $\overline{u^2}(x, 1)$ at $x = 4$ is only logarithmic. This is what creates the illusion of a jump discontinuity at $x = 4$ in the tabulated values of $\bar{u}$ and $\overline{u^2}$.

8. Arbitrary Negative Potentials

In Section 1 we have assumed for convenience that the potential has a single critical point. Here we indicate how arbitrary negative potentials can be handled by our method. A full exposition, going beyond what is described here, will be given by David Levermore.

Weyl's law, Theorem 1.1, is generally valid. For simplicity we take the case that there are three critical points: two minima, and between them a saddle point. We denote the value of the potential at the minima by $-\mu_1^2$, $-\mu_2^2$, the saddle point by $-\sigma^2$, see Figure 2.

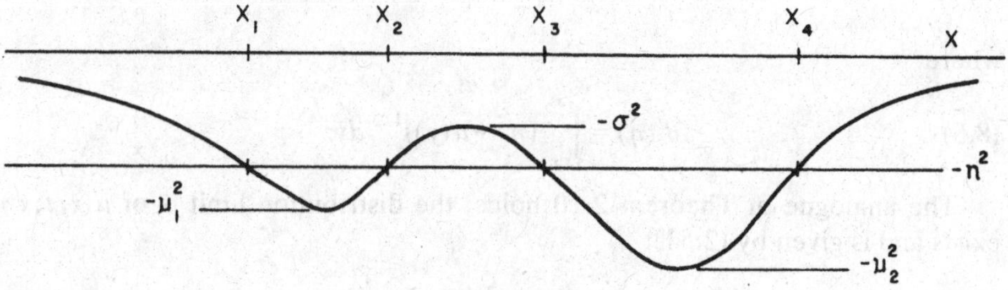

Figure 2

The equation

$$u(x) = -\eta^2 \tag{8.1}$$

has four, two or no solutions, depending on the relation of η^2 to the critical levels of u. We denote these, wherever defined, by $x_1(\eta)$, $x_2(\eta)$, $x_3(\eta)$, $x_4(\eta)$, arranged so that

$$x_1 < x_2 < x_3 < x_4.$$

We assume that the minimum on the right is deeper than on the left.

The number $N(\varepsilon, \eta)$ of eigenvalues of L, see formula (1.10), can be described with the aid of two functions Φ_1 and Φ_2 in place of the single function Φ defined in (1.13);

$$N(\varepsilon, \eta) \cong \frac{1}{\pi\varepsilon}[\Phi_1(\eta) + \Phi_2(\eta)], \tag{8.2}$$

where

$$\Phi_1(\eta)=\begin{cases}\displaystyle\int_{x_1}^{x_4}(-u(y)-\eta^2)^{1/2}\,dy & \text{for } \eta<\sigma,\\ \displaystyle\int_{x_3}^{x_4}(-u(y)-\eta^2)^{1/2}\,dy & \text{for } \sigma<\eta<\mu_2,\end{cases}\tag{8.3}$$

and

$$\Phi_2(\eta)=\begin{cases}0 & \text{for } \eta<\sigma,\\ \displaystyle\int_{x_1}^{x_2}(-u(y)-\eta^2)^{1/2}\,dy & \text{for } \sigma<\eta<\mu.\end{cases}\tag{8.4}$$

This divides the eigenvalues into two classes; the corresponding norming constants $c_1(\eta)$ and $c_2(\eta)$ are given by (1.17), with

$$\theta_{+1}(\eta)=\eta x_1(\eta)+\int_{x_1(\eta)}[\eta-(\eta^2+u(y))^{1/2}\,dy,\tag{8.4$_1$}$$

$$\theta_{+2}(\eta)=\theta_{+1}(\eta)-\theta(\eta),\tag{8.4$_2$}$$

where

$$\theta(\eta)=\int_{x_2}^{x_3}(\eta^2+u(y))^{1/2}\,dy.\tag{8.5}$$

The analogue of Theorem 2.10 holds; the distribution limit $\bar{u}$ of $u(x,t;\varepsilon)$ exists and is given by (2.51):

$$\bar{u}(x,t)=\partial_x^2Q^*(x,t),\tag{8.6}$$

where Q^* is characterized by the minimum problem

$$Q^*(x,t)=\min Q(\psi_1;\psi_2;x,t)\tag{8.7}$$

among all ψ_2, ψ_2 satisfying

$$0\leqq\psi_1\leqq\phi_1,\qquad 0\leqq\psi_2\leqq\phi_2.\tag{8.8}$$

Here

$$Q^*(\psi_1,\psi_2)=\frac{4}{\pi}[(a_1,\psi_1)+(a_2,\psi_2)]-\frac{2}{\pi}(L(\psi_1+\psi_2),\psi_1+\psi_2),\tag{8.9}$$

a_1 and a_2 being defined as in (1.23):

$$a_j(\eta,x,t)=\eta x-4\eta^3t-\theta_{+j}(\eta),\qquad j=1,2;\tag{8.10}$$

ϕ_1, ϕ_2 are given by

$$\phi_1(\eta)=\begin{cases}\displaystyle\int_{x_1}^{x_4}\frac{\eta}{(-u(y)-\eta^2)^{1/2}}\,dy & \text{for} \quad \eta<\sigma,\\ \displaystyle\int_{x_3}^{x_4}\frac{\eta}{(-u(y)-\eta^2)^{1/2}}\,dy & \text{for} \quad \sigma<\eta<\mu_2,\end{cases} \tag{8.11}$$

and

$$\phi_2(\eta)=\begin{cases}0 & \text{for} \quad \eta<\sigma,\\ \displaystyle\int_{x_1}^{x_2}\frac{\eta}{(-u(y)-\eta^2)^{1/2}}\,dy & \text{for} \quad \sigma<\eta<\mu_1.\end{cases} \tag{8.12}$$

We conclude with this observation: the KdV equation is the first of an infinite sequence of equations, such that the corresponding flows commute; all these so-called higher order KdV equations can be solved by the scattering transform. These explicit solutions can be used to determine the zero dispersion limit of the solutions of these equations; formulas similar to those in Section 5 hold.

Acknowledgment. The research of the first author was supported by the U.S. Department of Energy, Office of Basic Energy Sciences, Applied Mathematical Sciences Research Program under Contract DE-AC02-76ER03077, that of the second by the John and Fannie Hertz Foundation.

Bibliography

[1] Cole, J. D., *On a quasi-linear parabolic equation occurring in aerodynamics*, Quart. Appl. Math. 9, 1951, pp. 225–236.

[2] Flaschka, H., Forest, M. G., and McLaughlin, D. W., *Multiphase averaging and the inverse spectral solution of the Korteweg-de Vries equation*. Comm. Pure Appl. Math. 33, 1980, pp. 739–784.

[3] Gardner, C. S., Greene, J. M., Kruskal, M. D., and Miura, R. M., *Method for solving the Korteweg–de Vries equation*, Phys. Rev. Lett. 19, 1967, pp. 1095–1097.

[4] Gelfand, I. M., and Levitan, B. M., *On the determination of a differential equation by its spectral function*, Izv. Akad. Nauk SSR Ser. Mat. 15, 1951, pp. 309–360. (In Russian.)

[5] Hopf, E., *The partial differential equation* $u_1+uu_x=\mu_{xx}$, Comm. Pure Appl. Math. 3, 1950, pp. 201–230.

[6] Kay, I., and Moses, H. E., *Reflectionless transmission through dielectrics and scattering potentials*, J. Appl. Phys. 27, 1956, pp. 1503–1508.

[7] Koosis, P., *Introduction to* H^p *spaces*, LMS Lecture Notes 40, 1980.

[8] Lax, P. D., and Levermore, C. D., *The zero dispersion limit for KdV equation*, Proc. Nat. Acad. Sci. USA, 1979, pp. 3602–3606.

[9] Levermore, C. D., *The small dispersion limit of the Korteweg–de Vries equation*, Dissertation, FGAS, 1982, New York University.

[10] Miura, R. M., and Kruskal, M. D., *Application of a nonlinear WKB method to the Korteweg–de Vries equation*, SIAM J. Appl. Math. 26, 1974, pp. 376–395.

[11] Muskhelishvili, N. I., *Singular Integral Equations*, 1953, P. Noordhoff, Groningen.
[12] Tanaka, S., *On the N-tuple wave solutions of the Korteweg–de Vries equation*, Res. Inst. Math. Sci., Kyoto 8, 1973, pp. 419–427.
[13] Venakides, S., *The zero dispersion limit of the Korteweg–de Vries equation*, Dissertaion, FGAS, 1982, New York University.
[14] Whitham, G. B., *Linear and Nonlinear Waves*, Wiley-Interscience, New York, 1974.

Received June, 1982.

On Dispersive Difference Schemes. I

JONATHAN GOODMAN AND PETER D. LAX
Courant Institute

Introduction

It is well known that solutions of nonlinear hyperbolic equations, in one space variable, of which

$$u_t + uu_x = 0 \tag{1}$$

is the best known and best beloved example, break down after a finite elapse of time. It is also known that solutions of (1) can be continued beyond the time of the breakdown as solutions in the integral sense of the conservation law

$$u_t + \left(\tfrac{1}{2}u^2\right)_x = 0. \tag{1$'$}$$

These solutions in the integral sense contain discontinuities, the mathematical representation of shock waves; they are uniquely determined by their initial data provided that the discontinuities are constrained to satisfy an entropy condition.

It is further known that discontinuous solutions of (1)′ in the integral sense can be obtained as strong limits—L^1 and bounded a.e.—of smooth solutions of the viscous equation

$$u_t + uu_x = \mu u_{xx}, \qquad \mu > 0, \tag{2}$$

as μ tends to zero. They can also be obtained as strong limits of solutions of difference equations approximating (1)′ that contain a sufficient amount of artificial viscosity, such as the one proposed and used by von Neumann and Richtmyer [10], Friedrichs and Lax, Godunov, etc. The solutions constructed as such limits satisfy the entropy condition.

On the other hand, it is known that solutions of dispersive approximations to equation (1) behave quite differently. Specifically, it is known (see [4], [12]) that solutions $u(x, t; \varepsilon)$ of the KdV equation

$$u_t + uu_x + \varepsilon^2 u_{xxx} = 0, \tag{3}$$

$$u(x,0) = u_0(x), \tag{4}$$

behave as follows, as ε tends to 0:

As long as the solution of equation (1) with initial value (4) has a smooth solution, $u(x, t; \varepsilon)$ tends uniformly to that smooth solution. However, when t

Communications on Pure and Applied Mathematics, Vol. XLI 591–613 (1988)

CCC 0010-3640/88/050591-23$04.00

exceeds the time when the solution of (1), (4) breaks down, $u(x, t; \varepsilon)$ behaves in an oscillatory manner over some x-interval; as ε tends to zero, the amplitude of these oscillations remains bounded but does not tend to zero, and its wave length is of order ε. Clearly, $u(x, t; \varepsilon)$ does not tend strongly to a limit as ε tends to zero; however, $u(x, t; \varepsilon)$ converges weakly. It is easy to show that this weak limit —in the sense of distribution—which we denote as $\bar{u}$,

$$d\text{-}\lim_{\varepsilon \to 0} u(x, t; \varepsilon) = \bar{u}(x, t), \tag{5}$$

does *not* satisfy the conservation law (1)′ in the integral sense. For rewrite (3) in conservation form:

$$u_t + \left(\tfrac{1}{2}u^2\right)_x + \varepsilon^2 u_{xxx} = 0. \tag{3$'$}$$

It follows from (5) that the d-limit of the first term is $\bar{u}_t$, and that of the third term zero; therefore, the middle term also converges, and its limit is $(\tfrac{1}{2}\overline{u^2})_x$, where

$$\overline{u^2} = d\text{-}\lim_{u \to 0} u^2(x, t; \varepsilon). \tag{5$'$}$$

So we get from (3)′ that

$$\bar{u}_t + \left(\tfrac{1}{2}\overline{u^2}\right)_x = 0. \tag{6}$$

It is a well-known fact that if $u(\varepsilon)$ tends weakly but *not* strongly to $\bar{u}$, then the weak limit of $u^2(\varepsilon)$ is *not* the square of the weak limit of $u(\varepsilon)$:

$$\overline{u^2} \neq \bar{u}^2; \tag{7}$$

in fact,

$$\overline{u^2} > \bar{u}^2. \tag{7$'$}$$

The explicit formulas that have been derived in [4] for $\bar{u}$ and $\overline{u^2}$ bear out these inequalities. Setting this into (6) we deduce that

$$\bar{u}_t + \left(\tfrac{1}{2}\bar{u}^2\right)_x \neq 0, \tag{6$'$}$$

unless $\overline{u^2}$ happens to differ from $\bar{u}^2$ by a constant; the explicit formulas for $\bar{u}$ and $\overline{u^2}$ show that this is not the case.

The aim of this paper is to study the corresponding phenomena of oscillatory behavior and weak, but not strong, convergence of solutions of difference approximations to (1)′ that are *dispersive*. The particular approximation we study here is semi-discrete, i.e., continuous in time, discrete in x:

$$\frac{d}{dt}U_k + U_k\frac{U_{k+1} - U_{k-1}}{2\Delta} = 0, \tag{8}$$

$$U_k(0) = u_0(k\Delta). \tag{8$'$}$$

To see the dispersive nature of this approximation we use Taylor's expansion:

$$\frac{U_{k+1} - U_{k-1}}{2\Delta} \simeq u_x + \tfrac{1}{6}\Delta^2 u_{xxx}. \tag{9}$$

Setting this into (8) we obtain

$$u_t + uu_x + \tfrac{1}{6}\Delta^2 uu_{xxx} \simeq 0,$$

an equation very much like the KdV equation (3) as long as u does not change sign. If this analogy is valid, we can expect equation (8) to have solutions that oscillate with a wave length $O(\Delta)$, i.e., on mesh scale.

The purpose of this study is to show that the analogy is indeed valid, and that solutions of (8) behave analogously in their dependence on Δ as solutions of (3) do in their dependence on ε. As usual, the study of the difference equation is trickier than that of the differential equation.

The organization of this paper is as follows: in Section 1 we show that, for t less than the breakdown time of a solution of (1) with given initial data (4), the solution of (8) with the same initial data converges to the solution of (1), provided the initial data are sufficiently smooth.

In Section 2 we construct a special class of oscillatory solutions of (8) and show their weak convergence. These weak limits gratifyingly fail to satisfy the conservation law (1)′.

In Section 3 we show that beyond breakdown time for a solution of (1) the corresponding solution of (8) does not converge strongly. Using a simple transformation we change (8) to a system studied by Kac and van Moerbeke [3] and shown by them to be completely integrable; this system was reduced by Moser [8] to the Toda lattice, in the form due to Flaschka [2]. The integrals can be used to bound solutions of (8).

In Section 4 we present ample numerical evidence for the oscillatory character of solutions of (8) beyond breakdown time, and for their weak convergence to nonsolutions of (1).

In a subsequent publication we hope to use the complete integrability of (8) as discussed in [7] to furnish rigorous proofs for these experimental findings.

The oscillatory nature of solutions of dispersive difference schemes was discovered, accidentally, by von Neumann [9] in 1944 in a calculation of compressible flows with shocks in one space dimension, employing centered difference schemes. The solutions—calculated using punched card equipment—contained, as expected, a shock; but, unexpectedly, it contained post shock oscillations on mesh scale. Von Neumann suggested that these mesh-scale oscillations in velocity are to be interpreted as heat energy produced by the irreversible action of the shock wave, and conjectured that, as Δx, Δt tend to zero, the solutions of the difference equations converge weakly to exact discontinuous solutions of the equations governing the flow of compressible fluids. If the simplified model (1)′ is a guide to the equations of compressible flow, then there is reason to doubt the validity of von Neumann's surmise. A rigorous proof either way is a long way off; further analysis is contained in [5] and [6].

1. Convergence to Smooth Solutions

For convenience we take all functions of x to be periodic with period 2π, and choose $\Delta = 2\pi/N$, N integer. This makes U_k periodic in k with period N.

We rely on a basic convergence theorem of Strang. As stated in [11], the theorem deals with explicit difference schemes, but the proof applies to semi-discrete schemes as well.

Let

$$u_t + Cu_x = 0 \tag{1.1}$$

be a quasilinear hyperbolic system of partial differential equations. Here $C(u)$ is a matrix-valued function of u, depending smoothly on its arguments.

We consider a semi-discrete approximation to (1.1) of the form

$$\frac{d}{dt}U_k + F(U_{k-1}, U_k, U_{k+1}, \Delta) = 0 \tag{1.2}$$

that is consistent with (1.1). We assume that F is a smooth function of its arguments.

CONVERGENCE THEOREM OF STRANG.[1] *Let $u(x, t)$ be a smooth, x-periodic solution of the hyperbolic system* (1.1). *Assume that the approximation* (1.2) *to* (1.1) *is linearly l^2 stable around $u(x, t)$.*

Choose the initial data of U_k to coincide with those of u:

$$U_k(0) = u(k\Delta, 0), \tag{1.2$'$}$$

then $U_k(t)$ converges uniformly to $u(k\Delta, t)$ as $\Delta \to 0$, for $0 \leqq t \leqq T$. Furthermore,

$$|U_k(t) - u(k\Delta, t)| \leqq O(\Delta^p), \tag{1.3}$$

where p is the order of accuracy of the scheme (1.2) *for equation* (1.1).

The *linearization* of (1.2) is

$$\frac{d}{dt}\eta_k + \sum_{-1}^{1} F_j \eta_{k+j} = 0, \tag{1.4}$$

[1] Strang's theorem applies to any number of space variables, and to any finite stencil in the approximation (1.2).

where F_j is the gradient of the function F with respect to its $(j+2)$-nd argument, $j = -1, 0, 1$, evaluated at

$$u((k-1)\Delta, t), u(k\Delta, t), u((k+1)\Delta, t).$$

l^2 *stability* means that solutions of (1.4) are uniformly bounded in the l^2 norm:

$$\sum_1^N \eta_k^2(t) \leq K(t) \sum_1^N \eta_k^2(0), \tag{1.4'}$$

where $K(t)$ is *independent* of Δ.

We turn now to equation (1); let u be a smooth solution. We linearize (8) around u, obtaining the specialization of (1.4) to our case:

$$\frac{d}{dt}\eta_k + u_k \frac{\eta_{k+1} - \eta_{k-1}}{2\Delta} + \eta_k \frac{u_{k+1} - u_{k-1}}{2\Delta} = 0, \tag{1.5}$$

where

$$u_k = u(k\Delta, t). \tag{1.5'}$$

Multiply (1.5) by η_k and sum:

$$\frac{d}{dt}\frac{1}{2}\sum \eta_k^2 + \sum u_k \frac{\eta_k \eta_{k+1} - \eta_{k-1}\eta_k}{2\Delta} + \eta_k^2 \frac{u_{k+1} - u_{k-1}}{2\Delta} = 0.$$

Summing by parts the middle term, we get, using the abbreviation

$$H(t) = \frac{1}{2}\sum_1^N \eta_k^2, \tag{1.6}$$

that

$$\frac{d}{dt}H + \sum \frac{u_k - u_{k+1}}{2\Delta}\eta_k \eta_{k+1} + \sum \frac{u_{k+1} - u_{k-1}}{2\Delta}\eta_k^2 = 0. \tag{1.7}$$

Since $u(x, t)$ is a smooth function of x, it follows from (1.5)′ and (1.7) that

$$\frac{d}{dt}H \leq 2LH,$$

where L is a Lipschitz constant for $u(x, t)$. From this stability (1.4)′ follows with

$$K(t) = \exp 2Lt.$$

We appeal now to Strang's convergence theorem to conclude:

THEOREM 1.1. *Let $u(x, t)$ be a smooth x-periodic solution of* (1) *for* $0 \leqq t \leqq T$, $U_k(t)$ *the solution of* (8), (8)′, *with* $u_0(x) = u(x,0)$. *Then* $U_k(t)$ *converges uniformly to* $u(k\Delta, t)$ *as* $\Delta \to 0$, *for* $0 \leqq t \leqq T$.

We recall that the order of accuracy of a scheme (1.2) approximating (1.1) is the smallest integer p such that

$$|C(u)u_x - F(u(x-\Delta), u(x), u(x+\Delta), \Delta)| \leqq O(\Delta^p). \tag{1.8}$$

For the approximation (8) to equation (1) it follows from (9) that the order of accuracy is 2. Therefore, by (1.3),

$$|u(k\Delta, t) - U_k(t)| \leqq O(\Delta^2), \tag{1.9}$$

for $0 \leqq t \leqq T$. This is born out by the numerical calculations reported in Section 4.

2. A Class of Oscillatory Solutions

We choose the number N of mesh points per period to be even. We represent U_k as

$$U_k = V_k + (-1)^k W_k, \tag{2.1}$$

V_k and W_k of period N.

Inserting (2.1) into (8) gives

$$\frac{d}{dt}(V_k + W_k) + (V_k + W_k)\left(\frac{V_{k+1} - V_{k-1}}{2\Delta} - \frac{W_{k+1} - W_{k-1}}{2\Delta}\right) = 0 \tag{2.2$_+$}$$

for k even, and

$$\frac{d}{dt}(V_k - W_k) + (V_k - W_k)\left(\frac{V_{k+1} - V_{k-1}}{2\Delta} + \frac{W_{k+1} - W_{k-1}}{2\Delta}\right) = 0 \tag{2.2$_-$}$$

for k odd.

We require now that $(2.2)_+$ and $(2.2)_-$ hold for k both even and odd. The resulting equations can be interpreted as a semi-discrete approximation to the following system of partial differential equations:

$$\begin{aligned}(v + w)_t + (v + w)(v_x - w_x) &= 0,\\ (v - w)_t + (v - w)(v_x + w_x) &= 0.\end{aligned} \tag{2.3}$$

Introducing

$$v + w = a, \qquad v - w = b, \tag{2.4}$$

we can rewrite (2.3) as

$$\begin{aligned} a_t + ab_x &= 0, \\ b_t + ba_x &= 0. \end{aligned} \tag{2.5}$$

Clearly, (2.5) is a hyperbolic system if and only if $ab > 0$, in particular if a and b are positive. In view of (2.1), (2.4), this is related to $u > 0$. If this condition is satisfied by the initial data of a, b, then it will continue to be satisfied by the solution $a(x, t)$, $b(x, t)$, at least until the first shock forms.

We denote

$$V_k + W_k = A_k, \qquad V_k - W_k = B_k, \tag{2.6}$$

and rewrite $(2.2)_\pm$ as

$$\frac{d}{dt}A_k + A_k\frac{B_{k+1} - B_{k-1}}{2\Delta} = 0, \tag*{$(2.7)_+$}$$

$$\frac{d}{dt}B_k + B_k\frac{A_{k+1} - A_{k-1}}{2\Delta} = 0. \tag*{$(2.7)_-$}$$

We claim that if A_k and B_k are positive at $t = 0$, then they are positive at all future t. To see this we add the two equations $(2.7)_+$ and $(2.7)_-$; the sum of the second terms is a perfect difference:

$$\frac{A_kB_{k+1} + A_{k+1}B_k}{2\Delta} - \frac{A_{k-1}B_k + A_kB_{k-1}}{2\Delta}. \tag{2.8}$$

Therefore when we sum from $k = 1$ to N the sum of (2.8) is zero, and we get

$$\frac{d}{dt}\sum A_k + B_k = 0,$$

i.e.,

$$\sum A_k(t) + B_k(t) \tag{2.9}$$

is a conserved quantity.

If we divide $(2.7)_+$ by A_k and sum from $k = 1$ to N, we get likewise

$$\sum \frac{1}{A_k}\frac{d}{dt}A_k = 0,$$

i.e.,

$$\prod A_k(t) \tag*{$(2.10)_+$}$$

is a conserved quantity. Similarly,

$$(2.10)_- \qquad \prod B_k(t)$$

is a conserved quantity.

It follows from these conservation laws that if $A_k(t)$, $B_k(t)$ are initially positive, they can never change sign. For otherwise one or several of them would have to approach zero at some time t while all the others remain positive. It follows then from $(2.10)_\pm$ that another A_j or B_j would have to approach ∞; but this contradicts the conservation of (2.9). This argument shows that the system $(2.7)_\pm$ of ordinary differential equations has solution for all time.

THEOREM 2.1. *Let $a(x, t)$, $b(x, t)$ be a smooth, x-periodic solution of* (2.5) *for $0 \leqq t \leqq T$, a and b both positive. Let A_k, B_k be solutions of* (2.7) *whose initial values are equal to $a(k\Delta, 0)$, $b(k\Delta, 0)$. Then $A_k(t)$, $B_k(t)$ converge uniformly to $a(k\Delta, t)$, $b(k\Delta, t)$ for $0 \leqq t \leqq T$.*

Proof: We appeal to Strang's convergence theorem. We have to verify that the linearization of (2.7) around a, b is l^2 stable. The linearization is

$$(2.11)_+ \qquad \frac{d}{dt}\eta_k + a_k\frac{\zeta_{k+1} - \zeta_{k-1}}{2\Delta} + \eta_k\frac{b_{k+1} - b_{k-1}}{2\Delta} = 0,$$

$$(2.11)_- \qquad \frac{d}{dt}\zeta_k + b_k\frac{\eta_{k+1} - \eta_{k-1}}{2\Delta} + \zeta_k\frac{a_{k+1} - a_{k-1}}{2\Delta} = 0,$$

where

$$a_k = a(\Delta k, t), \qquad b_k = b(k\Delta, t).$$

Multiply $(2.11)_+$ by $b_k\eta_k$, $(2.11)_-$ by $a_k\zeta_k$, and add them; the sum of the first terms can be written as

$$(2.12)_1 \qquad \frac{1}{2}\frac{d}{dt}\left[b_k\eta_k^2 + a_k\zeta_k^2\right] - \frac{1}{2}\eta_k^2\frac{d}{dt}b_k - \frac{1}{2}\zeta_k^2\frac{d}{dt}a_k.$$

Using (2.8) we can write the sum of the second term as

$$(2.12)_2 \qquad \frac{1}{2\Delta}a_kb_k\left[\eta_k\zeta_{k+1} + \eta_{k+1}\zeta_k - \eta_{k-1}\zeta_k - \eta_k\zeta_{k-1}\right].$$

Since $a(x, t)$ and $b(x, t)$ are Lipschitz continuous in x, the sum of the third terms is less than

$$(2.12)_3 \qquad K\left(\eta_k^2 + \zeta_k^2\right),$$

independent of k, Δ and t. We sum (2.12) from $k = 1$ to N; in $(2.12)_2$ we sum by parts. Altogether we get

$$\frac{1}{2}\frac{d}{dt}\sum b_k\eta_k^2 + a_k\zeta_k^2 \leqq K'\sum \eta_k^2 + \zeta_k^2. \tag{2.13}$$

Introduce the abbreviation

$$H = \frac{1}{2}\sum b_k\eta_k^2 + a_k\zeta_k^2;$$

since for $0 \leqq t \leqq T$, a_k and b_k are bounded from below by a positive constant, we can rewrite (2.13) as

$$\frac{d}{dt}H \leqq K''H.$$

This shows that

$$H(t) \leqq \exp K''tH(0).$$

This proves l^2 stability of the linearized system, and so by Strang's theorem we conclude that Theorem 2.1 holds.

If we switch back to the original variables v, w, and V, W we see that $V_k(t)$ tends to $v(k\Delta, t)$, $W_k(t)$ to $w(k\Delta, t)$.

We extend now U_k, V_k and W_k to continuous arguments by setting

$$\begin{aligned} U_\Delta(x,t) &= U_k(t), \\ V_\Delta(x,t) &= V_k(t) \quad \text{for} \quad (k - \tfrac{1}{2}\Delta < x < (k + \tfrac{1}{2})\Delta, \\ W_\Delta(x,t) &= W_k(t). \end{aligned} \tag{2.14}$$

Similarly, we set

$$E_\Delta(x) = (-1)^k \quad \text{for} \quad (k - \tfrac{1}{2})\Delta < x < (k + \tfrac{1}{2})\Delta. \tag{2.14$'$}$$

It follows from (2.1) that

$$U_\Delta = V_\Delta + E_\Delta W_\Delta. \tag{2.15}$$

We have seen above that V_Δ and W_Δ converge uniformly to v and w, respectively. It follows from (2.1)′ that E_Δ tends weakly to zero. Setting these into (2.15) we conclude that U_Δ tends weakly to v:

$$d\text{-}\lim U_\Delta = v. \tag{2.16}$$

Denote by $\mathbf{T}_\Delta$ translation in x by Δ:

$$(2.17) \qquad (\mathbf{T}_\Delta U)(x) = U(x+\Delta).$$

It follows from the definition (2.14)′ of E_Δ that

$$(2.18) \qquad \mathbf{T}_\Delta E_\Delta = -E_\Delta.$$

Apply $\mathbf{T}_\Delta$ to (2.15); then from the above we get

$$\mathbf{T}_\Delta U_\Delta = \mathbf{T}_\Delta V_\Delta - E_\Delta \mathbf{T} W_\Delta.$$

Multiply this by (2.15); using (2.18), we obtain

$$\begin{aligned} U_\Delta \mathbf{T}_\Delta U_\Delta = {} & V_\Delta \mathbf{T}_\Delta V_\Delta - W_\Delta \mathbf{T}_\Delta W_\Delta \\ & + E_\Delta [W_\Delta \mathbf{T}_\Delta V_\Delta - V_\Delta \mathbf{T}_\Delta W_\Delta]. \end{aligned}$$

Taking the weak limit as $\Delta \to 0$ we conclude, as above, that

$$(2.19) \qquad d\text{-lim}\, U_\Delta \mathbf{T}_\Delta U_\Delta = v^2 - w^2.$$

Using the notation U_Δ we can rewrite (8) as an equation in conservation form:

$$\frac{d}{dt} U_\Delta + \frac{U_\Delta \mathbf{T}_\Delta U_\Delta - (\mathbf{T}_\Delta^{-1} U_\Delta) U_\Delta}{2\Delta} = 0.$$

We take the limit of this equation in the distribution sense; we obtain, taking (2.16) and (2.19) into account,

$$(2.20) \qquad v_t + \tfrac{1}{2}(v^2 - w^2)_x = 0.$$

Unless w is independent of x, it follows that

$$(2.20)' \qquad v_t + \left(\tfrac{1}{2} v^2\right)_x \neq 0.$$

This proves

THEOREM 2.1. *Equation* (8) *has solutions of the form* (2.1), *oscillating on mesh scale, whose weak limit does not satisfy the conservation law* (1)′.

3. Complete Integrability

Let N be an even positive integer, as in Section 2. Let A denote an N-vector of real numbers:

$$A = (A_1, \cdots, A_N).$$

We extend A_k to be defined for all k and be periodic with period N:

$$A_{k+N} = A_k.$$

The operator $\mathbf{T}$ is translation:

$$(\mathbf{T}A)_k = A_{k+1}. \tag{3.1}$$

Clearly, $\mathbf{T}$ maps periodic sequences into periodic sequences. We shall use the abbreviation

$$\mathbf{T}A = A_+. \tag{3.2$_+$}$$

The inverse of $\mathbf{T}$ is $\mathbf{T}^{-1}$, translation in the opposite direction. We abbreviate

$$\mathbf{T}^{-1}A = A_-. \tag{3.2$_-$}$$

Following Kac and van Moerbeke [2], we associate with each vector A the operators

$$\mathbf{L} = A\mathbf{T} + A_-\mathbf{T}^{-1}, \tag{3.3}$$

where A acts as the multiplication operator. L is a symmetric, cyclic Jacobi matrix with zero diagonal. We associate with A a second operator

$$\mathbf{B} = AA_+\mathbf{T}^2 - A_-A_{--}\mathbf{T}^{-2}, \tag{3.4}$$

which is antisymmetric. Here

$$A_{--} = \mathbf{T}^{-1}A_- = \mathbf{T}^{-2}A.$$

The commutator $[\mathbf{B}, \mathbf{L}] = \mathbf{BL} - \mathbf{LB}$ is

$$[\mathbf{B}, \mathbf{L}] = A\left(A_+^2 - A_-^2\right)\mathbf{T} + A_-(A - A_{--})\mathbf{T}^{-1}. \tag{3.5}$$

Consider the differential equation

$$\frac{d}{dt}\mathbf{L} + [\mathbf{B}, \mathbf{L}] = 0. \tag{3.6}$$

Componentwise this reads as

$$\frac{d}{dt}A_k + A_k\left(A_{k+1}^2 - A_{k-1}^2\right) = 0. \tag{3.6$'$}$$

Kac and van Moerbeke show that this system is completely integrable. In [8], Moser reduced this system to Flaschka's form of the Toda lattice, known to be completely integrable.

It follows from (3.6) that, for any positive integer p,

$$\frac{d}{dt}\mathbf{L}^p + [\mathbf{B},\mathbf{L}^p] = 0. \tag{3.6}''$$

This implies that

$$\operatorname{Tr}\mathbf{L}^p \tag{3.7}$$

are conserved quantities for solutions of (3.6). These quantities are nontrivial for p even: $2, 4, \cdots, N$.

In particular,

$$\mathbf{L}^2 = AA_+\mathbf{T}^2 + \left(A^2 + A_-^2\right)\mathbf{T} + A_-A_{--}\mathbf{T}^{-2}$$

and

$$\mathbf{L}^4 = AA_+A_{++}A_{+++}\mathbf{T}^4 + AA_+\left(A_-^2 + A^2 + A_+^2 + A_{++}^2\right)\mathbf{T}^2$$
$$+\left[A^2A_+^2 + A_-^2A_{--}^2 + \left(A^2 + A_-^2\right)^2\right]\mathbf{T} + \cdots.$$

Thus, by (3.7), the quantities

$$\operatorname{Tr}\mathbf{L}^2 = \sum A_k^2 + A_{k-1}^2 = 2\sum A_k^2 \tag{3.7}'$$

and

$$\operatorname{Tr}\mathbf{L}^4 = \sum 2A_k^2A_{k+1}^2 + \left(A_k^2 + A_{k-1}^2\right)^2 \tag{3.7}''$$

are conserved, etc.

Multiply (3.6)′ by $2A_k$; abbreviating

$$A_k^2 = U_k, \tag{3.8}$$

we obtain

$$\frac{d}{dt}U_k + 2U_k(U_{k+1} - U_{k-1}) = 0. \tag{3.8}'$$

This equation is, except for a numerical coefficient, the same as our equation (8). The correct numerical factor can be obtained by rescaling t or U_k.

Conversely, every positive solution $\{U_k\}$ can be reduced via a rescaling and the transformation (3.8) to a positive solution $\{A_k\}$ of (3.6)′.

It is easy to show that every solution of (8) whose initial values are positive remains positive for all t:

We rewrite (8) in conservation form,

$$\frac{d}{dt} U_k + \frac{U_k U_{k+1} - U_{k-1} U_k}{2\Delta} = 0; \tag{3.9}$$

we obtain, summing from $k = 1$ to N, and using the periodicity of U_k, that

$$\frac{d}{dt} \sum U_k = 0.$$

Hence

$$\sum U_k \tag{3.10}$$

is a conserved quantity. In view of (3.8), this is the same as (3.7)′.

If we divide (3.9) by U_k, we obtain another conservation form,

$$\frac{d}{dt} \log U_k + \frac{U_{k+1} - U_{k-1}}{2\Delta} = 0. \tag{3.11}$$

Summing from $k = 1$ to N, and using the periodicity of U_k, we deduce that

$$\frac{d}{dt} \sum \log U_k = 0;$$

hence,

$$\prod U_k \tag{3.12}$$

is a conserved quantity.

The positivity of $U_k(t)$ for all t can be deduced from these two conservation laws by an argument identical to the one used in Section 2 to show the positivity of $A_k(t)$ and $B_k(t)$ for all t.

The conservation of (3.10) and the positivity of $U_k(t)$ imply that all the $U_k(t)$ are bounded from above by M,

$$M = \sum U_k(0). \tag{3.13}$$

The uniform boundedness of $U_k(t)$ guarantees that every solution of (3.9) which has positive data exists for all time.

We turn now to the question of convergence of the approximate solutions. We use the notation introduced in Section 2:

$$U_\Delta(x, t) = U_k(t) \quad \text{for} \quad (k - \tfrac{1}{2}\Delta) < x < (k + \tfrac{1}{2})\Delta. \tag{3.14}$$

We denote the shift operator by $\mathbf{T}_\Delta$:

$$\mathbf{T}_\Delta U(x) = U(x + \Delta). \tag{3.15}$$

Using (3.10) and the above notation we conclude that

$$\int U_\Delta \, dx = \Delta \sum U_k \tag{3.16}$$

is a conserved quantity.

Using (3.8), the conservation of (3.7)″ can be restated in the above notation as the conservation of

$$\Delta \sum U_k U_{k+1} + \tfrac{1}{2}(U_k + U_{k-1})^2 = \int \left[U_\Delta \mathbf{T}_\Delta U_\Delta + \tfrac{1}{2}(U_\Delta + \mathbf{T}_\Delta U_\Delta)^2 \right] dx. \tag{3.17}$$

Since U_Δ is positive, it follows from the above that, for all t,

$$\int U_\Delta^2(x,t)\, dx \leqq 3 \int U_\Delta^2(x,0)\, dx. \tag{3.18}$$

Entirely similarly, we can deduce from the conservation of $\operatorname{Tr} \mathbf{L}^8$ that, for all t,

$$\int U_\Delta^4(x,t)\, dx \leqq \text{const.} \int U_\Delta^4(x,0)\, dx. \tag{3.18$'$}$$

LEMMA 3.1. *Let U_Δ denote the solution of* (3.9) *for given, smooth initial data. Then every sequence $\{\Delta\}$ tending to zero contains a subsequence such that U_Δ and $U_\Delta \mathbf{T}_\Delta U_\Delta$ both converge weakly with respect to the Hilbert space norm*

$$\int_0^T \int V^2(x,t)\, dx\, dt = \|V\|_T^2 \tag{3.19}$$

for any T.

Proof: For fixed T this is a standard consequence of the weak sequential compactness of bounded sets in Hilbert space, combined with inequalities (3.18) and (3.18)′. To get simultaneous weak convergence for all T we choose a denumerable sequence $T_n \to \infty$ and apply a diagonal process.

THEOREM 3.2. *Let U_Δ denote the solution of* (3.9) *for given, smooth initial data. Let T denote a time exceeding the breakdown time for the given initial data; then no subsequence of U_Δ can converge strongly in the norm* (3.19).

Proof: Suppose on the contrary that

$$s\text{-}\lim U_\Delta = \overline{U}. \tag{3.20}$$

From this it follows easily that

$$L^1\text{-lim}\, U_\Delta \mathrm{T}_\Delta U_\Delta = \overline{U}^2. \tag{3.21}$$

Using (3.18)′ we further conclude from (3.20) that

$$L^1\text{-lim}\, U_\Delta (\mathrm{T}_\Delta U_\Delta)^2 = \overline{U}^3, \tag{3.22}$$

and

$$L^1\text{-lim}\, U_\Delta (\mathrm{T}_\Delta U_\Delta)(\mathrm{T}_\Delta^2 U_\Delta) = \overline{U}^3. \tag{3.22$'$}$$

We write now the conservation law (3.9) in the notation (3.14), (3.15) as

$$\frac{d}{dt} U_\Delta + \frac{1}{2\Delta}\left[U_\Delta \mathrm{T}_\Delta U_\Delta - (\mathrm{T}_\Delta^{-1} U_\Delta) U_\Delta\right] = 0.$$

We take the limit of this equation in the sense of distributions; we get, using (3.20) and (3.21), that

$$\overline{U}_t + \left(\tfrac{1}{2}\overline{U}^2\right)_x = 0. \tag{3.23}$$

Next we derive another conservation law from (3.9), corresponding to the conserved quantity (3.17). A brief calculation shows that every solution of (3.9) satisfies

$$\begin{aligned}\frac{d}{dt}&\left[U_k^2 + U_{k+1}U_k + U_k U_{k-1}\right]\\ &+ \frac{1}{2\Delta}\left[U_{k+2}U_{k+1}U_k + U_{k+1}^2 U_k\right.\\ &\left.+ U_{k+1}U_k^2 - U_k U_{k-1} U_{k-2} - U_k^2 U_{k-1} - U_k U_{k-1}^2\right] = 0.\end{aligned}$$

In the notation of (3.14), (3.15) this can be written as

$$\begin{aligned}\frac{d}{dt}&\left[U_\Delta^2 + U_\Delta \mathrm{T}_\Delta U_\Delta + U_\Delta \mathrm{T}_\Delta^{-1} U_\Delta\right] + \frac{1}{2\Delta}\left[(\mathrm{T}_\Delta^2 U_\Delta)(\mathrm{T}_\Delta U_\Delta) U_\Delta - U_\Delta(\mathrm{T}_\Delta^{-1} U_\Delta)\mathrm{T}_\Delta^2 U_\Delta\right]\\ &+ \frac{1}{2\Delta}\left[(\mathrm{T}_\Delta U_\Delta)^2 U_\Delta - U_\Delta^2 \mathrm{T}_\Delta U_\Delta\right] + \frac{1}{2\Delta}\left[(\mathrm{T}_\Delta U_\Delta) U_\Delta^2 - U_\Delta(\mathrm{T}_\Delta U_\Delta)^2\right] = 0.\end{aligned}$$

We take now the limit in the sense of distributions of the above equation. Using (3.21), (3.22) and (3.22)′ we get

$$(3\overline{U}^2)_t + (2\overline{U}^3)_x = 0. \tag{3.24}$$

Equations (3.23) and (3.24) are equivalent as differential equations, i.e., they have the same smooth solutions; but they are incompatible for discontinuous solutions. For T exceeding breakdown time, $\overline{U}$ is a discontinuous solution of (3.23); therefore it cannot satisfy (3.24) in the distribution sense. This shows that (3.20) is untenable.

We remark that when U_Δ tends weakly but not strongly to $\overline{U}$, $U_\Delta \mathbf{T}_\Delta U_\Delta$ could, unlike U_Δ^2, very well tend weakly to $\overline{U}^2$. If that were the case, the weak limit $\overline{U}$ would satisfy (3.23) in the distribution sense. We believe that the weak limit of $U_\Delta \mathbf{T}_\Delta U_\Delta$ is in fact different from $\overline{U}^2$; the numerical calculations confirm this.

4. Numerical Experiments

We describe some numerical experiments with the difference scheme (8). The initial data $u_0(x)$ is taken to be the positive periodic function $u_0(x) = \cos(x) + \frac{3}{2}$. We solve the ordinary differential equations (8) using the four-stage Runge-Kutta method, (see [1], p. 346). The Runge-Kutta method, with fixed CFL ratio, is dissipative of high frequencies; we have to take quite a small time step to get full time accuracy.

For this problem a shock first forms at time $t = 1$. Table 1 illustrates the second-order accuracy of the scheme at time $t = \frac{2}{10}\pi \simeq 0.628$ before the breakdown time. Similar results hold in the L^2 and L^∞ norms. This agrees with the conclusion of Theorem 1.1, inequality (1.9).

Figure 1 illustrates the oscillations that have appeared by time $t = 0.8\pi \simeq 2.51$. These do not seem to be of the type constructed in Section 2. Each frame contains a plot of the "raw" solution U and a smoothed solution $V = \mathbf{S}_\Delta^k U$, where $\mathbf{S}_\Delta = \frac{1}{4}\mathbf{T}_\Delta^{-1} + \frac{1}{2}\mathbf{I} + \frac{1}{4}\mathbf{T}_\Delta$ in the notation of Section 3.

In both figures, k is increased with N. In view of the central limit theorem, the effective number of grid points over which smoothing is performed is proportional to $\sqrt{k}$.

Figure 3 contains plots of the discrete indefinite integral of the oscillatory solution. This is motivated by the work of Venakides [12]. Suppose that $u^\varepsilon(x,t)$ satisfies KdV: $u_t^\varepsilon + u^\varepsilon u_x^\varepsilon = \varepsilon u_{xxx}^\varepsilon$. Venakides proved that $u^\varepsilon \rightharpoonup u$ weakly (for some function $u(x,t)$) by showing that $u^\varepsilon(x,t) = (\partial/\partial x)v^\varepsilon(x,t)$, where $v^\varepsilon(x,t) \to v(x,t)$ strongly and $u = (\partial/\partial x)v$. This was an improvement over Lax

Table I. e_N is the l_1 error using N points at time $t = 2\pi/10$.

N	e_N	e_{2N}/e_N
10	2.95×10^{-2}	5.47
20	5.39×10^{-3}	3.66
40	1.47×10^{-3}	3.90
80	3.78×10^{-4}	4.03
160	9.36×10^{-5}	4.005
320	2.34×10^{-5}	

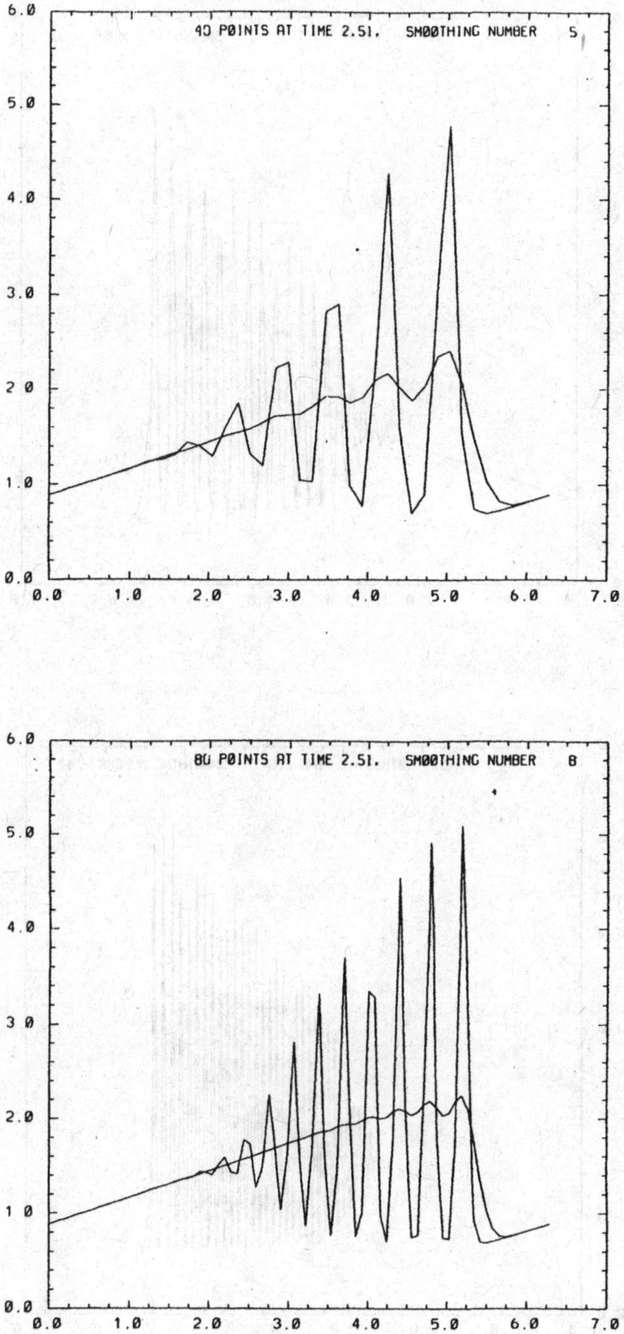

Figure 1. Oscillatory solutions after the breakdown time. The smoother curve is S^k operating on the raw solution where k is the smoothing number. All curves are piecewise linear interpolants of the N discrete values.

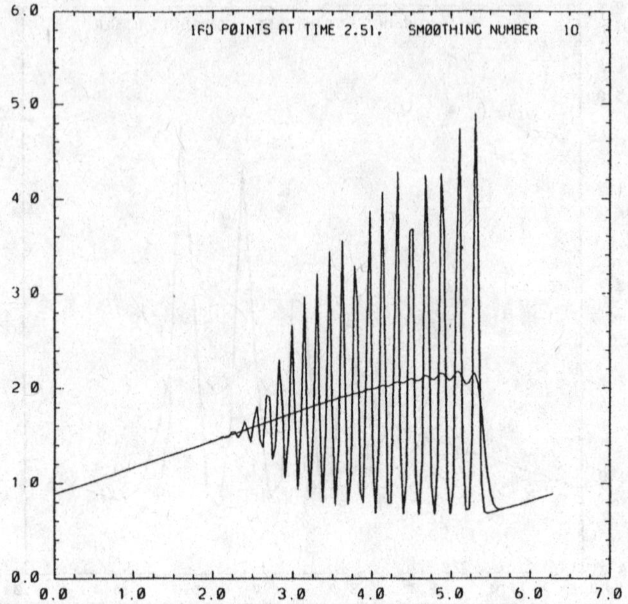

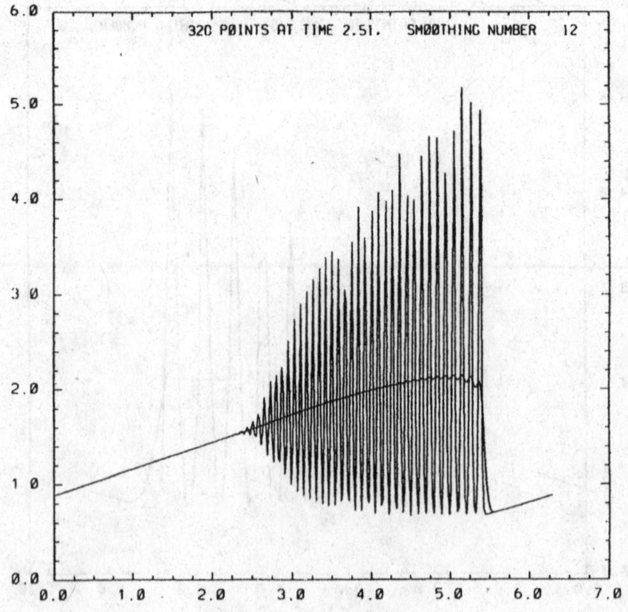

Figure 1. Continued

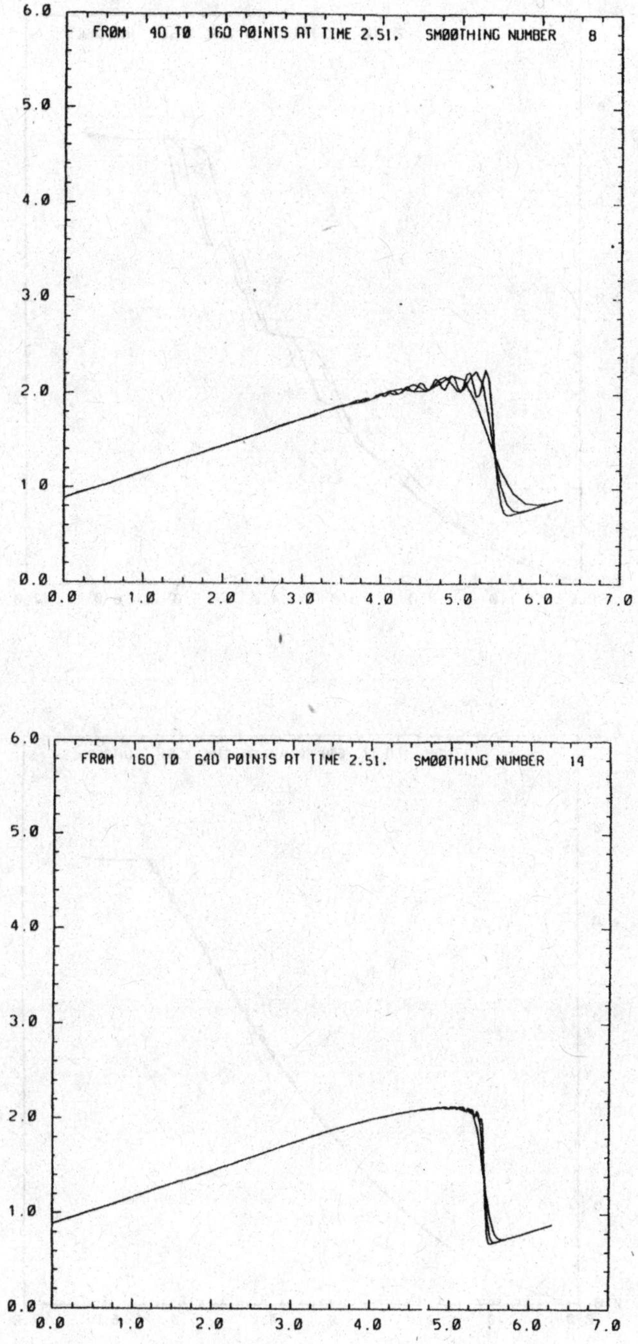

Figure 2. Superposition of several smoothed curves. The steeper and more oscillatory curves have larger N.

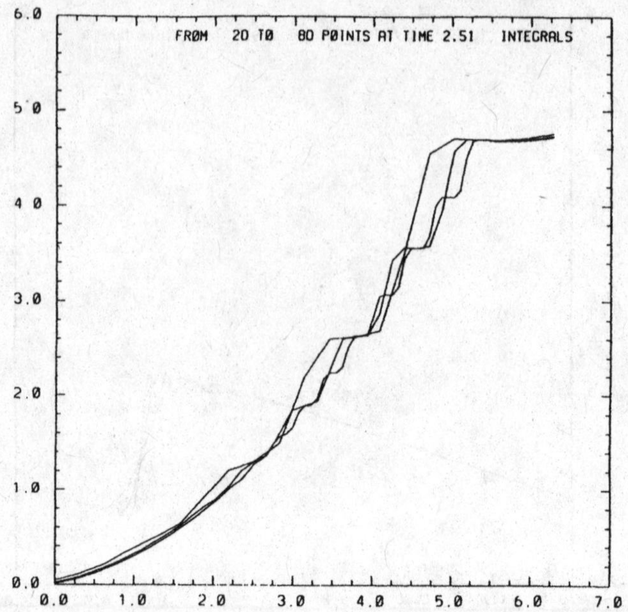

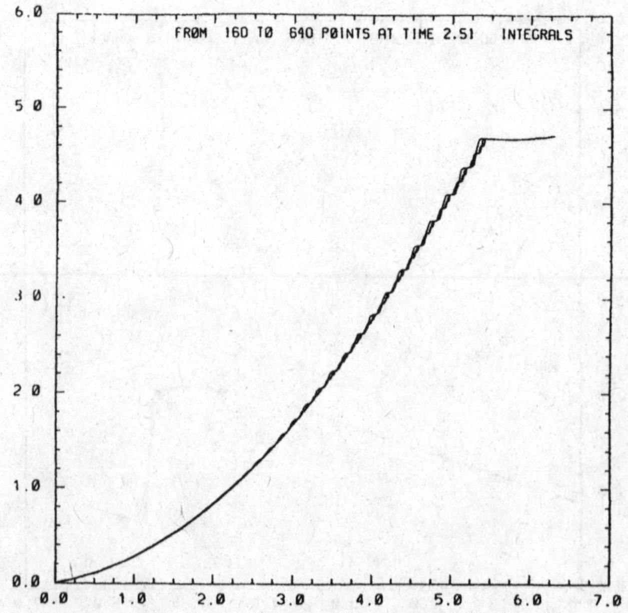

Figure 3. Indefinite integrals of the oscillatory solutions illustrating weak convergence (see text). The curves are piecewise linear interpolants of the data points. Curves with more corners correspond to larger N.

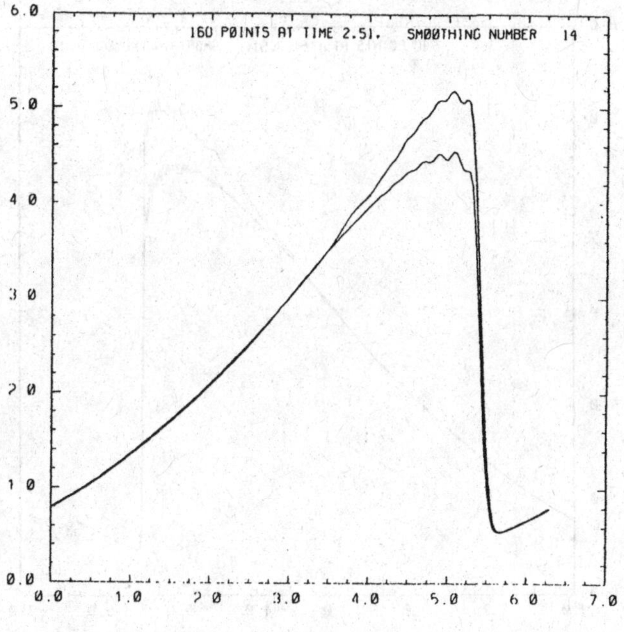

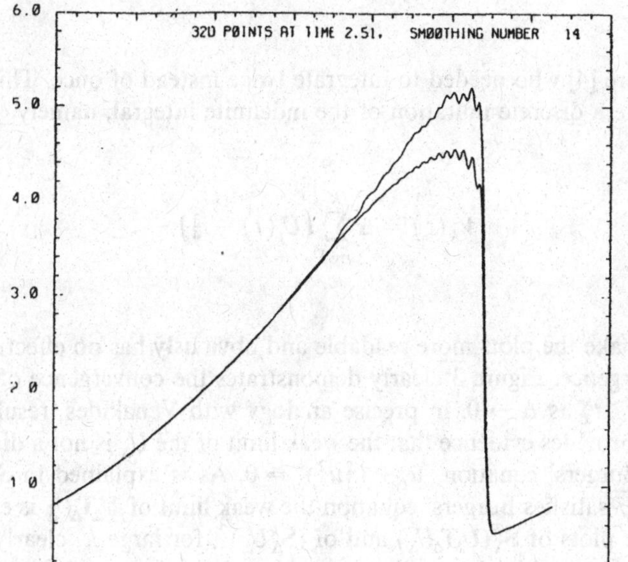

Figure 4. **Weak limits of oscillatory solutions are not weak solutions of Burgers' equation. Plots of smoothings of $U_\Delta T_\Delta U_\Delta$ and of the square of the smoothing of U_Δ for the same N and t. The two are clearly different in the oscillatory region.**

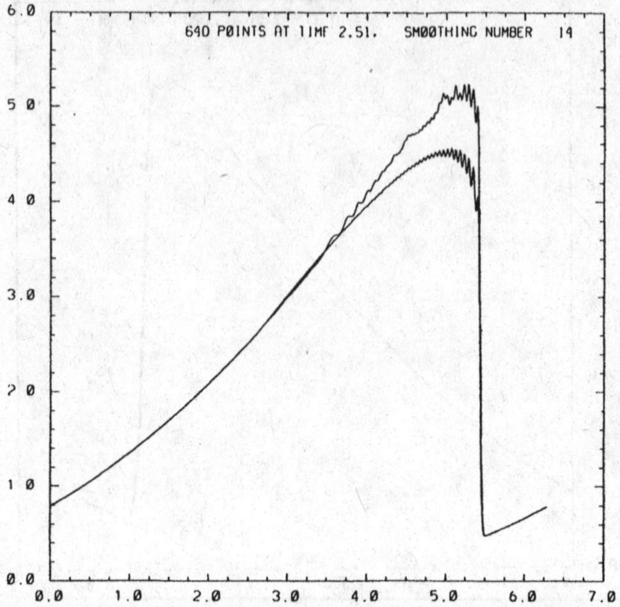

Figure 4. Continued

and Levermore [4] who needed to integrate twice instead of once. This motivates us to compute a discrete imitation of the indefinite integral, namely

$$V_k(t) = \Delta \sum_{j=1}^{k} \left(U_j(t) - \tfrac{3}{4}\right).$$

The $\frac{3}{4}$ is to make the plots more readable and obviously has no effect on weak or strong convergence. Figure 3 clearly demonstrates the convergence of $V_k(t)$ to a function $v(x_k, t)$ as $\Delta \to 0$, in precise analogy with Venakides' result for KdV.

Figure 4 provides evidence that the weak limit of the U_Δ is not a distributional solution of Burgers' equation, $u_t + (\frac{1}{2}u^2)_x = 0$. As is explained in Section 2, if $U_\Delta \rightharpoonup \overline{U}$ and $\overline{U}$ satisfies Burgers' equation the weak limit of $U_\Delta \mathbf{T}_\Delta U_\Delta$ is equal to $\overline{U}^2$. However, the plots of $\mathbf{S}_\Delta^k(U_\Delta \mathbf{T}_\Delta U_\Delta)$ and of $(\mathbf{S}_\Delta^k U_\Delta)^2$, for large k, clearly show that the two are not equal in the oscillatory region.

Acknowledgment. The research of the first author is supported by NSF Grant DMS-8553215, the AT&T Fund, and a Sloan Foundation Fellowship. The research of the second author is supported by DOE Grant DEAC 0276ER 03077.

Bibliography

[1] Dahlquist, G., and Björk, Å., *Numerical Methods*, Prentice-Hall, Englewood Cliffs, New Jersey, 1974.

[2] Flashchka, H., *The Toda lattice I*, Phys. Rev. B9, 1974, pp. 1924–1925; *II*, Progr. Theoret. Phys. 51, 1974, pp. 703–716.

[3] Kac, M. and van Moerbeke, P., *On an explicitly soluble system of nonlinear differential equations related to certain Toda lattices*, Adv. in Math. 16, 1975, pp. 160–169.

[4] Lax, P. D., and Levermore, C. D., *The small dispersion limit of the KdV equation*, Comm. Pure Appl. Math. 36, 1983, *I* pp. 253–290, *II* pp. 571–593, *III* pp. 809–830.

[5] Lax, P. D., *On dispersive difference schemes*, Physica 180, North-Holland, Amsterdam, 1986, pp. 250–252.

[6] Lax, P. D., *Oscillatory solutions of partial differential and difference equations* in *Mathematics Applied to Science*, Goldstein, Rosencrans, and Sod, eds., Academic Press, 1988, pp. 155–170.

[7] van Moerbeke, P., *The spectrum of Jacobi matrices*, Inventiones Math. 37, 1976, pp. 45–81.

[8] Moser, J., *Three integrable Hamiltonian systems connected with isospectral deformations*, Advances in Mathematics 16, 1975, 197–220.

[9] Neumann, J. v., *Proposal and analysis of a new numerical method for the treatment of hydrodynamical shock problems*, Collected Works, Pergamon, London.

[10] Neumann, J. v., and Richtmyer, R. D., *A method for the numerical calculation of hydrodynamical shocks*, J. Appl. Phys. 21, 1950, pp. 380–385.

[11] Strang, G., *Accurate partial difference methods, II. Nonlinear problems*, Num. Math. 6, 1964, pp. 37–46.

[12] Venakides, S., *The zero dispersion limit of the periodic KdV equation*, AMS Transaction, Vol. 301, 1987, pp. 189–226.

Received February, 1988.

Dispersive Approximations in Fluid Dynamics

THOMAS Y. HOU AND PETER D. LAX
Courant Institute

To Heinz Kreiss, with friendship and admiration

Historical Review

While working at Los Alamos in 1943–44, von Neumann became convinced that the calculation of the flows of compressible fluids containing strong shocks could be accomplished only by numerical methods. He conceived the idea of capturing shocks,* i.e., of ignoring the presence of a discontinuity. Employing a Lagrangian description of compressible flow, setting heat conduction and viscosity equal to zero, von Neumann replaced space and time derivatives by symmetric difference quotients. Calculations using this scheme were carried out at the Ballistics Research Laboratory at Aberdeen, using punched card equipment; the approximations resulting from these calculations (see [11]) showed oscillations on mesh scale behind the shock; see Figure 1. Von Neumann boldly conjectured that the oscillations in velocity represent the heat energy created by the irreversible action of the shock, and that as $\Delta\zeta$ and Δt tend to zero, the approximate solutions tend *in the weak sense* to the correct discontinuous solution of the equations of compressible flow. In his own words:

> These considerations suggest the surmise that (the difference equation) is always a valid approximation of the hydrodynamical motion, but with this qualification: it is not the $x_a(t)$ which approximate the $x(y, t)$ but the averages of the x_a over an interval of sufficient length of contiguous $a' - s$. The x_a themselves perform oscillations around these averages and these oscillations do not tend to zero, but they make finite contributions to the total energy.
>
> In the mathematical terminology, the surmise means that the quasi-molecular kinetic solutions converge to the hydrodynamical one but in the weak sense . . . A mathematical proof of this surmise would be most important, but it seems to be very difficult, even in the simplest special cases. The procedure to be followed will therefore be a different one: We shall test the surmise experimentally by carrying out the necessary computations. . . .

Garrett Birkhoff in [1] recalls the following scene:

> My desk in Bob Kent's** office made me privy to other fascinating interchanges. One of these occurred towards the end of the war in Germany. Von Neumann by this time had begun making pioneer numerical experiments on the leading digital computers of the time. He had observed that the kinetic theory of gas predicts essentially the same statistical behavior of a gas from a wide variety of force laws, and was trying to simulate plane shock wave propagation with a "gas" consisting of about 100 molecules in this spirit. He was presenting informally his ideas and results to a small but select group, which included Kent and von Karman, the famous aeronautical engineer. When von Neumann had finished

---

* The felicitous phrase *shock capturing* was coined by Harvard Lomax.

** Robert H. Kent was an outstanding ordinance expert and Chief Scientist of the Aberdeen Laboratory during and after the Second World War.

Communications on Pure and Applied Mathematics, Vol. XLIV, 1–40 (1991)
 CCC 0010-3640/91/010001-40$04.00

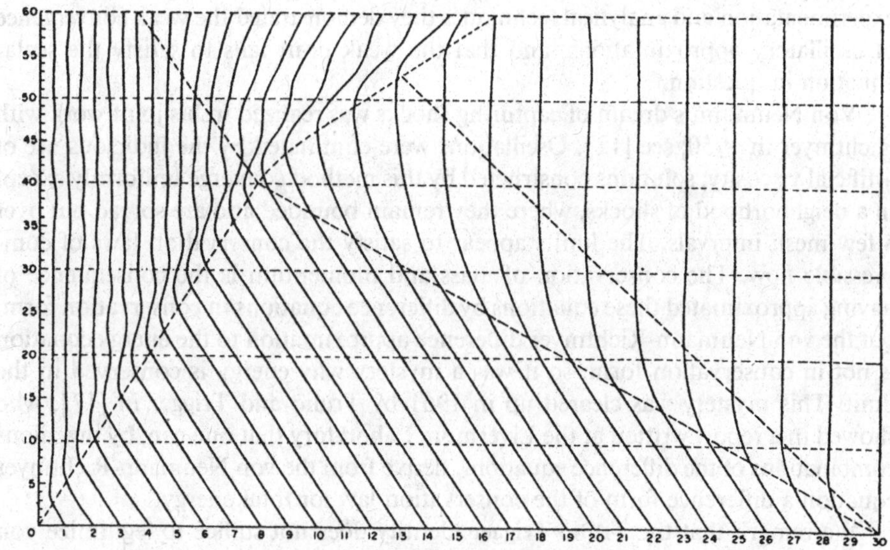

Figure 1. From John von Neumann, *Collected Works*, Vol. VI, p. 377.

his talk, with its dizzying prospects of glamorous future developments, von Karman said with a mischievous smile: "Of course you realize that Lagrange used the same model in his 'Méchanique Analytique' nearly two centuries ago."

The numerical method of von Neumann was analyzed in an interesting report (see [13]) by Sir Rudolf Peierls, written in 1946 while Peierls was a member of the British Mission to Los Alamos. We quote the abstract of his report:

> The method of allowing fluctuations in the mechanical solutions of hydrodynamic problems to take the place of the entropy increase in the shock is analyzed statistically. The model corresponds to the thermal behavior of a substance of anomalously low specific heat, and the effect of the pseudo thermal motion is greater than for any real substance. Therefore the model is a good approximation only when the contribution of the thermal pressure is negligible. Curves are given from which this contribution can be estimated. Other complicated features arise when the mass intervals are not equal and in the case of radial motion. A typical case, taken from I.B.M. calculations, is discussed by way of illustration.

In [8] it was counterconjectured that von Neumann was wrong in his surmise, i.e., that although the approximate solutions constructed by his method do converge weakly, the weak limit fails to satisfy the law of conservation of energy. Some plausible reasons were presented, including an analogy with the zero dispersion limit for the KdV equation; see [9]. Here we would like to remark that a statistical analysis such as Peierls used is probably not quite right; the oscillations that appear in these approximations are, after all, completely deterministic. Although they may appear random, they are at most pseudorandom.

A time-honored testing bed for the equations of compressible flow is the scalar equation $u_t + uu_x = 0$. In [5] Goodman and Lax investigated von Neumann's algorithm applied to this equation, in the semidiscrete case $\Delta t = 0$. Using numerical experimentation and analytical techniques they demonstrated the weak convergence of oscillatory approximations, and that the weak limit fails to satisfy the scalar equation in question.

Von Neumann's dream of capturing shocks was realized in his joint work with Richtmyer in 1950; see [12]. Oscillations were eliminated by the judicious use of artificial viscosity; solutions constructed by this method converge uniformly except in a neighborhood of shocks, where they remain bounded and are spread out over a few mesh intervals. The limits appear to satisfy the conservation laws of compressible flow. The conservation of mass and momentum is the consequence of having approximated these equations by difference equations in conservation form; but the von Neumann-Richtmyer difference approximation to the energy equation is not in conservation form; so it was a mystery why energy is conserved in the limit. This mystery was cleared up in 1961 by Trulio and Trigger in [17], who showed in a report written at the Livermore Laboratory that one can, by ingenious combinations of the difference equations, derive from the von Neumann-Richtmyer equation a difference form of the conservation law for total energy.

We remark that the Trulio-Trigger identity does not suffice to legitimize von Neumann's original method. On the contrary, it clarifies what goes wrong in the weak limit.

In this paper we have compared the results of a von Neumann-Richtmyer calculation with the weak limit of calculations performed by von Neumann's orginal method; although there are discernible differences, they are surprisingly small. We realize that it is not quite fair to subject to such searching scrutiny laboratory reports such as [11] and [13]. These reports, written in haste and under war-time pressure, were not meant for publication in refereed journals; they were merely to be used to design an atomic bomb. Actually, according to Edward Teller (see [16]), the implosion calculations of von Neumann were not used to estimate the performance of the bomb.

Introduction

The purpose of this paper is to analyze von Neumann's algorithm for solving the initial value problem for the Lagrangian equations of compressible flow; see equations (1.1)–(1.3). We largely follow the procedure recommended by von Neumann: detailed numerical experimentation.

Since the Lagrangian equations form a strictly hyperbolic nonlinear system of equations, it follows from the theory of such equations that for given smooth initial data the compressible flow equations have a uniquely determined smooth solution. Because of the propensity for shock formation, such smooth solutions are local, i.e., they do not exist for all space and time.

Von Neumann's difference scheme consists of replacing the time and space derivatives by centered difference quotients; see equations (1.5)–(1.7). This scheme, a variant of von Neumann's, is essentially a staggered leap-frog scheme; it is linearly stable in the l^2 sense when the CFL condition is satisfied. We shall study the convergence of approximations produced by this scheme as $\Delta\xi$ tends to zero, while $\Delta t/\Delta\xi = \lambda$ is kept fixed.

According to a classical result of Strang (see [14]), when one solves a smooth initial value problem for a nonlinear hyperbolic equation by a difference scheme, as $\Delta\xi$ tends to zero the approximations converge uniformly to the smooth solution of the differential equation in every domain of determinacy of the difference equation in which the smooth solution is defined, provided that the difference scheme is linearly stable. This result is applicable to the Lagrange equations of compressible flow and to von Neumann's difference scheme. Our numerical calculations demonstrate admirably the uniform convergence of the approximations produced by von Neumann's scheme in those regions of ξ, t space where Strang's theorem guarantees their convergence. The calculations also indicate uniform convergence to the smooth solutions in a region decidedly larger than the domain of determinacy of the scheme although contained within the domain of determinacy of the differential equation. This is very surprising, since outside this region the solutions of the difference equations are highly oscillatory with a wavelength approximately but not exactly equal to $\Delta\xi$, yet the boundary of the oscillatory region propagates with speed much smaller than $\Delta\xi/\Delta t$. The amplitude of the oscillations appears to be bounded away from zero and infinity, uniformly for all $\Delta\xi$. As $\Delta\xi$ tends to zero, the oscillatory regions seem to converge to a limiting region; within this region, the oscillatory functions u, p, and V appear to converge weakly to limits that we denote as $\bar{u}$, $\bar{p}$, and $\bar{V}$, respectively. The weak limit $\bar{u}$, $\bar{p}$, $\bar{V}$ contain shock-like structures, i.e., discontinuities or at least regions of very rapid change.

The Lagrangian laws of conservation of momentum and mass, equations (1.1) and (1.2), are linear in u, p, and V; so are their difference analogues, equations (1.5) and (1.6). Therefore we can take the weak limit of these equations as $\Delta\xi$, Δt tend to zero, and obtain the laws of conservation of momentum and mass for $\bar{u}$, $\bar{p}$, and $\bar{V}$.

It is quite otherwise with the law of conservation of energy; not only does a derivation of the above kind fail, but our numerical calculations indicate unequivocally that $\bar{u}$, $\bar{p}$, $\bar{V}$ do *not* satisfy the law of conservation of energy. This brings up the question, what equations do the weak limits satisfy? Some speculations are included in Section 3 of this paper.

As stated earlier, the ratio $\lambda = \Delta t/\Delta\xi$ is kept constant in the limiting procedure $\Delta\xi \to 0$. Since λ appears as a parameter in the algorithm, one would expect the weak limits to depend on λ; yet the numerical results suggest strongly that $\bar{u}$, $\bar{p}$, $\bar{V}$ are independent of λ even in the oscillatory region. Some speculations on why this might be so are given in the body of this paper.

We have found strong numerical evidence for the following: if the initial data have discontinuities whose time evolution leads to rarefaction waves, or contact discontinuities, not shocks, then the approximations constructed by von Neumann's method converge uniformly to the solution containing the rarefaction waves. If true, this would be a remarkable strengthening of Strang's theorem.

The organization of this paper is as follows: In Section 1 the differential and difference equations are stated and the applicability of Strang's theorem is spelled out. In Section 2 we discuss the numerical results, with three kinds of initial data: periodic, Riemann shock data, and Riemann rarefaction data. The main phenomena are pointed out: the existence of smooth and oscillatory regions and their propagation in space-time, weak convergence in the oscillatory region, the presence of discontinuities or near-discontinuities, and the seeming independence of λ of the weak limit. We present numerical evidence that for $\bar{u}, \bar{p}, \bar{V}$ total energy is not conserved. In Section 3 two versions of the Trulio-Trigger energy conservation law are derived and used to analyze the failure of the law of conservation of energy for $\bar{u}, \bar{p}, \bar{V}$.

Section 4 deals with the limiting case $\lambda = 0$, i.e., a semidiscrete approximation. Using the conservation law of energy and entropy we find a plausible reason for the development of oscillations. When the initial data are isentropic, so is the flow for all later times; in this case, it is possible to interpret the semidiscrete equations as describing a lattice vibration, about which there is a rich literature. The special case when p depends exponentially on V is the Toda lattice and it is completely integrable. We give a reinterpretation from our point of view of some results on shocks in lattices.

1. Strong Convergence to Smooth Solutions

The 1-D compressible gas dynamic equations in Lagrangian coordinates (see [2]) are

$$u_t + p_\xi = 0, \tag{1.1}$$

$$V_t - u_\xi = 0, \tag{1.2}$$

$$e_t + p u_\xi = 0, \tag{1.3}$$

where u is velocity, p is pressure, V is specific volume, and e is specific internal energy. Here we mostly study the case when the equation of state is given by the γ-law, with $\gamma = 1.4$:

$$e = \frac{pV}{\gamma - 1}. \tag{1.4}$$

Equations (1.1) and (1.2) are the conservation laws of momentum and mass. Multiplying (1.1) by u and adding to (1.3) yields the law of conservation of energy:

$$(e + u^2/2)_t + (up)_\xi = 0. \tag{1.3$'$}$$

Similarly, multiplying (1.2) by p and adding to (1.3) yields

$$e_t + pV_t = 0$$

which, when multiplied by an integrating factor becomes the law of conservation of entropy. For a γ-law gas this takes the form

$$(pV^{\gamma})_t = 0. \tag{1.4$'$}$$

What we study is a variant of von Neumann's difference scheme, a staggered, centered difference approximation to (1.1)–(1.3):

$$u_k^{n+1/2} = u_k^{n-1/2} - \lambda(p_{k+1/2}^n - p_{k-1/2}^n), \tag{1.5}$$

$$V_{k+1/2}^{n+1} = V_{k+1/2}^n + \lambda(u_{k+1}^{n+1/2} - u_k^{n+1/2}), \tag{1.6}$$

$$e_{k+1/2}^{n+1} = e_{k+1/2}^n - \lambda p_{k+1/2}^{n+1/2}(u_{k+1}^{n+1/2} - u_k^{n+1/2}), \tag{1.7}$$

where

$$p_{k+1/2}^{n+1/2} = \tfrac{1}{2}(p_{k+1/2}^{n+1} + p_{k+1/2}^n), \quad \text{and} \quad \lambda = \Delta t/\Delta\xi. \tag{1.8}$$

Here $u_k^{n+1/2}$ approximates $u(k\Delta\xi, (n+1/2)\Delta t)$ and $e_{k+1/2}^n$ approximates $e((k+1/2)\Delta\xi, n\Delta t)$, etc.

Given the initial values $u_k^{-1/2}$, $V_{k+1/2}^0$, and $p_{k+1/2}^0$, u, V, and p can be determined at all future times from the difference equations (1.5)–(1.8) and the equation of the state. These initial values are determined from the initial data $u(\xi, 0)$, $V(\xi, 0)$, and $p(\xi, 0)$.

If the initial data are smooth, equations (1.1)–(1.3) have a unique smooth solution up to some critical time t_{crit}. We shall prove that the numerical solution of von Neumann's scheme converges strongly to the solution of equations (1.1)–(1.3) for $t < t_{\text{crit}}$. This relies on a basic convergence theorem of Strang; see [14].

Let

$$u_t + C(u, x, t)u_x = 0 \tag{1.9}$$

be a quasilinear hyperbolic system. Here $C(u, x, t)$ is a matrix-valued function depending smoothly on its arguments, with real and distinct eigenvalues.

Consider an explicit finite difference approximation to (1.9) of the form

$$U_k^{n+1} + F(U_{k-1}^n, U_k^n, U_{k+1}^n) = 0 \tag{1.10}$$

that is consistent with (1.9). We assume that F is a smooth function of its arguments.

Convergence Theorem of Strang. *Let $u(x, t)$ be a smooth solution of the hyperbolic system (1.9) for $t < t_{\text{crit}}$. Assume that the approximation (1.10) to (1.9) is linearly l^2 stable around $u(x, t)$. Then for $t < t_{\text{crit}}$*

$$(1.11)\qquad |U_k^n - u(k\Delta x, n\Delta t)| \leqq C\Delta t^p,$$

where p is the order of accuracy of the scheme (1.10) *for equation* (1.9).

In order to prove linear stability of von Neumann's scheme when an appropriate CFL condition is satisfied, we extend the scheme from $\{u_k^{n-1/2}, V_{k+1/2}^n, e_{k+1/2}^n, p_{k+1/2}^n\}$ to $\{u_{k+1/2}^{n-1/2}, V_{k+1}^n, e_{k+1}^n, p_{k+1}^n\}$ and $\{u_k^n, V_{k+1/2}^{n+1/2}, e_{k+1/2}^{n+1/2}, p_{k+1/2}^{n+1/2}\}$. Then von Neumann's scheme is reduced to the following leap-frog scheme:

$$(1.12)\qquad u_k^{n+1} = u_k^{n-1} - \lambda(p_{k+1}^n - p_{k-1}^n),$$

$$(1.13)\qquad V_k^{n+1} = V_k^{n-1} + \lambda(u_{k+1}^n - u_{k-1}^n),$$

$$(1.14)\qquad e_k^{n+1} = e_k^{n-1} - \lambda\tfrac{1}{2}(p_k^{n+1} + p_k^{n-1})(u_{k+1}^n - u_{k-1}^n).$$

Substituting the γ-law $e = pV/(\gamma - 1)$ into (1.14) and using (1.13), we obtain

$$(1.15)\qquad p_k^{n+1} = \frac{p_k^{n-1} - \dfrac{\lambda(\gamma-1)}{2}\dfrac{p_k^{n-1}}{V_k^{n-1}}(u_{k+1}^n - u_{k-1}^n)}{1 + \dfrac{\lambda(\gamma+1)}{2V_k^{n-1}}(u_{k+1}^n - u_{k-1}^n)}.$$

Equations (1.12), (1.13), and (1.15) completely determine (u, V, p), e can be recovered through the γ-law.

Assume that $(\bar{u}, \bar{V}, \bar{p})$ is a smooth solution of the continuous equations. Let

$$\begin{cases} u = \bar{u} + \tilde{u} \\ V = \bar{V} + \tilde{V}. \\ p = \bar{p} + \tilde{p} \end{cases}$$

Linearizing (1.5), (1.6), and (1.15) around $(\bar{u}, \bar{V}, \bar{p})$, we obtain after some manipulations that

$$(1.16)\qquad \tilde{u}_k^{n+1} = \tilde{u}_k^{n-1} - \lambda(\tilde{p}_{k+1}^n - \tilde{p}_{k-1}^n),$$

$$(1.17)\qquad \tilde{V}_k^{n+1} = \tilde{V}_k^{n-1} + \lambda(\tilde{u}_{k+1}^n - \tilde{u}_{k-1}^n),$$

$$(1.18)\qquad \tilde{p}_k^{n+1} = \tilde{p}_k^{n-1} - \lambda\gamma\frac{\bar{p}_k^{n-1}}{\bar{V}_k^{n-1}}(\tilde{u}_{k-1}^n - \tilde{u}_{k-1}^n) + O(\Delta t)(\tilde{p}_k^{n-1} + \tilde{V}_k^{n-1}),$$

where $|O(\Delta t)| \leqq C\Delta t$ and C depends only on $(\bar{u}, \bar{V}, \bar{p})$. Denote

$$\tilde{\mathbf{W}} = \begin{pmatrix} \tilde{u} \\ \tilde{V} \\ \tilde{p} \end{pmatrix}, \quad \bar{\mathbf{W}} = \begin{pmatrix} \bar{u} \\ \bar{V} \\ \bar{p} \end{pmatrix},$$

and

$$A(\bar{\mathbf{W}}) = \lambda \begin{bmatrix} 0 & 0 & -1 \\ 1 & 0 & 0 \\ -\gamma\bar{p}/\bar{V} & 0 & 0 \end{bmatrix};$$

then (1.16)–(1.18) can be expressed as

$$\text{(1.19)} \quad \tilde{\mathbf{W}}_k^{n+1} = \tilde{\mathbf{W}}_k^{n-1} + A(\bar{\mathbf{W}}_k^{n-1})(\tilde{\mathbf{W}}_{k+1}^n - \tilde{\mathbf{W}}_{k-1}^n) + O(\Delta t)\tilde{\mathbf{W}}_k^{n-1}.$$

Note that matrix A has three distinct eigenvalues: $\mu_1 = -\lambda\sqrt{\gamma\bar{p}/\bar{V}}$, $\mu_2 = 0$, $\mu_3 = \lambda\sqrt{\gamma\bar{p}/\bar{V}}$. Therefore, there exists invertible matrix $X(\bar{\mathbf{W}})$ such that

$$X^{-1}AX = \begin{bmatrix} \mu_1 & 0 & 0 \\ 0 & \mu_2 & 0 \\ 0 & 0 & \mu_3 \end{bmatrix}.$$

Introduce change of variables

$$\tilde{\mathbf{U}}_k^n = X(\bar{\mathbf{W}}_k^n)^{-1}\tilde{\mathbf{W}}_k^n.$$

Then equation (1.19) reduces to

$$\text{(1.20)} \quad \tilde{\mathbf{U}}_k^{n+1} = \tilde{\mathbf{U}}_k^{n-1} + \begin{bmatrix} \mu_1 & 0 & 0 \\ 0 & \mu_2 & 0 \\ 0 & 0 & \mu_3 \end{bmatrix}(\tilde{\mathbf{U}}_{k+1}^n - \tilde{\mathbf{U}}_{k-1}^n) + O(\Delta t)(\tilde{\mathbf{U}}_k^{n-1} + \tilde{\mathbf{U}}_k^n),$$

where we have used the fact that $X(\bar{\mathbf{W}}(\xi_k, t_n))$ is a smooth function in ξ_k and t_n.

It is well known that leap-frog difference method is stable in discrete l^2 norm for linear symmetric hyperbolic system if $|\mu_i| < 1$, i.e., $\lambda\sqrt{\gamma\bar{p}/\bar{V}} < 1$. Thus (1.20) is stable. Consequently we prove the linear stability of von Neumann's scheme in l^2 norm.

We appeal now to Strang's convergence theorem to conclude:

THEOREM 1. *Let* $\mathbf{W} = (u, V, p, e)$ *be a smooth solution of* (1.1)–(1.4) *for* $0 \leqq t \leqq T$, $\mathbf{W}_k^n$ *be the solution of von Neumann's scheme* (1.5)–(1.8), *with* $\mathbf{W}_k^0 = \mathbf{W}(k\Delta\xi, 0)$. *Then* $\mathbf{W}_k^n$ *converges uniformly to* $\mathbf{W}(k\Delta\xi, n\Delta t)$ *as* $\Delta\xi, \Delta t \to 0$ *for* $0 \leqq t \leqq T$. *More precisely, we have*

$$|\mathbf{W}_k^n - \mathbf{W}(k\Delta\xi, n\Delta t)| \leqq C(\Delta\xi^2 + \Delta t^2) \quad \text{for} \quad 0 \leqq t \leqq T.$$

Remark. Typically, as t approaches t_{crit}, $\mathbf{W}(\xi, t)$ develops a finite number of singularities. We can still apply Strang's convergence theorem to some region where a smooth solution exists. This implies strong convergence of von Neumann's scheme in the region where the solution $\mathbf{W}$ is smooth and which is a domain of dependence for the difference scheme. Actually, numerical experiments indicate uniform convergence to the smooth solutions in a region decidedly larger than the domain of determinacy of the scheme. We also found that the numerical solutions of von Neumann's method converge uniformly to the solution containing rarefaction waves. (See Section 2.)

2. Description of Numerical Experiments

The graphs presented and analyzed in this section are taken from numerically computed solutions of the difference equations (1.5)–(1.7) for several sets of initial data. We have experimented with three kinds of initial values $u_0(\xi)$, $V_0(\xi)$, and $p_0(\xi)$.

I. Periodic initial data.

$$(2.1)\qquad \begin{aligned} u_0(\xi) &= 2 + 0.5\sin(2\pi\xi),\\ p_0(\xi) &\equiv 1,\\ V_0(\xi) &\equiv 1. \end{aligned}$$

There is a critical value t_{crit} of t such that equations (1.1)–(1.3) have a smooth solution for all ξ when $t < t_{\text{crit}}$, but for no longer value of t. For the initial data (2.1), $t_{\text{crit}} \sim 0.41$.

We have also studied Riemann initial data, i.e., piecewise constant data with a single discontinuity at $\xi = 0$:

$$(2.2)\qquad \begin{aligned} u_0(\xi) &= u_l, \quad V_0(\xi) = V_l, \quad p_0(\xi) = p_l, \quad \text{for} \quad \xi < 0,\\ u_0(\xi) &= u_r, \quad V_0(\xi) = V_r, \quad p_0(\xi) = p_r, \quad \text{for} \quad \xi > 0. \end{aligned}$$

Riemann initial data are *positive* homogeneous:

$$u_0(a\xi) = u_0(\xi) \quad \text{for} \quad a > 0, \quad \text{etc.}$$

It is convenient to extend the approximations constructed by solving (1.5)–(1.9) to all ξ and t as piecewise constant in rectangles centered at $(k\Delta\xi, (n + 1/2)\Delta t)$ and $((k + 1/2)\Delta\xi, n\Delta t)$, respectively. We denote these approximations thus extended as $u(\xi, t; \Delta\xi, \lambda)$, etc. It follows that the approximate solutions computed by the von Neumann scheme are also positive homogeneous:

$$u(a\xi, at; a\Delta\xi, \lambda) = u(\xi, t; \Delta\xi, \lambda), \quad \text{etc.} \tag{2.3}$$

for $a > 0$.

We shall study two kinds of Riemann data:

II. Riemann shock data.

$$\begin{aligned} &u_l = 1, \quad V_l = 1, \quad p_l = 1, \quad \text{for} \quad \xi < 0, \\ &u_r = 1 + \delta, \quad V_r = 1 - \delta, \quad p_r = 1 + \delta, \quad \text{for} \quad \xi > 0, \end{aligned} \tag{2.4}$$

where $\delta = -2(\gamma - 1)/(\gamma + 1)$. With these initial data the three conservation laws have a solution that consists of a single shock propagating to the right with velocity 1.

III. Riemann rarefaction data.

$$\begin{aligned} &u_l = 1, \quad V_l = 1, \quad p_l = 1, \quad \text{for} \quad \xi < 0, \\ &u_r = 1 - \delta, \quad V_r = 1 + \delta, \quad p_r = 1 - \delta, \quad \text{for} \quad \xi > 0, \end{aligned} \tag{2.5}$$

where $\delta = -2(\gamma - 1)/(\gamma + 1)$. The solution of the Lagrange equations of compressible flow, (1.1)–(1.4), with these initial rarefaction data consists of two rarefaction waves and one contact discontinuity issuing from the origin. The front of the rarefaction waves propagate with velocity C_r and $-C_l$ respectively, the sound speeds associated with states on the right and on the left. Between the two rarefaction waves there are two constant states separated by the contact discontinuity at $\xi = 0$:

$$u_m \sim 1.057, \quad V_m^+ \sim 0.86, \quad V_m^- \sim 1.05, \quad p_m \sim 0.93.$$

Figures 2 to 6 picture solutions whose initial data are periodic. Figures 2a, 2b, and 2c depict $u = u(\xi, t, \Delta\xi, \lambda)$ at $t = 0.7$ as function of the discretized space variable ξ, for three different values of $\Delta\xi : 0.01, 0.005, 0.0025$, and $\lambda = \Delta t/\Delta\xi = 0.2$. In these figures the smooth and oscillatory regions are clearly recognizable, and they appear to be independent of $\Delta\xi$. The number of peaks in the oscillatory region is proportional to $(\Delta\xi)^{-1}$, a clear indication that the oscillations are on mesh scale.

The calculations presented below strongly support the contentions that u, V, $p(\xi, t; \Delta\xi, \lambda)$, and polynomials formed of them, converge weakly as $\Delta\xi$ tends to zero. Here we want to point out that u and V converge weakly with respect to ξ and strongly with respect to t. That is, for all smooth test functions $\phi(\xi)$,

$$\int u\phi \, d\xi \equiv \sum u_k^{n+1/2} \phi_k^{n+1/2}$$

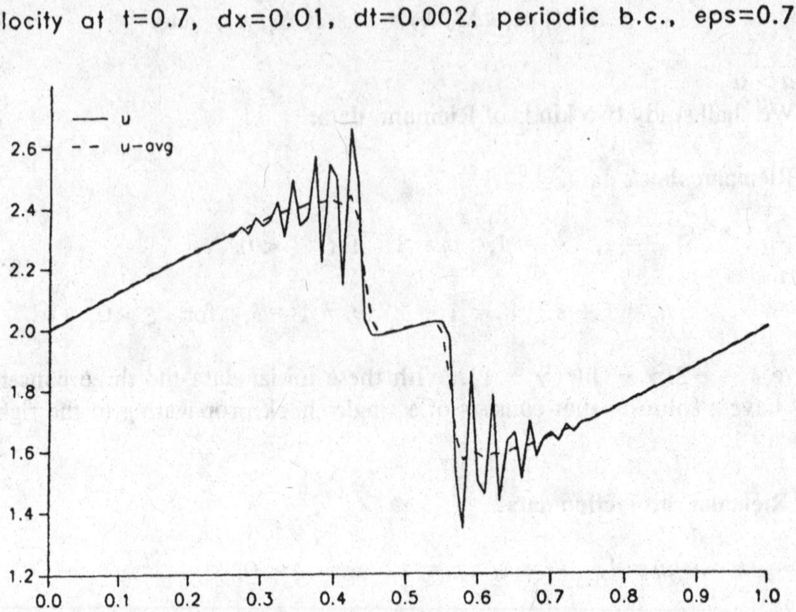

Figure 2a.

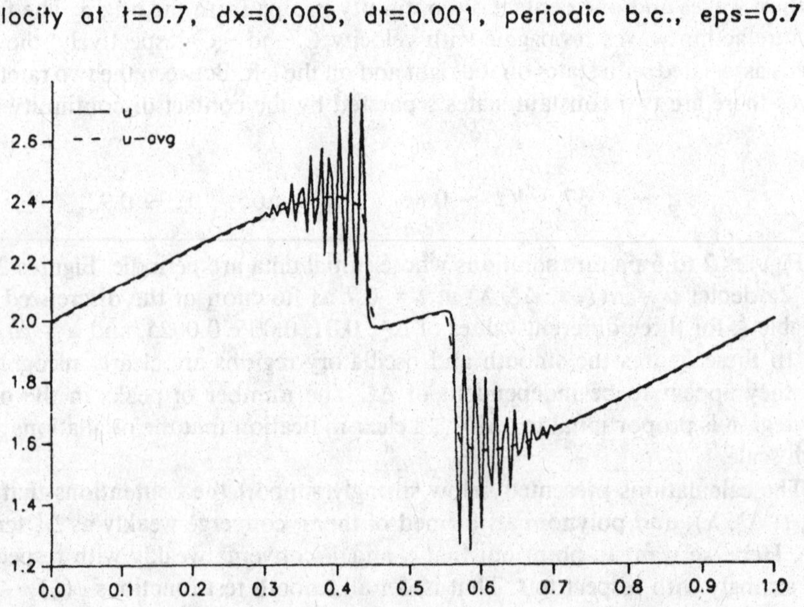

Figure 2b.

velocity at t=0.7, dx=0.0025, dt=0.0005, periodic b.c.,eps=0.65

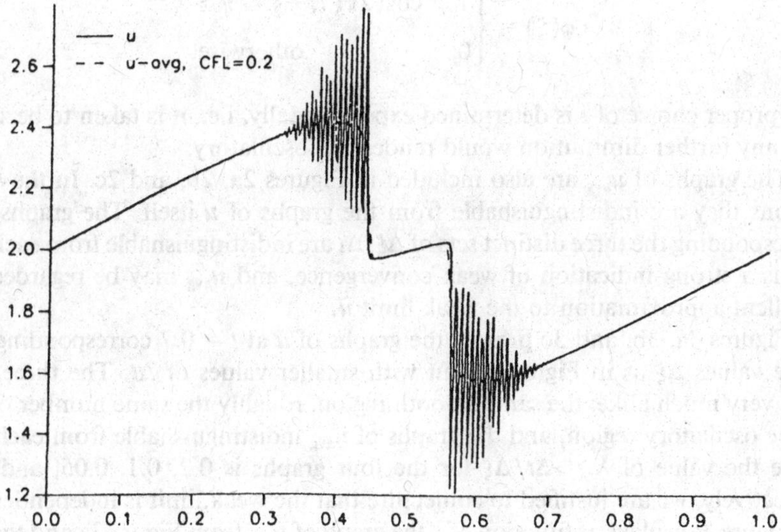

Figure 2c.

converges strongly as functions of t, where $\phi_k^{n+1/2} = \phi(k\Delta\xi, (n+1/2)\Delta t)$. To see this we multiply (1.5) by $\phi_k^{n+1/2}$ and sum over all k and n. Summation by parts on the right gives

$$\int u(\xi, T)\phi(\xi)\,d\xi - \int u(\xi, 0)\phi(\xi)\,d\xi = \int_0^T \int \phi_\xi(\xi) p(\xi, t; \Delta\xi, \lambda)\,d\xi\,dt.$$

Clearly, since p remains bounded, the right side is Lipschitz continuous as function of T: a similar argument applies to V. For the products of u, V, p, it is necessary to smooth both in space and time to obtain a weak limit, except for combinations such as $(3.11)_\pm$ that satisfy a conservation law $(3.10)_\pm$.

We can determine weak limits $\bar{u}$, $\bar{V}$, $\bar{p}$ by suitable space-time averaging. As pointed out above, for u and V, it is sufficient to average in ξ. The average value of u, denoted as u_{avg}, is obtained by convolving u as function of ξ with a mollifier whose support is small with respect to 1 and large with respect to $\Delta\xi$:

$$u_{\text{avg}} = u * \phi_\epsilon,$$

where

$$\phi_\epsilon(\xi) = \frac{1}{\epsilon}\phi\left(\frac{\xi}{\epsilon}\right),$$

and

$$\phi(\xi) = \begin{cases} 1 + \cos(2\pi\xi), -\frac{1}{2} \leq \xi \leq \frac{1}{2} \\ 0, \qquad \text{otherwise.} \end{cases}$$

The proper choice of ϵ is determined experimentally, i.e., it is taken to be so small that any further diminution would render u_{avg} oscillatory.

The graphs of u_{avg} are also included in Figures 2a, 2b, and 2c. In the smooth regions they are indistinguishable from the graphs of u itself. The graphs of u_{avg} corresponding the three distinct sets of $\Delta\xi$, Δt are indistinguishable from each other; this is a strong indication of weak convergence, and u_{avg} may be regarded as an excellent approximation to the weak limit $\bar{u}$.

Figures 3a, 3b, and 3c present the graphs of u at $t = 0.7$ corresponding to the same values $\Delta\xi$ as in Figure 2c but with smaller values of Δt. The three graphs look very much alike: the same smooth region, roughly the same number of peaks in the oscillatory region, and the graphs of u_{avg} indistinguishable from each other. Since the value of $\lambda = \Delta t/\Delta\xi$ for the four graphs is 0.2, 0.1, 0.05, and 0.025, respectively, we are justified to conjecture that the weak limit is independent of λ.

Figure 4 displays as function of ξ the graph of u_{avg} from Figure 3a and the graph of u_{NR} also at $t = 0.7$ computed using the von Neumann-Richtmyer artificial viscosity, using the same values of $\Delta\xi$ and λ that have been used to compute u_{avg}. The two graphs differ from each other, but not by much.

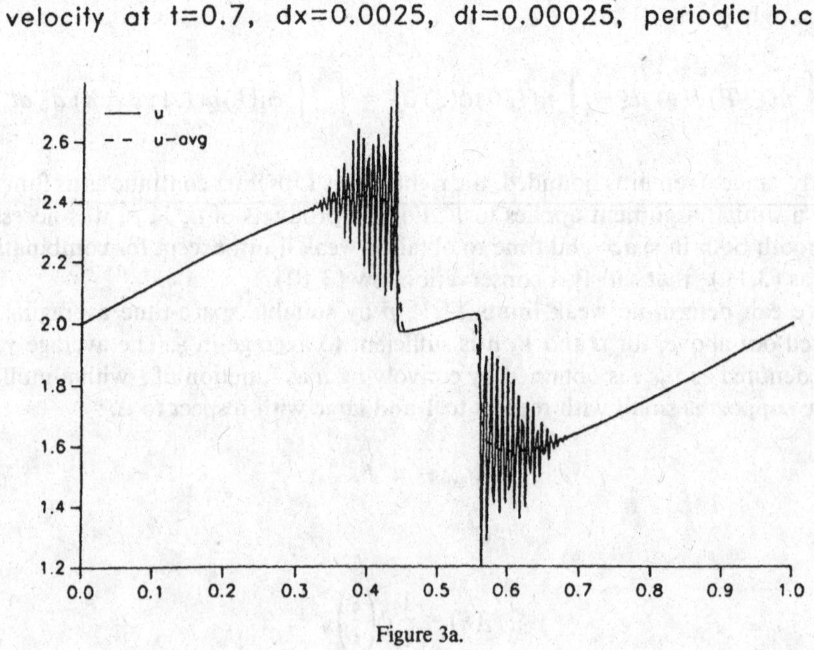

Figure 3a.

velocity at=0.7, dx=0.0025, dt=0.000125, periodic b.c.

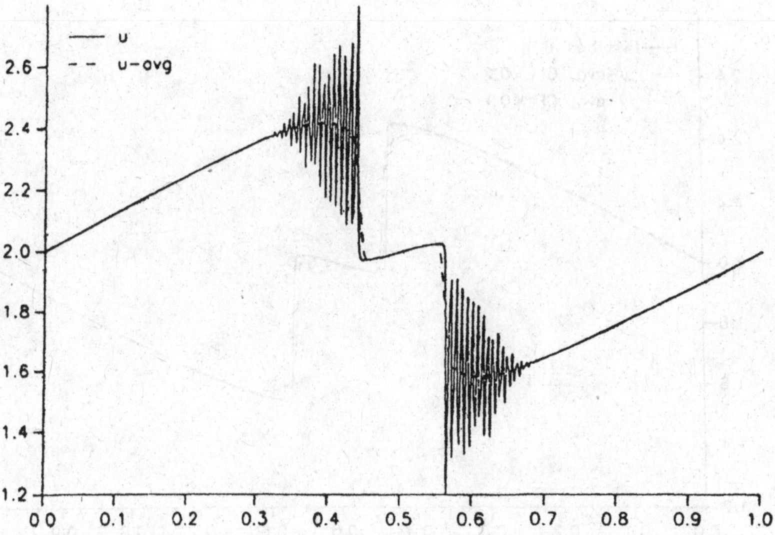

Figure 3b.

velocity at t=0.7, dx=0.0025, dt=0.0000625, periodic b.c.

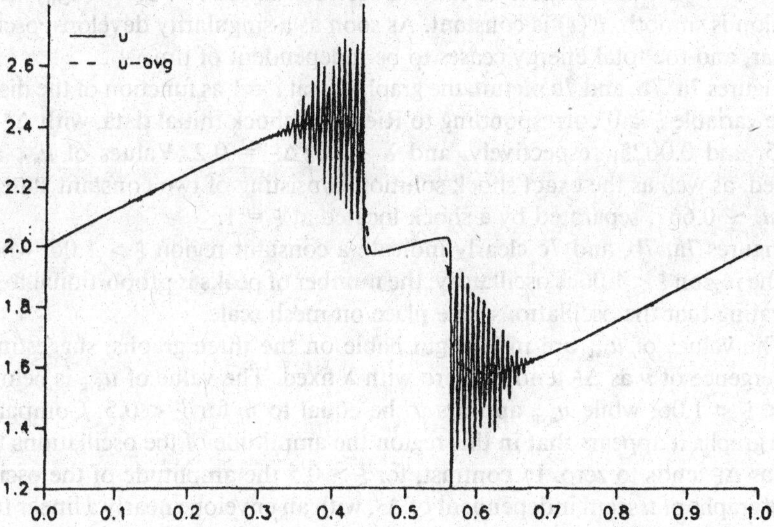

Figure 3c.

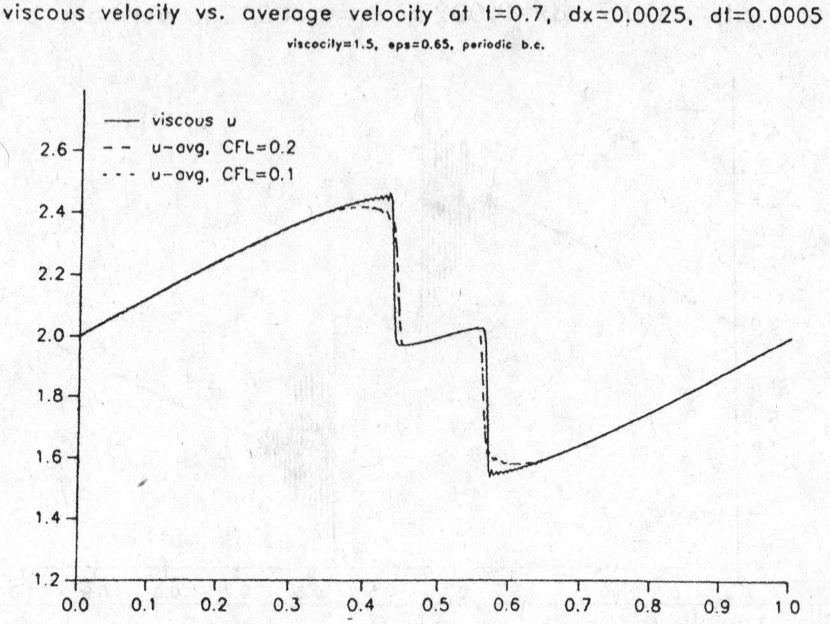

Figure 4.

Figure 5 pictures the propagation in space-time of the boundaries of the oscillatory regions.

Figure 6 pictures to total energy E, internal plus kinetic computed as $e(\bar{p}, \bar{V}) + \frac{1}{2}\bar{u}^2$, as function of time. Observe that for $t < 0.4 \sim t_{\text{crit}}$, where the solution is smooth, $E(t)$ is constant. As soon as a singularity develops, oscillations appear, and the total energy ceases to be independent of time.

Figures 7a, 7b, and 7c picture the graph of u at $t = 1$ as function of the discretized space variable $\xi \geqq 0$ corresponding to Riemann shock initial data, with $\Delta\xi = 0.01$, 0.005, and 0.0025, respectively, and $\lambda = \Delta t/\Delta\xi = 0.2$. Values of u_{avg} are also plotted, as well as the exact shock solution consisting of two constant states $u_l = 1$ and $u_r \sim 0.667$, separated by a shock located at $\xi = 1$.

Figures 7a, 7b, and 7c clearly indicate a constant region $\xi > 1.06$, where $u = u_r$. The region $\xi < 1.06$ is oscillatory; the number of peaks is proportional to $(\Delta\xi)^{-1}$, indicating that the oscillations take place on mesh scale.

The values of u_{avg} are indistinguishable on the three graphs, suggesting weak convergence of u as $\Delta\xi$ tends to zero with λ fixed. The value of u_{avg} is below u_l for $0.5 < \xi < 1.06$, while u_{avg} appears to be equal to u_l for $\xi < 0.5$. Comparing the three graphs it appears that in this region the amplitude of the oscillations tends to zero as $\Delta\xi$ tends to zero. In contrast, for $\xi > 0.5$ the amplitude of the oscillations of the graphs of u seem independent of $\Delta\xi$, with an envelope nearly a linear function of ξ. This gives the graph 7c of u, when turned by 90°, the appearance of a martini glass as viewed by one who has had one too many.

Figures 8a and 8b picture u at time $t = 1$ for $\Delta\xi = 0.005$, $\Delta\xi = 0.0025$, and

Inner boundary of oscillations, dx=0.0025, 0.00125, dt=dx/2

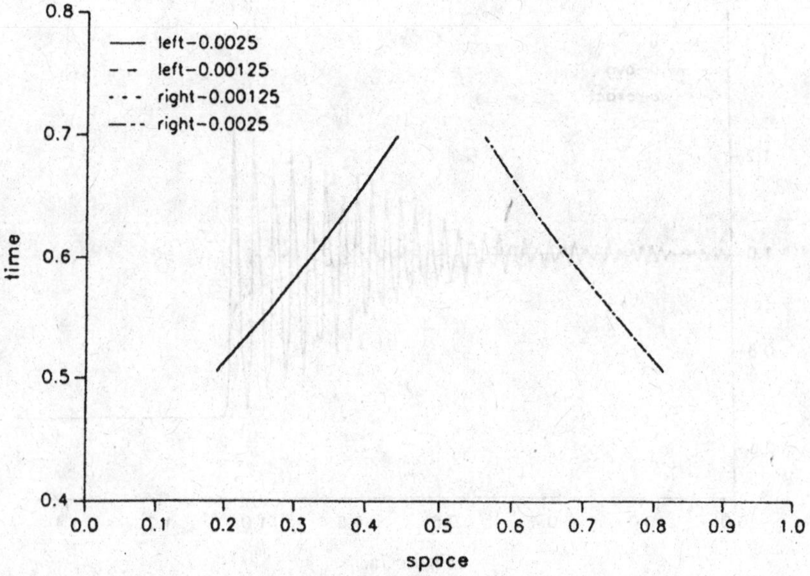

Figure 5.

total average numerical energy, dx=0.005, dt=0.001, eps=0.7

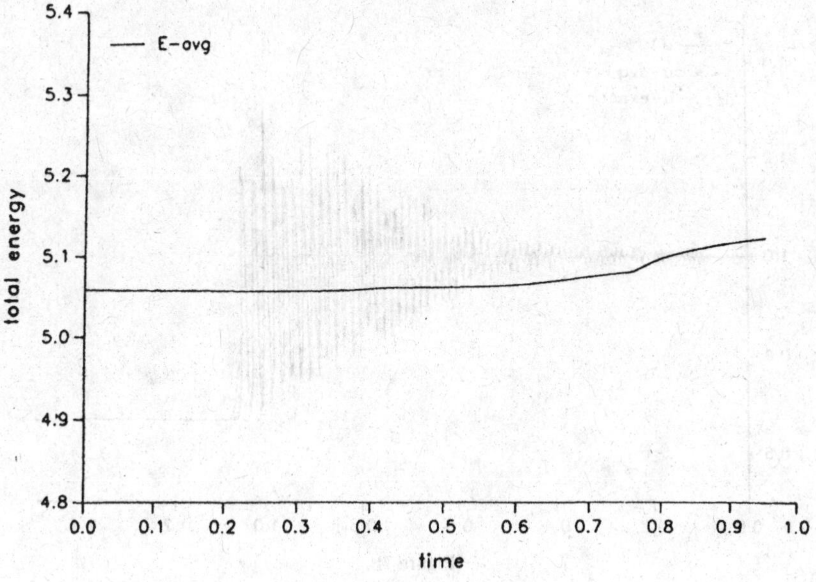

Figure 6.

velocity for shock Riemann data, t=1, dx=0.01, dt=0.002

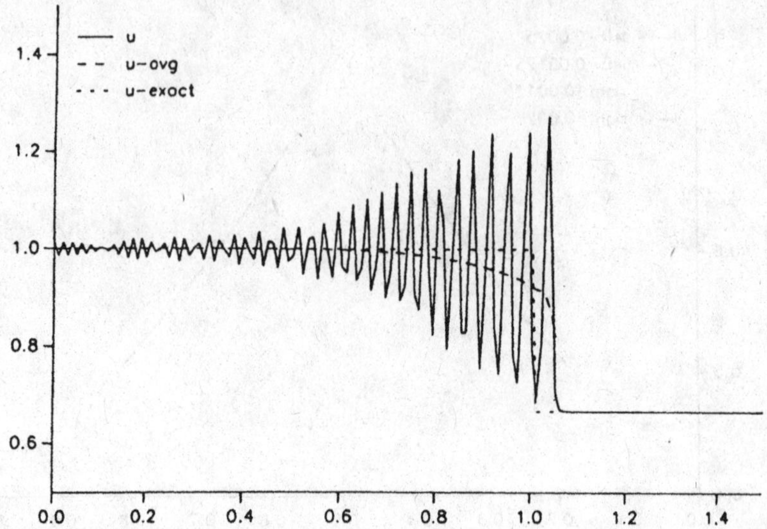

Figure 7a.

velocity for shock Riemann data, t=1, dx=0.005, dt=0.001

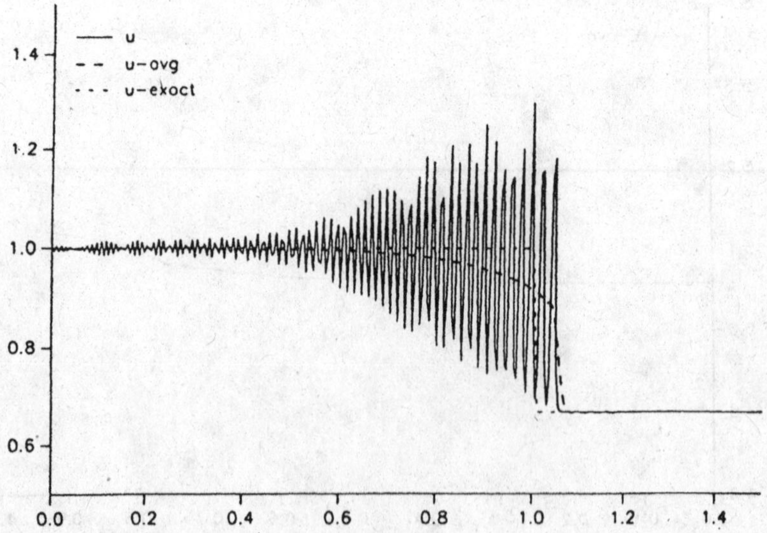

Figure 7b.

velocity for shock Riemann data, t=1, dx=0.0025, dt=0.0005

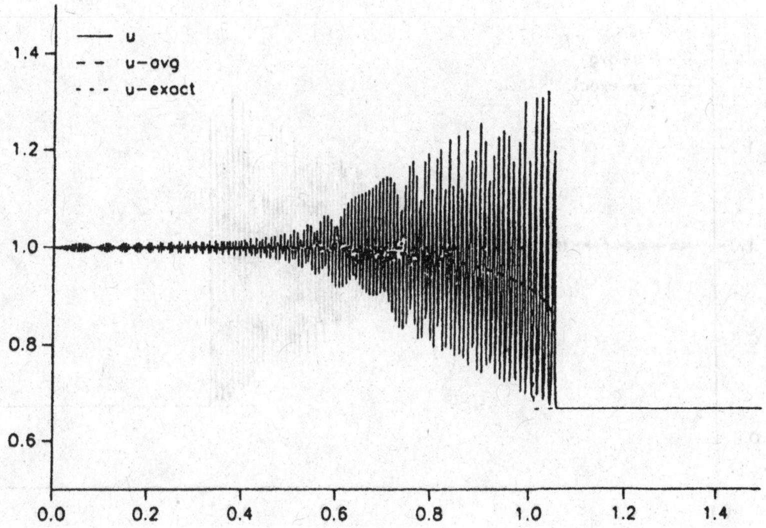

Figure 7c.

velocity for shock Riemann data, t=1, dx=0.005, dt=0.0005

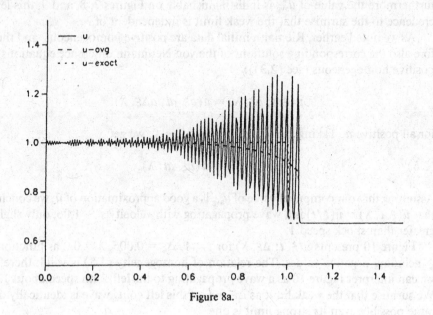

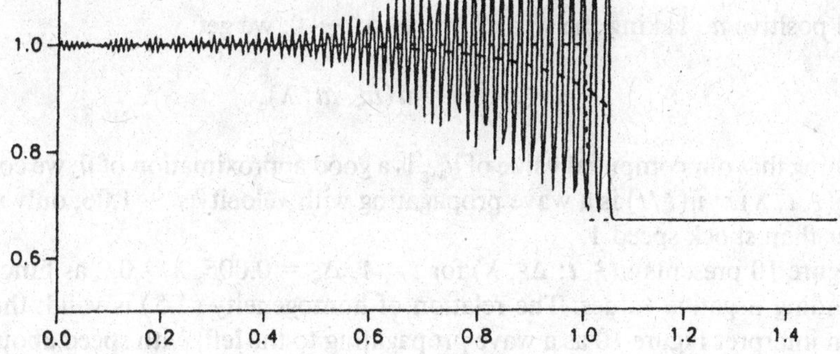

Figure 8a.

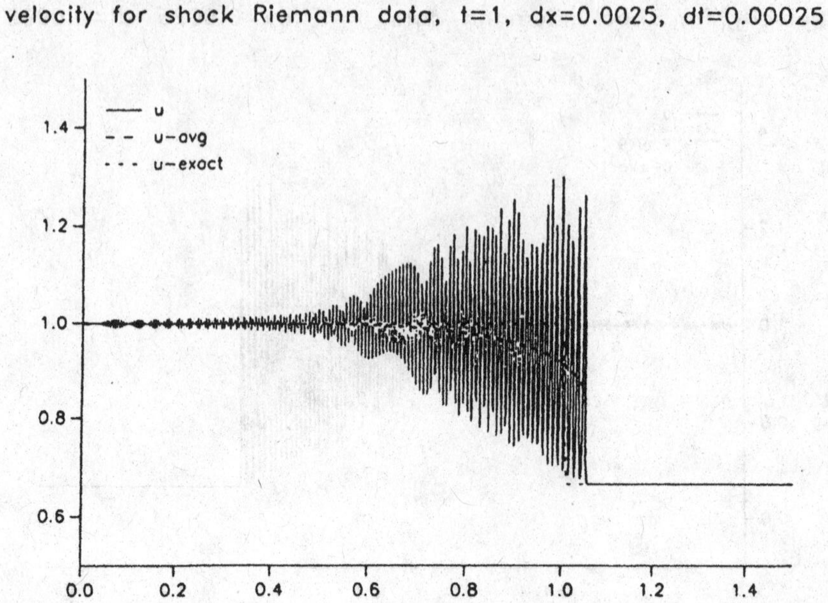

Figure 8b.

$\lambda = 0.1$. Figure 9 pictures u at time $t = 1$ for $\Delta\xi = 0.0025$, $\lambda = 0.4$. Although the details of the oscillations are different, their scale seems to depend only on $\Delta\xi$; furthermore the value of u_{avg} is indistinguishable on Figures 7, 8, and 9; this lends credence to the surmise that the weak limit is independent of λ.

As we noted earlier, Riemann initial data are positive homogeneous, and therefore also the corresponding solutions of the von Neumann difference equations are positive homogeneous (see (2.3)):

$$u(\xi, t; \Delta\xi, \lambda) \equiv u(a\xi, at; a\Delta\xi, \lambda)$$

for all positive a. Taking the weak limit as $\Delta\xi \to 0$, we get

$$\bar{u}(\xi, t; \lambda) = \bar{u}(a\xi, at; \lambda). \tag{2.6}$$

Assuming that our computed value of u_{avg} is a good approximation of $\bar{u}$, we conclude that $\bar{u}(\xi, t, \lambda) = \bar{u}(\xi/t)$ is a wave propagating with velocity $s \sim 1.06$, only slightly greater than shock speed 1.

Figure 10 presents $u(\xi, t; \Delta\xi, \lambda)$ for $t = 1$, $\Delta\xi = 0.005$, $\lambda = 0.2$ as function of ξ including negative values. The relation of homogeneity (2.5) is valid; therefore we can interpret Figure 10 as a wave propagating to the left, with speed about 1.25. We surmise that the weak limit as $\Delta\xi \to 0$ of this left going wave is identically one; quite possibly even its strong limit is one.

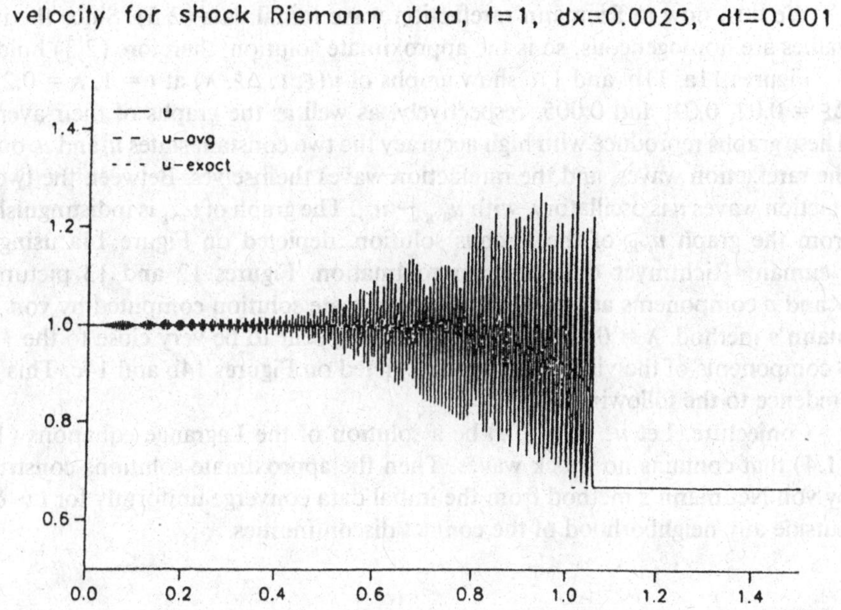

Figure 9.

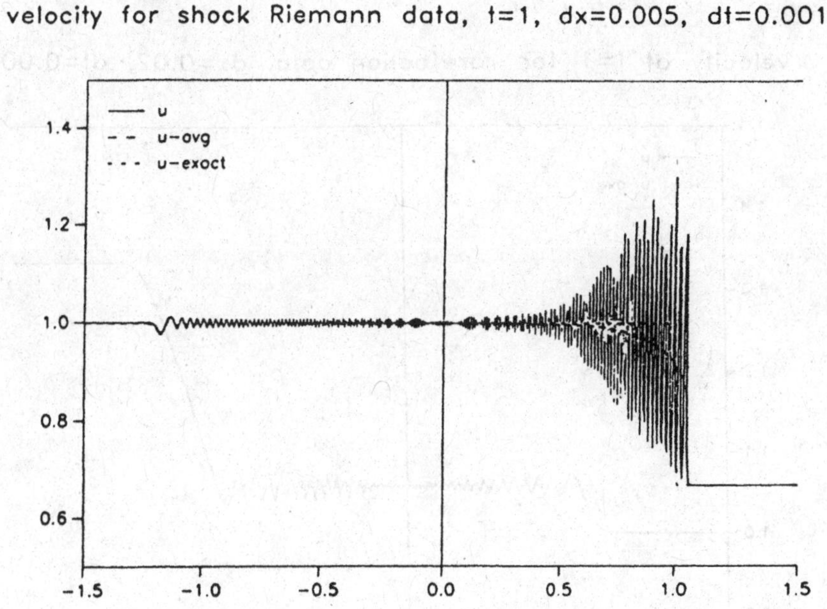

Figure 10.

We turn now to Riemann rarefaction wave initial data (2.5). Since the initial values are homogeneous, so is the approximate solution; therefore (2.3) holds.

Figures 11a, 11b, and 11c show graphs of $u(\xi, t; \Delta\xi, \lambda)$ at $t = 1$, $\lambda = 0.2$, and $\Delta\xi = 0.02$, 0.01, and 0.005, respectively, as well as the graphs of their averages. These graphs reproduce with high accuracy the two constant states u_l and u_r outside the rarefaction waves, and the rarefaction waves themselves. Between the two rarefaction waves u is oscillatory, with $u_{\text{avg}} \sim u_m$. The graph of u_{avg} is indistinguishable from the graph u_{NR} of the viscous solution, depicted on Figure 14a using von Neumann-Richtmyer difference approximation. Figures 12 and 13 picture the V and p components at $t = 1$ of the approximate solution computed by von Neumann's method, $\lambda = 0.2$, $\Delta\xi = 0.005$. They appear to be very close to the V and p components of the viscous solution depicted on Figures 14b and 14c. This gives credence to the following:

Conjecture. Let u, V, $p(\xi, t)$ be a solution of the Lagrange equations (1.1)–(1.4) that contains no shock waves. Then the approximate solutions constructed by von Neumann's method from the initial data converge uniformly for $t > \delta$ and outside any neighborhood of the contact discontinuities.

3. Analysis of the Weak Limit of von Neumann's Scheme

We assume at the outset that the solutions u, V, $p(\xi, t; \Delta\xi, \lambda)$ constructed by von Neumann's method converge weakly as $\Delta\xi$ tends to zero to limits that we

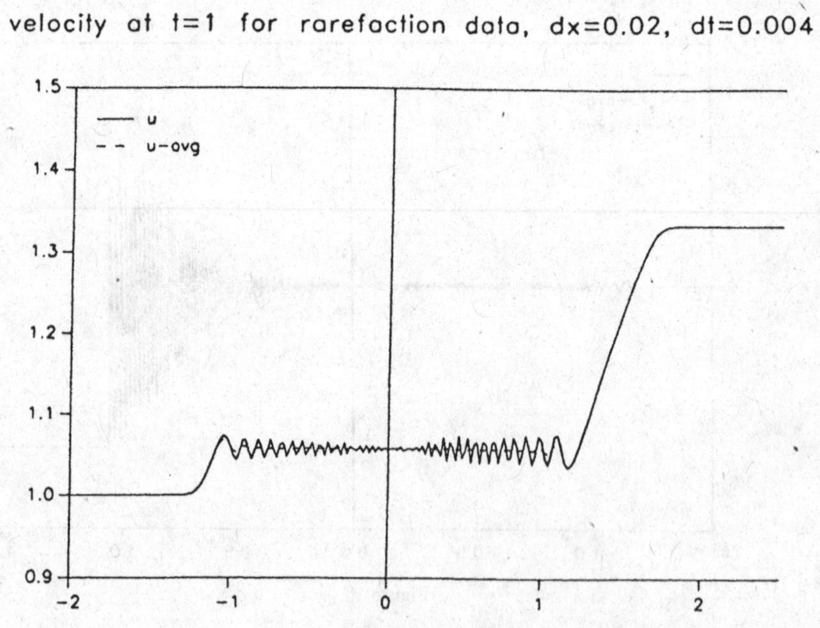

Figure 11a.

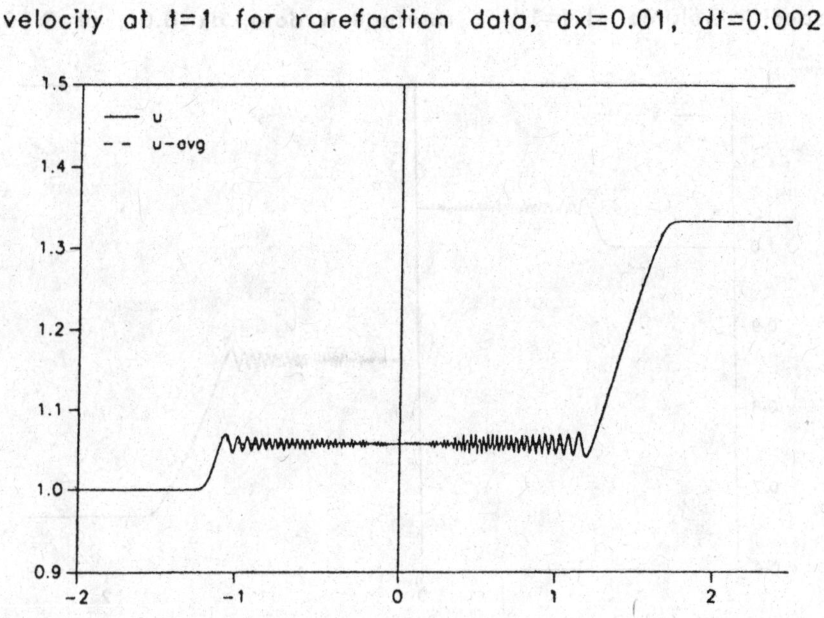

Figure 11b.

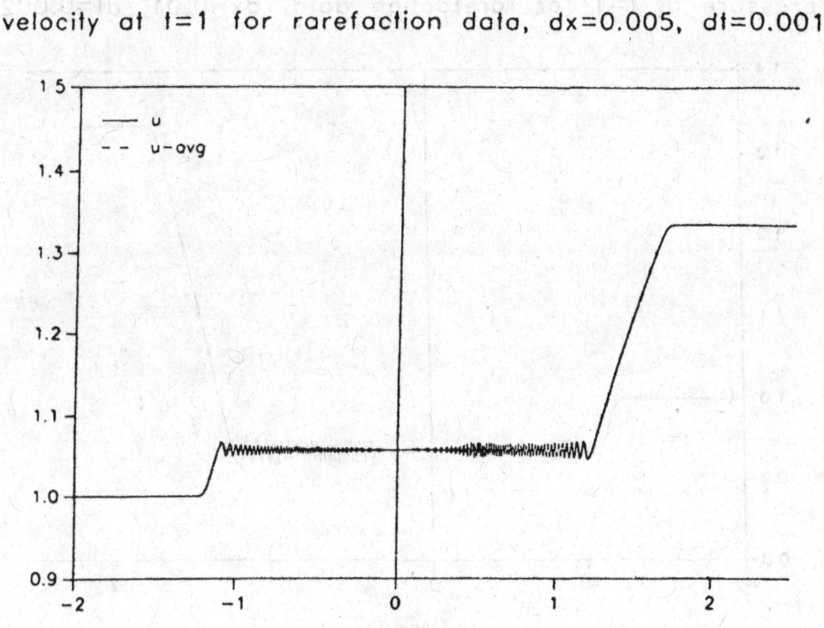

Figure 11c.

specific volume at t=1 for rarefaction data, dx=0.01, dt=0.002

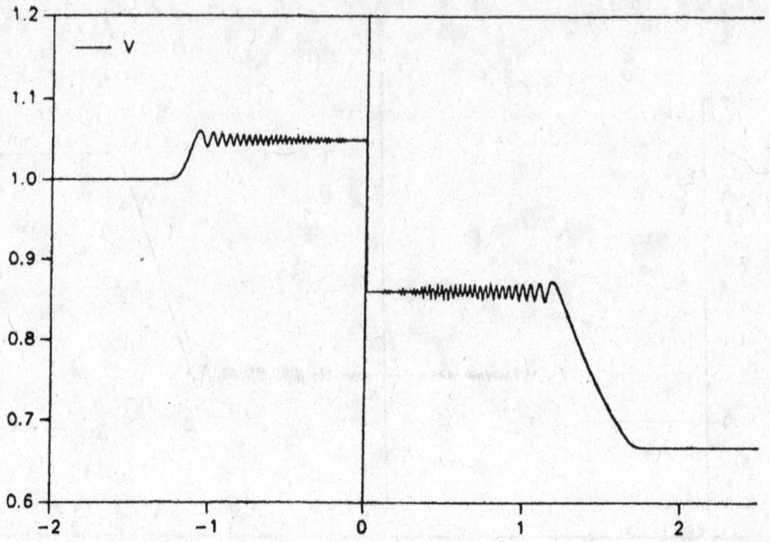

Figure 12.

pressure at t=1 for rarefaction data, dx=0.01, dt=0.002

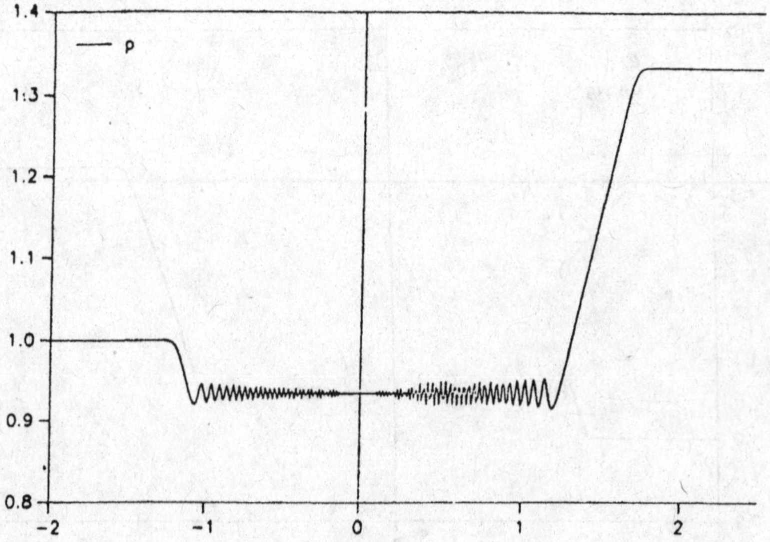

Figure 13.

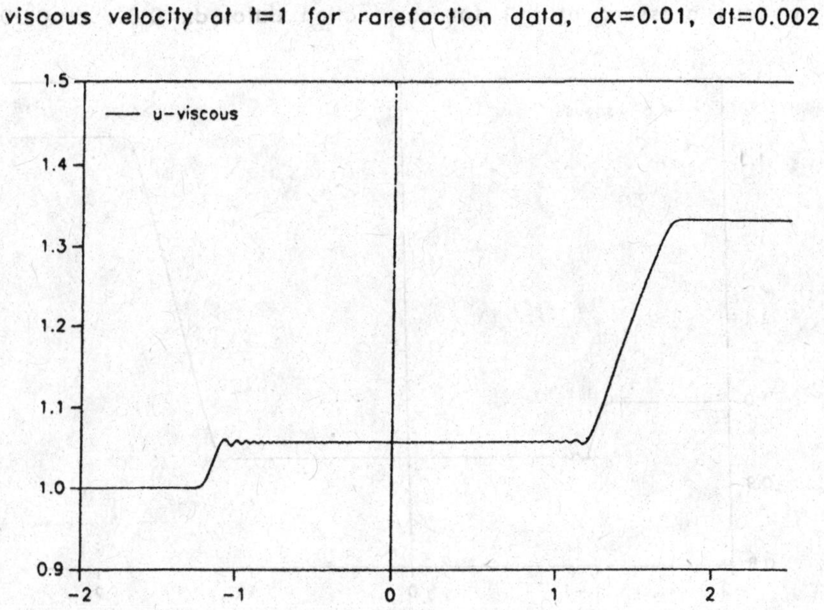

Figure 14a.

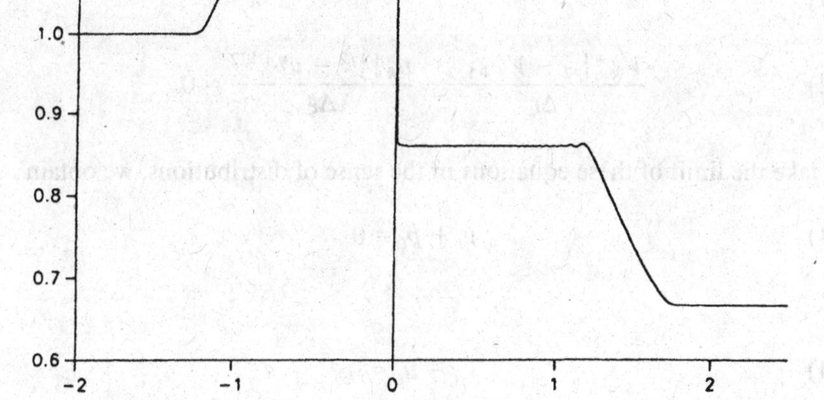

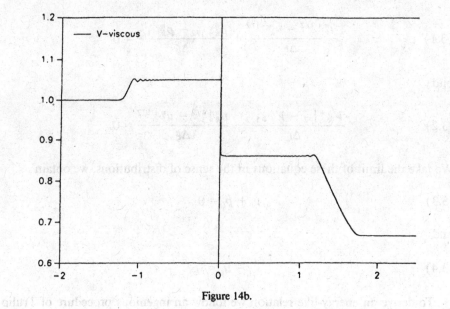

Figure 14b.

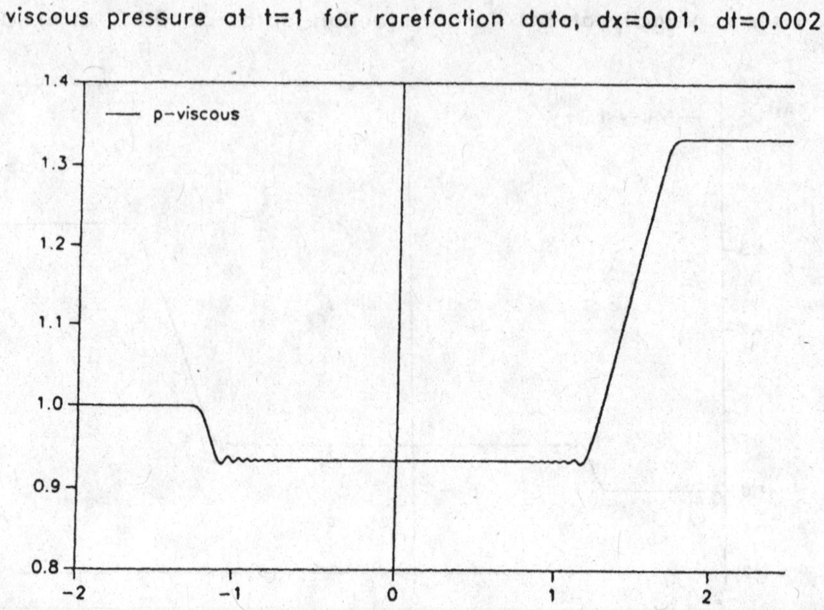

Figure 14c.

denote as $\bar{u}$, $\bar{V}$, $p(\xi, t)$. We do not indicate explicitly the possible dependence of the weak limit on λ.

Dividing equations (1.5) and (1.6) by Δt gives

$$\frac{u_k^{n+1/2} - u_k^{n-1/2}}{\Delta t} + \frac{p_{k+1/2}^n - p_{k-1/2}^n}{\Delta \xi} = 0 \tag{3.1}$$

and

$$\frac{V_{k+1/2}^{n+1} - V_{k+1/2}^n}{\Delta t} - \frac{u_{k+1}^{n+1/2} - u_k^{n+1/2}}{\Delta \xi} = 0. \tag{3.2}$$

We take the limit of these equations in the sense of distributions; we obtain

$$\bar{u}_t + \bar{p}_\xi = 0, \tag{3.3}$$

and

$$\bar{V}_t - \bar{u}_\xi = 0. \tag{3.4}$$

To derive an energy-like relation we follow an ingenious procedure of Trulio and Trigger; see [17]. Add equations $(3.1)_n$ and $(3.1)_{n+1}$ and divide by 2:

(3.5) $$\frac{u_k^{n+3/2}-u_k^{n-1/2}}{2\Delta t}+\frac{p_{k+1/2}^{n+1/2}-p_{k-1/2}^{n+1/2}}{\Delta\xi}=0,$$

where $p_k^{n+1/2}$ is defined in (1.8) as

(3.6) $$p_{k+1/2}^{n+1/2}=\tfrac{1}{2}p_{k+1/2}^{n+1}+p_{k+1/2}^{n}.$$

Multiply (3.5) by $u_k^{n+1/2}$; using the abbreviation

(3.7) $$K_k^n=\tfrac{1}{2}u_k^{n+1/2}u_k^{n-1/2},$$

we can write the resulting relation as

(3.8) $$\frac{K_k^{n+1}-K_k^n}{\Delta t}+u_k^{n+1/2}\frac{(p_{k+1/2}^{n+1/2}-p_{k-1/2}^{n+1/2})}{\Delta\xi}=0.$$

Next we divide equation (1.7) by Δt:

(3.9) $$\frac{e_{k+1/2}^{n+1}-e_{k+1/2}^{n}}{\Delta t}+p_{k+1/2}^{n+1/2}\frac{(u_{k+1}^{n+1/2}-u_k^{n+1/2})}{\Delta\xi}=0.$$

We add (3.8) and (3.9); the result can be written in the conservation form

$(3.10)_+$ $$\frac{E_k^{n+1}-E_k^n}{\Delta t}+\frac{(p_{k+1/2}u_{k+1}-p_{k-1/2}u_k)^{n+1/2}}{\Delta\xi}=0,$$

where

$(3.11)_+$ $$E_k^n=e_{k+1/2}^n+K_k^n.$$

Or we could replace k by $k-1$ in (3.9) and then add it to (3.8); the result is another, similar equation in conservation form

$(3.10)_-$ $$\frac{F_k^{n+1}-F_k^n}{\Delta t}+\frac{(p_{k+1/2}u_k-p_{k-1/2}u_{k-1})^{n+1/2}}{\Delta\xi}=0,$$

where

$(3.11)_-$ $$F_k^n=e_{k-1/2}^n+K_k^n.$$

We remark that the relation derived by Trulio and Trigger is the average of $(3.10)_+$ and $(3.10)_-$.

It is not possible to derive a difference analogue of conservation of entropy. Nevertheless, numerical experiments indicate that if entropy is constant initially, then it converges uniformly to that constant for $t>0$ as $\Delta\xi$, $\Delta t\to 0$; see Figure 19.

Suppose now—and this is born out by our numerical experiments—that not only u, V, and p but also $e(p, V)$, K, $p_{k+1/2}u_k$, and $p_{k-1/2}u_k$ have weak limits as $\Delta\xi$ tends to zero. We denote these weak limits as $\bar{e}$, $\bar{K}$, $\overline{p_+u}$, and $\overline{p_-u}$. Taking the limits of $(3.10)_\pm$ in the sense of distributions, we obtain two conservation laws:

$$(3.12)_+ \qquad (\bar{e} + \bar{K})_t + (\overline{p_-u})_\xi = 0,$$

and

$$(3.12)_- \qquad (\bar{e} + \bar{K})_t + (\overline{p_+u})_\xi = 0.$$

From $(3.12)_+$ and $(3.12)_-$ we obtain the following result:

$$(3.13) \qquad \overline{p_-u} = \overline{p_+u}\,.$$

This can be also deduced directly from equation (1.7).

Denoting the common value of (3.13) as $\overline{pu}$, we write $(3.12)_\pm$ as

$$(3.12) \qquad (\bar{e} + \bar{K})_t + (\overline{pu})_\xi = 0.$$

Equation (3.12) looks like the law of conservation of energy, but it is not. That law ought to say that

$$(3.14) \qquad (e(\bar{V}, \bar{p}) + \tfrac{1}{2}\bar{u}^2)_t + (\bar{p}\bar{u})_\xi = 0.$$

There is no theoretical reason to expect (3.12) and (3.14) to be identical, and numerical evidence indicates that

$$(3.15) \qquad \bar{e} + \bar{K} \neq e(\bar{V}, \bar{p}) + \tfrac{1}{2}\bar{u}^2,$$

and

$$(3.15)' \qquad \overline{pu} \neq \bar{p}\bar{u}.$$

In fact, as already pointed out in Section 2, the quantity

$$\int (e(\bar{V}, \bar{p}) + \tfrac{1}{2}\bar{u}^2)\, d\xi$$

is not independent of time for $t > t_{\text{crit}}$, whereas

$$\int (\bar{e} + \bar{K})\, d\xi$$

is.

We turn now to the question: what are the equations governing the time evolution of $\bar{u}$, $\bar{V}$, and $\bar{p}$? Equations (3.3) and (3.4) describe the time evolution of $\bar{u}$

and $\bar{V}$. Equation (3.12) does not furnish an equation for $\bar{p}$, because the quantities appearing in (3.12), $\bar{e}+\bar{K}$ and $\overline{up}$, cannot be expressed as functions of $\bar{u}$, $\bar{V}$, $\bar{p}$, nor can we expect to express $\bar{p}_t$ in terms of $\bar{u}$, $\bar{V}$, $\bar{p}$ and their ξ derivatives. It seems necessary to enlarge very much the set of functions for which we expect to derive an autonomous system of differential equations; we have to include in addition to $\bar{u}$, $\bar{V}$, and $\bar{p}$ all weak limits of polynomials of u, V, p and their translations. To be more precise, consider all polynomials q in the number of variables indicated below, and define

$$q_k^n = q(u_j^{m+1/2}, V_{i+1/2}^l, p_{g+1/2}^h) \tag{3.16}$$

where

$$|m-n|,\ |l-n|,\ |h-n|,\ |j-k|,\ |i-k|,\ |g-k| < d.$$

We take $u_k^{n+1/2}$, $V_{k+1/2}^n$, $p_{k+1/2}^n$ to be a solution of (1.5)–(1.7). We denote by $q(\xi, t; \Delta\xi, \lambda)$ the extension of the lattice function q_k^n to all real ξ and t.

Numerical evidence indicates that as $\Delta\xi$ tends to zero, q tends weakly to a limit; we denote this limit by $\bar{q}(\xi, t; \lambda)$. Note that (3.13) shows that two distinct polynomials q_1 and q_2 might very well have the same weak limit.

We define the *states* of our system to be the collection of functions $\{\bar{q}(\xi, t)\}$ at any particular time t for all initial data u_0, V_0, p_0. This is a useful concept only if the set of functions $\bar{q}$ are *complete* in the sense that if for two different sets of initial data $u^{(1)}$, $V^{(1)}$, $p^{(1)}$ and $u^{(2)}$, $V^{(2)}$, $p^{(2)}$ all the states $\{q^{(1)}(\xi, T)\}$ and $\{q^{(2)}(\xi, T)\}$ are identical at some time T, then they are identical at all future times $t > T$.

If the set of functions we have used to describe the state of our system are complete, then the time evolution of the state is a flow that forms a one-parameter semigroup of nonlinear operators. We can then form, or at least speculate about, the infinitesimal generator G of the semigroup, for clearly the time derivative of the flow at time t, if it exists, depends only on the state of the flow at t:

$$\frac{d}{dt} Q = G(Q), \tag{3.17}$$

where Q denotes the collection of all q. To be sure, even if the principle of completeness is true for the class of weak limits described above, we know very little about the generator G, because we know so little about the flow itself.

Another intriguing question is the seeming independence of the weak limits of λ. If we could give a constructive description of the generator G, and if we could show that it is independent of λ, and if we could prove uniqueness for the initial value problem for the resulting equation (3.17), then independence of λ would follow.

4. Semidiscrete Version of von Neumann's Scheme

In the case of $\lambda \to 0$, von Neumann's scheme reduces to the following semidiscrete system:

$$\frac{d}{dt} u_k = -\frac{1}{\Delta}(p_{k+1/2} - p_{k-1/2}), \tag{4.1}$$

$$\frac{d}{dt} V_{k+1/2} = \frac{1}{\Delta}(u_{k+1} - u_k), \tag{4.2}$$

$$\frac{d}{dt} e_{k+1/2} = -\frac{1}{\Delta} p_{k+1/2}(u_{k+1} - u_k), \tag{4.3}$$

where e, V, and p satisfy an equation of state

$$e_{k+1/2} = f(p_{k+1/2}, V_{k+1/2}). \tag{4.4}$$

Figures 15a and 15b depict the u component of the solution of the semidiscrete system (4.1)–(4.4) with a γ-law gas, whose initial data are periodic. The ODE was solved by a fourth-order Runge-Kutta method with $\Delta t = 0.002$ and 0.001 respectively. The graphs of $u(\xi, t; \Delta)$ are shown at $t = 0.7$ as functions of the discretized space variable ξ, for $\Delta = 0.01$ and $\Delta = 0.005$ respectively; the averages u_{avg} are also shown. Observe that in every respect—location of oscillatory region, average value, frequency, and amplitude of oscillations—these graphs are very similar to, well-nigh indistinguishable from, the graphs of $u(\xi, t; \Delta, \lambda)$ depicted in Figures 2a, 2b, 3a, 3b, and 3c respectively. This is evidence for the surmise that as Δ tends to zero the solutions u, V, $p(\xi, t; \Delta)$ of (4.1)–(4.4) tend weakly to a limit, and that this limit is the same as the weak limit $\bar{u}$ of the solutions of the fully discrete equations for every nonzero value of λ.

Figures 16a and 16b depict the u component of the solution of the semidiscrete equations for the same choice of parameters but with Riemann shock initial data. Again the similarity to graphs of u obtained as solutions of the fully discrete equations with the same initial data, depicted in Figures 7a, 7b, and 8a, is striking; this furnishes further evidence for the weak convergence as Δ tends to zero, and of the independence of λ of the weak limit.

In the semidiscrete case more analysis is possible than in the fully discrete case. Arguing as in the time discrete case, we can show that solutions of the semidiscrete von Neumann scheme converge strongly if the continuous equations admit a smooth solution. We shall show that this is no longer true in a region that contains points where the smooth solution breaks down.

Conservation of energy can be obtained by multiplying (4.1) by u_k and adding the resulting equation to (4.3). This gives

$$\frac{d}{dt}(\tfrac{1}{2}u_k^2 + e_{k+1/2}) = -\frac{1}{\Delta}(p_{k+1/2}u_{k+1} - p_{k-1/2}u_k). \tag{4.5$_+$}$$

semidiscrete velocity at t=0.7, dx=0.01, 4-th order Runge-Kutta

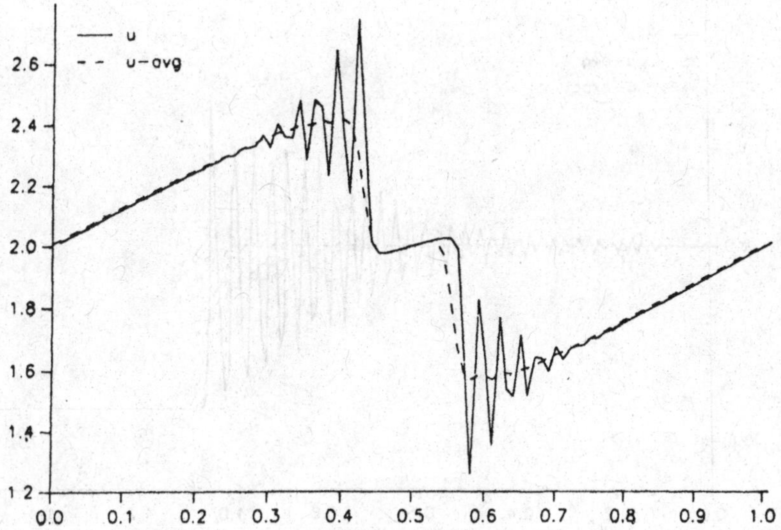

Figure 15a.

semidiscrete velocity at t=0.7, dx=0.005, 4-th order Runge-Kutta

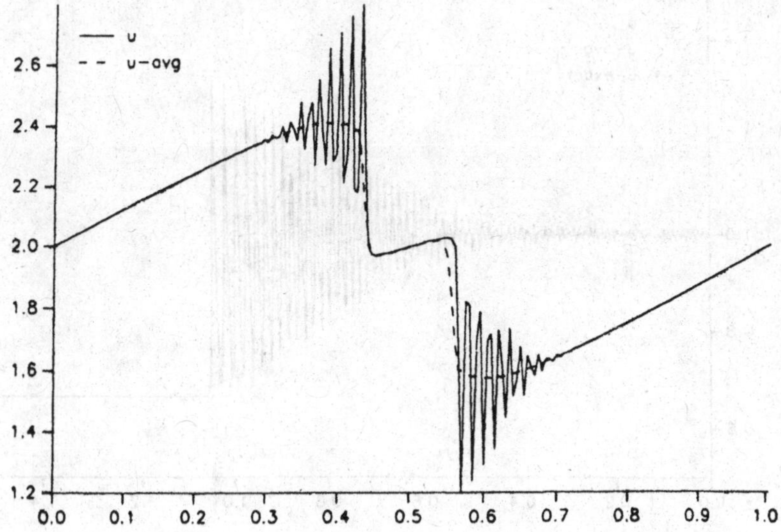

Figure 15b.

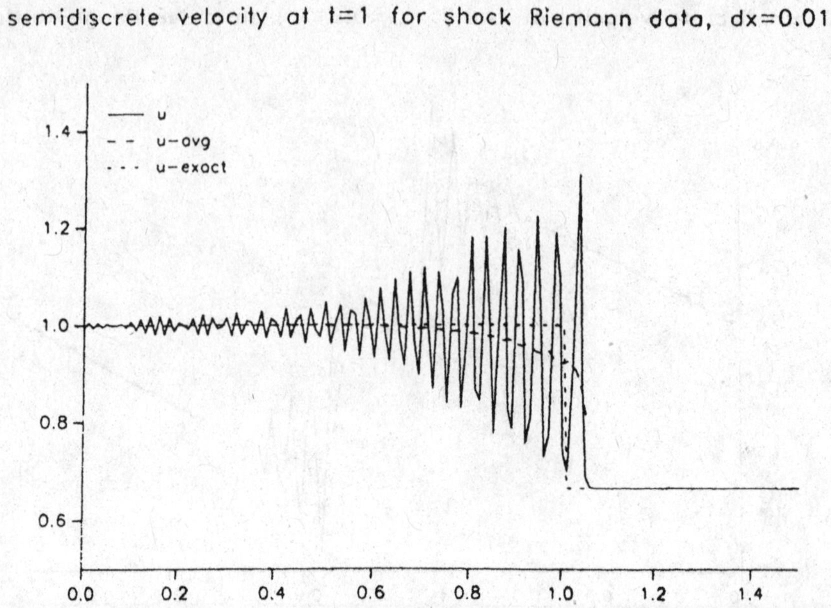

Figure 16a.

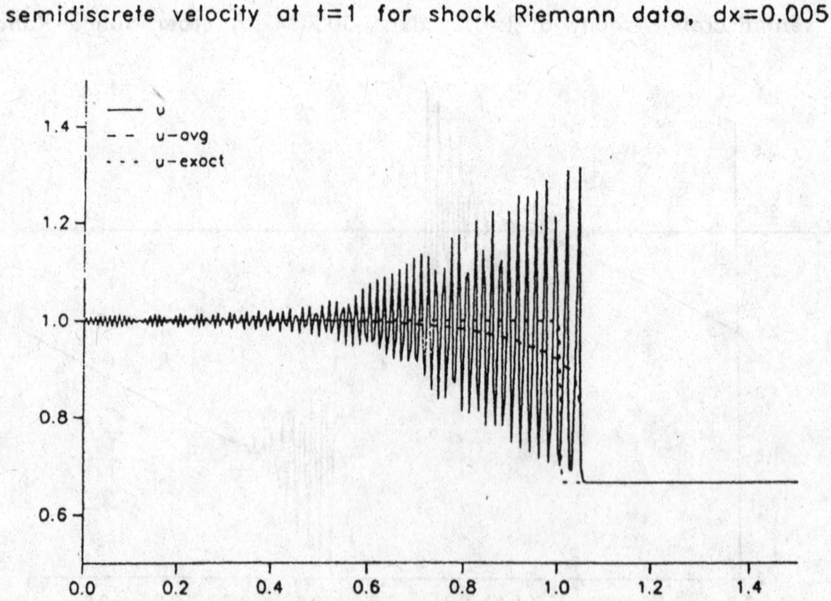

Figure 16b.

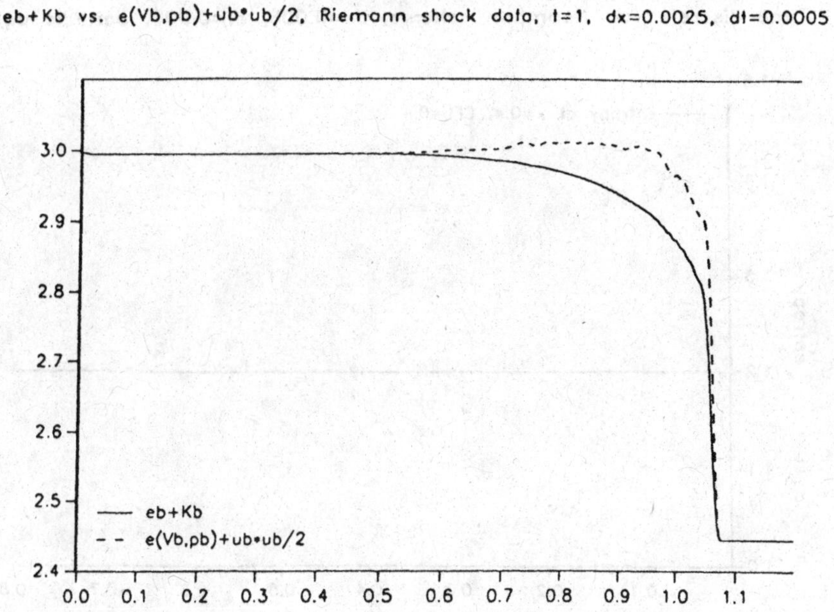

Figure 17.

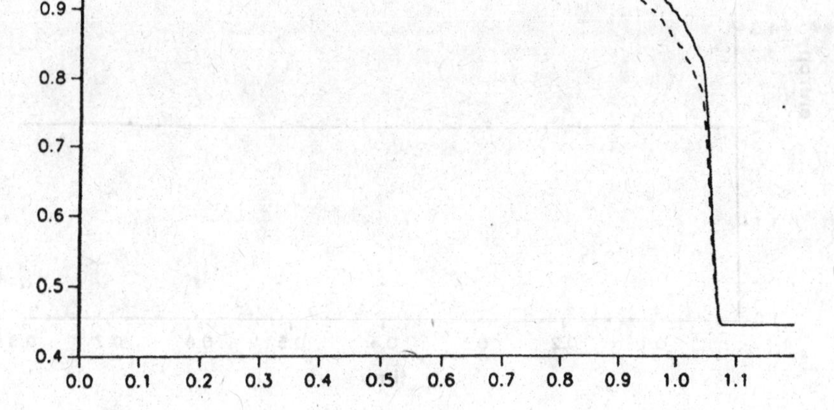

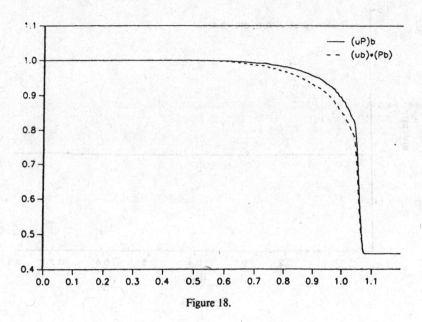

Figure 18.

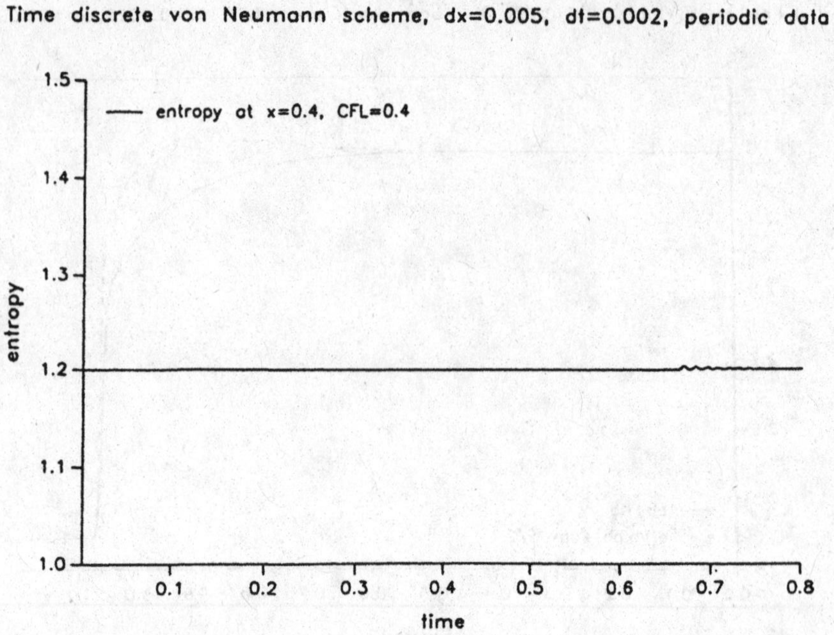

Figure 19a.

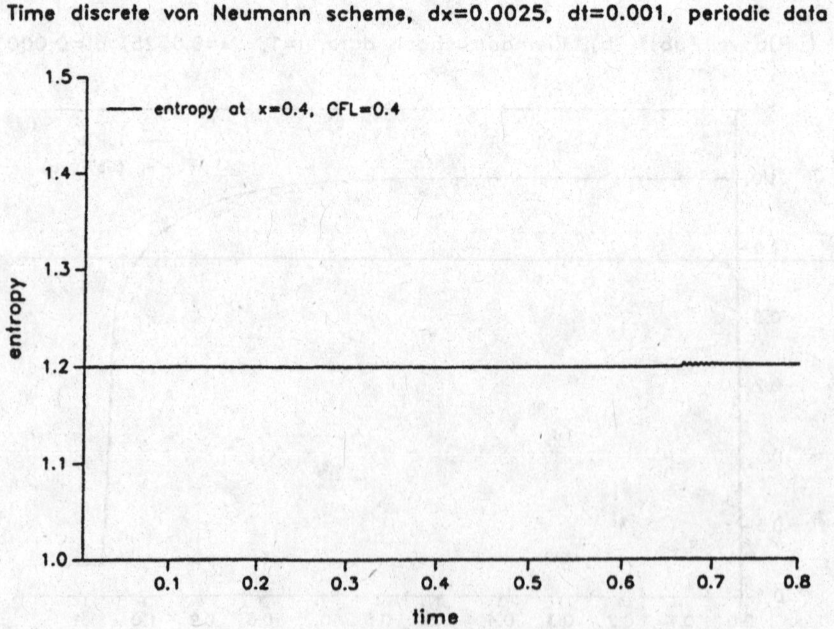

Figure 19b.

This equation is in *conservation form* in the sense that the right side is a perfect difference. Similarly, multiplying (4.1) by u_k and adding the resulting equation to $(4.3)_{k-1}$, we obtain

$$(4.5)_- \qquad \frac{d}{dt}\left(\frac{u_k^2}{2} + e_{k-1/2}\right) = -\frac{1}{\Delta}(p_{k+1/2}u_k - p_{k-1/2}u_{k-1}).$$

Assume—and this is supported by numerical evidence—that u_k^2, $e_{k+1/2}$, $p_{k+1/2}u_{k+1}$, and $p_{k+1/2}u_k$ have weak limits as $\Delta \to 0$. We deduce from $(4.5)_\pm$ that

$$(4.6)_+ \qquad (\tfrac{1}{2}\overline{u^2} + \bar{e})_t + (\overline{pu_+})_\xi = 0,$$

and

$$(4.6)_- \qquad (\tfrac{1}{2}\overline{u^2} + \bar{e})_t + (\overline{pu_-})_\xi = 0.$$

By comparing $(4.6)_+$ with $(4.6)_-$, we conclude that the weak limits of $p_{k+1/2}u_{k+1}$ and $p_{k+1/2}u_k$ are equal.

On the other hand, multiplying (4.2) by $p_{k+1/2}$ and adding the resulting equation to (4.3) gives

$$(4.7) \qquad p_{k+1/2}\frac{d}{dt}V_{k+1/2} + \frac{d}{dt}e_{k+1/2} = 0.$$

Substituting (4.4) into (4.7) yields a differential equation for $p_{k+1/2}$ and $V_{k+1/2}$,

$$(4.7)' \qquad \frac{d}{dt}S(p_{k+1/2}, V_{k+1/2}) = 0,$$

where S is entropy. Equation (4.7)′ is conservation of entropy.

We claim that in regions where the smooth solution breaks down, the semidiscrete solutions of (4.1)–(4.4) do not converge strongly in the sense of bounded, L^1 convergence. For if they did, then according to an observation of Lax and Wendroff (see [10]), since (4.1), (4.2), (4.5), and (4.7)′ are in conservation form, the strong limit of their solution would satisfy in the sense of distributions the corresponding continuum conservation laws of momentum, mass, energy, and entropy. On the other hand, it is known that in the region where the smooth solution breaks down, no solution exists, even in the sense of distributions, that satisfies all four conservation laws.

From this, we deduce

THEOREM 2. *In a region that contains points where the smooth solution breaks down, any subsequence* $u(\xi, t; \Delta)$, $\Delta \to 0$, *forms a noncompact set in the sense of bounded* L^1 *convergence.*

Proof: If not, we could select a strongly convergent subsequence.

Noncompactness implies that the total variation of $u(\xi, t; \Delta)$ must tend to infinity as Δ tends to zero. Thus Theorem 2 implies the observed oscillatory nature of the solutions of the difference equations.

Suppose now that initially the entropy S is independent of ξ; then by (4.7), S is independent of ξ and t, and p can be expressed as a function of V:

$$p = f(V). \tag{4.8}$$

This is the assumption von Neumann made in his calculations. In this case, the momentum and mass equations (4.1) and (4.2), together with the initial values of u and V, suffice to determine uniquely u and V for all t.

We show now how to transform (4.1) and (4.2) into a more familiar form, the analogue of introducing the Eulerian coordinate as a new unknown function. We recall that the Eulerian coordinate $X = X(\xi, t)$ satisfies

$$X_t = u. \tag{4.9}$$

Setting this into (1.2), we get

$$0 = V_t - X_{t\xi} = (V - X_\xi)_t. \tag{4.10}$$

Consequently if

$$V = X_\xi \tag{4.11}$$

at $t = 0$, then (4.11) holds for all t as well. Setting (4.8) into (1.1), we obtain

$$X_{tt} + p_\xi = 0.$$

Setting (4.11) into the equation of state (4.8), we can rewrite the above as the nonlinear wave equation

$$X_{tt} + f(X_\xi)_\xi = 0. \tag{4.12}$$

The introduction of Eulerian coordinates $X_k(t)$ can be carried out on the semi-discretized level as well. To simplify notation we denote differentiation with respect to t by a dot

$$\cdot = \frac{d}{dt}. \tag{4.13}$$

Then (4.1) and (4.2) can be rewritten as

$$\dot{u}_k = \frac{p_{k-1/2} - p_{k+1/2}}{\Delta}, \tag{4.14}$$

$$\dot{V}_{k+1/2} = \frac{u_{k+1} - u_k}{\Delta}. \tag{4.15}$$

We define $X_k(t)$ by

$$\dot{X}_k = u_k. \tag{4.16}$$

Set (4.16) into (4.15), obtaining,

$$\dot{V}_{k+1/2} = \frac{\dot{X}_{k+1} - \dot{X}_k}{\Delta}; \tag{4.17}$$

if the initial values of X_k are so chosen that

$$V_{k+1/2} = \frac{X_{k+1} - X_k}{\Delta} \tag{4.18}$$

at $t = 0$, then it follows from (4.17) that (4.18) holds for all t as well. In the recursion relation (4.18), $X_0(0)$ may be chosen arbitrarily, say zero.

Set (4.16) into equation (4.14); the result is

$$\ddot{X}_k = \frac{p_{k-1/2} - p_{k+1/2}}{\Delta}.$$

Setting (4.18) into the equation of state (4.8), we can rewrite the above as

$$\ddot{X}_k + \left[f\left(\frac{X_{k+1} - X_k}{\Delta}\right) - f\left(\frac{X_k - X_{k-1}}{\Delta}\right)\right] \Big/ \Delta = 0. \tag{4.19}$$

This can be looked upon as a semidiscrete approximation to the nonlinear wave equation (4.12). It can be given a different interpretation, after an appropriate change of scale:

Set

$$\frac{X_k}{\Delta} = q_k, \tag{4.20}$$

and define a rescaled time s that is related to t by

$$s = \frac{t}{\Delta}; \tag{4.21}$$

then

$$\frac{d}{dt} = \frac{1}{\Delta}\frac{d}{ds}. \tag{4.22}$$

Setting this into (4.19) and denoting d/ds by $'$, we get

$$q_k'' + f(q_{k+1} - q_k) - f(q_k - q_{k-1}) = 0. \tag{4.23}$$

We recognize (4.23) as the equation governing the vibrations of an anharmonic chain of particles with unit mass, linked by identical springs to their two nearest neighbors. Here $q_k(t)$ is the distance at time t of the k-th mass from the origin, and $f(d)$ the negative of the force exerted on each by two particles separated by distance d and linked by a spring. The force $-f$ is a monotonically increasing function of d, assumed to be zero when the two masses are a unit distance apart.

Equation (4.23) is the semidiscrete version of von Neumann's original formulation of his method; see [11].

There is an extensive literature in physics about lattice vibrations; see, e.g., those quoted in [6]. In [6] and [15] Holian and Straub investigated the so-called shock problem for lattices: particles are set in motion by giving them constant velocity directed toward origin:

$$q_k(0) = k, \quad q_k'(0) = \begin{cases} a & k < 0 \\ 0 & k = 0 \\ -a & k > 0; \end{cases} \tag{4.24}$$

here a is some positive quantity.

Using the relations (4.16)–(4.21) we can transform this initial value problem for lattice vibration governed by equation (4.23) into an initial value problem for the semidiscrete approximations (4.14), (4.15) of the isentropic flow equations in Lagrangian coordinates. The initial values corresponding to (4.24) are

$$V_{k+1/2}(0) = 1, \quad u_k(0) = \begin{cases} a & k < 0 \\ 0 & k = 0 \\ -a & k > 0; \end{cases} \tag{4.25}$$

this is the discretized version of the continuum Riemann initial value problem

$$V(\xi, 0) = 1, \quad u(\xi, 0) = \begin{cases} a & \xi < 0 \\ -a & \xi > 0. \end{cases} \tag{4.25$'$}$$

The rescalings (4.20) and (4.21) show that the large time limit $s \to \infty$ for the lattice shock problem (4.23), (4.24) is equivalent with the $\Delta \to 0$ limit for the semidiscrete approximation (4.14), (4.15) for the Riemann shock problem (4.25)$'$, provided that the spring force is identified as pressure.

In their numerical experiments, Holian and Straub solved equation (4.23) by a centered difference scheme in time; so effectively they carried out von Neumann's scheme for isentropic gas. They included in their studies a variety of physically

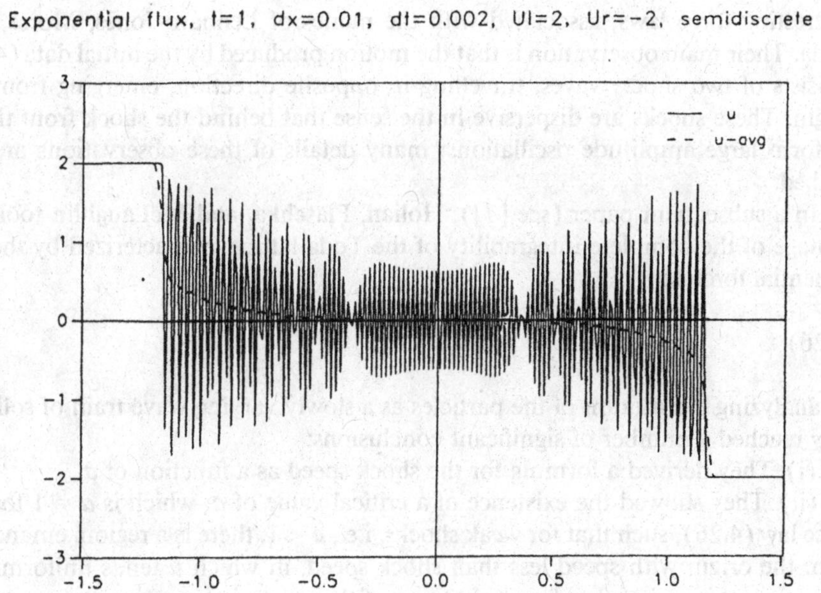

Figure 20a.

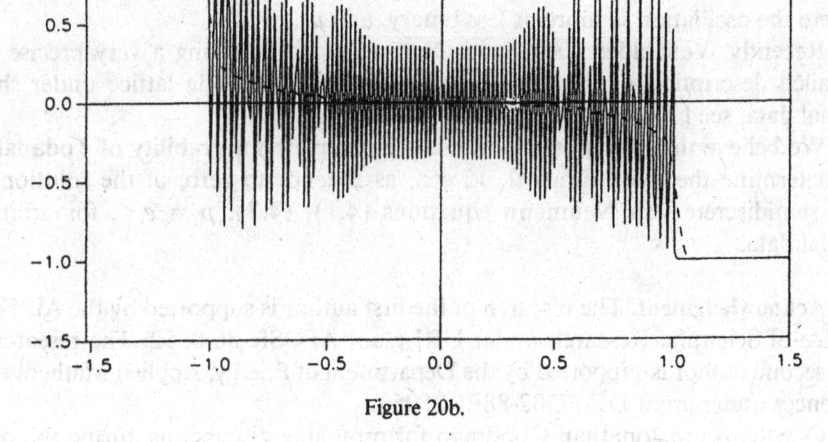

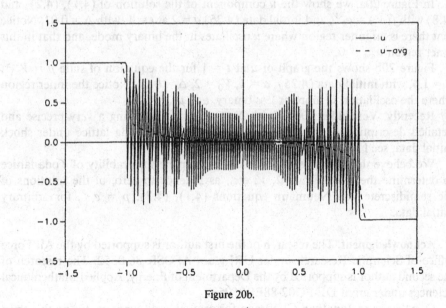

Figure 20b.

interesting force laws, associated with the names of Lennard-Jones, Morse, and Toda. Their main observation is that the motion produced by the initial data (4.24) consists of two shock waves, travelling in opposite direction, emerging from the origin. These shocks are dispersive in the sense that behind the shock front the q_k perform large amplitude oscillations; many details of these observations are described.

In a subsequent paper (see [7]), Holian, Flaschka, and McLaughlin took advantage of the complete integrability of the Toda lattice, characterized by the exponential force law

$$f(V) = e^{-V}. \tag{4.26}$$

By analyzing the motion of the particles as a slowly varying wave train of solitons they reached a number of significant conclusions:

(i) They derived a formula for the shock speed as a function of a.

(ii) They showed the existence of a critical value of a, which is $a = 1$ for the force law (4.26), such that for weak shocks, i.e., $a < 1$, there is a region, emanating from the origin with speed less than shock speed, in which u tends uniformly to zero—it is this region that forms the stem of the martini glass. As a approaches 1, the stem disappears. For $a > 1$, u is oscillatory everywhere behind the shock; there is a region in which the oscillation is in the binary mode, i.e., the displacement of adjacent particles is equal and opposite.

(iii) They observed that the qualitative features of the behavior described in (ii) hold for other anharmonic lattices.

In Figure 20a, we show the u component of the solution of (4.1), (4.2), and (4.8) with $f(V) = e^{-V}$, and initial data (4.25) $a = 2$ at $t = 1$, with $\Delta = 0.01$. Notice that there is an inner region where u oscillates in the binary mode, and that in this inner region $u_{\text{avg}} = 0$.

Figure 20b shows the graph of u at $t = 1$ for the equation of state $p = V^{-\gamma}$, $\gamma = 1.4$, with initial data (4.25), $a = 1$, $V_0 = 2$, $\Delta = 0.01$. Notice the inner region where the oscillation is more or less binary, and $u_{\text{avg}} = 0$.

Recently, Venakides, Deift, and Oba succeeded in giving a very precise and detailed description of the large time behavior of the Toda lattice under shock initial data; see [3].

We believe that it is possible to use the complete integrability of Toda lattice to determine the weak limits $\bar{u}$, $\bar{V}$, etc., as Δ tends to zero, of the solutions of the semidiscrete von Neumann equations (4.1), (4.2), $p = e^{-V}$, for arbitrary initial data.

Acknowledgment. The research of the first author is supported by the Air Force Office of Scientific Research under URI grant AFOSR 86-0352. The research of the second author is supported by the Department of Energy, Applied Mathematical Sciences under grant DE-FG02-88ER25053.

Our thanks to Jonathan Goodman for profitable discussions during the preparation of this paper.

Bibliography

[1] Birkhoff, G., *The rise of modern algebra*, 1936–1950, in *Men and Institutions in American Mathematics*, D. Tarwater, ed., Graduate Studies Texas Tech. Univ. 13, 1976, p. 74.
[2] Courant, R., and Friedrichs, K. O., *Supersonic Flow and Shock Waves*, Wiley, New York, 1948.
[3] Venakides, S., Deift, P., and Oba, R., *The Toda shock problem*, to appear.
[4] Flaschka, H., *The Toda lattice I*, Prog. Theor. Phys. 51, 1974, pp. 703–716.
[5] Goodman, J., and Lax, P. D., *On dispersive difference schemes I*, Comm. Pure Appl. Math. 41, 1988, pp. 591–613.
[6] Holian, B. L., and Straub, G. K., *Molecular dynamics of shock waves in one-dimensional chains*, Phys. Rev. B 18, 1978, pp. 1593–1608.
[7] Holian, B. L., Flaschka, H., and McLaughlin, D. W., *Shock waves in the Toda lattice: Analysis*, Phys. Rev. A 24, 1981, pp. 2595–2623.
[8] Lax, P. D., *On dispersive difference schemes*, Physica 18D, North-Holland, Amsterdam, 1986, pp. 250–254.
[9] Lax, P. D., and Levermore, C. D., *The small dispersion limit of the Korteweg-deVries equation I*, Comm. Pure Appl. Math. 36, 1983, pp. 253–290, II pp. 571–593, III pp. 809–829.
[10] Lax, P. D., and Wendroff, B., *Systems of conservation laws*, Comm. Pure Appl. Math. 13, 1960, pp. 217–237.
[11] Von Neumann, J., *Proposal and Analysis of a New Numerical Method in the Treatment of Hydrodynamical Shock Problems*, Collected Works VI, Pergamon, London.
[12] Von Neumann, J., and Richtmyer, R. D., *A method for the numerical calculation of hydrodynamical shocks*, J. Appl. Phys. 21, 1950, pp. 380–385.
[13] Peierls, R. E., *Theory of von Neumann's method of treating shocks*, Los Alamos Report No. 332, 1946.
[14] Strang, G., *Accurate partial differential methods II: Nonlinear problems*, Num. Math. 6, 1964, pp. 37–46.
[15] Straub, G. K., Holian, B. L., and Petschek, R. G., *Molecular dynamics of shock waves in one-dimensional chains II: Thermalization*, Phys. Rev. B, 19, 1979, pp. 4049–4055.
[16] Teller, E., talk at *Heritage of von Neumann Conference*, Hofstra, J. Impagliazzo, ed., Amer. Math. Soc., 1990.
[17] Trulio, J. G., and Trigger, K. R., *Numerical solution of one-dimensional hydrodynamical shock problem*, UCRL Report 6522, 1961.

Received September 1989.

Commentary on Part IV

49, 72, 74

The three papers in question investigate the phase portrait of the Korteweg–de Vries flow defined by $\partial u/\partial t + u\partial u/\partial x + \partial^3 u/\partial^3 x = 0$ in the class of functions on the line either subject to rapid vanishing at $\pm\infty$ or else of fixed period, $= 1$, say.

Kruskal and Zabusky [1965] had recognized from numerical experiments that the associated solitary waves $s(x) = 3csech^2(\frac{\sqrt{c}}{2}(x-ct))$ exhibit a very curious behavior: The speed $(c > 0)$ is proportional to the amplitude $(3c)$ so that big waves move fast. Let a big wave be placed far to the left of a little wave, far enough so they scarcely see each other, and let them run. Then the fast wave catches up to the slow wave, and a complicated interaction takes place, as you would expect in such a nonlinear problem. The surprise is that after some lapse of time, the big wave emerges, more or less in its original shape, far to the right of the reemergent small wave, with an error (radiation) that vanishes rapidly as time goes by; and not only this, but trains of three or more widely separated solitary waves do the same, the faster passing through the slower and resuming their original form. In short, the solitary waves behave somewhat like particles, whence the name "soliton."

Now, this kind of behavior suggests lots of memory, and lots of memory in mechanics can mean only many constants of motion. Indeed, Gardner–Kruskal–Miura [1968] confirmed the existence of an infinite series of constants of motion of the form $H =$ the integral of a polynomial in u, u', u'', etc., starting with $H_1 = \int u$, $H_2 = \int (u^2/2)$, $H_3 = \int [u^3/6 - (u')^2/2]$. The neat rule of A. Lenard $[D^3 + \frac{1}{3}(Du + uD)]\partial H^-/\partial u = D\partial H^+/\partial u$ leads from H^- to its successor H^+, producing the whole series, which is a (more or less) complete list of constants of motion; see [5].

Now for a further surprise: any reasonable solution of KdV vanishing $\pm\infty$ separates out for large time into a train of solitons, plus a little evanescent radiation, the speeds of these being predicted by the bound states of the Schrödinger operator $L - \lambda = D^2 + u/6 - \lambda$; to wit, $c = 4\lambda$. Naturally, these numbers also are constants of motion, prompting Lax to suggest that under the KdV flow of u, L moves in an isospectral manner, which is to say $L^\bullet = [P, L]$ with skew-symmetric P, the bracket being the commutator.* In fact, $P = D^3 + \frac{1}{8}(uD + Du)$. This is the celebrated Lax pair of which versions have been found in all the fashionable integrable partial differential equations known to date such as sine–Gordon, cubic Schrödinger, and Camassa–Holm. It is one of the most important tools in the subject, its appearance being more or less coextensive with (complete) integrability, on which now just a word.

Gardner [1971] discovered that KdV is a Hamiltonian system relative to the nonstandard Poisson bracket $[A, B] = \int (\partial A/\partial u) D (\partial B/\partial u) dx$ and that the series H_1, H_2, H_3, etc. and also the bound states of L all commute. This is, of course, favorable to integrability. In fact, Gardner–Greene–Kruskal–Miura [1967] discovered an explicit recipe for integrating KdV based on the scattering theory of L, and this is the basis of the proof of Faddeev–Zakharov

*Lax [1968] writes B for my P. I follow the dictum of Gelfand, so that the right side reads [P(eter), L(ax)].

[1971] that KdV is indeed "integrable" in the full technical sense of the word. In particular, the method shows that the space of nice functions vanishing at $\pm\infty$ (of which I am thinking now) is foliated into invariant manifolds of the full KdV hierarchy of (commuting) flows $\partial u/\partial t = D(\partial H_n/\partial u)(n = 1, 2, 3 \text{ etc.})$ cut out by fixing the bound states and transmission coefficient of L, each leaf being essentially an infinite-dimensional linear space (with a few exceptions). Thus far Lax [1968].

Lax [1975/1976] deals with the periodic case. Now the foliation is changed: each leaf is an algebraic torus of variable dimension $d \leq \infty$, $2d + 1$ being the number of simple periodic or antiperiodic eigenvalues of L. Lax elicits this picture starting from the (then unproven but correct) idea that functions producing $2d + 1 < \infty$ such simple eigenvalues are typical. Granting this, it follows that in the KdV hierarchy, only the first d of the vector fields $\mathbb{X}_n : u \to D(\partial H_n/\partial u)$ are independent: In particular, $\mathbb{X}_{d+1}$ is a combination of $\mathbb{X}_1, \ldots, \mathbb{X}_d$, expressing the fact that the value of H_{d+1} is critical, subject to certain fixed values of $H_1, \ldots, H_d$. Lax [1975] shows that this leads to the vanishing of d independent polynomials in the $2d$ variables $D^n u; 0 \leq n \leq 2d - 1$, cutting out an algebraic torus of dimension d in the phase space. Novikov [1974] and McKean–van Moerbeke [1975] recognized this torus as the (real) Jacobi variety of the (2-sheeted) projective curve based on the irrationality $\sqrt{\Delta^2 - 1}$, Δ being the Hill's discriminant of L, for which see also Mumford [1984]. The nice feature of Lax's method is its perfect directness: it employs only elementary calculus in function space, showing that the fancier method of projective curves and their Jacobi varieties is not at all necessary for understanding what is happening. Indeed, in these three papers, Lax brings to light, by the most uncluttered reasoning, many of the themes subsequently found to be quite general in the subject of integrable partial differential equations. For that reason, they repay close study.

The same type of invariant manifold is also to be found in sine–Gordon, cubic Schrödinger, Camassa-Holm, etc. Lax [1975/1976] is supplemented by numerical studies carried out by J.M. Hyman. Here you see an almost periodic behavior of KdV, as was inevitable for $d < \infty$: a finite-dimensional integrable flow confined to one of its invariant torus reduces, in action-angle variables, to straight line motion at constant speed. The fact persists for $d = \infty$ as well, as shown in McKean–Trubowitz [1976]. There also the method is fancy (projective curves of infinite genus), while Lax elicits the facts for $d < \infty$ with typical directness.

References

[1] Faddeev, L.; Zakharov, V. Korteweg–de Vries equation as completely integrable Hamiltonian system. *Funk. Anal. Priloz.* **5** (1971) 18–27.

[2] Gardner, C. Korteweg–de Vries equation and generalizations IV. The Korteweg–de Vries equation as a Hamiltonian system. *J. Math. Phys.* **12** (1971) 1548–1551.

[3] Gardner, C.; Kruskal, M.; Miura, R. Korteweg–de Vries equation and generalizations II. Existence of conservation laws and constants of motion: *J. Math. Phys.* **9** (1968) 1204–1209.

[4] ——— Method for solving the Korteweg–de Vries equation. *Phys. Rev. Letters* **19** (1967) 1095–1097.

[5] Gardner, C.; Greene, J.; Kruskal, M.; Miura, R. Korteweg–de Vries equation and generalizations VI. Methods for exact solutions. *CPAM* **27** (1974) 97–133.

[6] Kruskal, M.; Zabusky, N. Interaction of "solitons" in a collisionless plasma and the recurrence of initial states. *Phys. Rev. Letters* **15** (1965) 240–243.

[7] Lax, P. Integrals of nonlinear equations of evolution and solitary waves. *CPAM* **21** (1968) 467–490.

[8] ——— Periodic solutions of the KdV equation. *CPAM* **28** (1975) 141–188.

[9] ——— Almost periodic solutions of the KdV equation. *SIAM Review* **18** (1976) 351–375.

[10] McKean, H.P.; van Moerbeke, P. The spectrum of Hill's equation. *Invent. Math.* **30** (1975) 217–274.

[11] McKean, H.P.; Trubowitz, E. Hill's operator and hyperelliptic function theory in the presence of infinitely many branch points. *CPAM* **29** (1976) 143–226.

[12] Mumford, D. *Tata Lectures on Theta.* Birkhäuser, Boston/Basel/Stuttgart, 1984.

[13] Novikov, S. The periodic problem for the Korteweg–de Vries equation. *Funk. Anal. Priloz.* **8** (1974) 54–66.

H.P. McKean

103

In these remarkable papers (the PNAS paper is an announcement of the results in I, II, III) the authors consider the zero dispersion limit $\varepsilon \downarrow 0$ of solutions $u(x,t;\varepsilon)$ of the Cauchy problem for the Korteweg–de Vries (KdV) equation

$$u_t - 6uu_x + \varepsilon^2 u_{xxx} = 0, \quad u(x, 0 \cdot \varepsilon) = u(x), \tag{1}$$

where $u(x) \to 0$ as $|x| \to \infty$. The main result in I (see Theorem 2.10) is that $\bar{u}(x,t) = \lim_{\varepsilon\downarrow 0} u(x,t;\varepsilon)$ exists as a weak limit for all $t \geq 0$, and that $\bar{u}(x,t)$ can be evaluated in terms of a measure $\psi(\eta; x,t)d\eta$, which in turn is the (unique) solution of a quadratic minimization problem (see I (2.16)). In II, the authors make the assumption that $\psi(\eta; x,t)d\eta$ is supported on a finite number $N(x,t)$ of intervals, and then show that the endpoints $\{\beta_k = \beta_k(x,t)\}$ of these intervals satisfy (see II (5.32)) a distinguished system of multiphase Whitham equations discovered independently by Flaschka, Forest, and McLaughlin [6]. Paper III contains various results on the long-time behavior of $\bar{u}(x,t)$.

The analysis in these papers depends critically on the fact that the KdV equation (1) is completely integrable. Indeed, the results follow by a careful, and inspired, analysis of the explicit formulae of inverse scattering theory, and as such, the paper is a landmark in the theory of integrable systems. Ever since the seminal work of Gardner, Greene, Kruskal, and Miura in 1967, it was known that the Cauchy problem for KdV, and later for many other integrable systems, could be solved by the inverse scattering transform. However, the formulae are complicated, and with few exceptions, such as the work of Zakharov and Manakov [23] on the long-time behavior of solutions of the nonlinear Schrödinger (NLS) equation, very little information had been extracted prior to I, II, III from the inverse scattering transform about the detailed behavior of the solutions of the integrable system at hand. The papers of Lax and Levermore constitute one of the earliest successes in turning the inverse scattering method into a tool for detailed asymptotic analysis.

In I, II, III the authors restrict their attention mostly, but not exclusively, to initial data $u(x)$ that are negative. In this case, as $\varepsilon \downarrow 0$ a "gas" of order-ε^{-1} solitons gives the leading-order contribution to the inverse scattering transform. If $u(x)$ is positive, there are no solitons, and in [17] Venakides showed how to extend the Lax–Levermore method to this case by an ingenious device: he placed a barrier to the right of the initial data, thus creating a potential well bounded by the initial data to the left and by the barrier to the right. The resonances of this well then play the role of the solitons in I, II, III. In later publications [18, 19] Venakides also considered the zero dispersion limit for periodic initial data.

In their papers, Lax and Levermore show that the limit $\bar{u}(x,t) = \lim_{\varepsilon\downarrow 0} u(x,t;\varepsilon)$ is indeed weak, and they infer that high oscillations must emerge in the solution as ε becomes small. The original Lax–Levermore–Venakides theory computes only the weak limit of the solution, and new ideas are needed to derive these oscillations. In a series of papers culminating in [19], Venakides was able to describe these oscillations in terms of theta functions evaluated on the microscales x/ε and t/ε. The theta functions correspond to the hyperelliptic Riemann surface that arises naturally when the complex plane is slit along the support of $\psi(\eta; x, t)d\eta$. The key new step in [20] was the introduction of a "quantum condition" in the quadratic minimization problem for $\psi\ d\eta$.

It is interesting to consider the above results from a more physical point of view. As $\varepsilon \downarrow 0$, equation (1) becomes, at least formally, Burger's equation $u_t - 6uu_x = 0$. As is well known, solutions of this equation generally develop shocks. One way to investigate these shocks is to consider solutions of the dissipative regularization of Burger's equation $u_t - 6uu_x = \varepsilon^2 u_{xx}$ with ε small. Another way is through dispersive regularization, as in (1) above. With the dissipative regularization, the term $\varepsilon^2 u_{xx}$ "burns" away the shock energy and results in the generation of heat. On the other hand, with the dispersive regularization, the results of Lax, Levermore, and Venakides show that as $\varepsilon \downarrow 0$ the solution $u(x,t;\varepsilon)$ develops rapid oscillations that transport energy away from the shock in the form of highly oscillating, modulated wave trains.

If $\psi(\eta; x, t)d\eta$ is supported on $N = N(x,t)$ intervals, one says that one has an $(N-1)$-phase solution of the Whitham equations. As time evolves, these equations typically develop shocks, generating new phases. These phase changes make the analysis of the Cauchy problem for the Whitham equations extremely difficult. Using ideas of Tsarëv [16], who showed that the Whitham equations are in fact "integrable," and also Krichever [11], Tian [15] was the first to solve rigorously the Cauchy problem for a general class of 0-phase initial data, tracing the evolution of the solution through the singularity to the formation of a one-phase wave. Results for slightly more restricted initial data were also obtained by Wright [22] and K. McLaughlin [14]. Recent work of Grava and Tian [7] considers a class of initial data that leads, in different regions of space-time x, t, to 0-, 1-, and 2-phase solutions of the Whitham equations.

In the above papers, Lax, Levermore, and Venakides use the classical methods of inverse spectral theory introduced by Gel'fand and Levitan in the 1950s and then tailored to the scattering situation by Faddeev and Marchenko. As noted by Shabat, however, the inverse scattering transform can also be viewed as a Riemann–Hilbert problem. In 1993 Deift and Zhou [4] introduced a steepest-descent method for Riemann–Hilbert problems with oscillatory coefficients. In [3], Deift, Venakides, and Zhou extended the method of [4] in an essential way, and were able to use the Riemann–Hilbert steepest-descent formalism to derive the Lax–Levermore–Venakides theory in a natural way to (in principle) all orders of perturbation theory, together with explicit formulae for constants such as the phase shifts that arise in the theta function oscillations described above. In [3] the authors also derived integral representations for the solution of the Whitham equations.

The methods of Lax–Levermore–Venakides have also been used to solve a variety of asymptotic problems, such as the semiclassical limit of the defocusing NLS equation [9] , the Toda shock problem [21], and the continuum limit for the Toda lattice [2, 14]. A very useful review of the original work of Lax, Levermore, and Venakides, together with some of these more recent developments and related issues, is given in [12]. In later work, the extended steepest-descent method in [3] has provided the solution to a variety of problems in mathematics and mathematical physics, such as the proof of the so-called universality conjecture in random matrix theory given in [1]. Recently, Kamvissis, K. McLaughlin, and Miller [10] have extended the method in [3, 4] even further and have computed the semiclassical limit of the focusing NLS equation with special initial data. In the focusing case the associated spectral problem is not self-adjoint, and this leads to considerable new difficulties that do not arise in the defocusing case considered in [9].

The Whitham equations appear in many different mathematical and physical contexts and have been studied extensively in their own right. Further discussion of these developments would take us beyond the scope of this review. We do mention, however, that the earliest applications of the Whitham equations to the KdV equation were made by Gurevich and Pitaevskii, [8] who considered the problem of the long-time behavior of solutions of KdV with an initial discontinuity and also the problem of the "overturning" of the front for a simple wave. We also note the paper of Levermore [13] in which the author proves strict hyperbolicity and genuine nonlinearity for the Whitham modulation system.

References

[1] Deift, P.; Kriecherbauer, T.; McLaughlin, K.T.-R.; Venakides, S.; Zhou, X. Uniform asymptotics for polynomials orthogonal with respect to varying exponential weights and applications to universality questions in random matrix theory. *Comm. Pure Appl. Math.* **52** (1999), no. 11, 1335–1425.

[2] Deift, P.; McLaughlin, K.T.-R. A continuum limit of the Toda lattice. *Mem. Amer. Math. Soc.* **131** (1998), no. 624.

[3] Deift, P.; Venakides, S.; Zhou, X. New results in small dispersion KdV by an extension of the steepest descent method for Riemann–Hilbert problems. *Internat. Math. Res. Notices* **1997**, no. 6, 286–299.

[4] Deift, P.; Zhou, X. A steepest descent method for oscillatory Riemann–Hilbert problems. Asymptotics for the MKdV equation. *Ann. of Math. (2)* **137** (1993), no. 2, 294–368.

[5] Dubrovin, B.A.; Novikov, S.P. Hydrodynamics of weakly deformed soliton lattices. Differential geometry and Hamiltonian theory. *Uspekhi Mat. Nauk* **44** (1989), no. 6, 29–98. Translation in *Russian Math. Surveys* **44** (1989), no. 6, 35–124.

[6] Flaschka, H.; Forest, M.G.; McLaughlin, D.W. Multiphase averaging and the inverse spectral solution of the Korteweg–de Vries equation. *Comm. Pure Appl. Math.* **33** (1980), no. 6, 739–784.

[7] Grava, T.; Tian, F.R. The generation, propagation and extinction of multiphase in the KdV zero dispersion limit. Preprint, 2001.

[8] Gurevich, A.V.; Pitaevskii, L.P. Nonstationary structure of a collision-less shock wave, *Sov. Phys.-JETP*, **38** (1974), 291–297. *Zh. Eksp. Theor. Fiz.* **65** (1973), 590–604.

[9] Jin, S.; Levermore, C.D.; McLaughlin, D.W. The behavior of solutions of the NLS equation in the semiclassical limit. *Singular limits of dispersive waves (Lyon, 1991)*, 235–255. Edited by N.M. Ercolani, I.R. Gabitov, C.D. Levermore, and D. Serre. NATO Adv. Sci. Inst. Ser. B Phys., 320. Plenum, New York, 1994.

[10] Kamvissis, S.; McLaughlin, K.T.-R.; Miller, P.D. Semiclassical soliton ensembles for the focusing nonlinear Schrödinger equation. Preprint, 2000. Available online at `http://arXiv.org/abs/nlin.SI/0012034`.

[11] Krichever, I.M. The averaging method for two-dimensional "integrable" equations. *Funktsional. Anal. i Prilozhen.* **22** (1988), no. 3, 37–52. Translation in *Funct. Anal. Appl.* **22** (1988), no. 3, 200–213.

[12] Lax, P.D.; Levermore, C.D.; Venakides, S. The generation and propagation of oscillations in dispersive initial value problems and their limiting behavior. *Important Developments in Soliton Theory*, 205–241. Edited by A.S. Fokas and V.E. Zakharov. Springer Ser. Nonlinear Dynam. Springer, Berlin, 1993.

[13] Levermore, C.D. The hyperbolic nature of the zero dispersion KdV limit. *Comm. PDE* **13** (1988), no. 4, 495–514.

[14] McLaughlin, K.T.-R. A continuum limit of the Toda lattice. Ph.D. dissertation, New York University, 1996.

[15] Tian F.R. Oscillations of the zero dispersion limit of the Korteweg–de Vries equation. *Comm. Pure Appl. Math.* **46** (1993), no. 8, 1093–1129.

[16] Tsarëv, S.P. Poisson brackets and one-dimensional Hamiltonian systems of hydrodynamic type. *Dokl. Akad. Nauk SSSR* **282** (1985), no. 3, 534–537. Translation in *Soviet Math. Dokl.* **31** (1985), 488–491.

[17] Venakides, S. The zero dispersion limit of the Korteweg–de Vries equation for initial potentials with nontrivial reflection coefficient. *Comm. Pure Appl. Math.* 38 (1985), no. 2, 125–155.

[18] Venakides, S. The zero dispersion limit of the Korteweg–de Vries equation with periodic initial data. *Trans. Amer. Math. Soc.* **301** (1987), no. 1, 189–226.

[19] Venakides, S. The continuum limit of theta functions. *Comm. Pure Appl. Math.* **42** (1989), no. 6, 711–728.

[20] Venakides, S. The Korteweg–de Vries equation with small dispersion: higher order Lax–Levermore theory. *Comm. Pure Appl. Math.* **43** (1990), no. 3, 335–361.

[21] Venakides, S.; Deift, P.; Oba, R. The Toda shock problem. *Comm. Pure Appl. Math.* **44** (1991), no. 8–9, 1171–1242.

[22] Wright, O.C. Korteweg–de Vries zero dispersion limit: through first breaking for cubic-like analytic initial data. *Comm. Pure Appl. Math.* **46** (1993), no. 3, 423–440.

[23] Zakharov, V.E.; Manakov, S.V. Asymptotic behavior of non-linear wave systems integrated by the inverse scattering method. *soviet Physics JETP* **44** (1976), no. 1, 106–112. Translated from *Z. Èksper. Toeret. Fiz.* **71** (1976), no. 1, 203–215.

P. Deift

120

While working at Los Alamos in 1943–1944, von Neumann became interested in designing shock capturing schemes to calculate compressible fluid flows containing strong shocks. Employing a Lagrangian description of compressible flow, setting heat conduction and viscosity equal to zero, von Neumann replaced space and time derivatives by symmetric difference quotients. Approximations resulting from these calculations showed oscillations on mesh scale behind the shock. Von Neumann conjectured that the oscillations in velocity represent the heat energy created by the irreversible action of the shock, and as the mesh size tends to zero, the approximate solutions tend in the weak sense to the correct discontinuous solution of the equations of the compressible flow.

Based on his extensive research experience on zero dispersive limit of the KdV equation, Peter Lax counterconjectured in [1] that von Neumann was wrong in his surmise, i.e., that the dispersive approximations do not converge to the correct weak solution of the equations of the compressible flow. The work by Goodman and Lax on "Dispersive difference schemes I" provided the first systematic study of von Neumann's algorithm applied to the inviscid Burgers equation in the semidiscrete case. By expanding the symmetric difference quotient, they derived a modified equation for the dispersive approximation that resembles the small-dispersion KdV equation with the dispersion coefficient depending on the square of the mesh size. The dispersive scheme was shown to admit period-two oscillations on mesh scale. Further, they showed that before the shock forms, the dispersive approximation converges to the smooth solution of the Burgers equation. By using numerical experimentation and analytical techniques, they demonstrated the weak convergence of oscillatory approximations after the shock forms, and that the weak limit fails to satisfy the Burgers equation. Finally, Goodman and Lax showed that their semidiscrete dispersive approximation can be transformed into an integral system of Kac and van Moerbeke that allows one to carry out a rigorous study to characterize this weak limit. This is a beautiful piece of work.

The work of Goodman and Lax has inspired many subsequent numerical and analytical studies of dispersive approximations to nonlinear hyperbolic PDEs. In [3], Hou and Lax carried out a careful numerical study of von Neumann's original dispersive scheme to the Euler equations of gas dynamics and found convincing evidence that the weak solution fails to satisfy the gas dynamics equations. In [4], Levermore and Liu studied a nonintegrable semidiscrete scheme of the Burgers equation. Modulation equations were derived that describe period-two oscillations. However, those equations have an elliptic region that may be entered by its solutions in finite time, after which the corresponding period-two oscillations are seen to break down. This kind of phenomenon does not seem to occur for integrable schemes (see, e.g., the work of Deift and McLaughlin on a continuum limit of the Toda lattice [6]). This phenomenon was further analyzed in great depth by Turner and Rosales in [5] for another nonintegrable semidiscrete scheme of the Burgers equation. They showed that the modulation equations for the period-two oscillations have both a hyperbolic and an elliptic region. The period-two oscillations break down after they enter the elliptic region, and the solution blows up.

References

[[1] Lax, P.D. On dispersive difference schemes, *Physica D*, **18** (1986), 250–254.

[2] Goodman, J.; Lax, P.D. Dispersive difference schemes I, *CPAM*, **33** (1988), 591–613.

[3] Hou, T.Y.; Lax, P.D. Dispersive approximations in fluid dynamics, *CPAM*, **44** (1991), 1–40.

[4] Levermore, C.D.; Liu, J.G. Large oscillations arising in a dispersive numerical scheme, *Physica D*, **99** (1996), 191–216.

[5] Turner, C.V.; Rosales, R.R. The small dispersion limit for a nonlinear semidiscrete system of equations, *Studies in Applied Mathematics*, **99** (1997), 205–254.

[6] Deift, P.; McLaughlin, K.T.R. A continuum limit of the Toda lattice, *Memoirs of the American Mathematical Society*, **131** (624), 1998.

T. Hou

129

Working with Peter on "Dispersive approximations in fluid dynamics" was a very enlightening and pleasant experience for me. The subject of studying propagation of oscillations is a fascinating one. There has been a lot of work on this subject at the continuum level assuming periodic structure and scale separation (i.e., homogenization). But studying propagation of dispersive oscillations generated dynamically at the discrete level is much more difficult, and there are few tools available. Through careful numerical study, we found convincing evidence that the discrete oscillatory solution of von Neumann's scheme does not converge to the physical solution even in the weak sense, contrary to von Neumann's original conjecture. The reason why the weak limit fails to satisfy the gas dynamics equations is that nonlinear interaction of the high-frequency oscillations makes a finite contribution. Von Neumann conjectured that the oscillations in velocity represent the heat energy created by the irreversible action of the shock and that they make finite (and in the right amount) contributions to the total energy. Using an ingenious conservative reformulation of von Neumann's scheme by Trulio and Trigger (reference [17] in our paper), we demonstrated convincingly that the nonlinear interaction of high-frequency oscillations produces an order-one error term that is similar in spirit to the Reynolds stress term in turbulence modeling. For this reason, Peter also called this phenomenon *deterministic turbulence.*

It is clear that Peter had a much bigger picture in mind when he studied this problem. How should one quantify and characterize the continuum (weak) limit of lattice approximations, which are used frequently in the materials science, physics, and biology communities? There are many physically interesting applications (such as crack propagation) in which the continuum model is no longer valid. One has to rely on microscopic modeling (such as an atomistic model) to supplement the continuum model. How to model dynamic interaction across different length scales and to integrate the discrete and the continuum models is a great challenge. Many observations made by Peter in his study of dispersive approximations are still very relevant and inspiring in these studies.

Peter was very intrigued by the robustness of the discrete oscillatory solutions generated by von Neumann's dispersive approximation. These oscillations have a well-defined weak limit that seems to be independent of the time and space mesh sizes. In a special case, Peter observed that the semidiscrete version of von Neumann's scheme could be reduced to a Toda lattice problem. The continuum limit of the Toda lattice was subsequently studied in a beautiful piece of work by Percy Deift and Ken McLaughlin (Memoirs of AMS, 131: (624) 1-JAN 1998). Peter believed that the result should be true for the time discrete case and for the gas dynamics if one could generalize a KAM type of argument to this problem. It was a pity that we did not find the time and energy to carry this idea further.

Looking back, I felt extremely fortunate that I had such a valuable opportunity to work with Peter. I have benefited tremendously from his vision, his knowledge, and his approach to research. There were many happy moments throughout our collaborations. I found it most enjoyable when Peter told me stories about his personal interactions with various well-known mathematicians and scientists (von Neumann included) in the early stage of his career. As a postdoc and then a junior faculty member at the Courant Institute, I also witnessed at a close distance how Peter worked both as a mathematician and as a leader in science. All these have left me with unforgettable memories and have impacted the rest of my academic career. I wish Peter a very happy 75th birthday and many happy years to come.

T. Hou